Moments of Inertia of Common Geometric Shapes

Rectangle	
$\bar{I}_{x'} = \frac{1}{12}bh^3$ $\bar{I}_{y'} = \frac{1}{12}b^3h$ $I_x = \frac{1}{3}bh^3$ $I_y = \frac{1}{3}b^3h$ $J_C = \frac{1}{12}bh(b^2 + h^2)$	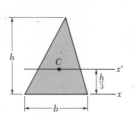
Triangle	
$\bar{I}_{x'} = \frac{1}{36}bh^3$ $I_x = \frac{1}{12}bh^3$	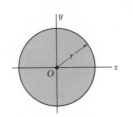
Circle	
$\bar{I}_x = \bar{I}_y = \frac{1}{4}\pi r^4$ $J_O = \frac{1}{2}\pi r^4$	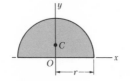
Semicircle	
$I_x = I_y = \frac{1}{8}\pi r^4$ $J_O = \frac{1}{4}\pi r^4$	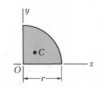
Quarter circle	
$I_x = I_y = \frac{1}{16}\pi r^4$ $J_O = \frac{1}{8}\pi r^4$	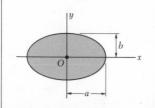
Ellipse	
$\bar{I}_x = \frac{1}{4}\pi ab^3$ $\bar{I}_y = \frac{1}{4}\pi a^3b$ $J_O = \frac{1}{4}\pi ab(a^2 + b^2)$	

Mass Moments of Inertia of Common Geometric Shapes

Slender rod	
$I_y = I_z = \frac{1}{12}mL^2$	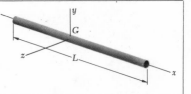
Thin rectangular plate	
$I_x = \frac{1}{12}m(b^2 + c^2)$ $I_y = \frac{1}{12}mc^2$ $I_z = \frac{1}{12}mb^2$	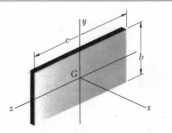
Rectangular prism	
$I_x = \frac{1}{12}m(b^2 + c^2)$ $I_y = \frac{1}{12}m(c^2 + a^2)$ $I_z = \frac{1}{12}m(a^2 + b^2)$	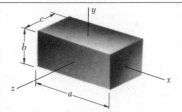
Thin disk	
$I_x = \frac{1}{2}mr^2$ $I_y = I_z = \frac{1}{4}mr^2$	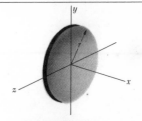
Circular cylinder	
$I_x = \frac{1}{2}ma^2$ $I_y = I_z = \frac{1}{12}m(3a^2 + L^2)$	
Circular cone	
$I_x = \frac{3}{10}ma^2$ $I_y = I_z = \frac{3}{5}m(\frac{1}{4}a^2 + h^2)$	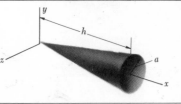
Sphere	
$I_x = I_y = I_z = \frac{2}{5}ma^2$	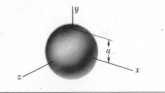

Vector Mechanics
for Engineers
Dynamics

Vector Mechanics for Engineers

Dynamics

Fifth Edition

Ferdinand P. Beer

Lehigh University

E. Russell Johnston, Jr.

University of Connecticut

McGraw-Hill Book Company

New York St. Louis San Francisco Auckland Bogotá Caracas
Colorado Springs Hamburg Lisbon London Madrid Mexico Milan
Montreal New Delhi Oklahoma City Panama Paris San Juan
São Paulo Singapore Sydney Tokyo Toronto

The cover photograph is of the high-speed train (TGV, Train à Grande Vitesse) designed and operated by the French National Railroads. Each train unit consists of eight passenger cars and two 3150-kW power cars running on 25-kV alternating current. Operated at speeds up to 270 km/h (168 mi/h) on a special track between Paris and Lyons, these trains may proceed at a reduced speed on normal tracks, thus providing fast, direct connections among more than 40 French and Swiss cities. *(Photograph by Richard Kalvar, Magnum.)*

Back cover photographs: (top) Lightscapes, The Stock Market. (bottom) Peter Vadnai, The Stock Market.

VECTOR MECHANICS FOR ENGINEERS: Dynamics

34567890 KGPKGP 8932109

P/N 004498-8
Part of
ISBN 0-07-079926-1

This book was set in Laurel by York Graphic Services, Inc.
The editors were John Corrigan and David A. Damstra;
the designer was Merrill Haber;
the production supervisor was Joe Campanella.
The drawings were done by Felix Cooper.
Arcata Graphics/Kingsport was printer and binder.

Library of Congress Cataloging-in-Publication Data

Beer, Ferdinand Pierre, (date).
   Vector mechanics for engineers: Dynamics/Ferdinand P. Beer,
E. Russell Johnston, Jr.—5th ed.
      p. cm.
   Bibliography: p.
   Includes index.
   ISBN 0-07-079926-1
   1. Mechanics, Applied.   2. Vector analysis.   3. Mechanics,
Applied—Problems, exercises, etc.   I. Johnston, E. Russell (Elwood
Russell) (date).   II. Title.
TA350.B3552 1988b
620. 1'05—dc19   87-35374

"How did you happen to write your books together, with one of you at Lehigh and the other at UConn, and how do you manage to keep collaborating on their successive revisions?" These are the two questions most often asked of our two authors.

The answer to the first question is simple. Russ Johnston's first teaching appointment was in the department of civil engineering and mechanics at Lehigh University. There he met Ferd Beer, who had joined that department two years earlier and was in charge of the courses in statics and dynamics. Born in France and educated in France and Switzerland (he holds an M.S. degree from the Sorbonne and a Sc.D. degree in the field of theoretical mechanics from the University of Geneva), Ferd had come to the United States after serving in the French army during the early part of World War II and had taught for four years at Williams College in the Williams-MIT joint arts and engineering program. Born in Philadelphia, Russ had obtained a B.S. degree in civil engineering from the University of Delaware and a Sc.D. degree in the field of structural engineering from MIT.

Ferd was delighted to discover that the young man who had been hired chiefly to teach graduate structural engineering courses was not only willing but eager to help him reorganize the courses in statics and dynamics. Both believed that these courses should be taught from a few basic principles and that the various concepts involved would be best understood and remembered by the students if they were presented in a graphic way. Together they wrote lecture notes, to which they later added problems they felt would appeal to future engineers, and soon they had produced the manuscript of the first edition of *Mechanics for Engineers*.

The second edition of their text found Russ Johnston at Worcester Polytechnic Institute and the third at the University of Connecticut. In the meantime, both Ferd and Russ had assumed administrative responsibilities in their departments, and both were involved in research, consulting, and the supervision of graduate students, Ferd in the area of stochastic processes and random vibrations, and Russ in the area of elastic stability and structural analysis and design. However, their interest in improving the teaching of the basic mechanics courses had not subsided and they both taught sections of these courses as they kept revising their texts.

This brings us to the second question: How did the authors manage to work together so effectively after Russ Johnston had left Lehigh? Part of the answer may be provided by their phone bills and the money they spend on postage. As the publication date of a new edition approaches, they call each other daily and rush to the post office with express-mail packages in order to double-check their work. There are also frequent visits between the two families. At one time there were even joint camping trips, with both families pitching their tents next to each other. The Beers were the first to graduate to a trailer, which was used to illustrate a problem in one of the early editions of their text, but was replaced by the Johnstons' trailer in the next one. Now this trailer has also been replaced, both authors preferring the comforts of a motel and its dining room to those of a camping ground and its fireplaces.

Contents

Preface

The main objective of a first course in mechanics should be to develop in the engineering student the ability to analyze any problem in a simple and logical manner and to apply to its solution a few, well-understood, basic principles. It is hoped that this text, as well as the preceding volume, *Vector Mechanics for Engineers: Statics,* will help the instructor achieve this goal.†

Vector algebra was introduced at the beginning of the first volume and used in the presentation of the basic principles of statics, as well as in the solution of many problems, particularly three-dimensional problems. Similarly, the concept of vector differentiation will be introduced early in this volume, and vector analysis will be used throughout the presentation of dynamics. This approach results in a more concise derivation of the fundamental principles. It also makes it possible to analyze many problems in kinematics and kinetics which could not be solved by scalar methods. The emphasis in this text, however, remains on the correct understanding of the principles of mechanics and on their application to the solution of engineering problems, and vector analysis is presented chiefly as a convenient tool.‡

One of the characteristics of the approach used in these volumes is that the mechanics of *particles* has been clearly separated from the mechanics of *rigid bodies.* This approach makes it possible to consider simple practical applications at an early stage and to postpone the introduction of more difficult concepts. In the volume on statics, the statics of particles was treated first, and the principle of equilibrium was immediately applied to practical situations involving only concurrent forces. The statics of rigid bodies was considered later, at which time the vector and scalar products of two vectors were introduced and used to define the moment of a force about a point and about an axis. In this volume, the same division is observed. The basic concepts of force, mass, and acceleration, of work and energy, and of impulse and momentum are introduced and first applied to problems involving only particles. Thus students may familiarize themselves with the three basic methods used in dynamics and learn their respective advantages before facing the difficulties associated with the motion of rigid bodies.

---

† Both texts are also available in a single volume, Vector Mechanics for Engineers: Statics and Dynamics, fifth edition.

‡ In a parallel text, *Mechanics for Engineers: Dynamics,* fourth edition, the use of vector algebra is limited to the addition and subtraction of vectors, and vector differentiation is omitted.

Since this text is designed for a first course in dynamics, new concepts have been presented in simple terms and every step explained in detail. On the other hand, by discussing the broader aspects of the problems considered and by stressing methods of general applicability, a definite maturity of approach has been achieved. For example, the concept of potential energy is discussed in the general case of a conservative force. Also, the study of the plane motion of rigid bodies has been designed to lead naturally to the study of their general motion in space. This is true in kinematics as well as in kinetics, where the principle of equivalence of external and effective forces is applied directly to the analysis of plane motion, thus facilitating the transition to the study of three-dimensional motion.

The fact that mechanics is essentially a *deductive* science based on a few fundamental principles has been stressed. Derivations have been presented in their logical sequence and with all the rigor warranted at this level. However, the learning process being largely *inductive*, simple applications have been considered first. Thus the dynamics of particles precedes the dynamics of rigid bodies; and, in the latter, the fundamental principles of kinetics are first applied to the solution of two-dimensional problems, which can be more easily visualized by the student (Chaps. 16 and 17), while three-dimensional problems are postponed until Chap. 18.

The fifth edition of *Vector Mechanics for Engineers* retains the unified presentation of the principles of kinetics which characterized the previous three editions. The concepts of linear and angular momentum are introduced in Chap. 12 so that Newton's second law of motion may be presented not only in its conventional form $\mathbf{F} = m\mathbf{a}$, but also as a law relating, respectively, the sum of the forces acting on a particle and the sum of their moments to the rates of change of the linear and angular momentum of the particle. This makes possible an earlier introduction of the principle of conservation of angular momentum and a more meaningful discussion of the motion of a particle under a central force (Sec. 12.9). More importantly this approach may be readily extended to the study of the motion of a system of particles (Chap. 14) and leads to a more concise and unified treatment of the kinetics of rigid bodies in two and three dimensions (Chaps. 16 through 18).

Free-body diagrams were introduced early in statics. They were used not only to solve equilibrium problems but also to express the equivalence of two systems of forces or, more generally, of two systems of vectors. The advantage of this approach becomes apparent in the study of the dynamics of rigid bodies, where it is used to solve three-dimensional as well as two-dimensional problems. By placing the emphasis on "free-body-diagram equations" rather than on the standard algebraic equations of motion, a more intuitive and more complete understanding of the fundamental principles of dynamics may be achieved. This approach, which was first introduced in 1962 in the first edition of *Vector Mechanics for Engineers,* has now gained wide acceptance among mechanics teachers in this country. It is, therefore, used in preference to the method of dynamic equilibrium and to the equations of motion in the solution of all sample problems in this edition.

Color has again been used in this edition to distinguish forces from other elements of the free-body diagrams. This makes it easier for the students to identify the forces acting on a given particle or rigid body and to follow the discussion of sample problems and other examples given in the text.

Because of the current trend among American engineers to adopt the international system of units (SI metric units), the SI units most frequently used in mechanics were introduced in Chap. 1 of *Statics*. They are discussed again in Chap. 12 of this volume and used throughout the text. Approximately half the sample problems and 60 percent of the problems to be assigned have been stated in these units, while the remainder retain U.S. customary units. The authors believe that this approach will best serve the needs of the students, who will be entering the engineering profession during the period of transition from one system of units to the other. It also should be recognized that the passage from one system to the other entails more than the use of conversion factors. Since the SI system of units is an absolute system based on the units of time, length, and mass, whereas the U.S. customary system is a gravitational system based on the units of time, length, and force, different approaches are required for the solution of many problems. For example, when SI units are used, a body is generally specified by its mass expressed in kilograms; in most problems of statics it was necessary to determine the weight of the body in newtons, and an additional calculation was required for this purpose. On the other hand, when U.S. customary units are used, a body is specified by its weight in pounds and, in dynamics problems, an additional calculation will be required to determine its mass in slugs (or $lb \cdot s^2/ft$). The authors, therefore, believe that problem assignments should include both systems of units. The actual distribution of assigned problems between the two systems of units, however, has been left to the instructor, and a sufficient number of problems of each type have been provided so that four complete lists of assignments may be selected with the proportion of problems stated in SI units set anywhere between 50 and 75 percent. If so desired, two complete lists of assignments may also be selected from problems stated in SI units only and two others from problems stated in U.S. customary units.

A number of optional sections have been included. These sections are indicated by asterisks and may thus easily be distinguished from those which form the core of the basic dynamics course. They may be omitted without prejudice to the understanding of the rest of the text. The topics covered in these additional sections include graphical methods for the solution of rectilinear-motion problems, the trajectory of a particle under a central force, the deflection of fluid streams, problems involving jet and rocket propulsion, the kinematics and kinetics of rigid bodies in three dimensions, damped mechanical vibrations, and electrical analogues. These topics will be found of particular interest when dynamics is taught in the junior year.

The material presented in this volume and most of the problems require no previous mathematical knowledge beyond algebra, trigonometry, elementary calculus, and the elements of vector algebra presented in

Chaps. 2 and 3 of the volume on statics.† However, special problems have been included, which make use of a more advanced knowledge of calculus, and certain sections, such as Secs. 19.8 and 19.9 on damped vibrations, should be assigned only if the students possess the proper mathematical background.

Each chapter begins with an introductory section setting the purpose and goals of the chapter and describing in simple terms the material to be covered and its application to the solution of engineering problems. The body of the text has been divided into units, each consisting of one or several theory sections, one or several sample problems, and a large number of problems to be assigned. Each unit corresponds to a well-defined topic and generally may be covered in one lesson. In a number of cases, however, the instructor will find it desirable to devote more than one lesson to a given topic. Each chapter ends with a review and summary of the material covered in that chapter. Marginal notes have been added to help the students organize their review work, and cross-references have been included to help them find the portions of material requiring their special attention.

The sample problems have been set up in much the same form that students will use in solving the assigned problems. They thus serve the double purpose of amplifying the text and demonstrating the type of neat and orderly work that students should cultivate in their own solutions. Most of the problems to be assigned are of a practical nature and should appeal to engineering students. They are primarily designed, however, to illustrate the material presented in the text and to help students understand the basic principles of mechanics. The problems have been grouped according to the portions of material they illustrate and have been arranged in order of increasing difficulty. Problems requiring special attention have been indicated by asterisks. Answers to all even-numbered problems are given at the end of the book.

The introduction in the engineering curriculum of instruction in computer programming and the increasing availability of personal computers or mainframe terminals on most campuses make it now possible for engineering students to solve a number of challenging dynamics problems. Only a few years ago these problems would have been considered inappropriate for an undergraduate course because of the large number of computations their solutions require. In this new edition of *Vector Mechanics for Engineers: Dynamics*, a group of four problems designed to be solved with a computer has been added to the review problems at the end of each chapter. These problems may involve the determination of the motion of a particle under various initial conditions, the kinematic or kinetic analysis of mechanisms in successive positions, or the numerical integration of various equations of motion. Developing the algorithm

---

† Some useful definitions and properties of vector algebra have been summarized in Appendix A at the end of this volume for the convenience of the reader. Also, Secs. 9.11 through 9.17 of the volume on statics, which deal with the moments of inertia of masses, have been reproduced in Appendix B.

required to solve a given dynamics problem will benefit the students in two different ways: (1) it will help them gain a better understanding of the mechanics principles involved; (2) it will provide them with an opportunity to apply the skills acquired in their computer programming course to the solution of a meaningful engineering problem.

A diskette containing interactive software designed for the IBM PC or any compatible computer has been placed in the pocket inside the rear cover of this text. This software includes twelve "Tutorials," each consisting of a problem in dynamics for which students are more likely to require help; thus, the software plays a role similar to that of the Sample Problems. However, being interactive, the Tutorials allow the students to vary the data, let them participate in the solution of the problem, and check the results obtained. By allowing the students to vary the data, the Tutorials provide them with an inexhaustible source of problems of the same type. They also allow the more inquisitive students to test the "sensitivity" of a given problem to a change in data or to find the conditions which lead to special or critical situations. The authors wish to thank Professor Raymond P. Canale of the University of Michigan for helping them prepare the Tutorials and Mr. Tad Slawecki of EnginComp Software, Inc., for programming them.

The authors wish to acknowledge gratefully the many helpful comments and suggestions offered by the users of the previous editions of *Mechanics for Engineers* and of *Vector Mechanics for Engineers*.

Ferdinand P. Beer
E. Russell Johnston, Jr.

List of Symbols

$\mathbf{a}, a$	Acceleration
a	Constant; radius; distance; semimajor axis of ellipse
$\overline{\mathbf{a}}, \overline{a}$	Acceleration of mass center
$\mathbf{a}_{B/A}$	Acceleration of B relative to frame in translation with A
$\mathbf{a}_{P/\mathcal{F}}$	Acceleration of P relative to rotating frame $\mathcal{F}$
$\mathbf{a}_c$	Coriolis acceleration
$\mathbf{A}, \mathbf{B}, \mathbf{C}, \ldots$	Reactions at supports and connections
$A, B, C, \ldots$	Points
A	Area
b	Width; distance; semiminor axis of ellipse
c	Constant; coefficient of viscous damping
C	Centroid; instantaneous center of rotation; capacitance
d	Distance
$\mathbf{e}_n, \mathbf{e}_t$	Unit vectors along normal and tangent
$\mathbf{e}_r, \mathbf{e}_\theta$	Unit vectors in radial and transverse directions
e	Coefficient of restitution; base of natural logarithms
E	Total mechanical energy; voltage
f	Frequency; scalar function
$\mathbf{F}$	Force; friction force
g	Acceleration of gravity
G	Center of gravity; mass center; constant of gravitation
h	Angular momentum per unit mass
$\mathbf{H}_O$	Angular momentum about point O
$\dot{\mathbf{H}}_G$	Rate of change of angular momentum $\mathbf{H}_G$ with respect to frame of fixed orientation
$(\dot{\mathbf{H}}_G)_{Gxyz}$	Rate of change of angular momentum $\mathbf{H}_G$ with respect to rotating frame $Gxyz$
$\mathbf{i}, \mathbf{j}, \mathbf{k}$	Unit vectors along coordinate axes
i	Current
$I, I_x, \ldots$	Moment of inertia
$\overline{I}$	Centroidal moment of inertia
$I_{xy}, \ldots$	Product of inertia
J	Polar moment of inertia
k	Spring constant
k_x, k_y, k_O	Radius of gyration
$\overline{k}$	Centroidal radius of gyration
l	Length
$\mathbf{L}$	Linear momentum

L	Length; inductance
m	Mass
m'	Mass per unit length
$\mathbf{M}$	Couple; moment
$\mathbf{M}_O$	Moment about point O
$\mathbf{M}_O^R$	Moment resultant about point O
M	Magnitude of couple or moment; mass of earth
M_{OL}	Moment about axis OL
n	Normal direction
$\mathbf{N}$	Normal component of reaction
O	Origin of coordinates
p	Circular frequency
$\mathbf{P}$	Force; vector
$\dot{\mathbf{P}}$	Rate of change of vector $\mathbf{P}$ with respect to frame of fixed orientation
q	Mass rate of flow; electric charge
$\mathbf{Q}$	Force; vector
$\dot{\mathbf{Q}}$	Rate of change of vector $\mathbf{Q}$ with respect to frame of fixed orientation
$(\dot{\mathbf{Q}})_{Oxyz}$	Rate of change of vector $\mathbf{Q}$ with respect to frame $Oxyz$
$\mathbf{r}$	Position vector
$\mathbf{r}_{B/A}$	Position vector of B relative to A
r	Radius; distance; polar coordinate
$\mathbf{R}$	Resultant force; resultant vector; reaction
R	Radius of earth; resistance
$\mathbf{s}$	Position vector
s	Length of arc
t	Time; thickness; tangential direction
$\mathbf{T}$	Force
T	Tension; kinetic energy
$\mathbf{u}$	Velocity
u	Variable
U	Work
$\mathbf{v}, v$	Velocity
v	Speed
$\bar{\mathbf{v}}, \bar{v}$	Velocity of mass center
$\mathbf{v}_{B/A}$	Velocity of B relative to frame in translation with A
$\mathbf{v}_{P/\mathscr{F}}$	Velocity of P relative to rotating frame $\mathscr{F}$
$\mathbf{V}$	Vector product
V	Volume; potential energy
w	Load per unit length
$\mathbf{W}, W$	Weight; load
x, y, z	Rectangular coordinates; distances
$\dot{x}, \dot{y}, \dot{z}$	Time derivatives of coordinates x, y, z
$\bar{x}, \bar{y}, \bar{z}$	Rectangular coordinates of centroid, center of gravity, or mass center
$\boldsymbol{\alpha}, \alpha$	Angular acceleration
α, β, γ	Angles

γ Specific weight

δ Elongation

ε Eccentricity of conic section or of orbit

$\boldsymbol{\lambda}$ Unit vector along a line

η Efficiency

θ Angular coordinate; Eulerian angle; angle; polar coordinate

μ Coefficient of friction

ρ Density; radius of curvature

τ Period; periodic time

ϕ Angle of friction; Eulerian angle; phase angle; angle

φ Phase difference

ψ Eulerian angle

$\boldsymbol{\omega}, \omega$ Angular velocity

ω Circular frequency of forced vibration

Ω Angular velocity of frame of reference

Vector Mechanics
for Engineers
Dynamics

Kinematics of Particles

11.1. Introduction to Dynamics. Chapters 1 to 10 were devoted to *statics*, i.e., to the analysis of bodies at rest. We shall now begin the study of *dynamics*, which is the part of mechanics that deals with the analysis of bodies in motion.

While the study of statics goes back to the time of the Greek philosophers, the first significant contribution to dynamics was made by Galileo (1564–1642). Galileo's experiments on uniformly accelerated bodies led Newton (1642–1727) to formulate his fundamental laws of motion.

Dynamics is divided into two parts: (1) *Kinematics*, which is the study of the geometry of motion; kinematics is used to relate displacement, velocity, acceleration, and time, without reference to the cause of the motion. (2) *Kinetics*, which is the study of the relation existing between the forces acting on a body, the mass of the body, and the motion of the body; kinetics is used to predict the motion caused by given forces or to determine the forces required to produce a given motion.

Chapters 11 to 14 are devoted to the *dynamics of particles*, and Chap. 11 more particularly to the *kinematics of particles*. The use of the word particles does not imply that we shall restrict our study to that of small corpuscles; it rather indicates that in these first chapters we shall study the motion of bodies—possibly as large as cars, rockets, or airplanes—without regard to their size. By saying that the bodies are analyzed as particles, we mean that only their motion as an entire unit will be considered; any rotation about their own mass center will be neglected. There are cases, however, when such a rotation is not negligible; the bodies may not then be considered as particles. The analysis of such motions will be carried out in later chapters dealing with the *dynamics of rigid bodies*.

In the first part of Chap. 11, we shall analyze the rectilinear motion of a particle; that is, we shall determine at every instant the position, velocity, and acceleration of a particle as it moves along a straight line. After first studying the motion of a particle by general methods of analysis, we shall consider two important particular cases, namely, the uniform motion and the uniformly accelerated motion of a particle (Secs. 11.4 and 11.5). We shall then consider in Sec. 11.6 the simultaneous motion of several particles and introduce the concept of the relative motion of one particle with respect to another. The first part of this chapter concludes with a study of graphical methods of analysis and their application to the solution of various problems involving the rectilinear motion of particles (Secs. 11.7 and 11.8).

In the second part of this chapter, we shall analyze the motion of a particle as it moves along a curved path. Since the position, velocity, and acceleration of a particle will be defined as vector quantities, the concept of the derivative of a vector function will be introduced in Sec. 11.10 and added to our mathematical tools. We shall then consider applications where the motion of a particle is defined by the rectangular components of its velocity and acceleration; at this point, the motion of a projectile will be analyzed (Sec. 11.11). In Sec. 11.12, we shall consider the motion of a particle relative to a reference frame in translation. Finally, we shall analyze the curvilinear motion of a particle in terms of components other than rectangular. In Sec. 11.13, we shall introduce the tangential and normal components of the velocity and acceleration of a particle; and in Sec. 11.14, the radial and transverse components of its velocity and acceleration.

RECTILINEAR MOTION OF PARTICLES

11.2. Position, Velocity, and Acceleration. A particle moving along a straight line is said to be in *rectilinear motion*. At any given instant t, the particle will occupy a certain position on the straight line. To define the position P of the particle, we choose a fixed origin O on the straight line and a positive direction along the line. We measure the distance x from O to P and record it with a plus or minus sign, according to whether P is reached from O by moving along the line in the positive or the negative direction. The distance x, with the appropriate sign, completely defines the position of the particle; it is called the *position coordinate* of the particle considered. For example, the position coordinate corresponding to P in Fig. 11.1a is $x = +5$ m, while the coordinate corresponding to P' in Fig. 11.1b is $x' = -2$ m.

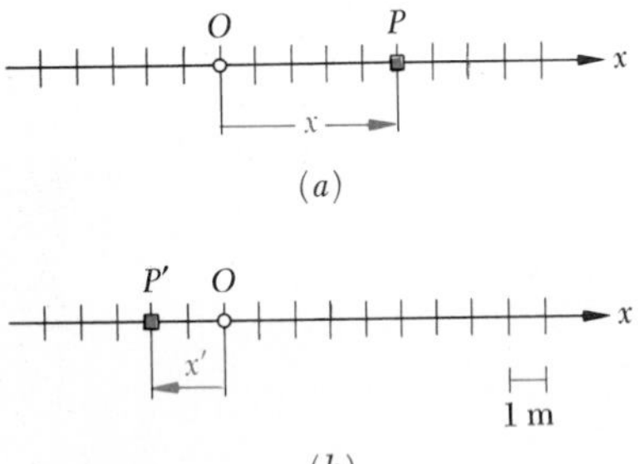

When the position coordinate x of a particle is known for every value of time t, we say that the motion of the particle is known. The "timetable" of the motion may be given in the form of an equation in x and t, such as $x = 6t^2 - t^3$, or in the form of a graph of x versus t as shown in Fig. 11.6. The units most generally used to measure the position coordinate x are the meter (m) in the SI system of units,† and the foot (ft) in the U.S. customary

(a)

(b)

Fig. 11.1

†Cf. Sec. 1.3.

system of units. Time t will generally be measured in seconds (s).

Consider the position P occupied by the particle at time t and the corresponding coordinate x (Fig. 11.2). Consider also the position P' occupied by the particle at a later time $t + \Delta t$; the position coordinate of P' may be obtained by adding to the coordinate x of P the small displacement Δx, which will be positive or negative according to whether P' is to the right or to the left of P. The *average velocity* of the particle over the time interval Δt is defined as the quotient of the displacement Δx and the time interval Δt:

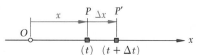

Fig. 11.2

$$\text{Average velocity} = \frac{\Delta x}{\Delta t}$$

If SI units are used, Δx is expressed in meters and Δt in seconds; the average velocity will thus be expressed in meters per second (m/s). If U.S. customary units are used, Δx is expressed in feet and Δt in seconds; the average velocity will then be expressed in feet per second (ft/s).

The *instantaneous velocity* v of the particle at the instant t is obtained from the average velocity by choosing shorter and shorter time intervals Δt and displacements Δx:

$$\text{Instantaneous velocity} = v = \lim_{\Delta t \to 0} \frac{\Delta x}{\Delta t}$$

The instantaneous velocity will also be expressed in m/s or ft/s. Observing that the limit of the quotient is equal, by definition, to the derivative of x with respect to t, we write

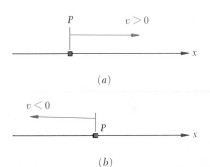

(a)

(b)

Fig. 11.3

$$v = \frac{dx}{dt} \tag{11.1}$$

The velocity v is represented by an algebraic number which may be positive or negative.[†] A positive value of v indicates that x increases, i.e., that the particle moves in the positive direction (Fig. 11.3a); a negative value of v indicates that x decreases, i.e., that the particle moves in the negative direction (Fig. 11.3b). The magnitude of v is known as the *speed* of the particle.

Consider the velocity v of the particle at time t and also its velocity $v + \Delta v$ at a later time $t + \Delta t$ (Fig. 11.4). The *average acceleration* of the particle over the time interval Δt is defined as the quotient of Δv and Δt:

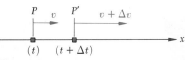

Fig. 11.4

$$\text{Average acceleration} = \frac{\Delta v}{\Delta t}$$

[†] As we shall see in Sec. 11.9, the velocity is actually a vector quantity. However, since we are considering here the rectilinear motion of a particle, where the velocity of the particle has a known and fixed direction, we need only specify the sense and magnitude of the velocity; this may be conveniently done by using a scalar quantity with a plus or minus sign. The same remark will apply to the acceleration of a particle in rectilinear motion.

If SI units are used, Δv is expressed in m/s and Δt in seconds; the average acceleration will thus be expressed in m/s². If U.S. customary units are used, Δv is expressed in ft/s and Δt in seconds; the average acceleration will then be expressed in ft/s².

The *instantaneous acceleration* a of the particle at the instant t is obtained from the average acceleration by choosing smaller and smaller values for Δt and Δv:

$$\text{Instantaneous acceleration} = a = \lim_{\Delta t \to 0} \frac{\Delta v}{\Delta t}$$

The instantaneous acceleration will also be expressed in m/s² or ft/s². The limit of the quotient is by definition the derivative of v with respect to t and measures the rate of change of the velocity. We write

$$a = \frac{dv}{dt} \tag{11.2}$$

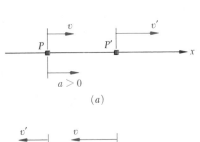

or, substituting for v from (11.1),

$$a = \frac{d^2x}{dt^2} \tag{11.3}$$

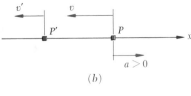

The acceleration a is represented by an algebraic number which may be positive or negative.† A positive value of a indicates that the velocity (i.e., the algebraic number v) increases. This may mean that the particle is moving faster in the positive direction (Fig. 11.5a) or that it is moving more slowly in the negative direction (Fig. 11.5b); in both cases, Δv is positive. A negative value of a indicates that the velocity decreases; either the particle is moving more slowly in the positive direction (Fig. 11.5c) or it is moving faster in the negative direction (Fig. 11.5d).

The term *deceleration* is sometimes used to refer to a when the speed of the particle (i.e., the magnitude of v) decreases; the particle is then moving more slowly. For example, the particle of Fig. 11.5 is decelerated in parts b and c, while it is truly accelerated (i.e., moves faster) in parts a and d.

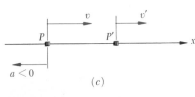

Another expression may be obtained for the acceleration by eliminating the differential dt in Eqs. (11.1) and (11.2). Solving (11.1) for dt, we obtain $dt = dx/v$; carrying into (11.2), we write

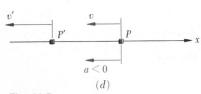

Fig. 11.5

$$a = v\frac{dv}{dx} \tag{11.4}$$

† See footnote, page 461.

Example. Consider a particle moving in a straight line, and assume that its position is defined by the equation

$$x = 6t^2 - t^3$$

where t is expressed in seconds and x in meters. The velocity v at any time t is obtained by differentiating x with respect to t:

$$v = \frac{dx}{dt} = 12t - 3t^2$$

The acceleration a is obtained by differentiating again with respect to t:

$$a = \frac{dv}{dt} = 12 - 6t$$

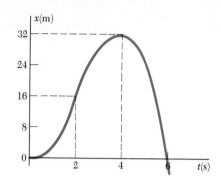

The position coordinate, the velocity, and the acceleration have been plotted against t in Fig. 11.6. The curves obtained are known as *motion curves*. It should be kept in mind, however, that the particle does not move along any of these curves; the particle moves in a straight line. Since the derivative of a function measures the slope of the corresponding curve, the slope of the x–t curve at any given time is equal to the value of v at that time and the slope of the v–t curve is equal to the value of a. Since $a = 0$ at $t = 2$ s, the slope of the v–t curve must be zero at $t = 2$ s; the velocity reaches a maximum at this instant. Also, since $v = 0$ at $t = 0$ and at $t = 4$ s, the tangent to the x–t curve must be horizontal for both these values of t.

A study of the three motion curves of Fig. 11.6 shows that the motion of the particle from $t = 0$ to $t = \infty$ may be divided into four phases:

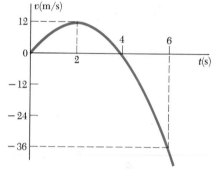

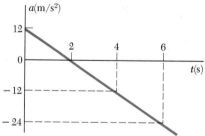

Fig. 11.6

1. The particle starts from the origin, $x = 0$, with no velocity but with a positive acceleration. Under this acceleration, the particle gains a positive velocity and moves in the positive direction. From $t = 0$ to $t = 2$ s, x, v, and a are all positive.

2. At $t = 2$ s, the acceleration is zero; the velocity has reached its maximum value. From $t = 2$ s to $t = 4$ s, v is positive, but a is negative; the particle still moves in the positive direction but more and more slowly; the particle is decelerated.

3. At $t = 4$ s, the velocity is zero; the position coordinate x has reached its maximum value. From then on, both v and a are negative; the particle is accelerated and moves in the negative direction with increasing speed.

4. At $t = 6$ s, the particle passes through the origin; its coordinate x is then zero, while the total distance traveled since the beginning of the motion is 64 m. For values of t larger than 6 s, x, v, and a will all be negative. The particle keeps moving in the negative direction, away from O, faster and faster.

11.3. Determination of the Motion of a Particle. We saw in the preceding section that the motion of a particle is said to be known if the position of the particle is known for every value of the time t. In practice, however, a motion is seldom defined by a relation between x and t. More often, the conditions of the motion will be specified by the type of acceleration that the particle possesses. For example, a freely falling body will have a constant acceleration, directed downward and equal to 9.81 m/s², or 32.2 ft/s²; a mass attached to a spring which has been stretched will have an acceleration proportional to the instantaneous elongation of the spring measured from the equilibrium position; etc. In general, the acceleration of the particle may be expressed as a function of one or more of the variables x, v, and t. In order to determine the position coordinate x in terms of t, it will thus be necessary to perform two successive integrations.

We shall consider three common classes of motion:

1. $a = f(t)$. *The Acceleration Is a Given Function of t.* Solving (11.2) for dv and substituting $f(t)$ for a, we write

$$dv = a\,dt$$
$$dv = f(t)\,dt$$

Integrating both members, we obtain the equation

$$\int dv = \int f(t)\,dt$$

which defines v in terms of t. It should be noted, however, that an arbitrary constant will be introduced as a result of the integration. This is due to the fact that there are many motions which correspond to the given acceleration $a = f(t)$. In order to uniquely define the motion of the particle, it is necessary to specify the *initial conditions* of the motion, i.e., the value v_0 of the velocity and the value x_0 of the position coordinate at $t = 0$. Replacing the indefinite integrals by *definite integrals* with lower limits corresponding to the initial conditions $t = 0$ and $v = v_0$ and upper limits corresponding to $t = t$ and $v = v$, we write

$$\int_{v_0}^{v} dv = \int_{0}^{t} f(t)\,dt$$

$$v - v_0 = \int_{0}^{t} f(t)\,dt$$

which yields v in terms of t.

We shall now solve (11.1) for dx:

$$dx = v\,dt$$

and substitute for v the expression just obtained. Both members are then integrated, the left-hand member with respect to x from x = x_0 to x = x, and the right-hand member with respect to t from t = 0 to t = t. The position coordinate x is thus obtained in terms of t; the motion is completely determined.

Two important particular cases will be studied in greater detail in Secs. 11.4 and 11.5: the case when $a = 0$, corresponding to a *uniform motion*, and the case when $a =$ constant, corresponding to a *uniformly accelerated motion*.

2. $a = f(x)$. *The Acceleration Is a Given Function of x*. Rearranging Eq. (11.4) and substituting $f(x)$ for a, we write

$$v \, dv = a \, dx$$
$$v \, dv = f(x) \, dx$$

Since each member contains only one variable, we may integrate the equation. Denoting again by v_0 and x_0, respectively, the initial values of the velocity and of the position coordinate, we obtain

$$\int_{v_0}^{v} v \, dv = \int_{x_0}^{x} f(x) \, dx$$

$$\tfrac{1}{2}v^2 - \tfrac{1}{2}v_0^2 = \int_{x_0}^{x} f(x) \, dx$$

which yields v in terms of x. We now solve (11.1) for dt:

$$dt = \frac{dx}{v}$$

and substitute for v the expression just obtained. Both members may be integrated to obtain the desired relation between x and t. However, in most cases this last integration cannot be performed analytically and one must resort to a numerical method of integration.

3. $a = f(v)$. *The Acceleration Is a Given Function of v*. We may then substitute $f(v)$ for a either in (11.2) or in (11.4) to obtain either of the following relations:

$$f(v) = \frac{dv}{dt} \qquad f(v) = v\frac{dv}{dx}$$

$$dt = \frac{dv}{f(v)} \qquad dx = \frac{v \, dv}{f(v)}$$

Integration of the first equation will yield a relation between v and t; integration of the second equation will yield a relation between v and x. Either of these relations may be used in conjunction with Eq. (11.1) to obtain the relation between x and t which characterizes the motion of the particle.

SAMPLE PROBLEM 11.1

The position of a particle which moves along a straight line is defined by the relation $x = t^3 - 6t^2 - 15t + 40$, where x is expressed in feet and t in seconds. Determine (a) the time at which the velocity will be zero, (b) the position and distance traveled by the particle at that time, (c) the acceleration of the particle at that time, (d) the distance traveled by the particle from $t = 4$ s to $t = 6$ s.

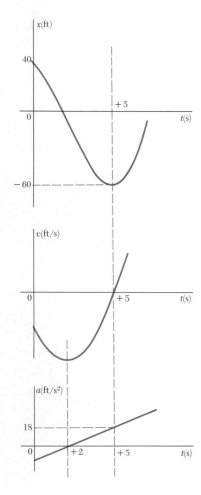

Solution. The equations of motion are

$$x = t^3 - 6t^2 - 15t + 40 \qquad (1)$$

$$v = \frac{dx}{dt} = 3t^2 - 12t - 15 \qquad (2)$$

$$a = \frac{dv}{dt} = 6t - 12 \qquad (3)$$

a. Time at Which $v = 0$. We make $v = 0$ in (2):

$$3t^2 - 12t - 15 = 0 \qquad t = -1\,\text{s} \qquad \text{and} \qquad t = +5\,\text{s} \blacktriangleleft$$

Only the root $t = +5$ s corresponds to a time after the motion has begun: for $t < 5$ s, $v < 0$, the particle moves in the negative direction; for $t > 5$ s, $v > 0$, the particle moves in the positive direction.

b. Position and Distance Traveled When $v = 0$. Carrying $t = +5$ s into (1), we have

$$x_5 = (5)^3 - 6(5)^2 - 15(5) + 40 \qquad\qquad x_5 = -60\,\text{ft} \blacktriangleleft$$

The initial position at $t = 0$ was $x_0 = +40$ ft. Since $v \neq 0$ during the interval $t = 0$ to $t = 5$ s, we have

$$\text{Distance traveled} = x_5 - x_0 = -60\,\text{ft} - 40\,\text{ft} = -100\,\text{ft}$$

$$\text{Distance traveled} = 100\,\text{ft in the negative direction} \blacktriangleleft$$

c. Acceleration When $v = 0$. We carry $t = +5$ s into (3):

$$a_5 = 6(5) - 12 \qquad\qquad a_5 = +18\,\text{ft/s}^2 \blacktriangleleft$$

d. Distance Traveled from $t = 4$ s to $t = 6$ s. Since the particle moves in the negative direction from $t = 4$ s to $t = 5$ s and in the positive direction from $t = 5$ s to $t = 6$ s, we shall compute separately the distance traveled during each of these time intervals.

From $t = 4$ s to $t = 5$ s: $\quad x_5 = -60$ ft

$$x_4 = (4)^3 - 6(4)^2 - 15(4) + 40 = -52\,\text{ft}$$

$$\text{Distance traveled} = x_5 - x_4 = -60\,\text{ft} - (-52\,\text{ft}) = -8\,\text{ft}$$
$$= 8\,\text{ft in the negative direction}$$

From $t = 5$ s to $t = 6$ s: $\quad x_5 = -60$ ft

$$x_6 = (6)^3 - 6(6)^2 - 15(6) + 40 = -50\,\text{ft}$$

$$\text{Distance traveled} = x_6 - x_5 = -50\,\text{ft} - (-60\,\text{ft}) = +10\,\text{ft}$$
$$= 10\,\text{ft in the positive direction}$$

Total distance traveled from $t = 4$ s to $t = 6$ s is

$$8\,\text{ft} + 10\,\text{ft} = 18\,\text{ft} \blacktriangleleft$$

SAMPLE PROBLEM 11.2

A ball is tossed with a velocity of 10 m/s directed vertically upward from a window located 20 m above the ground. Knowing that the acceleration of the ball is constant and equal to 9.81 m/s² downward, determine (*a*) the velocity v and elevation y of the ball above the ground at any time t, (*b*) the highest elevation reached by the ball and the corresponding value of t, (*c*) the time when the ball will hit the ground and the corresponding velocity. Draw the v–t and y–t curves.

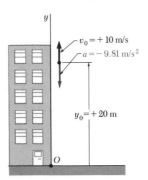

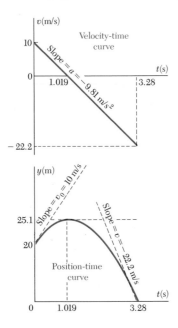

a. Velocity and Elevation. The y axis measuring the position coordinate (or elevation) is chosen with its origin O on the ground and its positive sense upward. The value of the acceleration and the initial values of v and y are as indicated. Substituting for a in $a = dv/dt$ and noting that at $t = 0$, $v_0 = +10$ m/s, we have

$$\frac{dv}{dt} = a = -9.81 \text{ m/s}^2$$

$$\int_{v_0=10}^{v} dv = -\int_{0}^{t} 9.81 \, dt$$

$$[v]_{10}^{v} = -[9.81t]_{0}^{t}$$

$$v - 10 = -9.81t$$

$$v = 10 - 9.81t \quad (1) \quad \blacktriangleleft$$

Substituting for v in $v = dy/dt$ and noting that at $t = 0$, $y_0 = 20$ m, we have

$$\frac{dy}{dt} = v = 10 - 9.81t$$

$$\int_{y_0=20}^{y} dy = \int_{0}^{t} (10 - 9.81t) \, dt$$

$$[y]_{20}^{y} = [10t - 4.905t^2]_{0}^{t}$$

$$y - 20 = 10t - 4.905t^2$$

$$y = 20 + 10t - 4.905t^2 \quad (2) \quad \blacktriangleleft$$

b. Highest Elevation. When the ball reaches its highest elevation, we have $v = 0$. Substituting into (1), we obtain

$$10 - 9.81t = 0 \qquad\qquad t = 1.019 \text{ s} \quad \blacktriangleleft$$

Carrying $t = 1.019$ s into (2), we have

$$y = 20 + 10(1.019) - 4.905(1.019)^2 \qquad y = 25.1 \text{ m} \quad \blacktriangleleft$$

c. Ball Hits the Ground. When the ball hits the ground, we have $y = 0$. Substituting into (2), we obtain

$$20 + 10t - 4.905t^2 = 0 \qquad t = -1.243 \text{ s} \quad \text{and} \quad t = +3.28 \text{ s} \quad \blacktriangleleft$$

Only the root $t = +3.28$ s corresponds to a time after the motion has begun. Carrying this value of t into (1), we have

$$v = 10 - 9.81(3.28) = -22.2 \text{ m/s} \qquad v = 22.2 \text{ m/s} \downarrow \quad \blacktriangleleft$$

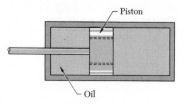

The brake mechanism used to reduce recoil in certain types of guns consists essentially of a piston which is attached to the barrel and may move in a fixed cylinder filled with oil. As the barrel recoils with an initial velocity v_0, the piston moves and oil is forced through orifices in the piston, causing the piston and the barrel to decelerate at a rate proportional to their velocity, that is, $a = -kv$. Express (a) v in terms of t, (b) x in terms of t, (c) v in terms of x. Draw the corresponding motion curves.

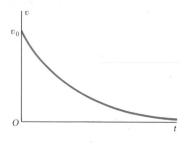

a. v in Terms of t. Substituting $-kv$ for a in the fundamental formula defining acceleration, $a = dv/dt$, we write

$$-kv = \frac{dv}{dt} \qquad \frac{dv}{v} = -k\,dt \qquad \int_{v_0}^{v} \frac{dv}{v} = -k \int_0^t dt$$

$$\ln\frac{v}{v_0} = -kt \qquad\qquad v = v_0 e^{-kt} \quad \blacktriangleleft$$

b. x in Terms of t. Substituting the expression just obtained for v into $v = dx/dt$, we write

$$v_0 e^{-kt} = \frac{dx}{dt}$$

$$\int_0^x dx = v_0 \int_0^t e^{-kt}\,dt$$

$$x = -\frac{v_0}{k}[e^{-kt}]_0^t = -\frac{v_0}{k}(e^{-kt} - 1)$$

$$x = \frac{v_0}{k}(1 - e^{-kt}) \quad \blacktriangleleft$$

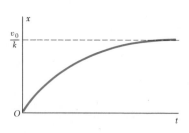

c. v in Terms of x. Substituting $-kv$ for a in $a = v\,dv/dx$, we write

$$-kv = v\frac{dv}{dx}$$

$$dv = -k\,dx$$

$$\int_{v_0}^{v} dv = -k \int_0^x dx$$

$$v - v_0 = -kx \qquad\qquad v = v_0 - kx \quad \blacktriangleleft$$

Check. Part c could have been solved by eliminating t from the answers obtained for parts a and b. This alternate method may be used as a check. From part a we obtain $e^{-kt} = v/v_0$; substituting in the answer of part b, we obtain

$$x = \frac{v_0}{k}(1 - e^{-kt}) = \frac{v_0}{k}\left(1 - \frac{v}{v_0}\right) \qquad v = v_0 - kx \qquad \text{(checks)}$$

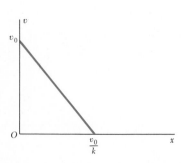

Problems

11.1 The motion of a particle is defined by the relation $x = t^4 - 12t^2 - 40$, where x is expressed in meters and t in seconds. Determine the position, velocity, and acceleration when $t = 2$ s.

11.2 The motion of a particle is defined by the relation $x = t^3 - 9t^2 + 24t - 6$, where x is expressed in meters and t in seconds. Determine the position, velocity, and acceleration when $t = 5$ s.

11.3 The motion of a particle is defined by the relation $x = 2t^3 - 8t^2 + 5t + 15$, where x is expressed in inches and t in seconds. Determine the position, velocity, and acceleration when $t = 3$ s.

11.4 The motion of a particle is defined by the relation $x = 2t^3 - 6t^2 + 10$, where x is expressed in feet and t in seconds. Determine the time, position, and acceleration when $v = 0$.

11.5 The motion of a particle is defined by the relation $x = t^3 - 12t^2 + 36t + 30$, where x is expressed in meters and t in seconds. Determine the time, position, and acceleration when $v = 0$.

11.6 The motion of a particle is defined by the relation $x = t^3 - 6t^2 + 9t + 5$, where x is expressed in meters and t in seconds. Determine (*a*) when the velocity is zero, (*b*) the position, acceleration, and total distance traveled when $t = 5$ s.

11.7 The motion of a particle is defined by the relation $x = 2t^3 - 15t^2 + 24t + 4$, where x is expressed in meters and t in seconds. Determine (*a*) when the velocity is zero, (*b*) the position and the total distance traveled when the acceleration is zero.

11.8 The acceleration of a particle is defined by the relation $a = -4$ ft/s^2. If $v = +24$ ft/s and $x = 0$ when $t = 0$, determine the velocity, position, and total distance traveled when $t = 8$ s.

11.9 The acceleration of a particle is directly proportional to the time t. At $t = 0$, the velocity of the particle is $v = -12$ m/s. Knowing that $v = 0$ and $x = 15$ m when $t = 4$ s, write the equations of motion of the particle.

11.10 The acceleration of a particle is defined by the relation $a = 9 - 3t^2$. The particle starts at $t = 0$ with $v = 0$ and $x = 5$ m. Determine (*a*) the time when the velocity is again zero, (*b*) the position and velocity when $t = 4$ s, (*c*) the total distance traveled by the particle from $t = 0$ to $t = 4$ s.

11.11 The acceleration of a particle is defined by the relation $a = kt^2$. (*a*) Knowing that $v = -32$ ft/s when $t = 0$ and that $v = +32$ ft/s when $t = 4$ s, determine the constant k. (*b*) Write the equations of motion, knowing also that $x = 0$ when $t = 4$ s.

11.12 The acceleration of a particle is defined by the relation $a = -kx^{-2}$. The particle starts with no initial velocity at $x = 800$ mm, and it is observed that its velocity is 6 m/s when $x = 500$ mm. Determine (a) the value of k, (b) the velocity of the particle when $x = 250$ mm.

11.13 The acceleration of a particle is defined by the relation $a = -k/x$. It has been experimentally determined that $v = 5$ m/s when $x = 200$ mm and that $v = 3$ m/s when $x = 400$ mm. Determine (a) the velocity of the particle when $x = 500$ mm, (b) the position of the particle at which its velocity is zero.

11.14 The acceleration of an oscillating particle is defined by the relation $a = -kx$. Determine (a) the value of k such that $v = 15$ in./s when $x = 0$ and $x = 3$ in. when $v = 0$, (b) the speed of the particle when $x = 2$ in.

11.15 The acceleration of a particle is defined by the relation $a = 25 - 3x^2$, where a is expressed in in./s^2 and x in inches. The particle starts with no initial velocity at the position $x = 0$. Determine (a) the velocity when $x = 2$ in., (b) the position where the velocity is again zero, (c) the position where the velocity is maximum.

11.16 The acceleration of a particle is defined by the relation $a = -4x(1 + kx^2)$, where a is expressed in m/s^2 and x in meters. Knowing that $v = 17$ m/s when $x = 0$, determine the velocity when $x = 4$ m, for (a) $k = 0$, (b) $k = 0.015$, (c) $k = -0.015$.

11.17 The acceleration of a particle is defined by the relation $a = -60x^{-1.5}$, where a is expressed in m/s^2 and x in meters. Knowing that the particle starts with no initial velocity at $x = 4$ m, determine the velocity of the particle when (a) $x = 2$ m, (b) $x = 1$ m, (c) $x = 100$ mm.

11.18 The acceleration of a particle is defined by the relation $a = -0.4v$, where a is expressed in in./s^2 and v in in./s. Knowing that at $t = 0$ the velocity is 30 in./s, determine (a) the distance the particle will travel before coming to rest, (b) the time required for the particle to come to rest, (c) the time required for the velocity of the particle to be reduced to 1 percent of its initial value.

11.19 The acceleration of a particle is defined by the relation $a = -kv^2$, where a is expressed in ft/s^2 and v in ft/s. The particle starts at $x = 0$ with a velocity of 25 ft/s and when $x = 40$ ft the velocity is found to be 20 ft/s. Determine the distance the particle will travel (a) before its velocity drops to 10 ft/s, (b) before it comes to rest.

11.20 Solve Prob. 11.19, assuming that the acceleration of the particle is defined by the relation $a = -kv^3$.

11.21 The acceleration of a particle falling through the atmosphere is defined by the relation $a = g(1 - k^2v^2)$. Knowing that the particle starts at $t = 0$ with no initial velocity, (a) show that the velocity at time t is $v = (1/k)$ tanh kgt, (b) write an equation defining the velocity of the particle for any value of the distance x through which it has fallen. (c) Why is $v_t = 1/k$ called the terminal velocity?

11.22 The acceleration of a particle is defined by the relation $a = -0.02v^{1.75}$, where a is expressed in m/s² and v in m/s. Knowing that the initial velocity of the particle is 15 m/s at $x = 0$, determine (*a*) the position where the velocity is 14 m/s, (*b*) the velocity of the particle when $x = 100$ m.

11.23 The acceleration of a particle is defined by the relation $a = -kv^{1.5}$. The particle starts at $t = 0$ and $x = 0$ with an initial velocity v_0. (*a*) Show that the velocity and position coordinate at any time t are related by the equation $x/t = \sqrt{v_0 v}$. (*b*) Knowing that for $v_0 = 25$ m/s the particle comes to rest after traveling 50 m, determine the velocity of the particle and the time when $x = 30$ m.

11.24 The velocity of a particle is defined by the relation $v = 8 - 0.02x$, where v is expressed in m/s and x in meters. Knowing that $x = 0$ at $t = 0$, determine (*a*) the distance traveled before the particle comes to rest, (*b*) the acceleration at $t = 0$, (*c*) the time when $x = 100$ m.

11.25 A projectile enters a resisting medium at $x = 0$ with an initial velocity $v_0 = 1200$ ft/s and travels 4 in. before coming to rest. Assuming that the velocity of the projectile was defined by the relation $v = v_0 - kx$, where v is expressed in ft/s and x in feet, determine (*a*) the initial acceleration of the projectile, (*b*) the time required for the projectile to penetrate 3.75 in. into the resisting medium.

11.26 The acceleration due to gravity at an altitude y above the surface of the earth may be expressed as

$$a = \frac{-9.81}{\left(1 + \dfrac{y}{6.37 \times 10^6}\right)^2}$$

where a is measured in m/s² and y in meters. Using this expression, compute the height reached by a bullet fired vertically upward from the surface of the earth with the following initial velocities: (*a*) 200 m/s, (*b*) 2000 m/s, (*c*) 11.18 km/s.

11.27 The acceleration due to gravity of a particle falling toward the earth is $a = -gR^2/r^2$, where r is the distance from the *center* of the earth to the particle, R is the radius of the earth, and g is the acceleration due to gravity at the surface of the earth. Derive an expression for the *escape velocity*, i.e., for the minimum velocity with which a particle should be projected vertically upward from the surface of the earth if it is not to return to the earth. (*Hint.* $v = 0$ for $r = \infty$.)

11.28 The acceleration of a particle is $a = k \sin (\pi t/T)$. Knowing that both the velocity and the position coordinate of the particle are zero when $t = 0$, determine (*a*) the equations of motion, (*b*) the maximum velocity, (*c*) the position at $t = 2T$, (*d*) the average velocity during the interval $t = 0$ to $t = 2T$.

11.29 The acceleration of a collar moving in a straight line is defined by the relation $a = 50 \sin \frac{1}{2}\pi t$, where a is expressed in mm/s² and t in seconds. Knowing that $x = 0$ and $v = 0$ when $t = 0$, determine (*a*) the maximum velocity of the collar, (*b*) its position at $t = 4$ s, (*c*) its average velocity over the interval $0 < t < 4$ s.

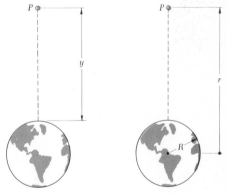

Fig. P11.26   **Fig. P11.27**

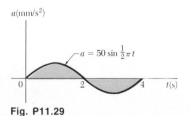

Fig. P11.29

11.4. Uniform Rectilinear Motion. Uniform rectilinear motion is a type of straight-line motion which is frequently encountered in practical applications. In this motion, the acceleration a of the particle is zero for every value of t. The velocity v is therefore constant, and Eq. (11.1) becomes

$$\frac{dx}{dt} = v = \text{constant}$$

The position coordinate x is obtained by integrating this equation. Denoting by x_0 the initial value of x, we write

$$\int_{x_0}^{x} dx = v \int_{0}^{t} dt$$

$$x - x_0 = vt$$

$$\boxed{x = x_0 + vt} \tag{11.5}$$

This equation may be used *only if the velocity of the particle is known to be constant.*

11.5. Uniformly Accelerated Rectilinear Motion. Uniformly accelerated rectilinear motion is another common type of motion. In this motion, the acceleration a of the particle is constant, and Eq. (11.2) becomes

$$\frac{dv}{dt} = a = \text{constant}$$

The velocity v of the particle is obtained by integrating this equation:

$$\int_{v_0}^{v} dv = a \int_{0}^{t} dt$$

$$v - v_0 = at$$

$$\boxed{v = v_0 + at} \tag{11.6}$$

where v_0 is the initial velocity. Substituting for v into (11.1), we write

$$\frac{dx}{dt} = v_0 + at$$

Denoting by x_0 the initial value of x and integrating, we have

$$\int_{x_0}^{x} dx = \int_{0}^{t} (v_0 + at)\, dt$$

$$x - x_0 = v_0 t + \tfrac{1}{2}at^2$$

$$\boxed{x = x_0 + v_0 t + \tfrac{1}{2}at^2} \tag{11.7}$$

We may also use Eq. (11.4) and write

$$v\frac{dv}{dx} = a = \text{constant}$$

$$v\, dv = a\, dx$$

Integrating both sides, we obtain

$$\int_{v_0}^{v} v \, dv = a \int_{x_0}^{x} dx$$

$$\tfrac{1}{2}(v^2 - v_0^2) = a(x - x_0)$$

$$v^2 = v_0^2 + 2a(x - x_0) \qquad (11.8)$$

The three equations we have derived provide useful relations among position coordinate, velocity, and time in the case of a uniformly accelerated motion, as soon as appropriate values have been substituted for a, v_0, and x_0. The origin O of the x axis should first be defined and a positive direction chosen along the axis; this direction will be used to determine the signs of a, v_0, and x_0. Equation (11.6) relates v and t and should be used when the value of v corresponding to a given value of t is desired, or inversely. Equation (11.7) relates x and t; Eq. (11.8) relates v and x. An important application of uniformly accelerated motion is the motion of a *freely falling body*. The acceleration of a freely falling body (usually denoted by g) is equal to 9.81 m/s^2, or 32.2 ft/s^2.

It is important to keep in mind that the three equations above may be used *only when the acceleration of the particle is known to be constant.* If the acceleration of the particle is variable, its motion should be determined from the fundamental equations (11.1) to (11.4), according to the methods outlined in Sec. 11.3.

11.6. Motion of Several Particles. When several particles move independently along the same line, independent equations of motion may be written for each particle. Whenever possible, time should be recorded from the same initial instant for all particles, and displacements should be measured from the same origin and in the same direction. In other words, a single clock and a single measuring tape should be used.

Relative Motion of Two Particles. Consider two particles A and B moving along the same straight line (Fig. 11.7). If the position coordinates x_A and x_B are measured from the same origin, the difference $x_B - x_A$ defines the *relative position coordinate of B with respect to A* and is denoted by $x_{B/A}$. We write

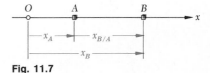

Fig. 11.7

$$x_{B/A} = x_B - x_A \qquad \text{or} \qquad x_B = x_A + x_{B/A} \qquad (11.9)$$

A positive sign for $x_{B/A}$ means that B is to the right of A; a negative sign means that B is to the left of A, regardless of the position of A and B with respect to the origin.

The rate of change of $x_{B/A}$ is known as the *relative velocity of B with respect to A* and is denoted by $v_{B/A}$. Differentiating (11.9), we write

$$v_{B/A} = v_B - v_A \qquad \text{or} \qquad v_B = v_A + v_{B/A} \qquad (11.10)$$

A positive sign for $v_{B/A}$ means that B is *observed from A* to move in the positive direction; a negative sign, that it is observed to move in the negative direction.

The rate of change of $v_{B/A}$ is known as the *relative acceleration of B with respect to A* and is denoted by $a_{B/A}$. Differentiating (11.10), we obtain†

$$a_{B/A} = a_B - a_A \qquad \text{or} \qquad \boxed{a_B = a_A + a_{B/A}} \tag{11.11}$$

Dependent Motions. Sometimes, the position of a particle will depend upon the position of another or of several other particles. The motions are then said to be dependent. For example, the position of block B in Fig. 11.8 depends upon the position of block A. Since the rope $ACDEFG$ is of constant length, and since the lengths of the portions of rope CD and EF wrapped around the pulleys remain constant, it follows that the sum of the lengths of the segments AC, DE, and FG is constant. Observing that the length of the segment AC differs from x_A only by a constant, and that, similarly, the lengths of the segments DE and FG differ from x_B only by a constant, we write

$$x_A + 2x_B = \text{constant}$$

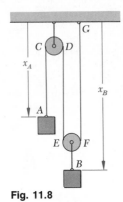

Fig. 11.8

Since only one of the two coordinates x_A and x_B may be chosen arbitrarily, we say that the system shown in Fig. 11.8 has *one degree of freedom*. From the relation between the position coordinates x_A and x_B, it follows that if x_A is given an increment Δx_A, that is, if block A is lowered by an amount Δx_A, the coordinate x_B will receive an increment $\Delta x_B = -\frac{1}{2}\Delta x_A$; that is, block B will rise by half the same amount; this may easily be checked directly from Fig. 11.8.

In the case of the three blocks of Fig. 11.9, we may again observe that the length of the rope which passes over the pulleys is constant, and thus that the following relation must be satisfied by the position coordinates of the three blocks:

$$2x_A + 2x_B + x_C = \text{constant}$$

Since two of the coordinates may be chosen arbitrarily, we say that the system shown in Fig. 11.9 has *two degrees of freedom*.

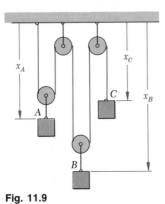

Fig. 11.9

When the relation existing between the position coordinates of several particles is *linear*, a similar relation holds between the velocities and between the accelerations of the particles. In the case of the blocks of Fig. 11.9, for instance, we differentiate twice the equation obtained and write

$$2\frac{dx_A}{dt} + 2\frac{dx_B}{dt} + \frac{dx_C}{dt} = 0 \qquad \text{or} \qquad 2v_A + 2v_B + v_C = 0$$

$$2\frac{dv_A}{dt} + 2\frac{dv_B}{dt} + \frac{dv_C}{dt} = 0 \qquad \text{or} \qquad 2a_A + 2a_B + a_C = 0$$

† Note that the product of the subscripts A and B/A used in the right-hand member of Eqs. (11.9), (11.10), and (11.11) is equal to the subscript B used in their left-hand member.

SAMPLE PROBLEM 11.4

A ball is thrown vertically upward from the 12-m level in an elevator shaft, with an initial velocity of 18 m/s. At the same instant an open-platform elevator passes the 5-m level, moving upward with a constant velocity of 2 m/s. Determine (a) when and where the ball will hit the elevator, (b) the relative velocity of the ball with respect to the elevator when the ball hits the elevator.

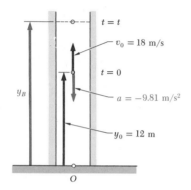

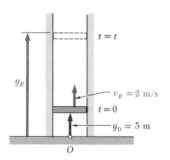

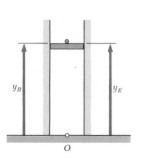

Motion of Ball. Since the ball has a constant acceleration, its motion is *uniformly accelerated*. Placing the origin O of the y axis at ground level and choosing its positive direction upward, we find that the initial position is $y_0 = +12$ m, the initial velocity is $v_0 = +18$ m/s, and the acceleration is $a = -9.81$ m/s². Substituting these values in the equations for uniformly accelerated motion, we write

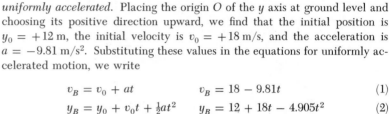

$$v_B = v_0 + at \qquad\qquad v_B = 18 - 9.81t \qquad\qquad (1)$$
$$y_B = y_0 + v_0 t + \tfrac{1}{2}at^2 \qquad y_B = 12 + 18t - 4.905t^2 \qquad (2)$$

Motion of Elevator. Since the elevator has a constant velocity, its motion is *uniform*. Again placing the origin O at the ground level and choosing the positive direction upward, we note that $y_0 = +5$ m and write

$$v_E = +2 \text{ m/s} \qquad\qquad (3)$$
$$y_E = y_0 + v_E t \qquad y_E = 5 + 2t \qquad\qquad (4)$$

Ball Hits Elevator. We first note that the same time t and the same origin O were used in writing the equations of motion of both the ball and the elevator. We see from the figure that when the ball hits the elevator,

$$y_E = y_B \qquad\qquad (5)$$

Substituting for y_E and y_B from (2) and (4) into (5), we have

$$5 + 2t = 12 + 18t - 4.905t^2$$
$$t = -0.39 \text{ s} \quad\text{and}\quad t = 3.65 \text{ s} \qquad \blacktriangleleft$$

Only the root $t = 3.65$ s corresponds to a time after the motion has begun. Substituting this value into (4), we have

$$y_E = 5 + 2(3.65) = 12.30 \text{ m}$$
Elevation from ground $= 12.30$ m $\qquad \blacktriangleleft$

The relative velocity of the ball with respect to the elevator is

$$v_{B/E} = v_B - v_E = (18 - 9.81t) - 2 = 16 - 9.81t$$

When the ball hits the elevator at time $t = 3.65$ s, we have

$$v_{B/E} = 16 - 9.81(3.65) \qquad v_{B/E} = -19.81 \text{ m/s} \qquad \blacktriangleleft$$

The negative sign means that the ball is observed from the elevator to be moving in the negative sense (downward).

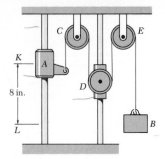

SAMPLE PROBLEM 11.5

Collar A and block B are connected by a cable passing over three pulleys C, D, and E as shown. Pulleys C and E are fixed, while D is attached to a collar which is pulled downward with a constant velocity of 3 in./s. At $t = 0$, collar A starts moving downward from position K with a constant acceleration and no initial velocity. Knowing that the velocity of collar A is 12 in./s as it passes through point L, determine the change in elevation, the velocity, and the acceleration of block B when collar A passes through L.

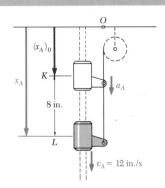

Motion of Collar A. We place the origin O at the upper horizontal surface and choose the positive direction downward. We observe that when $t = 0$, collar A is at position K and $(v_A)_0 = 0$. Since $v_A = 12$ in./s and $x_A - (x_A)_0 = 8$ in. when the collar passes through L, we write

$$v_A^2 = (v_A)_0^2 + 2a_A[x_A - (x_A)_0] \qquad (12)^2 = 0 + 2a_A(8)$$
$$a_A = 9 \text{ in./s}^2$$

The time at which collar A reaches point L is obtained by writing

$$v_A = (v_A)_0 + a_A t \qquad 12 = 0 + 9t \qquad t = 1.333 \text{ s}$$

Motion of Pulley D. Recalling that the positive direction is downward, we write

$$a_D = 0 \qquad v_D = 3 \text{ in./s} \qquad x_D = (x_D)_0 + v_D t = (x_D)_0 + 3t$$

When collar A reaches L, at $t = 1.333$ s, we have

$$x_D = (x_D)_0 + 3(1.333) = (x_D)_0 + 4$$

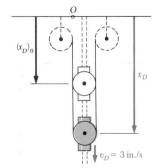

Thus, $\qquad\qquad\qquad\qquad x_D - (x_D)_0 = 4 \text{ in.}$

Motion of Block B. We note that the total length of cable $ACDEB$ differs from the quantity $(x_A + 2x_D + x_B)$ only by a constant. Since the cable length is constant during the motion, this quantity must also remain constant. Thus considering the times $t = 0$ and $t = 1.333$ s, we write

$$x_A + 2x_D + x_B = (x_A)_0 + 2(x_D)_0 + (x_B)_0 \qquad (1)$$
$$[x_A - (x_A)_0] + 2[x_D - (x_D)_0] + [x_B - (x_B)_0] = 0 \qquad (2)$$

But we know that $x_A - (x_A)_0 = 8$ in. and $x_D - (x_D)_0 = 4$ in.; substituting these values in (2), we find

$$8 + 2(4) + [x_B - (x_B)_0] = 0 \qquad x_B - (x_B)_0 = -16 \text{ in.}$$

Thus: Change in elevation of $B = 16$ in. ↑ ◀

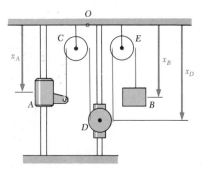

Differentiating (1) twice, we obtain equations relating the velocities and the accelerations of A, B, and D. Substituting for the velocities and accelerations of A and D at $t = 1.333$ s, we have

$$v_A + 2v_D + v_B = 0: \qquad 12 + 2(3) + v_B = 0$$
$$v_B = -18 \text{ in./s} \qquad v_B = 18 \text{ in./s} ↑ ◀$$

$$a_A + 2a_D + a_B = 0: \qquad 9 + 2(0) + a_B = 0$$
$$a_B = -9 \text{ in./s}^2 \qquad a_B = 9 \text{ in./s}^2 ↑ ◀$$

Problems

11.30 A stone is thrown vertically upward from a point on a bridge located 40 m above the water. Knowing that it strikes the water 4 s after release, determine (*a*) the speed with which the stone was thrown upward, (*b*) the speed with which the stone strikes the water.

11.31 A motorist is traveling at 54 km/h when she observes that a traffic light 240 m ahead of her turns red. The traffic light is timed to stay red for 24 s. If the motorist wishes to pass the light without stopping just as it turns green again, determine (*a*) the required uniform deceleration of the car, (*b*) the speed of the car as it passes the light.

11.32 An automobile travels 1200 ft in 30 s while being accelerated at a constant rate of 1.8 ft/s². Determine (*a*) its initial velocity, (*b*) its final velocity, (*c*) the distance traveled during the first 10 s.

11.33 A stone is released from an elevator moving up at a speed of 15 ft/s and reaches the bottom of the shaft in 3 s. (*a*) How high was the elevator when the stone was released? (*b*) With what speed does the stone strike the bottom of the shaft?

11.34 A bus is accelerated at the rate of 0.75 m/s² as it travels from *A* to *B*. Knowing that the speed of the bus was $v_0 = 27$ km/h as it passed *A*, determine (*a*) the time required for the bus to reach *B*, (*b*) the corresponding speed as it passes *B*.

11.35 Solve Prob. 11.34, assuming that the velocity of the bus as it passed *A* was 18 km/h.

11.36 Automobile *A* starts from *O* and accelerates at the constant rate of 0.75 m/s². A short time later it is passed by bus *B* which is traveling in the opposite direction at a constant speed of 6 m/s. Knowing that bus *B* passes point *O* 20 s after automobile *A* started from there, determine when and where the vehicles passed each other.

11.37 Two automobiles *A* and *B* are traveling in the same direction in adjacent highway lanes. Automobile *B* is stopped when it is passed by *A*, which travels at a constant speed of 15 mi/h. Two seconds later automobile *B* starts and accelerates at a constant rate of 3 ft/s². Determine (*a*) when and where *B* will overtake *A*, (*b*) the speed of *B* at that time.

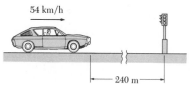

Fig. P11.31

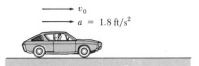

Fig. P11.32

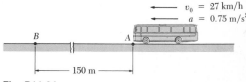

Fig. P11.34

Fig. P11.36

11.38 Automobiles A and B are traveling in adjacent highway lanes and at $t = 0$ have the positions and speeds shown. Knowing that automobile A has a constant acceleration of 1.8 ft/s^2 and that B has a constant deceleration of 1.2 ft/s^2, determine (a) when and where A will overtake B, (b) the speed of each automobile at that time.

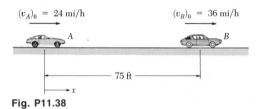

Fig. P11.38

11.39 An open-platform elevator is moving down a mine shaft with a constant velocity v_e when the elevator platform hits and dislodges a stone. (a) Assuming that the stone starts falling with no initial velocity, show that the stone will hit the platform with a relative velocity of magnitude v_e. (b) If $v_e = 7.5 \text{ m/s}$, determine when and where the stone will hit the elevator platform.

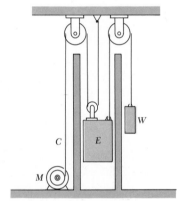

Fig. P11.40 and P11.41

11.40 The elevator shown in the figure moves downward with a constant velocity of 5 m/s. Determine (a) the velocity of the cable C, (b) the velocity of the counterweight W, (c) the relative velocity of the cable C with respect to the elevator, (d) the relative velocity of the counterweight W with respect to the elevator.

11.41 The elevator shown starts from rest and moves upward with a constant acceleration. If the counterweight W moves through 10 m in 5 s, determine (a) the accelerations of the elevator and the cable C, (b) the velocity of the elevator after 5 s.

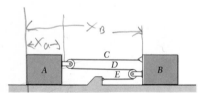

Fig. P11.42 and P11.43

11.42 The slider block B moves to the right with a constant velocity of 18 in./s. Determine (a) the velocity of block A, (b) the velocity of portion D of the cable, (c) the relative velocity of A with respect to B, (d) the relative velocity of portion C of the cable with respect to portion D.

11.43 The slider block A starts from rest and moves to the left with a constant acceleration. Knowing that the velocity of block B is 12 in./s after moving 24 in., determine (a) the accelerations of A and B, (b) the velocity and position of A after 5 s.

11.44 Collar A starts from rest and moves to the left with a constant acceleration. Knowing that after 6 s the relative velocity of collar B with respect to collar A is 450 mm/s, determine (a) the accelerations of A and B, (b) the position and velocity of B after 8 s.

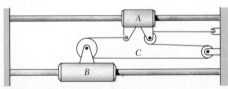

Fig. P11.44 and P11.45

11.45 In the position shown collar B moves to the left with a velocity of 100 mm/s. Determine (a) the velocity of collar A, (b) the velocity of portion C of the cable, (c) the relative velocity of portion C of the cable with respect to collar B.

11.46 Under normal operating conditions, tape is transferred between the reels shown at a speed of 720 mm/s. At $t = 0$, portion A of the tape is moving to the right at a speed of 600 mm/s and has a constant acceleration. Knowing that portion B of the tape has a constant speed of 720 mm/s and that the speed of portion A reaches 720 mm/s at $t = 6$ s, determine (a) the acceleration and velocity of the compensator C at $t = 4$ s, (b) the distance through which C will have moved at $t = 6$ s.

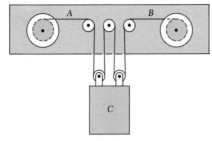

Fig. P11.46 and P11.47

11.47 Portions A and B of the tape shown start from rest at $t = 0$ and are each accelerated uniformly until a speed of 720 mm/s is reached; each portion of tape then moves at a constant speed of 720 mm/s. Knowing that portions A and B reached the speed of 720 mm/s in, respectively, 18 s and 16 s, determine (a) the acceleration and velocity of the compensator C at $t = 10$ s, (b) the distance through which C will have moved when both portions of tape reach their final speed.

11.48 Collar A starts from rest at $t = 0$ and moves upward with a constant acceleration of 3.6 in./s^2. Knowing that collar B moves downward with a constant velocity of 16 in./s, determine (a) the time at which the velocity of block C is zero, (b) the corresponding position of block C.

11.49 Collars A and B start from rest and move with the following accelerations: $a_A = 2.5t$ in./s^2 upward and $a_B = 15$ in./s^2 downward. Determine (a) the time at which the velocity of block C is again zero, (b) the distance through which block C will have moved at that time.

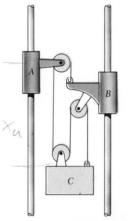

Fig. P11.48 and P11.49

*11.50 The three blocks shown are equally spaced horizontally and move vertically with constant velocities. Knowing that initially they are at the same level and that the relative velocity of C with respect to B is 200 mm/s downward, determine the velocity of each block so that the three blocks will remain aligned during their motion.

11.51 The three blocks shown move with constant velocities. Find the velocity of each block, knowing that the relative velocity of A with respect to C is 120 mm/s upward and that the relative velocity of B with respect to A is 40 mm/s upward.

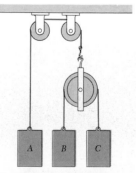

Fig. P11.50 and P11.51

* 11.7. Graphical Solution of Rectilinear-Motion Problems.

It was observed in Sec. 11.2 that the fundamental formulas

$$v = \frac{dx}{dt} \quad \text{and} \quad a = \frac{dv}{dt}$$

have a geometrical significance. The first formula expresses that the velocity at any instant is equal to the slope of the x–t curve at the same instant (Fig. 11.10). The second formula expresses that the acceleration is equal to

Fig. 11.10

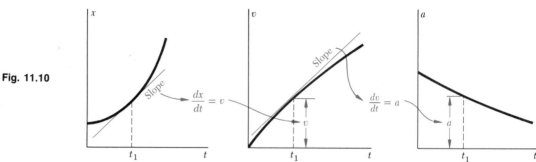

the slope of the v–t curve. These two properties may be used to derive graphically the v–t and a–t curves of a motion when the x–t curve is known.

Integrating the two fundamental formulas from a time t_1 to a time t_2, we write

$$x_2 - x_1 = \int_{t_1}^{t_2} v \, dt \quad \text{and} \quad v_2 - v_1 = \int_{t_1}^{t_2} a \, dt \qquad (11.12)$$

The first formula expresses that the area measured under the v–t curve from t_1 to t_2 is equal to the change in x during that time interval (Fig. 11.11). The second formula expresses similarly that the area measured under the a–t curve from t_1 to t_2 is equal to the change in v during that time interval. These two properties may be used to determine graphically the x–t curve of a motion when its v–t curve or its a–t curve is known (see Sample Prob. 11.6).

Graphical solutions are particularly useful when the motion considered is defined from experimental data and when x, v, and a are not analytical functions of t. They may also be used to advantage when the motion consists of distinct parts and when its analysis requires writing a different equation for each of its parts. When using a graphical solution, however, one should be careful to note (1) that the area under the v–t curve measures the *change in x*, not x itself, and, similarly, that the area under the a–t curve measures the change in v; (2) that while an area above the t axis corresponds to an increase in x or v, an area located below the t axis measures a decrease in x or v.

It will be useful to remember in drawing motion curves that if the velocity is constant, it will be represented by a horizontal straight line; the position coordinate x will then be a linear function of t and will be repre-

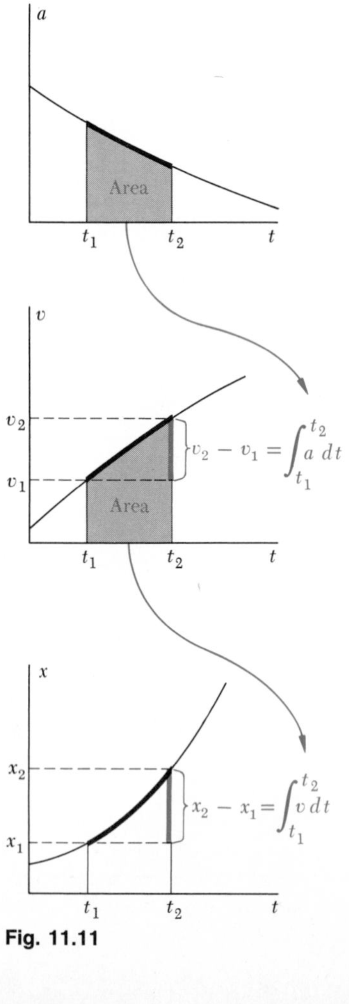

Fig. 11.11

sented by an oblique straight line. If the acceleration is constant and different from zero, it will be represented by a horizontal straight line; v will
then be a linear function of t, represented by an oblique straight line; and x
will be expressed as a second-degree polynomial in t, represented by a
parabola. If the acceleration is a linear function of t, the velocity and the
position coordinate will be equal, respectively, to second-degree and
third-degree polynomials; a will then be represented by an oblique straight
line, v by a parabola, and x by a cubic. In general, if the acceleration is a
polynomial of degree n in t, the velocity will be a polynomial of degree
$n + 1$ and the position coordinate a polynomial of degree $n + 2$; these
polynomials are represented by motion curves of a corresponding degree.

* **11.8. Other Graphical Methods.** An alternative graphical
solution may be used to determine directly from the a–t curve the position
of a particle at a given instant. Denoting, respectively, by x_0 and v_0 the
values of x and v at $t = 0$ and by x_1 and v_1 their values at $t = t_1$, and
observing that the area under the v–t curve may be divided into a rec-
tangle of area $v_0 t_1$ and horizontal differential elements of area $(t_1 - t)dv$
(Fig. 11.12a), we write

$$x_1 - x_0 = \text{area under } v\text{–}t \text{ curve} = v_0 t_1 + \int_{v_0}^{v_1} (t_1 - t)\, dv$$

Substituting $dv = a\, dt$ in the integral, we obtain

$$x_1 - x_0 = v_0 t_1 + \int_0^{t_1} (t_1 - t)a\, dt$$

Referring to Fig. 11.12b, we note that the integral represents the first moment of the area under the a–t curve with respect to the line $t = t_1$ bound-
ing the area on the right. This method of solution is known, therefore, as
the *moment-area method*. If the abscissa $\bar{t}$ of the centroid C of the area is
known, the position coordinate x_1 may be obtained by writing

$$x_1 = x_0 + v_0 t_1 + (\text{area under } a\text{–}t \text{ curve})(t_1 - \bar{t}) \qquad (11.13)$$

If the area under the a–t curve is a composite area, the last term in (11.13)
may be obtained by multiplying each component area by the distance from
its centroid to the line $t = t_1$. Areas above the t axis should be considered
as positive and areas below the t axis, as negative.

Another type of motion curve, the v–x curve, is sometimes used. If
such a curve has been plotted (Fig. 11.13), the acceleration a may be obtained at any time by drawing the normal to the curve and *measuring the
subnormal BC.* Indeed, observing that the angle between AC and AB is
equal to the angle θ between the horizontal and the tangent at A (the slope
of which is $\tan \theta = dv/dx$), we write

$$BC = AB \tan \theta = v\frac{dv}{dx}$$

and thus, recalling formula (11.4),

$$BC = a$$

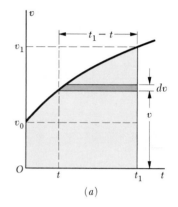

(a)

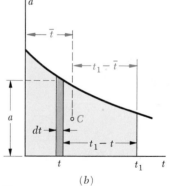

(b)

Fig. 11.12

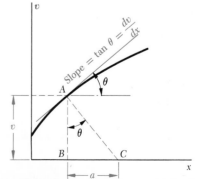

Fig. 11.13

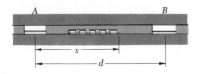

SAMPLE PROBLEM 11.6

A subway train leaves station A; it gains speed at the rate of 4 ft/s^2 for 6 s and then at the rate of 6 ft/s^2 until it has reached the speed of 48 ft/s. The train maintains the same speed until it approaches station B; brakes are then applied, giving the train a constant deceleration and bringing it to a stop in 6 s. The total running time from A to B is 40 s. Draw the a–t, v–t, and x–t curves, and determine the distance between stations A and B.

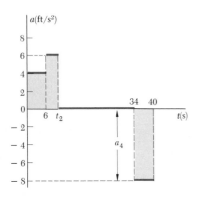

Acceleration-Time Curve. Since the acceleration is either constant or zero, the a–t curve is made of horizontal straight-line segments. The values of t_2 and a_4 are determined as follows:

$0 < t < 6$:   Change in v = area under a–t curve
$$v_6 - 0 = (6\text{ s})(4\text{ ft/s}^2) = 24\text{ ft/s}$$

$6 < t < t_2$:   Since the velocity increases from 24 to 48 ft/s,

Change in v = area under a–t curve
$$48 - 24 = (t_2 - 6)(6\text{ ft/s}^2) \qquad t_2 = 10\text{ s}$$

$t_2 < t < 34$:   Since the velocity is constant, the acceleration is zero.

$34 < t < 40$:   Change in v = area under a–t curve
$$0 - 48 = (6\text{ s})a_4 \qquad a_4 = -8\text{ ft/s}^2$$

The acceleration being negative, the corresponding area is below the t axis; this area represents a decrease in velocity.

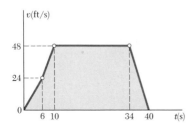

Velocity-Time Curve. Since the acceleration is either constant or zero, the v–t curve is made of segments of straight line connecting the points determined above.

$$\text{Change in } x = \text{area under } v\text{–}t \text{ curve}$$

$0 < t < 6$: $x_6 - 0 = \frac{1}{2}(6)(24) = 72$ ft
$6 < t < 10$: $x_{10} - x_6 = \frac{1}{2}(4)(24 + 48) = 144$ ft
$10 < t < 34$: $x_{34} - x_{10} = (24)(48) = 1152$ ft
$34 < t < 40$: $x_{40} - x_{34} = \frac{1}{2}(6)(48) = 144$ ft

Adding the changes in x, we obtain the distance from A to B:

$$d = x_{40} - 0 = 1512\text{ ft}$$

$$d = 1512\text{ ft} \quad \blacktriangleleft$$

Position-Time Curve. The points determined above should be joined by three arcs of parabola and one segment of straight line. The construction of the x–t curve will be performed more easily and more accurately if we keep in mind that for any value of t the slope of the tangent to the x–t curve is equal to the value of v at that instant.

Problems

11.52 A particle moves in a straight line with the acceleration shown in the figure. Knowing that it starts from the origin with $v_0 = -18$ ft/s, (a) plot the $v-t$ and $x-t$ curves for $0 < t < 20$ s, (b) determine its velocity, its position, and the total distance traveled after 12 s.

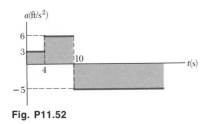

Fig. P11.52

11.53 For the particle and motion of Prob. 11.52, plot the $v-t$ and $x-t$ curves for $0 < t < 20$ s and determine (a) the maximum value of the velocity of the particle, (b) the maximum value of its position coordinate.

11.54 A particle moves in a straight line with the velocity shown in the figure. Knowing that $x = -16$ m at $t = 0$, draw the $a-t$ and $x-t$ curves for $0 < t < 40$ s and determine (a) the maximum value of the position coordinate of the particle, (b) the values of t for which the particle is at a distance of 36 m from the origin.

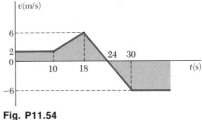

Fig. P11.54

11.55 For the particle and motion of Prob. 11.54, plot the $a-t$ and $x-t$ curves for $0 < t < 40$ s and determine (a) the total distance traveled by the particle during the period $t = 0$ to $t = 30$ s, (b) the two values of t for which the particle passes through the origin.

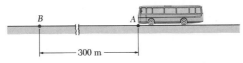

Fig. P11.56

11.56 A bus starts from rest at point A and accelerates at the rate of 0.8 m/s² until it reaches a speed of 12 m/s. It then proceeds at 12 m/s until the brakes are applied; it comes to rest at point B, 42 m beyond the point where the brakes were applied. Assuming uniform deceleration and knowing that the distance between A and B is 300 m, determine the time required for the bus to travel from A to B.

11.57 A series of city traffic signals is timed so that an automobile traveling at a constant speed of 45 km/h will reach each signal just as it turns green. A motorist misses a signal and is stopped at signal A. Knowing that the next signal B is 325 m ahead and that the maximum acceleration of the automobile is 1.5 m/s², determine what the motorist should do to keep the maximum speed as small as possible, yet reach signal B just as it turns green. What is the maximum speed reached?

11.58 Firing a howitzer causes the barrel to recoil 40 in. before a braking mechanism brings it to rest. From a high-speed photographic record, it is found that the maximum value of the recoil velocity is 270 in./s and that this is reached 0.02 s after firing. Assuming that the recoil period consists of two phases during which the acceleration has, respectively, a constant positive value a_1 and a constant negative value a_2, determine (a) the values of a_1 and a_2, (b) the position of the barrel 0.02 s after firing, (c) the time at which the velocity of the barrel is zero.

11.59 During a finishing operation the bed of an industrial planer moves alternately 30 in. to the right and 30 in. to the left. The velocity of the bed is limited to a maximum value of 6 in./s to the right and 12 in./s to the left; the acceleration is successively equal to 6 in./s² to the right, zero, 6 in./s² to the left, zero, etc. Determine the time required for the bed to complete a full cycle, and draw the v–t and x–t curves.

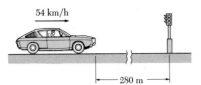

Fig. P11.60

11.60 A motorist is traveling at 54 km/h when she observes that a traffic signal 280 m ahead of her turns red. She knows that the signal is timed to stay red for 28 s. What should she do to pass the signal at 54 km/h just as it turns green again? Draw the v–t curve, selecting the solution which calls for the smallest possible deceleration and acceleration, and determine (a) the deceleration and acceleration in m/s², (b) the minimum speed reached in km/h.

11.61 Solve Prob. 11.60, knowing that the acceleration of the automobile cannot exceed 0.6 m/s².

11.62 An automobile at rest is passed by a truck traveling at a constant speed of 54 km/h. The automobile starts and accelerates for 10 s at a constant rate until it reaches a speed of 90 km/h. If the automobile then maintains a constant speed of 90 km/h, determine when and where it will overtake the truck, assuming that the automobile starts (a) just as the truck passes it, (b) 3 s after the truck has passed it.

11.63 A car and a truck are both traveling at the constant speed of 35 mi/h; the car is 40 ft behind the truck. The driver of the car wants to pass the truck, i.e., he wishes to place his car at B, 40 ft in front of the truck, and then resume the speed of 35 mi/h. The maximum acceleration of the car is 5 ft/s² and the maximum deceleration obtained by applying the brakes is 20 ft/s². What is the shortest time in which the driver of the car can complete the passing operation if he does not at any time exceed a speed of 50 mi/h? Draw the v–t curve.

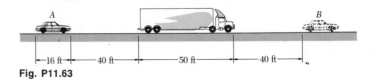

Fig. P11.63

11.64 Solve Prob. 11.63, assuming that the driver of the car does not pay any attention to the speed limit while passing and concentrates on reaching position B and resuming a speed of 35 mi/h in the shortest possible time. What is the maximum speed reached? Draw the v–t curve.

11.65 A car and a truck are both traveling at a constant speed v_0; the car is 30 ft behind the truck. The truck driver suddenly applies his brakes, causing the truck to decelerate at the constant rate of 9 ft/s^2. Two seconds later the driver of the car applies her brakes and barely manages to avoid a rear-end collision. Determine the constant rate at which the car decelerated if (a) $v_0 = 60$ mi/h, (b) $v_0 = 45$ mi/h.

11.66 Car A is traveling at the constant speed v_A. It approaches car B, which is traveling in the same direction at the constant speed of 72 km/h. The driver of car B notices car A when it is still 60 m behind him and then accelerates at the constant rate of 0.75 m/s^2 to avoid being passed or struck by car A. Knowing that the closest that A comes to B is 6 m, determine the speed v_A of car A.

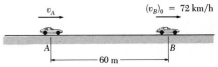

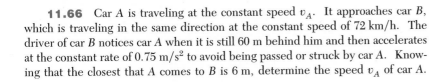

Fig. P11.66

11.67 A freight elevator F moving upward with a constant velocity of 5 m/s passes a passenger elevator P which is stopped. Three seconds later, the passenger elevator starts upward with an acceleration of 1.25 m/s^2. When the passenger elevator has reached a velocity of 10 m/s, it proceeds at constant speed. Draw the v–t and y–t curves, and from them determine the time and distance required by the passenger elevator to overtake the freight elevator.

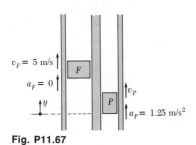

Fig. P11.67

11.68 The acceleration record shown was obtained for a small airplane traveling along a straight course. Knowing that $x = 0$ and $v = 60$ m/s when $t = 0$, determine (a) the velocity and position of the plane at $t = 20$ s, (b) its average velocity during the interval 6 s $< t <$ 14 s.

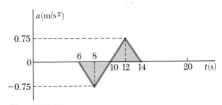

Fig. P11.68

11.69 The acceleration of a particle varies uniformly from $a = 90$ in./s^2 at $t = 0$, to $a = -90$ in./s^2 at $t = 6$ s. Knowing that $x = 0$ and $v = 0$ when $t = 0$, determine (a) the maximum velocity of the particle, (b) its position at $t = 6$ s, (c) its *average* velocity over the interval $0 < t < 6$ s. Draw the a–t, v–t, and x–t curves for the motion.

11.70 The rate of change of acceleration is known as the *jerk;* large or abrupt rates of change of acceleration cause discomfort to elevator passengers. If the jerk, or rate of change of acceleration, of an elevator is limited to ± 1.6 ft/s^2 per second, determine (a) the shortest time required for an elevator, starting from rest, to rise 50 ft and stop, (b) the corresponding average velocity of the elevator.

11.71 In order to maintain passenger comfort, the acceleration of an elevator is limited to ±1.5 m/s^2 and the jerk, or rate of change of acceleration, is limited to ±0.5 m/s^2 per second. If the elevator starts from rest, determine (*a*) the shortest time required for it to attain a constant velocity of 8 m/s, (*b*) the distance traveled in that time, (*c*) the corresponding average velocity of the elevator.

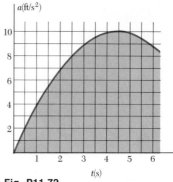

Fig. P11.72

11.72 The acceleration record shown was obtained for a truck traveling on a straight highway. Knowing that the initial velocity of the truck was 15 mi/h, determine (*a*) the velocity at $t = 6$ s, (*b*) the distance the truck traveled during the 6-s test.

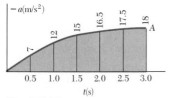

Fig. P11.73

11.73 A training airplane lands on an aircraft carrier and is brought to rest in 3 s by the arresting gear of the carrier. An accelerometer attached to the airplane provides the acceleration record shown. Determine by approximate means (*a*) the initial velocity of the airplane relative to the deck, (*b*) the distance the airplane travels along the deck before coming to rest.

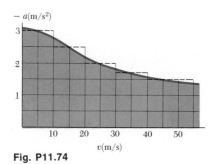

Fig. P11.74

11.74 The maximum possible deceleration of a passenger train under emergency conditions was determined experimentally; the results are shown (solid curve) in the figure. If the brakes are applied when the train is traveling at 90 km/h, determine by approximate means (*a*) the time required for the train to come to rest, (*b*) the distance traveled in that time.

11.75 The v–x curve shown was obtained experimentally during the motion of the bed of an industrial planer. Determine by approximate means the acceleration (a) when $x = 3$ in., (b) when $v = 10$ in./s.

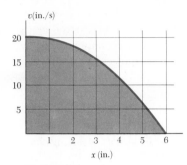

Fig. P11.75

11.76 Using the method of Sec. 11.8, derive the formula $x = x_0 + v_0 t + \frac{1}{2}at^2$ for the position coordinate of a particle in uniformly accelerated rectilinear motion.

11.77 The acceleration of an object subjected to the pressure wave of a large explosion is defined approximately by the curve shown. The object is initially at rest and is again at rest at time t_1. Using the method of Sec. 11.8, determine (a) the time t_1, (b) the distance through which the object is moved by the pressure wave.

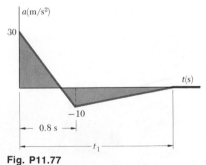

Fig. P11.77

11.78 For the particle of Prob. 11.54, draw the a–t curve and, using the method of Sec. 11.8, determine (a) the position of the particle when $t = 20$ s, (b) the maximum value of its position coordinate.

11.79 Using the method of Sec. 11.8, determine the position of the particle of Prob. 11.52 when $t = 14$ s.

CURVILINEAR MOTION OF PARTICLES

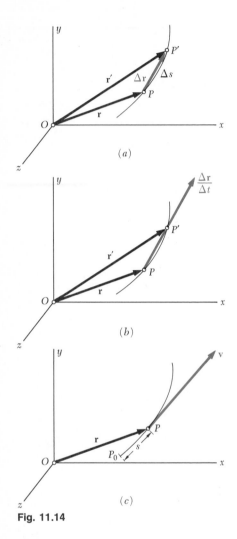

Fig. 11.14

11.9. Position Vector, Velocity, and Acceleration. When a particle moves along a curve other than a straight line, we say that the particle is in *curvilinear motion*. To define the position P occupied by the particle at a given time t, we select a fixed reference system, such as the x, y, z axes shown in Fig. 11.14a, and draw the vector $\mathbf{r}$ joining the origin O and point P. Since the vector $\mathbf{r}$ is characterized by its magnitude r and its direction with respect to the reference axes, it completely defines the position of the particle with respect to those axes; the vector $\mathbf{r}$ is referred to as the *position vector* of the particle at time t.

Consider now the vector $\mathbf{r}'$ defining the position P' occupied by the same particle at a later time $t + \Delta t$. The vector $\Delta \mathbf{r}$ joining P and P' represents the change in the position vector during the time interval Δt since, as we may easily check from Fig. 11.14a, the vector $\mathbf{r}'$ is obtained by adding the vectors $\mathbf{r}$ and $\Delta \mathbf{r}$ according to the triangle rule. We note that $\Delta \mathbf{r}$ represents a change in *direction* as well as a change in *magnitude* of the position vector $\mathbf{r}$. The *average velocity* of the particle over the time interval Δt is defined as the quotient of $\Delta \mathbf{r}$ and Δt. Since $\Delta \mathbf{r}$ is a vector and Δt is a scalar, the quotient $\Delta \mathbf{r} / \Delta t$ is a vector attached at P, of the same direction as $\Delta \mathbf{r}$, and of magnitude equal to the magnitude of $\Delta \mathbf{r}$ divided by Δt (Fig. 11.14b).

The *instantaneous velocity* of the particle at time t is obtained by choosing shorter and shorter time intervals Δt and, correspondingly, shorter and shorter vector increments $\Delta \mathbf{r}$. The instantaneous velocity is thus represented by the vector

$$\mathbf{v} = \lim_{\Delta t \to 0} \frac{\Delta \mathbf{r}}{\Delta t} \tag{11.14}$$

As Δt and $\Delta \mathbf{r}$ become shorter, the points P and P' get closer; the vector $\mathbf{v}$ obtained at the limit must therefore be tangent to the path of the particle (Fig. 11.14c).

Since the position vector $\mathbf{r}$ depends upon the time t, we may refer to it as a *vector function* of the scalar variable t and denote it by $\mathbf{r}(t)$. Extending the concept of derivative of a scalar function introduced in elementary calculus, we shall refer to the limit of the quotient $\Delta \mathbf{r} / \Delta t$ as the *derivative* of the vector function $\mathbf{r}(t)$. We write

$$\mathbf{v} = \frac{d\mathbf{r}}{dt} \tag{11.15}$$

The magnitude v of the vector $\mathbf{v}$ is called the *speed* of the particle. It may be obtained by substituting for the vector $\Delta \mathbf{r}$ in formula (11.14) its magnitude represented by the straight-line segment PP'. But the length of the segment PP' approaches the length Δs of the arc PP' as Δt decreases (Fig. 11.14a), and we may write

$$v = \lim_{\Delta t \to 0} \frac{PP'}{\Delta t} = \lim_{\Delta t \to 0} \frac{\Delta s}{\Delta t} \qquad v = \frac{ds}{dt} \tag{11.16}$$

The speed v may thus be obtained by differentiating with respect to t the length s of the arc described by the particle.

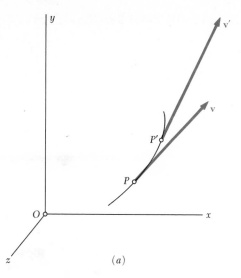

(a)

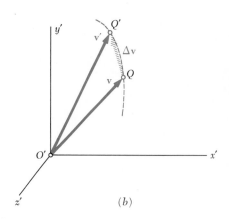

(b)

Consider the velocity **v** of the particle at time t and also its velocity **v′** at a later time $t + \Delta t$ (Fig. 11.15a). Let us draw both vectors **v** and **v′** from the same origin O' (Fig. 11.15b). The vector Δ**v** joining Q and Q' represents the change in the velocity of the particle during the time interval Δt, since the vector **v′** may be obtained by adding the vectors **v** and Δ**v**. We should note that Δ**v** represents a change in the *direction* of the velocity as well as a change in *speed*. The *average acceleration* of the particle over the time interval Δt is defined as the quotient of Δ**v** and Δt. Since Δ**v** is a vector and Δt a scalar, the quotient Δ**v**$/\Delta t$ is a vector of the same direction as Δ**v**.

The *instantaneous acceleration* of the particle at time t is obtained by choosing smaller and smaller values for Δt and Δ**v**. The instantaneous acceleration is thus represented by the vector

$$\mathbf{a} = \lim_{\Delta t \to 0} \frac{\Delta \mathbf{v}}{\Delta t} \tag{11.17}$$

Noting that the velocity **v** is a vector function **v**(t) of the time t, we may refer to the limit of the quotient Δ**v**$/\Delta t$ as the derivative of **v** with respect to t. We write

$$\mathbf{a} = \frac{d\mathbf{v}}{dt} \tag{11.18}$$

We observe that the acceleration **a** is tangent to the curve described by the tip Q of the vector **v** when the latter is drawn from a fixed origin O' (Fig. 11.15c) and that, in general, the acceleration is *not* tangent to the path of the particle (Fig. 11.15d). The curve described by the tip of **v** and shown in Fig. 11.15c is called the *hodograph* of the motion.

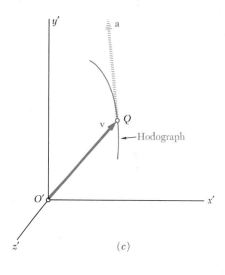

(c)

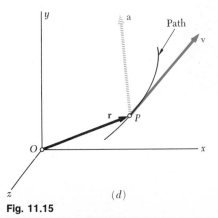

(d)

Fig. 11.15

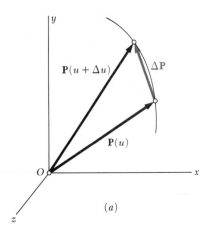

(a)

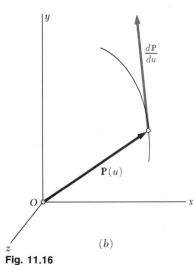

(b)

Fig. 11.16

11.10. Derivatives of Vector Functions. We saw in the preceding section that the velocity **v** of a particle in curvilinear motion may be represented by the derivative of the vector function **r**(t) characterizing the position of the particle. Similarly, the acceleration **a** of the particle may be represented by the derivative of the vector function **v**(t). In this section, we shall give a formal definition of the derivative of a vector function and establish a few rules governing the differentiation of sums and products of vector functions.

Let **P**(u) be a vector function of the scalar variable u. By that we mean that the scalar u completely defines the magnitude and direction of the vector **P**. If the vector **P** is drawn from a fixed origin O and the scalar u is allowed to vary, the tip of **P** will describe a given curve in space. Consider the vectors **P** corresponding, respectively, to the values u and u + Δu of the scalar variable (Fig. 11.16a). Let Δ**P** be the vector joining the tips of the two given vectors; we write

$$\Delta \mathbf{P} = \mathbf{P}(u + \Delta u) - \mathbf{P}(u)$$

Dividing through by Δu and letting Δu approach zero, *we define the derivative of the vector function* **P**(u):

$$\frac{d\mathbf{P}}{du} = \lim_{\Delta u \to 0} \frac{\Delta \mathbf{P}}{\Delta u} = \lim_{\Delta u \to 0} \frac{\mathbf{P}(u + \Delta u) - \mathbf{P}(u)}{\Delta u} \qquad (11.19)$$

As Δu approaches zero, the line of action of Δ**P** becomes tangent to the curve of Fig. 11.16a. Thus, the derivative d**P**/du of the vector function **P**(u) *is tangent to the curve described by the tip of* **P**(u) (Fig. 11.16b).

We shall now show that the standard rules for the differentiation of the sums and products of scalar functions may be extended to vector functions. Consider first the *sum of two vector functions* **P**(u) and **Q**(u) of the same scalar variable u. According to the definition given in (11.19), the derivative of the vector **P** + **Q** is

$$\frac{d(\mathbf{P} + \mathbf{Q})}{du} = \lim_{\Delta u \to 0} \frac{\Delta(\mathbf{P} + \mathbf{Q})}{\Delta u} = \lim_{\Delta u \to 0} \left(\frac{\Delta \mathbf{P}}{\Delta u} + \frac{\Delta \mathbf{Q}}{\Delta u} \right)$$

or since the limit of a sum is equal to the sum of the limits of its terms,

$$\frac{d(\mathbf{P} + \mathbf{Q})}{du} = \lim_{\Delta u \to 0} \frac{\Delta \mathbf{P}}{\Delta u} + \lim_{\Delta u \to 0} \frac{\Delta \mathbf{Q}}{\Delta u}$$

$$\frac{d(\mathbf{P} + \mathbf{Q})}{du} = \frac{d\mathbf{P}}{du} + \frac{d\mathbf{Q}}{du} \qquad (11.20)$$

Next we shall consider the *product of a scalar function f(u) and of a vector function* **P**(u) of the same scalar variable u. The derivative of the vector f**P** is

$$\frac{d(f\mathbf{P})}{du} = \lim_{\Delta u \to 0} \frac{(f + \Delta f)(\mathbf{P} + \Delta \mathbf{P}) - f\mathbf{P}}{\Delta u} = \lim_{\Delta u \to 0} \left(\frac{\Delta f}{\Delta u} \mathbf{P} + f \frac{\Delta \mathbf{P}}{\Delta u} \right)$$

or recalling the properties of the limits of sums and products,

$$\frac{d(f\mathbf{P})}{du} = \frac{df}{du}\mathbf{P} + f\frac{d\mathbf{P}}{du} \qquad (11.21)$$

The derivatives of the *scalar product* and of the *vector product* of two vector functions $\mathbf{P}(u)$ and $\mathbf{Q}(u)$ may be obtained in a similar way. We have

$$\frac{d(\mathbf{P} \cdot \mathbf{Q})}{du} = \frac{d\mathbf{P}}{du} \cdot \mathbf{Q} + \mathbf{P} \cdot \frac{d\mathbf{Q}}{du} \qquad (11.22)$$

$$\frac{d(\mathbf{P} \times \mathbf{Q})}{du} = \frac{d\mathbf{P}}{du} \times \mathbf{Q} + \mathbf{P} \times \frac{d\mathbf{Q}}{du} \qquad (11.23)\dagger$$

We shall use the properties established above to determine the *rectangular components of the derivative of a vector function* $\mathbf{P}(u)$. Resolving $\mathbf{P}$ into components along fixed rectangular axes x, y, z, we write

$$\mathbf{P} = P_x\mathbf{i} + P_y\mathbf{j} + P_z\mathbf{k} \qquad (11.24)$$

where P_x, P_y, P_z are the rectangular scalar components of the vector $\mathbf{P}$, and $\mathbf{i}$, $\mathbf{j}$, $\mathbf{k}$ the unit vectors corresponding, respectively, to the x, y, and z axes (Sec. 2.12). By (11.20), the derivative of $\mathbf{P}$ is equal to the sum of the derivatives of the terms in the right-hand member. Since each of these terms is the product of a scalar and a vector function, we should use (11.21). But the unit vectors $\mathbf{i}$, $\mathbf{j}$, $\mathbf{k}$ have a constant magnitude (equal to 1) and fixed directions. Their derivatives are therefore zero, and we write

$$\frac{d\mathbf{P}}{du} = \frac{dP_x}{du}\mathbf{i} + \frac{dP_y}{du}\mathbf{j} + \frac{dP_z}{du}\mathbf{k} \qquad (11.25)$$

Noting that the coefficients of the unit vectors are, by definition, the scalar components of the vector $d\mathbf{P}/du$, we conclude that *the rectangular scalar components of the derivative $d\mathbf{P}/du$ of the vector function $\mathbf{P}(u)$ are obtained by differentiating the corresponding scalar components of $\mathbf{P}$.*

Rate of Change of a Vector. When the vector $\mathbf{P}$ is a function of the time t, its derivative $d\mathbf{P}/dt$ represents the *rate of change* of $\mathbf{P}$ with respect to the frame $Oxyz$. Resolving $\mathbf{P}$ into rectangular components, we have, by (11.25),

$$\frac{d\mathbf{P}}{dt} = \frac{dP_x}{dt}\mathbf{i} + \frac{dP_y}{dt}\mathbf{j} + \frac{dP_z}{dt}\mathbf{k}$$

or, using dots to indicate differentiation with respect to t,

$$\dot{\mathbf{P}} = \dot{P}_x\mathbf{i} + \dot{P}_y\mathbf{j} + \dot{P}_z\mathbf{k} \qquad (11.25')$$

†Since the vector product is not commutative (Sec. 3.4), the order of the factors in (11.23) must be maintained.

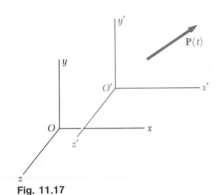

Fig. 11.17

As we shall see in Sec. 15.10, the rate of change of a vector, as observed from a *moving frame of reference*, is, in general, different from its rate of change as observed from a fixed frame of reference. However, if the moving frame $O'x'y'z'$ is in *translation*, i.e., if its axes remain parallel to the corresponding axes of the fixed frame $Oxyz$ (Fig. 11.17), the same unit vectors, **i**, **j**, **k** are used in both frames, and the vector **P** has, at any given instant, the same components P_x, P_y, P_z in both frames. It follows from (11.25′) that the rate of change $\dot{\mathbf{P}}$ is the same with respect to the frames $Oxyz$ and $O'x'y'z'$. We state, therefore: *The rate of change of a vector is the same with respect to a fixed frame and with respect to a frame in translation.* This property will greatly simplify our work, since we shall deal mainly with frames in translation.

11.11. Rectangular Components of Velocity and Acceleration.
When the position of a particle P is defined at any instant by its rectangular coordinates x, y, and z, it is convenient to resolve the velocity **v** and the acceleration **a** of the particle into rectangular components (Fig. 11.18).

Resolving the position vector **r** of the particle into rectangular components, we write

$$\mathbf{r} = x\mathbf{i} + y\mathbf{j} + z\mathbf{k} \tag{11.26}$$

where the coordinates x, y, z are functions of t. Differentiating twice, we obtain

$$\mathbf{v} = \frac{d\mathbf{r}}{dt} = \dot{x}\mathbf{i} + \dot{y}\mathbf{j} + \dot{z}\mathbf{k} \tag{11.27}$$

$$\mathbf{a} = \frac{d\mathbf{v}}{dt} = \ddot{x}\mathbf{i} + \ddot{y}\mathbf{j} + \ddot{z}\mathbf{k} \tag{11.28}$$

where $\dot{x}$, $\dot{y}$, $\dot{z}$ and $\ddot{x}$, $\ddot{y}$, $\ddot{z}$ represent, respectively, the first and second derivatives of x, y, and z with respect to t. It follows from (11.27) and (11.28) that the scalar components of the velocity and acceleration are

$$v_x = \dot{x} \qquad v_y = \dot{y} \qquad v_z = \dot{z} \tag{11.29}$$

$$a_x = \ddot{x} \qquad a_y = \ddot{y} \qquad a_z = \ddot{z} \tag{11.30}$$

A positive value for v_x indicates that the vector component $\mathbf{v}_x$ is directed to the right, a negative value that it is directed to the left; the sense of each of the other vector components may be determined in a similar way from the sign of the corresponding scalar component. If desired, the magnitudes and directions of the velocity and acceleration may be obtained from their scalar components by the methods of Secs. 2.7 and 2.12.

The use of rectangular components to describe the position, the velocity, and the acceleration of a particle is particularly effective when the

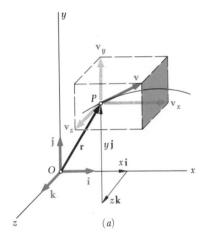

(a)

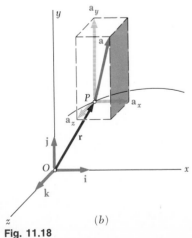

(b)

Fig. 11.18

component a_x of the acceleration depends only upon t, x, and/or v_x, and when, similarly, a_y depends only upon t, y, and/or v_y, and a_z upon t, z, and/or v_z. Equations (11.30) may then be integrated independently, and so may Eqs. (11.29). In other words, the motion of the particle in the x direction, its motion in the y direction, and its motion in the z direction may be considered separately.

In the case of the *motion of a projectile*, for example, it may be shown (see Sec. 12.5) that the components of the acceleration are

$$a_x = \ddot{x} = 0 \qquad a_y = \ddot{y} = -g \qquad a_z = \ddot{z} = 0$$

if the resistance of the air is neglected. Denoting by x_0, y_0, z_0 the coordinates of the gun, and by $(v_x)_0$, $(v_y)_0$, $(v_z)_0$ the components of the initial velocity $\mathbf{v}_0$ of the projectile, we integrate twice in t and obtain

$$
\begin{aligned}
v_x = \dot{x} = (v_x)_0 && v_y = \dot{y} = (v_y)_0 - gt && v_z = \dot{z} = (v_z)_0 \\
x = x_0 + (v_x)_0 t && y = y_0 + (v_y)_0 t - \tfrac{1}{2}gt^2 && z = z_0 + (v_z)_0 t
\end{aligned}
$$

If the projectile is fired in the xy plane from the origin O, we have $x_0 = y_0 = z_0 = 0$ and $(v_z)_0 = 0$, and the equations of motion reduce to

$$
\begin{aligned}
v_x = (v_x)_0 && v_y = (v_y)_0 - gt && v_z = 0 \\
x = (v_x)_0 t && y = (v_y)_0 t - \tfrac{1}{2}gt^2 && z = 0
\end{aligned}
$$

These equations show that the projectile remains in the xy plane and that its motion in the horizontal direction is uniform, while its motion in the vertical direction is uniformly accelerated. The motion of a projectile may thus be replaced by two independent rectilinear motions, which are easily visualized if we assume that the projectile is fired vertically with an initial velocity $(\mathbf{v}_y)_0$ from a platform moving with a constant horizontal velocity $(\mathbf{v}_x)_0$ (Fig. 11.19). The coordinate x of the projectile is equal at any instant to the distance traveled by the platform, while its coordinate y may be computed as if the projectile were moving along a vertical line.

It may be observed that the equations defining the coordinates x and y of a projectile at any instant are the parametric equations of a parabola. Thus, the trajectory of a projectile is *parabolic*. This result, however, ceases to be valid when the resistance of the air or the variation with altitude of the acceleration of gravity is taken into account.

11.12. Motion Relative to a Frame in Translation.

In the preceding section, a single frame of reference was used to describe the motion of a particle. In most cases this frame was attached to the earth and was considered as fixed. We shall now analyze situations in which it is convenient to use simultaneously several frames of reference. If one of the frames is attached to the earth, we shall call it a *fixed frame of reference*, and the other frames will be referred to as *moving frames of reference*. It should be understood, however, that the selection of a fixed frame of reference is purely arbitrary. Any frame may be designated as "fixed"; all other frames not rigidly attached to this frame will then be described as "moving."

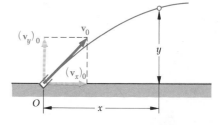

(a) Motion of a projectile

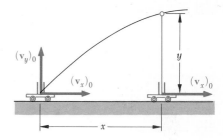

(b) Equivalent rectilinear motions

Fig. 11.19

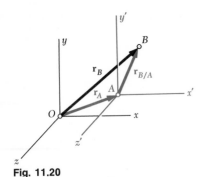

Fig. 11.20

Consider two particles A and B moving in space (Fig. 11.20); the vectors $\mathbf{r}_A$ and $\mathbf{r}_B$ define their positions at any given instant with respect to the fixed frame of reference $Oxyz$. Consider now a system of axes x', y', z' centered at A and parallel to the x, y, z axes. While the origin of these axes moves, their orientation remains the same; the frame of reference $Ax'y'z'$ is in *translation* with respect to $Oxyz$. The vector $\mathbf{r}_{B/A}$ joining A and B defines *the position of B relative to the moving frame $Ax'y'z'$* (or, for short, *the position of B relative to A*).

We note from Fig. 11.20 that the position vector $\mathbf{r}_B$ of particle B is the sum of the position vector $\mathbf{r}_A$ of particle A and of the position vector $\mathbf{r}_{B/A}$ of B relative to A; we write

$$\mathbf{r}_B = \mathbf{r}_A + \mathbf{r}_{B/A} \tag{11.31}$$

Differentiating (11.31) with respect to t within the fixed frame of reference, and using dots to indicate time derivatives, we have

$$\dot{\mathbf{r}}_B = \dot{\mathbf{r}}_A + \dot{\mathbf{r}}_{B/A} \tag{11.32}$$

The derivatives $\dot{\mathbf{r}}_A$ and $\dot{\mathbf{r}}_B$ represent, respectively, the velocities $\mathbf{v}_A$ and $\mathbf{v}_B$ of the particles A and B. The derivative $\dot{\mathbf{r}}_{B/A}$ represents the rate of change of $\mathbf{r}_{B/A}$ with respect to the frame $Ax'y'z'$, as well as with respect to the fixed frame, since $Ax'y'z'$ is in translation (Sec. 11.10). This derivative therefore defines *the velocity $\mathbf{v}_{B/A}$ of B relative to the frame $Ax'y'z'$* (or, for short, *the velocity $\mathbf{v}_{B/A}$ of B relative to A*). We write

$$\mathbf{v}_B = \mathbf{v}_A + \mathbf{v}_{B/A} \tag{11.33}$$

Differentiating Eq. (11.33) with respect to t, and using the derivative $\dot{\mathbf{v}}_{B/A}$ to define *the acceleration $\mathbf{a}_{B/A}$ of B relative to the frame $Ax'y'z'$* (or, for short, *the acceleration $\mathbf{a}_{B/A}$ of B relative to A*), we write

$$\mathbf{a}_B = \mathbf{a}_A + \mathbf{a}_{B/A} \tag{11.34}$$

The motion of B with respect to the fixed frame $Oxyz$ is referred to as the *absolute motion of B*. The equations derived in this section show that *the absolute motion of B may be obtained by combining the motion of A and the relative motion of B with respect to the moving frame attached to A.* Equation (11.33), for example, expresses that the absolute velocity $\mathbf{v}_B$ of particle B may be obtained by adding vectorially the velocity of A and the velocity of B relative to the frame $Ax'y'z'$. Equation (11.34) expresses a similar property in terms of the accelerations.† We should keep in mind, however, that *the frame $Ax'y'z'$ is in translation;* that is, while it moves with A, it maintains the same orientation. As we shall see later (Sec. 15.14), different relations must be used in the case of a rotating frame of reference.

---

† Note that the product of the subscripts A and B/A used in the right-hand member of Eqs. (11.31) through (11.34) is equal to the subscript B used in their left-hand member.

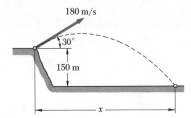

A projectile is fired from the edge of a 150-m cliff with an initial velocity of 180 m/s, at an angle of 30° with the horizontal. Neglecting air resistance, find (*a*) the horizontal distance from the gun to the point where the projectile strikes the ground, (*b*) the greatest elevation above the ground reached by the projectile.

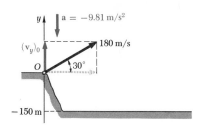

Solution. We shall consider separately the vertical and the horizontal motion.

Vertical Motion. Uniformly accelerated motion. Choosing the positive sense of the y axis upward and placing the origin O at the gun, we have

$$(v_y)_0 = (180 \text{ m/s}) \sin 30° = +90 \text{ m/s}$$
$$a = -9.81 \text{ m/s}^2$$

Substituting into the equations of uniformly accelerated motion, we have

$$
\begin{array}{llr}
v_y = (v_y)_0 + at & v_y = 90 - 9.81t & (1) \\
y = (v_y)_0 t + \tfrac{1}{2}at^2 & y = 90t - 4.90t^2 & (2) \\
v_y^2 = (v_y)_0^2 + 2ay & v_y^2 = 8100 - 19.62y & (3)
\end{array}
$$

Horizontal Motion. Uniform motion. Choosing the positive sense of the x axis to the right, we have

$$(v_x)_0 = (180 \text{ m/s}) \cos 30° = +155.9 \text{ m/s}$$

Substituting into the equation of uniform motion, we obtain

$$x = (v_x)_0 t \qquad x = 155.9t \qquad (4)$$

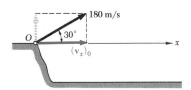

a. **Horizontal Distance.** When the projectile strikes the ground, we have

$$y = -150 \text{ m}$$

Carrying this value into Eq. (2) for the vertical motion, we write

$$-150 = 90t - 4.90t^2 \qquad t^2 - 18.37t - 30.6 = 0 \qquad t = 19.91 \text{ s}$$

Carrying $t = 19.91$ s into Eq. (4) for the horizontal motion, we obtain

$$x = 155.9(19.91) \qquad\qquad x = 3100 \text{ m} \quad \blacktriangleleft$$

b. **Greatest Elevation.** When the projectile reaches its greatest elevation, we have $v_y = 0$; carrying this value into Eq. (3) for the vertical motion, we write

$$0 = 8100 - 19.62y \qquad y = 413 \text{ m}$$
$$\text{Greatest elevation above ground} = 150 \text{ m} + 413 \text{ m}$$
$$= 563 \text{ m} \quad \blacktriangleleft$$

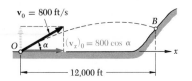

A projectile is fired with an initial velocity of 800 ft/s at a target B located 2000 ft above the gun A and at a horizontal distance of 12,000 ft. Neglecting air resistance, determine the value of the firing angle α.

Solution. We shall consider separately the horizontal and the vertical motion.

Horizontal Motion. Placing the origin of coordinates at the gun, we have

$$(v_x)_0 = 800 \cos \alpha$$

Substituting into the equation of uniform horizontal motion, we obtain

$$x = (v_x)_0 t \qquad x = (800 \cos \alpha)t$$

The time required for the projectile to move through a horizontal distance of 12,000 ft is obtained by making x equal to 12,000 ft.

$$12{,}000 = (800 \cos \alpha)t$$

$$t = \frac{12{,}000}{800 \cos \alpha} = \frac{15}{\cos \alpha}$$

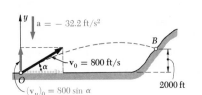

Vertical Motion

$$(v_y)_0 = 800 \sin \alpha \qquad a = -32.2 \text{ ft/s}^2$$

Substituting into the equation of uniformly accelerated vertical motion, we obtain

$$y = (v_y)_0 t + \tfrac{1}{2}at^2 \qquad y = (800 \sin \alpha)t - 16.1t^2$$

Projectile Hits Target. When $x = 12{,}000$ ft, we must have $y = 2000$ ft. Substituting for y and making t equal to the value found above, we write

$$2000 = 800 \sin \alpha \frac{15}{\cos \alpha} - 16.1\left(\frac{15}{\cos \alpha}\right)^2$$

Since $1/\cos^2 \alpha = \sec^2 \alpha = 1 + \tan^2 \alpha$, we have

$$2000 = 800(15) \tan \alpha - 16.1(15^2)(1 + \tan^2 \alpha)$$
$$3622 \tan^2 \alpha - 12{,}000 \tan \alpha + 5622 = 0$$

Solving this quadratic equation for $\tan \alpha$, we have

$$\tan \alpha = 0.565 \qquad \text{and} \qquad \tan \alpha = 2.75$$

$$\alpha = 29.5° \qquad \text{and} \qquad \alpha = 70.0° \quad \blacktriangleleft$$

The target will be hit if either of these two firing angles is used (see figure).

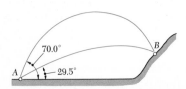

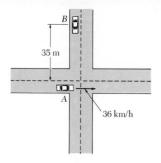

Automobile A is traveling east at the constant speed of 36 km/h. As automobile A crosses the intersection shown, automobile B starts from rest 35 m north of the intersection and moves south with a constant acceleration of 1.2 m/s². Determine the position, velocity, and acceleration of B relative to A five seconds after A crosses the intersection.

Solution. We choose x and y axes with origin at the intersection of the two streets and with positive senses directed respectively east and north.

Motion of Automobile A. First the speed is expressed in m/s:

$$v_A = \left(36 \frac{\text{km}}{\text{h}}\right)\left(\frac{1000 \text{ m}}{1 \text{ km}}\right)\left(\frac{1 \text{ h}}{3600 \text{ s}}\right) = 10 \text{ m/s}$$

Noting that the motion of A is uniform, we write, for any time t,

$$a_A = 0$$
$$v_A = +10 \text{ m/s}$$
$$x_A = (x_A)_0 + v_A t = 0 + 10t$$

For $t = 5$ s, we have

$$a_A = 0 \qquad\qquad \mathbf{a}_A = 0$$
$$v_A = +10 \text{ m/s} \qquad\qquad \mathbf{v}_A = 10 \text{ m/s} \rightarrow$$
$$x_A = +(10 \text{ m/s})(5 \text{ s}) = +50 \text{ m} \qquad \mathbf{r}_A = 50 \text{ m} \rightarrow$$

Motion of Automobile B. We note that the motion of B is uniformly accelerated, and write

$$a_B = -1.2 \text{ m/s}^2$$
$$v_B = (v_B)_0 + at = 0 - 1.2t$$
$$y_B = (y_B)_0 + (v_B)_0 t + \tfrac{1}{2}a_B t^2 = 35 + 0 - \tfrac{1}{2}(1.2)t^2$$

For $t = 5$ s, we have

$$a_B = -1.2 \text{ m/s}^2 \qquad\qquad \mathbf{a}_B = 1.2 \text{ m/s}^2 \downarrow$$
$$v_B = -(1.2 \text{ m/s}^2)(5 \text{ s}) = -6 \text{ m/s} \qquad \mathbf{v}_B = 6 \text{ m/s} \downarrow$$
$$y_B = 35 - \tfrac{1}{2}(1.2 \text{ m/s}^2)(5 \text{ s})^2 = +20 \text{ m} \qquad \mathbf{r}_B = 20 \text{ m} \uparrow$$

Motion of B Relative to A. We draw the triangle corresponding to the vector equation $\mathbf{r}_B = \mathbf{r}_A + \mathbf{r}_{B/A}$ and obtain the magnitude and direction of the position vector of B relative to A.

$$r_{B/A} = 53.9 \text{ m} \qquad \alpha = 21.8° \qquad \mathbf{r}_{B/A} = 53.9 \text{ m} \; \diagdown \; 21.8° \quad \blacktriangleleft$$

Proceeding in a similar fashion, we find the velocity and acceleration of B relative to A.

$$\mathbf{v}_B = \mathbf{v}_A + \mathbf{v}_{B/A}$$
$$v_{B/A} = 11.66 \text{ m/s} \qquad \beta = 31.0° \qquad \mathbf{v}_{B/A} = 11.66 \text{ m/s} \; \diagup \; 31.0° \quad \blacktriangleleft$$
$$\mathbf{a}_B = \mathbf{a}_A + \mathbf{a}_{B/A}$$
$$\mathbf{a}_{B/A} = 1.2 \text{ m/s}^2 \downarrow \quad \blacktriangleleft$$

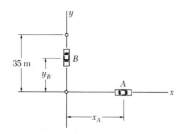

498

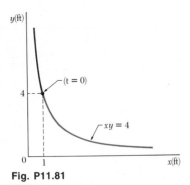

Fig. P11.81

Problems

Note. Neglect air resistance in problems concerning projectiles.

11.80 The motion of a particle is defined by the equations $x = 2t^3 - 8t^2$ and $y = 3t^2 - 12t$, where x and y are expressed in inches and t in seconds. Determine the velocity and acceleration when (*a*) $t = 1$ s, (*b*) $t = 2$ s, (*c*) $t = 3$ s.

11.81 The motion of a particle is defined by the equations $x = (t + 1)^2$ and $y = 4(t + 1)^{-2}$, where x and y are expressed in feet and t in seconds. Show that the path of the particle is part of the rectangular hyperbola shown and determine the velocity and acceleration when (*a*) $t = 0$, (*b*) $t = \frac{1}{2}$ s.

11.82 The motion of a particle is defined by the equations $x = t^2 - 8t + 7$ and $y = 0.5t^2 + 2t - 4$, where x and y are expressed in meters and t in seconds. Determine (*a*) the magnitude of the smallest velocity reached by the particle, (*b*) the corresponding time, position, and direction of the velocity.

11.83 The motion of a particle is defined by the equations $x = 48 \sin \frac{1}{2}\pi t$ and $y = 15t^2$, where x and y are expressed in millimeters and t in seconds. Determine the velocity and acceleration of the particle when (*a*) $t = 1$ s, (*b*) $t = 2$ s.

11.84 A particle moves in an elliptic path defined by the position vector $\mathbf{r} = (A \cos pt)\mathbf{i} + (B \sin pt)\mathbf{j}$. Show that the acceleration (*a*) is directed toward the origin, (*b*) is proportional to the distance from the origin to the particle.

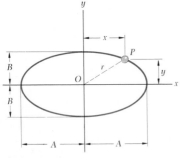

Fig. P11.84

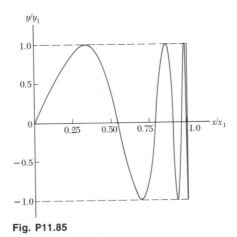

Fig. P11.85

11.85 The motion of a vibrating particle is defined by the position vector $\mathbf{r} = [x_1(1 - e^{-\pi t})]\mathbf{i} + (y_1 \sin 4\pi t)\mathbf{j}$, where t is expressed in seconds. For $x_1 = 16$ in. and $y_1 = 3$ in., determine the velocity and acceleration when (*a*) $t = 0$, (*b*) $t = 0.125$ s.

11.86 The three-dimensional motion of a particle is defined by the position vector $\mathbf{r} = (R \sin pt)\mathbf{i} + ct\mathbf{j} + (R \cos pt)\mathbf{k}$. Determine the magnitudes of the velocity and acceleration of the particle. (The space curve described by the particle is a helix.)

11.87 The three-dimensional motion of a particle is defined by the position vector $\mathbf{r} = At\mathbf{i} + ABt^3\mathbf{j} + Bt^2\mathbf{k}$, where r is expressed in meters and t in seconds. Show that the space curve described by the particle lies on the hyperbolic paraboloid $y = xz$. For $A = 1$, $B = 2$, determine the magnitudes of the velocity and acceleration when (a) $t = 0$, (b) $t = 2$ s.

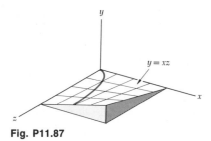

Fig. P11.87

11.88 A man standing on a bridge 30 m above the water throws a stone in a horizontal direction. Knowing that the stone hits the water 40 m from a point on the water directly below the man, determine (a) the initial velocity of the stone, (b) the distance at which the stone would hit the water if it were thrown with the same velocity from a bridge 10 m lower.

11.89 A handball player throws a ball from A with a horizontal velocity v_0. Knowing that $d = 15$ ft, determine (a) the value of v_0 for which the ball will strike the corner C, (b) the range of values of v_0 for which the ball will strike the corner region BCD.

11.90 Solve Prob. 11.89, assuming that the distance from point A to the wall is $d = 25$ ft.

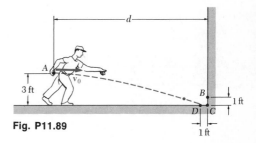

Fig. P11.89

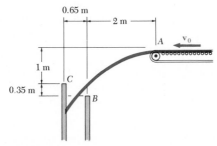

Fig. P11.91

11.91 Sand is discharged at A from a horizontal conveyor belt with an initial velocity v_0. Determine the range of values of v_0 for which the sand will enter the vertical chute shown.

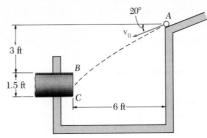

Fig. P11.92

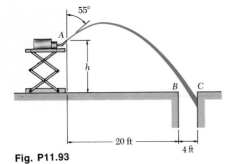

Fig. P11.93

11.92 After rolling down a 20° incline a sphere has a velocity v_0 at point A. Determine the range of values of v_0 for which the sphere will enter the horizontal pipe shown.

11.93 A pump is located near the edge of the horizontal platform shown. The nozzle at A discharges water with an initial velocity of 25 ft/s at an angle of 55° with the vertical. Determine the range of values of the height h for which the water enters the opening BC.

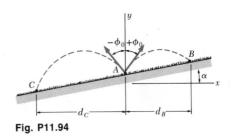

Fig. P11.94

11.94 An oscillating water sprinkler is operated at point A on an incline which forms an angle α with the horizontal. The sprinkler discharges water with an initial velocity v_0 at an angle ϕ with the vertical which varies from $-\phi_0$ to $+\phi_0$. Knowing that $v_0 = 8$ m/s, $\phi_0 = 40°$, and $\alpha = 10°$, determine the horizontal distance between the sprinkler and points B and C which define the watered area.

11.95 A nozzle at A discharges water with an initial velocity of 12 m/s at an angle of 60° with the horizontal. Determine where the stream of water strikes the roof. Check that the stream will clear the edge of the roof.

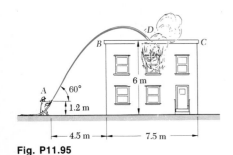

Fig. P11.95

11.96 In Prob. 11.95, determine the range of values of the initial velocity for which the water will fall on the roof.

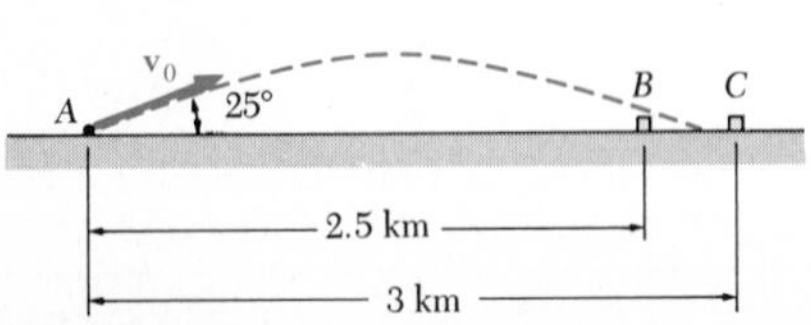

11.97 A projectile is fired with an initial velocity v_0 at an angle of 25° with the horizontal. Determine the required value of v_0 if the projectile is to hit (a) point B, (b) point C.

Fig. P11.97

11.98 Sand is discharged at A from a conveyor belt and falls onto the top of a stockpile at B. Knowing that the conveyor belt forms an angle $\alpha = 20°$ with the horizontal, determine the speed v_0 of the belt.

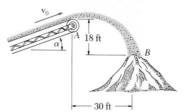

Fig. P11.98, P11.100, and P11.101

11.99 A ball is dropped onto a pad at A and rebounds with a velocity $\mathbf{v}_0$ at an angle of 70° with the horizontal. Determine the range of values of v_0 for which the ball will enter the opening BC.

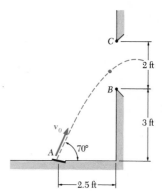

Fig. P11.99

11.100 Knowing that the conveyor belt moves at the constant speed $v_0 = 25$ ft/s, determine the angle α for which the sand is deposited on the stockpile at B.

*11.101 Knowing that the conveyor belt moves at the constant speed v_0, determine (a) the smallest value of v_0 for which sand can be deposited on the stockpile at B, (b) the corresponding value of α.

*11.102 A player throws a ball with an initial velocity $\mathbf{v}_0$ of 18 m/s from point A. (a) Determine the maximum height h at which the ball can strike the wall, (b) the corresponding angle α.

11.103 A player throws a ball with an initial velocity $\mathbf{v}_0$ of 18 m/s from point A. Knowing that the ball strikes the wall at a height $h = 13.5$ m above the ground, determine the angle α.

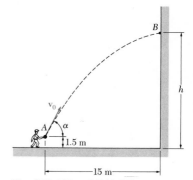

Fig. P11.102 and P11.103

11.104 A player throws a ball with an initial velocity $\mathbf{v}_0$ of 16 m/s from a point A located 1.5 m above the floor. Knowing that $h = 3.5$ m, determine the angle α for which the ball will strike the wall at point B.

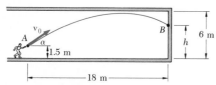

Fig. P11.104 and P11.105

11.105 A player throws a ball with an initial velocity $\mathbf{v}_0$ of 16 m/s from a point A located 1.5 m above the floor. Knowing that the ceiling of the gymnasium is 6 m high, determine the highest point B at which the ball can strike the wall 18 m away.

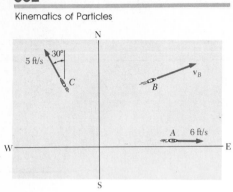

Fig. P11.106

11.106 The velocities of boats A and C are as shown and the relative velocity of boat B with respect to A is $v_{B/A} = 4$ ft/s $\angle 50°$. Determine (a) $v_{A/C}$, (b) $v_{C/B}$, (c) the change in position of B with respect to C during a 10-s interval. Also show that for any motion, $v_{B/A} + v_{C/B} + v_{A/C} = 0$.

11.107 Instruments in an airplane indicate that with respect to the air, the plane is moving north at a speed of 500 km/h. At the same time ground-based radar indicates that the plane is moving at a speed of 530 km/h in a direction 5° east of north. Determine the magnitude and direction of the velocity of the air.

11.108 Airplane A is flying due east at 700 km/h, while airplane B is flying at 500 km/h at the same altitude and in a direction to the west of south. Knowing that the speed of B with respect to A is 1125 km/h, determine the direction of the flight path of B.

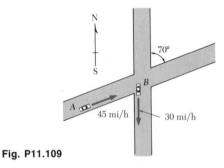

Fig. P11.109

11.109 Three seconds after automobile B passes through the intersection shown, automobile A passes through the same intersection. Knowing that the speed of each automobile is constant, determine (a) the relative velocity of B with respect to A, (b) the change in position of B with respect to A during a 4-s interval, (c) the distance between the two automobiles 2 s after A has passed through the intersection.

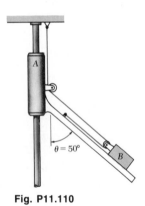

Fig. P11.110

11.110 Knowing that at the instant shown assembly A has a velocity of 9 in./s and an acceleration of 15 in./s² both directed downward, determine (a) the velocity of block B, (b) the acceleration of block B.

11.111 Pin P moves at a constant speed of 7.5 in./s in a counterclockwise sense along a circular slot which has been milled in the slider block A shown. Knowing that the block moves downward at a constant speed of 5 in./s, determine the velocity of pin P when (a) $\theta = 30°$, (b) $\theta = 120°$.

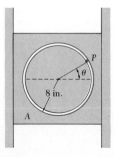

Fig. P11.111

11.112 and 11.113   At $t = 0$, wedge A starts moving to the left with a constant acceleration of 80 mm/s² and block B starts moving along the wedge toward the right with a constant acceleration of 120 mm/s² relative to the wedge. Determine (a) the acceleration of block B, (b) the velocity of block B when $t = 3$ s.

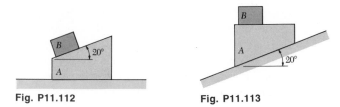

Fig. P11.112

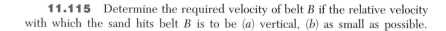

Fig. P11.113

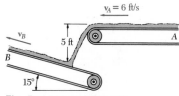

Fig. P11.114 and P11.115

11.114   The conveyor belt A moves with a constant velocity and discharges sand onto belt B as shown. Knowing that the velocity of belt B is 8 ft/s, determine the velocity of the sand relative to belt B as it lands on belt B.

11.115   Determine the required velocity of belt B if the relative velocity with which the sand hits belt B is to be (a) vertical, (b) as small as possible.

11.116   As observed from a ship moving due east at 9 km/h, the wind appears to blow from the south. After the ship has changed course and speed, and as it is moving due north at 6 km/h, the wind appears to blow from the southwest. Assuming that the wind velocity is constant during the period of observation, determine the magnitude and direction of the true wind velocity.

11.117   A small motorboat maintains a constant speed of 1.5 m/s relative to the water as it is maneuvering in a tidal current. When the boat is directed due east, it is observed from shore to move due south and when it is directed toward the northeast it is observed to move due west. Determine the speed and direction of the current.

11.118   During a rainstorm the paths of the raindrops appear to form an angle of 30° with the vertical and to be directed to the left when observed from a side window of a train moving at a speed of 10 mi/h. A short time later, after the speed of the train has increased to 16 mi/h, the angle between the vertical and the paths of the drops appears to be 45°. If the train were stopped, at what angle and with what velocity would the drops be observed to fall?

11.119   As the speed of the train of Prob. 11.118 increases, the angle between the vertical and the paths of the drops becomes equal to 60°. Determine the speed of the train at that time.

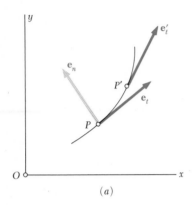

(a)

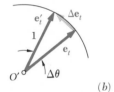

(b)

Fig. 11.21

11.13. Tangential and Normal Components. We saw in Sec. 11.9 that the velocity of a particle is a vector tangent to the path of the particle but that, in general, the acceleration is not tangent to the path. It is sometimes convenient to resolve the acceleration into components directed, respectively, along the tangent and the normal to the path of the particle.

Plane Motion of a Particle. We shall first consider a particle which moves along a curve contained in the plane of the figure. Let P be the position of the particle at a given instant. We attach at P a unit vector $\mathbf{e}_t$ tangent to the path of the particle and pointing toward the direction of motion (Fig. 11.21a). Let $\mathbf{e}_t'$ be the unit vector corresponding to the position P' of the particle at a later instant. Drawing both vectors from the same origin O', we define the vector $\Delta\mathbf{e}_t = \mathbf{e}_t' - \mathbf{e}_t$ (Fig. 11.21b). Since $\mathbf{e}_t$ and $\mathbf{e}_t'$ are of unit length, their tips lie on a circle of radius 1. Denoting by $\Delta\theta$ the angle formed by $\mathbf{e}_t$ and $\mathbf{e}_t'$, we find that the magnitude of $\Delta\mathbf{e}_t$ is $2\sin(\Delta\theta/2)$. Considering now the vector $\Delta\mathbf{e}_t/\Delta\theta$, we note that as $\Delta\theta$ approaches zero, this vector becomes tangent to the unit circle of Fig. 11.21b, i.e., perpendicular to $\mathbf{e}_t$, and that its magnitude approaches

$$\lim_{\Delta\theta\to 0}\frac{2\sin(\Delta\theta/2)}{\Delta\theta} = \lim_{\Delta\theta\to 0}\frac{\sin(\Delta\theta/2)}{\Delta\theta/2} = 1$$

Thus, the vector obtained at the limit is a unit vector along the normal to the path of the particle, in the direction toward which $\mathbf{e}_t$ turns. Denoting this vector by $\mathbf{e}_n$, we write

$$\mathbf{e}_n = \lim_{\Delta\theta\to 0}\frac{\Delta\mathbf{e}_t}{\Delta\theta}$$

$$\mathbf{e}_n = \frac{d\mathbf{e}_t}{d\theta} \tag{11.35}$$

Since the velocity $\mathbf{v}$ of the particle is tangent to the path, we may express it as the product of the scalar v and the unit vector $\mathbf{e}_t$. We have

$$\mathbf{v} = v\mathbf{e}_t \tag{11.36}$$

To obtain the acceleration of the particle, we shall differentiate (11.36) with respect to t. Applying the rule for the differentiation of the product of a scalar and a vector function (Sec. 11.10), we write

$$\mathbf{a} = \frac{d\mathbf{v}}{dt} = \frac{dv}{dt}\mathbf{e}_t + v\frac{d\mathbf{e}_t}{dt} \tag{11.37}$$

But

$$\frac{d\mathbf{e}_t}{dt} = \frac{d\mathbf{e}_t}{d\theta}\frac{d\theta}{ds}\frac{ds}{dt}$$

Recalling from (11.16) that $ds/dt = v$, from (11.35) that $d\mathbf{e}_t/d\theta = \mathbf{e}_n$, and from elementary calculus that $d\theta/ds$ is equal to $1/\rho$, where ρ is the radius of curvature of the path at P (Fig. 11.22), we have

$$\frac{d\mathbf{e}_t}{dt} = \frac{v}{\rho}\mathbf{e}_n \qquad (11.38)$$

Substituting into (11.37), we obtain

$$\mathbf{a} = \frac{dv}{dt}\mathbf{e}_t + \frac{v^2}{\rho}\mathbf{e}_n \qquad (11.39)$$

Thus, the scalar components of the acceleration are

$$a_t = \frac{dv}{dt} \qquad a_n = \frac{v^2}{\rho} \qquad (11.40)$$

The relations obtained express that the *tangential component* of the acceleration is equal to the *rate of change of the speed of the particle*, while the *normal component* is equal to the *square of the speed divided by the radius of curvature of the path at P*. Depending upon whether the speed of the particle increases or decreases, a_t is positive or negative, and the vector component $\mathbf{a}_t$ points in the direction of motion or against the direction of motion. The vector component $\mathbf{a}_n$, on the other hand, *is always directed toward the center of curvature C of the path* (Fig. 11.23).

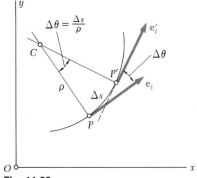

Fig. 11.22

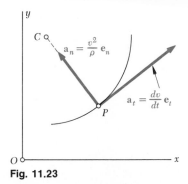

Fig. 11.23

It appears from the above that the tangential component of the acceleration reflects a change in the speed of the particle, while its normal component reflects a change in the direction of motion of the particle. The acceleration of a particle will be zero only if both its components are zero. Thus, the acceleration of a particle moving with constant speed along a curve will not be zero, unless the particle happens to pass through a point of inflection of the curve (where the radius of curvature is infinite) or unless the curve is a straight line.

The fact that the normal component of the acceleration depends upon the radius of curvature of the path followed by the particle is taken into account in the design of structures or mechanisms as widely different as airplane wings, railroad tracks, and cams. In order to avoid sudden changes in the acceleration of the air particles flowing past a wing, wing profiles are designed without any sudden change in curvature. Similar care is taken in designing railroad curves, to avoid sudden changes in the acceleration of the cars (which would be hard on the equipment and unpleasant for the passengers). A straight section of track, for instance, is never directly followed by a circular section. Special transition sections are used, to help pass smoothly from the infinite radius of curvature of the straight section to the finite radius of the circular track. Likewise, in the design of high-speed cams, abrupt changes in acceleration are avoided by using transition curves which produce a continuous change in acceleration.

Motion of a Particle in Space. The relations (11.39) and (11.40) still hold in the case of a particle moving along a space curve. However, since there is an infinite number of straight lines which are perpendicular to the tangent at a given point P of a space curve, it is then necessary to define more precisely the direction of the unit vector $\mathbf{e}_n$.

Let us consider again the unit vectors $\mathbf{e}_t$ and $\mathbf{e}'_t$ tangent to the path of the particle at two neighboring points P and P' (Fig. 11.24a) and the vector $\Delta\mathbf{e}_t$ representing the difference between $\mathbf{e}_t$ and $\mathbf{e}'_t$ (Fig. 11.24b). Let us now

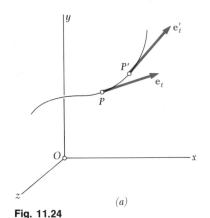

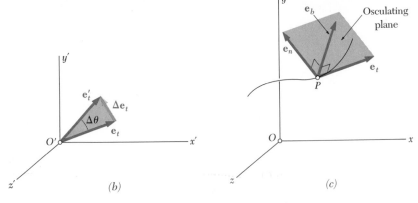

Fig. 11.24

imagine a plane through P (Fig. 11.24a) parallel to the plane defined by the vectors $\mathbf{e}_t$, $\mathbf{e}'_t$, and $\Delta\mathbf{e}_t$ (Fig. 11.24b). This plane contains the tangent to the curve at P and is parallel to the tangent at P'. If we let P' approach P, we obtain at the limit the plane which fits the curve most closely in the neighborhood of P. This plane is called the *osculating plane* at P.† It follows from this definition that the osculating plane contains the unit vector $\mathbf{e}_n$, since this vector represents the limit of the vector $\Delta\mathbf{e}_t/\Delta\theta$. The normal defined by $\mathbf{e}_n$ is thus contained in the osculating plane; it is called the *principal normal* at P. The unit vector $\mathbf{e}_b = \mathbf{e}_t \times \mathbf{e}_n$ which completes the right-handed triad $\mathbf{e}_t$, $\mathbf{e}_n$, $\mathbf{e}_b$ (Fig. 11.24c) defines the *binormal* at P. The binormal is thus perpendicular to the osculating plane. We conclude that as stated in (11.39), the acceleration of the particle at P may be resolved into two components, one along the tangent, the other along the principal normal at P. The acceleration has no component along the binormal.

---

† From the Latin *osculari*, to embrace.

11.14. Radial and Transverse Components.

In certain problems of plane motion, the position of the particle P is defined by its polar coordinates r and θ (Fig. 11.25a). It is then convenient to resolve the velocity and acceleration of the particle into components parallel and perpendicular, respectively, to the line OP. These components are called *radial* and *transverse components*.

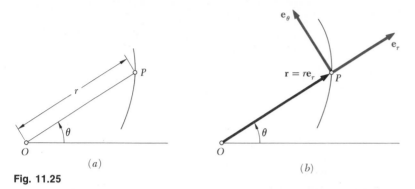

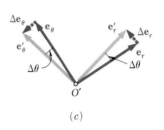

Fig. 11.25

We attach at P two unit vectors, $\mathbf{e}_r$ and $\mathbf{e}_\theta$ (Fig. 11.25b). The vector $\mathbf{e}_r$ is directed along OP and the vector $\mathbf{e}_\theta$ is obtained by rotating $\mathbf{e}_r$ through 90° counterclockwise. The unit vector $\mathbf{e}_r$ defines the *radial* direction, i.e., the direction in which P would move if r were increased and θ kept constant; the unit vector $\mathbf{e}_\theta$ defines the *transverse* direction, i.e., the direction in which P would move if θ were increased and r kept constant. A derivation similar to the one we used in Sec. 11.13 to determine the derivative of the unit vector $\mathbf{e}_t$ leads to the relations

$$\frac{d\mathbf{e}_r}{d\theta} = \mathbf{e}_\theta \qquad \frac{d\mathbf{e}_\theta}{d\theta} = -\mathbf{e}_r \qquad (11.41)$$

where $-\mathbf{e}_r$ denotes a unit vector of sense opposite to that of $\mathbf{e}_r$ (Fig. 11.25c). Using the chain rule of differentiation, we express the time derivatives of the unit vectors $\mathbf{e}_r$ and $\mathbf{e}_\theta$ as follows:

$$\frac{d\mathbf{e}_r}{dt} = \frac{d\mathbf{e}_r}{d\theta}\frac{d\theta}{dt} = \mathbf{e}_\theta\frac{d\theta}{dt} \qquad \frac{d\mathbf{e}_\theta}{dt} = \frac{d\mathbf{e}_\theta}{d\theta}\frac{d\theta}{dt} = -\mathbf{e}_r\frac{d\theta}{dt}$$

or, using dots to indicate differentiation with respect to t,

$$\dot{\mathbf{e}}_r = \dot{\theta}\mathbf{e}_\theta \qquad \dot{\mathbf{e}}_\theta = -\dot{\theta}\mathbf{e}_r \qquad (11.42)$$

To obtain the velocity $\mathbf{v}$ of the particle P, we express the position vector $\mathbf{r}$ of P as the product of the scalar r and the unit vector $\mathbf{e}_r$ and differentiate with respect to t:

$$\mathbf{v} = \frac{d}{dt}(r\mathbf{e}_r) = \dot{r}\mathbf{e}_r + r\dot{\mathbf{e}}_r$$

or, recalling the first of the relations (11.42),

$$\mathbf{v} = \dot{r}\mathbf{e}_r + r\dot{\theta}\mathbf{e}_\theta \qquad (11.43)$$

Differentiating again with respect to t to obtain the acceleration, we write

$$\mathbf{a} = \frac{d\mathbf{v}}{dt} = \ddot{r}\mathbf{e}_r + \dot{r}\dot{\mathbf{e}}_r + \dot{r}\dot{\theta}\mathbf{e}_\theta + r\ddot{\theta}\mathbf{e}_\theta + r\dot{\theta}\dot{\mathbf{e}}_\theta$$

or, substituting for $\dot{\mathbf{e}}_r$ and $\dot{\mathbf{e}}_\theta$ from (11.42) and factoring $\mathbf{e}_r$ and $\mathbf{e}_\theta$,

$$\mathbf{a} = (\ddot{r} - r\dot{\theta}^2)\mathbf{e}_r + (r\ddot{\theta} + 2\dot{r}\dot{\theta})\mathbf{e}_\theta \qquad (11.44)$$

The scalar components of the velocity and acceleration in the radial and transverse directions are therefore

$$v_r = \dot{r} \qquad\qquad v_\theta = r\dot{\theta} \qquad (11.45)$$

$$a_r = \ddot{r} - r\dot{\theta}^2 \qquad a_\theta = r\ddot{\theta} + 2\dot{r}\dot{\theta} \qquad (11.46)$$

It is important to note that a_r is *not* equal to the time derivative of v_r, and that a_θ is *not* equal to the time derivative of v_θ.

In the case of a particle moving along a circle of center O, we have $r = $ constant, $\dot{r} = \ddot{r} = 0$, and the formulas (11.43) and (11.44) reduce, respectively, to

$$\mathbf{v} = r\dot{\theta}\mathbf{e}_\theta \qquad \mathbf{a} = -r\dot{\theta}^2\mathbf{e}_r + r\ddot{\theta}\mathbf{e}_\theta \qquad (11.47)$$

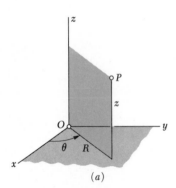

Extension to the Motion of a Particle in Space: Cylindrical Coordinates. The position of a particle P in space is sometimes defined by its cylindrical coordinates R, θ, and z (Fig. 11.26a). It is then convenient to use the unit vectors $\mathbf{e}_R$, $\mathbf{e}_\theta$, and $\mathbf{k}$ shown in Fig. 11.26b. Resolving the position vector $\mathbf{r}$ of the particle P into components along the unit vectors, we write

$$\mathbf{r} = R\mathbf{e}_R + z\mathbf{k} \qquad (11.48)$$

Observing that $\mathbf{e}_R$ and $\mathbf{e}_\theta$ define, respectively, the radial and transverse direction in the horizontal xy plane, and that the vector $\mathbf{k}$, which defines the *axial* direction, is constant in direction as well as in magnitude, we easily verify that

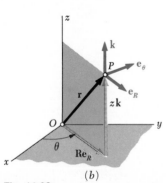

$$\mathbf{v} = \frac{d\mathbf{r}}{dt} = \dot{R}\mathbf{e}_R + R\dot{\theta}\mathbf{e}_\theta + \dot{z}\mathbf{k} \qquad (11.49)$$

$$\mathbf{a} = \frac{d\mathbf{v}}{dt} = (\ddot{R} - R\dot{\theta}^2)\mathbf{e}_R + (R\ddot{\theta} + 2\dot{R}\dot{\theta})\mathbf{e}_\theta + \ddot{z}\mathbf{k} \qquad (11.50)$$

Fig. 11.26

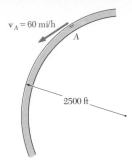

$\mathbf{v}_A = 60$ mi/h

A

2500 ft

SAMPLE PROBLEM 11.10

A motorist is traveling on a curved section of highway of radius 2500 ft at the speed of 60 mi/h. The motorist suddenly applies the brakes, causing the automobile to slow down at a constant rate. Knowing that after 8 s the speed has been reduced to 45 mi/h, determine the acceleration of the automobile immediately after the brakes have been applied.

Tangential Component of Acceleration. First the speeds are expressed in ft/s.

$$60 \text{ mi/h} = \left(60 \frac{\text{mi}}{\text{h}}\right)\left(\frac{5280 \text{ ft}}{1 \text{ mi}}\right)\left(\frac{1 \text{ h}}{3600 \text{ s}}\right) = 88 \text{ ft/s}$$

$$45 \text{ mi/h} = 66 \text{ ft/s}$$

Since the automobile slows down at a constant rate, we have

$$a_t = \text{average } a_t = \frac{\Delta v}{\Delta t} = \frac{66 \text{ ft/s} - 88 \text{ ft/s}}{8 \text{ s}} = -2.75 \text{ ft/s}^2$$

Normal Component of Acceleration. Immediately after the brakes have been applied, the speed is still 88 ft/s, and we have

$$a_n = \frac{v^2}{\rho} = \frac{(88 \text{ ft/s})^2}{2500 \text{ ft}} = 3.10 \text{ ft/s}^2$$

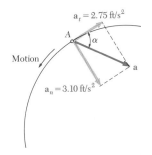

$a_t = 2.75$ ft/s²

A

α

Motion

$a_n = 3.10$ ft/s²

a

Magnitude and Direction of Acceleration. The magnitude and direction of the resultant **a** of the components $\mathbf{a}_n$ and $\mathbf{a}_t$ are

$$\tan \alpha = \frac{a_n}{a_t} = \frac{3.10 \text{ ft/s}^2}{2.75 \text{ ft/s}^2} \qquad \alpha = 48.4° \quad \blacktriangleleft$$

$$a = \frac{a_n}{\sin \alpha} = \frac{3.10 \text{ ft/s}^2}{\sin 48.4°} \qquad a = 4.14 \text{ ft/s}^2 \quad \blacktriangleleft$$

SAMPLE PROBLEM 11.11

Determine the minimum radius of curvature of the trajectory described by the projectile considered in Sample Prob. 11.7.

Solution. Since $a_n = v^2/\rho$, we have $\rho = v^2/a_n$. The radius will be small when v is small or when a_n is large. The speed v is minimum at the top of the trajectory since $v_y = 0$ at that point; a_n is maximum at that same point, since the direction of the vertical coincides with the direction of the normal. Therefore, the minimum radius of curvature occurs at the top of the trajectory. At this point, we have

$$v = v_x = 155.9 \text{ m/s} \qquad a_n = a = 9.81 \text{ m/s}^2$$

$$\rho = \frac{v^2}{a_n} = \frac{(155.9 \text{ m/s})^2}{9.81 \text{ m/s}^2} \qquad \rho = 2480 \text{ m} \quad \blacktriangleleft$$

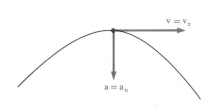

$v = v_x$

$a = a_n$

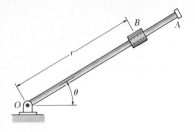

SAMPLE PROBLEM 11.12

The rotation of the 0.9-m arm OA about O is defined by the relation $\theta = 0.15t^2$, where θ is expressed in radians and t in seconds. Collar B slides along the arm in such a way that its distance from O is $r = 0.9 - 0.12t^2$, where r is expressed in meters and t in seconds. After the arm OA has rotated through 30°, determine (a) the total velocity of the collar, (b) the total acceleration of the collar, (b) the relative acceleration of the collar with respect to the arm.

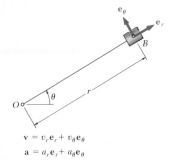

$$\mathbf{v} = v_r \mathbf{e}_r + v_\theta \mathbf{e}_\theta$$
$$\mathbf{a} = a_r \mathbf{e}_r + a_\theta \mathbf{e}_\theta$$

Solution. We first find the time t at which $\theta = 30°$. Substituting $\theta = 30° = 0.524$ rad into the expression for θ, we obtain

$$\theta = 0.15t^2 \qquad 0.524 = 0.15t^2 \qquad t = 1.869 \text{ s}$$

Equations of Motion. Substituting $t = 1.869$ s in the expressions for r, θ, and their first and second derivatives, we have

$$r = 0.9 - 0.12t^2 = 0.481 \text{ m} \qquad \theta = 0.15t^2 = 0.524 \text{ rad}$$
$$\dot{r} = -0.24t = -0.449 \text{ m/s} \qquad \dot{\theta} = 0.30t = 0.561 \text{ rad/s}$$
$$\ddot{r} = -0.24 = -0.240 \text{ m/s}^2 \qquad \ddot{\theta} = 0.30 = 0.300 \text{ rad/s}^2$$

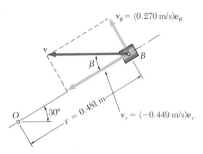

a. Velocity of B. Using Eqs. (11.45), we obtain the values of v_r and v_θ when $t = 1.869$ s.

$$v_r = \dot{r} = -0.449 \text{ m/s}$$
$$v_\theta = r\dot{\theta} = 0.481(0.561) = 0.270 \text{ m/s}$$

Solving the right triangle shown, we obtain the magnitude and direction of the velocity,

$$v = 0.524 \text{ m/s} \qquad \beta = 31.0° \quad \blacktriangleleft$$

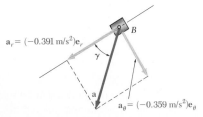

b. Acceleration of B. Using Eqs. (11.46), we obtain

$$a_r = \ddot{r} - r\dot{\theta}^2$$
$$\quad = -0.240 - 0.481(0.561)^2 = -0.391 \text{ m/s}^2$$
$$a_\theta = r\ddot{\theta} + 2\dot{r}\dot{\theta}$$
$$\quad = 0.481(0.300) + 2(-0.449)(0.561) = -0.359 \text{ m/s}^2$$
$$a = 0.531 \text{ m/s}^2 \qquad \gamma = 42.6° \quad \blacktriangleleft$$

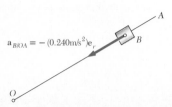

c. Acceleration of B with Respect to Arm OA. We note that the motion of the collar with respect to the arm is rectilinear and defined by the coordinate r. We write

$$a_{B/OA} = \ddot{r} = -0.240 \text{ m/s}^2$$
$$a_{B/OA} = 0.240 \text{ m/s}^2 \text{ toward } O. \quad \blacktriangleleft$$

Problems

11.120 What is the smallest radius which should be used for a highway curve if the normal component of the acceleration of a car traveling at 45 mi/h is not to exceed 2.4 ft/s^2?

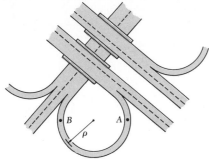

Fig. P11.121

11.121 A motorist drives along the circular exit ramp of a turnpike at the constant speed v_0. Knowing that the odometer indicates a distance of 0.8 km between point A where the automobile is going due south and B where it is going due north, determine the speed v_0 for which the normal component of the acceleration is $0.1g$.

11.122 Determine the peripheral speed of the centrifuge test cab A for which the normal component of the acceleration is $10g$.

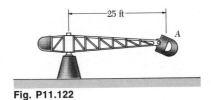

Fig. P11.122

11.123 A computer tape moves over two drums at a constant speed v_0. Knowing that the normal component of the acceleration of the portion of tape in contact with drum B is 480 ft/s^2, determine (*a*) the speed v_0, (*b*) the normal component of the acceleration of the portion of tape in contact with drum A.

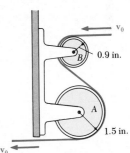

Fig. P11.123

11.124 A motorist is traveling on a curved portion of highway of radius 350 m at a speed of 72 km/h. The brakes are suddenly applied, causing the speed to decrease at a constant rate of 1.25 m/s^2. Determine the magnitude of the total acceleration of the automobile (*a*) immediately after the brakes have been applied, (*b*) 4 s later.

11.125 A bus starts from rest on a curve of 300-m radius and accelerates at the constant rate $a_t = 0.75 \text{ m/s}^2$. Determine the distance and time that the bus will travel before the magnitude of its total acceleration is 0.9 m/s^2.

11.126 The speed of a racing car is increased at a constant rate from 60 mi/h to 75 mi/h over a distance of 600 ft along a curve of 800-ft radius. Determine the magnitude of the total acceleration of the car after it has traveled 400 ft along the curve.

11.127 A monorail train starts from rest on a curve of radius 1200 ft and accelerates at the constant rate a_t. If the maximum total acceleration of the train must not exceed 4 ft/s², determine (a) the shortest distance in which the train can reach a speed of 45 mi/h, (b) the corresponding constant rate of acceleration a_t.

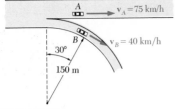

Fig. P11.128

11.128 Automobile A is traveling along a straight highway, while B is moving along a circular exit ramp of 150-m radius. The speed of A is being increased at the rate of 1.5 m/s² and the speed of B is being decreased at the rate of 0.9 m/s². For the position shown, determine (a) the velocity of A relative to B, (b) the acceleration of A relative to B.

11.129 Solve Prob. 11.128, assuming that the velocity of B is 20 km/h and is being decreased at the rate of 0.9 m/s².

11.130 A nozzle discharges a stream of water in the direction shown with an initial velocity of 25 ft/s. Determine the radius of curvature of the stream (a) as it leaves the nozzle, (b) at the maximum height of the stream.

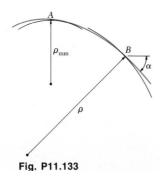

Fig. P11.130

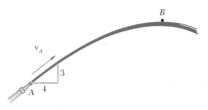

Fig. P11.131

11.131 From measurements of a photograph it has been found that as the stream of water shown left the nozzle at A, it had a radius of curvature of 35 m. Determine (a) the initial velocity $\mathbf{v}_A$ of the stream, (b) the radius of curvature of the stream at its maximum height.

11.132 Determine the radius of curvature of the trajectory described by the projectile of Sample Prob. 11.7 (a) as the projectile leaves the gun, (b) at the maximum elevation of the projectile.

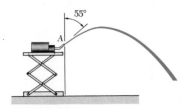

Fig. P11.133

11.133 (a) Show that the radius of curvature of the trajectory of a projectile reaches its minimum value at the highest point A of the trajectory. (b) Denoting by α the angle formed by the trajectory and the horizontal at a given point B, show that the radius of curvature of the trajectory at B is $\rho = \rho_{min}/\cos^3 \alpha$.

*11.134 Determine the radius of curvature of the helix of Prob. 11.86.

*11.135 Determine the radius of curvature of the path described by the particle of Prob. 11.87 when (a) $t = 0$, (b) $t = 2$ s.

11.136 A satellite will travel indefinitely in a circular orbit around the earth if the normal component of its acceleration is equal to $g(R/r)^2$, where $g = 32.2$ ft/s^2, R = radius of the earth = 3960 mi, and r = distance from the center of the earth to the satellite. Determine the height above the surface of the earth at which a satellite will travel indefinitely around the earth at a speed of 16,000 mi/h.

11.137 Determine the speed of a space shuttle traveling in a circular orbit 140 mi above the surface of the earth. (See information given in Prob. 11.136.)

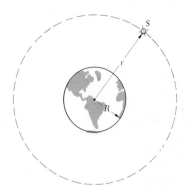

Fig. P11.136

11.138 Show that the speed of an earth satellite traveling in a circular orbit is inversely proportional to the square root of the radius of its orbit. Knowing that the radius of the earth is 6370 km and that $g = 9.81$ m/s^2, determine the minimum time required for a satellite to circle the earth. (See information given in Prob. 11.136.)

11.139 Assuming that the orbit of the moon is a circle of radius 384×10^3 km, determine the speed of the moon relative to the earth. (See information given in Probs. 11.136 and 11.138.)

11.140 The rotation of rod OA about O is defined by the relation $\theta = 2t^2$, where θ is expressed in radians and t in seconds. Collar B slides along the rod in such a way that its distance from O is $r = 60t^2 - 20t^3$, where r is expressed in inches and t in seconds. When $t = 1$ s, determine (a) the velocity of the collar, (b) the total acceleration of the collar, (c) the acceleration of the collar relative to the rod.

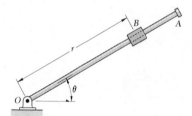

Fig. P11.140 and P11.141

11.141 The rotation of rod OA about O is defined by the relation $\theta = \frac{1}{2}\pi(4t - 3t^2)$, where θ is expressed in radians and t in seconds. Collar B slides along the rod in such a way that its distance from O is $r = 1.25t^2 - 0.9t^3$, where r is expressed in meters and t in seconds. When $t = 1$ s, determine (a) the velocity of the collar, (b) the total acceleration of the collar, (c) the acceleration of the collar relative to the rod.

11.142 The two-dimensional motion of a particle is defined by the relations $r = 2b \cos \omega t$ and $\theta = \omega t$, where b and ω are constants. Determine (a) the velocity and acceleration of the particle at any instant, (b) the radius of curvature of its path. What conclusion can you draw regarding the path of the particle?

11.143 The path of a particle P is an Archimedean spiral. The motion of the particle is defined by the relations $r = 10t$ and $\theta = 2\pi t$, where r is expressed in inches, t in seconds, and θ in radians. Determine the velocity and acceleration of the particle when (a) $t = 0$, (b) $t = 0.25$ s.

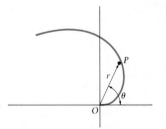

Fig. P11.143

11.144 As circle B rolls on the fixed circle A, point P describes a cardioid defined by the relations $r = 2b(1 + \cos \frac{1}{2}\pi t)$ and $\theta = \frac{1}{2}\pi t$, where t is expressed in seconds. Determine the velocity and acceleration of P when (a) $t = 0$, (b) $t = 1$ s.

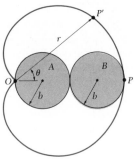

Fig. P11.144

11.145 Solve Prob. 11.144, when (a) $t = 0.5$ s, (b) $t = 2$ s.

11.146 A rocket is fired vertically from a launching pad at B. Its flight is tracked by radar from point A. Determine the velocity of the rocket in terms of b, θ, and $\dot{\theta}$.

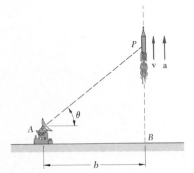

Fig. P11.146 and P11.150

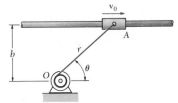

Fig. P11.147

11.147 A wire OA connects the collar A and a reel located at O. Knowing that the collar moves to the right at a constant speed v_0, determine $\dot{\theta}$ in terms of v_0, b, and θ.

11.148 Determine the acceleration of the rocket of Prob. 11.146 in terms of b, θ, $\dot{\theta}$, and $\ddot{\theta}$.

11.149 In Prob. 11.147, determine $\ddot{\theta}$ in terms of v_0, b, and θ.

11.150 A test rocket is fired vertically from a launching pad at B. When the rocket is at P the angle of elevation is $\theta = 47.0°$, and 0.5 s later it is $\theta = 48.0°$. Knowing that $b = 4$ km, determine approximately the speed of the rocket during the 0.5-s interval.

11.151 An airplane passes over a radar tracking station at A and continues to fly due east. When the airplane is at P, the distance and angle of elevation of the plane are, respectively, $r = 12{,}600$ ft and $\theta = 31.2°$. Two seconds later the radar station sights the plane at $r = 13{,}600$ ft and $\theta = 28.3°$. Determine approximately the speed and the angle of dive α of the plane during the 2-s interval.

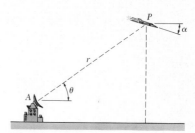

Fig. P11.151

11.152 and 11.153 A particle moves along the spiral shown; determine the magnitude of the velocity of the particle in terms of b, θ, and $\dot{\theta}$.

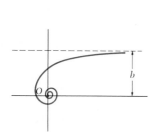

Hyperbolic spiral $r\theta = b$
Fig. P11.152 and P11.154

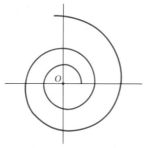

Logarithmic spiral $r = e^{b\theta}$
Fig. P11.153 and P11.155

11.154 and 11.155 A particle moves along the spiral shown. Knowing that $\dot{\theta}$ is constant and denoting this constant by ω, determine the magnitude of the acceleration of the particle in terms of b, θ, and ω.

11.156 As rod OA rotates, pin P moves along the parabola BCD. Knowing that the equation of this parabola is $r = 2b/(1 + \cos \theta)$ and that $\theta = kt$, determine the velocity and acceleration of P when (a) $\theta = 0$, (b) $\theta = 90°$.

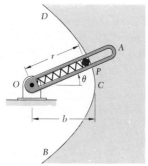

Fig. P11.156

11.157 The pin at B is free to slide along the circular slot DE and along the rotating rod OC. Assuming that the rod OC rotates at a constant rate $\dot{\theta}$, (a) show that the acceleration of pin B is of constant magnitude, (b) determine the direction of the acceleration of pin B.

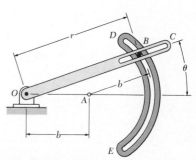

Fig. P11.157

11.158 The motion of a particle on the surface of a right circular cylinder is defined by the relations $R = A$, $\theta = 2\pi t$, and $z = B \sin 2\pi nt$, where A and B are constants and n is an integer. Determine the magnitudes of the velocity and acceleration of the particle at any time t.

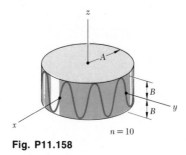

Fig. P11.158

11.159 For the case when $n = 1$ in Prob. 11.158, (*a*) show that the path of the particle is contained in a plane, (*b*) determine the maximum and minimum radii of curvature of the path.

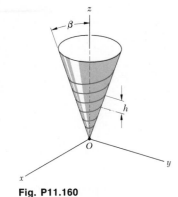

Fig. P11.160

11.160 The motion of a particle on the surface of a right circular cone is defined by the relations $R = ht \tan \beta$, $\theta = 2\pi t$, and $z = ht$, where β is the apex angle of the cone and h is the distance the particle rises in one passage around the cone. Determine the magnitudes of the velocity and acceleration at any time t.

11.161 The three-dimensional motion of a particle is defined by the relations $R = A(1 - e^{-t})$, $\theta = 2\pi t$, and $z = B(1 - e^{-t})$. Determine the magnitudes of the velocity and acceleration when (*a*) $t = 0$, (*b*) $t = \infty$.

11.162 The three-dimensional motion of a particle is defined by the relations $R = A$, $\theta = 2\pi t$, and $z = A \sin^2 \pi t$. Determine the magnitudes of the velocity and acceleration at any time t.

11.163 The three-dimensional motion of a particle is defined by the relations $R = 2k \cos t$, $\theta = t$, and $z = pt$. Determine (*a*) the path of the particle, (*b*) the magnitudes of the velocity and acceleration at any time t, (*c*) the radius of curvature of the path at any time t.

*11.164 For the helix of Prob. 11.86, determine the angle that the osculating plane forms with the y axis.

*11.165 Determine the direction of the binormal of the path described by the particle of Prob. 11.87 when (*a*) $t = 0$, (*b*) $t = 2$ s.

Review and Summary

In the first half of the chapter, we analyzed the *rectilinear motion of a particle*, i.e., the motion of a particle along a straight line. To define the position P of the particle on that line, we chose a fixed origin O and a positive direction (Fig. 11.27). The distance x from O to P, with the appropriate sign, completely defines the position of the particle on the line and is called the *position coordinate* of the particle [Sec. 11.2].

Position coordinate of a particle in rectilinear motion

Fig. 11.27

The *velocity v* of the particle was shown to be equal to the time derivative of the position coordinate x,

$$v = \frac{dx}{dt} \tag{11.1}$$

Velocity and acceleration in rectilinear motion

and the *acceleration a* was obtained by differentiating v with respect to t,

$$a = \frac{dv}{dt} \tag{11.2}$$

or

$$a = \frac{d^2x}{dt^2} \tag{11.3}$$

We also noted that a could be expressed as

$$a = v\frac{dv}{dx} \tag{11.4}$$

We observed that the velocity v and the acceleration a were represented by algebraic numbers which may be positive or negative. A positive value for v indicates that the particle moves in the positive direction, and a negative value that it moves in the negative direction. A positive value for a, however, may mean that the particle is truly accelerated (i.e., moves faster) in the positive direction, or that it is decelerated (i.e., moves more slowly) in the negative direction. A negative value for a is subject to a similar interpretation [Sample Prob. 11.1].

In most problems, the conditions of motion of a particle are defined by the type of acceleration that the particle possesses and by the initial conditions [Sec. 11.3]. The velocity and position of the particle may then be obtained by integrating two of the equations (11.1) to (11.4). Which of these equations should be selected depends upon the type of acceleration involved [Sample Probs. 11.2 and 11.3].

Determination of the velocity and acceleration by integration

Two types of motion are frequently encountered: the *uniform recti-linear motion* [Sec. 11.4], in which the velocity v of the particle is constant and

Uniform rectilinear motion

$$x = x_0 + vt \tag{11.5}$$

and the *uniformly accelerated rectilinear motion* [Sec. 11.5], in which the acceleration a of the particle is constant and we have

Uniformly accelerated rectilinear motion

$$v = v_0 + at \tag{11.6}$$
$$x = x_0 + v_0 t + \tfrac{1}{2}at^2 \tag{11.7}$$
$$v^2 = v_0^2 + 2a(x - x_0) \tag{11.8}$$

Relative motion of two particles

When two particles A and B move along the same straight line, we may wish to consider the *relative motion* of B with respect to A [Sec. 11.6].

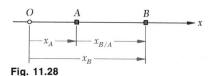

Fig. 11.28

Denoting by $x_{B/A}$ the *relative position coordinate* of B with respect to A (Fig. 11.28), we had

$$x_B = x_A + x_{B/A} \tag{11.9}$$

Differentiating Eq. (11.9) twice with respect to t, we obtained successively

$$v_B = v_A + v_{B/A} \tag{11.10}$$
$$a_B = a_A + a_{B/A} \tag{11.11}$$

where $v_{B/A}$ and $a_{B/A}$ represent, respectively, the *relative velocity* and the *relative acceleration* of B with respect to A.

Blocks connected by inextensible cords

When several blocks are *connected by inextensible cords*, it is possible to write a *linear relation* between their position coordinates. Similar relations may then be written between their velocities and between their accelerations and used to analyze their motion [Sample Prob. 11.5].

Graphical solutions

It is sometimes convenient to use a *graphical solution* for problems involving the rectilinear motion of a particle [Secs. 11.7 and 11.8]. The graphical solution most commonly used involves the x–t, v–t, and a–t curves [Sec. 11.7; Sample Prob. 11.6]. It was shown that, at any given time t,

$$v = \text{slope of } x\text{–}t \text{ curve}$$
$$a = \text{slope of } v\text{–}t \text{ curve}$$

while, over any given time interval from t_1 to t_2,

$$v_2 - v_1 = \text{area under } a\text{–}t \text{ curve}$$
$$x_2 - x_1 = \text{area under } v\text{–}t \text{ curve}$$

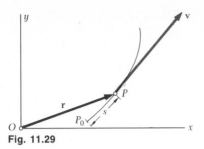

Fig. 11.29

In the second half of the chapter, we analyzed the *curvilinear motion of a particle*, i.e., the motion of a particle along a curved path. The position P of the particle at a given time [Sec. 11.9] was defined by the *position vector* **r** joining the origin O of the coordinates and point P (Fig. 11.29). The *velocity* **v** of the particle was defined by the relation

$$\mathbf{v} = \frac{d\mathbf{r}}{dt} \qquad (11.15)$$

and was found to be a *vector tangent to the path of the particle* and of magnitude v (called the *speed* of the particle) equal to the time derivative of the length s of the arc described by the particle:

$$v = \frac{ds}{dt} \qquad (11.16)$$

The *acceleration* **a** of the particle was defined by the relation

$$\mathbf{a} = \frac{d\mathbf{v}}{dt} \qquad (11.18)$$

and we noted that, in general, *the acceleration is not tangent to the path of the particle*.

Before proceeding to the consideration of the components of velocity and acceleration, we reviewed the formal definition of the derivative of a vector function and established a few rules governing the differentiation of sums and products of vector functions. We then showed that the rate of change of a vector is the same with respect to a fixed frame and with respect to a frame in translation [Sec. 11.10].

Denoting by x, y, and z the rectangular coordinates of a particle P, we found that the rectangular components of the velocity and acceleration of P equal, respectively, the first and second derivatives with respect to t of the corresponding coordinates:

$$v_x = \dot{x} \qquad v_y = \dot{y} \qquad v_z = \dot{z} \qquad (11.29)$$
$$a_x = \ddot{x} \qquad a_y = \ddot{y} \qquad a_z = \ddot{z} \qquad (11.30)$$

Position vector and velocity in curvilinear motion

Acceleration in curvilinear motion

Derivative of a vector function

Rectangular components of velocity and acceleration

Component motions

Relative motion of two particles

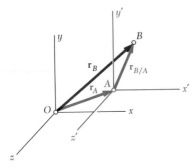

Fig. 11.30

Tangential and normal components

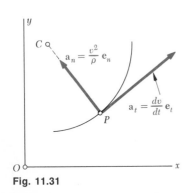

Fig. 11.31

Motion along a space curve

When the component a_x of the acceleration depends only upon t, x, and/or v_x, and when similarly a_y depends only upon t, y, and/or v_y, and a_z upon t, z, and/or v_z, Eqs. (11.30) may be integrated independently. The analysis of the given curvilinear motion may thus be reduced to the analysis of three independent rectilinear component motions [Sec. 11.11]. This approach is particularly effective in the study of the motion of projectiles [Sample Probs. 11.7 and 11.8].

For two particles A and B moving in space (Fig. 11.30), we considered the relative motion of B with respect to A, or more precisely, with respect to a moving frame attached to A and in translation with A [Sec. 11.12]. Denoting by $\mathbf{r}_{B/A}$ the *relative position vector* of B with respect to A (Fig. 11.30), we had

$$\mathbf{r}_B = \mathbf{r}_A + \mathbf{r}_{B/A} \qquad (11.31)$$

Denoting by $\mathbf{v}_{B/A}$ and $\mathbf{a}_{B/A}$, respectively, the *relative velocity* and the *relative acceleration* of B with respect to A, we also showed that

$$\mathbf{v}_B = \mathbf{v}_A + \mathbf{v}_{B/A} \qquad (11.33)$$

and

$$\mathbf{a}_B = \mathbf{a}_A + \mathbf{a}_{B/A} \qquad (11.34)$$

It is sometimes convenient to resolve the velocity and acceleration of a particle P into components other than the rectangular x, y, and z components. For a particle P moving along a path contained in a plane, we attached to P unit vectors $\mathbf{e}_t$ tangent to the path and $\mathbf{e}_n$ normal to the path and directed toward the center of curvature of the path [Sec. 11.13]. We then expressed the velocity and acceleration of the particle in terms of tangential and normal components. We wrote

$$\mathbf{v} = v\mathbf{e}_t \qquad (11.36)$$

and

$$\mathbf{a} = \frac{dv}{dt}\mathbf{e}_t + \frac{v^2}{\rho}\mathbf{e}_n \qquad (11.39)$$

where v is the speed of the particle and ρ the radius of curvature of its path [Sample Probs. 11.10 and 11.11]. We observed that while the velocity $\mathbf{v}$ is directed along the tangent to the path, the acceleration $\mathbf{a}$ consists of a component $\mathbf{a}_t$ directed along the tangent to the path and a component $\mathbf{a}_n$ directed toward the center of curvature of the path (Fig. 11.31).

For a particle P moving along a space curve we defined the plane which most closely fits the curve in the neighborhood of P as the *osculating plane*. This plane contains the unit vectors $\mathbf{e}_t$ and $\mathbf{e}_n$ which define, respectively, the tangent and principal normal to the curve. The unit vector $\mathbf{e}_b$ which is perpendicular to the osculating plane defines the *binormal*.

When the position of a particle P moving in a plane is defined by its polar coordinates r and θ, it is convenient to use radial and transverse components directed, respectively, along the position vector $\mathbf{r}$ of the particle and in the direction obtained by rotating $\mathbf{r}$ through 90° counterclockwise [Sec. 11.14]. We attached to P unit vectors $\mathbf{e}_r$ and $\mathbf{e}_\theta$ directed, respectively, in the radial and transverse directions (Fig. 11.32). We then expressed the velocity and acceleration of the particle in terms of radial and transverse components

$$\mathbf{v} = \dot{r}\mathbf{e}_r + r\dot{\theta}\mathbf{e}_\theta \qquad (11.43)$$

$$\mathbf{a} = (\ddot{r} - r\dot{\theta}^2)\mathbf{e}_r + (r\ddot{\theta} + 2\dot{r}\dot{\theta})\mathbf{e}_\theta \qquad (11.44)$$

where dots are used to indicate differentiation with respect to time. The scalar components of the velocity and acceleration in the radial and transverse directions are therefore

$$\begin{aligned} v_r &= \dot{r} & v_\theta &= r\dot{\theta} & (11.45) \\ a_r &= \ddot{r} - r\dot{\theta}^2 & a_\theta &= r\ddot{\theta} + 2\dot{r}\dot{\theta} & (11.46) \end{aligned}$$

It is important to note that a_r is *not* equal to the time derivative of v_r, and that a_θ is *not* equal to the time derivative of v_θ [Sample Prob. 11.12].

The chapter ended with a discussion of the use of cylindrical coordinates to define the position and motion of a particle in space.

Radial and transverse components

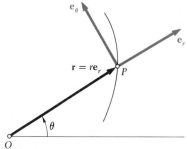

Fig. 11.32

Review Problems

11.166 The speed of a racing car is increased at a constant rate from 54 km/h to 90 km/h over a distance of 150 m along a curve of 240-m radius. Determine the magnitude of the total acceleration of the car after it has traveled 100 m along the curve.

11.167 The magnitude in m/s² of the deceleration due to air resistance of the nose cone of a small experimental rocket is known to be $0.0005v^2$, where v is expressed in m/s. If the nose cone is projected vertically from the ground with an initial velocity of 150 m/s, determine the maximum height that it will reach. [*Hint.* The total acceleration is $-(g + 0.0005v^2)$, where $g = 9.81$ m/s².]

11.168 The motion of a vibrating particle is defined by the position vector $\mathbf{r} = (8 \sin \pi t)\mathbf{i} - (2 \cos 2\pi t)\mathbf{j}$, where r is expressed in inches and t in seconds. (*a*) Determine the velocity and acceleration when $t = 1$ s. (*b*) Show that the path of the particle is parabolic.

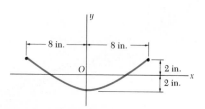

Fig. P11.168

11.169 A nozzle discharges a stream of water with an initial velocity v_0 of 40 ft/s into the end of a horizontal pipe of inside diameter $d = 4$ ft. Determine the largest distance x that the stream can reach.

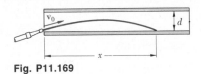

Fig. P11.169

11.170 Two airplanes A and B are flying at the same altitude; plane A is flying due east at a constant speed of 800 km/h, while plane B is flying southwest at a constant speed of 500 km/h. Determine the change in position of plane B relative to plane A, which takes place during a 3-min interval.

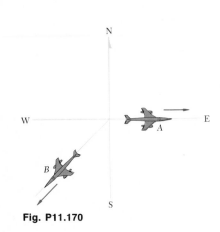

Fig. P11.170

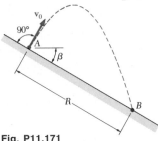

Fig. P11.171

11.171 A ball is projected from point A with a velocity v_0 which is perpendicular to the incline shown. Knowing that the ball strikes the incline at B, determine the range R in terms of v_0 and β.

11.172 In Prob. 11.171, determine the range R when $v_0 = 10$ m/s and $\beta = 30°$.

11.173 Drops of water are observed to drip from a faucet at constant intervals of time. As any drop B begins to fall freely, the preceding drop A has already fallen 250 mm. Determine the distance drop A will have fallen by the time the distance between A and B will have increased to 750 mm.

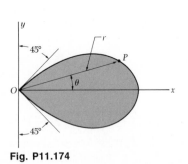

Fig. P11.174

11.174 The motion of the particle P is defined by the relations $r = 6 \cos 2\pi t$ and $\theta = \pi t$, where r is expressed in inches, t in seconds, and θ in radians. Determine the velocity and acceleration of the particle when (*a*), $t = 0$, (*b*) $t = 0.25$ s.

11.175 A motorcycle and an automobile are both traveling at the constant speed of 40 mi/h; the motorcycle is 60 ft behind the automobile. The motorcyclist wants to pass the automobile, i.e., he wishes to place his motorcycle at B, 60 ft in front of the automobile, and then resume the speed of 40 mi/h. The maximum acceleration of the motorcycle is 5 ft/s^2 and the maximum deceleration obtained by applying the brakes is 15 ft/s^2. What is the shortest time in which the motorcyclist can complete the passing operation if he does not at any time exceed a speed of 55 mi/h? Draw the v–t curve.

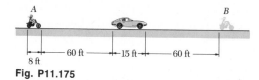

Fig. P11.175

11.176 Knowing that block A moves down to the left with a constant velocity of 80 mm/s, determine (a) the velocity of block B, (b) the change in position of block A relative to block B, which takes place in 4 s.

11.177 At $t = 0$, block A starts from rest and moves down to the left with a constant acceleration of 48 mm/s^2. Determine (a) the acceleration of block B, (b) the velocity of block B relative to block A when $t = 5$ s.

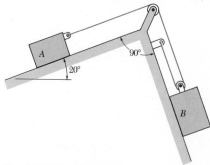

Fig. P11.176 and P11.177

The following problems are designed to be solved with a computer.

11.C1 Several test firings are to be made of the experimental rocket of Prob. 11.167. Write a computer program and calculate, for values of the initial velocity from 0 to 400 m/s at 25-m/s intervals, the maximum height reached by the nose cone of the rocket and its velocity as it returns to the ground.

11.C2 A nozzle at A discharges water with an initial velocity v_0 at an angle of 60° with the horizontal. Write a computer program and use it to determine where the stream of water strikes the ground or the building for values of v_0 from 4 m/s to 16 m/s at 1-m/s intervals.

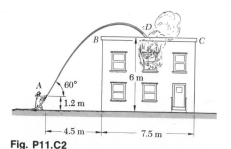

Fig. P11.C2

11.C3 A gun located on a plain fires a projectile with a velocity v_0 at an angle α with the horizontal. After being fired, the projectile is subjected to the acceleration of gravity, $g = -32.2 \text{ ft/s}^2$, directed vertically downward and to a deceleration due to air resistance, $a_t = -kv^2$, in a direction opposite to that of its velocity $\mathbf{v}$. Express the horizontal and vertical components a_x and a_y of the acceleration of the projectile in terms of g, k, and the components v_x and v_y of its velocity. Considering successive time intervals Δt and assuming the acceleration to remain constant over each time interval, use the expressions obtained for a_x and a_y to write a computer program to calculate at any instant the position of the projectile on its trajectory. Knowing that $v_0 = 1200 \text{ ft/s}$, $\alpha = 40°$, and using time intervals $\Delta t = 0.5 \text{ s}$, determine the range of the projectile and print the coordinates of the projectile along its trajectory at 5-s intervals, assuming (*a*) $k = 0$, (*b*) $k = 2 \times 10^{-6} \text{ ft}^{-1}$, (*c*) $k = 10 \times 10^{-6} \text{ ft}^{-1}$, (*d*) $k = 20 \times 10^{-6} \text{ ft}^{-1}$.

11.C4 As a train enters a curve of radius 4000 ft at a speed of 60 mi/h, the brakes are applied sufficiently to cause the magnitude of the total acceleration of the train to be 2.5 ft/s². After 1 s the brakes are more fully applied so that the magnitude of the total acceleration is again 2.5 ft/s². This brake setting is maintained for 1 s when the brakes are again more fully applied so that the total acceleration is again 2.5 ft/s². If this procedure is repeated each second, determine the total time required to bring the train to a stop.

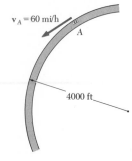

$v_A = 60 \text{ mi/h}$

A

4000 ft

Fig. P11.C4

Kinetics of Particles: Newton's Second Law

12.1. Introduction. Newton's first and third laws of motion were used extensively in statics to study bodies at rest and the forces acting upon them. These two laws are also used in dynamics; in fact, they are sufficient for the study of the motion of bodies which have no acceleration. However, when bodies are accelerated, i.e., when the magnitude or the direction of their velocity changes, it is necessary to use Newton's second law of motion to relate the motion of the body with the forces acting on it.

In this chapter we shall discuss Newton's second law and apply it to the analysis of the motion of particles. As we state in Sec. 12.2, if the resultant of the forces acting on a particle is not zero, the particle will have an acceleration proportional to the magnitude of the resultant and in the direction of this resultant force. Moreover, the ratio of the magnitudes of the resultant force and of the acceleration may be used to define the *mass* of the particle.

In Sec. 12.3, we shall define the *linear momentum* of a particle as the product $\mathbf{L} = m\mathbf{v}$ of the mass m and velocity $\mathbf{v}$ of the particle and show that Newton's second law may be expressed in an alternative form relating the rate of change of the linear momentum with the resultant of the forces acting on that particle.

Section 12.4 stresses the need for consistent units in the solution of dynamics problems and provides a review of the two systems used in this text, namely, the International System of Units (SI units) and the U.S. customary units.

In Secs. 12.5 and 12.6 and in the Sample Problems which follow, Newton's second law is applied to the solution of engineering problems, using either rectangular components or tangential and normal components of the forces and accelerations involved. We recall that an actual body—possibly as large as a car, a rocket, or an airplane—may be considered as a particle for the purpose of analyzing its motion, as long as the effect of a rotation of the body about its mass center may be ignored.

The second part of the chapter is devoted to the solution of problems in terms of radial and transverse components, with particular emphasis on the motion of a particle under a central force. In Sec. 12.7, we shall define the *angular momentum* $\mathbf{H}_O$ of a particle about a point O as the moment about O of the linear momentum of the particle: $\mathbf{H}_O = \mathbf{r} \times m\mathbf{v}$. It then follows from Newton's second law that the rate of change of the angular momentum $\mathbf{H}_O$ of a particle is equal to the sum of the moments about O of the forces acting on that particle.

Section 12.9 deals with the motion of a particle under a *central force*, i.e., under a force directed toward or away from a fixed point O. Since such a force has zero moment about O, it follows that the angular momentum of the particle about O is conserved. This property greatly simplifies the analysis of the motion of a particle under a central force and will be applied to the solution of problems involving the orbital motion of bodies under gravitational attraction (Sec. 12.10).

Sections 12.11 through 12.13 are optional. They present a more extensive discussion of orbital motion and contain a number of problems related to space mechanics.

12.2. Newton's Second Law of Motion.

Newton's second law may be stated as follows:

If the resultant force acting on a particle is not zero, the particle will have an acceleration proportional to the magnitude of the resultant and in the direction of this resultant force.

Newton's second law of motion may best be understood if we imagine the following experiment: A particle is subjected to a force F_1 of constant direction and constant magnitude F_1. Under the action of that force, the particle will be observed to move in a straight line and *in the direction of the force* (Fig. 12.1a). By determining the position of the particle at various instants, we find that its acceleration has a constant magnitude a_1. If the experiment is repeated with forces F_2, F_3, etc., of different magnitude or direction (Fig. 12.1b and c), we find each time that the particle moves in the direction of the force acting on it and that the magnitudes a_1, a_2, a_3, etc., of the accelerations are proportional to the magnitudes F_1, F_2, F_3, etc., of the corresponding forces,

$$\frac{F_1}{a_1} = \frac{F_2}{a_2} = \frac{F_3}{a_3} = \cdots = \text{constant}$$

The constant value obtained for the ratio of the magnitudes of the forces and accelerations is a characteristic of the particle under consideration. It is called the *mass* of the particle and is denoted by m. When a particle of mass m is acted upon by a force $\mathbf{F}$, the force $\mathbf{F}$ and the acceleration $\mathbf{a}$ of the particle must therefore satisfy the relation

(a)

(b)

(c)

Fig. 12.1

$$\mathbf{F} = m\mathbf{a} \tag{12.1}$$

This relation provides a complete formulation of Newton's second law; it expresses not only that the magnitudes of **F** and **a** are proportional but also (since m is a positive scalar) that the vectors **F** and **a** have the same direction (Fig. 12.2). We should note that Eq. (12.1) still holds when **F** is not constant but varies with t in magnitude or direction. The magnitudes of **F** and **a** remain proportional, and the two vectors have the same direction at any given instant. However, they will not, in general, be tangent to the path of the particle.

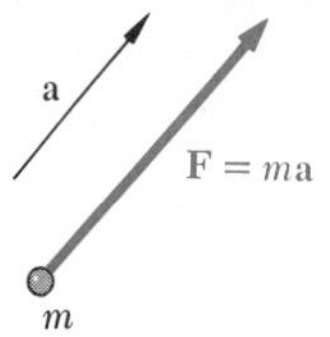

When a particle is subjected simultaneously to several forces, Eq. (12.1) should be replaced by

Fig. 12.2

$$\Sigma \mathbf{F} = m\mathbf{a} \qquad (12.2)$$

where $\Sigma\mathbf{F}$ represents the sum, or resultant, of all the forces acting on the particle.

It should be noted that the system of axes with respect to which the acceleration **a** is determined is not arbitrary. These axes must have a constant orientation with respect to the stars, and their origin must either be attached to the sun† or move with a constant velocity with respect to the sun. Such a system of axes is called a *newtonian frame of reference.*‡ A system of axes attached to the earth does *not* constitute a newtonian frame of reference, since the earth rotates with respect to the stars and is accelerated with respect to the sun. However, in most engineering applications, the acceleration **a** may be determined with respect to axes attached to the earth and Eqs. (12.1) and (12.2) used without any appreciable error. On the other hand, these equations do not hold if **a** represents a relative acceleration measured with respect to moving axes, such as axes attached to an accelerated car or to a rotating piece of machinery.

We may observe that if the resultant $\Sigma\mathbf{F}$ of the forces acting on the particle is zero, it follows from Eq. (12.2) that the acceleration **a** of the particle is also zero. If the particle is initially at rest ($\mathbf{v}_0 = 0$) with respect to the newtonian frame of reference used, it will thus remain at rest ($\mathbf{v} = 0$). If originally moving with a velocity $\mathbf{v}_0$, the particle will maintain a constant velocity $\mathbf{v} = \mathbf{v}_0$; that is, it will move with the constant speed v_0 in a straight line. This, we recall, is the statement of Newton's first law (Sec. 2.10). Thus, Newton's first law is a particular case of Newton's second law and may be omitted from the fundamental principles of mechanics.

† More accurately, to the mass center of the solar system.

‡ Since the stars are not actually fixed, a more rigorous definition of a newtonian frame of reference (also called *inertial system*) is *one with respect to which Eq. (12.2) holds.*

12.3. Linear Momentum of a Particle. Rate of Change of Linear Momentum.

Replacing the acceleration **a** by the derivative $d\mathbf{v}/dt$ in Eq. (12.2), we write

$$\Sigma \mathbf{F} = m\frac{d\mathbf{v}}{dt}$$

or, since the mass m of the particle is constant,

$$\Sigma \mathbf{F} = \frac{d}{dt}(m\mathbf{v}) \tag{12.3}$$

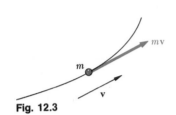

Fig. 12.3

The vector $m\mathbf{v}$ is called the *linear momentum*, or simply the *momentum*, of the particle. It has the same direction as the velocity of the particle and its magnitude is equal to the product of the mass m and the speed v of the particle (Fig. 12.3). Equation (12.3) expresses that *the resultant of the forces acting on the particle is equal to the rate of change of the linear momentum of the particle.* It is in this form that the second law of motion was originally stated by Newton. Denoting by **L** the linear momentum of the particle,

$$\mathbf{L} = m\mathbf{v} \tag{12.4}$$

and by $\dot{\mathbf{L}}$ its derivative with respect to t, we may write Eq. (12.3) in the alternative form

$$\Sigma \mathbf{F} = \dot{\mathbf{L}} \tag{12.5}$$

It should be noted that the mass m of the particle is assumed to be constant in Eqs. (12.3) to (12.5). Equation (12.3) or (12.5) should therefore not be used to solve problems involving the motion of bodies, such as rockets, which gain or lose mass. Problems of that type will be considered in Sec. 14.12.†

It follows from Eq. (12.3) that the rate of change of the linear momentum $m\mathbf{v}$ is zero when $\Sigma \mathbf{F} = 0$. Thus, *if the resultant force acting on a particle is zero, the linear momentum of the particle remains constant, both in magnitude and direction.* This is the principle of *conservation of linear momentum* for a particle, which we may recognize as just an alternative statement of Newton's first law (Sec. 2.10).

12.4. Systems of Units.

In using the fundamental equation $\mathbf{F} = m\mathbf{a}$, the units of force, mass, length, and time cannot be chosen arbitrarily. If they are, the magnitude of the force **F** required to give an acceleration **a** to the mass m will *not* be numerically equal to the product ma; it will be only proportional to this product. Thus, we may choose three of the four units arbitrarily but must choose the fourth unit so that the equation $\mathbf{F} = m\mathbf{a}$ is satisfied. The units are then said to form a system of consistent kinetic units.

† On the other hand, Eqs. (12.3) and (12.5) do hold in *relativistic mechanics*, where the mass m of the particle is assumed to vary with the speed of the particle.

Two systems of consistent kinetic units are currently used by American engineers, the International System of Units (SI units†), and the U.S. customary units. Since both systems have been discussed in detail in Sec. 1.3, we shall describe them only briefly in this section.

International System of Units (SI Units). In this system, the base units are the units of length, mass, and time, and are called, respectively, the *meter* (m), the *kilogram* (kg), and the *second* (s). All three are arbitrarily defined (Sec. 1.3). The unit of force is a derived unit. It is called the *newton* (N) and is defined as the force which gives an acceleration of 1 m/s² to a mass of 1 kg (Fig. 12.4). From Eq. (12.1) we write

$$1 \text{ N} = (1 \text{ kg})(1 \text{ m/s}^2) = 1 \text{ kg} \cdot \text{m/s}^2$$

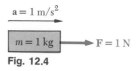

Fig. 12.4

The SI units are said to form an *absolute* system of units. This means that the three base units chosen are independent of the location where measurements are made. The meter, the kilogram, and the second may be used anywhere on the earth; they may even be used on another planet. They will always have the same significance.

The *weight* **W** of a body, or *force of gravity* exerted on that body, should, like any other force, be expressed in newtons. Since a body subjected to its own weight acquires an acceleration equal to the acceleration of gravity *g*, it follows from Newton's second law that the magnitude *W* of the weight of a body of mass *m* is

$$W = mg \tag{12.6}$$

Recalling that $g = 9.81$ m/s², we find that the weight of a body of mass 1 kg (Fig. 12.5) is

$$W = (1 \text{ kg})(9.81 \text{ m/s}^2) = 9.81 \text{ N}$$

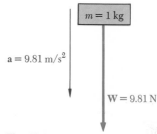

Fig. 12.5

Multiples and submultiples of the units of length, mass, and force are frequently used in engineering practice. They are, respectively, the *kilometer* (km) and the *millimeter* (mm); the *megagram‡* (Mg) and the *gram* (g); and the *kilonewton* (kN). By definition,

$$\begin{array}{ll} 1 \text{ km} = 1000 \text{ m} & 1 \text{ mm} = 0.001 \text{ m} \\ 1 \text{ Mg} = 1000 \text{ kg} & 1 \text{ g} = 0.001 \text{ kg} \\ \multicolumn{2}{c}{1 \text{ kN} = 1000 \text{ N}} \end{array}$$

The conversion of these units to meters, kilograms, and newtons, respectively, can be effected by simply moving the decimal point three places to the right or to the left.

Units other than the units of mass, length, and time may all be expressed in terms of these three base units. For example, the unit of linear momentum may be obtained by recalling the definition of linear momentum and writing

$$mv = (\text{kg})(\text{m/s}) = \text{kg} \cdot \text{m/s}$$

† SI stands for *Système International d'Unités* (French).

‡ Also known as a *metric ton.*

U.S. Customary Units. Most practicing American engineers still commonly use a system in which the base units are the units of length, force, and time. These units are, respectively, the *foot* (ft), the *pound* (lb), and the *second* (s). The second is the same as the corresponding SI unit. The foot is defined as 0.3048 m. The pound is defined as the *weight* of a platinum standard, called the *standard pound* and kept at the National Bureau of Standards in Washington, the mass of which is 0.453 592 43 kg. Since the weight of a body depends upon the gravitational attraction of the earth, which varies with location, it is specified that the standard pound should be placed at sea level and at the latitude of 45° to properly define a force of 1 lb. Clearly the U.S. customary units do not form an absolute system of units. Because of their dependence upon the gravitational attraction of the earth, they are said to form a *gravitational* system of units.

While the standard pound also serves as the unit of mass in commercial transactions in the United States, it cannot be so used in engineering computations since such a unit would not be consistent with the base units defined in the preceding paragraph. Indeed, when acted upon by a force of 1 lb, that is, when subjected to its own weight, the standard pound receives the acceleration of gravity, $g = 32.2$ ft/s^2 (Fig. 12.6), not the unit acceleration required by Eq. (12.1). The unit of mass consistent with the foot, the pound, and the second is the mass which receives an acceleration of 1 ft/s^2 when a force of 1 lb is applied to it (Fig. 12.7). This unit, sometimes called a *slug*, can be derived from the equation $F = ma$ after substituting 1 lb and 1 ft/s^2 for F and a, respectively. We write

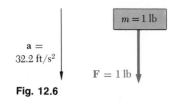

a =
32.2 ft/s²

F = 1 lb

Fig. 12.6

$$F = ma \qquad 1 \text{ lb} = (1 \text{ slug})(1 \text{ ft/s}^2)$$

and obtain

$$1 \text{ slug} = \frac{1 \text{ lb}}{1 \text{ ft/s}^2} = 1 \text{ lb} \cdot \text{s}^2/\text{ft}$$

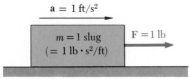

a = 1 ft/s²

m = 1 slug
(= 1 lb · s²/ft)

F = 1 lb

Fig. 12.7

Comparing Figs. 12.6 and 12.7, we conclude that the slug is a mass 32.2 times larger than the mass of the standard pound.

The fact that bodies are characterized in the U.S. customary system of units by their weight in pounds rather than by their mass in slugs was a convenience in the study of statics, where we were dealing constantly with weights and other forces and only seldom with masses. However, in the study of kinetics, where forces, masses, and accelerations are involved, we repeatedly shall have to express in slugs the mass m of a body, the weight W of which has been given in pounds. Recalling Eq. (12.6), we shall write

$$m = \frac{W}{g} \qquad (12.7)$$

where g is the acceleration of gravity ($g = 32.2$ ft/s^2).

Units other than the units of force, length, and time may all be expressed in terms of these three base units. For example, the unit of linear momentum may be obtained by recalling the definition of linear momentum and writing

$$mv = (\text{lb} \cdot \text{s}^2/\text{ft})(\text{ft/s}) = \text{lb} \cdot \text{s}$$

The conversion from U.S. customary units to SI units, and vice versa, has been discussed in Sec. 1.4. We shall recall the conversion factors obtained, respectively, for the units of length, force, and mass:

Length: 1 ft = 0.3048 m
Force: 1 lb = 4.448 N
Mass: 1 slug = 1 lb·s²/ft = 14.59 kg

Although it cannot be used as a consistent unit of mass, we also recall that the mass of the standard pound is, by definition,

$$1 \text{ pound-mass} = 0.4536 \text{ kg}$$

This constant may be used to determine the *mass* in SI units (kilograms) of a body which has been characterized by its *weight* in U.S. customary units (pounds).

12.5. Equations of Motion. Consider a particle of mass m acted upon by several forces. We recall from Sec. 12.2 that Newton's second law may be expressed by writing the equation

$$\Sigma \mathbf{F} = m\mathbf{a} \tag{12.2}$$

which relates the forces acting on the particle and the vector $m\mathbf{a}$ (Fig. 12.8). In order to solve problems involving the motion of a particle, however, it will be found more convenient to replace Eq. (12.2) by equivalent equations involving scalar quantities.

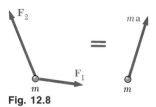

Fig. 12.8

Rectangular Components. Resolving each force $\mathbf{F}$ and the acceleration $\mathbf{a}$ into rectangular components, we write

$$\Sigma(F_x\mathbf{i} + F_y\mathbf{j} + F_z\mathbf{k}) = m(a_x\mathbf{i} + a_y\mathbf{j} + a_z\mathbf{k})$$

from which it follows that

$$\Sigma F_x = ma_x \qquad \Sigma F_y = ma_y \qquad \Sigma F_z = ma_z \tag{12.8}$$

Recalling from Sec. 11.11 that the components of the acceleration are equal to the second derivatives of the coordinates of the particle, we have

$$\Sigma F_x = m\ddot{x} \qquad \Sigma F_y = m\ddot{y} \qquad \Sigma F_z = m\ddot{z} \tag{12.8'}$$

Consider, as an example, the motion of a projectile. If the resistance of the air is neglected, the only force acting on the projectile after it has been fired is its weight $\mathbf{W} = -W\mathbf{j}$. The equations defining the motion of the projectile are therefore

$$m\ddot{x} = 0 \qquad m\ddot{y} = -W \qquad m\ddot{z} = 0$$

and the components of the acceleration of the projectile are

$$\ddot{x} = 0 \qquad \ddot{y} = -\frac{W}{m} = -g \qquad \ddot{z} = 0$$

where g is 9.81 m/s², or 32.2 ft/s². The equations obtained may be integrated independently, as was shown in Sec. 11.11, to obtain the velocity and displacement of the projectile at any instant.

When a problem involves two or more bodies, equations of motion should be written for each of the bodies (see Sample Probs. 12.3 and 12.4). We also recall from Sec. 12.2 that all accelerations should be measured with respect to a newtonian frame of reference. In most engineering applications accelerations may be determined with respect to axes attached to the earth, but relative accelerations measured with respect to moving axes, such as axes attached to an accelerated body, cannot be substituted for $\mathbf{a}$ in the equations of motion.

Tangential and Normal Components. Resolving the forces and the acceleration of the particle into components along the tangent to the path (in the direction of motion) and the normal (toward the inside of the

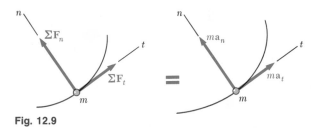

Fig. 12.9

path) (Fig. 12.9), and substituting into Eq. (12.2), we obtain the two scalar equations

$$\Sigma F_t = ma_t \qquad \Sigma F_n = ma_n \qquad (12.9)$$

Substituting for a_t and a_n from Eqs. (11.40), we have

$$\Sigma F_t = m\frac{dv}{dt} \qquad \Sigma F_n = m\frac{v^2}{\rho} \qquad (12.9')$$

The equations obtained may be solved for two unknowns.

12.6. Dynamic Equilibrium. Returning to Eq. (12.2) and transposing the right-hand member, we write Newton's second law in the alternative form

$$\Sigma \mathbf{F} - m\mathbf{a} = 0 \qquad (12.10)$$

which expresses that if we add the vector $-m\mathbf{a}$ to the forces acting on the particle, *we obtain a system of vectors equivalent to zero* (Fig. 12.10). The vector $-m\mathbf{a}$, of magnitude ma and of *direction opposite* to that of the acceleration, is called an *inertia vector*. The particle may thus be considered to be in equilibrium under the given forces and the inertia vector. The particle is said to be in *dynamic equilibrium,* and the problem under consideration may be solved by the methods developed earlier in statics.

In the case of coplanar forces, we may draw in tip-to-tail fashion all the vectors shown in Fig. 12.10, *including the inertia vector,* to form a closed-vector polygon. Or we may write that the sums of the components of all the vectors in Fig. 12.10, including again the inertia vector, are zero. Using rectangular components, we therefore write

$$\Sigma F_x = 0 \qquad \Sigma F_y = 0 \qquad \textit{including inertia vector} \quad (12.11)$$

When tangential and normal components are used, it is more convenient to represent the inertia vector by its two components $-m\mathbf{a}_t$ and $-m\mathbf{a}_n$ in the sketch itself (Fig. 12.11). The tangential component of the inertia vector provides a measure of the resistance the particle offers to a change in speed, while its normal component (also called *centrifugal force*) represents the tendency of the particle to leave its curved path. We should note that either of these two components may be zero under special conditions: (1) if the particle starts from rest, its initial velocity is zero and the normal component of the inertia vector is zero at $t = 0$; (2) if the particle moves at constant speed along its path, the tangential component of the inertia vector is zero and only its normal component needs to be considered.

Because they measure the resistance that particles offer when we try to set them in motion or when we try to change the conditions of their motion, inertia vectors are often called *inertia forces.* The inertia forces, however, are not forces like the forces found in statics, which are either contact forces or gravitational forces (weights). Many people, therefore, object to the use of the word "force" when referring to the vector $-m\mathbf{a}$ or even avoid altogether the concept of dynamic equilibrium. Others point out that inertia forces and actual forces, such as gravitational forces, affect our senses in the same way and cannot be distinguished by physical measurements. A man riding in an elevator which is accelerated upward will have the feeling that his weight has suddenly increased; and no measurement made within the elevator could establish whether the elevator is truly accelerated or whether the force of attraction exerted by the earth has suddenly increased.

Sample problems have been solved in this text by the direct application of Newton's second law, as illustrated in Figs. 12.8 and 12.9, rather than by the method of dynamic equilibrium.

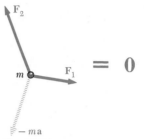

Fig. 12.10

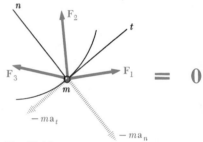

Fig. 12.11

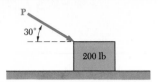

SAMPLE PROBLEM 12.1

A 200-lb block rests on a horizontal plane. Find the magnitude of the force **P** required to give the block an acceleration of 10 ft/s^2 to the right. The coefficient of kinetic friction between the block and the plane is $\mu_k = 0.25$.

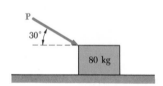

Solution. The mass of the block is

$$m = \frac{W}{g} = \frac{200\ \text{lb}}{32.2\ \text{ft/s}^2} = 6.21\ \text{lb}\cdot\text{s}^2/\text{ft}$$

We note that $F = \mu_k N = 0.25N$ and that $a = 10$ ft/s^2. Expressing that the forces acting on the block are equivalent to the vector $m\mathbf{a}$, we write

$$\xrightarrow{+} \Sigma F_x = ma: \qquad P \cos 30° - 0.25N = (6.21\ \text{lb}\cdot\text{s}^2/\text{ft})(10\ \text{ft/s}^2)$$
$$P \cos 30° - 0.25N = 62.1\ \text{lb} \tag{1}$$
$$+\uparrow \Sigma F_y = 0: \qquad N - P \sin 30° - 200\ \text{lb} = 0 \tag{2}$$

Solving (2) for N and carrying the result into (1), we obtain

$$N = P \sin 30° + 200\ \text{lb}$$

$$P \cos 30° - 0.25(P \sin 30° + 200\ \text{lb}) = 62.1\ \text{lb} \qquad P = 151\ \text{lb} \quad \blacktriangleleft$$

SAMPLE PROBLEM 12.2

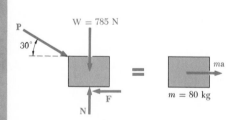

An 80-kg block rests on a horizontal plane. Find the magnitude of the force **P** required to give the block an acceleration of 2.5 m/s^2 to the right. The coefficient of kinetic friction between the block and the plane is $\mu_k = 0.25$.

Solution. The weight of the block is

$$W = mg = (80\ \text{kg})(9.81\ \text{m/s}^2) = 785\ \text{N}$$

We note that $F = \mu_k N = 0.25N$ and that $a = 2.5$ m/s^2. Expressing that the forces acting on the block are equivalent to the vector $m\mathbf{a}$, we write

$$\xrightarrow{+} \Sigma F_x = ma: \qquad P \cos 30° - 0.25N = (80\ \text{kg})(2.5\ \text{m/s}^2)$$
$$P \cos 30° - 0.25N = 200\ \text{N} \tag{1}$$
$$+\uparrow \Sigma F_y = 0: \qquad N - P \sin 30° - 785\ \text{N} = 0 \tag{2}$$

Solving (2) for N and carrying the result into (1), we obtain

$$N = P \sin 30° + 785\ \text{N}$$

$$P \cos 30° - 0.25(P \sin 30° + 785\ \text{N}) = 200\ \text{N} \qquad P = 535\ \text{N} \quad \blacktriangleleft$$

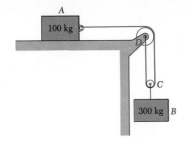

SAMPLE PROBLEM 12.3

The two blocks shown start from rest. The horizontal plane and the pulley are frictionless, and the pulley is assumed to be of negligible mass. Determine the acceleration of each block and the tension in each cord.

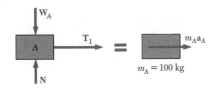

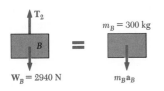

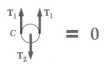

Kinematics. We note that if block A moves through x_A to the right, block B moves down through

$$x_B = \tfrac{1}{2}x_A$$

Differentiating twice with respect to t, we have

$$a_B = \tfrac{1}{2}a_A \tag{1}$$

Kinetics. We shall apply Newton's second law successively to block A, block B, and pulley C.

Block A. Denoting by T_1 the tension in cord ACD, we write

$$\xrightarrow{+} \Sigma F_x = m_A a_A: \qquad\qquad T_1 = 100a_A \tag{2}$$

Block B. Observing that the weight of block B is

$$W_B = m_B g = (300 \text{ kg})(9.81 \text{ m/s}^2) = 2940 \text{ N}$$

and denoting by T_2 the tension in cord BC, we write

$$+{\downarrow}\Sigma F_y = m_B a_B: \qquad\qquad 2940 - T_2 = 300a_B$$

or, substituting for a_B from (1),

$$2940 - T_2 = 300(\tfrac{1}{2}a_A)$$
$$T_2 = 2940 - 150a_A \tag{3}$$

Pulley C. Since m_C is assumed to be zero, we have

$$+{\downarrow}\Sigma F_y = m_C a_C = 0: \qquad T_2 - 2T_1 = 0 \tag{4}$$

Substituting for T_1 and T_2 from (2) and (3), respectively, into (4) we write

$$2940 - 150a_A - 2(100a_A) = 0$$
$$2940 - 350a_A = 0 \qquad\qquad a_A = 8.40 \text{ m/s}^2 \quad \blacktriangleleft$$

Substituting the value obtained for a_A into (1) and (2), we have

$$a_B = \tfrac{1}{2}a_A = \tfrac{1}{2}(8.40 \text{ m/s}^2) \qquad\qquad a_B = 4.20 \text{ m/s}^2 \quad \blacktriangleleft$$
$$T_1 = 100a_A = (100 \text{ kg})(8.40 \text{ m/s}^2) \qquad\qquad T_1 = 840 \text{ N} \quad \blacktriangleleft$$

Recalling (4), we write

$$T_2 = 2T_1 \qquad T_2 = 2(840 \text{ N}) \qquad T_2 = 1680 \text{ N} \quad \blacktriangleleft$$

We note that the value obtained for T_2 is *not* equal to the weight of block B.

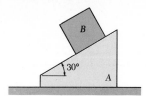

SAMPLE PROBLEM 12.4

The 12-lb block B starts from rest and slides on the 30-lb wedge A, which is supported by a horizontal surface. Neglecting friction, determine (a) the acceleration of the wedge, (b) the acceleration of the block relative to the wedge.

Kinematics. We shall first examine the accelerations of the wedge and of the block.

Wedge A. Since the wedge is constrained to move on the horizontal surface, its acceleration $\mathbf{a}_A$ is horizontal. We shall assume that it is directed to the right.

Block B. The acceleration $\mathbf{a}_B$ of block B may be expressed as the sum of the acceleration of A and of the acceleration of B relative to A. We have

$$\mathbf{a}_B = \mathbf{a}_A + \mathbf{a}_{B/A}$$

where $\mathbf{a}_{B/A}$ is directed along the sloping surface of the wedge.

Kinetics. We shall draw the free-body diagrams of the wedge and of the block and apply Newton's second law.

Wedge A. We denote respectively by $\mathbf{N}_1$ and $\mathbf{N}_2$ the forces exerted by the block and the horizontal surface on wedge A.

$$\xrightarrow{+} \Sigma F_x = m_A a_A: \qquad N_1 \sin 30° = m_A a_A$$

$$0.5 N_1 = (W_A/g) a_A \qquad (1)$$

Block B. Using the coordinate axes shown and resolving $\mathbf{a}_B$ into its components $\mathbf{a}_A$ and $\mathbf{a}_{B/A}$, we write

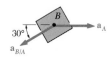

$$+ \nearrow \Sigma F_x = m_B a_x: \qquad -W_B \sin 30° = m_B a_A \cos 30° - m_B a_{B/A}$$

$$-W_B \sin 30° = (W_B/g)(a_A \cos 30° - a_{B/A})$$

$$a_{B/A} = a_A \cos 30° + g \sin 30° \qquad (2)$$

$$+ \nwarrow \Sigma F_y = m_B a_y: \qquad N_1 - W_B \cos 30° = -m_B a_A \sin 30°$$

$$N_1 - W_B \cos 30° = -(W_B/g) a_A \sin 30° \qquad (3)$$

a. Acceleration of Wedge A. Substituting for N_1 from Eq. (1) into Eq. (3), we have

$$2(W_A/g) a_A - W_B \cos 30° = -(W_B/g) a_A \sin 30°$$

Solving for a_A and substituting the numerical data, we write

$$a_A = \frac{W_B \cos 30°}{2 W_A + W_B \sin 30°} g = \frac{(12 \text{ lb}) \cos 30°}{2(30 \text{ lb}) + (12 \text{ lb}) \sin 30°} (32.2 \text{ ft/s}^2)$$

$$a_A = +5.07 \text{ ft/s}^2 \qquad \mathbf{a}_A = 5.07 \text{ ft/s}^2 \rightarrow \quad \blacktriangleleft$$

b. Acceleration of Block B relative to A. Substituting the value obtained for a_A into Eq. (2), we have

$$a_{B/A} = (5.07 \text{ ft/s}^2) \cos 30° + (32.2 \text{ ft/s}^2) \sin 30°$$

$$a_{B/A} = +20.5 \text{ ft/s}^2 \qquad \mathbf{a}_{B/A} = 20.5 \text{ ft/s}^2 \,\triangledown\, 30° \quad \blacktriangleleft$$

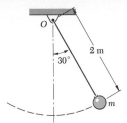

SAMPLE PROBLEM 12.5

The bob of a 2-m pendulum describes an arc of circle in a vertical plane. If the tension in the cord is 2.5 times the weight of the bob for the position shown, find the velocity and acceleration of the bob in that position.

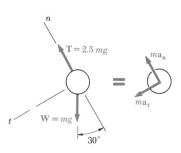

Solution. The weight of the bob is $W = mg$; the tension in the cord is thus $2.5\ mg$. Recalling that $\mathbf{a}_n$ is directed toward O and assuming $\mathbf{a}_t$ as shown, we apply Newton's second law and obtain

$$+\swarrow \Sigma F_t = ma_t: \qquad mg \sin 30° = ma_t$$
$$a_t = g \sin 30° = +4.90 \text{ m/s}^2 \qquad \mathbf{a}_t = 4.90 \text{ m/s}^2 \swarrow \quad \blacktriangleleft$$

$$+\nwarrow \Sigma F_n = ma_n: \qquad 2.5\ mg - mg \cos 30° = ma_n$$
$$a_n = 1.634\ g = +16.03 \text{ m/s}^2 \qquad \mathbf{a}_n = 16.03 \text{ m/s}^2 \nwarrow \quad \blacktriangleleft$$

Since $a_n = v^2/\rho$, we have $v^2 = \rho a_n = (2 \text{ m})(16.03 \text{ m/s}^2)$
$$v = \pm 5.66 \text{ m/s} \qquad \mathbf{v} = 5.66 \text{ m/s} \nearrow \text{(up or down)} \quad \blacktriangleleft$$

SAMPLE PROBLEM 12.6

Determine the rated speed of a highway curve of radius $\rho = 400$ ft banked through an angle $\theta = 18°$. The rated speed of a banked curved road is the speed at which a car should travel if no lateral friction force is to be exerted on its wheels.

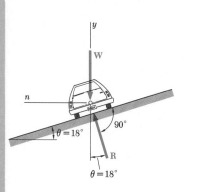

Solution. The car travels in a *horizontal* circular path of radius ρ. The normal component $\mathbf{a}_n$ of the acceleration is directed toward the center of the path; its magnitude is $a_n = v^2/\rho$, where v is the speed of the car in ft/s. The mass m of the car is W/g, where W is the weight of the car. Since no lateral friction force is to be exerted on the car, the reaction $\mathbf{R}$ of the road is shown perpendicular to the roadway. Applying Newton's second law, we write

$$+\uparrow \Sigma F_y = 0: \qquad R \cos \theta - W = 0 \qquad R = \frac{W}{\cos \theta} \qquad (1)$$

$$\xleftrightarrow{+} \Sigma F_n = ma_n: \qquad R \sin \theta = \frac{W}{g} a_n \qquad (2)$$

Substituting for R from (1) into (2), and recalling that $a_n = v^2/\rho$,

$$\frac{W}{\cos \theta} \sin \theta = \frac{W}{g} \frac{v^2}{\rho} \qquad v^2 = g\rho \tan \theta$$

Substituting the given data, $\rho = 400$ ft and $\theta = 18°$, into this equation, we obtain

$$v^2 = (32.2 \text{ ft/s}^2)(400 \text{ ft}) \tan 18°$$
$$v = 64.7 \text{ ft/s} \qquad v = 44.1 \text{ mi/h} \quad \blacktriangleleft$$

Problems

12.1 The value of g at any latitude ϕ may be obtained from the formula

$$g = 32.09(1 + 0.0053 \sin^2 \phi)\, \text{ft/s}^2$$

which takes into account the effect of the rotation of the earth, as well as the fact that the earth is not truly spherical. Determine to four significant figures (*a*) the weight in pounds, (*b*) the mass in pounds, (*c*) the mass in lb·s²/ft, at the latitudes of 0°, 45°, and 90°, of a silver bar, the mass of which has been officially designated as 5 lb.

12.2 The acceleration due to gravity on the moon is 1.62 m/s². Determine (*a*) the weight in newtons, (*b*) the mass in kilograms, on the moon, of a gold bar, the mass of which has been officially designated as 2 kg.

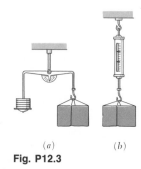

(*a*)   (*b*)

Fig. P12.3

12.3 Two boxes are weighed on the scales shown: scale *a* is a lever scale; scale *b* is a spring scale. The scales are attached to the roof of an elevator. When the elevator is at rest, each scale indicates a mass of 15 kg. If the spring scale indicates a mass of 20 kg, determine (*a*) the acceleration of the elevator, (*b*) the mass indicated by the lever scale.

12.4 A 500-kg satellite has been placed in a circular orbit 1200 km above the surface of the earth. The acceleration of gravity at this elevation is 6.95 m/s². Determine the linear momentum of the satellite, knowing that its orbital speed is 26.1×10^3 km/h.

12.5 Determine the maximum theoretical speed that may be achieved over a distance of 60 m by a car starting from rest, knowing that the coefficient of static friction is 0.80 between the tires and the pavement and that 60 percent of the weight of the car is distributed over its front wheels and 40 percent over its rear wheels. Assume (*a*) four-wheel drive, (*b*) front-wheel drive, (*c*) rear-wheel drive.

12.6 A motorist traveling at a speed of 108 km/h suddenly applies the brakes and comes to a stop after skidding 75 m. Determine (*a*) the time required for the car to stop, (*b*) the coefficient of friction between the tires and the pavement.

2.5% grade

Fig. P12.7

12.7 A truck is proceeding up a long 2.5 percent grade at a constant speed of 45 mi/h. If the driver does not change the setting of his throttle or shift gears, what will be the acceleration of the truck as it starts moving on the level section of the road?

12.8 A 2800-lb automobile is driven down a 4° incline at a speed of 55 mi/h when the brakes are applied, causing a total braking force of 1500 lb to be applied to the automobile. Determine the distance traveled by the automobile before it comes to a stop.

12.9 The two blocks shown are originally at rest. Neglecting the masses of the pulleys and the effect of friction in the pulleys and between the blocks and the inclines, determine (a) the acceleration of each block, (b) the tension in the cable.

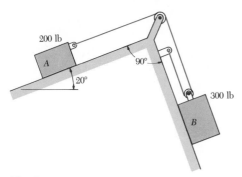

Fig. P12.9

12.10 Solve Prob. 12.9, assuming that the coefficients of friction between the blocks and the inclines are $\mu_s = 0.25$ and $\mu_k = 0.20$.

12.11 A light train made up of two cars is traveling at 90 km/h when the brakes are applied to both cars. Knowing that car A has a mass of 25 Mg and car B a mass of 20 Mg, and that the braking force is 30 kN on each car, determine (a) the distance traveled by the train before it comes to a stop, (b) the force in the coupling between the cars while the train is slowing down.

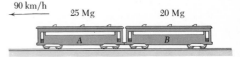

Fig. P12.11

12.12 Solve Prob. 12.11, assuming that the brakes of car B fail to operate.

12.13 Block A has a mass of 25 kg and block B a mass of 15 kg. The coefficients of friction between all surfaces of contact are $\mu_s = 0.20$ and $\mu_k = 0.15$. Knowing that $\theta = 25°$ and that the magnitude of the force **P** applied to block A is 250 N, determine (a) the acceleration of block A, (b) the tension in the cord.

12.14 Solve Prob. 12.13, assuming that the 250-N force **P** is applied to block B instead of block A.

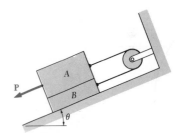

Fig. P12.13

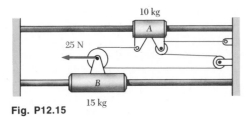

Fig. P12.15

12.15 Knowing that the system shown starts from rest, find the velocity at $t = 1.2$ s of (a) collar A, (b) collar B. Neglect the masses of the pulleys and the effect of friction.

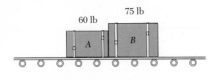

Fig. P12.16

12.16 Two packages are placed on a conveyor belt which is at rest. The coefficient of kinetic friction is 0.20 between the belt and package A, and 0.10 between the belt and package B. If the belt is suddenly started to the right and slipping occurs between the belt and the packages, determine (a) the acceleration of the packages, (b) the force exerted by package A on package B.

12.17 Each of the systems shown is initially at rest. Assuming the pulleys to be weightless and neglecting axle friction, determine for each system (a) the acceleration of block A, (b) the velocity of block A after 2 s, (c) the velocity of block A after it has moved through 8 ft.

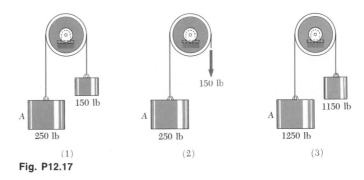

Fig. P12.17

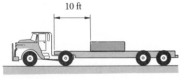

Fig. P12.18

12.18 The coefficients of friction between the load and the flat-bed trailer shown are $\mu_s = 0.40$ and $\mu_k = 0.30$. Knowing that the speed of the rig is 45 mi/h, determine the shortest distance in which the rig can be brought to a stop if the load is not to shift.

12.19 The tractor-trailer of Prob. 12.18 is traveling at 60 mi/h when the driver makes an emergency stop, causing the rig to skid to rest in 4 s. Determine (a) whether the load will shift, (b) if the load does shift, the relative velocity with which it will reach the forward edge of the trailer.

12.20 In a manufacturing process, disks are moved from level A to level B by the lifting arm shown. The arm starts from level A with no initial velocity, moves first with a constant acceleration $\mathbf{a}_1$ as shown, then with a constant deceleration $\mathbf{a}_2$, and comes to a stop at level B. Knowing that the coefficient of static friction between the disks and the arm is 0.30, determine the largest allowable acceleration $\mathbf{a}_1$ and the largest allowable deceleration $\mathbf{a}_2$ if the disks are not to slide.

12.21 If the disks of Prob. 12.20 are to be moved from level A to level B in the shortest possible time without sliding, determine (a) the maximum velocity reached by the lifting arm, (b) the time required to move each disk.

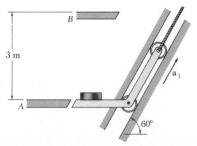

Fig. P12.20

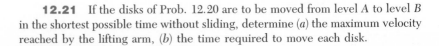

12.22 A ship of total mass m is anchored in the middle of a river which is flowing with a constant velocity $\mathbf{v}_0$. The horizontal component of the force exerted on the ship by the anchor chain is $\mathbf{T}_0$. If the anchor chain suddenly breaks, determine the time required for the ship to attain a velocity equal to $\frac{1}{2}\mathbf{v}_0$. Assume that the frictional resistance of the water is proportional to the velocity of the ship relative to the water.

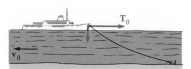

Fig. P12.22

12.23 An airplane has a mass of 25 Mg and its engines develop a total thrust of 40 kN during take-off. If the drag $\mathbf{D}$ exerted on the plane has a magnitude $D = 2.25 \, v^2$, where v is expressed in meters per second and D in newtons, and if the plane becomes airborne at a speed of 240 km/h, determine the length of runway required for the plane to take off.

12.24 For the airplane of Prob. 12.23, determine the time required to take off.

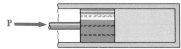

Fig. P12.25

12.25 A constant force $\mathbf{P}$ is applied to a piston and rod of total mass m in order to make them move in a cylinder filled with oil. As the piston moves, the oil is forced through orifices in the piston and exerts on the piston an additional force of magnitude kv, proportional to the speed v of the piston and in a direction opposite to its motion. Express the distance x traveled by the piston as a function of the time t, assuming that the piston starts from rest at $t = 0$ and $x = 0$.

12.26 Block A weighs 30 lb, and blocks B and C weigh 15 lb each. Knowing that $P = 3$ lb and that the blocks are initially at rest, determine at $t = 4$ s the velocity (a) of B relative to A, (b) of C relative to A. Neglect the weights of the pulleys and the effect of friction.

12.27 Block A weighs 30 lb, and blocks B and C weigh 15 lb each. Knowing that the blocks are initially at rest and that B moves through 12 ft in 2 s, determine (a) the magnitude of the force $\mathbf{P}$, (b) the tension in the cord AD. Neglect the weights of the pulleys and the effect of friction.

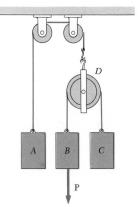

Fig. P12.26 and P12.27

12.28 The coefficients of friction between blocks A and C and the horizontal surfaces are $\mu_s = 0.24$ and $\mu_k = 0.20$. Knowing that $m_A = 5$ kg, $m_B = 10$ kg, and $m_C = 10$ kg, determine (a) the tension in the cord, (b) the acceleration of each block.

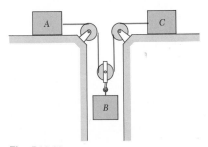

Fig. P12.28

12.29 Solve Prob. 12.28, assuming that $m_A = 5$ kg, $m_B = 10$ kg, and $m_C = 20$ kg.

12.30 A 15-lb block B rests as shown on the upper surface of a 25-lb wedge A. Neglecting friction, determine (a) the acceleration of A, (b) the acceleration of B relative to A, immediately after the system is released from rest.

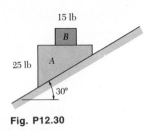

Fig. P12.30

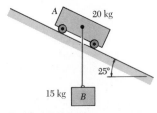

Fig. P12.31

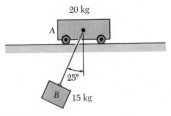

Fig. 12.32

12.31 and 12.32 A 15-kg block B is suspended from a 2.5-m cord attached to a 20-kg cart A. Neglecting friction, determine (a) the acceleration of the cart, (b) the tension in the cord, immediately after the system is released from rest in the position shown.

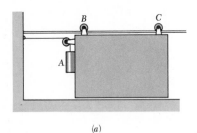

(a)

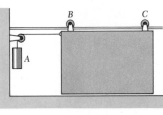

(b)

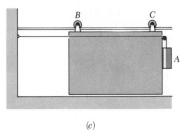

(c)

Fig. P12.33

12.33 A 40-lb sliding panel is supported by rollers at B and C. A 25-lb counterweight A is attached to a cable as shown and, in cases a and c, is initially in contact with a vertical edge of the panel. Neglecting friction, determine in each case shown the acceleration of the panel and the tension in the cord immediately after the system is released from rest.

Fig. P12.34 and P12.35

12.34 A 4-lb ball revolves in a horizontal circle as shown. Knowing that $L = 3$ ft and that the maximum allowable tension in the cord is 10 lb, determine (a) the maximum allowable speed, (b) the corresponding value of the angle θ.

12.35 A 2-kg ball revolves in a horizontal circle as shown at a constant speed of 1.5 m/s. Knowing that $L = 600$ mm, determine (a) the angle θ that the cord forms with the vertical, (b) the tension in the cord.

12.36 and 12.37 A single wire ACB passes through a ring at C attached to a sphere which revolves at a constant speed v in the horizontal circle shown. Knowing that the tension is the same in both portions of the wire, determine the speed v.

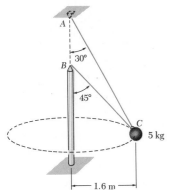

Fig. P12.36, P12.38, P12.39 and P12.41

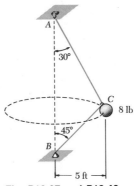

Fig. P12.37 and P12.40

12.38 Two wires AC and BC are tied at C to a sphere which revolves at a constant speed v in the horizontal circle shown. Determine the range of values of v for which both wires remain taut.

12.39 Two wires AC and BC are tied at C to a sphere which revolves at a constant speed v in the horizontal circle shown. Determine the range of the allowable values of v if the tension in either of the wires is not to exceed 40 N.

*12.40 Two wires AC and BC are tied at C to a sphere which revolves at a constant speed v in the horizontal circle shown. Determine the range of the allowable values of v if both wires are to remain taut and if the tension in either of the wires is not to exceed 15 lb.

*12.41 Two wires AC and BC are tied at C to a sphere which revolves at a constant speed v in the horizontal circle shown. Determine the range of the allowable values of v if both wires are to remain taut and if the tension in either of the wires is not to exceed 60 N.

12.42 A small sphere of weight W is held as shown by two wires AB and CD. If wire AB is cut, determine the tension in the other wire (a) before AB is cut, (b) immediately after AB has been cut.

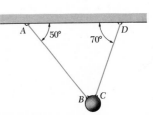

Fig. P12.42

12.43 Solve Prob. 12.42, assuming that wire CD is cut instead of wire AB.

Fig. P12.44

12.44 Since the French high-speed (TGV) train running from Paris to Lyons uses a special track from which freight trains are barred, the designers of the track could use steeper grades than are usually permissible for railroads. On the other hand, because of the high speeds involved, they had to avoid sudden changes in grade and set a lower limit for the radius of curvature ρ of the vertical profile of the track. (*a*) Knowing that the train's maximum speed in test runs was 382 km/h, determine the smallest allowable value of ρ if the train is not to leave the track at that speed. (*b*) Assuming the value of ρ found in part *a*, determine the force exerted by his seat on a 75-kg passenger at the top and at the bottom of a hill when the train is traveling at 300 km/h.

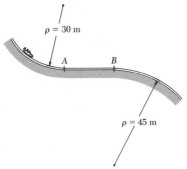

Fig. P12.45

12.45 The roller-coaster track shown is contained in a vertical plane. The portion of track between *A* and *B* is straight and horizontal, while the portions to the left of *A* and to the right of *B* have radii of curvature as indicated. A car is traveling at a speed of 72 km/h when the brakes are suddenly applied, causing the wheels of the car to slide on the track ($\mu_k = 0.20$). Determine the initial deceleration of the car if the brakes are applied as the car (*a*) has almost reached *A*, (*b*) is traveling between *A* and *B*, (*c*) has just passed *B*.

12.46 A 160-lb pilot flies a small plane in a vertical loop of 500-ft radius. Determine the speed of the plane at points *A* and *B*, knowing that at point *A* the pilot experiences weightlessness and that at point *B* the pilot's apparent weight is 550 lb.

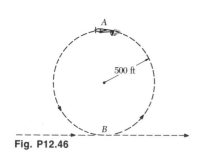

Fig. P12.46

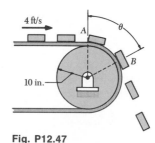

Fig. P12.47

12.47 A series of small packages, each weighing $\frac{3}{4}$ lb, are discharged from a conveyor belt as shown. Knowing that the coefficient of static friction between each package and the conveyor belt is 0.40, determine (*a*) the force exerted by the belt on a package just after it has passed point *A*, (*b*) the angle θ defining the point *B* where the packages first *slip* relative to the belt.

12.48 A 250-g block B fits inside a small cavity cut in arm OA, which rotates in the vertical plane at a constant rate such that $v = 3$ m/s. Knowing that the spring exerts on block B a force of magnitude $P = 1.5$ N and neglecting the effect of friction, determine the range of values of θ for which block B is in contact with the face of the cavity closest to the axis of rotation O.

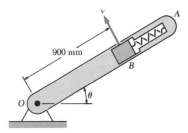

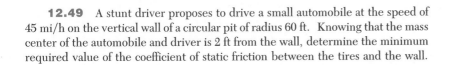

900 mm

Fig. P12.48

12.49 A stunt driver proposes to drive a small automobile at the speed of 45 mi/h on the vertical wall of a circular pit of radius 60 ft. Knowing that the mass center of the automobile and driver is 2 ft from the wall, determine the minimum required value of the coefficient of static friction between the tires and the wall.

12.50 Express the minimum and maximum safe speeds with respect to skidding of a car traveling on a banked road, in terms of the radius r of the curve, the banking angle θ, and the friction angle ϕ_s between the tires and the pavement.

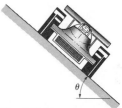

Fig. P12.51

12.51 A curve in a speed track has a radius of 200 m and a rated speed of 180 km/h. (See Sample Prob. 12.6 for the definition of rated speed.) Knowing that a racing car starts skidding on the curve when traveling at a speed of 320 km/h, determine (a) the banking angle θ, (b) the coefficient of static friction between the tires and the track under the prevailing conditions, (c) the minimum speed at which the same car could negotiate that curve.

12.52 For the highway curve of Sample Prob. 12.6, determine the maximum safe speed, assuming that the coefficient of static friction between the tires and the pavement is 0.75.

200 mm

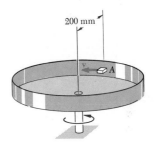

12.53 The assembly shown rotates about a vertical axis at a constant rate. Knowing that the coefficient of static friction between the small block A and the cylindrical wall is 0.25, determine the lowest speed v for which the block will remain in contact with the wall.

Fig. P12.53

12.54 A small 250-g collar C may slide on a semicircular rod which is made to rotate about the vertical AB at a constant rate of 7.5 rad/s. Determine the three values of θ for which the collar will not slide on the rod, assuming no friction between the collar and the rod.

12.55 Determine the minimum required value of the coefficient of static friction between the collar and the rod of Prob. 12.54 if the collar is not to slide when (a) $\theta = 90°$, (b) $\theta = 60°$, (c) $\theta = 45°$. Indicate in each case the direction of the impending motion.

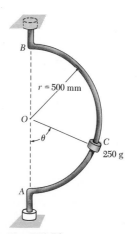

$r = 500$ mm

C
250 g

12.56 For the collar and rod of Prob. 12.54, and assuming that the coefficients of friction are $\mu_s = 0.25$ and $\mu_k = 0.20$, indicate whether the collar will slide on the rod if it is released in the position corresponding to (a) $\theta = 75°$, (b) $\theta = 40°$. Also determine the magnitude and direction of the friction force exerted on the collar immediately after release.

Fig. P12.54

Fig. P12.57 and P12.58

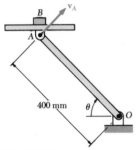

Fig. P12.59

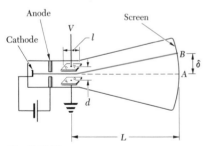

Fig. P12.61

12.57 A small block B is supported by a turntable which starting from rest is rotated in such a way that the block undergoes a constant tangential acceleration $a_t = 4.5$ ft/s^2. Knowing that the coefficient of static friction between the block and the turntable is 0.50, determine (*a*) how long it will take for the block to start slipping on the turntable, (*b*) the speed v of the block at that instant.

12.58 A small block B is supported by a turntable which starting from rest is rotated in such a way that the block undergoes a constant tangential acceleration. Knowing that the coefficient of static friction between the block and the turntable is 0.60, determine the smallest interval of time in which the block can reach the speed $v = 5$ ft/s without slipping.

12.59 A small block B is supported by a platform connected at A to rod OA. Point A describes a circle in a vertical plane at the constant speed v_A, while the platform is constrained to remain horizontal throughout its motion by the use of a special linkage (not shown in the figure). The coefficients of friction between the block and the platform are $\mu_s = 0.40$ and $\mu_k = 0.30$. Determine (*a*) the maximum allowable speed v_A if the block is not to slide on the platform, (*b*) the values of θ for which sliding is impending.

*12.60 For the system of Prob. 12.59, it is observed that block B slides intermittently on the platform when point A moves in a vertical circle at the constant speed $v_A = 1.22$ m/s. Determine the values of θ for which the block starts sliding.

12.61 In the cathode-ray tube shown, electrons emitted by the cathode and attracted by the anode pass through a small hole in the anode and keep traveling in a straight line with a speed v_0 until they strike the screen at A. However, if a difference of potential V is established between the two parallel plates, each electron will be subjected to a force $\mathbf{F}$ perpendicular to the plates while it travels between the plates and will strike the screen at point B at a distance δ from A. The magnitude of the force $\mathbf{F}$ is $F = eV/d$, where $-e$ is the charge of the electron and d is the distance between the plates. Derive an expression for the deflection δ in terms of V, v_0, the charge $-e$ of the electron, its mass m, and the dimensions d, l, and L.

12.62 In Prob. 12.61, determine the smallest allowable value of the ratio d/l in terms of e, m, v_0, and V if the electrons are not to strike the positive plate.

12.63 A manufacturer wishes to design a new cathode-ray tube which will be only half as long as his current model. If the size of the screen is to remain the same, how should the length l of the plates be modified if all the other characteristics of the circuit are to remain unchanged? (See Prob. 12.61 for a description of a cathode-ray tube.)

12.7. Angular Momentum of a Particle. Rate of Change of Angular Momentum.

Consider a particle P of mass m moving with respect to a newtonian frame of reference $Oxyz$. As we saw in Sec. 12.3, the linear momentum of the particle at a given instant is defined as the vector $m\mathbf{v}$ obtained by multiplying the velocity $\mathbf{v}$ of the particle by its mass m. The moment about O of the vector $m\mathbf{v}$ is called the *moment of momentum*, or the *angular momentum*, of the particle about O at that instant and is denoted by $\mathbf{H}_O$. Recalling the definition of the moment of a vector (Sec. 3.6), and denoting by $\mathbf{r}$ the position vector of P, we write

$$\mathbf{H}_O = \mathbf{r} \times m\mathbf{v} \qquad (12.12)$$

and note that $\mathbf{H}_O$ is a vector perpendicular to the plane containing $\mathbf{r}$ and $m\mathbf{v}$, and of magnitude

$$H_O = rmv \sin \phi \qquad (12.13)$$

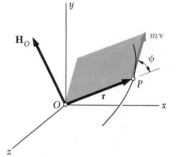

Fig. 12.12

where ϕ is the angle between $\mathbf{r}$ and $m\mathbf{v}$ (Fig. 12.12). The sense of $\mathbf{H}_O$ may be determined from the sense of $m\mathbf{v}$ by applying the right-hand rule. The unit of angular momentum is obtained by multiplying the units of length and of linear momentum (Sec. 12.4). With SI units we have

$$(\text{m})(\text{kg} \cdot \text{m/s}) = \text{kg} \cdot \text{m}^2/\text{s}$$

while, with U.S. customary units, we write

$$(\text{ft})(\text{lb} \cdot \text{s}) = \text{ft} \cdot \text{lb} \cdot \text{s}$$

Resolving the vectors $\mathbf{r}$ and $m\mathbf{v}$ into components, and applying formula (3.10), we write

$$\mathbf{H}_O = \begin{vmatrix} \mathbf{i} & \mathbf{j} & \mathbf{k} \\ x & y & z \\ mv_x & mv_y & mv_z \end{vmatrix} \qquad (12.14)$$

The components of $\mathbf{H}_O$, which also represent the moments of the linear momentum $m\mathbf{v}$ about the coordinate axes, may be obtained by expanding the determinant in (12.14). We have

$$\begin{aligned} H_x &= m(yv_z - zv_y) \\ H_y &= m(zv_x - xv_z) \\ H_z &= m(xv_y - yv_x) \end{aligned} \qquad (12.15)$$

In the case of a particle moving in the xy plane, we have $z = v_z = 0$ and the components H_x and H_y reduce to zero. The angular momentum is thus perpendicular to the xy plane; it is then completely defined by the scalar

$$H_O = H_z = m(xv_y - yv_x) \qquad (12.16)$$

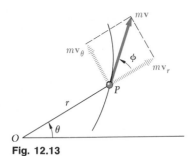

Fig. 12.13

which will be positive or negative, according to the sense in which the particle is observed to move from O. If polar coordinates are used, we resolve the linear momentum of the particle into radial and transverse components (Fig. 12.13) and write

$$H_O = rmv \sin \phi = rmv_\theta \tag{12.17}$$

or, recalling from (11.45) that $v_\theta = r\dot{\theta}$,

$$H_O = mr^2\dot{\theta} \tag{12.18}$$

We shall now compute the derivative with respect to t of the angular momentum $\mathbf{H}_O$ of a particle P moving in space. Differentiating both members of Eq. (12.12), and recalling the rule for the differentiation of a vector product (Sec. 11.10), we write

$$\dot{\mathbf{H}}_O = \dot{\mathbf{r}} \times m\mathbf{v} + \mathbf{r} \times m\dot{\mathbf{v}} = \mathbf{v} \times m\mathbf{v} + \mathbf{r} \times m\mathbf{a}$$

Since the vectors $\mathbf{v}$ and $m\mathbf{v}$ are collinear, the first term of the expression obtained is zero; and, by Newton's second law, $m\mathbf{a}$ is equal to the sum $\Sigma\mathbf{F}$ of the forces acting on P. Noting that $\mathbf{r} \times \Sigma\mathbf{F}$ represents the sum $\Sigma\mathbf{M}_O$ of the moments about O of these forces, we write

$$\Sigma\mathbf{M}_O = \dot{\mathbf{H}}_O \tag{12.19}$$

Equation (12.19), which results directly from Newton's second law, states that *the sum of the moments about O of the forces acting on the particle is equal to the rate of change of the moment of momentum, or angular momentum, of the particle about O.*

12.8. Equations of Motion in Terms of Radial and Transverse Components.

Consider a particle P, of polar coordinates r and θ, which moves in a plane under the action of several forces. Resolving the forces and the acceleration of the particle into radial and transverse components (Fig. 12.14), and substituting into Eq. (12.2), we obtain the two scalar equations

$$\Sigma F_r = ma_r \qquad \Sigma F_\theta = ma_\theta \tag{12.20}$$

Substituting for a_r and a_θ from Eqs. (11.46), we have

$$\Sigma F_r = m(\ddot{r} - r\dot{\theta}^2) \tag{12.21}$$

$$\Sigma F_\theta = m(r\ddot{\theta} + 2\dot{r}\dot{\theta}) \tag{12.22}$$

The equations obtained may be solved for two unknowns.

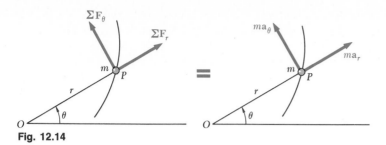

Fig. 12.14

Equation (12.22) could have been derived from Eq. (12.19). Recalling (12.18) and noting that $\Sigma M_O = r\Sigma F_\theta$, Eq. (12.19) yields

$$r\Sigma F_\theta = \frac{d}{dt}(mr^2\dot{\theta})$$

$$= m(r^2\ddot{\theta} + 2r\dot{r}\dot{\theta})$$

and, after dividing both members by r,

$$\Sigma F_\theta = m(r\ddot{\theta} + 2\dot{r}\dot{\theta}) \tag{12.22}$$

12.9. Motion under a Central Force. Conservation of Angular Momentum.

When the only force acting on a particle P is a force **F** directed toward or away from a fixed point O, the particle is said to be moving *under a central force*, and the point O is referred to as the *center of force* (Fig. 12.15). Since the line of action of **F** passes through O, we must have $\Sigma \mathbf{M}_O = 0$ at any given instant. Substituting into Eq. (12.19), we therefore obtain

$$\dot{\mathbf{H}}_O = 0$$

for all values of t or, integrating in t,

$$\mathbf{H}_O = \text{constant} \tag{12.23}$$

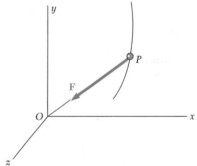

Fig. 12.15

We thus conclude that *the angular momentum of a particle moving under a central force is constant, both in magnitude and direction.*

Recalling the definition of the angular momentum of a particle (Sec. 12.7), we write

$$\mathbf{r} \times m\mathbf{v} = \mathbf{H}_O = \text{constant} \tag{12.24}$$

from which it follows that the position vector **r** of the particle P must be perpendicular to the constant vector $\mathbf{H}_O$. Thus, a particle under a central

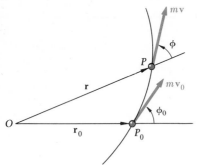

Fig. 12.16

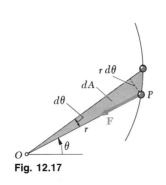

Fig. 12.17

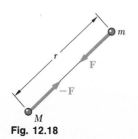

Fig. 12.18

force moves in a fixed plane perpendicular to $\mathbf{H}_O$. The vector $\mathbf{H}_O$ and the fixed plane are defined by the initial position vector $\mathbf{r}_0$ and the initial velocity $\mathbf{v}_0$ of the particle. For convenience, we shall assume that the plane of the figure coincides with the fixed plane of motion (Fig. 12.16).

Since the magnitude H_O of the angular momentum of the particle P is constant, the right-hand member in Eq. (12.13) must be constant. We therefore write

$$rmv \sin \phi = r_0 mv_0 \sin \phi_0 \qquad (12.25)$$

This relation applies to the motion of any particle under a central force. Since the gravitational force exerted by the sun on a planet is a central force directed toward the center of the sun, Eq. (12.25) is fundamental to the study of planetary motion. For a similar reason, it is also fundamental to the study of the motion of space vehicles in orbit about the earth.

Recalling Eq. (12.18), we may alternatively express the fact that the magnitude H_O of the angular momentum of the particle P is constant by writing

$$mr^2\dot{\theta} = H_O = \text{constant} \qquad (12.26)$$

or, dividing by m and denoting by h the angular momentum per unit mass H_O/m,

$$r^2\dot{\theta} = h \qquad (12.27)$$

Equation (12.27) may be given an interesting geometric interpretation. Observing from Fig. 12.17 that the radius vector OP sweeps an infinitesimal area $dA = \frac{1}{2}r^2\, d\theta$ as it rotates through an angle $d\theta$, and defining the *areal velocity* of the particle as the quotient dA/dt, we note that the left-hand member of Eq. (12.27) represents twice the areal velocity of the particle. We thus conclude that *when a particle moves under a central force, its areal velocity is constant.*

12.10. Newton's Law of Gravitation. As we saw in the preceding section, the gravitational force exerted by the sun on a planet, or by the earth on an orbiting satellite, is an important example of a central force. In this section we shall learn how to determine the magnitude of a gravitational force.

In his *law of universal gravitation*, Newton states that two particles at a distance r from each other and, respectively, of mass M and m attract each other with equal and opposite forces $\mathbf{F}$ and $-\mathbf{F}$ directed along the line joining the particles (Fig. 12.18). The common magnitude F of the two forces is

$$F = G\frac{Mm}{r^2} \qquad (12.28)$$

where G is a universal constant, called the *constant of gravitation*. Experiments show that the value of G is $(66.73 \pm 0.03) \times 10^{-12}$ m³/kg·s² in SI units, or approximately 34.4×10^{-9} ft⁴/lb·s⁴ in U.S. customary units. While gravitational forces exist between any pair of bodies, their effect is appreciable only when one of the bodies has a very large mass. The effect of gravitational forces is apparent in the case of the motion of a planet about the sun, of satellites orbiting about the earth, or of bodies falling on the surface of the earth.

Since the force exerted by the earth on a body of mass m located on or near its surface is defined as the weight $\mathbf{W}$ of the body, we may substitute the magnitude $W = mg$ of the weight for F, and the radius R of the earth for r, in Eq. (12.28). We obtain

$$W = mg = \frac{GM}{R^2}m \qquad \text{or} \qquad g = \frac{GM}{R^2} \qquad (12.29)$$

where M is the mass of the earth. Since the earth is not truly spherical, the distance R from the center of the earth depends upon the point selected on its surface, and the values of W and g will thus vary with the altitude and latitude of the point considered. Another reason for the variation of W and g with the latitude is that a system of axes attached to the earth does not constitute a newtonian frame of reference (see Sec. 12.2). A more accurate definition of the weight of a body should therefore include a component representing the centrifugal force due to the rotation of the earth. Values of g at sea level vary from 9.781 m/s², or 32.09 ft/s², at the equator to 9.833 m/s², or 32.26 ft/s², at the poles.†

The force exerted by the earth on a body of mass m located in space at a distance r from its center may be found from Eq. (12.28). The computations will be somewhat simplified if we note that according to Eq. (12.29), the product of the constant of gravitation G and of the mass M of the earth may be expressed as

$$GM = gR^2 \qquad (12.30)$$

where g and the radius R of the earth will be given their average values $g = 9.81$ m/s² and $R = 6.37 \times 10^6$ m in SI units,‡ or $g = 32.2$ ft/s² and $R = (3960 \text{ mi})(5280 \text{ ft/mi})$ in U.S. customary units.

The discovery of the law of universal gravitation has often been attributed to the fact that, after observing an apple falling from a tree, Newton had reflected that the earth must attract an apple and the moon in much the same way. While it is doubtful that this incident actually took place, it may be said that Newton would not have formulated his law if he had not first perceived that the acceleration of a falling body must have the same cause as the acceleration which keeps the moon in its orbit. This basic concept of continuity of the gravitational attraction is more easily understood now, when the gap between the apple and the moon is being filled with long-range ballistic missiles and artificial earth satellites.

---

† A formula expressing g in terms of the latitude ϕ was given in Prob. 12.1.

‡ The value of R is easily found if one recalls that the circumference of the earth is $2\pi R = 40 \times 10^6$ m.

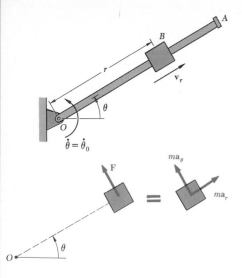

$\dot{\theta} = \dot{\theta}_0$

SAMPLE PROBLEM 12.7

A block B of mass m may slide freely on a frictionless arm OA which rotates in a horizontal plane at a constant rate $\dot{\theta}_0$. Knowing that B is released at a distance r_0 from O, express as a function of r, (a) the component v_r of the velocity of B along OA, (b) the magnitude of the horizontal force $\mathbf{F}$ exerted on B by the arm OA.

Solution. Since all other forces are perpendicular to the plane of the figure, the only force shown acting on B is the force $\mathbf{F}$ perpendicular to OA.
Equations of Motion. Using radial and transverse components,

$$+\nearrow \Sigma F_r = ma_r: \qquad\qquad 0 = m(\ddot{r} - r\dot{\theta}^2) \qquad\qquad (1)$$
$$+\nwarrow \Sigma F_\theta = ma_\theta: \qquad\qquad F = m(r\ddot{\theta} + 2\dot{r}\dot{\theta}) \qquad\qquad (2)$$

a. **Component $\mathbf{v}_r$ of Velocity.** Since $v_r = \dot{r}$, we have

$$\ddot{r} = \dot{v}_r = \frac{dv_r}{dt} = \frac{dv_r}{dr}\frac{dr}{dt} = v_r\frac{dv_r}{dr}$$

Substituting for $\ddot{r}$ into (1), recalling that $\dot{\theta} = \dot{\theta}_0$, and separating the variables,

$$v_r\, dv_r = \dot{\theta}_0^2 r\, dr$$

Multiplying by 2, and integrating form 0 to v_r and from r_0 to r,

$$v_r^2 = \dot{\theta}_0^2(r^2 - r_0^2) \qquad\qquad v_r = \dot{\theta}_0(r^2 - r_0^2)^{1/2} \quad \blacktriangleleft$$

b. **Horizontal Force F.** Making $\dot{\theta} = \dot{\theta}_0$, $\ddot{\theta} = 0$, $\dot{r} = v_r$ in Eq. (2), and substituting for v_r the expression obtained in part a,

$$F = 2m\dot{\theta}_0(r^2 - r_0^2)^{1/2}\dot{\theta}_0 \qquad F = 2m\dot{\theta}_0^2(r^2 - r_0^2)^{1/2} \quad \blacktriangleleft$$

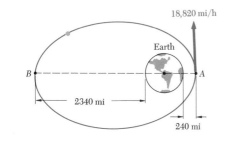

18,820 mi/h

Earth

B ———————— A

2340 mi

240 mi

SAMPLE PROBLEM 12.8

A satellite is launched in a direction parallel to the surface of the earth with a velocity of 18,820 mi/h from an altitude of 240 mi. Determine the velocity of the satellite as it reaches its maximum altitude of 2340 mi. It is recalled that the radius of the earth is 3960 mi.

Solution. Since the satellite is moving under a central force directed toward the center O of the earth, its angular momentum $\mathbf{H}_O$ is constant. From Eq. (12.13) we have

$$rmv \sin \phi = H_O = \text{constant}$$

which shows that v is minimum at B, where both r and $\sin \phi$ are maximum. Expressing conservation of angular momentum between A and B,

$$r_A mv_A = r_B mv_B$$

$$v_B = v_A \frac{r_A}{r_B} = (18,820 \text{ mi/h})\frac{3960 \text{ mi} + 240 \text{ mi}}{3960 \text{ mi} + 2340 \text{ mi}}$$

$$v_B = 12,550 \text{ mi/h} \quad \blacktriangleleft$$

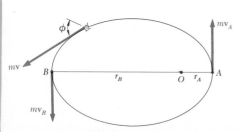

Problems

12.64 The motion of a 4-lb block B in a horizontal plane is defined by the relations $r = 3t^2 - t^3$ and $\theta = 2t^2$, where r is expressed in feet, t in seconds, and θ in radians. Determine the radial and transverse components of the force exerted on the block when (*a*) $t = 0$, (*b*) $t = 1$ s.

12.65 For the motion defined in Prob. 12.64, determine the radial and transverse components of the force exerted on the 4-lb block as it passes again through the origin at $t = 3$ s.

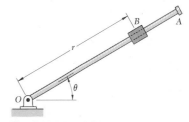

Fig. P12.64 and P12.66

12.66 The motion of a 500-g block B in a horizontal plane is defined by the relations $r = 2(1 + \cos 2\pi t)$ and $\theta = 2\pi t$, where r is expressed in meters, t in seconds, and θ in radians. Determine the radial and transverse components of the force exerted on the block when (*a*) $t = 0$, (*b*) $t = 0.75$ s.

12.67 A block B of mass m may slide on the frictionless arm OA which rotates in a horizontal plane at a constant rate $\dot{\theta}_0$. As the arm rotates, the cord wraps around a *fixed* drum of radius b and pulls the block toward O with a speed $b\dot{\theta}_0$. Express as a function of m, r, b, and $\dot{\theta}_0$, (*a*) the tension T in the cord, (*b*) the magnitude of the horizontal force $\mathbf{Q}$ exerted on B by arm OA.

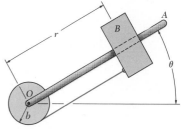

Fig. P12.67

12.68 Solve Prob. 12.67, knowing that the mass of the block is 2 kg and that $r = 500$ mm, $b = 60$ mm, and $\dot{\theta}_0 = 8$ rad/s.

12.69 Slider C has a mass of 200 g and may move in a slot cut in arm AB, which rotates at the constant rate $\dot{\theta}_0 = 12$ rad/s in a horizontal plane. The slider is attached to a spring of constant $k = 36$ N/m, which is unstretched when $r = 0$. Knowing that the slider passes through the position $r = 400$ mm with a radial velocity $v_r = +1.8$ m/s, determine at that instant (*a*) the radial and transverse components of its acceleration, (*b*) its acceleration relative to arm AB, (*c*) the horizontal force exerted on the slider by arm AB.

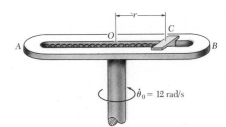

Fig. P12.69 and P12.70

*12.70 Slider C has a mass of 200 g and may move in a slot cut in arm AB, which rotates at the constant rate $\dot{\theta}_0 = 12$ rad/s in a horizontal plane. The slider is attached to a spring of constant $k = 36$ N/m, which is unstretched when $r = 0$. Knowing that the slider is released with no radial velocity in the position $r = 500$ mm and neglecting friction, determine for the position $r = 300$ mm (*a*) the radial and transverse components of the velocity of the slider, (*b*) the radial and transverse components of its acceleration, (*c*) the horizontal force exerted on the slider by arm AB.

*12.71 Solve Prob. 12.70, assuming that the spring is unstretched when slider C is located 45 mm to the left of the midpoint O of arm AB ($r = -45$ mm).

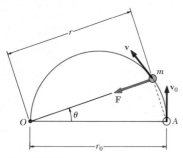

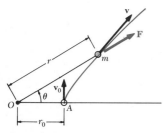

Fig. P12.72

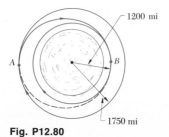

Fig. P12.73

12.72 A particle of mass m is projected from point A with an initial velocity $\mathbf{v}_0$ perpendicular to line OA and moves under a central force $\mathbf{F}$ along a semicircular path of diameter OA. Observing that $r = r_0 \cos \theta$ and using Eq. (12.27), show that the speed v of the particle is inversely proportional to the square of the distance r from the particle to the center of force O.

12.73 A particle of mass m is projected from point A with an initial velocity $\mathbf{v}_0$ perpendicular to OA and moves under a central force $\mathbf{F}$ directed away from the center of force O. Knowing that the particle follows a path defined by the equation $r = r_0/\cos 2\theta$, and using Eq. (12.27), express the radial and transverse components of the velocity $\mathbf{v}$ of the particle as functions of θ.

12.74 For the particle of Prob. 12.72, (a) show that the central force $\mathbf{F}$ is inversely proportional to the fifth power of the distance r from the particle to the center of force O, (b) determine the magnitude of $\mathbf{F}$ for $\theta = 0$.

12.75 For the particle of Prob. 12.73, (a) show that the central force $\mathbf{F}$ is inversely proportional to the cube of the distance r from the particle to the center of force O, (b) determine the magnitude of $\mathbf{F}$ for $\theta = 0$.

12.76 Denoting by ρ the mean density of a planet, show that the minimum time required by a satellite to complete one full revolution about the planet is $(3\pi/G\rho)^{1/2}$, where G is the constant of gravitation.

12.77 Show that the radius r of the moon's orbit may be determined from the radius R of the earth, the acceleration of gravity g at the surface of the earth, and the time τ required by the moon to complete one full revolution about the earth. Compute r knowing that $\tau = 27.3$ days, giving the answer in both SI and U.S. customary units.

12.78 Communication satellites have been placed in a geosynchronous orbit, i.e., in a circular orbit such that they complete one full revolution about the earth in one sidereal day (23 h 56 min), and thus appear stationary with respect to the ground. Determine (a) the altitude of the satellites above the surface of the earth, (b) the velocity with which they describe their orbit. Give the answers in both SI and U.S. customary units.

12.79 The periodic times of two of the planet Jupiter's satellites, Io and Callisto, have been observed to be, respectively, 1 day 18 h 28 min and 16 days 16 h 32 min. Knowing that the radius of Callisto's orbit is 1.884×10^6 km, determine (a) the mass of the planet Jupiter, (b) the radius of Io's orbit. (The periodic time of a satellite is the time it requires to complete one full revolution about the planet.)

12.80 An Apollo spacecraft on a lunar landing mission was describing a circular orbit of 1750-mi radius about the moon with a velocity of 2950 mi/h. To transfer it to a smaller circular orbit of 1200-mi radius, the spacecraft was first placed on an elliptic path AB by reducing its velocity by 290 mi/h as it passed through A. Determine (a) the velocity of the spacecraft as it approached B on the elliptic path, (b) the amount by which its velocity was reduced at B to insert it into the smaller circular orbit.

Fig. P12.80

12.81 Plans for an unmanned landing mission on the planet Mars called for the earth-return vehicle to first describe a circular orbit at an altitude $d_A = 2200$ km above the surface of the planet with a velocity of 2771 m/s. As it passed through point A, the vehicle was to be inserted into an elliptic transfer orbit by firing its engine and increasing its speed by $\Delta v_A = 1046$ m/s. As it passed through point B, at an altitude $d_B = 100\,000$ km, the vehicle was to be inserted into a second transfer orbit located in a slightly different plane, by changing the direction of its velocity and reducing its speed by $\Delta v_B = -22.0$ m/s. Finally, as the vehicle passed through point C, at an altitude $d_C = 1000$ km, its speed was to be increased by $\Delta v_C = 660$ m/s to insert it into its return trajectory. Knowing that the radius of the planet Mars is $R = 3400$ km, determine the velocity of the vehicle after completion of the last maneuver.

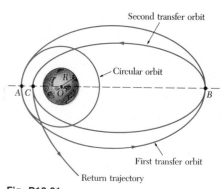

12.82 Solve Prob. 12.81, assuming that $\Delta v_B = -37.2$ m/s and $d_C = 300$ km, and that all other data remain unchanged.

Fig. P12.81

12.83 A space tug is used to place communication satellites into a geosynchronous orbit (see Prob. 12.78) at an altitude of 22,230 mi above the surface of the earth. The tug initially describes a circular orbit at an altitude of 340 mi and is inserted in an elliptic transfer orbit by firing its engine as it passes through A, thus increasing its velocity by 7740 ft/s. By how much should its velocity be increased as it reaches B to insert it in the geosynchronous orbit?

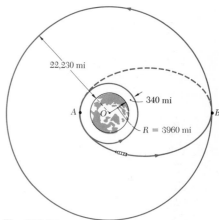

12.84 A 4-lb ball is mounted on a horizontal rod which is free to rotate about a vertical shaft. In the position shown, the speed of the ball is $v_1 = 30$ in./s and the ball is held by a cord attached to the shaft. The cord is suddenly cut and the ball moves to position B as the rod rotates. Neglecting the mass of the rod, determine (a) the radial and transverse components of the acceleration of the ball immediately after the cord has been cut, (b) the acceleration of the ball relative to the rod at the same instant, (c) the speed of the ball after it has reached the stop B.

Fig. P12.83

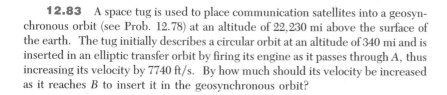

Fig. P12.84

12.85 A small ball swings in a horizontal circle at the end of a cord of length l_1, which forms an angle θ_1 with the vertical. The cord is then slowly drawn through the support at O until the length of the free end is l_2. (a) Derive a relation between l_1, l_2, θ_1, and θ_2. (b) If the ball is set in motion so that initially $l_1 = 30$ in. and $\theta_1 = 30°$, determine the angle θ_2 when $l_2 = 20$ in.

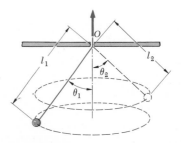

Fig. P12.85

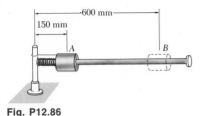

12.86 A 300-g collar may slide on a horizontal rod which is free to rotate about a vertical shaft. The collar is initially held at A by a cord attached to the shaft and compresses a spring of constant 5 N/m, which is undeformed when the collar is located 750 mm from the shaft. As the rod rotates at the rate $\dot{\theta} = 12$ rad/s, the cord is cut and the collar moves out along the rod. Neglecting friction and the mass of the rod, determine for position B of the collar (a) the transverse component of the velocity of the collar, (b) the radial and transverse components of its acceleration, (c) the acceleration of the collar relative to the rod.

*12.87 In Prob. 12.86, determine for position B of the collar, (a) the radial component of the velocity of the collar, (b) the value of $\ddot{\theta}$.

* 12.11. Trajectory of a Particle under a Central Force.

Consider a particle P moving under a central force $\mathbf{F}$. We propose to obtain the differential equation which defines its trajectory.

Assuming that the force $\mathbf{F}$ is directed toward the center of force O, we note that ΣF_r and ΣF_θ reduce, respectively, to $-F$ and zero in Eqs. (12.21) and (12.22). We therefore write

$$m(\ddot{r} - r\dot{\theta}^2) = -F \tag{12.31}$$

$$m(r\ddot{\theta} + 2\dot{r}\dot{\theta}) = 0 \tag{12.32}$$

These equations define the motion of P. We shall, however, replace Eq. (12.32) by Eq. (12.27), which is more convenient to use and which is equivalent to Eq. (12.32), as we may easily check by differentiating it with respect to t. We write

$$r^2\dot{\theta} = h \qquad \text{or} \qquad r^2\frac{d\theta}{dt} = h \tag{12.33}$$

Equation (12.33) may be used to eliminate the independent variable t from Eq. (12.31). Solving Eq. (12.33) for $\dot{\theta}$ or $d\theta/dt$, we have

$$\dot{\theta} = \frac{d\theta}{dt} = \frac{h}{r^2} \tag{12.34}$$

from which it follows that

$$\dot{r} = \frac{dr}{dt} = \frac{dr}{d\theta}\frac{d\theta}{dt} = \frac{h}{r^2}\frac{dr}{d\theta} = -h\frac{d}{d\theta}\left(\frac{1}{r}\right) \tag{12.35}$$

$$\ddot{r} = \frac{d\dot{r}}{dt} = \frac{d\dot{r}}{d\theta}\frac{d\theta}{dt} = \frac{h}{r^2}\frac{d\dot{r}}{d\theta}$$

or, substituting for $\dot{r}$ from (12.35),

$$\ddot{r} = \frac{h}{r^2}\frac{d}{d\theta}\left[-h\frac{d}{d\theta}\left(\frac{1}{r}\right)\right]$$

$$\ddot{r} = -\frac{h^2}{r^2}\frac{d^2}{d\theta^2}\left(\frac{1}{r}\right) \tag{12.36}$$

Substituting for $\dot{\theta}$ and $\ddot{r}$ from (12.34) and (12.36), respectively, into Eq. (12.31), and introducing the function $u = 1/r$, we obtain after reductions

$$\frac{d^2u}{d\theta^2} + u = \frac{F}{mh^2u^2} \tag{12.37}$$

In deriving Eq. (12.37), the force $\mathbf{F}$ was assumed directed toward O. The magnitude F should therefore be positive if $\mathbf{F}$ is actually directed toward O (attractive force) and negative if $\mathbf{F}$ is directed away from O (repulsive force). If F is a known function of r and thus of u, Eq. (12.37) is a differential equation in u and θ. This differential equation defines the trajectory followed by the particle under the central force $\mathbf{F}$. The equation of the trajectory will be obtained by solving the differential equation (12.37) for u as a function of θ and determining the constants of integration from the initial conditions.

* 12.12. Application to Space Mechanics. After the last stage of their launching rockets has burned out, earth satellites and other space vehicles are subjected to only the gravitational pull of the earth. Their motion may therefore be determined from Eqs. (12.33) and (12.37), which govern the motion of a particle under a central force, after F has been replaced by the expression obtained for the force of gravitational attraction.† Setting in Eq. (12.37)

$$F = \frac{GMm}{r^2} = GMmu^2$$

where M = mass of earth
m = mass of space vehicle
r = distance from center of earth to vehicle
$u = 1/r$

we obtain the differential equation

$$\frac{d^2u}{d\theta^2} + u = \frac{GM}{h^2} \tag{12.38}$$

where the right-hand member is observed to be a constant.

The solution of the differential equation (12.38) is obtained by adding the particular solution $u = GM/h^2$ to the general solution $u = C\cos(\theta - \theta_0)$ of the corresponding homogeneous equation (i.e., the equation obtained by setting the right-hand member equal to zero). Choosing

---

† It is assumed that the space vehicles considered here are attracted by only the earth and that their mass is negligible compared with the mass of the earth. If a vehicle moves very far from the earth, its path may be affected by the attraction of the sun, the moon, or another planet.

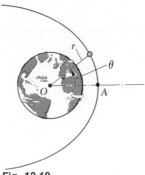

Fig. 12.19

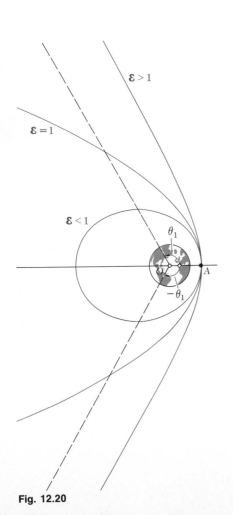

Fig. 12.20

the polar axis so that $\theta_0 = 0$, we write

$$\frac{1}{r} = u = \frac{GM}{h^2} + C \cos \theta \tag{12.39}$$

Equation (12.39) is the equation of a *conic section* (ellipse, parabola, or hyperbola) in the polar coordinates r and θ. The origin O of the coordinates, which is located at the center of the earth, is a *focus* of this conic section, and the polar axis is one of its axes of symmetry (Fig. 12.19).

The ratio of the constants C and GM/h^2 defines the *eccentricity* ε of the conic section; setting

$$\varepsilon = \frac{C}{GM/h^2} = \frac{Ch^2}{GM} \tag{12.40}$$

we may write Eq. (12.39) in the form

$$\frac{1}{r} = \frac{GM}{h^2}(1 + \varepsilon \cos \theta) \tag{12.39'}$$

Three cases may be distinguished:

1. $\varepsilon > 1$, or $C > GM/h^2$: There are two values θ_1 and $-\theta_1$ of the polar angle, defined by $\cos \theta_1 = -GM/Ch^2$, for which the right-hand member of Eq. (12.39) becomes zero. For both these values, the radius vector r becomes infinite; the conic section is a *hyperbola* (Fig. 12.20).
2. $\varepsilon = 1$, or $C = GM/h^2$: The radius vector becomes infinite for $\theta = 180°$; the conic section is a *parabola*.
3. $\varepsilon < 1$, or $C < GM/h^2$: The radius vector remains finite for every value of θ; the conic section is an *ellipse*. In the particular case when $\varepsilon = C = 0$, the length of the radius vector is constant; the conic section is a *circle*.

We shall see now how the constants C and GM/h^2 which characterize the trajectory of a space vehicle may be determined from the position and the velocity of the space vehicle at the beginning of its free flight. We shall assume, as is generally the case, that the powered phase of its flight has been programmed in such a way that as the last stage of the launching rocket burns out, the vehicle has a velocity parallel to the surface of the earth (Fig. 12.21). In other words, we shall assume that the space vehicle begins its free flight at the vertex A of its trajectory.†

Denoting, respectively, by r_0 and v_0 the radius vector and speed of the vehicle at the beginning of its free flight, we observe, since the velocity reduces to its transverse component, that $v_0 = r_0 \dot\theta_0$. Recalling Eq. (12.27), we express the angular momentum per unit mass h as

$$h = r_0^2 \dot\theta_0 = r_0 v_0 \qquad (12.41)$$

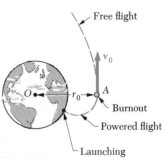

Free flight

v_0

O r_0 A

Burnout

Powered flight

Launching

Fig. 12.21

The value obtained for h may be used to determine the constant GM/h^2. We also note that the computation of this constant will be simplified if we use the relation indicated in Sec. 12.10:

$$GM = gR^2 \qquad (12.30)$$

where R is the radius of the earth ($R = 6.37 \times 10^6$ m or 3960 mi) and g the acceleration of gravity at the surface of the earth.

The constant C will be determined by setting $\theta = 0$, $r = r_0$ in Eq. (12.39); we obtain

$$C = \frac{1}{r_0} - \frac{GM}{h^2} \qquad (12.42)$$

Substituting for h from (12.41), we may then easily express C in terms of r_0 and v_0.

Let us now determine the initial conditions corresponding to each of the three fundamental trajectories indicated above. Considering first the parabolic trajectory, we set C equal to GM/h^2 in Eq. (12.42) and eliminate h between Eqs. (12.41) and (12.42). Solving for v_0, we obtain

$$v_0 = \sqrt{\frac{2GM}{r_0}}$$

We may easily check that a larger value of the initial velocity corresponds to a hyperbolic trajectory and a smaller value, to an elliptic orbit. Since the value of v_0 obtained for the parabolic trajectory is the smallest value for which the space vehicle does not return to its starting point, it is called the *escape velocity*. We write therefore

$$v_{\text{esc}} = \sqrt{\frac{2GM}{r_0}} \qquad \text{or} \qquad v_{\text{esc}} = \sqrt{\frac{2gR^2}{r_0}} \qquad (12.43)$$

if we make use of Eq. (12.30). We note that the trajectory will be (1) hyperbolic if $v_0 > v_{\text{esc}}$; (2) parabolic if $v_0 = v_{\text{esc}}$; (3) elliptic if $v_0 < v_{\text{esc}}$.

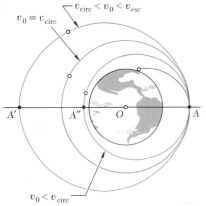

Fig. 12.22

Among the various possible elliptic orbits, one is of special interest, the *circular orbit*, which is obtained when $C = 0$. The value of the initial velocity corresponding to a circular orbit is easily found to be

$$v_{\text{circ}} = \sqrt{\frac{GM}{r_0}} \qquad \text{or} \qquad v_{\text{circ}} = \sqrt{\frac{gR^2}{r_0}} \qquad (12.44)$$

if Eq. (12.30) is taken into account. We may note from Fig. 12.22 that for values of v_0 comprised between v_{circ} and v_{esc}, point A where free flight begins is the point of the orbit closest to the earth; this point is called the *perigee*, while point A', which is farthest away from the earth, is known as the *apogee*. For values of v_0 smaller than v_{circ}, point A becomes the apogee, while point A'', on the other side of the orbit, becomes the perigee. For values of v_0 much smaller than v_{circ}, the trajectory of the space vehicle intersects the surface of the earth; in such a case, the vehicle does not go into orbit.

Ballistic missiles, which are designed to hit the surface of the earth, also travel along elliptic trajectories. In fact, we should now realize that any object projected in vacuum with an initial velocity v_0 smaller than v_{esc} will move along an elliptic path. It is only when the distances involved are small that the gravitational field of the earth may be assumed uniform, and that the elliptic path may be approximated by a parabolic path, as was done earlier (Sec. 11.11) in the case of conventional projectiles.

Periodic Time. An important characteristic of the motion of an earth satellite is the time required by the satellite to describe its orbit. This time is known as the *periodic time* of the satellite and is denoted by τ. We first observe, in view of the definition of the areal velocity (Sec. 12.9), that τ may be obtained by dividing the area inside the orbit by the areal velocity. Since the area of an ellipse is equal to πab, where a and b denote, respectively, the semimajor and semiminor axes, and since the areal velocity is equal to $h/2$, we write

$$\tau = \frac{2\pi ab}{h} \qquad (12.45)$$

While h may be readily determined from r_0 and v_0 in the case of a satellite launched in a direction parallel to the surface of the earth, the semiaxes a and b are not directly related to the initial conditions. Since, on the other hand, the values r_0 and r_1 of r corresponding to the perigee and apogee of the orbit may easily be determined from Eq. (12.39), we shall express the semiaxes a and b in terms of r_0 and r_1.

Consider the elliptic orbit shown in Fig. 12.23. The earth's center is located at O and coincides with one of the two foci of the ellipse, while the points A and A' represent, respectively, the perigee and apogee of the orbit. We easily check that

$$r_0 + r_1 = 2a$$

and thus

$$a = \tfrac{1}{2}(r_0 + r_1) \tag{12.46}$$

Recalling that the sum of the distances from each of the foci to any point of the ellipse is constant, we write

$$O'B + BO = O'A + OA = 2a \qquad \text{or} \qquad BO = a$$

On the other hand, we have $CO = a - r_0$. We may therefore write

$$b^2 = (BC)^2 = (BO)^2 - (CO)^2 = a^2 - (a - r_0)^2$$
$$b^2 = r_0(2a - r_0) = r_0 r_1$$

and thus

$$b = \sqrt{r_0 r_1} \tag{12.47}$$

Formulas (12.46) and (12.47) indicate that the semimajor and semiminor axes of the orbit are equal, respectively, to the arithmetic and geometric means of the maximum and minimum values of the radius vector. Once r_0 and r_1 have been determined, the lengths of the semiaxes may thus be easily computed and substituted for a and b in formula (12.45).

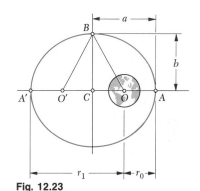

Fig. 12.23

12.13. Kepler's Laws of Planetary Motion. The equations governing the motion of an earth satellite may be used to describe the motion of the moon around the earth. In that case, however, the mass of the moon is not negligible compared with the mass of the earth, and the results obtained are not entirely accurate.

The theory developed in the preceding sections may also be applied to the study of the motion of the planets around the sun. While another error is introduced by neglecting the forces exerted by the planets on one another, the approximation obtained is excellent. Indeed, the properties expressed by Eq. (12.39), where M now represents the mass of the sun, and by Eq. (12.33) had been discovered by the German astronomer Johann Kepler (1571–1630) from astronomical observations of the motion of the planets, even before Newton had formulated his fundamental theory.

Kepler's three *laws of planetary motion* may be stated as follows:

1. Each planet describes an ellipse, with the sun located at one of its foci.
2. The radius vector drawn from the sun to a planet sweeps equal areas in equal times.
3. The squares of the periodic times of the planets are proportional to the cubes of the semimajor axes of their orbits.

The first law states a particular case of the result established in Sec. 12.12, while the second law expresses that the areal velocity of each planet is constant (see Sec. 12.9). Kepler's third law may also be derived from the results obtained in Sec. 12.12.†

† See Prob. 12.114.

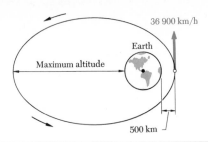

36 900 km/h

Earth

Maximum altitude

500 km

SAMPLE PROBLEM 12.9

A satellite is launched in a direction parallel to the surface of the earth with a velocity of 36 900 km/h from an altitude of 500 km. Determine (a) the maximum altitude reached by the satellite, (b) the periodic time of the satellite.

a. Maximum Altitude. After launching, the satellite is subjected only to the gravitational attraction of the earth; its motion is thus governed by Eq. (12.39),

$$\frac{1}{r} = \frac{GM}{h^2} + C \cos \theta \qquad (1)$$

Since the radial component of the velocity is zero at the point of launching A, we have $h = r_0 v_0$. Recalling that the radius of the earth is $R = 6370$ km, we compute

$$r_0 = 6370 \text{ km} + 500 \text{ km} = 6870 \text{ km} = 6.87 \times 10^6 \text{ m}$$

$$v_0 = 36\,900 \text{ km/h} = \frac{36.9 \times 10^6 \text{ m}}{3.6 \times 10^3 \text{ s}} = 10.25 \times 10^3 \text{ m/s}$$

$$h = r_0 v_0 = (6.87 \times 10^6 \text{ m})(10.25 \times 10^3 \text{ m/s}) = 70.4 \times 10^9 \text{ m}^2/\text{s}$$

$$h^2 = 4.96 \times 10^{21} \text{ m}^4/\text{s}^2$$

Since $GM = gR^2$, where R is the radius of the earth, we have

$$GM = gR^2 = (9.81 \text{ m/s}^2)(6.37 \times 10^6 \text{ m})^2 = 398 \times 10^{12} \text{ m}^3/\text{s}^2$$

$$\frac{GM}{h^2} = \frac{398 \times 10^{12} \text{ m}^3/\text{s}^2}{4.96 \times 10^{21} \text{ m}^4/\text{s}^2} = 80.3 \times 10^{-9} \text{ m}^{-1}$$

Substituting this value into (1), we obtain

$$\frac{1}{r} = 80.3 \times 10^{-9} \text{ m}^{-1} + C \cos \theta \qquad (2)$$

Noting that at point A we have $\theta = 0$ and $r = r_0 = 6.87 \times 10^6$ m, we compute the constant C:

$$\frac{1}{6.87 \times 10^6 \text{ m}} = 80.3 \times 10^{-9} \text{ m}^{-1} + C \cos 0° \qquad C = 65.3 \times 10^{-9} \text{ m}^{-1}$$

At A', the point on the orbit farthest from the earth, we have $\theta = 180°$. Using (2), we compute the corresponding distance r_1:

$$\frac{1}{r_1} = 80.3 \times 10^{-9} \text{ m}^{-1} + (65.3 \times 10^{-9} \text{ m}^{-1}) \cos 180°$$

$$r_1 = 66.7 \times 10^6 \text{ m} = 66\,700 \text{ km}$$

$$\textit{Maximum altitude} = 66\,700 \text{ km} - 6370 \text{ km} = 60\,300 \text{ km} \quad \blacktriangleleft$$

b. Periodic Time. Since A and A' are the perigee and apogee, respectively, of the elliptic orbit, we use Eqs. (12.46) and (12.47) and compute the semimajor and semiminor axes of the orbit:

$$a = \tfrac{1}{2}(r_0 + r_1) = \tfrac{1}{2}(6.87 + 66.7)(10^6) \text{ m} = 36.8 \times 10^6 \text{ m}$$

$$b = \sqrt{r_0 r_1} = \sqrt{(6.87)(66.7)} \times 10^6 \text{ m} = 21.4 \times 10^6 \text{ m}$$

$$\tau = \frac{2\pi a b}{h} = \frac{2\pi (36.8 \times 10^6 \text{ m})(21.4 \times 10^6 \text{ m})}{70.4 \times 10^9 \text{ m}^2/\text{s}}$$

$$\tau = 70.3 \times 10^3 \text{ s} = 1171 \text{ min} = 19 \text{ h } 31 \text{ min} \quad \blacktriangleleft$$

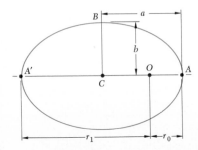

Problems

12.88 For the particle of Prob. 12.72, and using Eq. (12.37), show that the central force **F** is inversely proportional to the fifth power of the distance r from the particle to the center of force O.

12.89 For the particle of Prob. 12.73, and using Eq. (12.37), show that the central force **F** is inversely proportional to the cube of the distance r from the particle to the center of force O

12.90 A particle of mass m describes the hyperbolic spiral $r = b/\theta$ under a central force **F** directed toward the center of force O. Using Eq. (12.37), show that **F** is inversely proportional to the cube of the distance r from the particle to O.

12.91 A particle of mass m describes the logarithmic spiral $r = r_0 e^{b\theta}$ under a central force **F** directed toward the center of force O. Using Eq. (12.37), show that **F** is inversely proportional to the cube of the distance r from the particle to O.

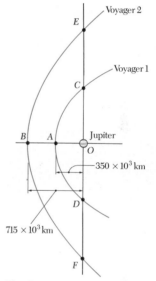

12.92 It was observed that as the spacecraft Voyager 1 reached the point on its trajectory closest to the planet Jupiter, it was at a distance of 350×10^3 km from the center of the planet and had a velocity of 26.9 km/s. Determine the mass of Jupiter, assuming that the trajectory of the spacecraft was parabolic.

12.93 It was observed that as the spacecraft Voyager 1 reached the point of its trajectory closest to the planet Saturn, it was at a distance of 115×10^3 mi from the center of the planet and had a velocity of 68.8×10^3 ft/s. Knowing that Tethys, one of Saturn's satellites, describes a circular orbit of radius 183×10^3 mi at a speed of 37.2×10^3 ft/s, determine the eccentricity of the trajectory of Voyager 1 on its approach to Saturn.

Fig. P12.92 and P12.94

12.94 It was observed that as the spacecraft Voyager 2 reached the point on its trajectory closest to the planet Jupiter, it was at a distance of 715×10^3 km from the center of the planet. Assuming the trajectory of the spacecraft to be parabolic and using the data given in Prob. 12.92 for Voyager 1, determine the maximum velocity of Voyager 2 on its approach to Jupiter.

12.95 Solve Prob. 12.94, assuming that the eccentricity of the trajectory of the spacecraft Voyager 2 was $\varepsilon = 1.50$.

12.96 A satellite describes an elliptic orbit about a planet of mass M. Denoting by r_0 and r_1, respectively, the minimum and maximum values of the distance r from the satellite to the center of the planet, derive the relation

$$\frac{1}{r_0} + \frac{1}{r_1} = \frac{2GM}{h^2}$$

where h is the angular momentum per unit mass of the satellite.

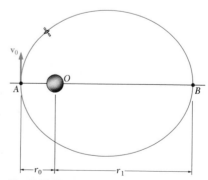

Fig. P12.96

12.97 At engine burnout on one of its early missions, the space shuttle Columbia had reached point A at an altitude of 40 mi and had a horizontal velocity v_0. Knowing that its first orbit was elliptic and that the shuttle was transferred to a circular orbit as it passed through point B at an altitude of 240 mi above the surface of the earth, determine (a) the speed v_0 of the shuttle at engine burnout, (b) the increase in speed required at B to insert the shuttle on the circular orbit.

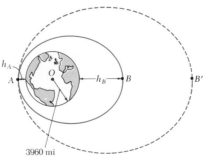

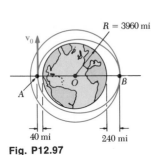

R = 3960 mi

40 mi   240 mi

Fig. P12.97

3960 mi

Fig. P12.98

12.98 A spacecraft is describing an elliptic orbit of minimum altitude $h_A = 1500$ mi and maximum altitude $h_B = 6000$ mi above the surface of the earth. (a) Determine the speed of the spacecraft at A. (b) If its engine is fired as the spacecraft passes through point A and its speed is increased by 8 percent, determine the maximum altitude reached by the spacecraft on its new orbit.

12.99 A space probe is describing a circular orbit about a planet of radius R. The altitude of the probe above the surface of the planet is αR and its velocity is v_0. In order to place the probe on an elliptic orbit which will bring it closer to the planet, its velocity is reduced from v_0 to βv_0, where $\beta < 1$, by firing its engine for a short interval of time. Determine the smallest permissible value of β if the probe is not to crash on the surface of the planet.

12.100 A space probe is to be placed in a circular orbit of 9000-km radius about the planet Venus in a specified plane. As the probe reaches A, the point of its original trajectory closest to Venus, it is inserted in a first elliptic transfer orbit by reducing its speed by Δv_A. This orbit brings it to point B with a much reduced velocity. There the probe is inserted in a second transfer orbit located in the specified plane by changing the direction of its velocity and further reducing its speed by Δv_B. Finally, as the probe reaches point C, it is inserted in the desired circular orbit by reducing its speed by Δv_C. Knowing that the mass of Venus is 0.82 times the mass of the earth, that $r_A = 15 \times 10^3$ km and $r_B = 300 \times 10^3$ km, and that the probe approaches A on a parabolic trajectory, determine by how much the velocity of the probe should be reduced (a) at A, (b) at B, (c) at C.

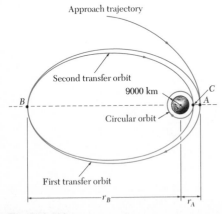

Approach trajectory

Second transfer orbit

9000 km

C

B

Circular orbit

A

First transfer orbit

r_B r_A

Fig. P12.100

12.101 For the space probe of Prob. 12.100, it is known that $r_A = 15 \times 10^3$ km and that the velocity of the probe is reduced to 6500 m/s as it passes through A. Determine (a) the distance from the center of Venus to point B, (b) the amounts by which the velocity of the probe should be reduced at B and C, respectively.

12.102 Determine the time needed for the space probe of Prob. 12.100 to travel from A to B on its first transfer orbit.

12.103 Determine the time needed for the space probe of Prob. 12.100 to travel from B to C on its second transfer orbit.

12.104 For the shuttle Columbia of Prob. 12.97, determine (*a*) the periodic time of the shuttle on its final circular orbit, (*b*) the time needed for the shuttle to travel from A to B on its original elliptic orbit.

12.105 Halley's comet travels in an elongated elliptic orbit for which the minimum distance from the sun is approximately $\frac{1}{2}r_E$, where $r_E = 92.9 \times 10^6$ mi is the mean distance from the sun to the earth. Knowing that the periodic time of Halley's comet is about 76 years, determine the maximum distance from the sun reached by the comet.

12.106 Determine the time needed for the spacecraft Voyager 1 of Prob. 12.92 to travel from C to D on its parabolic trajectory.

12.107 Determine the time needed for the spacecraft Voyager 2 of Prob. 12.94 to travel from E to F on its parabolic trajectory.

12.108 A space probe is describing a circular orbit of radius nR with a velocity v_0 about a planet of radius R and center O. As the probe passes through point A, its velocity is reduced from v_0 to βv_0, where $\beta < 1$, to place the probe on a crash trajectory. Show that the angle AOB, where B denotes the point of impact of the probe on the planet, depends only upon the values of n and β.

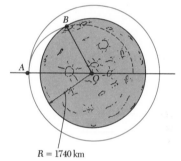

$R = 1740$ km

Fig. P12.109

12.109 Upon completion of their moon-exploration mission, the two astronauts forming the crew of an Apollo lunar excursion module (LEM) would rejoin the command module which had remained in a circular orbit around the moon. Before their return to earth, the astronauts would position their craft so that the LEM faced to the rear. As the command module passed through A, the LEM would be cast adrift and crash on the moon's surface at point B. Knowing that the command module was orbiting the moon at an altitude of 120 km and that the angle AOB was 50°, determine the velocity of the LEM relative to the command module as it was cast adrift. (*Hint.* Point A is the apogee of the elliptic crash trajectory. It is also recalled that the mass of the moon is 0.01230 times the mass of the earth.)

12.110 In Prob. 12.109, determine the angle AOB defining the point of impact of the LEM on the surface of the moon, assuming that the command module was orbiting the moon at an altitude of 150 km and that the LEM was cast adrift with a velocity of 80 m/s relative to the command module.

12.111 A space shuttle is describing a circular orbit at an altitude of 200 mi above the surface of the earth. As it passes through A it fires its engine for a short interval of time to reduce its speed by 6 percent and begin its descent toward the earth. Determine the altitude of the shuttle at point B, knowing that the angle AOB is equal to 50°. (*Hint.* Point A is the apogee of the elliptic descent trajectory.)

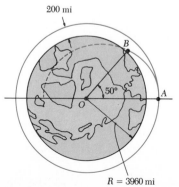

200 mi

$R = 3960$ mi

Fig. P12.111

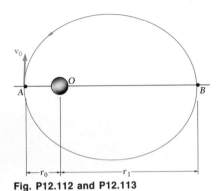

Fig. P12.112 and P12.113

12.112 A satellite describes an elliptic orbit about a planet. Denoting by r_0 and r_1 the distances corresponding, respectively, to the perigee and apogee of the orbit, show that the curvature of the orbit at each of these two points may be expressed as

$$\frac{1}{\rho} = \frac{1}{2}\left(\frac{1}{r_0} + \frac{1}{r_1}\right)$$

12.113 (*a*) Express the eccentricity ε of the elliptic orbit described by a satellite about a planet in terms of the distances r_0 and r_1 corresponding, respectively, to the perigee and apogee of the orbit. (*b*) Use the result obtained in part *a* to determine the eccentricities of the two transfer orbits described in Prob. 12.100.

12.114 Derive Kepler's third law of planetary motion from Eqs. (12.39) and (12.45).

12.115 Show that the angular momentum per unit mass h of a satellite describing an elliptic orbit of semimajor axis a and eccentricity ε about a planet of mass M may be expressed as

$$h = \sqrt{GMa(1 - \varepsilon^2)}$$

Review and Summary

This chapter was devoted to Newton's second law and its application to the analysis of the motion of particles.

Newton's second law

Denoting by m the mass of a particle, by $\Sigma\mathbf{F}$ the sum, or resultant, of the forces acting on the particle, and by $\mathbf{a}$ the acceleration of the particle relative to a *newtonian frame of reference* [Sec. 12.2], we wrote

$$\Sigma\mathbf{F} = m\mathbf{a} \qquad (12.2)$$

Linear momentum

Introducing the *linear momentum* of a particle, $\mathbf{L} = m\mathbf{v}$ [Sec. 12.3], we saw that Newton's second law may also be written in the form

$$\Sigma\mathbf{F} = \dot{\mathbf{L}} \qquad (12.5)$$

which expresses that *the resultant of the forces acting on a particle is equal to the rate of change of the linear momentum of the particle.*

Consistent systems of units

Equation (12.2) holds only if a consistent system of units is used. With SI units, the forces should be expressed in newtons, the masses in kilograms, and the accelerations in m/s²; with U.S. customary units, the forces should be expressed in pounds, the masses in lb·s²/ft (also referred to as *slugs*), and the accelerations in ft/s² [Sec. 12.4].

To solve a problem involving the motion of a particle, Eq. (12.2) should be replaced by equations containing scalar quantities [Sec. 12.5]. Using *rectangular components* of **F** and **a**, we wrote

$$\Sigma F_x = ma_x \qquad \Sigma F_y = ma_y \qquad \Sigma F_z = ma_z \qquad (12.8)$$

Using *tangential and normal components*, we had

$$\Sigma F_t = m\frac{dv}{dt} \qquad \Sigma F_n = m\frac{v^2}{\rho} \qquad (12.9')$$

We also noted [Sec. 12.6] that the equations of motion of a particle may be replaced by equations similar to the equilibrium equations used in statics if a vector $-m\mathbf{a}$ of magnitude ma but of sense opposite to that of the acceleration is added to the forces applied to the particle; the particle is then said to be in *dynamic equilibrium*. For the sake of uniformity, however, all the Sample Problems were solved from the equations of motion, first using rectangular components [Sample Probs. 12.1 through 12.4], then tangential and normal components [Sample Probs. 12.5 and 12.6].

Dynamic equilibrium

In the second part of the chapter, we defined the *angular momentum* $\mathbf{H}_O$ of a particle about a point O as the moment about O of the linear momentum $m\mathbf{v}$ of that particle [Sec. 12.7]. We wrote

Angular momentum

$$\mathbf{H}_O = \mathbf{r} \times m\mathbf{v} \qquad (12.12)$$

and noted that $\mathbf{H}_O$ is a vector perpendicular to the plane containing $\mathbf{r}$ and $m\mathbf{v}$ (Fig. 12.24) and of magnitude

$$H_O = rmv \sin \phi \qquad (12.13)$$

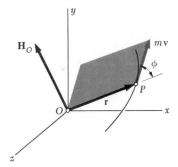

Resolving the vectors $\mathbf{r}$ and $m\mathbf{v}$ into rectangular components, we expressed the angular momentum $\mathbf{H}_O$ in the determinant form

$$\mathbf{H}_O = \begin{vmatrix} \mathbf{i} & \mathbf{j} & \mathbf{k} \\ x & y & z \\ mv_x & mv_y & mv_z \end{vmatrix} \qquad (12.14)$$

Fig. 12.24

In the case of a particle moving in the xy plane, we have $z = v_z = 0$. The angular momentum is perpendicular to the xy plane and is completely defined by its magnitude. We wrote

$$H_O = H_z = m(xv_y - yv_x) \qquad (12.16)$$

Computing the rate of change $\dot{\mathbf{H}}_O$ of the angular momentum $\mathbf{H}_O$, and applying Newton's second law, we wrote the equation

Rate of change of angular momentum

$$\Sigma \mathbf{M}_O = \dot{\mathbf{H}}_O \qquad (12.19)$$

which states that *the sum of the moments about O of the forces acting on a particle is equal to the rate of change of the angular momentum of the particle about O.*

In many problems involving the plane motion of a particle, it is found convenient to use *radial and transverse components* [Sec. 12.8, Sample Prob. 12.7] and to write the equations

$$\Sigma F_r = m(\ddot{r} - r\dot{\theta}^2) \qquad (12.21)$$
$$\Sigma F_\theta = m(r\ddot{\theta} + 2\dot{r}\dot{\theta}) \qquad (12.22)$$

Motion under a central force

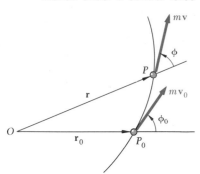

Fig. 12.25

When the only force acting on a particle P is a force $\mathbf{F}$ directed toward or away from a fixed point O, the particle is said to be moving *under a central force* [Sec. 12.9]. Since $\Sigma \mathbf{M}_O = 0$ at any given instant, it follows from Eq. (12.19) that $\dot{\mathbf{H}}_O = 0$ for all values of t and, thus, that

$$\mathbf{H}_O = \text{constant} \qquad (12.23)$$

We concluded that *the angular momentum of a particle moving under a central force is constant, both in magnitude and direction,* and that the particle moves in a plane perpendicular to the vector $\mathbf{H}_O$.

Recalling Eq. (12.13), we wrote the relation

$$rmv \sin \phi = r_0 m v_0 \sin \phi_0 \qquad (12.25)$$

which applies to the motion of any particle under a central force (Fig. 12.25). Using polar coordinates and recalling Eq. (12.18), we also had

$$r^2 \dot{\theta} = h \qquad (12.27)$$

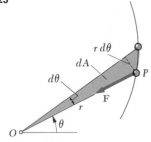

Fig. 12.26

where h is a constant representing the angular momentum per unit mass, H_O/m, of the particle. We observed (Fig. 12.26) that the infinitesimal area dA swept by the radius vector OP as it rotates through $d\theta$ is equal to $\frac{1}{2}r^2 \, d\theta$ and, thus, that the left-hand member of Eq. (12.27) represents twice the *areal velocity dA/dt* of the particle. Therefore, *the areal velocity of a particle moving under a central force is constant.*

Newton's law of universal gravitation

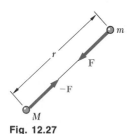

Fig. 12.27

An important application of the motion under a central force is provided by the orbital motion of bodies under gravitational attraction [Sec. 12.10]. According to *Newton's law of universal gravitation,* two particles at a distance r from each other and, respectively, of mass M and m attract each other with equal and opposite forces $\mathbf{F}$ and $-\mathbf{F}$ directed along the line joining the particles (Fig. 12.27). The common magnitude F of the two forces is

$$F = G\frac{Mm}{r^2} \qquad (12.28)$$

where G is the *constant of gravitation.* In the case of a body of mass m subjected to the gravitational attraction of the earth, the product GM, where M is the mass of the earth, may be expressed as

$$GM = gR^2 \qquad (12.30)$$

where $g = 9.81 \text{ m/s}^2 = 32.2 \text{ ft/s}^2$, and where R represents the radius of the earth.

It was shown in Sec. 12.11 that a particle moving under a central force describes a trajectory defined by the differential equation

$$\frac{d^2u}{d\theta^2} + u = \frac{F}{mh^2u^2} \tag{12.37}$$

where $F > 0$ corresponds to an attractive force and where $u = 1/r$. In the case of a particle moving under a force of gravitational attraction [Sec. 12.12], we substituted for F the expression given in Eq. (12.28). Measuring θ from the axis OA joining the focus O to the point A of the trajectory closest to O (Fig. 12.28), we found that the solution to Eq. (12.37) was

$$\frac{1}{r} = u = \frac{GM}{h^2} + C \cos \theta \tag{12.39}$$

This is the equation of a conic of eccentricity $\varepsilon = Ch^2/GM$. The conic is an *ellipse* if $\varepsilon < 1$, a *parabola* if $\varepsilon = 1$, and a *hyperbola* if $\varepsilon > 1$. The constants C and h may be determined from the initial conditions; if the particle is projected from point A ($\theta = 0$, $r = r_0$) with an initial velocity v_0 perpendicular to OA, we have $h = r_0v_0$ [Sample Prob. 12.9].

It was also shown that the values of the initial velocity corresponding, respectively, to a parabolic and a circular trajectory were

$$v_{\text{esc}} = \sqrt{\frac{2GM}{r_0}} \tag{12.43}$$

$$v_{\text{circ}} = \sqrt{\frac{GM}{r_0}} \tag{12.44}$$

and that the first of these values, called the *escape velocity*, is the smallest value of v_0 for which the particle will not return to its starting point.

The *periodic time* τ of a planet or satellite was defined as the time required by that body to describe its orbit. It was shown that

$$\tau = \frac{2\pi ab}{h} \tag{12.45}$$

where $h = r_0v_0$ and where a and b represent the semimajor and semiminor axes of the orbit. It was further shown that these semiaxes are respectively equal to the arithmetic and geometric means of the maximum and minimum values of the radius vector r.

The last section of the chapter [Sec. 12.13] presented *Kepler's laws of planetary motion* and showed that these empirical laws, obtained from early astronomical observations, confirm Newton's laws of motion as well as his law of gravitation.

Orbital motion

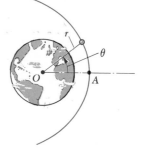

Fig. 12.28

Escape velocity

Periodic time

Kepler's laws

Review Problems

2% grade ⟍ ⟋ 3% grade

Fig. P12.116

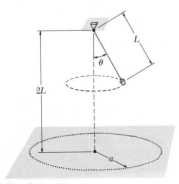

Fig. P12.117

12.116 A car has been traveling up a long 2 percent grade at a constant speed of 85 km/h. If the driver does not change the setting of the throttle or shift gears as the car reaches the top of the hill, what will be the acceleration of the car as it starts moving down the 3 percent grade?

12.117 A bucket is attached to a rope of length $L = 1.2$ m and is made to revolve in a horizontal circle. Drops of water leaking from the bucket fall and strike the floor along the perimeter of a circle of radius a. Determine the radius a when $\theta = 30°$.

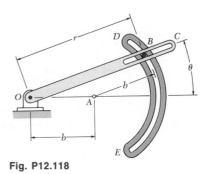

Fig. P12.118

12.118 Pin B weighs 4 oz and is free to slide along the rotating arm OC and along the circular slot DE of radius $b = 18$ in. Neglecting friction and assuming that rod OC is made to rotate at the constant rate $\dot\theta_0 = 12$ rad/s in a horizontal plane, determine for any given value of θ (a) the radial and transverse components of the resultant force exerted on pin B, (b) the forces $\mathbf{P}$ and $\mathbf{Q}$ exerted on pin B, respectively, by rod OC and the wall of slot DE.

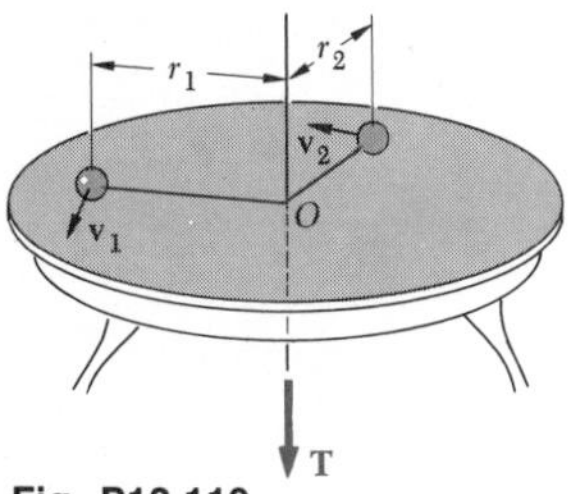

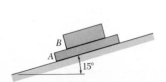

Fig. P12.119

12.119 A 5-oz ball slides on a smooth horizontal table at the end of a string which passes through a small hole in the table at O. When the length of the portion of string above the table is $r_1 = 24$ in., the speed of the ball is $v_1 = 4$ ft/s. Knowing that the breaking strength of the string is 6.00 lb, determine (a) the smallest distance r_2 which can be achieved by slowly drawing the string through the hole, (b) the corresponding speed v_2.

Fig. P12.120

12.120 Two plates A and B, each of mass 50 kg, are placed as shown on a 15° incline. The coefficient of friction between A and B is 0.10; the coefficient of friction between A and the incline is 0.20. (a) If the plates are released from rest, determine the acceleration of each plate. (b) Solve part a assuming that plates A and B are welded together and act as a single rigid body.

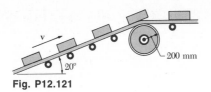

Fig. P12.121

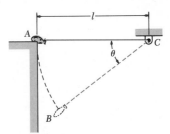

Fig. P12.122

12.121 A series of small packages, being moved by a conveyor belt at a constant speed v, passes over an idler roller as shown. Knowing that the coefficient of friction between the packages and the belt is 0.75, determine the maximum value of v for which the packages do not slip with respect to the belt.

12.122 A bag is gently pushed off the top of a wall at A and swings in a vertical plane at the end of a rope of length l. (a) For any position B of the bag, determine the tangential component a_t of its acceleration and obtain its velocity v by integration. (b) Determine the value of θ for which the rope will break, knowing that it can withstand a maximum tension equal to twice the weight of the bag.

12.123 Two space stations S_1 and S_2 are describing coplanar circular counterclockwise orbits of radius r_0 and $8r_0$, respectively, around the earth. It is desired to send a vehicle from S_1 to S_2. The vehicle is to be launched in a direction tangent to the orbit of S_1 and is to reach S_2 with a velocity tangent to the orbit of S_2. After a short powered phase, the vehicle will travel in free flight from S_1 to S_2. (a) Determine the launching velocity (velocity of the vehicle relative to S_1), in terms of the velocity $\mathbf{v}_0$ of S_1. (b) Determine the angle θ defining the required position of S_2 relative to S_1 at the time of launching.

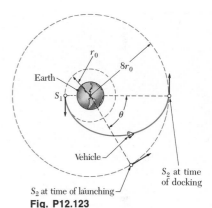

Fig. P12.123

12.124 The two blocks shown are originally at rest. Neglecting the masses of the pulleys and the effect of friction in the pulleys and between block A and the incline, determine (a) the acceleration of each block, (b) the tension in the cable.

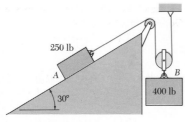

Fig. P12.124

12.125 Solve Prob. 12.124, assuming that the coefficients of friction between block A and the incline are $\mu_s = 0.25$ and $\mu_k = 0.20$.

12.126 At engine burnout an Explorer satellite was 170 mi above the surface of the earth and had a horizontal velocity $\mathbf{v}_0$ of magnitude 32.6×10^3 ft/s. Determine (a) the highest altitude reached by the satellite, (b) its velocity at its apogee B.

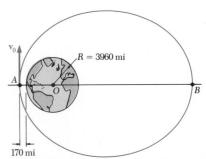

Fig. P12.126

12.127 A 6-kg block B rests as shown on a 10-kg bracket A. The coefficients of friction are $\mu_s = 0.30$ and $\mu_k = 0.25$ between block B and bracket A, and there is no friction in the pulley or between the bracket and the horizontal surface. Determine (a) the maximum force $\mathbf{P}$ which may be exerted on the cord if block B is not to slide on bracket A, (b) the corresponding acceleration of the bracket.

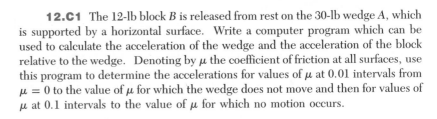

Fig. P12.127

The following problems are designed to be solved with a computer.

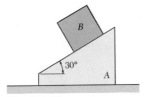

Fig. P12.C1

12.C1 The 12-lb block B is released from rest on the 30-lb wedge A, which is supported by a horizontal surface. Write a computer program which can be used to calculate the acceleration of the wedge and the acceleration of the block relative to the wedge. Denoting by μ the coefficient of friction at all surfaces, use this program to determine the accelerations for values of μ at 0.01 intervals from $\mu = 0$ to the value of μ for which the wedge does not move and then for values of μ at 0.1 intervals to the value of μ for which no motion occurs.

12.C2 In Prob. 12.54, determine with an accuracy of $0.1°$ the three ranges of values of θ for which the collar will not slide on the rod, assuming a coefficient of static friction of 0.25.

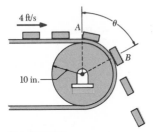

Fig. P12.C3

12.C3 A series of small packages are discharged from a conveyor belt as shown. The coefficients of friction are $\mu_s = 0.40$ and $\mu_k = 0.35$ between the belt and the packages and it is known that the packages first slip with respect to the belt when $\theta = 9.01°$ (see answer to Prob. 12.47). Write a computer program and use it to determine by numerical integration at intervals of time $\Delta t = 0.001$ s the angle θ defining the point where the packages will leave the belt.

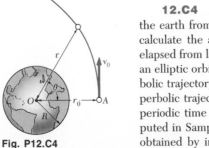

Fig. P12.C4

12.C4 A spacecraft is launched with a velocity v_0 parallel to the surface of the earth from an altitude of 500 km. Write a computer program and use it to calculate the altitude $d = r - R$ of the spacecraft and the time t (in minutes) elapsed from launching for values of θ at $5°$ intervals, (a) from 0 to $180°$, assuming an elliptic orbit with $v_0 = 36.9 \times 10^3$ km/h, (b) from 0 to $170°$, assuming a parabolic trajectory with $v_0 = 38.75 \times 10^3$ km/h, (c) from 0 to $115°$, assuming a hyperbolic trajectory with $v_0 = 47.5 \times 10^3$ km/h. For case a, also determine the periodic time of the spacecraft and compare the value obtained with that computed in Sample Prob. 12.9. [*Hint.* The time t elapsed from launching may be obtained by integrating Eq. (12.27) numerically, using increments $\Delta\theta = 0.1°$.]

Kinetics of Particles: Energy and Momentum Methods

13.1. Introduction. In the preceding chapter, most problems dealing with the motion of particles were solved through the use of the fundamental equation of motion $\mathbf{F} = m\mathbf{a}$. Given a particle acted upon by a force $\mathbf{F}$, we could solve this equation for the acceleration $\mathbf{a}$; then, by applying the principles of kinematics, we could determine from $\mathbf{a}$ the velocity and position of the particle at any time.

If the equation $\mathbf{F} = m\mathbf{a}$ and the principles of kinematics are combined, two additional methods of analysis may be obtained, the *method of work and energy* and the *method of impulse and momentum*. The advantage of these methods lies in the fact that they make the determination of the acceleration unnecessary. Indeed, the method of work and energy relates directly force, mass, velocity, and displacement, while the method of impulse and momentum relates force, mass, velocity, and time.

The method of work and energy will be considered first. In Secs. 13.2 through 13.4, we shall discuss the *work of a force* and the *kinetic energy of a particle* and apply the principle of work and energy to the solution of engineering problems. The concepts of *power* and *efficiency* of a machine will be introduced in Sec. 13.5.

Sections 13.6 through 13.8 are devoted to the concept of *potential energy* of a conservative force and to the application of the principle of conservation of energy to various problems of practical interest. In Sec. 13.9, we shall see how the principles of conservation of energy and of conservation of angular momentum may be used jointly to solve problems of space mechanics.

The second part of the chapter is devoted to the *principle of impulse and momentum* and to its application to the study of the motion of a particle. As we shall see in Sec. 13.11, this principle is particularly effective in the study of the *impulsive motion* of a particle, where very large forces are applied for a very short time interval.

In Secs. 13.12 through 13.14, we shall consider the *central impact* of two bodies. It will be shown that a certain relation exists between the relative velocities of the two colliding bodies before and after impact. This relation may be used together with the fact that the total momentum of the two bodies is conserved to solve a number of problems of practical interest.

Finally, in Sec. 13.15, we shall learn to select from the three fundamental methods presented in Chaps. 12 and 13 the method best suited for the solution of a given problem. We shall also see how the principle of conservation of energy and the method of impulse and momentum may be combined to solve problems involving only conservative forces, except for a short impact phase during which impulsive forces must also be taken into consideration.

13.2. Work of a Force. We shall first define the terms *displacement* and *work* as they are used in mechanics.† Consider a particle which moves from a point A to a neighboring point A' (Fig. 13.1). If $\mathbf{r}$ denotes the position vector corresponding to point A, the small vector joining A and A' may be denoted by the differential $d\mathbf{r}$; the vector $d\mathbf{r}$ is called the *displacement* of the particle. Now, let us assume that a force $\mathbf{F}$ is acting on the particle. The *work of the force $\mathbf{F}$ corresponding to the displacement $d\mathbf{r}$* is defined as the quantity

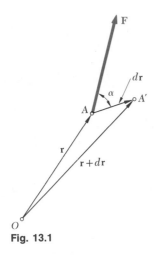

Fig. 13.1

$$dU = \mathbf{F} \cdot d\mathbf{r} \tag{13.1}$$

obtained by forming the scalar product of the force $\mathbf{F}$ and of the displacement $d\mathbf{r}$. Denoting, respectively, by F and ds the magnitudes of the force and of the displacement, and by α the angle formed by $\mathbf{F}$ and $d\mathbf{r}$, and recalling the definition of the scalar product of two vectors (Sec. 3.9), we write

$$dU = F \, ds \cos \alpha \tag{13.1'}$$

Using formula (3.30), we may also express the work dU in terms of the rectangular components of the force and of the displacement:

$$dU = F_x \, dx + F_y \, dy + F_z \, dz \tag{13.1''}$$

Being a *scalar quantity*, work has a magnitude and a sign but no direction. We also note that work should be expressed in units obtained by multiplying units of length by units of force. Thus, if U.S. customary units are used, work should be expressed in ft·lb or in·lb. If SI units are used, work should

---

† The definition of work was given in Sec. 10.2, and the basic properties of the work of a force were outlined in Secs. 10.2 and 10.6. For convenience, we repeat here the portions of this material which relate to the kinetics of particles.

be expressed in N·m. The unit of work N·m is called a *joule* (J).† Recalling the conversion factors indicated in Sec. 12.4, we write

$$1 \text{ ft·lb} = (1 \text{ ft})(1 \text{ lb}) = (0.3048 \text{ m})(4.448 \text{ N}) = 1.356 \text{ J}$$

It follows from (13.1′) that the work dU is positive if the angle α is acute and negative if α is obtuse. Three particular cases are of special interest. If the force $\mathbf{F}$ has the same direction as $d\mathbf{r}$, the work dU reduces to $F \, ds$. If $\mathbf{F}$ has a direction opposite to that of $d\mathbf{r}$, the work is $dU = -F \, ds$. Finally, if $\mathbf{F}$ is perpendicular to $d\mathbf{r}$, the work dU is zero.

The work of $\mathbf{F}$ during a *finite* displacement of the particle from A_1 to A_2 (Fig. 13.2a) is obtained by integrating Eq. (13.1) along the path described by the particle. This work, denoted by $U_{1\to2}$, is

$$U_{1\to2} = \int_{A_1}^{A_2} \mathbf{F} \cdot d\mathbf{r} \qquad (13.2)$$

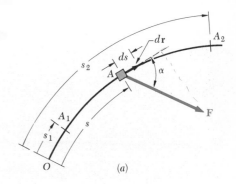

Using the alternate expression (13.1′) for the elementary work dU, and observing that $F \cos \alpha$ represents the tangential component F_t of the force, we may also express the work $U_{1\to2}$ as

$$U_{1\to2} = \int_{s_1}^{s_2} (F \cos \alpha) \, ds = \int_{s_1}^{s_2} F_t \, ds \qquad (13.2')$$

where the variable of integration s measures the distance traveled by the particle along the path. The work $U_{1\to2}$ is represented by the area under the curve obtained by plotting $F_t = F \cos \alpha$ against s (Fig. 13.2b).

When the force $\mathbf{F}$ is defined by its rectangular components, the expression (13.1″) may be used for the elementary work. We write then

$$U_{1\to2} = \int_{A_1}^{A_2} (F_x \, dx + F_y \, dy + F_z \, dz) \qquad (13.2'')$$

where the integration is to be performed along the path described by the particle.

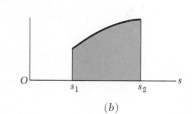

Fig. 13.2

Work of a Constant Force in Rectilinear Motion. When a particle moving in a straight line is acted upon by a force $\mathbf{F}$ of constant magnitude and of constant direction (Fig. 13.3), formula (13.2′) yields

$$U_{1\to2} = (F \cos \alpha) \Delta x \qquad (13.3)$$

where α = angle the force forms with direction of motion
    Δx = displacement from A_1 to A_2

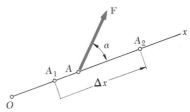

Fig. 13.3

---

† The joule (J) is the SI unit of *energy*, whether in mechanical form (work, potential energy, kinetic energy) or in chemical, electrical, or thermal form. We should note that even though N·m = J, the moment of a force must be expressed in N·m and not in joules, since the moment of a force is not a form of energy.

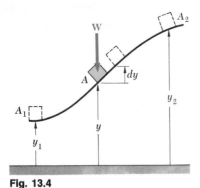

W

A_2

dy

A

y_2

A_1

y

y_1

Fig. 13.4

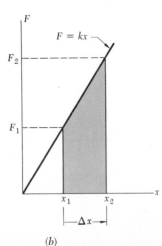

Work of the Force of Gravity. The work of the weight $\mathbf{W}$ of a body, i.e., of the force of gravity exerted on that body, is obtained by substituting the components of $\mathbf{W}$ into (13.1″) and (13.2″). With the y axis chosen upward (Fig. 13.4), we have $F_x = 0$, $F_y = -W$, $F_z = 0$, and we write

$$dU = -W \, dy$$

$$U_{1 \rightarrow 2} = -\int_{y_1}^{y_2} W \, dy = Wy_1 - Wy_2 \tag{13.4}$$

or

$$U_{1 \rightarrow 2} = -W(y_2 - y_1) = -W \, \Delta y \tag{13.4'}$$

where Δy is the vertical displacement from A_1 to A_2. The work of the weight $\mathbf{W}$ is thus equal to *the product of W and of the vertical displacement of the center of gravity of the body.* The work is *positive* when $\Delta y < 0$, that is, *when the body moves down.*

Work of the Force Exerted by a Spring. Consider a body A attached to a fixed point B by a spring; it is assumed that the spring is undeformed when the body is at A_0 (Fig. 13.5a). Experimental evidence shows that the magnitude of the force $\mathbf{F}$ exerted by the spring on body A is proportional to the deflection x of the spring measured from the position A_0. We have

$$F = kx \tag{13.5}$$

where k is the *spring constant,* expressed in N/m or kN/m if SI units are used and in lb/ft or lb/in. if U.S. customary units are used.†

The work of the force $\mathbf{F}$ exerted by the spring during a finite displacement of the body from $A_1(x = x_1)$ to $A_2(x = x_2)$ is obtained by writing

$$dU = -F \, dx = -kx \, dx$$

$$U_{1 \rightarrow 2} = -\int_{x_1}^{x_2} kx \, dx = \tfrac{1}{2}kx_1^2 - \tfrac{1}{2}kx_2^2 \tag{13.6}$$

Care should be taken to express k and x in consistent units. For example, if U.S. customary units are used, k should be expressed in lb/ft and x in feet, or k in lb/in. and x in inches; in the first case, the work is obtained in ft·lb, in the second case, in in·lb. We note that the work of the force $\mathbf{F}$ exerted by the spring on the body is *positive* when $x_2 < x_1$, that is, *when the spring is returning to its undeformed position.*

Since Eq.(13.5) is the equation of a straight line of slope k passing through the origin, the work $U_{1 \rightarrow 2}$ of $\mathbf{F}$ during the displacement from A_1 to

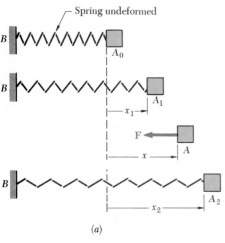

Spring undeformed

B

A_0

B

A_1

x_1

F

A

x

B

A_2

x_2

(a)

F

$F = kx$

F_2

F_1

x_1

x_2

x

Δx

(b)

Fig. 13.5

† The relation $F = kx$ is correct under static conditions only. Under dynamic conditions, formula (13.5) should be modified to take the inertia of the spring into account. However, the error introduced by using the relation $F = kx$ in the solution of kinetics problems is small if the mass of the spring is small compared with the other masses in motion.

A_2 may be obtained by evaluating the area of the trapezoid shown in Fig. 13.5b. This is done by computing F_1 and F_2 and multiplying the base Δx of the trapezoid by its mean height $\frac{1}{2}(F_1 + F_2)$. Since the work of the force $\mathbf{F}$ exerted by the spring is positive for a negative value of Δx, we write

$$U_{1\to2} = -\tfrac{1}{2}(F_1 + F_2)\,\Delta x \tag{13.6'}$$

Formula (13.6') is usually more convenient to use than (13.6) and affords fewer chances of confusing the units involved.

Work of a Gravitational Force. We saw in Sec. 12.10 that two particles at distance r from each other and, respectively, of mass M and m, attract each other with equal and opposite forces $\mathbf{F}$ and $-\mathbf{F}$ directed along the line joining the particles, and of magnitude

$$F = G\frac{Mm}{r^2}$$

Let us assume that the particle M occupies a fixed position O while the particle m moves along the path shown in Fig. 13.6. The work of the force $\mathbf{F}$ exerted on the particle m during an infinitesimal displacement of the particle from A to A' may be obtained by multiplying the magnitude F of the force by the radial component dr of the displacement. Since $\mathbf{F}$ is directed toward O, the work is negative and we write

$$dU = -F\,dr = -G\frac{Mm}{r^2}\,dr$$

The work of the gravitational force $\mathbf{F}$ during a finite displacement from $A_1(r = r_1)$ to $A_2(r = r_2)$ is therefore

$$U_{1\to2} = -\int_{r_1}^{r_2} \frac{GMm}{r^2}\,dr = \frac{GMm}{r_2} - \frac{GMm}{r_1} \tag{13.7}$$

The formula obtained may be used to determine the work of the force exerted by the earth on a body of mass m at a distance r from the center of the earth, when r is larger than the radius R of the earth. The letter M represents then the mass of the earth; recalling the first of the relations (12.29), we may thus replace the product GMm in Eq. (13.7) by WR^2, where R is the radius of the earth ($R = 6.37 \times 10^6$ m or 3960 mi) and W the value of the weight of the body at the surface of the earth.

A number of forces frequently encountered in problems of kinetics *do no work.* They are forces applied to fixed points ($ds = 0$) or acting in a direction perpendicular to the displacement ($\cos\alpha = 0$). Among the forces which do no work are the following: the reaction at a frictionless pin when the body supported rotates about the pin, the reaction at a frictionless surface when the body in contact moves along the surface, the reaction at a roller moving along its track, and the weight of a body when its center of gravity moves horizontally.

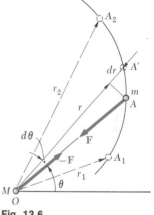

Fig. 13.6

13.3. Kinetic Energy of a Particle. Principle of Work and Energy. Consider a particle of mass m acted upon by a force $\mathbf{F}$ and moving along a path which is either rectilinear or curved (Fig. 13.7). Expressing Newton's second law in terms of the tangential components of the

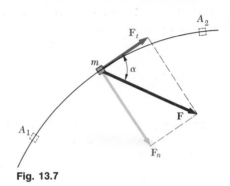

Fig. 13.7

force and of the acceleration (see Sec. 12.5), we write

$$F_t = ma_t \qquad \text{or} \qquad F_t = m\frac{dv}{dt}$$

where v is the speed of the particle. Recalling from Sec. 11.9 that $v = ds/dt$, we obtain

$$F_t = m\frac{dv}{ds}\frac{ds}{dt} = mv\frac{dv}{ds}$$

$$F_t\,ds = mv\,dv$$

Integrating from A_1, where $s = s_1$ and $v = v_1$, to A_2, where $s = s_2$ and $v = v_2$, we write

$$\int_{s_1}^{s_2} F_t\,ds = m\int_{v_1}^{v_2} v\,dv = \tfrac{1}{2}mv_2^2 - \tfrac{1}{2}mv_1^2 \qquad (13.8)$$

The left-hand member of Eq. (13.8) represents the work $U_{1\to2}$ of the force $\mathbf{F}$ exerted on the particle during the displacement from A_1 to A_2; as indicated in Sec. 13.2, the work $U_{1\to2}$ is a scalar quantity. The expression $\tfrac{1}{2}mv^2$ is also a scalar quantity; it is defined as the kinetic energy of the particle and is denoted by T. We write

$$T = \tfrac{1}{2}mv^2 \qquad (13.9)$$

Substituting into (13.8), we have

$$U_{1\to2} = T_2 - T_1 \qquad (13.10)$$

which expresses that, when a particle moves from A_1 to A_2 under the action of a force $\mathbf{F}$, *the work of the force $\mathbf{F}$ is equal to the change in kinetic energy of the particle.* This is known as the *principle of work and energy.* Rearranging the terms in (13.10), we write

$$T_1 + U_{1\to2} = T_2 \qquad (13.11)$$

Thus, *the kinetic energy of the particle at A_2 may be obtained by adding to its kinetic energy at A_1 the work done during the displacement from A_1 to A_2 by the force $\mathbf{F}$ exerted on the particle.* As Newton's second law from which it is derived, the principle of work and energy applies only with respect to a newtonian frame of reference (Sec. 12.2). The speed v used to determine the kinetic energy T should therefore be measured with respect to a newtonian frame of reference.

Since both work and kinetic energy are scalar quantities, their sum may be computed as an ordinary algebraic sum, the work $U_{1\to2}$ being considered as positive or negative according to the direction of $\mathbf{F}$. When several forces act on the particle, the expression $U_{1\to2}$ represents the total work of the forces acting on the particle; it is obtained by adding algebraically the work of the various forces.

As noted above, the kinetic energy of a particle is a scalar quantity. It further appears from the definition $T = \frac{1}{2}mv^2$ that the kinetic energy is always positive, regardless of the direction of motion of the particle. Considering the particular case when $v_1 = 0$, $v_2 = v$, and substituting $T_1 = 0$, $T_2 = T$ into (13.10), we observe that the work done by the forces acting on the particle is equal to T. Thus, the kinetic energy of a particle moving with a speed v represents the work which must be done to bring the particle from rest to the speed v. Substituting $T_1 = T$ and $T_2 = 0$ into (13.10), we also note that when a particle moving with a speed v is brought to rest, the work done by the forces acting on the particle is $-T$. Assuming that no energy is dissipated into heat, we conclude that the work done by the forces exerted *by the particle* on the bodies which cause it to come to rest is equal to T. Thus, the kinetic energy of a particle also represents *the capacity to do work associated with the speed of the particle.*

The kinetic energy is measured in the same units as work, i.e., in joules if SI units are used, and in ft·lb if U.S. customary units are used. We check that, in SI units,

$$T = \tfrac{1}{2}mv^2 = \text{kg}(\text{m/s})^2 = (\text{kg·m/s}^2)\text{m} = \text{N·m} = \text{J}$$

while, in customary units,

$$T = \tfrac{1}{2}mv^2 = (\text{lb·s}^2/\text{ft})(\text{ft/s})^2 = \text{lb·ft}$$

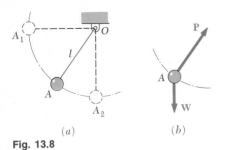

(a)

(b)

Fig. 13.8

13.4. Applications of the Principle of Work and Energy.
The application of the principle of work and energy greatly simplifies the solution of many problems involving forces, displacements, and velocities. Consider, for example, the pendulum OA consisting of a bob A of weight W attached to a cord of length l (Fig. 13.8a). The pendulum is released with no initial velocity from a horizontal position OA_1 and allowed to swing in a vertical plane. We wish to determine the speed of the bob as it passes through A_2, directly under O.

We first determine the work done during the displacement from A_1 to A_2 by the forces acting on the bob. We draw a free-body diagram of the bob, showing all the *actual* forces acting on it, i.e., the weight $\mathbf{W}$ and the force $\mathbf{P}$ exerted by the cord (Fig. 13.8b). (An inertia vector is not an actual force and *should not* be included in the free-body diagram.) We note that the force $\mathbf{P}$ does no work, since it is normal to the path; the only force which does work is thus the weight $\mathbf{W}$. The work of $\mathbf{W}$ is obtained by multiplying its magnitude W by the vertical displacement l (Sec. 13.2); since the displacement is downward, the work is positive. We therefore write $U_{1\rightarrow2} = Wl$.

Considering, now, the kinetic energy of the bob, we find $T_1 = 0$ at A_1 and $T_2 = \frac{1}{2}(W/g)v_2^2$ at A_2. We may now apply the principle of work and energy; recalling formula (13.11), we write

$$T_1 + U_{1\rightarrow2} = T_2 \qquad 0 + Wl = \frac{1}{2}\frac{W}{g}v_2^2$$

Solving for v_2, we find $v_2 = \sqrt{2gl}$. We note that the speed obtained is that of a body falling freely from a height l.

The example we have considered illustrates the following advantages of the method of work and energy:

1. In order to find the speed at A_2, there is no need to determine the acceleration in an intermediate position A and to integrate the expression obtained from A_1 to A_2.
2. All quantities involved are scalars and may be added directly, without using x and y components.
3. Forces which do no work are eliminated from the solution of the problem.

What is an advantage in one problem, however, may become a disadvantage in another. It is evident, for instance, that the method of work and energy cannot be used to directly determine an acceleration. We also note that it should be supplemented by the direct application of Newton's second law in order to determine a force which is normal to the path of the particle, since such a force does no work. Suppose, for example, that we wish to determine the tension in the cord of the pendulum of Fig. 13.8a as the bob passes through A_2. We draw a free-body diagram of the bob in that position (Fig. 13.9) and express Newton's second law in terms of tangential and normal components. The equations $\Sigma F_t = ma_t$ and $\Sigma F_n = ma_n$ yield, respectively, $a_t = 0$ and

$$P - W = ma_n = \frac{W}{g}\frac{v_2^2}{l}$$

But the speed at A_2 was determined earlier by the method of work and energy. Substituting $v_2^2 = 2gl$ and solving for P, we write

$$P = W + \frac{W}{g}\frac{2gl}{l} = 3W$$

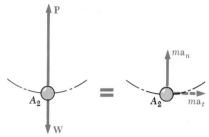

Fig. 13.9

When a problem involves two particles or more, each particle may be considered separately and the principle of work and energy may be applied to each particle. Adding the kinetic energies of the various particles, and considering the work of all the forces acting on them, we may also write a single equation of work and energy for all the particles involved. We have

$$T_1 + U_{1\to 2} = T_2 \tag{13.11}$$

where T represents the arithmetic sum of the kinetic energies of the particles involved (all terms are positive) and $U_{1\to 2}$ the work of all the forces acting on the particles, *including the forces of action and reaction exerted by the particles on each other.* In problems involving bodies connected by *inextensible cords or links,* however, the work of the forces exerted by a given cord or link on the two bodies it connects cancels out, since the points of application of these forces move through equal distances (see Sample Prob. 13.2).†

Since friction forces have a direction opposite to that of the displacement of the body on which they act, *the work of friction forces is always negative.* This work represents energy dissipated into heat and always results in a decrease in the kinetic energy of the body involved (see Sample Prob. 13.3).

13.5. Power and Efficiency. *Power* is defined as the time rate at which work is done. In the selection of a motor or engine, power is a much more important criterion than is the actual amount of work to be performed. A small motor or a large power plant may both be used to do a given amount of work; but the small motor may require a month to do the work done by the power plant in a matter of minutes. If ΔU is the work done during the time interval Δt, then the average power during this time interval is

$$\text{Average power} = \frac{\Delta U}{\Delta t}$$

Letting Δt approach zero, we obtain at the limit

$$\text{Power} = \frac{dU}{dt} \tag{13.12}$$

† The application of the method of work and energy to a system of particles is discussed in detail in Chap. 14.

Substituting the scalar product $\mathbf{F} \cdot d\mathbf{r}$ for dU, we may also write

$$\text{Power} = \frac{dU}{dt} = \frac{\mathbf{F} \cdot d\mathbf{r}}{dt}$$

and, recalling that $d\mathbf{r}/dt$ represents the velocity $\mathbf{v}$ of the point of application of $\mathbf{F}$,

$$\text{Power} = \mathbf{F} \cdot \mathbf{v} \tag{13.13}$$

Since power was defined as the time rate at which work is done, it should be expressed in units obtained by dividing units of work by the unit of time. Thus, if SI units are used, power should be expressed in J/s; this unit is called a *watt* (W). We have

$$1\,\text{W} = 1\,\text{J/s} = 1\,\text{N} \cdot \text{m/s}$$

If U.S. customary units are used, power should be expressed in ft·lb/s or in *horsepower* (hp), with the latter defined as

$$1\,\text{hp} = 550\,\text{ft} \cdot \text{lb/s}$$

Recalling from Sec. 13.2 that $1\,\text{ft} \cdot \text{lb} = 1.356\,\text{J}$, we verify that

$$1\,\text{ft} \cdot \text{lb/s} = 1.356\,\text{J/s} = 1.356\,\text{W}$$
$$1\,\text{hp} = 550(1.356\,\text{W}) = 746\,\text{W} = 0.746\,\text{kW}$$

The *mechanical efficiency* of a machine was defined in Sec. 10.5 as the ratio of the output work to the input work:

$$\eta = \frac{\text{output work}}{\text{input work}} \tag{13.14}$$

This definition is based on the assumption that work is done at a constant rate. The ratio of the output to the input work is therefore equal to the ratio of the rates at which output and input work are done, and we have

$$\eta = \frac{\text{power output}}{\text{power input}} \tag{13.15}$$

Because of energy losses due to friction, the output work is always smaller than the input work, and, consequently, the power output is always smaller than the power input. The mechanical efficiency of a machine is therefore always less than 1.

When a machine is used to transform mechanical energy into electric energy, or thermal energy into mechanical energy, its *overall efficiency* may be obtained from formula (13.15). The overall efficiency of a machine is always less than 1; it provides a measure of all the various energy losses involved (losses of electric or thermal energy as well as frictional losses). We should note that it is necessary, before using formula (13.15), to express the power output and the power input in the same units.

SAMPLE PROBLEM 13.1

An automobile weighing 4000 lb is driven down a 5° incline at a speed of 60 mi/h when the brakes are applied, causing a constant total braking force (applied by the road on the tires) of 1500 lb. Determine the distance traveled by the automobile as it comes to a stop.

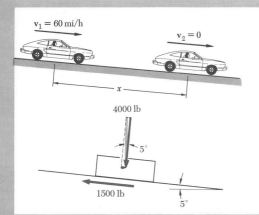

Solution. *Kinetic Energy*

Position 1: $\quad v_1 = \left(60\dfrac{\text{mi}}{\text{h}}\right)\left(\dfrac{5280\text{ ft}}{1\text{ mi}}\right)\left(\dfrac{1\text{ h}}{3600\text{ s}}\right) = 88\text{ ft/s}$

$$T_1 = \tfrac{1}{2}mv_1^2 = \tfrac{1}{2}(4000/32.2)(88)^2 = 481{,}000\text{ ft}\cdot\text{lb}$$

Position 2: $\qquad\qquad v_2 = 0 \qquad T_2 = 0$

Work $\qquad U_{1\to2} = -1500x + (4000\sin 5°)x = -1151x$

Principle of Work and Energy

$$T_1 + U_{1\to2} = T_2$$
$$481{,}000 - 1151x = 0 \qquad\qquad x = 418\text{ ft} \quad \blacktriangleleft$$

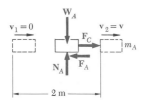

SAMPLE PROBLEM 13.2

Two blocks are joined by an inextensible cable as shown. If the system is released from rest, determine the velocity of block A after it has moved 2 m. Assume that the coefficient of kinetic friction between block A and the plane is $\mu_k = 0.25$ and that the pulley is weightless and frictionless.

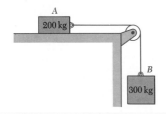

Solution. *Work and Energy for Block A.* We denote by $\mathbf{F}_A$ the friction force, by $\mathbf{F}_C$ the force exerted by the cable, and write

$$m_A = 200\text{ kg} \qquad W_A = (200\text{ kg})(9.81\text{ m/s}^2) = 1962\text{ N}$$

$$F_A = \mu_k N_A = \mu_k W_A = 0.25(1962\text{ N}) = 490\text{ N}$$

$T_1 + U_{1\to2} = T_2:\qquad 0 + F_C(2\text{ m}) - F_A(2\text{ m}) = \tfrac{1}{2}m_A v^2$

$$F_C(2\text{ m}) - (490\text{ N})(2\text{ m}) = \tfrac{1}{2}(200\text{ kg})v^2 \qquad (1)$$

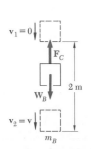

Work and Energy for Block B. We write

$$m_B = 300\text{ kg} \qquad W_B = (300\text{ kg})(9.81\text{ m/s}^2) = 2940\text{ N}$$

$T_1 + U_{1\to2} = T_2:\qquad 0 + W_B(2\text{ m}) - F_C(2\text{ m}) = \tfrac{1}{2}m_B v^2$

$$(2940\text{ N})(2\text{ m}) - F_C(2\text{ m}) = \tfrac{1}{2}(300\text{ kg})v^2 \qquad (2)$$

Adding the left-hand and right-hand members of (1) and (2), we observe that the work of the forces exerted by the cable on A and B cancels out:

$$(2940\text{ N})(2\text{ m}) - (490\text{ N})(2\text{ m}) = \tfrac{1}{2}(200\text{ kg} + 300\text{ kg})v^2$$
$$4900\text{ J} = \tfrac{1}{2}(500\text{ kg})v^2 \qquad v = 4.43\text{ m/s} \quad \blacktriangleleft$$

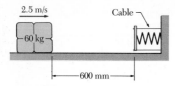

2.5 m/s

60 kg

Cable

600 mm

A spring is used to stop a 60-kg package which is sliding on a horizontal surface. The spring has a constant $k = 20$ kN/m and is held by cables so that it is initially compressed 120 mm. Knowing that the package has a velocity of 2.5 m/s in the position shown and that the maximum additional deflection of the spring is 40 mm, determine (a) the coefficient of kinetic friction between the package and the surface, (b) the velocity of the package as it passes again through the position shown.

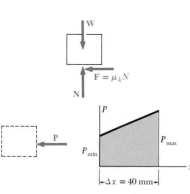

v_1 $v_2 = 0$

1   2

600 mm   40 mm

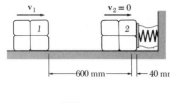

W

$F = \mu_k N$

N

P

P_{min} P_{max}

$\Delta x = 40$ mm

a. Motion from Position 1 to Position 2
Kinetic Energy. *Position 1:* $v_1 = 2.5$ m/s

$$T_1 = \tfrac{1}{2}mv_1^2 = \tfrac{1}{2}(60 \text{ kg})(2.5 \text{ m/s})^2 = 187.5 \text{ N·m} = 187.5 \text{ J}$$

Position 2 (maximum spring deflection): $v_2 = 0$ $T_2 = 0$

Work
Friction Force **F.**   We have
$$F = \mu_k N = \mu_k W = \mu_k mg = \mu_k(60 \text{ kg})(9.81 \text{ m/s}^2) = 588.6 \,\mu_k$$

The work of **F** is negative and equal to

$$(U_{1\to2})_f = -Fx = -(588.6 \,\mu_k)(0.600 \text{ m} + 0.040 \text{ m}) = -377 \,\mu_k$$

Spring Force **P.**   The variable force **P** exerted by the spring does an amount of negative work equal to the area under the force-deflection curve of the spring force.   We have

$$P_{min} = kx_0 = (20 \text{ kN/m})(120 \text{ m}) = (20\ 000 \text{ N/m})(0.120 \text{ m}) = 2400 \text{ N}$$
$$P_{max} = P_{min} + k\,\Delta x = 2400 \text{ N} + (20 \text{ kN/m})(40 \text{ mm}) = 3200 \text{ N}$$
$$(U_{1\to2})_e = -\tfrac{1}{2}(P_{min} + P_{max})\,\Delta x = -\tfrac{1}{2}(2400 \text{ N} + 3200 \text{ N})(0.040 \text{ m}) = -112.0 \text{ J}$$

The total work is thus

$$U_{1\to2} = (U_{1\to2})_f + (U_{1\to2})_e = -377 \,\mu_k - 112.0 \text{ J}$$

Principle of Work and Energy

$$T_1 + U_{1\to2} = T_2:\quad 187.5 \text{ J} - 377 \,\mu_k - 112.0 \text{ J} = 0 \qquad \mu_k = 0.20 \quad \blacktriangleleft$$

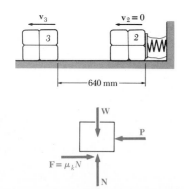

v_3 $v_2 = 0$

3   2

640 mm

W

P

$F = \mu_k N$

N

b. Motion from Position 2 to Position 3
Kinetic Energy. *Position 2:* $v_2 = 0$ $T_2 = 0$

Position 3: $T_3 = \tfrac{1}{2}mv_3^2 = \tfrac{1}{2}(60 \text{ kg})v_3^2$

Work.   Since the distances involved are the same, the numerical values of the work of the friction force **F** and of the spring force **P** are the same as above. However, while the work of **F** is still negative, the work of **P** is now positive.

$$U_{2\to3} = -377\mu_k + 112.0 \text{ J} = -75.5 \text{ J} + 112.0 \text{ J} = +36.5 \text{ J}$$

Principle of Work and Energy

$$T_2 + U_{2\to3} = T_3:\quad 0 + 36.5 \text{ J} = \tfrac{1}{2}(60 \text{ kg})v_3^2$$
$$v_3 = 1.103 \text{ m/s} \qquad v_3 = 1.103 \text{ m/s} \leftarrow \quad \blacktriangleleft$$

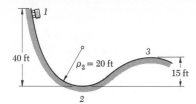

A 2000-lb car starts from rest at point *1* and moves without friction down the track shown. (*a*) Determine the force exerted by the track on the car at point 2, where the radius of curvature of the track is 20 ft. (*b*) Determine the minimum safe value of the radius of curvature at point 3.

a. **Force Exerted by the Track at Point 2.** The principle of work and energy is used to determine the velocity of the car as it passes through point 2.

Kinetic Energy: $\qquad T_1 = 0 \qquad T_2 = \tfrac{1}{2}mv_2^2 = \dfrac{1}{2}\dfrac{W}{g}v_2^2$

Work. The only force which does work is the weight **W**. Since the vertical displacement from point *1* to point *2* is 40 ft downward, the work of the weight is

$$U_{1\to2} = +W(40 \text{ ft})$$

Principle of Work and Energy

$$T_1 + U_{1\to2} = T_2 \qquad 0 + W(40 \text{ ft}) = \dfrac{1}{2}\dfrac{W}{g}v_2^2$$

$$v_2^2 = 80g = 80(32.2) \qquad v_2 = 50.8 \text{ ft/s}$$

Newton's Second Law at Point 2. The acceleration $\mathbf{a}_n$ of the car at point 2 has a magnitude $a_n = v_2^2/\rho$ and is directed upward. Since the external forces acting on the car are **W** and **N**, we write

$$+\uparrow\Sigma F_n = ma_n: \qquad -W + N = ma_n$$

$$= \dfrac{W}{g}\dfrac{v_2^2}{\rho}$$

$$= \dfrac{W}{g}\dfrac{80g}{20}$$

$$N = 5W \qquad N = 10{,}000 \text{ lb} \uparrow \quad \blacktriangleleft$$

b. **Minimum Value of ρ at Point 3.** *Principle of Work and Energy.* Applying the principle of work and energy between point *1* and point 3, we obtain

$$T_1 + U_{1\to3} = T_3 \qquad 0 + W(25 \text{ ft}) = \dfrac{1}{2}\dfrac{W}{g}v_3^2$$

$$v_3^2 = 50g = 50(32.2) \qquad v_3 = 40.1 \text{ ft/s}$$

Newton's Second Law at Point 3. The minimum safe value of ρ occurs when **N** = 0. In this case, the acceleration $\mathbf{a}_n$, of magnitude $a_n = v_3^2/\rho$, is directed downward, and we write

$$+\downarrow\Sigma F_n = ma_n: \qquad W = \dfrac{W}{g}\dfrac{v_3^2}{\rho}$$

$$= \dfrac{W}{g}\dfrac{50g}{\rho} \qquad\qquad \rho = 50 \text{ ft} \quad \blacktriangleleft$$

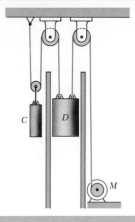

The dumb-waiter D and its load have a combined weight of 600 lb, while the counterweight C weighs 800 lb. Determine the power delivered by the electric motor M when the dumb-waiter (a) is moving up at a constant speed of 8 ft/s, (b) has an instantaneous velocity of 8 ft/s and an acceleration of 2.5 ft/s², both directed upward.

Solution. Since the force $\mathbf{F}$ exerted by the motor cable has the same direction as the velocity $\mathbf{v}_D$ of the dumb-waiter, the power is equal to Fv_D, where $v_D = 8$ ft/s. To obtain the power, we must first determine $\mathbf{F}$ in each of the two given situations.

a. Uniform Motion. We have $\mathbf{a}_C = \mathbf{a}_D = 0$; both bodies are in equilibrium.

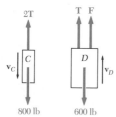

Free Body C: $+\!\uparrow\Sigma F_y = 0$: $2T - 800\ \text{lb} = 0$ $T = 400\ \text{lb}$
Free Body D: $+\!\uparrow\Sigma F_y = 0$: $F + T - 600\ \text{lb} = 0$

$$F = 600\ \text{lb} - T = 600\ \text{lb} - 400\ \text{lb} = 200\ \text{lb}$$

$$Fv_D = (200\ \text{lb})(8\ \text{ft/s}) = 1600\ \text{ft}\cdot\text{lb/s}$$

$$\text{Power} = (1600\ \text{ft}\cdot\text{lb/s})\frac{1\ \text{hp}}{550\ \text{ft}\cdot\text{lb/s}} = 2.91\ \text{hp} \quad\blacktriangleleft$$

b. Accelerated Motion. We have

$$\mathbf{a}_D = 2.5\ \text{ft/s}^2\!\uparrow \qquad \mathbf{a}_C = -\tfrac{1}{2}\mathbf{a}_D = 1.25\ \text{ft/s}^2\!\downarrow$$

The equations of motion are

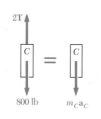

Free Body C: $+\!\downarrow\Sigma F_y = m_C a_C$: $800 - 2T = \dfrac{800}{32.2}(1.25)$ $T = 384.5\ \text{lb}$

Free Body D: $+\!\uparrow\Sigma F_y = m_D a_D$: $F + T - 600 = \dfrac{600}{32.2}(2.5)$

$$F + 384.5 - 600 = 46.6 \qquad F = 262.1\ \text{lb}$$

$$Fv_D = (262.1\ \text{lb})(8\ \text{ft/s}) = 2097\ \text{ft}\cdot\text{lb/s}$$

$$\text{Power} = (2097\ \text{ft}\cdot\text{lb/s})\frac{1\ \text{hp}}{550\ \text{ft}\cdot\text{lb/s}} = 3.81\ \text{hp} \quad\blacktriangleleft$$

Problems

13.1 A 100-lb satellite was placed in a circular orbit 1548 mi above the surface of the earth. At this elevation the acceleration of gravity is 16.7 ft/s². Determine the kinetic energy of the satellite, knowing that its orbital speed is 15,000 mi/h.

13.2 A 3-kg stone is dropped from a height h and strikes the ground with a velocity of 30 m/s. (*a*) Determine the kinetic energy of the stone as it strikes the ground and the height h from which it was dropped. (*b*) Solve part *a*, assuming that the same stone is dropped on the moon. (Acceleration of gravity on the moon = 1.62 m/s².)

13.3 Determine the maximum theoretical speed that may be achieved over a distance of 100 m by a car starting from rest, knowing that the coefficient of static friction is 0.80 between the tires and the pavement and that 60 percent of the weight of the car is distributed over its front wheels and 40 percent over its rear wheels. Assume (*a*) front-wheel drive, (*b*) rear-wheel drive, (*c*) four-wheel drive.

13.4 In an iron-ore mixing operation, a bucket full of ore is suspended from a traveling crane which is moving with a speed $v = 3$ m/s along a stationary bridge. If the crane suddenly stops, determine the additional horizontal distance through which the bucket will move.

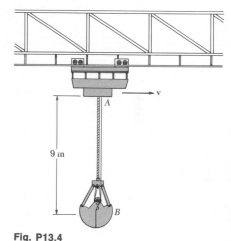

Fig. P13.4

13.5 Determine the speed v with which the crane of Prob. 13.4 was moving if the cables supporting the bucket B swing through an angle of 16° after the crane is brought to a sudden stop.

13.6 Packages are thrown down an incline at A with a velocity of 4 ft/s. The packages slide along the surface ABC to a conveyor belt which moves with a velocity of 8 ft/s. Knowing that $\mu_k = 0.25$ between the packages and the surface ABC, determine the distance d if the packages are to arrive at C with a velocity of 8 ft/s.

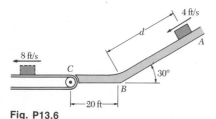

Fig. P13.6

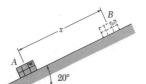

Fig. P13.7

13.7 A 50-lb package is projected up a 20° incline with an initial velocity of 40 ft/s. Knowing that the coefficient of kinetic friction between the package and the incline is 0.15, determine (*a*) the maximum distance x that the package will move up the incline, (*b*) the velocity of the package as it returns to its original position, (*c*) the total amount of energy dissipated due to friction.

13.8 A 1200-kg trailer is hitched to a 1400-kg car. The car and trailer are traveling at 72 km/h when the driver applies the brakes on both the car and the trailer. Knowing that the braking forces exerted on the car and the trailer are 5000 N and 4000 N, respectively, determine (*a*) the distance traveled by the car and trailer before they come to a stop, (*b*) the horizontal component of the force exerted by the trailer hitch on the car.

Fig. P13.8

13.9 The subway train shown is traveling at a speed of 30 mi/h when the brakes are fully applied on the wheels of cars *B* and *C*, causing them to slide on the track, but are not applied on the wheels of car *A*. Knowing that the coefficient of kinetic friction is 0.35 between the wheels and the track, determine (*a*) the distance required to bring the train to a stop, (*b*) the force in each coupling.

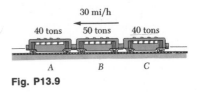

Fig. P13.9

13.10 Solve Prob. 13.9, assuming that the brakes are applied only on the wheels of car *A*.

13.11 Solve Prob. 13.8, assuming that the trailer brakes are inoperative.

13.12 The two blocks shown are originally at rest. Neglecting the masses of the pulleys and the effect of friction in the pulleys and between the blocks and the incline, determine (*a*) the velocity of block *A* after it has moved through 6 ft, (*b*) the tension in the cable.

13.13 Solve Prob. 12.17*c*, using the method of work and energy.

13.14 Solve Prob. 13.12, assuming that the coefficients of friction between the blocks and the inclines are $\mu_s = 0.25$ and $\mu_k = 0.20$.

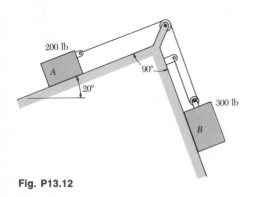

Fig. P13.12

13.15 The system shown is at rest when a constant 150-N force is applied to collar B. (*a*) If the force acts through the entire motion, determine the speed of collar B as it strikes the support at C. (*b*) After what distance d should the 150-N force be removed if the collar is to reach support C with zero velocity?

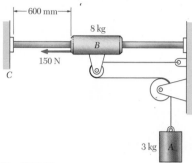

Fig. P13.15

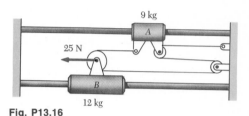

Fig. P13.16

13.16 Knowing that the system shown starts from rest, determine (*a*) the velocity of collar A after it has moved through 400 mm, (*b*) the corresponding velocity of collar B, (*c*) the tension in the cable. Neglect the masses of the pulleys and the effect of friction.

13.17 Two blocks A and B, of mass 8 kg and 12 kg, respectively, hang from a cable which passes over a pulley of negligible mass. Knowing that the blocks are released from rest and that the energy dissipated by axle friction in the pulley is 10 J, determine (*a*) the velocity of block B as it strikes the ground, (*b*) the force exerted by the cable on each of the two blocks during the motion.

13.18 Two blocks A and B, of mass 12 kg and 15 kg, respectively, hang from a cable which passes over a pulley of negligible mass. The blocks are released from rest in the positions shown and block B is observed to strike the ground with a velocity of 1.4 m/s. Determine (*a*) the energy dissipated due to axle friction in the pulley, (*b*) the force exerted by the cable on each of the two blocks during the motion.

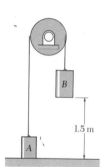

Fig. P13.17 and P13.18

13.19 Two blocks A and D, weighing, respectively, 100 lb and 250 lb, are attached to a rope which passes over two fixed pipes B and C as shown. It is observed that when the system is released from rest, block A acquires a velocity of 6 ft/s after moving 3 ft up. Determine (*a*) the force exerted by the rope on each of the two blocks during the motion, (*b*) the coefficient of kinetic friction between the rope and the pipes, (*c*) the energy dissipated due to friction.

13.20 Two blocks A and D are attached to a rope which passes over two fixed pipes B and C as shown. The coefficients of friction between the rope and the pipes are $\mu_s = 0.30$ and $\mu_k = 0.25$. Knowing that the masses of blocks A and D are, respectively, 20 kg and 55 kg and that the system is released from rest, determine (*a*) the velocity of A after it has moved 1.5 m up, (*b*) the force exerted by the rope on each of the two blocks during the motion, (*c*) the energy dissipated due to friction.

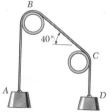

Fig. P13.19 and P13.20

13.21 Two blocks A and B, weighing 9 lb and 10 lb, respectively, are connected by a cord which passes over pulleys as shown. A collar C is placed on block A and the system is released from rest. After the blocks have moved 3 ft, collar C is removed and the blocks continue to move. Knowing that collar C weighs 5 lb, determine the speed of block A just before it strikes the ground.

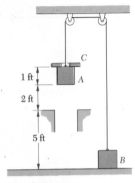

Fig. P13.21

13.22 Four packages, each having a mass of 15 kg, are placed as shown on a conveyor belt which is disengaged from its drive motor. Package 1 is just to the right of the horizontal portion of the belt. If the system is released from rest, determine the velocities of packages 1 and 2 as they fall off the belt at point A. Assume that the mass of the belt and rollers is small compared with the mass of the packages.

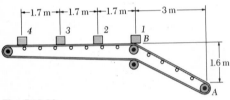

Fig. P13.22

13.23 Knowing that package 2 in Prob. 13.22 has a velocity of 4.72 m/s as it falls off the belt, determine the velocities of packages 3 and 4 as they fall off the belt at point A.

13.24 Guide angles have been attached to a conveyor belt at equal distances $d = 300$ mm. Four packages, each having a mass of 4 kg, are placed as shown on the belt which is at rest. If a constant force **P** of magnitude 60 N is applied to the belt, determine the velocities of packages 1 and 2 as they fall off the belt at point A. Assume that the mass of the belt and pulleys is small compared with the mass of the packages.

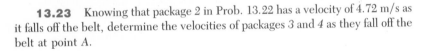

Fig. P13.24

13.25 Knowing that package 2 in Prob. 13.24 has a velocity of 2.29 m/s as it falls off the belt, determine the velocities of packages 3 and 4 as they fall off the belt at point A.

13.26 In Prob. 13.21, determine the smallest weight of collar C for which block A will reach the ground.

13.27 A 6-lb block is attached to a cable and to a spring as shown. The constant of the spring is $k = 8$ lb/in. and the tension in the cable is 3 lb. If the cable is cut, determine (a) the maximum displacement of the block, (b) the maximum velocity of the block.

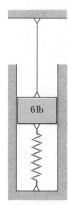

Fig. P13.27

13.28 An 8-kg plunger is released from rest in the position shown and is stopped by two nested springs; the constant of the outer spring is $k_1 = 3$ kN/m and the constant of the inner spring is $k_2 = 10$ kN/m. If the maximum deflection of the outer spring is observed to be 150 mm, determine the height h from which the plunger was released.

13.29 An 8-kg plunger is released from rest in the position shown and is stopped by two nested springs; the constant of the outer spring is $k_1 = 3$ kN/m and the constant of the inner spring is $k_2 = 10$ kN/m. If the plunger is released from a height $h = 600$ mm, determine the maximum deflection of the outer spring.

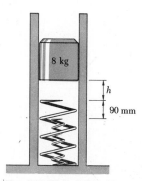

Fig. P13.28 and P13.29

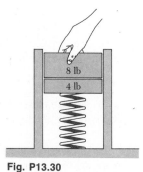

Fig. P13.30

13.30 A 4-lb block is at rest on a spring of constant 2 lb/in. An 8-lb block is held above the 4-lb block so that it just touches it, and then is released. Determine (a) the maximum velocity attained by the blocks, (b) the maximum force exerted on the blocks by the spring.

13.31 Two types of energy-absorbing fenders designed to be used on a pier are statically loaded. The force-deflection curve for each type of fender is given in the graph. Determine the maximum deflection of each fender when a 90-ton ship moving at 1 mi/h strikes the fender and is brought to rest.

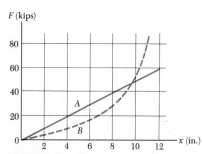

Fig. P13.31

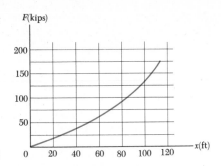

Fig. P13.32

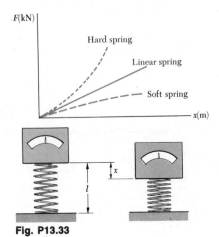

Fig. P13.33

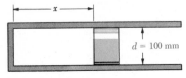

Fig. P13.35

13.32 A 15,000-lb airplane lands on an aircraft carrier and is caught by an arresting cable which is characterized by the force-deflection diagram shown. Knowing that the landing speed of the plane is 90 mi/h, determine (*a*) the distance required for the plane to come to rest, (*b*) the maximum rate of deceleration of the plane.

13.33 Nonlinear springs are classified as hard or soft, depending upon the curvature of their force-deflection curve (see figure). If a delicate instrument having a mass of 5 kg is placed on a spring of length *l* so that its base is just touching the undeformed spring and then inadvertently released from that position, determine the maximum deflection x_m of the spring and the maximum force F_m exerted by the spring, assuming (*a*) a linear spring of constant $k = 3$ kN/m, (*b*) a hard, nonlinear spring, for which $F = (3 \text{ kN/m})x(1 + 160x^2)$.

13.34 Solve Prob. 13.33, assuming that in part *b*, the hard spring has been replaced by a soft, nonlinear spring, for which $F = (3 \text{ kN/m})x(1 - 160x^2)$.

13.35 A 100-mm-diameter piston having a mass of 4 kg slides without friction in a cylinder. The pressure *p* within the cylinder varies inversely as the volume of the cylinder and equals the atmospheric pressure $p_a = 101.3$ kPa when $x = 250$ mm. If the piston is moved to the left and released with no velocity when $x = 120$ mm, determine the maximum velocity reached by the piston in the ensuing motion.

13.36 In Prob. 13.35, determine the maximum value of the coordinate *x* after the piston has been released with no velocity in the position $x = 120$ mm.

13.37 A bullet is fired straight up from the surface of the moon with an initial velocity of 600 m/s. Determine the maximum elevation reached by the bullet, (*a*) assuming a uniform gravitational field with $g = 1.62$ m/s², (*b*) using Newton's law of gravitation. (Radius of moon = 1740 km.)

13.38 Determine the maximum velocity with which an object dropped from a very large distance will hit the surface (*a*) of the earth, (*b*) of the moon. Neglect the effect of the earth's atmosphere. (The radius of the moon is 1740 km and its mass is 0.01230 times the mass of the earth.)

13.39 A rocket is fired vertically from the ground. Knowing that at burnout the rocket is 60 mi above the ground and has a velocity of 16,000 ft/s, determine the highest altitude it will reach.

13.40 A rocket is fired vertically from the ground. What should be its velocity v_B at burnout, 60 mi above the ground, if it is to reach an altitude of 900 mi?

13.41 Sphere C and block A are both moving to the left with a velocity v_0 when the block is suddenly stopped by the wall. Determine the smallest velocity v_0 for which the sphere C will swing in a full circle about the pivot B (a) if BC is a slender rod of negligible mass, (b) if BC is a cord.

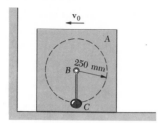

Fig. P13.41

13.42 A bag is gently pushed off the top of a wall at A and swings in a vertical plane at the end of a rope of length l. Determine the angle θ for which the rope will break, knowing that it can withstand a maximum tension equal to twice the weight of the bag.

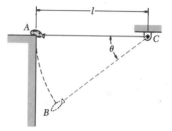

Fig. P13.42

13.43 A roller coaster starts from rest at A and rolls down the track shown. Assuming no energy loss and knowing that the radius of curvature of the track at B is 50 ft, determine the apparent weight of a 150-lb passenger at B.

13.44 A roller coaster is released with no velocity at A and rolls down the track shown. The brakes are suddenly applied as the car passes through point B, causing the wheels of the car to slide on the track ($\mu_k = 0.25$). Assuming no energy loss between A and B and knowing that the radius of curvature of the track at B is 50 ft, determine the normal and tangential components of the acceleration of the car just after the brakes have been applied.

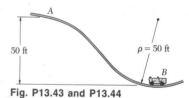

Fig. P13.43 and P13.44

13.45 A small block slides with velocity v_0 on the horizontal surface AB. Neglecting friction and knowing that $v_0 = 0.5\sqrt{gr}$, express in terms of r (a) the elevation h of point C where the block will leave the cylindrical surface BD, (b) the distance d from D to point E where it will hit the ground.

13.46 For the block of Prob. 13.45, and knowing that $r = 800$ mm, determine (a) the smallest value of v_0 for which the block leaves the surface ABD at point B and the corresponding value of d, (b) the smallest possible values of h and d (when v_0 approaches zero).

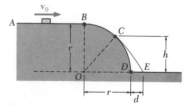

Fig. P13.45

13.47 An 18-g bullet leaves a fixed rifle barrel 2 ms after being fired. Knowing that the muzzle velocity is 750 m/s and neglecting friction, determine the average power developed by the rifle.

13.48 A 70-kg man runs up a 5-m-high flight of stairs in 6 s. (a) What is the average power developed by the man? (b) If a 55-kg woman can develop 80 percent as much power, how long will it take her to run up a 4-m-high flight of stairs?

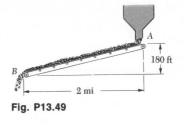

Fig. P13.49

13.49 Crushed stone is being moved from a quarry at *A* to a loading dock at *B* at the rate of 600 tons/h. An electric generator is attached to the system in order to maintain a constant belt speed. Knowing that the efficiency of the belt-generator system is 0.65, determine the average power in kilowatts developed by the generator if the belt speed is (*a*) 5 ft/s, (*b*) 12 ft/s.

13.50 A chair-lift is designed to transport 900 skiers per hour from the base *A* to the summit *B*. The average weight of a skier is 160 lb and the average speed of the lift is 250 ft/min. Determine (*a*) the average power required, (*b*) the required capacity of the motor if the mechanical efficiency is 85 percent and if a 300 percent overload is to be allowed.

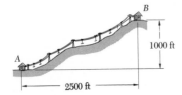

Fig. P13.50

13.51 A train of total mass equal to 500 Mg starts from rest and accelerates uniformly to a speed of 90 km/h in 50 s. After reaching this speed, the train travels with a constant velocity. The track is horizontal and axle friction and rolling resistance result in a total force of 15 kN in a direction opposite to the direction of motion. Determine the power required as a function of time.

13.52 Solve Prob. 13.51, assuming that during the entire motion, the train is traveling up a 1.5 percent grade.

13.53 Elevator *E* weighs 5000 lb when fully loaded and is connected to a 2200-lb counterweight *W*. Determine the power delivered by the electric motor when the elevator (*a*) is moving up at a constant speed of 15 ft/s, (*b*) has an instantaneous velocity of 15 ft/s and an acceleration of 3 ft/s², both directed upward.

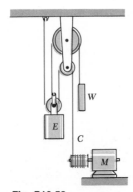

Fig. P13.53

13.54 For the dumb-waiter system of Sample Prob. 13.5, it is known that the power delivered by the motor is 3 hp at the instant the speed of the dumb-waiter is 6 ft/s upward. Determine the acceleration of the dumb-waiter at that instant.

13.55 The fluid transmission of a 15-Mg truck allows the engine to deliver an essentially constant power of 50 kW to the driving wheels. Determine the time required and the distance traveled as the speed of the truck is increased (*a*) from 36 km/h to 54 km/h, (*b*) from 54 km/h to 72 km/h.

13.56 The frictional resistance of a ship is known to vary directly as the 1.75 power of the speed *v* of the ship. A single tugboat at full power can tow the ship at a constant speed of 4.5 km/h by exerting a constant force of 300 kN. Determine (*a*) the power developed by the tugboat, (*b*) the maximum speed at which two tugboats, capable of delivering the same power, can tow the ship.

13.6. Potential Energy.† Let us consider again a body of weight **W** which moves along a curved path from a point A_1 of elevation y_1 to a point A_2 of elevation y_2 (Fig. 13.4). We recall from Sec. 13.2 that the work of the force of gravity **W** during this displacement is

$$U_{1\to2} = Wy_1 - Wy_2 \qquad (13.4)$$

The work of **W** may thus be obtained by subtracting the value of the function Wy corresponding to the second position of the body from its value corresponding to the first position. The work of **W** is independent of the actual path followed; it depends only upon the initial and final values of the function Wy. This function is called the *potential energy* of the body with respect to the *force of gravity* **W** and is denoted by V_g. We write

$$U_{1\to2} = (V_g)_1 - (V_g)_2 \qquad \text{with } V_g = Wy \qquad (13.16)$$

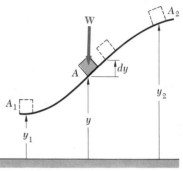

Fig. 13.4 (repeated)

We note that if $(V_g)_2 > (V_g)_1$, that is, *if the potential energy increases* during the displacement (as in the case considered here), *the work $U_{1\to2}$ is negative*. If, on the other hand, the work of **W** is positive, the potential energy decreases. Therefore, the potential energy V_g of the body provides a measure of the work which may be done by its weight **W**. Since only the *change* in potential energy, and not the actual value of V_g, is involved in formula (13.16), an arbitrary constant may be added to the expression obtained for V_g. In other words, the level, or datum, from which the elevation y is measured may be chosen arbitrarily. Note that potential energy is expressed in the same units as work, i.e., in joules if SI units are used, and in ft·lb or in·lb if U.S. customary units are used.

It should be noted that the expression just obtained for the potential energy of a body with respect to gravity is valid only as long as the weight **W** of the body may be assumed to remain constant, i.e., as long as the displacements of the body are small compared with the radius of the earth. In the case of a space vehicle, however, we should take into consideration the variation of the force of gravity with the distance r from the center of the earth. Using the expression obtained in Sec. 13.2 for the work of a gravitational force, we write (Fig. 13.6)

$$U_{1\to2} = \frac{GMm}{r_2} - \frac{GMm}{r_1} \qquad (13.7)$$

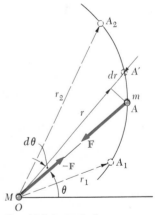

Fig. 13.6 (repeated)

The work of the force of gravity may therefore be obtained by subtracting the value of the function $-GMm/r$ corresponding to the second position of the body from its value corresponding to the first position. Thus, the expression which should be used for the potential energy V_g when the variation in the force of gravity cannot be neglected is

$$V_g = -\frac{GMm}{r} \qquad (13.17)$$

†Some of the material in this section has already been considered in Sec. 10.7.

Taking the first of the relations (12.29) into account, we write V_g in the alternate form

$$V_g = -\frac{WR^2}{r} \qquad (13.17')$$

where R is the radius of the earth and W the value of the weight of the body at the surface of the earth. When either of the relations (13.17) or (13.17') is used to express V_g, the distance r should, of course, be measured from the center of the earth.† Note that V_g is always negative and that it approaches zero for very large values of r.

Consider, now, a body attached to a spring and moving from a position A_1, corresponding to a deflection x_1 of the spring, to a position A_2, corresponding to a deflection x_2 (Fig. 13.5). We recall from Sec. 13.2 that the work of the force $\mathbf{F}$ exerted by the spring on the body is

$$U_{1\to2} = \tfrac{1}{2}kx_1^2 - \tfrac{1}{2}kx_2^2 \qquad (13.6)$$

The work of the elastic force is thus obtained by subtracting the value of the function $\tfrac{1}{2}kx^2$ corresponding to the second position of the body from its value corresponding to the first position. This function is denoted by V_e and is called the *potential energy* of the body with respect to the *elastic force* $\mathbf{F}$. We write

$$U_{1\to2} = (V_e)_1 - (V_e)_2 \qquad \text{with} \quad V_e = \tfrac{1}{2}kx^2 \qquad (13.18)$$

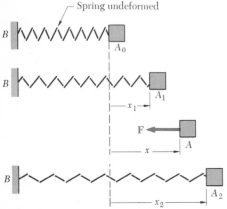

Spring undeformed

Fig. 13.5 (repeated)

and observe that during the displacement considered, the work of the force $\mathbf{F}$ exerted by the spring on the body is negative and the potential energy V_e increases. We should note that the expression obtained for V_e is valid only if the deflection of the spring is measured from its undeformed position. On the other hand, formula (13.18) may be used even when the spring is rotated about its fixed end (Fig. 13.10*a*). The work of the elastic force depends only upon the initial and final deflections of the spring (Fig. 13.10*b*).

†The expressions given for V_g in (13.17) and (13.17') are valid only when $r \geqq R$, that is, when the body considered is above the surface of the earth.

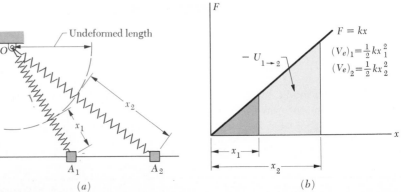

Fig. 13.10

The concept of potential energy may be used when forces other than gravity forces and elastic forces are involved. Indeed, it remains valid as long as the work of the force considered is independent of the path followed by its point of application as this point moves from a given position A_1 to a given position A_2. Such forces are said to be *conservative forces;* the general properties of conservative forces are studied in the following section.

* 13.7. Conservative Forces. As indicated in the preceding section, a force **F** acting on a particle A is said to be conservative *if its work $U_{1\rightarrow2}$ is independent of the path followed by the particle A as it moves from A_1 to A_2* (Fig. 13.11a). We may then write

$$U_{1\rightarrow2} = V(x_1, y_1, z_1) - V(x_2, y_2, z_2) \tag{13.19}$$

or, for short,

$$U_{1\rightarrow2} = V_1 - V_2 \tag{13.19'}$$

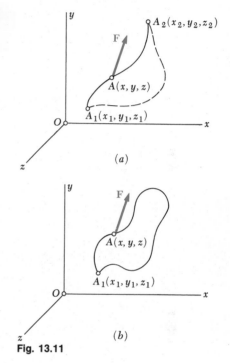

Fig. 13.11

The function $V(x, y, z)$ is called the potential energy, or *potential function,* of **F**.

We note that if A_2 is chosen to coincide with A_1, that is, if the particle describes a closed path (Fig. 13.11b), we have $V_1 = V_2$ and the work is zero. We may thus write for any conservative force **F**

$$\oint \mathbf{F} \cdot d\mathbf{r} = 0 \tag{13.20}$$

where the circle on the integral sign indicates that the path is closed.

Let us now apply (13.19) between two neighboring points $A(x, y, z)$ and $A'(x + dx, y + dy, z + dz)$. The elementary work dU corresponding to the displacement $d\mathbf{r}$ from A to A' is

$$dU = V(x, y, z) - V(x + dx, y + dy, z + dz)$$

or

$$dU = -dV(x, y, z) \tag{13.21}$$

Thus, the elementary work of a conservative force is an *exact differential.*

Substituting for dU in (13.21) the expression obtained in (13.1''), and recalling the definition of the differential of a function of several variables, we write

$$F_x\, dx + F_y\, dy + F_z\, dz = -\left(\frac{\partial V}{\partial x}dx + \frac{\partial V}{\partial y}dy + \frac{\partial V}{\partial z}dz\right)$$

from which it follows that

$$F_x = -\frac{\partial V}{\partial x} \qquad F_y = -\frac{\partial V}{\partial y} \qquad F_z = -\frac{\partial V}{\partial z} \qquad (13.22)$$

It is clear that the components of $\mathbf{F}$ must be functions of the coordinates x, y, z. Thus, a *necessary* condition for a conservative force is that it depend only upon the position of its point of application. The relations (13.22) may be expressed more concisely if we write

$$\mathbf{F} = F_x\mathbf{i} + F_y\mathbf{j} + F_z\mathbf{k} = -\left(\frac{\partial V}{\partial x}\mathbf{i} + \frac{\partial V}{\partial y}\mathbf{j} + \frac{\partial V}{\partial z}\mathbf{k}\right)$$

The vector in parentheses is known as the *gradient of the scalar function* V and is denoted by **grad** V. We thus write for any conservative force

$$\mathbf{F} = -\mathbf{grad}\ V \qquad (13.23)$$

The relations (13.19) to (13.23) were shown to be satisfied by any conservative force. It may also be shown that if a force $\mathbf{F}$ satisfies one of these relations, $\mathbf{F}$ must be a conservative force.

13.8. Conservation of Energy. We saw in the preceding two sections that the work of a conservative force, such as the weight of a particle or the force exerted by a spring, may be expressed as a change in potential energy. When a particle moves under the action of conservative forces, the principle of work and energy stated in Sec. 13.3 may be expressed in a modified form. Substituting for $U_{1 \to 2}$ from (13.19') into (13.10), we write

$$V_1 - V_2 = T_2 - T_1$$

$$T_1 + V_1 = T_2 + V_2 \qquad (13.24)$$

Formula (13.24) indicates that when a particle moves under the action of conservative forces, *the sum of the kinetic energy and of the potential energy of the particle remains constant.* The sum $T + V$ is called the *total mechanical energy* of the particle and is denoted by E.

Consider, for example, the pendulum analyzed in Sec. 13.4, which is released with no velocity from A_1 and allowed to swing in a vertical plane (Fig. 13.12). Measuring the potential energy from the level of A_2, we have, at A_1,

$$T_1 = 0 \qquad V_1 = Wl \qquad T_1 + V_1 = Wl$$

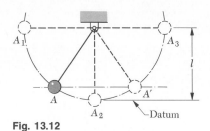

Fig. 13.12

Recalling that at A_2, the speed of the pendulum is $v_2 = \sqrt{2gl}$, we have

$$T_2 = \tfrac{1}{2}mv_2^2 = \frac{1}{2}\frac{W}{g}(2gl) = Wl \qquad V_2 = 0$$

$$T_2 + V_2 = Wl$$

We thus check that the total mechanical energy $E = T + V$ of the pendulum is the same at A_1 and A_2. While the energy is entirely potential at A_1, it becomes entirely kinetic at A_2 and, as the pendulum keeps swinging to the right, the kinetic energy is transformed back into potential energy. At A_3, we shall have $T_3 = 0$ and $V_3 = Wl$.

Since the total mechanical energy of the pendulum remains constant and since its potential energy depends only upon its elevation, the kinetic energy of the pendulum will have the same value at any two points located on the same level. Thus, the speed of the pendulum is the same at A and at A' (Fig. 13.12). This result may be extended to the case of a particle moving along any given path, regardless of the shape of the path, as long as the only forces acting on the particle are its weight and the normal reaction of the path. The particle of Fig. 13.13, for example, which slides in a vertical plane along a frictionless track, will have the same speed at A, A', and A''.

While the weight of a particle and the force exerted by a spring are conservative forces, *friction forces are nonconservative forces.* In other words, *the work of a friction force cannot be expressed as a change in potential energy.* The work of a friction force depends upon the path followed by its point of application; and while the work $U_{1 \to 2}$ defined by (13.19) is positive or negative according to the sense of motion, *the work of a friction force,* as we noted in Sec. 13.4, *is always negative.* It follows that when a mechanical system involves friction, its total mechanical energy does not remain constant but decreases. The energy of the system, however, is not lost; it is transformed into heat, and the sum of the *mechanical energy* and of the *thermal energy* of the system remains constant.

Other forms of energy may also be involved in a system. For instance, a generator converts mechanical energy into *electric energy;* a gasoline engine converts *chemical energy* into mechanical energy; a nuclear reactor converts *mass* into thermal energy. If all forms of energy are considered, the energy of any system may be considered as constant and the principle of conservation of energy remains valid under all conditions.

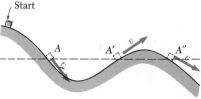

Fig. 13.13

13.9. Motion under a Conservative Central Force. Application to Space Mechanics. We saw in Sec. 12.9 that when a particle P moves under a central force $\mathbf{F}$, the angular momentum $\mathbf{H}_O$ of the particle about the center of force O is constant. If the force $\mathbf{F}$ is also conservative, there exists a potential energy V associated with $\mathbf{F}$, and the total energy $E = T + V$ of the particle is constant (Sec. 13.8). Thus, when a particle moves under a conservative central force, both the principle of conservation of angular momentum and the principle of conservation of energy may be used to study its motion.

Consider, for example, a space vehicle moving under the earth's gravitational force. We shall assume that it begins its free flight at point P_0 at a distance r_0 from the center of the earth, with a velocity $\mathbf{v}_0$ forming an angle ϕ_0 with the radius vector OP_0 (Fig. 13.14). Let P be a point of the trajectory described by the vehicle; we denote by r the distance from O to P, by $\mathbf{v}$ the velocity of the vehicle at P, and by ϕ the angle formed by $\mathbf{v}$ and the radius vector OP. Applying the principle of conservation of angular momentum about O between P_0 and P (Sec. 12.9), we write

$$r_0 m v_0 \sin \phi_0 = r m v \sin \phi \qquad (13.25)$$

Recalling the expression (13.17) obtained for the potential energy due to a gravitational force, we apply the principle of conservation of energy between P_0 and P and write

$$T_0 + V_0 = T + V$$

$$\tfrac{1}{2}mv_0^2 - \frac{GMm}{r_0} = \tfrac{1}{2}mv^2 - \frac{GMm}{r} \qquad (13.26)$$

where M is the mass of the earth.

Equation (13.26) may be solved for the magnitude v of the velocity of the vehicle at P when the distance r from O to P is known; Eq. (13.25) may then be used to determine the angle ϕ that the velocity forms with the radius vector OP.

Equations (13.25) and (13.26) may also be used to determine the maximum and minimum values of r in the case of a satellite launched from P_0 in a direction forming an angle ϕ_0 with the vertical OP_0 (Fig. 13.15). The desired values of r are obtained by making $\phi = 90°$ in (13.25) and eliminating v between Eqs. (13.25) and (13.26).

It should be noted that the application of the principles of conservation of energy and of conservation of angular momentum leads to a more fundamental formulation of the problems of space mechanics than does the method indicated in Sec. 12.12. In all cases involving oblique launchings, it will also result in much simpler computations. And while the method of Sec. 12.12 must be used when the actual trajectory or the periodic time of a space vehicle is to be determined, the calculations will be simplified if the conservation principles are first used to compute the maximum and minimum values of the radius vector r.

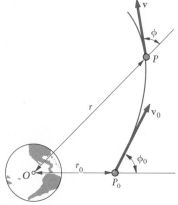

Fig. 13.14

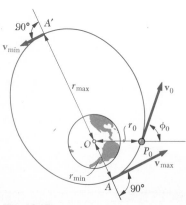

Fig. 13.15

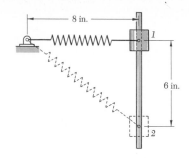

SAMPLE PROBLEM 13.6

A 20-lb collar slides without friction along a vertical rod as shown. The spring attached to the collar has an undeformed length of 4 in. and a constant of 3 lb/in. If the collar is released from rest in position *1*, determine its velocity after it has moved 6 in. to position *2*.

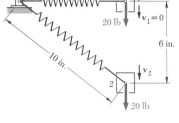

Position *1*. ***Potential Energy.*** The elongation of the spring is

$$x_1 = 8 \text{ in.} - 4 \text{ in.} = 4 \text{ in.}$$

and we have

$$V_e = \tfrac{1}{2}kx_1^2 = \tfrac{1}{2}(3 \text{ lb/in.})(4 \text{ in.})^2 = 24 \text{ in} \cdot \text{lb}$$

Choosing the datum as shown, we have $V_g = 0$. Therefore,

$$V_1 = V_e + V_g = 24 \text{ in} \cdot \text{lb} = 2 \text{ ft} \cdot \text{lb}$$

Kinetic Energy. Since the velocity in position *1* is zero, $T_1 = 0$.

Position *2*. ***Potential Energy.*** The elongation of the spring is

$$x_2 = 10 \text{ in.} - 4 \text{ in.} = 6 \text{ in.}$$

and we have

$$V_e = \tfrac{1}{2}kx_2^2 = \tfrac{1}{2}(3 \text{ lb/in.})(6 \text{ in.})^2 = 54 \text{ in} \cdot \text{lb}$$
$$V_g = Wy = (20 \text{ lb})(-6 \text{ in.}) = -120 \text{ in} \cdot \text{lb}$$

Therefore,

$$V_2 = V_e + V_g = 54 - 120 = -66 \text{ in} \cdot \text{lb}$$
$$= -5.5 \text{ ft} \cdot \text{lb}$$

Kinetic Energy

$$T_2 = \tfrac{1}{2}mv_2^2 = \frac{1}{2}\frac{20}{32.2}v_2^2 = 0.311v_2^2$$

Conservation of Energy. Applying the principle of conservation of energy between positions *1* and *2*, we write

$$T_1 + V_1 = T_2 + V_2$$
$$0 + 2 \text{ ft} \cdot \text{lb} = 0.311v_2^2 - 5.5 \text{ ft} \cdot \text{lb}$$
$$v_2 = \pm 4.91 \text{ ft/s}$$

$$\mathbf{v}_2 = 4.91 \text{ ft/s} \downarrow \quad \blacktriangleleft$$

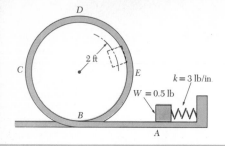

The 0.5-lb pellet is pushed against the spring at A and released from rest. Neglecting friction, determine the smallest deflection of the spring for which the pellet will travel around the loop $ABCDE$ and remain at all times in contact with the loop.

Required Speed at Point D. As the pellet passes through the highest point D, its potential energy with respect to gravity is maximum; thus, at the same point its kinetic energy and its speed are minimum. Since the pellet must remain in contact with the loop, the force $\mathbf{N}$ exerted on the pellet by the loop must be equal to, or greater than, zero. Setting $\mathbf{N} = 0$, we compute the smallest possible speed v_D.

$$+\downarrow \Sigma F_n = ma_n: \qquad W = ma_n \qquad mg = ma_n \qquad a_n = g$$

$$a_n = \frac{v_D^2}{r}: \qquad v_D^2 = ra_n = rg = (2 \text{ ft})(32.2 \text{ ft/s}^2) = 64.4 \text{ ft}^2/\text{s}^2$$

Position 1. Potential Energy. Denoting by x the deflection of the spring and noting that $k = 3 \text{ lb/in.} = 36 \text{ lb/ft}$, we write

$$V_e = \tfrac{1}{2}kx^2 = \tfrac{1}{2}(36 \text{ lb/ft})x^2 = 18x^2$$

Choosing the datum at A, we have $V_g = 0$; therefore

$$V_1 = V_e + V_g = 18x^2$$

Kinetic Energy. Since the pellet is released from rest, $v_A = 0$ and we have $T_1 = 0$.

Position 2. Potential Energy. The spring is now undeformed; thus $V_e = 0$. Since the pellet is 4 ft above the datum, we have

$$V_g = Wy = (0.5 \text{ lb})(4 \text{ ft}) = 2 \text{ ft} \cdot \text{lb}$$
$$V_2 = V_e + V_g = 2 \text{ ft} \cdot \text{lb}$$

Kinetic Energy. Using the value of v_D^2 obtained above, we write

$$T_2 = \tfrac{1}{2}mv_D^2 = \frac{1}{2}\frac{0.5 \text{ lb}}{32.2 \text{ ft/s}^2}(64.4 \text{ ft}^2/\text{s}^2) = 0.5 \text{ ft} \cdot \text{lb}$$

Conservation of Energy. Applying the principle of conservation of energy between positions 1 and 2, we write

$$T_1 + V_1 = T_2 + V_2$$
$$0 + 18x^2 = 0.5 \text{ ft} \cdot \text{lb} + 2 \text{ ft} \cdot \text{lb}$$
$$x = 0.3727 \text{ ft} \qquad\qquad x = 4.47 \text{ in.} \blacktriangleleft$$

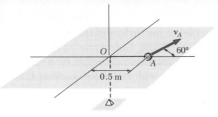

SAMPLE PROBLEM 13.8

A sphere of mass $m = 0.6\,\text{kg}$ is attached to an elastic cord of constant $k = 100\,\text{N/m}$, which is undeformed when the sphere is located at the origin O. Knowing that the sphere may slide without friction on the horizontal surface and that in the position shown its velocity $\mathbf{v}_A$ has a magnitude of 20 m/s, determine (a) the maximum and minimum distances from the sphere to the origin O, (b) the corresponding values of its speed.

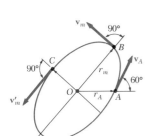

Solution. The force exerted by the cord on the sphere passes through the fixed point O, and its work may be expressed as a change in potential energy. It is therefore a conservative central force, and both the total energy of the sphere and its angular momentum about O are conserved.

Conservation of Angular Momentum About O. At point B, where the distance from O is maximum, the velocity of the sphere is perpendicular to OB and the angular momentum is $r_m v_m$. A similar property holds at point C, where the distance from O is minimum. Expressing conservation of angular momentum between A and B, we write

$$r_A m v_A \sin 60° = r_m m v_m$$
$$(0.5\,\text{m})(0.6\,\text{kg})(20\,\text{m/s}) \sin 60° = r_m(0.6\,\text{kg})v_m$$
$$v_m = \frac{8.66}{r_m} \qquad (1)$$

Conservation of Energy

At point A: $\quad T_A = \tfrac{1}{2}mv_A^2 = \tfrac{1}{2}(0.6\,\text{kg})(20\,\text{m/s})^2 = 120\,\text{J}$

$\qquad\qquad\quad V_A = \tfrac{1}{2}kr_A^2 = \tfrac{1}{2}(100\,\text{N/m})(0.5\,\text{m})^2 = 12.5\,\text{J}$

At point B: $\quad T_B = \tfrac{1}{2}mv_m^2 = \tfrac{1}{2}(0.6\,\text{kg})v_m^2 = 0.3v_m^2$

$\qquad\qquad\quad V_B = \tfrac{1}{2}kr_m^2 = \tfrac{1}{2}(100\,\text{N/m})r_m^2 = 50r_m^2$

Applying the principle of conservation of energy between points A and B, we write

$$T_A + V_A = T_B + V_B$$
$$120 + 12.5 = 0.3v_m^2 + 50r_m^2 \qquad (2)$$

a. Maximum and Minimum Values of Distance. Substituting for v_m from Eq. (1) into Eq. (2) and solving for r_m^2, we obtain

$$r_m^2 = 2.468, \text{ or } 0.1824 \qquad r_m = 1.571\,\text{m}, \ r_m' = 0.427\,\text{m} \blacktriangleleft$$

b. Corresponding Values of Speed. Substituting the values obtained for r_m and r_m' into Eq. (1), we have

$$v_m = \frac{8.66}{1.571} \qquad\qquad v_m = 5.51\,\text{m/s} \blacktriangleleft$$

$$v_m' = \frac{8.66}{0.427} \qquad\qquad v_m' = 20.3\,\text{m/s} \blacktriangleleft$$

Note. It may be shown that the path of the sphere is an ellipse of center O.

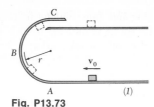
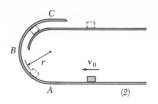

Fig. P13.73

13.73 A small package of weight W is projected into a vertical return loop at A with a velocity $\mathbf{v}_0$. The package travels without friction along a circle of radius r and is deposited on a horizontal surface at C. For each of the two loops shown, determine (a) the smallest velocity $\mathbf{v}_0$ for which the package will reach the horizontal surface at C, (b) the corresponding force exerted by the loop on the package as it passes point B.

13.74 In Prob. 13.73, it is desired to have the package deposited on the horizontal surface at C with a speed of 1.5 m/s. Knowing that $r = 0.3$ m, (a) show that this requirement cannot be fulfilled by the first loop, (b) determine the required initial velocity $\mathbf{v}_0$ when the second loop is used.

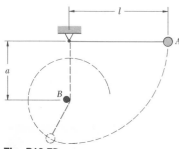

Fig. P13.75

13.75 The pendulum shown is released from rest at A and swings through 90° before the cord touches the fixed peg B. Determine the smallest value of a for which the pendulum bob will describe a circle about the peg.

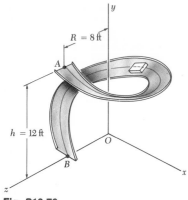

Fig. P13.76

*13.76 Packages are moved from point A on the upper floor of a warehouse to point B on the lower floor, 12 ft directly below A, by means of a chute, the centerline of which is in the shape of a helix of vertical axis y and radius $R = 8$ ft. The cross section of the chute is to be banked in such a way that each package, after being released at A with no velocity, will slide along the centerline of the chute, without ever touching its edges Neglecting friction, (a) express as a function of the elevation y of a given point P of the centerline the angle ϕ formed by the normal to the surface of the chute at P and the principal normal of the centerline at that point, (b) determine the magnitude and direction of the force exerted by the chute on a 20-lb package as it reaches point B. *Hint.* The principal normal to the helix at any point P is horizontal and directed toward the y axis, and the radius of curvature of the helix is $\rho = R[1 + (h/2\pi R)^2]$.

*13.77 Prove that a force $\mathbf{F}(x, y, z)$ is conservative if, and only if, the following relations are satisfied:

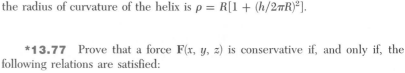

$$\frac{\partial F_x}{\partial y} = \frac{\partial F_y}{\partial x} \qquad \frac{\partial F_y}{\partial z} = \frac{\partial F_z}{\partial y} \qquad \frac{\partial F_z}{\partial x} = \frac{\partial F_x}{\partial z}$$

*13.78 The force $\mathbf{F} = (x\mathbf{i} + y\mathbf{j} + z\mathbf{k})/(x^2 + y^2 + z^2)^{3/2}$ acts on the particle $P(x, y, z)$ which moves in space. (a) Using the relations derived in Prob. 13.77, prove that $\mathbf{F}$ is a conservative force. (b) Determine the potential function $V(x, y, z)$ associated with $\mathbf{F}$.

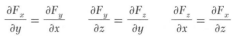

*13.79 A force **F** acts on a particle $P(x, y)$ which moves in the xy plane. Determine whether **F** is a conservative force and compute the work of **F** when P describes in counterclockwise sense the square of vertices A, B, C, and D, (a) when $\mathbf{F} = ky\mathbf{i}$, (b) when $\mathbf{F} = k(y\mathbf{i} + x\mathbf{j})$.

*13.80 A force $\mathbf{F} = (x\mathbf{i} + y\mathbf{j})(x^2 + y^2)$ acts on a particle $P(x, y)$ which moves in the xy plane. (a) Using the first of the relations derived in Prob. 13.77, prove that **F** is a conservative force. (b) Determine the potential function $V(x, y)$ associated with **F**.

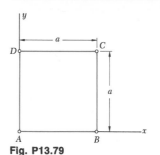

Fig. P13.79

13.81 (a) Determine the kinetic energy per unit mass which a missile must have after being fired from the surface of the earth if it is to reach an infinite distance from the earth. (b) What is the initial velocity of the missile (called the *escape velocity*)? Give your answers in SI units and show that the answer to part b is independent of the firing angle.

13.82 A satellite of mass m describes a circular orbit of radius r about the earth. Express (a) its potential energy, (b) its kinetic energy, (c) its total energy, as a function of r. Denote by R the radius of the earth, by g the acceleration of gravity at the surface of the earth, and assume that the potential energy of the satellite is zero on its launching pad.

13.83 (a) Show by setting $r = R + y$ in formula (13.17′) and expanding in a power series in y/R that the expression obtained in (13.16) for the potential energy V_g due to gravity is a first-order approximation for the expression given in (13.17′). (b) Using the same expansion, derive a second-order approximation for V_g.

13.84 A lunar excursion module (LEM) was used in the Apollo moon-landing missions to save fuel by making it unnecessary to launch the entire Apollo spacecraft from the moon's surface on its return trip to the earth. Check the effectiveness of this approach by computing the energy per kilogram required for a spacecraft to escape the gravitational field of the moon if the spacecraft starts (a) from the moon's surface, (b) from a circular orbit 100 km above the moon's surface. Neglect the effect of the earth's gravitational field. (The radius of the moon is 1740 km and its mass is 0.01230 times the mass of the earth.)

13.85 During its fifth mission, the space shuttle Columbia ejected two communication satellites while describing a circular orbit, 185 mi above the surface of the earth. Knowing that one of these satellites weighed 8000 lb, determine (a) the additional energy required to place the satellite in a geosynchronous orbit (see Prob. 12.78) at an altitude of 22,230 mi above the surface of the earth, (b) the energy required to place it in the same orbit by launching it from the surface of the earth.

13.86 Show that the ratio of the potential and kinetic energies of an electron as it enters the plates of the cathode-ray tube of Prob. 12.61 is equal to $d\delta/lL$. (Place the datum at the surface of the positive plate.)

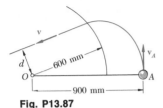

Fig. P13.87

13.87 A 200-g ball may slide on a horizontal frictionless surface and is attached to a fixed point O by means of an elastic cord of constant $k = 150$ N/m and undeformed length equal to 600 mm. The ball is placed at point A, 900 mm from O, and is given an initial velocity $\mathbf{v}_A$ in a direction perpendicular to OA. Knowing that the ball passes at a distance $d = 100$ mm from point O, determine (a) the initial speed v_A of the ball, (b) its speed v after the cord has become slack.

13.88 Knowing that the initial speed of the ball of Prob. 13.87 is $v_A = 2.5$ m/s, determine (a) the speed v of the ball after the cord has become slack, (b) the closest distance d that the ball will come to O.

13.89 For the ball of Prob. 13.87, determine (a) the smallest magnitude of the initial velocity $\mathbf{v}_A$ for which the elastic cord remains taut at all times, (b) the corresponding maximum speed reached by the ball.

13.90 In Sample Prob. 13.8, determine the required magnitude of the velocity $\mathbf{v}_A$ if the maximum distance from the origin O reached by the sphere is to be 1.2 m.

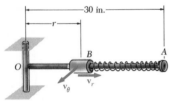

Fig. P13.91

13.91 Collar B weighs 10 lb and is attached to a spring of constant 50 lb/ft and of undeformed length equal to 18 in. The system is set in motion with $r = 12$ in., $v_\theta = 16$ ft/s, and $v_r = 0$. Neglecting the mass of the rod and the effect of friction, determine the radial and transverse components of the velocity of the collar when $r = 21$ in.

13.92 For the motion described in Prob. 13.91, determine (a) the maximum distance between the origin and the collar, (b) the corresponding velocity. (*Hint.* Solve by trial and error the equation obtained for r.)

13.93 through 13.95 Using the principles of conservation of energy and conservation of angular momentum, solve the following problems:

13.93 Prob. 12.97.
13.94 Prob. 12.98.
13.95 Prob. 12.101a

13.96 Using the principles of conservation of energy and conservation of angular momentum, determine the velocity of the probe of Prob. 12.100 (a) just after it has been placed on its second transfer orbit at B, (b) as it reaches point C.

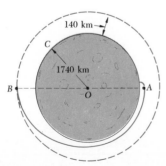

Fig. P13.97

13.97 After completing their moon-exploration mission, the two astronauts forming the crew of an Apollo lunar excursion module (LEM) would prepare to rejoin the command module which was orbiting the moon at an altitude of 140 km. They would fire the LEM's engine, bring it along a curved path to a point A, 8 km above the moon's surface, and shut off the engine. Knowing that the LEM was moving at that time in a direction parallel to the moon's surface and that it then coasted along an elliptic path to a rendezvous at B with the command module, determine (a) the speed of the LEM at engine shutoff, (b) the relative velocity with which the command module approached the LEM at B. (The radius of the moon is 1740 km and its mass is 0.01230 times the mass of the earth.)

13.98 While describing a circular orbit, 185 mi above the surface of the earth, a space shuttle ejects at point A an inertial upper stage (IUS) carrying a communication satellite to be placed in a geosynchronous orbit (see Prob. 12.78) at an altitude of 22,230 mi above the surface of the earth. Determine (a) the velocity of the IUS relative to the shuttle after its engine has been fired at A, (b) the increase in velocity required at B to place the satellite in its final orbit.

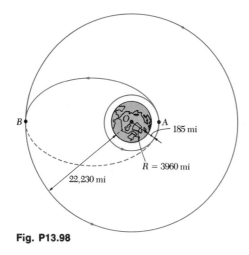

Fig. P13.98

13.99 Upon the LEM's return to the command module, the Apollo space-craft of Prob. 13.97 was turned around so that the LEM faced to the rear. The LEM was then cast adrift with a velocity of 200 m/s relative to the command module. Determine the magnitude and direction (angle ϕ formed with the vertical OC) of the velocity $\mathbf{v}_C$ of the LEM just before it crashed at C on the moon's surface.

13.100 A space shuttle is to rendezvous with an orbiting laboratory which circles the earth at a constant altitude of 360 km. The shuttle has reached an altitude of 60 km when its engine is shut off, and its velocity $\mathbf{v}_0$ forms an angle $\phi_0 = 50°$ with the vertical OB at that time. What magnitude should $\mathbf{v}_0$ have if the shuttle's trajectory is to be tangent at A to the orbit of the laboratory?

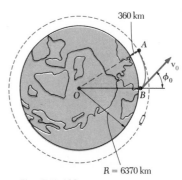

Fig. P13.100

13.101 A space shuttle is describing a circular orbit at an altitude of 150 mi above the surface of the earth. As it passes through A, it fires its engine for a short interval of time to reduce its speed by 5 percent and begin its descent toward the earth. Determine the magnitude and direction (angle ϕ formed with the vertical OB) of the velocity $\mathbf{v}_B$ of the shuttle as it reaches point B at an altitude of 30 mi.

13.102 A space shuttle is describing a circular orbit at an altitude of 150 mi above the surface of the earth. As it passes through A, it fires its engine to reduce its speed and begin its descent toward the earth. As it reaches point B at an altitude of 30 mi, its velocity $\mathbf{v}_B$ is observed to form an angle $\phi = 80°$ with the vertical. What is the magnitude of its velocity at that instant?

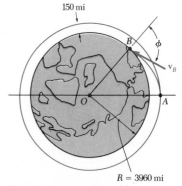

Fig. P13.101 and P13.102

13.103 At engine burnout, an experimental rocket has reached an altitude of 300 km and has a velocity v_0 of magnitude 8000 m/s forming an angle of 40° with the vertical. Determine the maximum altitude reached by the rocket.

13.104 At engine burnout, an experimental rocket has reached an altitude of 400 km and has a velocity v_0 of magnitude 7500 m/s. What angle should v_0 form with the vertical if the rocket is to reach a maximum altitude of 3000 km?

13.105 Show that the values v_1 and v_2 of the speed of an earth satellite at the perigee A and the apogee A' of an elliptic orbit are defined by the relations

$$v_1^2 = \frac{2GM}{r_1 + r_2}\frac{r_2}{r_1} \qquad v_2^2 = \frac{2GM}{r_1 + r_2}\frac{r_1}{r_2}$$

where M is the mass of the earth, and r_1 and r_2 represent, respectively, the minimum and maximum distances of the orbit to the center of the earth.

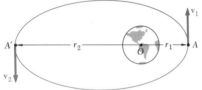

Fig. P13.105 and P13.106

13.106 Show that the total energy E of an earth satellite of mass m describing an elliptic orbit is $E = -GMm/(r_1 + r_2)$, where M is the mass of the earth, and r_1 and r_2 represent, respectively, the minimum and maximum distances of the orbit to the center of the earth. (It is recalled that the gravitational potential energy of a satellite was defined as being zero at an infinite distance from the earth.)

13.107 A spacecraft of mass m describes a circular orbit of radius r_1 around the earth. (a) Show that the additional energy ΔE which must be imparted to the spacecraft to transfer it to a circular orbit of larger radius r_2 is

$$\Delta E = \frac{GMm(r_2 - r_1)}{2r_1r_2}$$

where M is the mass of the earth. (b) Further show that if the transfer from one circular orbit to the other is executed by placing the spacecraft on a transitional semielliptic path AB, the amounts of energy ΔE_A and ΔE_B which must be imparted at A and B are, respectively, proportional to r_2 and r_1:

$$\Delta E_A = \frac{r_2}{r_1 + r_2}\Delta E \qquad \Delta E_B = \frac{r_1}{r_1 + r_2}\Delta E$$

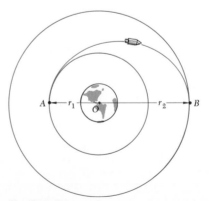

Fig. P13.107

13.108   A satellite is projected into space with a velocity v_0 at a distance r_0 from the center of the earth by the last stage of its launching rocket. The velocity v_0 was designed to send the satellite into a circular orbit of radius r_0. However, owing to a malfunction of control, the satellite is not projected horizontally but at an angle α with the horizontal and, as a result, is propelled into an elliptic orbit. Determine the maximum and minimum values of the distance from the center of the earth to the satellite.

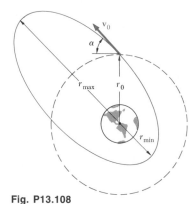

Fig. P13.108

13.109   A missile is fired from the ground with an initial velocity v_0 forming an angle ϕ_0 with the vertical. If the missile is to reach a maximum altitude equal to αR, where R is the radius of the earth, (a) show that the required angle ϕ_0 is defined by the relation

$$\sin \phi_0 = (1 + \alpha)\sqrt{1 - \frac{\alpha}{1 + \alpha}\left(\frac{v_{esc}}{v_0}\right)^2}$$

where v_{esc} is the escape velocity, (b) determine the range of allowable values of v_0.

*13.110   Using the answers obtained in Prob. 13.108, show that the intended circular orbit and the resulting elliptic orbit intersect at the ends of the minor axis of the elliptic orbit.

*13.111   (a) Express in terms of r_{min} and v_{max} the angular momentum per unit mass, h, and the total energy per unit mass, E/m, of a space vehicle moving under the gravitational attraction of a planet of mass M (Fig. 13.15). (b) Eliminating v_{max} between the equations obtained, derive the formula

$$\frac{1}{r_{min}} = \frac{GM}{h^2}\left[1 + \sqrt{1 + \frac{2E}{m}\left(\frac{h}{GM}\right)^2}\right]$$

(c) Show that the eccentricity ε of the trajectory of the vehicle may be expressed as

$$\varepsilon = \sqrt{1 + \frac{2E}{m}\left(\frac{h}{GM}\right)^2}$$

(d) Further show that the trajectory of the vehicle is a hyperbola, an ellipse, or a parabola, depending on whether E is positive, negative, or zero.

13.10. Principle of Impulse and Momentum.   A third basic method for the solution of problems dealing with the motion of particles will be considered now. This method is based on the principle of impulse and momentum and may be used to solve problems involving force, mass, velocity, and time. It is of particular interest in the solution of problems involving impulsive motion or impact (Secs. 13.11 and 13.12).

Consider a particle of mass m acted upon by a force $\mathbf{F}$. As we saw in Sec. 12.3, Newton's second law may be expressed in the form

$$\mathbf{F} = \frac{d}{dt}(m\mathbf{v}) \tag{13.27}$$

where $m\mathbf{v}$ is the linear momentum of the particle. Multiplying both sides of Eq. (13.27) by dt and integrating from a time t_1 to a time t_2, we write

$$\mathbf{F}\, dt = d(m\mathbf{v})$$

$$\int_{t_1}^{t_2} \mathbf{F}\, dt = m\mathbf{v}_2 - m\mathbf{v}_1$$

or, transposing the last term,

$$m\mathbf{v}_1 + \int_{t_1}^{t_2} \mathbf{F}\, dt = m\mathbf{v}_2 \tag{13.28}$$

The integral in Eq. (13.28) is a vector known as the *linear impulse,* or simply the *impulse,* of the force $\mathbf{F}$ during the interval of time considered. Resolving $\mathbf{F}$ into rectangular components, we write

$$\mathbf{Imp}_{1\to2} = \int_{t_1}^{t_2} \mathbf{F}\, dt$$

$$= \mathbf{i}\int_{t_1}^{t_2} F_x\, dt + \mathbf{j}\int_{t_1}^{t_2} F_y\, dt + \mathbf{k}\int_{t_1}^{t_2} F_z\, dt \tag{13.29}$$

and note that the components of the impulse of the force $\mathbf{F}$ are, respectively, equal to the areas under the curves obtained by plotting the components F_x, F_y, and F_z against t (Fig. 13.16). In the case of a force $\mathbf{F}$ of constant magnitude and direction, the impulse is represented by the vector $\mathbf{F}(t_2 - t_1)$, which has the same direction as $\mathbf{F}$.

If SI units are used, the magnitude of the impulse of a force is expressed in N·s. But, recalling the definition of the newton, we have

$$\text{N·s} = (\text{kg·m/s}^2)\text{·s} = \text{kg·m/s}$$

which is the unit obtained in Sec. 12.4 for the linear momentum of a particle. We thus check that Eq. (13.28) is dimensionally correct. If U.S. customary units are used, the impulse of a force is expressed in lb·s, which is also the unit obtained in Sec. 12.4 for the linear momentum of a particle.

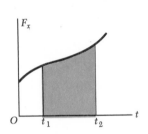

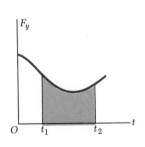

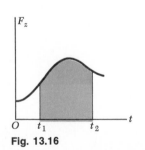

Fig. 13.16

Equation (13.28) expresses that when a particle is acted upon by a force **F** during a given time interval, *the final momentum* mv_2 *of the particle may be obtained by adding vectorially its initial momentum* mv_1 *and the impulse of the force* **F** *during the time interval considered* (Fig. 13.17).

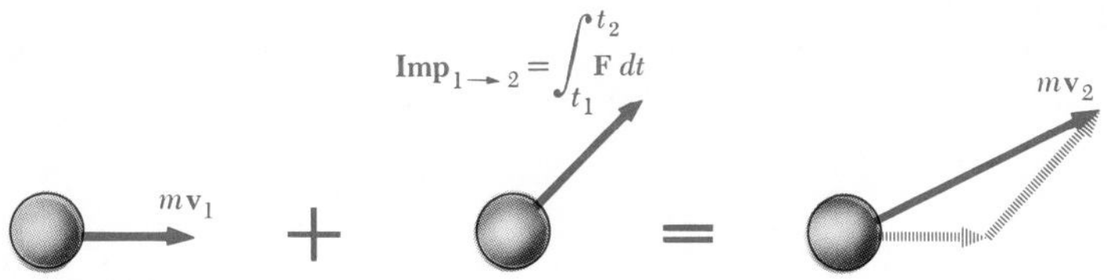

$$\text{Imp}_{1 \to 2} = \int_{t_1}^{t_2} \mathbf{F}\, dt$$

mv_1 $\quad + \quad$ $\quad = \quad$ mv_2

Fig. 13.17

We write

$$mv_1 + \text{Imp}_{1 \to 2} = mv_2 \tag{13.30}$$

We note that while kinetic energy and work are scalar quantities, momentum and impulse are vector quantities. To obtain an analytic solution, it is thus necessary to replace Eq. (13.30) by the equivalent component equations

$$(mv_x)_1 + \int_{t_1}^{t_2} F_x\, dt = (mv_x)_2$$
$$(mv_y)_1 + \int_{t_1}^{t_2} F_y\, dt = (mv_y)_2 \tag{13.31}$$
$$(mv_z)_1 + \int_{t_1}^{t_2} F_z\, dt = (mv_z)_2$$

When several forces act on a particle, the impulse of each of the forces must be considered. We have

$$mv_1 + \Sigma\, \text{Imp}_{1 \to 2} = mv_2 \tag{13.32}$$

Again, the equation obtained represents a relation between vector quantities; in the actual solution of a problem, it should be replaced by the corresponding component equations.

When a problem involves two particles or more, each particle may be considered separately and Eq. (13.32) may be written for each particle. We may also add vectorially the momenta of all the particles and the impulses of all the forces involved. We write then

$$\Sigma mv_1 + \Sigma\, \text{Imp}_{1 \to 2} = \Sigma mv_2 \tag{13.33}$$

Since the forces of action and reaction exerted by the particles on each other form pairs of equal and opposite forces, and since the time interval from t_1 to t_2 is common to all the forces involved, the impulses of the forces of action and reaction cancel out, and only the impulses of the external forces need be considered.†

If no external force is exerted on the particles or, more generally, if the sum of the external forces is zero, the second term in Eq. (13.33) vanishes, and Eq. (13.33) reduces to

$$\Sigma m\mathbf{v}_1 = \Sigma m\mathbf{v}_2 \tag{13.34}$$

which expresses that *the total momentum of the particles is conserved.* Consider, for example, two boats, of mass m_A and m_B, initially at rest, which are being pulled together (Fig. 13.18). If the resistance of the water

Fig. 13.18

is neglected, the only external forces acting on the boats are their weights and the buoyant forces exerted on them. Since these forces are balanced, we write

$$\Sigma m\mathbf{v}_1 = \Sigma m\mathbf{v}_2$$
$$0 = m_A\mathbf{v}'_A + m_B\mathbf{v}'_B$$

where $\mathbf{v}'_A$ and $\mathbf{v}'_B$ represent the velocities of the boats after a finite interval of time. The equation obtained indicates that the boats move in opposite directions (toward each other) with velocities inversely proportional to their masses.‡

13.11. Impulsive Motion. In some problems, a very large force may act during a very short time interval on a particle and produce a definite change in momentum. Such a force is called an *impulsive force* and the resulting motion an *impulsive motion.* For example, when a baseball is struck, the contact between bat and ball takes place during a very short time interval Δt. But the average value of the force $\mathbf{F}$ exerted by the bat on

---

†We should note the difference between this statement and the corresponding statement made in Sec. 13.4 regarding the work of the forces of action and reaction between several particles. While the sum of the impulses of these forces is always zero, the sum of their work is zero only under special circumstances, e.g., when the various bodies involved are connected by inextensible cords or links and are thus constrained to move through equal distances.

‡The application of the method of impulse and momentum to a system of particles and the concept of conservation of momentum for a system of particles are discussed in detail in Chap. 14.

the ball is very large, and the resulting impulse $\mathbf{F} \Delta t$ is large enough to change the sense of motion of the ball (Fig. 13.19).

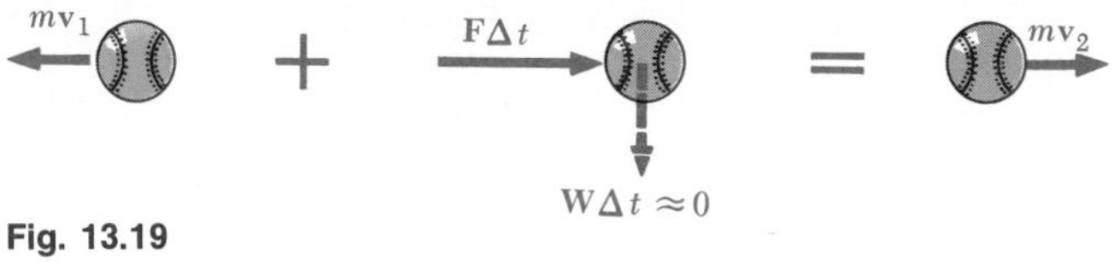

Fig. 13.19

When impulsive forces act on a particle, Eq. (13.32) becomes

$$m\mathbf{v}_1 + \Sigma \mathbf{F}\, \Delta t = m\mathbf{v}_2 \tag{13.35}$$

Any force which is not an impulsive force may be neglected, since the corresponding impulse $\mathbf{F} \Delta t$ is very small. *Nonimpulsive forces* include the weight of the body, the force exerted by a spring, or any other force which is *known* to be small compared with an impulsive force. Unknown reactions may or may not be impulsive; their impulse should therefore be included in Eq. (13.35) as long as it has not been proved negligible. The impulse of the weight of the baseball considered above, for example, may be neglected. If the motion of the bat is analyzed, the impulse of the weight of the bat may also be neglected. The impulses of the reactions of the player's hands on the bat, however, should be included; these impulses will not be negligible if the ball is incorrectly hit.

We note that the method of impulse and momentum is particularly effective in the analysis of the impulsive motion of a particle, since it involves only the initial and final velocities of the particle and the impulses of the forces exerted on the particle. The direct application of Newton's second law, on the other hand, would require the determination of the forces as functions of the time and the integration of the equations of motion over the time interval Δt.

In the case of the impulsive motion of several particles, Eq. (13.33) may be used. It reduces to

$$\Sigma m\mathbf{v}_1 + \Sigma \mathbf{F}\, \Delta t = \Sigma m\mathbf{v}_2 \tag{13.36}$$

where the second term involves only impulsive, external forces. If all the external forces acting on the various particles are nonimpulsive, the second term in Eq. (13.36) vanishes and this equation reduces to Eq. (13.34). We write

$$\Sigma m\mathbf{v}_1 = \Sigma m\mathbf{v}_2 \tag{13.34}$$

which expresses that the total momentum of the particles is conserved. This situation occurs, for example, when two particles which are moving freely collide with one another. We should note, however, that while the total momentum of the particles is conserved, their total energy is generally *not* conserved. Problems involving the collision or *impact* of two particles will be discussed in detail in Secs. 13.12 through 13.14.

SAMPLE PROBLEM 13.10

An automobile weighing 4000 lb is driven down a 5° incline at a speed of 60 mi/h when the brakes are applied, causing a constant total braking force (applied by the road on the tires) of 1500 lb. Determine the time required for the automobile to come to a stop.

Solution. We apply the principle of impulse and momentum. Since each force is constant in magnitude and direction, each corresponding impulse is equal to the product of the force and of the time interval t.

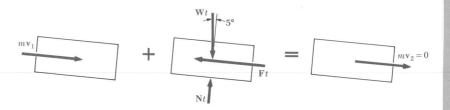

$$m\mathbf{v}_1 + \Sigma\,\mathbf{Imp}_{1\rightarrow2} = m\mathbf{v}_2$$

$+\searrow x$ components: $\quad mv_1 + (W \sin 5°)t - Ft = 0$

$(4000/32.2)(88 \text{ ft/s}) + (4000 \sin 5°)t - 1500t = 0 \qquad t = 9.49 \text{ s} \quad \blacktriangleleft$

SAMPLE PROBLEM 13.11

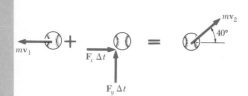

A 4-oz baseball is pitched with a velocity of 80 ft/s toward a batter. After the ball is hit by the bat B, it has a velocity of 120 ft/s in the direction shown. If the bat and ball are in contact 0.015 s, determine the average impulsive force exerted on the ball during the impact.

Solution. We apply the principle of impulse and momentum to the ball. Since the weight of the ball is a nonimpulsive force, we shall neglect it.

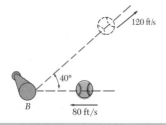

$$m\mathbf{v}_1 + \Sigma\,\mathbf{Imp}_{1\rightarrow2} = m\mathbf{v}_2$$

$\xrightarrow{+} x$ components: $\qquad -mv_1 + F_x\,\Delta t = mv_2 \cos 40°$

$$-\frac{\frac{4}{16}}{32.2}(80 \text{ ft/s}) + F_x(0.015 \text{ s}) = \frac{\frac{4}{16}}{32.2}(120 \text{ ft/s}) \cos 40°$$

$$F_x = +89.0 \text{ lb}$$

$+\uparrow y$ components: $\qquad 0 + F_y\,\Delta t = mv_2 \sin 40°$

$$F_y(0.015 \text{ s}) = \frac{\frac{4}{16}}{32.2}(120 \text{ ft/s}) \sin 40°$$

$$F_y = +39.9 \text{ lb}$$

From its components F_x and F_y we determine the magnitude and direction of the force $\mathbf{F}$:

$$\mathbf{F} = 97.5 \text{ lb} \measuredangle 24.2° \quad \blacktriangleleft$$

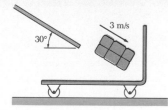

SAMPLE PROBLEM 13.12

A 10-kg package drops from a chute into a 25-kg cart with a velocity of 3 m/s. Knowing that the cart is initially at rest and may roll freely, determine (a) the final velocity of the cart, (b) the impulse exerted by the cart on the package, (c) the fraction of the initial energy lost in the impact.

Solution. We first apply the principle of impulse and momentum to the package-cart system to determine the velocity v_2 of the cart and package. We then apply the same principle to the package alone to determine the impulse $\mathbf{F}\,\Delta t$ exerted on it.

a. Impulse-Momentum Principle: Package and Cart

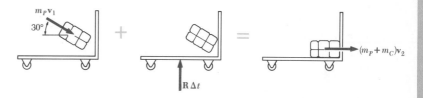

$$m_P \mathbf{v}_1 + \Sigma\,\mathbf{Imp}_{1\to 2} = (m_P + m_C)\mathbf{v}_2$$

$\xrightarrow{+} x$ components:
$$m_P v_1 \cos 30^\circ + 0 = (m_P + m_C)v_2$$
$$(10\text{ kg})(3\text{ m/s}) \cos 30^\circ = (10\text{ kg} + 25\text{ kg})v_2$$
$$\mathbf{v}_2 = 0.742\text{ m/s} \rightarrow \blacktriangleleft$$

We note that the equation used expresses conservation of momentum in the x direction.

b. Impulse-Momentum Principle: Package

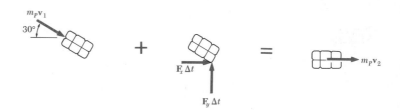

$$m_P \mathbf{v}_1 + \Sigma\,\mathbf{Imp}_{1\to 2} = m_P \mathbf{v}_2$$

$\xrightarrow{+} x$ components:
$$m_P v_1 \cos 30^\circ + F_x\,\Delta t = (10\text{ kg})(0.742\text{ m/s})$$
$$F_x\,\Delta t = -18.56\text{ N}\cdot\text{s}$$

$+\uparrow y$ components:
$$-m_P v_1 \sin 30^\circ + F_y\,\Delta t = 0$$
$$-(10\text{ kg})(3\text{ m/s}) \sin 30^\circ + F_y\,\Delta t = 0$$
$$F_y\,\Delta t = +15\text{ N}\cdot\text{s}$$

The impulse exerted on the package is
$$\mathbf{F}\,\Delta t = 23.9\text{ N}\cdot\text{s} \;\measuredangle\; 38.9^\circ \quad \blacktriangleleft$$

c. Fraction of Energy Lost.

The initial and final energies are

$$T_1 = \tfrac{1}{2}m_P v_1^2 = \tfrac{1}{2}(10\text{ kg})(3\text{ m/s})^2 = 45\text{ J}$$
$$T_2 = \tfrac{1}{2}(m_P + m_C)v_2^2 = \tfrac{1}{2}(10\text{ kg} + 25\text{ kg})(0.742\text{ m/s})^2 = 9.63\text{ J}$$

The fraction of energy lost is
$$\frac{T_1 - T_2}{T_1} = \frac{45\text{ J} - 9.63\text{ J}}{45\text{ J}} = 0.786 \quad \blacktriangleleft$$

Problems

13.112 A 40,000-ton ocean liner has an initial velocity of 2.5 mi/h. Neglecting the frictional resistance of the water, determine the time required to bring the liner to rest by using a single tugboat which exerts a constant force of 35 kips.

13.113 A 1200-kg automobile is moving at a speed of 90 km/h when the brakes are fully applied, causing all four wheels to skid. Determine the time required to stop the automobile (a) on dry pavement ($\mu_k = 0.75$), (b) on an icy road ($\mu_k = 0.10$).

13.114 The coefficients of friction between the load and the flatbed trailer shown are $\mu_s = 0.40$ and $\mu_k = 0.35$. Knowing that the speed of the rig is 90 km/h, determine the shortest time in which the rig can be brought to a stop if the load is not to shift.

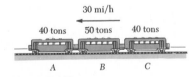

Fig. P13.114

13.115 A 10-lb particle is acted upon by the force, expressed in pounds, $\mathbf{F} = 2t\mathbf{i} + (3 - t)\mathbf{j} + t^3\mathbf{k}$. Knowing that the velocity of the particle at $t = 0$ is $\mathbf{v} = -(24 \text{ ft/s})\mathbf{i} + (15 \text{ ft/s})\mathbf{j} - (60 \text{ ft/s})\mathbf{k}$, determine the velocity of the particle at $t = 4$ s.

13.116 A 2-kg particle is acted upon by the force, expressed in newtons, $\mathbf{F} = (8 - 6t)\mathbf{i} + (4 - t^2)\mathbf{j} + (4 + t)\mathbf{k}$. Knowing that the velocity of the particle is $\mathbf{v} = (150 \text{ m/s})\mathbf{i} + (100 \text{ m/s})\mathbf{j} - (250 \text{ m/s})\mathbf{k}$ at $t = 0$, determine (a) the time at which the velocity of the particle is parallel to the yz plane, (b) the corresponding velocity of the particle.

13.117 Using the principle of impulse and momentum, solve Prob. 12.15.

13.118 The subway train shown is traveling at a speed of 30 mi/h when the brakes are fully applied on the wheels of cars B and C, causing them to slide on the track, but are not applied on the wheels of car A. Knowing that the coefficient of kinetic friction is 0.35 between the wheels and the track, determine (a) the time required to bring the train to a stop, (b) the force in each coupling.

13.119 Solve Prob. 13.118, assuming that the brakes are applied only on the wheels of car A.

13.120 The system shown is at rest when a constant 150-N force is applied to collar B. Neglecting the effect of friction, determine (a) the time at which the velocity of collar B will be 2.5 m/s to the left, (b) the corresponding tension in the cable.

13.121 Using the principle of impulse and momentum, solve Prob. 12.17b.

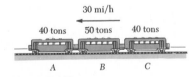

Fig. P13.118

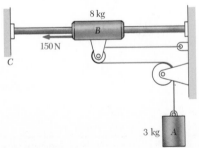

Fig. P13.120

Problems

13.122 Two packages are placed on an incline as shown. The coefficients of friction are $\mu_s = 0.30$ and $\mu_k = 0.25$ between the incline and package A, and $\mu_s = 0.20$ and $\mu_k = 0.15$ between the incline and package B. Knowing that the packages are in contact when released, determine (a) the velocity of each package after 3 s, (b) the force exerted by package A on package B.

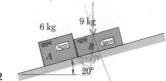

Fig. P13.122

13.123 The 3-kg collar is initially at rest and is acted upon by the force $\mathbf{Q}$ which varies as shown. Knowing that $\mu_k = 0.25$, determine the velocity of the collar at (a) $t = 1$ s, (b) $t = 2$ s.

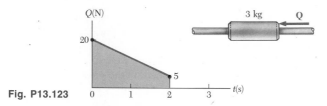

Fig. P13.123

13.124 In Prob. 13.123, determine (a) the maximum velocity reached by the collar and the corresponding time, (b) the time at which the collar comes to rest.

13.125 A 125-lb block initially at rest is acted upon by a force **P** which varies as shown. Knowing that the coefficients of friction between the block and the horizontal surface are $\mu_s = 0.50$ and $\mu_k = 0.40$, determine (a) the time at which the block will start moving, (b) the maximum velocity reached by the block, (c) the time at which the block will stop moving.

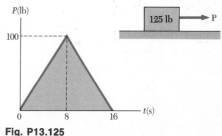

Fig. P13.125

13.126 Solve Prob. 13.125, assuming that the weight of the block is 175 lb.

13.127 A 25-g bullet is fired with a velocity of 600 m/s into a wooden block which rests against a solid vertical wall. Knowing that the bullet is brought to rest in 0.75 ms, determine the average impulsive force exerted by the bullet on the block.

13.128 A 3000-lb car moving with a velocity of 2.5 mi/h hits a garage wall and is brought to rest in 75 ms. Determine the average impulsive force exerted by the wall on the car bumper.

13.129 A 1-oz steel-jacketed bullet is fired with a velocity of 2000 ft/s toward a steel plate and ricochets along path CD with a velocity of 1600 ft/s. Knowing that the bullet leaves a 2-in. scratch on the surface of the plate and assuming that it has an average speed of 1800 ft/s while in contact with the plate, determine the magnitude and direction of the impulsive force exerted by the plate on the bullet.

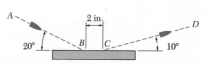

Fig. P13.129

13.130   After scaling a wall, a man lets himself drop 3 m to the ground. If his body comes to a complete stop 0.15 s after his feet first touch the ground, determine the vertical component of the average impulsive force exerted by the ground on his feet.

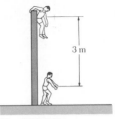

Fig. P13.130

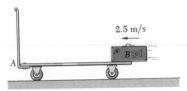

Fig. P13.131

13.131   An airline employee tosses a 12-kg suitcase with a horizontal velocity of 2.5 m/s onto a 30-kg baggage carrier. Knowing that the carrier is initially at rest and can roll freely, determine (*a*) the velocity of the carrier after the suitcase has slid to a relative stop on the carrier, (*b*) the ratio of the final kinetic energy of the carrier and suitcase to the initial kinetic energy of the suitcase.

13.132   A 45-Mg railroad car moving with a velocity of 3 km/h is to be coupled to a 25-Mg car which is at rest. Determine (*a*) the final velocity of the coupled cars, (*b*) the average impulsive force acting on each car if the coupling is completed in 0.3 s.

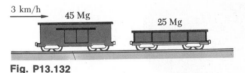

Fig. P13.132

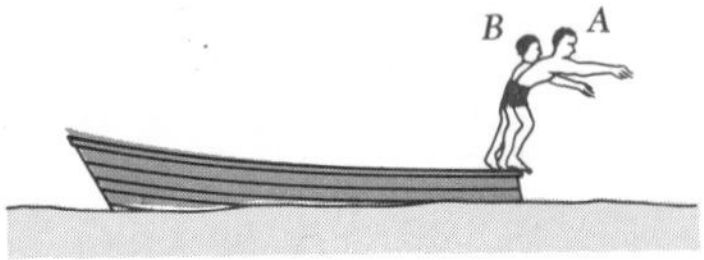

13.133   Two swimmers *A* and *B*, of mass 75 kg and 50 kg, respectively, dive off the end of a 200-kg boat. Each swimmer dives so that his relative horizontal velocity with respect to the boat is 3 m/s. If the boat is initially at rest, determine its final velocity, assuming that (*a*) the two swimmers dive simultaneously, (*b*) swimmer *A* dives first, (*c*) swimmer *B* dives first.

Fig. P13.133

Fig. P13.134

13.134   A 3-Mg barge is initially at rest and carries a 500-kg crate. The barge is equipped with a winch which exerts a constant 1200-N force on the crate for 6 s. The crate then slides along the deck until it comes to rest ($\mu_k = 0.20$). (*a*) Draw the *v–t* curve for the barge. (*b*) Determine the final position of the barge. (*c*) Determine the final position of the crate on the deck of the barge.

13.135 Car A was traveling due north through an intersection when it was hit broadside by car B which was traveling due east. While both drivers admitted having ignored the four-way stop signs at the intersection, each claimed that he was traveling at the 35-mi/h speed limit and that the other car was traveling much faster. Knowing that car A weighs 2000 lb, car B 3600 lb, and that inspection of the scene of the accident showed that as a result of the impact, the two cars got stuck together and skidded in a direction 40° north of east, determine (a) which of the two cars was actually traveling at 35 mi/h, (b) how fast the other car was moving.

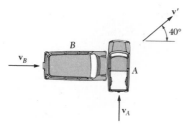

Fig. P13.135

13.136 An old 3000-lb gun fires an 18-lb shell with an initial velocity of 1500 ft/s at an angle of 30°. The gun rests on a horizontal surface and is free to move horizontally. Assuming that the barrel of the gun is rigidly attached to the frame (no recoil mechanism) and that the shell leaves the barrel 5 ms after firing, determine (a) the recoil velocity of the gun, (b) the resultant **R** of the vertical impulsive forces exerted by the ground on the gun.

Fig. P13.136

13.137 A small rivet connecting two pieces of sheet metal is being clinched by hammering. Determine the impulse exerted on the rivet and the energy absorbed by the rivet under each blow, knowing that the head of the hammer has a mass of 750 g and that it strikes the rivet with a velocity of 6 m/s. Assume that the hammer does not rebound and that the anvil is supported by springs and (a) has an infinite mass (rigid support), (b) has a mass of 4 kg.

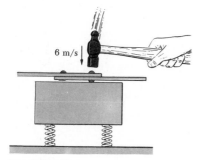

Fig. P13.137

13.138 A machine part is forged in a small drop forge. The hammer weighs 400 lb and is dropped from a height of 5 ft. Determine the initial impulse exerted on the machine part, assuming that the 1000-lb anvil (a) is resting directly on hard ground, (b) is supported by springs.

13.12. Impact. A collision between two bodies which occurs in a very small interval of time, and during which the two bodies exert on each other relatively large forces, is called an *impact*. The common normal to the surfaces in contact during the impact is called the *line of impact*. If the mass centers of the two colliding bodies are located on this line, the impact is a *central impact*. Otherwise, the impact is said to be *eccentric*. We shall limit our present study to that of the central impact of two particles and postpone until later the analysis of the eccentric impact of two rigid bodies (Sec. 17.12).

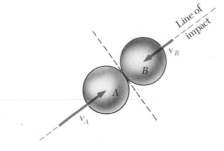

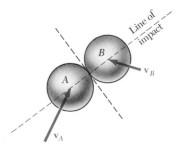

Fig. 13.20   (*a*) Direct central impact   (*b*) Oblique central impact

If the velocities of the two particles are directed along the line of impact, the impact is said to be a *direct impact* (Fig. 13.20*a*). If, on the other hand, either or both particles move along a line other than the line of impact, the impact is said to be an *oblique impact* (Fig. 13.20*b*).

13.13. Direct Central Impact. Consider two particles *A* and *B*, of mass m_A and m_B, which are moving in the same straight line and to the right with known velocities $\mathbf{v}_A$ and $\mathbf{v}_B$ (Fig. 13.21*a*). If $\mathbf{v}_A$ is larger than $\mathbf{v}_B$, particle *A* will eventually strike particle *B*. Under the impact, the two particles, will *deform* and, at the end of the period of deformation, they will have the same velocity $\mathbf{u}$ (Fig. 13.21*b*). A period of *restitution* will then take place, at the end of which, depending upon the magnitude of the impact forces and upon the materials involved, the two particles either will have regained their original shape or will stay permanently deformed. Our purpose here is to determine the velocities $\mathbf{v}'_A$ and $\mathbf{v}'_B$ of the particles at the end of the period of restitution (Fig. 13.21*c*).

Considering first the two particles as a single system, we note that there is no impulsive, external force. Thus, the total momentum of the two particles is conserved, and we write

$$m_A\mathbf{v}_A + m_B\mathbf{v}_B = m_A\mathbf{v}'_A + m_B\mathbf{v}'_B$$

Since all the velocities considered are directed along the same axis, we may replace the equation obtained by the following relation involving only scalar components:

$$m_A v_A + m_B v_B = m_A v'_A + m_B v'_B \tag{13.37}$$

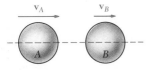

(*a*) Before impact

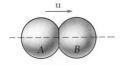

(*b*) At maximum deformation

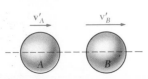

(*c*) After impact

Fig. 13.21

A positive value for any of the scalar quantities v_A, v_B, v'_A, or v'_B means that the corresponding vector is directed to the right; a negative value indicates that the corresponding vector is directed to the left.

To obtain the velocities $\mathbf{v}'_A$ and $\mathbf{v}'_B$, it is necessary to establish a second relation between the scalars v'_A and v'_B. For this purpose, we shall consider now the motion of particle A during the period of deformation and apply the principle of impulse and momentum. Since the only impulsive force acting on A during this period is the force $\mathbf{P}$ exerted by B (Fig. 13.22a), we write, using again scalar components,

$$m_A v_A - \int P\,dt = m_A u \tag{13.38}$$

where the integral extends over the period of deformation. Considering now the motion of A during the period of restitution, and denoting by $\mathbf{R}$ the force exerted by B on A during this period (Fig. 13.22b), we write

$$m_A u - \int R\,dt = m_A v'_A \tag{13.39}$$

where the integral extends over the period of restitution.

(a) Period of deformation

(b) Period of restitution

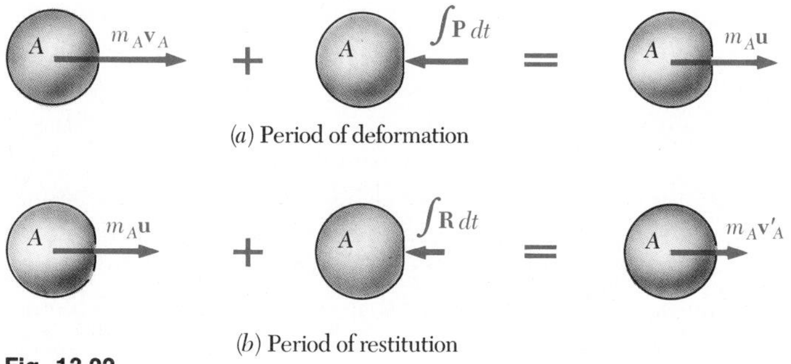

Fig. 13.22

In general, the force $\mathbf{R}$ exerted on A during the period of restitution differs from the force $\mathbf{P}$ exerted during the period of deformation, and the magnitude $\int R\,dt$ of its impulse is smaller than the magnitude $\int P\,dt$ of the impulse of $\mathbf{P}$. The ratio of the magnitudes of the impulses corresponding, respectively, to the period of restitution and to the period of deformation is called the *coefficient of restitution* and is denoted by e. We write

$$e = \frac{\int R\,dt}{\int P\,dt} \tag{13.40}$$

The value of the coefficient e is always between 0 and 1 and depends to a large extent on the two materials involved. However, it also varies considerably with the impact velocity and the shape and size of the two colliding bodies.

Solving Eqs. (13.38) and (13.39) for the two impulses and substituting into (13.40), we write

$$e = \frac{u - v'_A}{v_A - u} \tag{13.41}$$

A similar analysis of particle B leads to the relation

$$e = \frac{v_B' - u}{u - v_B} \tag{13.42}$$

Since the quotients in (13.41) and (13.42) are equal, they are also equal to the quotient obtained by adding, respectively, their numerators and their denominators. We have, therefore,

$$e = \frac{(u - v_A') + (v_B' - u)}{(v_A - u) + (u - v_B)} = \frac{v_B' - v_A'}{v_A - v_B}$$

and

$$\boxed{v_B' - v_A' = e(v_A - v_B)} \tag{13.43}$$

Since $v_B' - v_A'$ represents the relative velocity of the two particles after impact and $v_A - v_B$ their relative velocity before impact, formula (13.43) expresses that *the relative velocity of the two particles after impact may be obtained by multiplying their relative velocity before impact by the coefficient of restitution.* This property is used to determine experimentally the value of the coefficient of restitution of two given materials.

The velocities of the two particles after impact may now be obtained by solving Eqs. (13.37) and (13.43) simultaneously for v_A' and v_B'. It is recalled that the derivation of Eqs. (13.37) and (13.43) was based on the assumption that particle B is located to the right of A, and that both particles are initially moving to the right. If particle B is initially moving to the left, the scalar v_B should be considered negative. The same sign convention holds for the velocities after impact: a positive sign for v_A' will indicate that particle A moves to the right after impact and a negative sign, that it moves to the left.

Two particular cases of impact are of special interest:

1. $e = 0$, *Perfectly Plastic Impact.* When $e = 0$, Eq. (13.43) yields $v_B' = v_A'$. There is no period of restitution, and both particles stay together after impact. Substituting $v_B' = v_A' = v'$ into Eq. (13.37), which expresses that the total momentum of the particles is conserved, we write

$$m_A v_A + m_B v_B = (m_A + m_B)v' \tag{13.44}$$

This equation may be solved for the common velocity v' of the two particles after impact.

2. $e = 1$, *Perfectly Elastic Impact.* When $e = 1$, Eq. (13.43) reduces to

$$v_B' - v_A' = v_A - v_B \tag{13.45}$$

which expresses that the relative velocities before and after impact are equal. The impulses received by each particle during the period of deformation and during the period of restitution are equal.

The particles move away from each other after impact with the same velocity with which they approached each other before impact. The velocities v'_A and v'_B may be obtained by solving Eqs. (13.37) and (13.45) simultaneously.

It is worth noting that *in the case of a perfectly elastic impact, the total energy of the two particles,* as well as their total momentum, *is conserved.* Equations (13.37) and (13.45) may be written as follows:

$$m_A(v_A - v'_A) = m_B(v'_B - v_B) \qquad (13.37')$$
$$v_A + v'_A = v_B + v'_B \qquad (13.45')$$

Multiplying (13.37') and (13.45') member by member, we have

$$m_A(v_A - v'_A)(v_A + v'_A) = m_B(v'_B - v_B)(v'_B + v_B)$$
$$m_A v_A^2 - m_A(v'_A)^2 = m_B(v'_B)^2 - m_B v_B^2$$

Rearranging the terms in the equation obtained, and multiplying by $\frac{1}{2}$, we write

$$\tfrac{1}{2}m_A v_A^2 + \tfrac{1}{2}m_B v_B^2 = \tfrac{1}{2}m_A(v'_A)^2 + \tfrac{1}{2}m_B(v'_B)^2 \qquad (13.46)$$

which expresses that the kinetic energy of the particles is conserved. It should be noted, however, that *in the general case of impact,* i.e., when e is not equal to 1, *the total energy of the particles is not conserved.* This may be shown in any given case by comparing the kinetic energies before and after impact. The lost kinetic energy is in part transformed into heat and in part spent in generating elastic waves within the two colliding bodies.

13.14. Oblique Central Impact. Let us now consider the case when the velocities of the two colliding particles are *not* directed along the line of impact (Fig. 13.23). As indicated in Sec. 13.12, the impact is said to

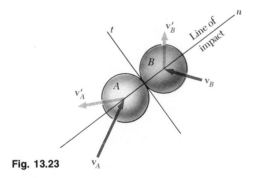

Fig. 13.23

be *oblique.* Since the velocities $\mathbf{v}'_A$ and $\mathbf{v}'_B$ of the particles after impact are unknown in direction as well as in magnitude, their determination will require the use of four independent equations.

We choose as coordinate axes the n axis along the line of impact, i.e., along the common normal to the surfaces in contact, and the t axis along their common tangent. Assuming that the particles are perfectly *smooth and frictionless,* we observe that the only impulses exerted on the particles

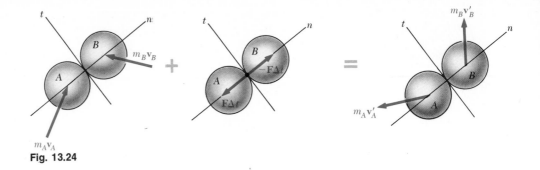

Fig. 13.24

during the impact are due to internal forces directed along the line of impact, i.e., along the n axis (Fig. 13.24). It follows that

1. The component along the t axis of the momentum of each particle, considered separately, is conserved; hence the t component of the velocity of each particle remains unchanged. We write

$$(v_A)_t = (v'_A)_t \qquad (v_B)_t = (v'_B)_t \qquad (13.37)$$

2. The component along the n axis of the total momentum of the two particles is conserved. We write

$$m_A(v_A)_n + m_B(v_B)_n = m_A(v'_A)_n + m_B(v'_B)_n \qquad (13.48)$$

3. The component along the n axis of the relative velocity of the two particles after impact is obtained by multiplying the n component of their relative velocity before impact by the coefficient of restitution. Indeed, a derivation similar to that given in Sec. 13.13 for direct central impact yields

$$(v'_B)_n - (v'_A)_n = e[(v_A)_n - (v_B)_n] \qquad (13.49)$$

We have thus obtained four independent equations which may be solved for the components of the velocities of A and B after impact. This method of solution is illustrated in Sample Prob. 13.15.

Our analysis of the oblique central impact of two particles has been based so far on the assumption that both particles moved freely before and after the impact. We shall now examine the case when one or both of the colliding particles is constrained in its motion. Consider, for instance, the collision between block A, which is constrained to move on a horizontal surface, and ball B, which is free to move in the plane of the figure (Fig. 13.25). Assuming no friction between the block and the ball, or between the block and the horizontal surface, we note that the impulses exerted on the system consist of the impulses of the internal forces $\mathbf{F}$ and $-\mathbf{F}$ directed along the line of impact, i.e., along the n axis, and of the impulse of the external force $\mathbf{F}_{ext}$ exerted by the horizontal surface on block A and directed along the vertical (Fig. 13.26).

The velocities of block A and ball B immediately after the impact are represented by three unknowns, namely, the magnitude of the velocity $\mathbf{v}'_A$ of block A, which is known to be horizontal, and the magnitude and direc-

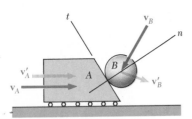

Fig. 13.25

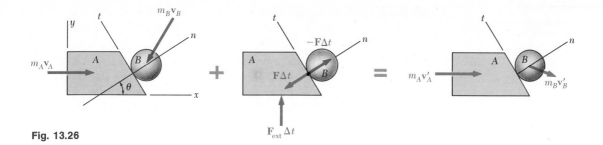

Fig. 13.26

tion of the velocity $\mathbf{v}'_B$ of ball B. We shall therefore write three equations by expressing that

1. The component along the t axis of the momentum of ball B is conserved; hence the t component of the velocity of ball B remains unchanged. We write

$$(v_B)_t = (v'_B)_t \qquad (13.50)$$

2. The component along the horizontal x axis of the total momentum of block A and ball B is conserved. We write

$$m_A v_A + m_B(v_B)_x = m_A v'_A + m_B(v'_B)_x \qquad (13.51)$$

3. The component along the n axis of the relative velocity of block A and ball B after impact is obtained by multiplying the n component of their relative velocity before impact by the coefficient of restitution. We write again

$$(v'_B)_n - (v'_A)_n = e[(v_A)_n - (v_B)_n] \qquad (13.49)$$

We should note, however, that in the case considered here, the validity of Eq. (13.49) cannot be established through a mere extension of the derivation given in Sec. 13.13 for the direct central impact of two particles moving in a straight line. Indeed, these particles were not subjected to any external impulse, while block A in the present analysis is subjected to the impulse exerted by the horizontal surface. To prove that Eq. (13.49) is still valid, we shall first apply the principle of impulse and momentum to block A over the period of deformation (Fig. 13.27). Considering only the horizontal components, we write

$$m_A v_A - (\smallint P \, dt) \cos \theta = m_A u \qquad (13.52)$$

where the integral extends over the period of deformation and where $\mathbf{u}$ represents the velocity of block A at the end of that period. Considering

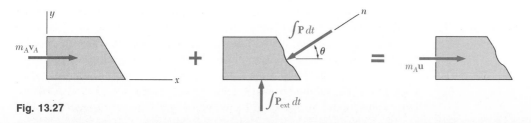

Fig. 13.27

now the period of restitution, we write in a similar way

$$m_A u - (\textstyle\int R\, dt) \cos \theta = m_A v'_A \qquad (13.53)$$

where the integral extends over the period of restitution.

Recalling from Sec. 13.13 the definition of the coefficient of restitution, we write

$$e = \frac{\int R\, dt}{\int P\, dt} \qquad (13.40)$$

Solving Eqs. (13.52) and (13.53) for the integrals $\int P\, dt$ and $\int R\, dt$, and substituting into Eq. (13.40), we have, after reductions,

$$e = \frac{u - v'_A}{v_A - u}$$

or, multiplying all velocities by $\cos \theta$ to obtain their projections on the line of impact,

$$e = \frac{u_n - (v'_A)_n}{(v_A)_n - u_n} \qquad (13.54)$$

We note that Eq. (13.54) is identical to Eq. (13.41) of Sec. 13.13, except for the subscripts n which are used here to indicate that we are considering velocity components along the line of impact. Since the motion of ball B is unconstrained, the proof of Eq. (13.49) may be completed in the same manner as the derivation of Eq. (13.43) of Sec. 13.13. Thus, we conclude that the relation (13.49) between the components along the line of impact of the relative velocities of two colliding particles remains valid when one of the particles is constrained in its motion. The validity of this relation may easily be extended to the case when both particles are constrained in their motion.

13.15. Problems Involving Energy and Momentum.
We have now at our disposal three different methods for the solution of kinetics problems: the direct application of Newton's second law, $\Sigma \mathbf{F} = m\mathbf{a}$, the method of work and energy, and the method of impulse and momentum. To derive maximum benefit from these three methods, we should be able to choose the method best suited for the solution of a given problem. We should also be prepared to use different methods for solving the various parts of a problem when such a procedure seems advisable.

We have already seen that the method of work and energy is in many cases more expeditious than the direct application of Newton's second law. As indicated in Sec. 13.4, however, the method of work and energy has limitations, and it must sometimes be supplemented by the use of $\Sigma \mathbf{F} = m\mathbf{a}$. This is the case, for example, when we wish to determine an acceleration or a normal force.

There is generally no great advantage in using the method of impulse and momentum for the solution of problems involving no impulsive forces. It will usually be found that the equation $\Sigma \mathbf{F} = m\mathbf{a}$ yields a solution just as fast and that the method of work and energy, if it applies, is more rapid and more convenient. However, the method of impulse and momentum is the only practicable method in problems of impact. A solution based on the direct application of $\Sigma F = ma$ would be unwieldy, and the method of work and energy cannot be used since impact (unless perfectly elastic) involves a loss of mechanical energy.

Many problems involve only conservative forces, except for a short impact phase during which impulsive forces act. The solution of such problems may be divided into several parts. While the part corresponding to the impact phase calls for the use of the method of impulse and momentum and of the relation between relative velocities, the other parts may usually be solved by the method of work and energy. The use of the equation $\Sigma F = ma$ will be necessary, however, if the problem involves the determination of a normal force.

Consider, for example, a pendulum A, of mass m_A and length l, which is released with no velocity from a position A_1 (Fig. 13.28a). The pendulum swings freely in a vertical plane and hits a second pendulum B, of mass m_B and same length l, which is initially at rest. After the impact (with coefficient of restitution e), pendulum B swings through an angle θ that we wish to determine.

The solution of the problem may be divided into three parts:

1. *Pendulum A Swings from A_1 to A_2.* The principle of conservation of energy may be used to determine the velocity $(\mathbf{v}_A)_2$ of the pendulum at A_2 (Fig. 13.28b).
2. *Pendulum A Hits Pendulum B.* Using the fact that the total momentum of the two pendulums is conserved and the relation between their relative velocities, we determine the velocities $(\mathbf{v}_A)_3$ and $(\mathbf{v}_B)_3$ of the two pendulums after impact (Fig. 13.28c).
3. *Pendulum B Swings from B_3 to B_4.* Applying the principle of conservation of energy to pendulum B, we determine the maximum elevation y_4 reached by that pendulum (Fig. 13.28d). The angle θ may then be determined by trigonometry.

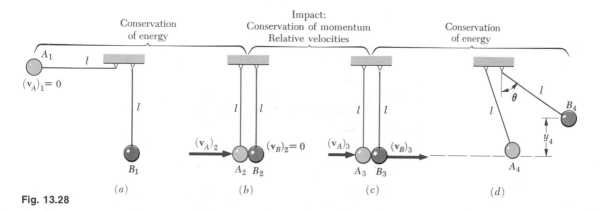

Fig. 13.28

We note that the method of solution just described should be supplemented by the use of $\Sigma F = ma$ if the tensions in the cords holding the pendulums are to be determined.

SAMPLE PROBLEM 13.13

A 20-Mg railroad car moving at a speed of 0.5 m/s to the right collides with a 35-Mg car which is at rest. If after the collision the 35-Mg car is observed to move to the right at a speed of 0.3 m/s, determine the coefficient of restitution between the two cars.

Solution. We express that the total momentum of the two cars is conserved.

$$m_A \mathbf{v}_A + m_B \mathbf{v}_B = m_A \mathbf{v}'_A + m_B \mathbf{v}'_B$$
$$(20 \text{ Mg})(+0.5 \text{ m/s}) + (35 \text{ Mg})(0) = (20 \text{ Mg})v'_A + (35 \text{ Mg})(+0.3 \text{ m/s})$$
$$v'_A = -0.025 \text{ m/s} \qquad \mathbf{v}'_A = 0.025 \text{ m/s} \leftarrow$$

The coefficient of restitution is obtained by writing

$$e = \frac{v'_B - v'_A}{v_A - v_B} = \frac{+0.3 - (-0.025)}{+0.5 - 0} = \frac{0.325}{0.5} \qquad e = 0.65 \blacktriangleleft$$

SAMPLE PROBLEM 13.14

A ball is thrown against a frictionless, vertical wall. Immediately before the ball strikes the wall, its velocity has a magnitude v and forms an angle of 30° with the horizontal. Knowing that $e = 0.90$, determine the magnitude and direction of the velocity of the ball as it rebounds from the wall.

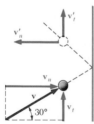

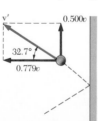

Solution. We resolve the initial velocity of the ball into components, respectively, perpendicular and parallel to the wall:

$$v_n = v \cos 30° = 0.866v \qquad v_t = v \sin 30° = 0.500v$$

Motion Parallel to the Wall. Since the wall is frictionless, the impulse it exerts on the ball is perpendicular to the wall. Thus, the component parallel to the wall of the momentum of the ball is conserved and we have

$$\mathbf{v}'_t = \mathbf{v}_t = 0.500v \uparrow$$

Motion Perpendicular to the Wall. Since the mass of the wall (and earth) is essentially infinite, expressing that the total momentum of the ball and wall is conserved would yield no useful information. Using the relation (13.49) between relative velocities, we write

$$0 - v'_n = e(v_n - 0)$$
$$v'_n = -0.90(0.866v) = -0.779v \qquad \mathbf{v}'_n = 0.779v \leftarrow$$

Resultant Motion. Adding vectorially the components $\mathbf{v}'_n$ and $\mathbf{v}'_t$,

$$\mathbf{v}' = 0.926v \; \diagdown \; 32.7° \blacktriangleleft$$

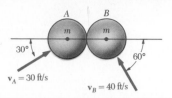

SAMPLE PROBLEM 13.15

The magnitude and direction of the velocities of two identical frictionless balls before they strike each other are as shown. Assuming $e = 0.90$, determine the magnitude and direction of the velocity of each ball after the impact.

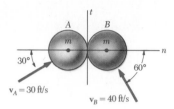

Solution. The impulsive forces acting between the balls during the impact are directed along a line joining the centers of the balls called the *line of impact*. Resolving the velocities into components directed, respectively, along the line of impact and along the common tangent to the surfaces in contact, we write

$$(v_A)_n = v_A \cos 30° = +26.0 \text{ ft/s}$$
$$(v_A)_t = v_A \sin 30° = +15.0 \text{ ft/s}$$
$$(v_B)_n = -v_B \cos 60° = -20.0 \text{ ft/s}$$
$$(v_B)_t = v_B \sin 60° = +34.6 \text{ ft/s}$$

Principle of Impulse and Momentum. In the adjoining sketches we show in turn the initial momenta, the impulses, and the final momenta.

Motion along the Common Tangent. Considering only the t components, we apply the principle of impulse and momentum to each ball *separately*. Since the impulsive forces are directed along the line of impact, the t component of the momentum, and hence the t component of the velocity of each ball, is unchanged. We have

$$(v'_A)_t = 15.0 \text{ ft/s} \uparrow \qquad (v'_B)_t = 34.6 \text{ ft/s} \uparrow$$

Motion along the Line of Impact. In the n direction, we consider the two balls as a single system and note that by Newton's third law, the internal impulses are, respectively, $\mathbf{F}\,\Delta t$ and $-\mathbf{F}\,\Delta t$ and cancel. We thus write that the total momentum of the balls is conserved:

$$m_A(v_A)_n + m_B(v_B)_n = m_A(v'_A)_n + m_B(v'_B)_n$$
$$m(26.0) + m(-20.0) = m(v'_A)_n + m(v'_B)_n$$
$$(v'_A)_n + (v'_B)_n = 6.0 \quad (1)$$

Using the relation (13.49) between relative velocities, we write

$$(v'_B)_n - (v'_A)_n = e[(v_A)_n - (v_B)_n]$$
$$(v'_B)_n - (v'_A)_n = (0.90)[26.0 - (-20.0)]$$
$$(v'_B)_n - (v'_A)_n = 41.4 \quad (2)$$

Solving Eqs. (1) and (2) simultaneously, we obtain

$$(v'_A)_n = -17.7 \qquad (v'_B)_n = +23.7$$
$$(v'_A)_n = 17.7 \text{ ft/s} \leftarrow \qquad (v'_B)_n = 23.7 \text{ ft/s} \rightarrow$$

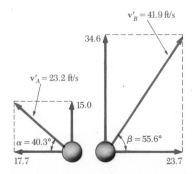

Resultant Motion. Adding vectorially the velocity components of each ball, we obtain

$$\mathbf{v}'_A = 23.2 \text{ ft/s} \seardown 40.3° \qquad \mathbf{v}'_B = 41.9 \text{ ft/s} \nearrow 55.6° \quad \blacktriangleleft$$

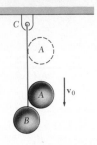

SAMPLE PROBLEM 13.16

Ball B is hanging from an inextensible cord BC. An identical ball A is released from rest when it is just touching the cord and acquires a velocity $\mathbf{v}_0$ before striking ball B. Assuming perfectly elastic impact ($e = 1$) and no friction, determine the velocity of each ball immediately after impact.

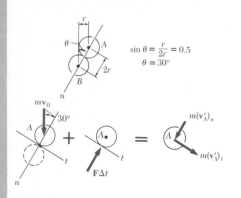

$\sin \theta = \dfrac{r}{2r} = 0.5$
$\theta = 30°$

Solution. Since ball B is constrained to move in a circle of center C, its velocity $\mathbf{v}'_B$ after impact must be horizontal. Thus the problem involves three unknowns: the magnitude v'_B of the velocity of B, and the magnitude and direction of the velocity $\mathbf{v}'_A$ of A after impact.

Impulse-Momentum Principle: Ball A

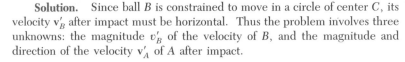

$$mv_A + F \Delta t = mv'_A$$
$+\searrow t$ components: $mv_0 \sin 30° + 0 = m(v'_A)_t$
$$(v'_A)_t = 0.5v_0 \qquad (1)$$

We note that the equation used expresses conservation of the momentum of ball A along the common tangent to balls A and B.

Impulse-Momentum Principle: Balls A and B

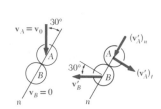

$$m\mathbf{v}_A + \mathbf{T} \Delta t = m\mathbf{v}'_A + m\mathbf{v}'_B$$
$\xrightarrow{+} x$ components: $0 = m(v'_A)_t \cos 30° - m(v'_A)_n \sin 30° - mv'_B$

We note that the equation obtained expresses conservation of the total momentum in the x direction. Substituting for $(v'_A)_t$ from Eq. (1) and rearranging terms, we write

$$0.5(v'_A)_n + v'_B = 0.433v_0 \qquad (2)$$

Relative Velocities along the Line of Impact. Since $e = 1$, Eq. (13.49) yields

$$(v'_B)_n - (v'_A)_n = (v_A)_n - (v_B)_n$$
$$v'_B \sin 30° - (v'_A)_n = v_o \cos 30° - 0$$
$$0.5v'_B - (v'_A)_n = 0.866v_0 \qquad (3)$$

Solving Eqs. (2) and (3) simultaneously, we obtain

$$(v'_A)_n = -0.520v_0 \qquad v'_B = 0.693v_0$$
$$\mathbf{v}'_B = 0.693v_0 \leftarrow \quad \blacktriangleleft$$

Recalling Eq. (1) we draw the adjoining sketch and obtain by trigonometry

$$v'_A = 0.721v_0 \qquad \beta = 46.1° \qquad \alpha = 46.1° - 30° = 16.1°$$
$$\mathbf{v}'_A = 0.721v_0 \measuredangle 16.1° \quad \blacktriangleleft$$

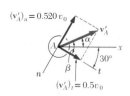

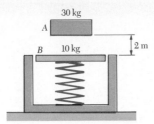

30 kg
A

B   10 kg   2 m

SAMPLE PROBLEM 13.17

A 30-kg block is dropped from a height of 2 m onto the 10-kg pan of a spring scale. Assuming the impact to be perfectly plastic, determine the maximum deflection of the pan. The constant of the spring is $k = 20$ kN/m.

Solution. The impact between the block and the pan *must* be treated separately; therefore we divide the solution into three parts.

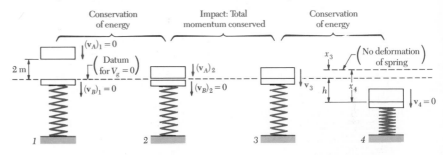

Conservation of energy · Impact: Total momentum conserved · Conservation of energy

$(v_A)_1 = 0$
Datum for $V_g = 0$
$(v_B)_1 = 0$
$(v_A)_2$
$(v_B)_2 = 0$
v_3
x_3 (No deformation of spring)
h · x_4
$v_4 = 0$

1   2   3   4

Conservation of Energy. Block: $W_A = (30 \text{ kg})(9.81 \text{ m/s}^2) = 294$ N

$$T_1 = \tfrac{1}{2}m_A(v_A)_1^2 = 0 \qquad V_1 = W_A y = (294 \text{ N})(2 \text{ m}) = 588 \text{ J}$$
$$T_2 = \tfrac{1}{2}m_A(v_A)_2^2 = \tfrac{1}{2}(30 \text{ kg})(v_A)_2^2 \qquad V_2 = 0$$
$$T_1 + V_1 = T_2 + V_2: \qquad 0 + 588 \text{ J} = \tfrac{1}{2}(30 \text{ kg})(v_A)_2^2 + 0$$
$$(v_A)_2 = +6.26 \text{ m/s} \qquad (\mathbf{v}_A)_2 = 6.26 \text{ m/s} \downarrow$$

Impact: Conservation of Momentum. Since the impact is perfectly plastic, $e = 0$; the block and pan move together after the impact.

$$m_A(v_A)_2 + m_B(v_B)_2 = (m_A + m_B)v_3$$
$$(30 \text{ kg})(6.26 \text{ m/s}) + 0 = (30 \text{ kg} + 10 \text{ kg})v_3$$
$$v_3 = +4.70 \text{ m/s} \qquad \mathbf{v}_3 = 4.70 \text{ m/s} \downarrow$$

Conservation of Energy. Initially the spring supports the weight W_B of the pan; thus the initial deflection of the spring is

$$F = ky$$

$$x_3 = \frac{W_B}{k} = \frac{(10 \text{ kg})(9.81 \text{ m/s}^2)}{20 \times 10^3 \text{ N/m}} = \frac{98.1 \text{ N}}{20 \times 10^3 \text{ N/m}} = 4.91 \times 10^{-3} \text{ m}$$

Denoting by x_4 the total maximum deflection of the spring, we write

$$T_3 = \tfrac{1}{2}(m_A + m_B)v_3^2 = \tfrac{1}{2}(30 \text{ kg} + 10 \text{ kg})(4.70 \text{ m/s})^2 = 442 \text{ J}$$
$$V_3 = V_g + V_e = 0 + \tfrac{1}{2}kx_3^2 = \tfrac{1}{2}(20 \times 10^3)(4.91 \times 10^{-3})^2 = 0.241 \text{ J}$$
$$T_4 = 0$$
$$V_4 = V_g + V_e = (W_A + W_B)(-h) + \tfrac{1}{2}kx_4^2 = -(392)h + \tfrac{1}{2}(20 \times 10^3)x_4^2$$

Noting that the displacement of the pan is $h = x_4 - x_3$, we write

$$T_3 + V_3 = T_4 + V_4:$$
$$442 + 0.241 = 0 - 392(x_4 - 4.91 \times 10^{-3}) + \tfrac{1}{2}(20 \times 10^3)x_4^2$$
$$x_4 = 0.230 \text{ m} \qquad h = x_4 - x_3 = 0.230 \text{ m} - 4.91 \times 10^{-3} \text{ m}$$
$$h = 0.225 \text{ m} \qquad\qquad h = 225 \text{ mm} \blacktriangleleft$$

Problems

13.139 The coefficient of restitution between the two collars is known to be 0.80. Determine (a) their velocities after impact, (b) the energy loss during impact.

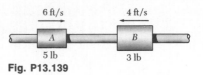

Fig. P13.139

13.140 Solve Prob. 13.139, assuming that the velocity of collar B is 2 ft/s to the right.

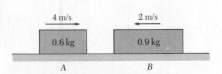

Fig. P13.141 and P13.142

13.141 Two steel blocks slide without friction on a horizontal surface; immediately before impact their velocities are as shown. Knowing that $e = 0.75$, determine (a) their velocities after impact, (b) the energy loss during impact.

13.142 The velocities of two steel blocks before impact are as shown. If after impact the velocity of block B is observed to be 2.5 m/s to the right, determine the coefficient of restitution between the two blocks.

13.143 Two identical cars B and C are at rest on a loading dock with their brakes released. Car A of the same model, which has been pushed by dock workers, hits car B with a velocity of 1.5 m/s, causing a series of collisions among the three cars. Assuming a coefficient of restitution $e = 0.75$ between the bumpers, determine the velocity of each car after *all* collisions have taken place.

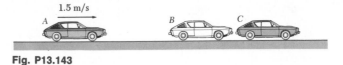

Fig. P13.143

13.144 Solve Prob. 13.143, assuming that the coefficients of restitution between the cars are $e = 0.50$ between A and B and $e = 1.00$ between B and C.

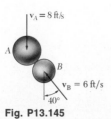

Fig. P13.145

13.145 A 2.5-lb ball A is falling vertically with a velocity of magnitude $v_A = 8$ ft/s when it is hit as shown by a 1.5-lb ball B which has a velocity of magnitude $v_B = 6$ ft/s. Knowing that the coefficient of restitution between the two balls is $e = 0.75$ and assuming no friction, determine the velocity of each ball immediately after impact.

13.146   Two identical pucks A and B, of 3-in. diameter, may move freely on a hockey rink.   Puck B is at rest and puck A has an initial velocity $\mathbf{v} = v_0\mathbf{i}$. (*a*) Knowing that $b = 1.5$ in. and $e = 0.80$, determine the velocity of each puck after impact. (*b*) Show that if $e = 1$, the final velocities of the pucks form a right angle for all values of b.

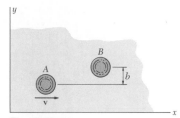

Fig. P13.146

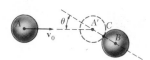

Fig. P13.147

13.147   Two identical billiard balls may move freely on a horizontal table. Ball A has a velocity $\mathbf{v}_0$ as shown and hits ball B, which is at rest, at a point C defined by $\theta = 45°$.   Knowing that the coefficient of restitution between the two balls is $e = 0.80$ and assuming no friction, determine the velocity of each ball after impact.

13.148   A billiard player wishes to have ball A hit ball B obliquely and then ball C squarely. Assuming perfectly elastic impact ($e = 1$) and denoting by r the radius of the balls and by d the distance between the centers of B and C, determine (*a*) the angle θ defining point D where ball B should be hit, (*b*) the ranges of values of angles ABC and ACB for which this play is possible.

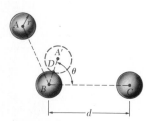

Fig. P13.148

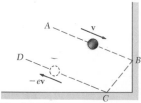

Fig. P13.149

13.149   A ball is thrown into a 90° corner with an initial velocity $\mathbf{v}$. Denoting the coefficient of restitution by e, show that the final velocity is of magnitude ev and that the initial and final paths AB and CD are parallel.

13.150   One of the requirements for tennis balls to be used in official competition is that, when dropped onto a rigid surface from a height of 100 in., the height of the first bounce of the ball must be in the range 53 in. $\leq h \leq$ 58 in. Determine the range of the coefficients of restitution of the tennis balls satisfying this requirement.

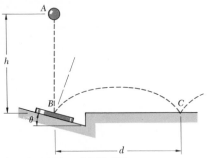

Fig. P13.151 and P13.152

13.151 A small ball A is dropped from a height h onto a rigid, frictionless plate at B and bounces to point C at the same elevation as B. Knowing that $\theta = 20°$ and that the coefficient of restitution between the ball and the plate is $e = 0.40$, determine the distance d.

13.152 A small ball A is dropped from a height h onto a rigid, frictionless plate at B and bounces to point C at the same elevation as B. Determine the value of θ for which the distance d is maximum and the corresponding value of d, assuming that the coefficient of restitution between the ball and the plate is (a) $e = 1$, (b) $e = 0.50$.

13.153 A ball moving with a horizontal velocity $\mathbf{v}_0$ drops from A through the vertical distance $h_0 = 40$ in. to a frictionless floor. Knowing that the ball hits the floor at a distance $d_0 = 6$ in. from B and that the coefficient of restitution between the ball and the floor is $e = 0.85$, determine (a) the height h_1 and length d_1 of the first bounce, (b) the height h_2 and length d_2 of the second bounce.

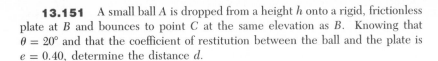

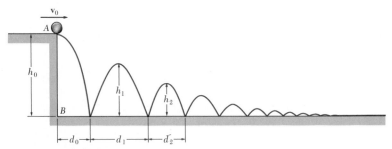

Fig. P13.153, P13.154, and P13.155

13.154 A ball moving with a horizontal velocity $\mathbf{v}_0$ drops from A through the vertical distance $h_0 = 25$ in. to a frictionless floor. Knowing that the ball hits the floor at a distance $d_0 = 5$ in. from B and that the height of its first bounce is $h_1 = 16$ in., determine (a) the coefficient of restitution between the ball and the floor, (b) the length d_1 of the first bounce.

13.155 A ball moving with a horizontal velocity of magnitude $v_0 = 0.4$ m/s drops from A to a frictionless floor. Knowing that the ball hits the floor at a distance $d_0 = 80$ mm from B and that the length of its first bounce is $d_1 = 140$ mm, determine (a) the coefficient of restitution between the ball and the floor, (b) the height h_1 of the first bounce.

13.156 In Prob. 13.155, determine how long the ball will keep bouncing after first hitting the floor.

13.157 A 4-lb sphere moving to the right with a velocity of 5 ft/s strikes at D the frictionless surface of a 10-lb half cylinder which is at rest. The half cylinder rests on rollers and may move freely in the horizontal direction. Knowing that $e = 0.75$ and $\theta = 60°$, determine the velocities of the half cylinder and of the sphere immediately after impact.

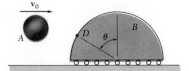

Fig. P13.157

13.158 Solve Prob. 13.157, assuming that $\theta = 30°$.

13.159 A sphere A of mass $m_A = 2$ kg is released from rest in the position shown and strikes the frictionless, inclined surface of a wedge B of mass $m_B = 6$ kg with a velocity of magnitude $v_0 = 3$ m/s. The wedge, which is supported by rollers and may move freely in the horizontal direction, is initially at rest. Knowing that $\theta = 30°$ and that the coefficient of restitution between the sphere and the wedge is $e = 0.80$, determine the velocities of the sphere and of the wedge immediately after impact.

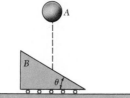

Fig. P13.159

13.160 Solve Prob. 13.159, assuming that $\theta = 60°$.

13.161 For the sphere and wedge of Prob. 13.159, determine the angle θ of the wedge for which the sphere bounces off horizontally to the right, assuming that the wedge (a) is free to move in the horizontal direction, (b) is rigidly attached to the ground.

13.162 A 1.5-lb sphere A is released from rest when $\theta_A = 45°$ and strikes a 3-lb sphere B which is at rest. Knowing that the coefficient of restitution is $e = 0.75$, determine the values of θ_A and θ_B corresponding to the highest positions to which the spheres will rise after impact.

13.163 A 1.5-lb sphere A is released from rest when $\theta_A = 70°$ and strikes a 3-lb sphere B which is at rest. Knowing that the velocity of sphere A is zero after impact, determine (a) the coefficient of restitution e, (b) the value of θ_B corresponding to the highest position to which sphere B will rise.

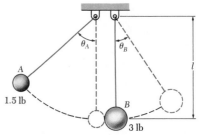

Fig. P13.162 and P13.163

13.164 A 6-kg block A is released from rest when $l = 900$ mm, moves down the 30° incline, and strikes a 3-kg block B which is at rest. Small rollers are attached to the incline and friction between block A and the incline may be neglected. After the impact each block slides to the left and comes to rest. Knowing that $e = 0.60$ and that $\mu_k = 0.40$ between each block and the horizontal surface, determine the distance through which each block will slide.

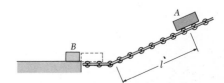

Fig. P13.164 and P13.165

13.165 A 5-kg block A is released from rest when $l = 800$ mm, moves down the 30° incline, and strikes a 3-kg block B which is at rest. Small rollers are attached to the incline and friction between block A and the incline may be neglected. Knowing that after the impact block A slides 240 mm to the left and block B slides 1.2 m to the left, determine (a) the coefficient of restitution between the two blocks, (b) the coefficient of friction between the blocks and the horizontal surface.

13.166 A 20-g bullet is fired with a velocity of magnitude $v_0 = 600$ m/s into a 4.5-kg block of wood. Knowing that the coefficient of kinetic friction between the block and the floor is 0.40, determine (a) how far the block will move, (b) the percentage of the initial energy lost in friction between the block and the floor.

Fig. P13.166

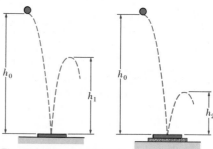

Fig. P13.167

13.167 Cylinder A is dropped 2.4 m onto cylinder B, which is resting on a spring of constant $k = 3$ kN/m. Assuming a perfectly plastic impact, determine (a) the maximum deflection of cylinder B, (b) the energy loss during the impact.

13.168 It is desired to drive the 300-lb pile into the ground until the resistance to its penetration is 20,000 lb. Each blow of the 1200-lb hammer is the result of a 3-ft free fall onto the top of the pile. Determine how far the pile will be driven into the ground by a single blow as the 20,000-lb resistance is being achieved. Assume that the impact is perfectly plastic.

13.169 The 1200-lb hammer of a drop-hammer pile driver falls from a height of 3 ft onto the top of a 300-lb pile. The pile is driven 4 in. into the ground. Assuming perfectly plastic impact, determine the average resistance of the ground to penetration.

13.170 A 35-g ball is dropped from a height h_0 onto a 140-g plate. The ball is observed to rebound to a height $h_1 = 580$ mm when the plate rests directly on hard ground and to a height $h_2 = 160$ mm when a foam-rubber mat is placed between the plate and the ground. Determine (a) the coefficient of restitution between the ball and the plate, (b) the height h_0 from which the ball was dropped.

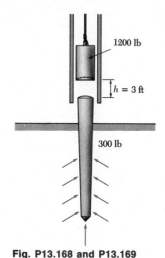

1200 lb

$h = 3$ ft

300 lb

Fig. P13.168 and P13.169

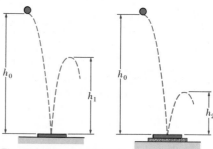

Fig. P13.170 and P13.171

13.171 A 35-g ball is dropped from a height $h_0 = 600$ mm onto a small plate. The ball is observed to rebound to a height $h_1 = 360$ mm when the plate rests directly on hard ground and to a height $h_2 = 200$ mm when a foam-rubber mat is placed between the plate and the ground. Determine (a) the coefficient of restitution between the ball and the plate, (b) the mass of the plate.

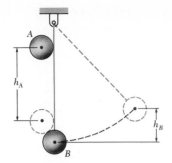

Fig. P13.172

13.172 Ball B is hanging from an inextensible cord. An identical ball A is released from rest when it is just touching the cord and drops through the vertical distance $h_A = 200$ mm before striking ball B. Assuming perfectly elastic impact ($e = 1$) and no friction, determine the resulting maximum vertical displacement h_B of ball B.

13.173 A 1.5-lb sphere A is moving to the left with a velocity of 50 ft/s when it strikes the inclined surface of a 4-lb block B which is at rest. The block is supported by rollers and is attached to a spring of constant $k = 15$ lb/in. Knowing that the coefficient of restitution between the sphere and the block is $e = 0.75$ and neglecting friction, determine the maximum deflection of the spring.

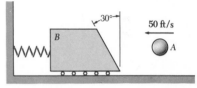

Fig. P13.173

13.174 A bumper is designed to protect a 2500-lb automobile from damage when it hits a rigid wall at speeds up to 5 mi/h. Assuming perfectly plastic impact, determine (*a*) the energy absorbed by the bumper during the impact, (*b*) the speed at which the automobile can hit another 2500-lb automobile without incurring any damage if the other automobile is similarly protected and is at rest with its brakes released.

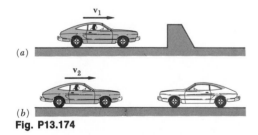

Fig. P13.174

*13.175 A ball of mass m_A moving with a velocity $\mathbf{v}_A$ strikes squarely a second ball of mass m_B which is at rest. Denoting by e the coefficient of restitution between the two balls, show that the percentage of energy lost during the impact is

$$\frac{100(1 - e^2)}{1 + (m_A/m_B)}$$

Fig. P13.175

Review and Summary

This chapter was devoted to the method of work and energy and to the method of impulse and momentum. In the first half of the chapter we studied the method of work and energy and its application to the analysis of the motion of particles.

Work of a force

We first considered a force $\mathbf{F}$ acting on a particle A and defined the *work of $\mathbf{F}$ corresponding to the small displacement $d\mathbf{r}$* [Sec. 13.2] as the quantity

$$dU = \mathbf{F} \cdot d\mathbf{r} \qquad (13.1)$$

or, recalling the definition of the scalar product of two vectors,

$$dU = F \, ds \cos \alpha \qquad (13.1')$$

where α is the angle between $\mathbf{F}$ and $d\mathbf{r}$ (Fig. 13.29). The work of $\mathbf{F}$ during a finite displacement from A_1 to A_2 was denoted by $U_{1\to2}$ and was obtained by integrating Eq. (13.1) along the path described by the particle:

$$U_{1\to2} = \int_{A_1}^{A_2} \mathbf{F} \cdot d\mathbf{r} \qquad (13.2)$$

For a force defined by its rectangular components, we wrote

$$U_{1\to2} = \int_{A_1}^{A_2} (F_x \, dx + F_y \, dy + F_z \, dz) \qquad (13.2'')$$

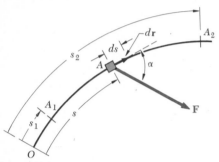

Fig. P13.29

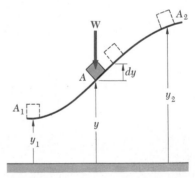

Fig. 13.30

Work of a weight

The work of the weight $\mathbf{W}$ of a body as its center of gravity moves from the elevation y_1 to y_2 (Fig. 13.30) was obtained by substituting $F_x = F_z = 0$ and $F_y = -W$ into Eq. (13.2'') and integrating. We found

$$U_{1\to2} = -\int_{y_1}^{y_2} W \, dy = Wy_1 - Wy_2 \qquad (13.4)$$

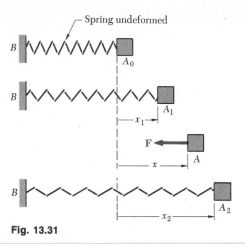

Spring undeformed

Fig. 13.31

The work of a force $\mathbf{F}$ exerted by a spring on a body A during a finite displacement of the body (Fig. 13.31) from $A_1(x = x_1)$ to $A_2(x = x_2)$ was obtained by writing

$$dU = -F\,dx = -kx\,dx$$

$$U_{1\to2} = -\int_{x_1}^{x_2} kx\,dx = \tfrac{1}{2}kx_1^2 - \tfrac{1}{2}kx_2^2 \qquad (13.6)$$

The work of $\mathbf{F}$ is therefore positive *when the spring is returning to its undeformed position.*

Work of the force exerted by a spring

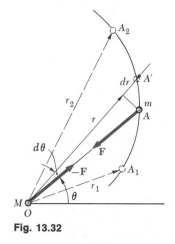

Fig. 13.32

The *work of the gravitational force* $\mathbf{F}$ exerted by a particle of mass M located at O on a particle of mass m as the latter moves from A_1 to A_2 (Fig. 13.32) was obtained by recalling from Sec. 12.10 the expression for the magnitude of $\mathbf{F}$ and writing

$$U_{1\to2} = -\int_{r_1}^{r_2} \frac{GMm}{r^2}\,dr = \frac{GMm}{r_2} - \frac{GMm}{r_1} \qquad (13.7)$$

Work of the gravitational force

The *kinetic energy of a particle* of mass m moving with a velocity $\mathbf{v}$ [Sec. 13.3] was defined as the scalar quantity

$$T = \tfrac{1}{2}mv^2 \tag{13.9}$$

Principle of work and energy

From Newton's second law we derived the *principle of work and energy*, which states that *the kinetic energy of a particle at A_2 may be obtained by adding to its kinetic energy at A_1 the work done during the displacement from A_1 to A_2 by the force $\mathbf{F}$ exerted on the particle:*

$$T_1 + U_{1 \rightarrow 2} = T_2 \tag{13.11}$$

Method of work and energy

The method of work and energy simplifies the solution of many problems dealing with forces, displacements, and velocities, since it does not require the determination of accelerations [Sec. 13.4]. We also note that it involves only scalar quantities and that forces which do no work need not be considered [Sample Probs. 13.1 and 13.3]. However, this method should be supplemented by the direct application of Newton's second law to determine a force normal to the path of the particle [Sample Prob. 13.4].

Power and mechanical efficiency

The power developed by a machine and its mechanical efficiency were discussed in Sec. 13.5. Power was defined as the time rate at which work is done

$$\text{Power} = \frac{dU}{dt} = \mathbf{F} \cdot \mathbf{v} \tag{13.12, 13.13}$$

where $\mathbf{F}$ is the force exerted on the particle and $\mathbf{v}$ the velocity of the particle [Sample Prob. 13.5]. The *mechanical efficiency*, denoted by η, was expressed as

$$\eta = \frac{\text{power output}}{\text{power input}} \tag{13.15}$$

Conservative force. Potential energy

When the work of a force $\mathbf{F}$ is independent of the path followed [Secs. 13.6 and 13.7], the force $\mathbf{F}$ is said to be a *conservative force*, and its work is equal to *minus the change in the potential energy V associated with $\mathbf{F}$:*

$$U_{1 \rightarrow 2} = V_1 - V_2 \tag{13.19'}$$

The following expressions were obtained for the potential energy associated with each of the forces considered earlier:

Force of gravity (weight): $\qquad V_g = Wy \tag{13.16}$

Gravitational force: $\qquad V_g = -\dfrac{GMm}{r} \tag{13.17}$

Elastic force exerted by a spring: $\quad V_e = \tfrac{1}{2}kx^2 \tag{13.18}$

Substituting for $U_{1\rightarrow2}$ from Eq. (13.19') into Eq. (13.11) and rearranging the terms [Sec. 13.8], we obtained

$$T_1 + V_1 = T_2 + V_2 \qquad (13.24)$$

This is the *principle of conservation of energy*, which states that when a particle moves under the action of conservative forces, *the sum of its kinetic and potential energies remains constant*. The application of this principle facilitates the solution of problems involving only conservative forces [Sample Probs. 13.6 and 13.7].

Recalling from Sec. 12.9 that, when a particle moves under a central force **F**, its angular momentum about the center of force O remains constant, we observed [Sec. 13.9] that, if the central force **F** is also conservative, the principles of conservation of angular momentum and of conservation of energy may be used jointly to analyze the motion of the particle [Sample Prob. 13.8]. Since the gravitational force exerted by the earth on a space vehicle is both central and conservative, this approach was used to study the motion of such vehicles [Sample Prob. 13.9] and was found particularly effective in the case of an *oblique launching*. Considering the initial position P_0 and an arbitrary position P of the vehicle (Fig. 13.33), we wrote

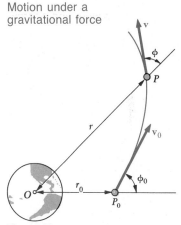

$(H_O)_0 = H_O:$ 　　　 $r_0 m v_0 \sin \phi_0 = r m v \sin \phi \qquad (13.25)$

$T_0 + V_0 = T + V:$ 　 $\frac{1}{2}mv_0^2 - \frac{GMm}{r_0} = \frac{1}{2}mv^2 - \frac{GMm}{r} \qquad (13.26)$

where m was the mass of the vehicle and M the mass of the earth.

Fig. 13.33

The second half of the chapter was devoted to the method of impulse and momentum and to its application to the solution of various types of problems involving the motion of particles.

The *linear momentum of a particle* was defined [Sec. 13.10] as the product $m\mathbf{v}$ of the mass m of the particle and its velocity **v**. From Newton's second law, $\mathbf{F} = m\mathbf{a}$, we derived the relation

$$m\mathbf{v}_1 + \int_{t_1}^{t_2} \mathbf{F}\,dt = m\mathbf{v}_2 \qquad (13.28)$$

where $m\mathbf{v}_1$ and $m\mathbf{v}_2$ represent the momentum of the particle at a time t_1 and a time t_2, respectively, and where the integral defines the *linear impulse of the force* **F** during the corresponding time interval. We wrote therefore

$$m\mathbf{v}_1 + \mathbf{Imp}_{1\rightarrow2} = m\mathbf{v}_2 \qquad (13.30)$$

which expresses the principle of impulse and momentum for a particle. When the particle considered is subjected to several forces, the sum of the impulses of these forces should be used; we had

$$mv_1 + \Sigma\, \mathbf{Imp}_{1\rightarrow2} = mv_2 \qquad (13.32)$$

Since Eqs. (13.30) and (13.32) involve *vector quantities,* it is necessary to consider separately their x and y components when applying them to the solution of a given problem [Sample Probs. 13.10 and 13.11].

Impulsive motion

The method of impulse and momentum is particularly effective in the study of the *impulsive motion* of a particle, when very large forces, called *impulsive forces,* are applied for a very short interval of time Δt, since this method involves the impulses $\mathbf{F}\,\Delta t$ of the forces, rather than the forces themselves [Sec. 13.11]. Neglecting the impulse of any nonimpulsive force, we wrote

$$m\mathbf{v}_1 + \Sigma \mathbf{F}\,\Delta t = m\mathbf{v}_2 \qquad (13.35)$$

In the case of the impulsive motion of several particles, we had

$$\Sigma m\mathbf{v}_1 + \Sigma \mathbf{F}\,\Delta t = \Sigma m\mathbf{v}_2 \qquad (13.36)$$

where the second term involves only impulsive, external forces [Sample Prob. 13.12].

In the particular case *when the sum of the impulses of the external forces is zero,* Eq. (13.36) reduces to $\Sigma m\mathbf{v}_1 = \Sigma m\mathbf{v}_2$, i.e., *the total momentum of the particles is conserved.*

Direct central impact

In Secs. 13.12 through 13.14, we considered the *central impact* of two colliding bodies. In the case of a *direct central impact* [Sec. 13.13], the two colliding bodies A and B were moving along the *line of impact*

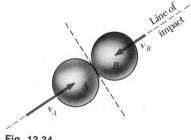

Fig. 13.34

with velocities $\mathbf{v}_A$ and $\mathbf{v}_B$, respectively (Fig. 13.34). Two equations could be used to determine their velocities $\mathbf{v}'_A$ and $\mathbf{v}'_B$ after the impact. The first expressed conservation of the total momentum of the two bodies,

$$m_A v_A + m_B v_B = m_A v'_A + m_B v'_B \qquad (13.37)$$

where a positive sign indicates that the corresponding velocity is directed to the right, while the second related the *relative velocities* of the two bodies before and after the impact,

$$v'_B - v'_A = e(v_A - v_B) \qquad (13.43)$$

The constant e is known as the *coefficient of restitution;* its value lies between 0 and 1 and depends in a large measure on the materials involved. When $e = 0$, the impact is said to be *perfectly plastic;* when $e = 1$, it is said to be *perfectly elastic* [Sample Prob. 13.13].

In the case of an *oblique central impact* [Sec. 13.14], the velocities of the two colliding bodies before and after the impact were resolved into n components along the line of impact and t components along the common tangent to the surfaces in contact (Fig. 13.35). We observed that the t

Oblique central impact

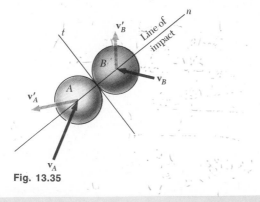

Fig. 13.35

component of the velocity of each body remained unchanged, while the n components satisfied equations similar to Eqs. (13.37) and (13.43) [Sample Probs. 13.14 and 13.15]. While this method was developed for bodies moving freely before and after the impact, it was shown that it could be extended to the case when one or both of the colliding bodies is constrained in its motion [Sample Prob. 13.16].

In Sec. 13.15, we discussed the relative advantages of the three fundamental methods presented in this chapter and the preceding one, namely, Newton's second law, work and energy, and impulse and momentum. We noted that the method of work and energy and the method of impulse and momentum may be combined to solve problems involving a short impact phase during which impulsive forces must be taken into consideration [Sample Prob. 13.17].

Using the three fundamental methods of kinetic analysis

Review Problems

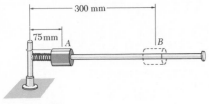

Fig. P13.176

13.176 A 200-g collar may slide on a horizontal rod which is free to rotate about a vertical shaft. The collar is initially held at A by a cord attached to the shaft and compresses a spring of constant 40 N/m, which is undeformed when the collar is located 225 mm from the shaft. As the rod rotates at the rate $\dot{\theta}_0 = 12$ rad/s, the cord is cut and the collar moves along the rod. Neglecting friction and the mass of the rod, and assuming that the spring is attached to the collar and to the shaft, determine the radial and transverse components of the velocity of the collar as it passes through B.

13.177 A weight W attached to an inextensible cord rotates in a vertical circle of radius r. Show that the difference between the maximum value T_{max} of the tension in the cord and its minimum value T_{min} is independent of the speed at which the weight rotates, and determine $T_{max} - T_{min}$.

13.178 An elevator travels upward at a constant speed of 2 m/s. A child riding the elevator throws a 0.6-kg stone upward with a speed of 5 m/s *relative* to the elevator. Determine (a) the work done by the child in throwing the stone, (b) the difference in the values of the kinetic energy of the stone before and after it was thrown. (c) Why are the values obtained in parts a and b not the same?

13.179 A 32,000-lb airplane lands on an aircraft carrier and is caught by an arresting cable. The cable is inextensible and is paid out at A and B from mechanisms located below deck and consisting of pistons moving in long oil-filled cylinders. Knowing that the piston-cylinder system maintains a constant tension of 85 kips in the cable during the entire landing, determine the landing speed of the airplane if it travels a distance $d = 95$ ft after being caught by the cable.

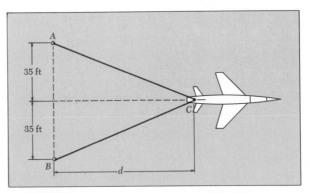

Fig. P13.179

13.180 For the airplane and aircraft carrier of Prob. 13.179, determine the value of the tension in the arresting cable, knowing that the airplane lands with a speed of 120 mi/h and travels a distance $d = 110$ ft after being caught by the cable.

13.181 A ball is dropped from a height h above the landing and bounces down a flight of stairs. Denoting by e the coefficient of restitution, determine the value of h for which the ball will bounce the same height above each step.

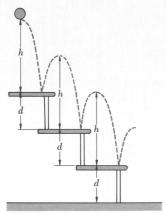

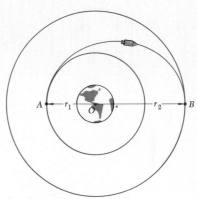

Fig. P13.181

Fig. P13.182

13.182 A spacecraft describes a circular orbit of radius $r_1 = 4500$ mi around the earth. As it passes through A its engine is fired for 180 s. This has the effect of placing the spacecraft on an elliptic orbit with apogee at B, at a distance of 8000 mi from the center of the earth. How long should the engine be fired at B if the spacecraft is to be inserted in a circular orbit of radius $r_2 = 8000$ mi?

13.183 A 12-Mg truck and a 30-Mg railroad flatcar are both at rest with their brakes released. An engine bumps the flatcar and causes the flatcar alone to start moving with a velocity of 1.4 m/s to the right. Assuming $e = 1$ between the truck and the ends of the flatcar and neglecting friction, determine the velocities of the truck and of the flatcar after (*a*) end A strikes the truck, (*b*) the truck strikes end B.

Fig. P13.183

13.184 A chain of length L is suspended from a strip of rubber, of natural length h, and is in equilibrium in the position shown. The chain is then cut at point A. Determine the length x, knowing that the remaining portion of the chain will rise sufficiently (*a*) to allow the rubber strip to become slack, (*b*) to touch the ceiling.

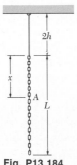

Fig. P13.184

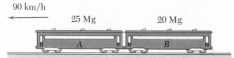

Fig. P13.186

13.185 A dime which is at rest on a *rough* surface is struck squarely by a half dollar moving to the right. After the impact, each coin slides and comes to rest; the dime slides 480 mm to the right, and the half dollar slides 95 mm to the right. Assuming that the coefficient of friction is the same for each coin, determine the value of the coefficient of restitution between the coins. (*Masses:* half dollar, 12.50 g; quarter dollar, 6.25 g; dime, 2.50 g.)

13.186 A light train made up of two cars travels at 90 km/h. When the brakes are applied, a constant braking force of 30 kN is applied to each car. Determine (*a*) the time required for the train to stop after the brakes are applied, (*b*) the force in the coupling between the cars while the train is slowing down.

13.187 Collar A is dropped 3 ft onto collar B, which is resting on a spring of constant $k = 5$ lb/in. Assuming perfectly plastic impact, determine (*a*) the maximum deflection of collar B, (*b*) the energy lost during the impact.

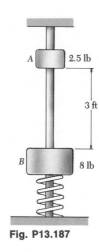

Fig. P13.187

The following problems are designed to be solved with a computer.

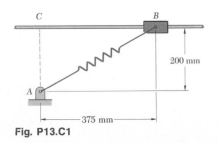

Fig. P13.C1

13.C1 The spring AB is of constant 60 N/m and is attached to the 1.8-kg collar B which may move freely along the horizontal rod. The unstretched length of the spring is 125 mm. Knowing that the collar is released from rest in the position shown, (*a*) write a computer program and use it to calculate the speed of the collar at 25-mm intervals from point B to point C, (*b*) expand this program to determine the approximate time required for the collar to move from B to C, using 5-mm intervals. [*Hint:* The time Δt_i required for the collar to move through Δx_i may be obtained by dividing Δx_i by the average velocity $\frac{1}{2}(v_i + v_{i+1})$ of the collar over Δt_i if the acceleration of the collar is assumed to remain constant over Δt_i.]

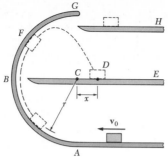

Fig. P13.C2

13.C2 A small package of mass m is projected into a vertical return loop at A with a velocity v_0. The package travels without friction along a circle of radius r and, if $v_0^2 > 5gr$, it is deposited on a horizontal shelf H. If $v_0^2 = 2gr$, the package reaches point B and slides back along the circular surface. For $2gr < v_0^2 < 5gr$, the package leaves the circular surface and strikes the horizontal shelf E at point D. Write a computer program and use it to determine where the package strikes shelf E for values of v_0^2 from $2gr$ to $5gr$ at intervals of $0.2gr$.

13.C3 At engine burnout a satellite has reached an altitude of 2400 km and has a velocity v_0 of magnitude 8100 m/s forming an angle ϕ_0 with the vertical. Write a computer program and use it to determine the maximum and minimum heights reached by the satellite for values of ϕ_0 from 60 to 120° at 5° intervals. Assuming that if the satellite gets closer than 300 km from the earth's surface, it will soon burn up, indicate those values of ϕ_0 for which a permanent orbit is not achieved.

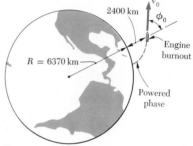

Fig. P13.C3

13.C4 In a game of billiards, a player wishes to hit ball B squarely with ball A at a distance nd, where d is the diameter of the balls. However, the player projects ball A with a velocity v_0 forming a small angle θ_0 with line AB. Denoting by e the coefficient of restitution, write a computer program to calculate the magnitude of the velocity v'_A of ball A after the impact and the angle θ'_A that v'_A will form with line AB. Using this program and knowing that $n = 10$ and $v_0 = 4$ m/s, determine v'_A and θ'_A for values of θ_0 from 0 to 0.5° at 0.05° intervals, and from 0.5 to 5.5° at 0.5° intervals, assuming (a) $e = 0.95$, (b) $e = 0.90$, (c) $e = 0.70$.

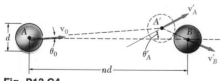

Fig. P13.C4

Systems of Particles

14.1. Introduction. We shall be concerned in this chapter with the motion of *systems of particles,* i.e., with the motion of a large number of particles considered together. In the first part of the chapter, we shall discuss systems consisting of well-defined particles, while in the second part we shall analyze the motion of variable systems, i.e., systems which are continually gaining or losing particles, or doing both at the same time.

In Sec. 14.2, we shall first apply Newton's second law to each particle of the system. Defining the *effective force* of a particle as the product $m_i\mathbf{a}_i$ of its mass m_i and its acceleration $\mathbf{a}_i$, we shall show that the *external forces* acting on the various particles form a system equipollent to the system of the effective forces, i.e., both systems have the same resultant and same moment resultant about any given point. In Sec. 14.3, we shall further show that the resultant and moment resultant of the external forces are equal, respectively, to the rate of change of the total linear momentum and of the total angular momentum of the particles of the system.

In Sec. 14.4, we shall define the *mass center* of a system of particles and describe the motion of that point, while in Sec. 14.5 we shall discuss the motion of the particles about their mass center. We shall be concerned in Sec. 14.6 with the conditions under which the linear momentum and the angular momentum of a system of particles are conserved, and we shall apply the results obtained to the solution of various problems.

Sections 14.7 and 14.8 deal with the application of the work-energy principle to a system of particles, and Sec. 14.9 with the application of the impulse-momentum principle. These sections also contain a number of problems of practical interest.

It should be noted that while the derivations given in the first part of this chapter are carried out for a system of independent particles, they remain valid when the particles of the system are rigidly connected, i.e., when they form a rigid body. In fact, the results obtained here will form the foundation of our discussion of the kinetics of rigid bodies in Chaps. 16 through 18.

The second part of this chapter is devoted to the study of variable systems of particles. In Sec. 14.11 we shall consider steady streams of particles, such as a stream of water diverted by a fixed vane, or the flow of air through a jet engine, and we shall learn to determine the force exerted by the stream on the vane and the thrust developed by the engine. Finally, in Sec. 14.12, we shall analyze systems which gain mass by continually absorbing particles or lose mass by continually expelling particles. Among the various practical applications of this analysis will be the determination of the thrust developed by a rocket engine.

14.2. Application of Newton's Laws to the Motion of a System of Particles. Effective Forces. In order to derive the equations of motion for a system of n particles, we shall begin by writing Newton's second law for each individual particle of the system. Consider the particle P_i, where $1 \leq i \leq n$. Let m_i be the mass of P_i and $\mathbf{a}_i$ its acceleration with respect to the newtonian frame of reference $Oxyz$. We shall denote by $\mathbf{f}_{ij}$ the force exerted on P_i by another particle P_j of the system (Fig. 14.1); this force is called an *internal force*. The resultant of the

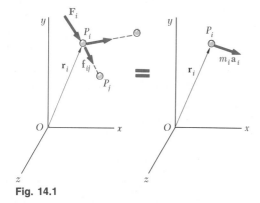

Fig. 14.1

internal forces exerted on P_i by all the other particles of the system is thus $\sum\limits_{j=1}^{n} \mathbf{f}_{ij}$ (where $\mathbf{f}_{ii}$ has no meaning and is assumed to be equal to zero). Denoting, on the other hand, by $\mathbf{F}_i$ the resultant of all the *external forces* acting on P_i, we write Newton's second law for the particle P_i as follows:

$$\mathbf{F}_i + \sum_{j=1}^{n} \mathbf{f}_{ij} = m_i \mathbf{a}_i \tag{14.1}$$

Denoting by $\mathbf{r}_i$ the position vector of P_i and taking the moments about O of the various terms in Eq. (14.1), we also write

$$\mathbf{r}_i \times \mathbf{F}_i + \sum_{j=1}^{n} (\mathbf{r}_i \times \mathbf{f}_{ij}) = \mathbf{r}_i \times m_i \mathbf{a}_i \tag{14.2}$$

Repeating this procedure for each particle P_i of the system, we obtain n equations of the type (14.1) and n equations of the type (14.2), where i takes successively the values 1, 2, . . . , n. The vectors $m_i\mathbf{a}_i$ are referred to as the *effective forces* of the particles.† Thus the equations obtained express the fact that the external forces $\mathbf{F}_i$ and the internal forces $\mathbf{f}_{ij}$ acting on the various particles form a system equivalent to the system of the effective forces $m_i\mathbf{a}_i$ (i.e., one system may be replaced by the other) (Fig. 14.2).

Before proceeding further with our derivation, let us examine the internal forces $\mathbf{f}_{ij}$. We note that these forces occur in pairs $\mathbf{f}_{ij}$, $\mathbf{f}_{ji}$, where $\mathbf{f}_{ij}$ represents the force exerted by the particle P_j on the particle P_i and $\mathbf{f}_{ji}$ the force exerted by P_i on P_j (Fig. 14.2). Now, according to Newton's third law

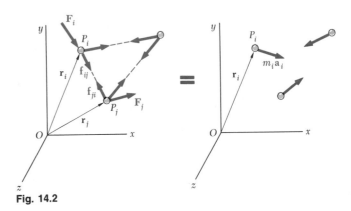

Fig. 14.2

(Sec. 6.1), as extended by Newton's law of gravitation to particles acting at a distance (Sec. 12.10), the forces $\mathbf{f}_{ij}$ and $\mathbf{f}_{ji}$ are equal and opposite and have the same line of action. Their sum is therefore $\mathbf{f}_{ij} + \mathbf{f}_{ji} = 0$ and the sum of their moments about O is

$$\mathbf{r}_i \times \mathbf{f}_{ij} + \mathbf{r}_j \times \mathbf{f}_{ji} = \mathbf{r}_i \times (\mathbf{f}_{ij} + \mathbf{f}_{ji}) + (\mathbf{r}_j - \mathbf{r}_i) \times \mathbf{f}_{ji} = 0$$

since the vectors $\mathbf{r}_j - \mathbf{r}_i$ and $\mathbf{f}_{ji}$ in the last term are collinear. Adding all the internal forces of the system, and summing their moments about O, we obtain the equations

$$\sum_{i=1}^{n}\sum_{j=1}^{n}\mathbf{f}_{ij} = 0 \qquad \sum_{i=1}^{n}\sum_{j=1}^{n}(\mathbf{r}_i \times \mathbf{f}_{ij}) = 0 \qquad (14.3)$$

which express the fact that the resultant and the moment resultant of the internal forces of the system are zero.

Returning now to the n equations (14.1), where $i = 1, 2, . . . , n$, we add them member by member. Taking into account the first of Eqs. (14.3), we obtain

$$\sum_{i=1}^{n}\mathbf{F}_i = \sum_{i=1}^{n}m_i\mathbf{a}_i \qquad (14.4)$$

---

† Since these vectors represent the resultants of the forces acting on the various particles of the system, they may truly be considered as forces.

Proceeding similarly with Eqs. (14.2) and taking into account the second of Eqs. (14.3), we have

655

14.2. Newton's Laws of Motion. Effective Forces

$$\sum_{i=1}^{n} (\mathbf{r}_i \times \mathbf{F}_i) = \sum_{i=1}^{n} (\mathbf{r}_i \times m_i \mathbf{a}_i) \tag{14.5}$$

Equations (14.4) and (14.5) express the fact that the system of the external forces $\mathbf{F}_i$ and the system of the effective forces $m_i\mathbf{a}_i$ have the same resultant and the same moment resultant. Referring to the definition given in Sec. 3.19 for two equipollent systems of vectors, we may therefore state that *the system of the external forces acting on the particles and the system of the effective forces of the particles are equipollent*† (Fig. 14.3).

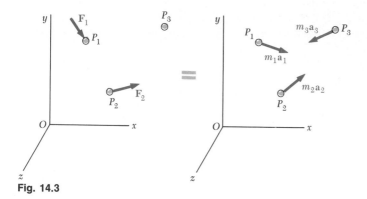

Fig. 14.3

We may note that Eqs. (14.3) express the fact that the system of the internal forces $\mathbf{f}_{ij}$ is equipollent to zero. It does *not* follow, however, that the internal forces have no effect on the particles under consideration. Indeed, the gravitational forces that the sun and the planets exert on one another are internal to the solar system and equipollent to zero. Yet these forces are alone responsible for the motion of the planets about the sun.

Similarly, it does not follow from Eqs. (14.4) and (14.5) that two systems of external forces which have the same resultant and the same moment resultant will have the same effect on a given system of particles. Clearly, the systems shown in Figs. 14.4a and 14.4b have the same resultant and the same moment resultant; yet the first system accelerates particle A and leaves particle B unaffected, while the second accelerates B and does not affect A. It is important to recall that when we stated in Sec. 3.19 that

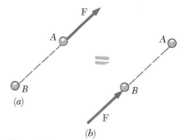

Fig. 14.4

†The result just obtained is often referred to as *d'Alembert's principle*, after the French mathematician Jean le Rond d'Alembert (1717–1783). However, d'Alembert's original statement refers to the motion of a system of connected bodies, with $\mathbf{f}_{ij}$ representing constraint forces which if applied by themselves, will not cause the system to move. Since, as it will now be shown, this is in general not the case for the internal forces acting on a system of free particles, we shall postpone the consideration of d'Alembert's principle until the study of the motion of rigid bodies (Chap. 16).

two equipollent systems of forces acting on a rigid body are also equivalent, we specifically noted that this property could *not* be extended to a system of forces acting on a set of independent particles such as those considered in this chapter.

In order to avoid any confusion, we shall use gray equals signs to connect equipollent systems of vectors, such as those shown in Figs. 14.3 and 14.4. These signs will indicate that the two systems of vectors have the same resultant and the same moment resultant. Colored equals signs will continue to be used to indicate that two systems of vectors are equivalent, i.e., that one system may actually be replaced by the other (Fig. 14.2).

14.3. Linear and Angular Momentum of a System of Particles. Equations (14.4) and (14.5), obtained in the preceding section for the motion of a system of particles, may be expressed in a more condensed form if we introduce the linear and the angular momentum of the system of particles. Defining the linear momentum $\mathbf{L}$ of the system of particles as the sum of the linear momenta of the various particles of the system (Sec. 12.3), we write

$$\mathbf{L} = \sum_{i=1}^{n} m_i \mathbf{v}_i \tag{14.6}$$

Defining the angular momentum $\mathbf{H}_O$ about O of the system of particles in a similar way (Sec. 12.7), we have

$$\mathbf{H}_O = \sum_{i=1}^{n} (\mathbf{r}_i \times m_i \mathbf{v}_i) \tag{14.7}$$

Differentiating both members of Eqs. (14.6) and (14.7) with respect to t, we write

$$\dot{\mathbf{L}} = \sum_{i=1}^{n} m_i \dot{\mathbf{v}}_i = \sum_{i=1}^{n} m_i \mathbf{a}_i \tag{14.8}$$

and

$$\dot{\mathbf{H}}_O = \sum_{i=1}^{n} (\dot{\mathbf{r}}_i \times m_i \mathbf{v}_i) + \sum_{i=1}^{n} (\mathbf{r}_i \times m_i \dot{\mathbf{v}}_i)$$

$$= \sum_{i=1}^{n} (\mathbf{v}_i \times m_i \mathbf{v}_i) + \sum_{i=1}^{n} (\mathbf{r}_i \times m_i \mathbf{a}_i)$$

which reduces to

$$\dot{\mathbf{H}}_O = \sum_{i=1}^{n} (\mathbf{r}_i \times m_i \mathbf{a}_i) \tag{14.9}$$

since the vectors $\mathbf{v}_i$ and $m_i \mathbf{v}_i$ are collinear.

We observe that the right-hand members of Eqs. (14.8) and (14.9) are, respectively, identical with the right-hand members of Eqs. (14.4) and (14.5). It follows that the left-hand members of these equations are respectively equal. Recalling that the left-hand member of Eq. (14.5) represents the sum of the moments $\mathbf{M}_O$ about O of the external forces acting on the particles of the system, and omitting the subscript i from the sums, we write

$$\Sigma\mathbf{F} = \dot{\mathbf{L}} \tag{14.10}$$

$$\Sigma\mathbf{M}_O = \dot{\mathbf{H}}_O \tag{14.11}$$

These equations express that *the resultant and the moment resultant about the fixed point O of the external forces are respectively equal to the rates of change of the linear momentum and of the angular momentum about O of the system of particles.*

14.4. Motion of the Mass Center of a System of Particles.

Equation (14.10) may be written in an alternate form if the *mass center* of the system of particles is considered. The mass center of the system is the point G defined by the position vector $\bar{\mathbf{r}}$, which satisfies the relation

$$m\bar{\mathbf{r}} = \sum_{i=1}^{n} m_i \mathbf{r}_i \tag{14.12}$$

where m represents the total mass $\sum_{i=1}^{n} m_i$ of the particles. Resolving the position vectors $\bar{\mathbf{r}}$ and $\mathbf{r}_i$ into rectangular components, we obtain the following three scalar equations, which may be used to determine the coordinates $\bar{x}$, $\bar{y}$, $\bar{z}$ of the mass center:

$$m\bar{x} = \sum_{i=1}^{n} m_i x_i \qquad m\bar{y} = \sum_{i=1}^{n} m_i y_i \qquad m\bar{z} = \sum_{i=1}^{n} m_i z_i \tag{14.12'}$$

Since $m_i g$ represents the weight of the particle P_i, and mg the total weight of the particles, we note that G is also the center of gravity of the system of particles. However, in order to avoid any confusion, we shall call G the *mass center* of the system of particles when discussing properties of the system associated with the *mass* of the particles, and we shall refer to it as the *center of gravity* of the system when considering properties associated with the *weight* of the particles. Particles located outside the gravitational field of the earth, for example, have a mass but no weight. We may then properly refer to their mass center, but obviously not to their center of gravity.†

---

†It may also be pointed out that the mass center and the center of gravity of a system of particles do not exactly coincide, since the weights of the particles are directed toward the center of the earth and thus do not truly form a system of parallel forces.

Differentiating both members of Eq. (14.12) with respect to t, we write

$$m\dot{\bar{\mathbf{r}}} = \sum_{i=1}^{n} m_i \dot{\mathbf{r}}_i$$

or

$$m\bar{\mathbf{v}} = \sum_{i=1}^{n} m_i \mathbf{v}_i \qquad (14.13)$$

where $\bar{\mathbf{v}}$ represents the velocity of the mass center G of the system of particles. But the right-hand member of Eq. (14.13) is, by definition, the linear momentum $\mathbf{L}$ of the system (Sec. 14.3). We therefore have

$$\mathbf{L} = m\bar{\mathbf{v}} \qquad (14.14)$$

and, differentiating both members with respect to t,

$$\dot{\mathbf{L}} = m\bar{\mathbf{a}} \qquad (14.15)$$

where $\bar{\mathbf{a}}$ represents the acceleration of the mass center G. Substituting for $\dot{\mathbf{L}}$ from (14.15) into (14.10), we write the equation

$$\Sigma\mathbf{F} = m\bar{\mathbf{a}} \qquad (14.16)$$

which defines the motion of the mass center G of the system of particles.

We note that Eq. (14.16) is identical with the equation we would obtain for a particle of mass m equal to the total mass of the particles of the system, acted upon by all the external forces. We state therefore: *The mass center of a system of particles moves as if the entire mass of the system and all the external forces were concentrated at that point.*

This principle is best illustrated by the motion of an exploding shell. We know that if the resistance of the air is neglected, a shell may be assumed to travel along a parabolic path. After the shell has exploded, the mass center G of the fragments of shell will continue to travel along the same path. Indeed, point G must move as if the mass and the weight of all fragments were concentrated at G; it must move, therefore, as if the shell had not exploded.

It should be noted that the preceding derivation does not involve the moments of the external forces. Therefore, *it would be wrong to assume* that the external forces are equipollent to a vector $m\bar{\mathbf{a}}$ attached at the mass center G. This is, in general, not the case, since, as we shall see in the next section, the sum of the moments about G of the external forces is, in general, not equal to zero.

14.5. Angular Momentum of a System of Particles about Its Mass Center.

In some applications (for example, in the analysis of the motion of a rigid body) it is convenient to consider the motion of the particles of the system with respect to a centroidal frame of reference $Gx'y'z'$ which translates with respect to the newtonian frame of reference $Oxyz$ (Fig. 14.5). While such a frame is not, in general, a newtonian frame of reference, we shall see that the fundamental relation (14.11) still holds when the frame $Oxyz$ is replaced by $Gx'y'z'$.

Denoting, respectively, by $\mathbf{r}_i'$ and $\mathbf{v}_i'$ the position vector and the velocity of the particle P_i relative to the moving frame of reference $Gx'y'z'$, we define the *angular momentum* $\mathbf{H}_G'$ of the system of particles *about the mass center G* as follows:

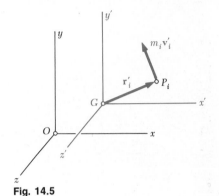

Fig. 14.5

$$\mathbf{H}_G' = \sum_{i=1}^{n} (\mathbf{r}_i' \times m_i \mathbf{v}_i') \qquad (14.17)$$

We now differentiate both members of Eq. (14.17) with respect to t. This operation being similar to that performed in Sec. 14.3 on Eq. (14.7), we write immediately

$$\dot{\mathbf{H}}_G' = \sum_{i=1}^{n} (\mathbf{r}_i' \times m_i \mathbf{a}_i') \qquad (14.18)$$

where $\mathbf{a}_i'$ denotes the acceleration of P_i relative to the moving frame of reference. Referring to Sec. 11.12, we write

$$\mathbf{a}_i = \overline{\mathbf{a}} + \mathbf{a}_i'$$

where $\mathbf{a}_i$ and $\overline{\mathbf{a}}$ denote, respectively, the accelerations of P_i and G relative to the frame $Oxyz$. Solving for $\mathbf{a}_i'$ and substituting into (14.18), we have

$$\dot{\mathbf{H}}_G' = \sum_{i=1}^{n} (\mathbf{r}_i' \times m_i \mathbf{a}_i) - \left(\sum_{i=1}^{n} m_i \mathbf{r}_i' \right) \times \overline{\mathbf{a}} \qquad (14.19)$$

But, by (14.12), the second sum in Eq. (14.19) is equal to $m\overline{\mathbf{r}}'$ and, thus, to zero, since the position vector $\overline{\mathbf{r}}'$ of G relative to the frame $Gx'y'z'$ is clearly zero. On the other hand, since $\mathbf{a}_i$ represents the acceleration of P_i relative to a newtonian frame, we may use Eq. (14.1) and replace $m_i \mathbf{a}_i$ by the sum of the internal forces $\mathbf{f}_{ij}$ and of the resultant $\mathbf{F}_i$ of the external forces acting on P_i. But a reasoning similar to that used in Sec. 14.2 shows that the moment resultant about G of the internal forces $\mathbf{f}_{ij}$ of the entire system is zero. The first sum in Eq. (14.19) reduces therefore to the moment resultant about G of the external forces acting on the particles of the system, and we write

$$\Sigma \mathbf{M}_G = \dot{\mathbf{H}}_G' \qquad (14.20)$$

which expresses that *the moment resultant about G of the external forces is* *equal to the rate of change of the angular momentum about G of the system of particles.*

It should be noted that in Eq. (14.17), we defined the angular momentum $\mathbf{H}'_G$ as the sum of the moments about G of the momenta of the particles $m_i\mathbf{v}'_i$ *in their motion relative to the centroidal frame of reference* $Gx'y'z'$. We may sometimes want to compute the sum $\mathbf{H}_G$ of the moments about G of the momenta of the particles $m_i\mathbf{v}_i$ *in their absolute motion*, i.e., in their

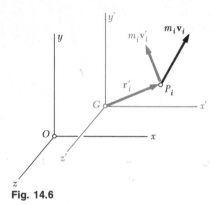

Fig. 14.6

motion as observed from the newtonian frame of reference $Oxyz$ (Fig. 14.6):

$$\mathbf{H}_G = \sum_{i=1}^{n} (\mathbf{r}'_i \times m_i\mathbf{v}_i) \tag{14.21}$$

Remarkably, the angular momenta $\mathbf{H}'_G$ and $\mathbf{H}_G$ are identically equal. This may be verified by referring to Sec. 11.12 and writing

$$\mathbf{v}_i = \bar{\mathbf{v}} + \mathbf{v}'_i \tag{14.22}$$

Substituting for $\mathbf{v}_i$ from (14.22) into Eq. (14.21), we have

$$\mathbf{H}_G = \left(\sum_{i=1}^{n} m_i\mathbf{r}'_i\right) \times \bar{\mathbf{v}} + \sum_{i=1}^{n} (\mathbf{r}'_i \times m_i\mathbf{v}'_i)$$

But, as observed earlier, the first sum is equal to zero. Thus $\mathbf{H}_G$ reduces to the second sum which, by definition, is equal to $\mathbf{H}'_G$.†

---

† Note that this property is peculiar to the centroidal frame $Gx'y'z'$ and does not hold, in general, for other frames of reference (see Prob. 14.24).

Taking advantage of the property we have just established, we shall simplify our notation by dropping the prime (′) from Eq. (14.20). We therefore write

$$\Sigma \mathbf{M}_G = \dot{\mathbf{H}}_G \qquad (14.23)$$

where it is understood that the angular momentum $\mathbf{H}_G$ may be computed by forming the moments about G of the momenta of the particles in their motion with respect to either the newtonian frame $Oxyz$ or the centroidal frame $Gx'y'z'$:

$$\mathbf{H}_G = \sum_{i=1}^{n} (\mathbf{r}_i' \times m_i \mathbf{v}_i) = \sum_{i=1}^{n} (\mathbf{r}_i' \times m_i \mathbf{v}_i') \qquad (14.24)$$

14.6. Conservation of Momentum for a System of Particles. If no external force acts on the particles of a system, the left-hand members of Eqs. (14.10) and (14.11) are equal to zero and these equations reduce to $\dot{\mathbf{L}} = 0$ and $\dot{\mathbf{H}}_O = 0$. We conclude that

$$\mathbf{L} = \text{constant} \qquad \mathbf{H}_O = \text{constant} \qquad (14.25)$$

The equations obtained express that the linear momentum of the system of particles and its angular momentum about the fixed point O are conserved.

In some applications, such as problems involving central forces, the moment about a fixed point O of each of the external forces may be zero, without any of the forces being zero. In such cases, the second of Eqs. (14.25) still holds; the angular momentum of the system of particles about O is conserved.

The concept of conservation of momentum may also be applied to the analysis of the motion of the mass center G of a system of particles and to the analysis of the motion of the system about G. For example, if the sum of the external forces is zero, the first of Eqs. (14.25) applies. Recalling Eq. (14.14), we write

$$\bar{\mathbf{v}} = \text{constant} \qquad (14.26)$$

which expresses that the mass center G of the system moves in a straight line and at a constant speed. On the other hand, if the sum of the moments about G of the external forces is zero, it follows from Eq. (14.23) that the angular momentum of the system about its mass center is conserved:

$$\mathbf{H}_G = \text{constant} \qquad (14.27)$$

SAMPLE PROBLEM 14.1

A 200-kg space vehicle is observed at $t = 0$ to pass through the origin of a newtonian reference frame $Oxyz$ with the velocity $\mathbf{v}_0 = (150 \text{ m/s})\mathbf{i}$ relative to the frame. Following the detonation of explosive charges, the vehicle separates into three parts, A, B, and C, of mass 100 kg, 60 kg, and 40 kg, respectively. Knowing that, at $t = 2.5$ s, the positions of parts A and B are observed to be $A(555, -180, 240)$ and $B(255, 0, -120)$, where the coordinates are expressed in meters, determine the position of part C at that time.

Solution. Since there is no external force, the mass center G of the system moves with the constant velocity $\mathbf{v}_0 = (150 \text{ m/s})\mathbf{i}$. At $t = 2.5$ s, its position is

$$\bar{\mathbf{r}} = \mathbf{v}_0 t = (150 \text{ m/s})\mathbf{i}(2.5 \text{ s}) = (375 \text{ m})\mathbf{i}$$

Recalling Eq. (14.12), we write

$$m\bar{\mathbf{r}} = m_A \mathbf{r}_A + m_B \mathbf{r}_B + m_C \mathbf{r}_C$$
$$(200 \text{ kg})(375 \text{ m})\mathbf{i} = (100 \text{ kg})[(555 \text{ m})\mathbf{i} - (180 \text{ m})\mathbf{j} + (240 \text{ m})\mathbf{k}]$$
$$+ (60 \text{ kg})[(255 \text{ m})\mathbf{i} - (120 \text{ m})\mathbf{k}] + (40 \text{ kg})\mathbf{r}_C$$

$$\mathbf{r}_C = (105 \text{ m})\mathbf{i} + (450 \text{ m})\mathbf{j} - (420 \text{ m})\mathbf{k} \quad \blacktriangleleft$$

SAMPLE PROBLEM 14.2

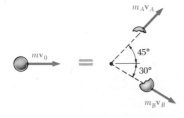

A 20-lb projectile is moving with a velocity of 100 ft/s when it explodes into two fragments A and B, weighing 5 lb and 15 lb, respectively. Knowing that immediately after the explosion fragments A and B travel in directions defined, respectively, by $\theta_A = 45°$ and $\theta_B = 30°$, determine the velocity of each fragment.

Solution. Since there is no external force, the linear momentum of the system is conserved, and we write

$$m_A \mathbf{v}_A + m_B \mathbf{v}_B = m \mathbf{v}_0$$
$$(5/g)\mathbf{v}_A + (15/g)\mathbf{v}_B = (20/g)\mathbf{v}_0$$

$\xrightarrow{+}$ x components: $\qquad 5v_A \cos 45° + 15v_B \cos 30° = 20(100)$

$+\uparrow y$ components: $\qquad 5v_A \sin 45° - 15v_B \sin 30° = 0$

Solving simultaneously the two equations for v_A and v_B, we have

$$v_A = 207 \text{ ft/s} \qquad v_B = 97.6 \text{ ft/s}$$

$$\mathbf{v}_A = 207 \text{ ft/s} \measuredangle 45° \qquad \mathbf{v}_B = 97.6 \text{ ft/s} \searrow 30° \quad \blacktriangleleft$$

Problems

14.1 A $\frac{3}{4}$-oz bullet is fired in a horizontal direction through block A and becomes embedded in block B. The bullet causes A and B to start moving with velocities of 7.5 and 5 ft/s, respectively. Determine (a) the initial velocity v_0 of the bullet, (b) the velocity of the bullet as it travels from block A to block B.

14.2 A $\frac{3}{4}$-oz bullet is fired in a horizontal direction through block A and becomes embedded in block B. Knowing that $v_0 = 1800$ ft/s and that block B starts moving with a velocity of 6 ft/s, determine (a) the velocity with which block A starts moving, (b) the velocity of the bullet as it travels from block $Å$ to block B.

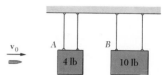

Fig. P14.1 and P14.2

14.3 A 60-Mg engine coasting at 7 km/h strikes a 10-Mg flatcar carrying a 25-Mg load which may slide along the floor of the car ($\mu_k = 0.20$). Knowing that the car was at rest with its brakes released and that it is automatically coupled with the engine upon impact, determine the velocity of the car (a) immediately after impact, (b) after the load has slid to a stop relative to the car.

14.4 Solve Prob. 14.3, assuming that the car fails to couple with the engine and that the impact is perfectly elastic ($e = 1$).

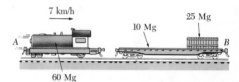

Fig. P14.3

14.5 Three identical freight cars have the velocities indicated. Assuming that car B is first hit by car A, determine the velocity of each car after all collisions have taken place, if (a) all three cars get automatically coupled, (b) cars A and B get automatically and tightly coupled, while cars B and C bounce off each other with a coefficient of restitution $e = 1$.

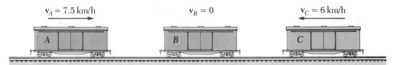

Fig. P14.5

14.6 Solve Prob. 14.5, assuming that car B is first hit by car C.

14.7 A system consists of three particles A, B, and C. We know that $W_A = 5$ lb, $W_B = 4$ lb, and $W_C = 3$ lb and that the velocities of the particles expressed in ft/s are, respectively, $v_A = 2i + 3j - 2k$, $v_B = v_x i + 2j + v_z k$, and $v_C = -3i - 2j + k$. Determine (a) the components v_x and v_z of the velocity of particle B for which the angular momentum H_O of the system about O is parallel to the x axis, (b) the value of H_O.

14.8 For the system of particles of Prob. 14.7, determine (a) the components v_x and v_z of the velocity of particle B for which the angular momentum H_O of the system about O is parallel to the z axis, (b) the value of H_O.

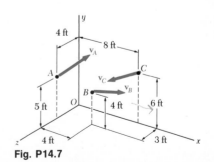

Fig. P14.7

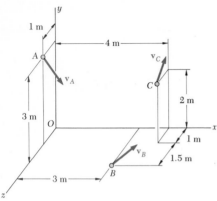

Fig. P14.9

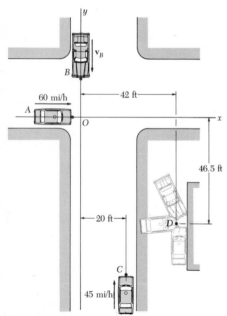

Fig. P14.12

14.9 A system consists of three particles A, B, and C. We know that $m_A = 1$ kg, $m_B = 2$ kg, and $m_C = 3$ kg and that the velocities of the particles expressed in m/s are, respectively, $v_A = 3i - 2j + 4k$, $v_B = 4i + 3j$, and $v_C = 2i + 5j - 3k$. (a) Determine the angular momentum $\mathbf{H}_O$ of the system about O. (b) Using the result of part a and the answers to Prob. 14.10, check that the relation given in Prob. 14.22 is satisfied.

14.10 For the system of particles of Prob. 14.9, determine (a) the position vector $\bar{r}$ of the mass center G of the system, (b) the linear momentum $m\bar{v}$ of the system, (c) the angular momentum $\mathbf{H}_G$ of the system about G.

14.11 A 240-kg space vehicle traveling with the velocity $v_0 = (500$ m/s$)k$ passes through the origin O at $t = 0$. Explosive charges then separate the vehicle into three parts, A, B, and C, of mass 40 kg, 80 kg, and 120 kg, respectively. Knowing that at $t = 3$ s the positions of parts B and C are observed to be $B(375, 825, 2025)$ and $C(-300, -600, 1200)$, where the coordinates are expressed in meters, determine the corresponding position of part A. Neglect the effect of gravity.

14.12 Cruiser A was traveling east at 60 mi/h on a police emergency call when it was hit at an intersection by car B which was traveling south at high speed. After sliding together on the wet pavement the two cars hit cruiser C which was traveling north at 45 mi/h. Stuck together, the three cars hit a wall and came to a stop at D. Knowing that car B weighed 3600 lb, that each of the cruisers weighed 3000 lb, and that the officer driving cruiser C observed that she was 63 ft from the intersection when the first collision occurred, determine (a) the speed of car B, (b) the time elapsed from the first collision to the stop at D. (Neglect the forces exerted on the cars by the wet pavement.)

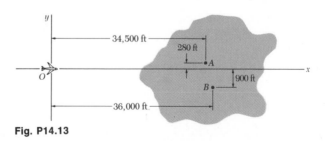

Fig. P14.13

14.13 An airliner weighing 60 tons and flying due east at an altitude of 30,000 ft and a speed of 540 mi/h suddenly explodes, breaking into three fragments A, B, and C. Fragment A, weighing 30 tons, and fragment B, weighing 20 tons, are found in a wooded area at the points shown in the figure. Knowing that the explosion occurred when the airliner was directly above point O and assuming that all three fragments hit the ground at the same time, determine the location of fragment C. Neglect the resistance of the air.

14.14 Two 15-kg cannon balls are chained together and fired horizontally with a velocity of 165 m/s from the top of a 15-m wall. The chain breaks during the flight of the cannon balls and one of them strikes the ground at $t = 1.5$ s, at a distance of 240 m from the foot of the wall, and 7 m to the right of the line of fire. Determine the position of the other cannon ball at that instant. Neglect the resistance of the air.

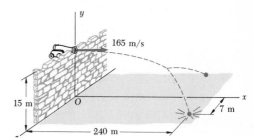

14.15 Solve Prob. 14.14, if the cannon ball which first strikes the ground has a mass of 12 kg, and the other a mass of 18 kg. Assume that the time of flight and the point of impact of the first cannon ball remain the same.

Fig. P14.14

14.16 Knowing that fragment A of Sample Prob. 14.2 has a velocity of magnitude $v_A = 250$ ft/s and travels in the direction $\theta_A = 60°$, determine the magnitude and direction of the velocity of fragment B immediately after the explosion.

14.17 In a game of billiards, ball A is moving with the velocity $v_0 = (10 \text{ ft/s})\mathbf{i}$ when it strikes balls B and C which are at rest side by side. After the collision, A is observed to move with the velocity $v_A = (3.92 \text{ ft/s})\mathbf{i} - (4.56 \text{ ft/s})\mathbf{j}$, while B and C move in the directions shown. Determine the magnitudes of the velocities v_B and v_C.

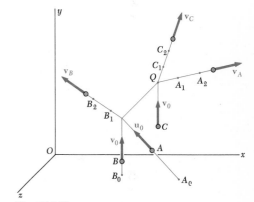

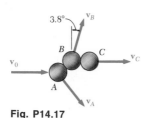

Fig. P14.17

Fig. P14.18

14.18 In a scattering experiment, an alpha particle A is projected with the velocity $u_0 = -(600 \text{ m/s})\mathbf{i} + (750 \text{ m/s})\mathbf{j} - (800 \text{ m/s})\mathbf{k}$ into a stream of oxygen nuclei moving with the common velocity $v_0 = (600 \text{ m/s})\mathbf{j}$. After colliding successively with nuclei B and C, particle A is observed to move along the path defined by the points $A_1(280, 240, 120)$, $A_2(360, 320, 160)$, while nuclei B and C are observed to move along paths defined, respectively, by $B_1(147, 220, 130)$, $B_2(114, 290, 120)$ and by $C_1(240, 232, 90)$, $C_2(240, 280, 75)$. All paths are along straight lines and all coordinates are expressed in millimeters. Knowing that the mass of an oxygen nucleus is four times that of an alpha particle, determine the speed of each of the three particles after the collisions.

14.19 A 1.2-kg shell is moving in a direction perpendicular to a wall when at point D, it explodes into three fragments A, B, and C, respectively, of mass 300 g, 400 g, and 500 g. Knowing that the fragments hit the wall at the points indicated and that $v_A = 490$ m/s, determine the velocity v_0 of the shell before it exploded.

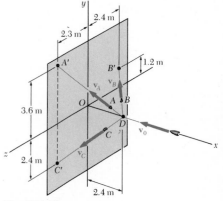

Fig. P14.19

14.20 A 4500-kg helicopter A was traveling due east at a speed of 120 km/h and an altitude of 800 m when it was hit by a 6000-kg helicopter B. As a result of the collision, both helicopters lost their lift, and their entangled wreckage fell to the ground in 12 s at a point located 525 m east and 135 m south of the point of impact. Neglecting air resistance, determine the velocity components of helicopter B just before the collision.

14.21 An archer hits a game bird flying in a horizontal straight line 9 m above the ground with a 40-g wooden arrow. Knowing that the arrow strikes the bird from behind with a velocity of 110 m/s at an angle of 30° with the vertical, and that the bird falls to the ground in 1.5 s and 18 m beyond the point where it was hit, determine (a) the mass of the bird, (b) the speed at which it was flying when it was hit.

14.22 Derive the relation

$$\mathbf{H}_O = \bar{\mathbf{r}} \times m\bar{\mathbf{v}} + \mathbf{H}_G$$

between the angular momenta $\mathbf{H}_O$ and $\mathbf{H}_G$ defined in Eqs. (14.7) and (14.24), respectively. The vectors $\bar{\mathbf{r}}$ and $\bar{\mathbf{v}}$ define, respectively, the position and velocity of the mass center G of the system of particles relative to the newtonian frame of reference $Oxyz$, and m represents the total mass of the system.

14.23 Show that Eq. (14.23) may be derived directly from Eq. (14.11) by substituting for $\mathbf{H}_O$ the expression given in Prob. 14.22.

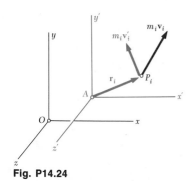

Fig. P14.24

14.24 Consider the frame of reference $Ax'y'z'$ in translation with respect to the newtonian frame of reference $Oxyz$. We define the angular momentum $\mathbf{H}'_A$ of a system of n particles about A as the sum

$$\mathbf{H}'_A = \sum_{i=1}^{n} \mathbf{r}'_i \times m_i \mathbf{v}'_i \tag{1}$$

of the moments about A of the momenta $m_i \mathbf{v}'_i$ of the particles in their motion relative to the frame $Ax'y'z'$. Denoting by $\mathbf{H}_A$ the sum

$$\mathbf{H}_A = \sum_{i=1}^{n} \mathbf{r}'_i \times m_i \mathbf{v}_i \tag{2}$$

of the moments about A of the momenta $m_i \mathbf{v}_i$ of the particles in their motion relative to the newtonian frame $Oxyz$, show that $\mathbf{H}_A = \mathbf{H}'_A$ at a given instant if, and only if, one of the following conditions is satisfied at that instant: (a) A has zero velocity with respect to the frame $Oxyz$, (b) A coincides with the mass center G of the system, (c) the velocity $\mathbf{v}_A$ relative to $Oxyz$ is directed along the line AG.

14.25 Show that the relation $\Sigma \mathbf{M}_A = \dot{\mathbf{H}}'_A$, where $\mathbf{H}'_A$ is defined by Eq. (1) of Prob. 14.24 and where $\Sigma \mathbf{M}_A$ represents the sum of the moments about A of the external forces acting on the system of particles, is valid if, and only if, one of the following conditions is satisfied: (a) the frame $Ax'y'z'$ is itself a newtonian frame of reference, (b) A coincides with the mass center G, (c) the acceleration $\mathbf{a}_A$ of A relative to $Oxyz$ is directed along the line AG.

14.7. Kinetic Energy of a System of Particles.

The kinetic energy T of a system of particles is defined as the sum of the kinetic energies of the various particles of the system. Referring to Sec. 13.3, we therefore write

$$T = \frac{1}{2} \sum_{i=1}^{n} m_i v_i^2 \tag{14.28}$$

Using a Centroidal Frame of Reference. It is often convenient when computing the kinetic energy of a system comprising a large number of particles (as in the case of a rigid body) to consider separately the motion of the mass center G of the system and the motion of the system relative to a moving frame of reference attached to G.

Let P_i be a particle of the system, $\mathbf{v}_i$ its velocity relative to the newtonian frame of reference $Oxyz$, and $\mathbf{v}'_i$ its velocity relative to the moving frame $Gx'y'z'$ which is in translation with respect to $Oxyz$ (Fig. 14.7). We recall from the preceding section that

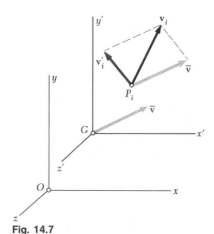

Fig. 14.7

$$\mathbf{v}_i = \overline{\mathbf{v}} + \mathbf{v}'_i \tag{14.22}$$

where $\overline{\mathbf{v}}$ denotes the velocity of the mass center G relative to the newtonian frame $Oxyz$. Observing that v_i^2 is equal to the scalar product $\mathbf{v}_i \cdot \mathbf{v}_i$, we express as follows the kinetic energy T of the system relative to the newtonian frame $Oxyz$:

$$T = \frac{1}{2} \sum_{i=1}^{n} m_i v_i^2 = \frac{1}{2} \sum_{i=1}^{n} (m_i \mathbf{v}_i \cdot \mathbf{v}_i)$$

or, substituting for $\mathbf{v}_i$ from (14.22),

$$T = \frac{1}{2} \sum_{i=1}^{n} [m_i (\overline{\mathbf{v}} + \mathbf{v}'_i) \cdot (\overline{\mathbf{v}} + \mathbf{v}'_i)]$$

$$= \frac{1}{2} \left(\sum_{i=1}^{n} m_i \right) \overline{v}^2 + \overline{\mathbf{v}} \cdot \sum_{i=1}^{n} m_i \mathbf{v}'_i + \frac{1}{2} \sum_{i=1}^{n} m_i v_i'^2$$

The first sum represents the total mass m of the system. Recalling Eq. (14.13), we note that the second sum is equal to $m\overline{\mathbf{v}}'$ and thus to zero, since $\overline{\mathbf{v}}'$, which represents the velocity of G relative to the frame $Gx'y'z'$, is clearly zero. We therefore write

$$T = \tfrac{1}{2} m \overline{v}^2 + \frac{1}{2} \sum_{i=1}^{n} m_i v_i'^2 \tag{14.29}$$

This equation shows that the kinetic energy T of a system of particles may be obtained *by adding the kinetic energy of the mass center G* (assuming the entire mass concentrated at G) *and the kinetic energy of the system in its motion relative to the frame $Gx'y'z'$.*

14.8. Work-Energy Principle. Conservation of Energy for a System of Particles.

The principle of work and energy may be applied to each particle P_i of a system of particles. We write

$$T_1 + U_{1 \to 2} = T_2 \tag{14.30}$$

for each particle P_i, where $U_{1 \to 2}$ represents the work done by the internal forces $\mathbf{f}_{ij}$ and the resultant external force $\mathbf{F}_i$ acting on P_i. Adding the kinetic energies of the various particles of the system, and considering the work of all the forces involved, we may apply Eq. (14.30) to the entire system. The quantities T_1 and T_2 now represent the kinetic energy of the entire system and may be computed from either Eq. (14.28) or Eq. (14.29). The quantity $U_{1 \to 2}$ represents the work of all the forces acting on the particles of the system. We should note that while the internal forces $\mathbf{f}_{ij}$ and $\mathbf{f}_{ji}$ are equal and opposite, the work of these forces, in general, will not cancel out, since the particles P_i and P_j on which they act will, in general, undergo different displacements. Therefore, in computing $U_{1 \to 2}$, *we should consider the work of the internal forces $\mathbf{f}_{ij}$ as well as the work of the external forces $\mathbf{F}_i$.*

If all the forces acting on the particles of the system are conservative, Eq. (14.30) may be replaced by

$$T_1 + V_1 = T_2 + V_2 \tag{14.31}$$

where V represents the potential energy associated with the internal and external forces acting on the particles of the system. Equation (14.31) expresses the principle of *conservation of energy* for the system of particles.

14.9. Principle of Impulse and Momentum for a System of Particles.

Integrating Eqs. (14.10) and (14.11) in t from a time t_1 to a time t_2, we write

$$\sum \int_{t_1}^{t_2} \mathbf{F} \, dt = \mathbf{L}_2 - \mathbf{L}_1 \tag{14.32}$$

$$\sum \int_{t_1}^{t_2} \mathbf{M}_O \, dt = (\mathbf{H}_O)_2 - (\mathbf{H}_O)_1 \tag{14.33}$$

Recalling the definition of the linear impulse of a force given in Sec. 13.10, we observe that the integrals in Eq. (14.32) represent the linear impulses of the external forces acting on the particles of the system. We shall refer in a similar way to the integrals in Eq. (14.33) as the *angular impulses* about O of the external forces. Thus, Eq. (14.32) expresses that the sum of the linear impulses of the external forces acting on the system is equal to the change in linear momentum of the system. Similarly, Eq. (14.33) expresses that the sum of the angular impulses about O of the external forces is equal to the change in angular momentum about O of the system.

In order to understand the physical significance of Eqs. (14.32) and (14.33), we shall rearrange the terms in these equations and write

$$\mathbf{L}_1 + \sum \int_{t_1}^{t_2} \mathbf{F}\, dt = \mathbf{L}_2 \tag{14.34}$$

$$(\mathbf{H}_O)_1 + \sum \int_{t_1}^{t_2} \mathbf{M}_O\, dt = (\mathbf{H}_O)_2 \tag{14.35}$$

We have sketched in parts a and c of Fig. 14.8 the momenta of the particles of the system at times t_1 and t_2, respectively, and we have shown in part b of the same figure a vector equal to the sum of the linear impulses of the external forces and a couple of moment equal to the sum of the angular impulses about O of the external forces. For simplicity, the particles have

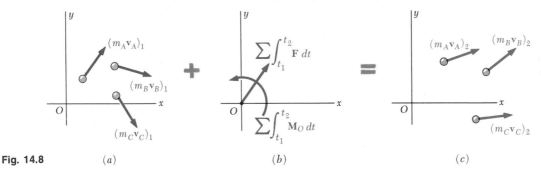

Fig. 14.8   (a)   (b)   (c)

been assumed to move in the plane of the figure, but the present discussion remains valid in the case of particles moving in space. Recalling from Eq. (14.6) that $\mathbf{L}$, by definition, is the resultant of the momenta $m_i\mathbf{v}_i$, we note that Eq. (14.34) expresses that the resultant of the vectors shown in parts a and b of Fig. 14.8 is equal to the resultant of the vectors shown in part c of the same figure. Recalling from Eq. (14.7) that $\mathbf{H}_O$ is the moment resultant of the momenta $m_i\mathbf{v}_i$, we note that Eq. (14.35) similarly expresses that the moment resultant of the vectors in parts a and b of Fig. 14.8 is equal to the moment resultant of the vectors in part c. Together, Eqs. (14.34) and (14.35) thus express that *the momenta of the particles at time t_1 and the impulses of the external forces from t_1 to t_2 form a system of vectors equipollent to the system of the momenta of the particles at time t_2.* This has been indicated in Fig. 14.8 by the use of gray plus and equals signs.

If no external force acts on the particles of the system, the integrals in Eqs. (14.34) and (14.35) are zero, and these equations reduce to

$$\mathbf{L}_1 = \mathbf{L}_2 \tag{14.36}$$

$$(\mathbf{H}_O)_1 = (\mathbf{H}_O)_2 \tag{14.37}$$

We thus check the result obtained in Sec. 14.6: If no external force acts on the particles of a system, the linear momentum and the angular momentum about O of the system of particles are conserved. The system of the initial momenta is equipollent to the system of the final momenta, and it follows that the angular momentum of the system of particles about *any* fixed point is conserved.

SAMPLE PROBLEM 14.3

For the 200-kg space vehicle of Sample Prob. 14.1, it is known that at $t = 2.5$ s, the velocity of part A is $\mathbf{v}_A = (270 \text{ m/s})\mathbf{i} - (120 \text{ m/s})\mathbf{j} + (160 \text{ m/s})\mathbf{k}$ and the velocity of part B is parallel to the xz plane. Determine the velocity of part C.

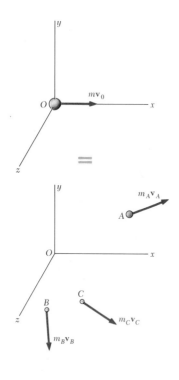

Solution. Since there is no external force, the initial momentum $m\mathbf{v}_0$ is equipollent to the system of the final momenta. Equating first the sums of the vectors in both parts of the adjoining sketch, and then the sums of their moments about O, we write

$\mathbf{L}_1 = \mathbf{L}_2$: $\qquad m\mathbf{v}_0 = m_A\mathbf{v}_A + m_B\mathbf{v}_B + m_C\mathbf{v}_C$ (1)

$(\mathbf{H}_O)_1 = (\mathbf{H}_O)_2$: $\quad 0 = \mathbf{r}_A \times m_A\mathbf{v}_A + \mathbf{r}_B \times m_B\mathbf{v}_B + \mathbf{r}_C \times m_C\mathbf{v}_C$ (2)

Recalling from Sample Prob. 14.1 that $\mathbf{v}_0 = (150 \text{ m/s})\mathbf{i}$,

$$m_A = 100 \text{ kg} \qquad m_B = 60 \text{ kg} \qquad m_C = 40 \text{ kg}$$
$$\mathbf{r}_A = (555 \text{ m})\mathbf{i} - (180 \text{ m})\mathbf{j} + (240 \text{ m})\mathbf{k}$$
$$\mathbf{r}_B = (255 \text{ m})\mathbf{i} - (120 \text{ m})\mathbf{k}$$
$$\mathbf{r}_C = (105 \text{ m})\mathbf{i} + (450 \text{ m})\mathbf{j} - (420 \text{ m})\mathbf{k}$$

and using the information given in the statement of this problem, we rewrite Eqs. (1) and (2) as follows:

$$200(150\mathbf{i}) = 100(270\mathbf{i} - 120\mathbf{j} + 160\mathbf{k}) + 60[(v_B)_x\mathbf{i} + (v_B)_z\mathbf{k}]$$
$$+ 40[(v_C)_x\mathbf{i} + (v_C)_y\mathbf{j} + (v_C)_z\mathbf{k}] \quad (1')$$

$$0 = 100\begin{vmatrix} \mathbf{i} & \mathbf{j} & \mathbf{k} \\ 555 & -180 & 240 \\ 270 & -120 & 160 \end{vmatrix} + 60\begin{vmatrix} \mathbf{i} & \mathbf{j} & \mathbf{k} \\ 255 & 0 & -120 \\ (v_B)_x & 0 & (v_B)_z \end{vmatrix}$$

$$+ 40\begin{vmatrix} \mathbf{i} & \mathbf{j} & \mathbf{k} \\ 105 & 450 & -420 \\ (v_C)_x & (v_C)_y & (v_C)_z \end{vmatrix} \quad (2')$$

Equating to zero the coefficient of $\mathbf{j}$ in (1') and the coefficients of $\mathbf{i}$ and $\mathbf{k}$ in (2'), we write, after reductions, the three scalar equations

$$(v_C)_y - 300 = 0$$
$$450(v_C)_z + 420(v_C)_y = 0$$
$$105(v_C)_y - 450(v_C)_x - 45\,000 = 0$$

which yield, respectively,

$$(v_C)_y = 300 \qquad (v_C)_z = -280 \qquad (v_C)_x = -30$$

The velocity of part C is thus

$$\mathbf{v}_C = -(30 \text{ m/s})\mathbf{i} + (300 \text{ m/s})\mathbf{j} - (280 \text{ m/s})\mathbf{k} \quad \blacktriangleleft$$

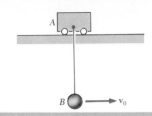

B $\rightarrow v_0$

SAMPLE PROBLEM 14.4

Ball B, of mass m_B, is suspended from a cord of length l attached to cart A, of mass m_A, which may roll freely on a frictionless horizontal track. If the ball is given an initial horizontal velocity $\mathbf{v}_0$ while the cart is at rest, determine (a) the velocity of B as it reaches its maximum elevation, (b) the maximum vertical distance h through which B will rise. (We assume that $v_0^2 < 2gl$.)

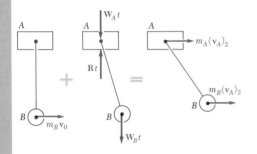

Solution We shall apply the impulse-momentum principle and the principle of conservation of energy to the cart-ball system between its initial position 1 and the position 2 when B reaches its maximum elevation.

Velocities. **Position 1:** $(\mathbf{v}_A)_1 = 0$ $(\mathbf{v}_B)_1 = \mathbf{v}_0$ (1)

Position 2: When ball B reaches its maximum elevation, its velocity $(\mathbf{v}_{B/A})_2$ relative to its support A is zero Thus, at that instant, its absolute velocity is

$$(\mathbf{v}_B)_2 = (\mathbf{v}_A)_2 + (\mathbf{v}_{B/A})_2 = (\mathbf{v}_A)_2 \qquad (2)$$

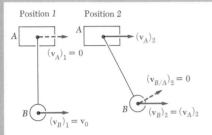

Impulse-Momentum Principle. Noting that the external impulses consist of $\mathbf{W}_A t$, $\mathbf{W}_B t$, and $\mathbf{R} t$, where $\mathbf{R}$ is the reaction of the track on the cart, and recalling (1) and (2), we draw the impulse-momentum diagram and write

$$\Sigma m\mathbf{v}_1 + \Sigma \ \mathbf{Ext}\ \mathbf{Imp}_{1\rightarrow 2} = \Sigma m\mathbf{v}_2$$

$\xrightarrow{+}$ x components: $m_B v_0 = (m_A + m_B)(v_A)_2$

which expresses that the linear momentum of the system is conserved in the horizontal direction. Solving for $(v_A)_2$:

$$(v_A)_2 = \frac{m_B}{m_A + m_B} v_0 \qquad (v_B)_2 = (v_A)_2 = \frac{m_B}{m_A + m_B} v_0 \rightarrow \ \blacktriangleleft$$

Conservation of Energy

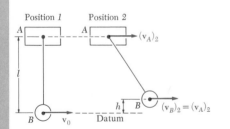

Position 1. *Potential Energy:*	$V_1 = m_A gl$
Kinetic Energy:	$T_1 = \frac{1}{2} m_B v_0^2$
Position 2. *Potential Energy:*	$V_2 = m_A gl + m_B gh$
Kinetic Energy:	$T_2 = \frac{1}{2}(m_A + m_B)(v_A)_2^2$

$T_1 + V_1 = T_2 + V_2$: $\frac{1}{2} m_B v_0^2 + m_A gl = \frac{1}{2}(m_A + m_B)(v_A)_2^2 + m_A gl + m_B gh$

Solving for h, we have

$$h = \frac{v_0^2}{2g} - \frac{m_A + m_B}{m_B} \frac{(v_A)_2^2}{2g}$$

or, substituting for $(v_A)_2$ the expression found above,

$$h = \frac{v_0^2}{2g} - \frac{m_B}{m_A + m_B} \frac{v_0^2}{2g} \qquad h = \frac{m_A}{m_A + m_B} \frac{v_0^2}{2g} \ \blacktriangleleft$$

Remarks. (1) Recalling that $v_0^2 < 2gl$, it follows from the last equation that $h < l$; we thus check that B stays below A as assumed in our solution.

(2) For $m_A \gg m_B$, the answers obtained reduce to $(\mathbf{v}_B)_2 = (\mathbf{v}_A)_2 = 0$ and $h = v_0^2/2g$; B oscillates as a simple pendulum with A fixed. For $m_A \ll m_B$, they reduce to $(\mathbf{v}_B)_2 = (\mathbf{v}_A)_2 = \mathbf{v}_0$ and $h = 0$; A and B move with the same constant velocity $\mathbf{v}_0$.

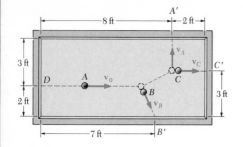

SAMPLE PROBLEM 14.5

In a game of billiards, ball A is given an initial velocity $\mathbf{v}_0$ of magnitude $v_0 = 10$ ft/s along line DA parallel to the axis of the table. It hits ball B and then ball C, which are both at rest. Knowing that A and C hit the sides of the table squarely at points A' and C', respectively, that B hits the side obliquely at B', and assuming frictionless surfaces and perfectly elastic impacts, determine the velocities $\mathbf{v}_A$, $\mathbf{v}_B$, and $\mathbf{v}_C$ with which the balls hit the sides of the table. (*Remark.* In this sample problem and in several of the problems which follow, the billiard balls are assumed to be particles moving freely in a horizontal plane, rather than the rolling and sliding spheres they actually are.)

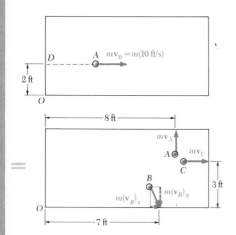

Solution. *Conservation of Momentum.* Since there is no external force, the initial momentum $m\mathbf{v}_0$ is equipollent to the system of momenta after the two collisions (and before any of the balls hits the side of the table). Referring to the adjoining sketch, we write

$\rightarrow x$ components:
$$m(10 \text{ ft/s}) = m(v_B)_x + mv_C \qquad (1)$$

$+\uparrow y$ components:
$$0 = mv_A{}' - m(v_B)_y \qquad (2)$$

$+\uparrow$ moments about O:
$$-(2 \text{ ft})m(10 \text{ ft/s}) = (8 \text{ ft})mv_A$$
$$-(7 \text{ ft})m(v_B)_y - (3 \text{ ft})mv_C \qquad (3)$$

Solving the three equations for v_A, $(v_B)_x$, and $(v_B)_y$ in terms of v_C,

$$v_A = (v_B)_y = 3v_C - 20 \qquad (v_B)_x = 10 - v_C \qquad (4)$$

Conservation of Energy. Since the surfaces are frictionless and the impacts are perfectly elastic, the initial kinetic energy $\frac{1}{2}mv_0^2$ is equal to the final kinetic energy of the system:

$$\tfrac{1}{2}mv_0^2 = \tfrac{1}{2}m_A v_A^2 + \tfrac{1}{2}m_B v_B^2 + \tfrac{1}{2}m_C v_C^2$$
$$v_A^2 + (v_B)_x^2 + (v_B)_y^2 + v_C^2 = (10 \text{ ft/s})^2 \qquad (5)$$

Substituting for v_A, $(v_B)_x$, and $(v_B)_y$ from (4) into (5), we have

$$2(3v_C - 20)^2 + (10 - v_C)^2 + v_C^2 = 100$$
$$20v_C^2 - 260v_C + 800 = 0$$

Solving for v_C, we find $v_C = 5$ ft/s and $v_C = 8$ ft/s. Since only the second root yields a positive value for v_A after substitution into Eqs. (4), we conclude that $v_C = 8$ ft/s and

$$v_A = (v_B)_y = 3(8) - 20 = 4 \text{ ft/s} \qquad (v_B)_x = 10 - 8 = 2 \text{ ft/s}$$

$$\mathbf{v}_A = 4 \text{ ft/s} \uparrow \qquad \mathbf{v}_B = 4.47 \text{ ft/s} \searrow 63.4° \qquad \mathbf{v}_C = 8 \text{ ft/s} \rightarrow \quad \blacktriangleleft$$

Problems

14.26 In Prob. 14.1, determine the energy lost due to friction as the bullet (a) moves through block A, (b) penetrates block B.

14.27 In Prob. 14.3, determine the energy lost (a) in the coupling of the engine and the flatcar, (b) while the load slides on the car.

14.28 Each of the three freight cars of Prob. 14.5 has a mass of 36 Mg. Assuming that car B is first hit by car A and gets automatically coupled with it and then is hit by car C and gets automatically coupled with it, determine the energy lost (a) in the first coupling, (b) in the second coupling.

use vel. results from Prob #5 —

14.29 In Prob. 14.19, determine the work done by the internal forces during the explosion of the shell.

They strick together after colision

14.30 Two automobiles A and B, of mass m_A and m_B, respectively, are traveling in opposite directions when they collide head on. The impact is assumed perfectly plastic, and it is further assumed that the energy absorbed by each automobile is equal to its loss of kinetic energy with respect to a moving frame of reference attached to the mass center of the two-vehicle system. Denoting by E_A and E_B, respectively, the energy absorbed by automobile A and by automobile B, (a) show that $E_A/E_B = m_B/m_A$, that is, the amount of energy absorbed by each vehicle is inversely proportional to its mass, (b) compute E_A and E_B, knowing that automobile A weighs 3500 lb and had a speed of 55 mi/h before the crash, while automobile B weighs 2000 lb and had a speed of 35 mi/h.

$G = $ mass center

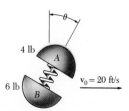

Fig. P14.30

14.31 It is assumed that each of the two automobiles involved in the collision described in Prob. 14.30 had been designed to safely withstand a test in which it crashed into a solid, immovable wall at the speed v_0. The severity of the collision of Prob. 14.30 may then be measured for each vehicle by the ratio of the energy it absorbed in the collision to the energy it absorbed in the test. On that basis, show that the collision described in Prob. 14.30 is $(m_A/m_B)^2$ times more severe for automobile B than for automobile A.

14.32 When the cord connecting particles A and B is severed, the compressed spring causes the particles to fly apart (the spring is not connected to the particles). The potential energy of the compressed spring is known to be 40 ft·lb and the assembly has an initial velocity v_0 as shown. If the cord is severed when $\theta = 25°$, determine the resulting velocity of each particle.

Fig. P14.32

14.33 A 20-kg block B is suspended from a 2-m cord attached to a 30-kg cart A, which may roll freely on a frictionless, horizontal track. If the system is released from rest in the position shown, determine the velocities of A and B as B passes directly under A.

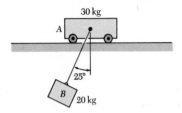

30 kg

A

25°

B 20 kg

Fig. P14.33

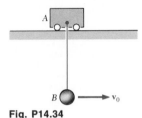

A

B v_0

Fig. P14.34

14.34 Ball B is suspended from a cord of length l attached to cart A, which may roll freely on a frictionless, horizontal track. The ball and the cart have the same mass m. If the ball is given an initial horizontal velocity v_0 while the cart is at rest, describe the subsequent motion of the system, specifying the velocities of A and B for the following successive values of the angle θ (assumed positive counterclockwise) that the cord will form with the vertical: (a) $\theta = \theta_{max}$, (b) $\theta = 0$, (c) $\theta = \theta_{min}$.

v_0

A

B

C

Fig. P14.35

14.35 In a game of billiards, ball A is moving with the velocity $v_0 = v_0 i$ when it strikes balls B and C which are at rest side by side. Assuming frictionless surfaces and perfectly elastic impact (i.e., conservation of energy), determine the final velocity of each ball, assuming that the path of A is (a) perfectly centered and that A strikes B and C simultaneously, (b) not perfectly centered and that A strikes B slightly before it strikes C.

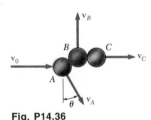

v_B

B C

v_0 v_C

A

θ v_A

Fig. P14.36

14.36 In a game of billiards, ball A is moving with the velocity $v_0 = (5 \text{ m/s})i$ when it strikes balls B and C which are at rest side by side. After the collision, the three balls are observed to move in the directions shown, with $\theta = 20°$. Assuming frictionless surfaces and perfectly elastic impacts (i.e., conservation of energy), determine the magnitudes of the velocities v_A, v_B, and v_C.

14.37 A 15-lb block B starts from rest and slides on the 25-lb wedge A, which is supported by a horizontal surface. Neglecting friction, determine (a) the velocity of B relative to A after it has slid 3 ft down the inclined surface of the wedge, (b) the corresponding velocity of A.

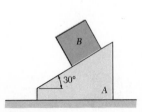

B

30°

A

Fig. P14.37

14.38 Three spheres, each of mass m, may slide freely on a frictionless, horizontal surface. Spheres A and B are attached to an inextensible, inelastic cord of length l and are at rest in the position shown when sphere B is struck squarely by sphere C which is moving to the right with a velocity $\mathbf{v}_0$. Knowing that the cord is taut when sphere B is struck by sphere C and assuming perfectly elastic impact between B and C, and thus conservation of energy for the entire system, determine the velocity of each sphere immediately after impact.

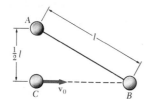

Fig. P14.38

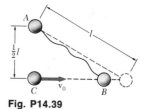

Fig. P14.39

14.39 Three spheres, each of mass m, may slide freely on a frictionless, horizontal surface. Spheres A and B are attached to an inextensible, inelastic cord of length l and are at rest in the position shown when sphere B is struck squarely by sphere C which is moving to the right with a velocity $\mathbf{v}_0$. Knowing that the cord is slack when sphere B is struck by sphere C and assuming perfectly elastic impact between B and C, determine (a) the velocity of each sphere immediately after the cord becomes taut, (b) the fraction of the initial kinetic energy of the system which is dissipated when the cord becomes taut.

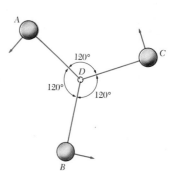

Fig. P14.40

14.40 Three small spheres A, B, and C, each of mass m, are connected to a small ring D by means of three inextensible, inelastic cords of length l which are equally spaced. The spheres may slide freely on a frictionless horizontal surface and are rotating initially at a speed v_0 about ring D which is at rest. Suddenly cord CD breaks. After the other two cords have again become taut, determine (a) the speed of ring D, (b) the relative speed at which spheres A and B rotate about D, (c) the fraction of the energy of the original system which is dissipated when cords AD and BD again become taut.

14.41 Two small spheres A and B, respectively of mass m and $2m$, are connected by a rigid rod of length l and negligible mass. The two spheres are resting on a horizontal, frictionless surface when A is suddenly given the velocity $\mathbf{v}_0 = v_0\mathbf{i}$. Determine (a) the linear momentum of the system and its angular momentum about its mass center G, (b) the velocities of A and B after the rod AB has rotated through $90°$, (c) the velocities of A and B after the rod AB has rotated through $180°$.

Fig. P14.41

14.42 In the scattering experiment of Prob. 14.18, it is known that the alpha particle is projected from $A_0(300, 0, 300)$ and that it collides with the oxygen nucleus C at $Q(240, 200, 100)$, where all coordinates are expressed in millimeters. Determine the coordinates of point B_0 where the original path of nucleus B intersects the xz plane. (*Hint.* Express that the angular momentum of the three particles about Q is conserved.)

14.43 A 750-lb space vehicle traveling with a velocity $\mathbf{v}_0 = (1500/\text{ft/s})\mathbf{k}$ passes through the origin O. Explosive charges then separate the vehicle into three parts A, B, and C, weighing, respectively, 125 lb, 250 lb, and 375 lb. Knowing that shortly thereafter the positions of the three parts are, respectively, $A(240, 240, 2160)$, $B(600, 1320, 3240)$, and $C(-480, -960, 1920)$, where the coordinates are expressed in feet, that the velocity of B is $\mathbf{v}_B = (500 \text{ ft/s})\mathbf{i} + (1100 \text{ ft/s})\mathbf{j} + (2200 \text{ ft/s})\mathbf{k}$, and that the x component of the velocity of C is -400 ft/s, determine the velocity of part A.

14.44 Three small identical spheres A, B, and C, which may slide on a horizontal, frictionless surface, are attached to three strings of length l which are tied to a ring G. Initially the spheres rotate about the ring which moves along the x axis with a velocity $\mathbf{v}_0$. Suddenly the ring breaks and the three spheres move freely in the xy plane. Knowing that $\mathbf{v}_A = (0.732 \text{ m/s})\mathbf{j}$, $\mathbf{v}_C = (1.2 \text{ m/s})\mathbf{i}$, $a = 208$ mm, and $d = 120$ mm, determine (*a*) the initial velocity of the ring, (*b*) the length l of the strings, (*c*) the rate in rad/s at which the spheres were rotating about G.

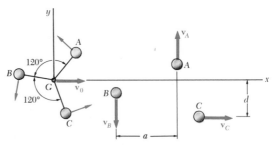

Fig. P14.44 and P14.45

14.45 Three small identical spheres A, B, and C, which may slide on a horizontal, frictionless surface, are attached to three strings, 150 mm long, which are tied to a ring G. Initially the spheres rotate counterclockwise about the ring with a relative velocity of 0.6 m/s and the ring moves along the x axis with a velocity $\mathbf{v}_0 = (0.3 \text{ m/s})\mathbf{i}$. Suddenly the ring breaks and the three spheres move freely in the xy plane with A and B following paths parallel to the y axis at a distance $a = 260$ mm from each other and C a path parallel to the x axis. Determine (*a*) the velocity of each sphere, (*b*) the distance d.

14.46 Two small disks A and B, of mass 3 kg and 1.5 kg, respectively, may slide on a horizontal, frictionless surface. They are connected by a cord, 600 mm long, and spin counterclockwise about their mass center G at the rate of 10 rad/s. At $t = 0$, the coordinates of G are $\bar{x}_0 = 0$, $\bar{y}_0 = 2$ m, and its velocity is $\bar{v}_0 = (1.2 \text{ m/s})\mathbf{i} + (0.96 \text{ m/s})\mathbf{j}$. Shortly thereafter the cord breaks; disk A is then observed to move along a path parallel to the y axis and disk B along a path which intersects the x axis at a distance $b = 7.5$ m from O. Determine (a) the velocities of A and B after the cord breaks, (b) the distance a from the y axis to the path of A.

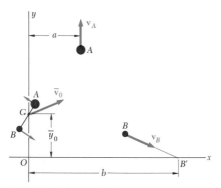

Fig. P14.46 and P14.47

14.47 Two small disks A and B, of mass 2 kg and 1 kg, respectively, may slide on a horizontal and frictionless surface. They are connected by a cord of negligible mass and spin about their mass center G. At $t = 0$, G is moving with the velocity $\bar{v}_0$ and its coordinates are $\bar{x}_0 = 0$, $\bar{y}_0 = 1.89$ m. Shortly thereafter, the cord breaks and disk A is observed to move with a velocity $\mathbf{v}_A = (5 \text{ m/s})\mathbf{j}$ in a straight line and at a distance $a = 2.56$ m from the y axis, while B moves with a velocity $\mathbf{v}_B = (7.2 \text{ m/s})\mathbf{i} - (4.6 \text{ m/s})\mathbf{j}$ along a path intersecting the x axis at a distance $b = 7.48$ m from the origin O. Determine (a) the initial velocity $\bar{v}_0$ of the mass center G of the two disks, (b) the length of the cord initially connecting the two disks, (c) the rate in rad/s at which the disks were spinning about G.

14.48 In a game of billiards, ball A is given an initial velocity $\mathbf{v}_0$ along line DA parallel to the axis of the table. It hits ball B and then ball C, which are both at rest. Balls A and C are observed to hit the sides of the table squarely at points A' and C', respectively, and ball B to hit the side obliquely at B'. Knowing that $v_0 = 10$ ft/s, $v_A = 4$ ft/s, and $a = 7$ ft, determine (a) the velocities $\mathbf{v}_B$ and $\mathbf{v}_C$ of balls B and C, (b) the point C' where ball C hits the side of the table. Assume frictionless surfaces and perfectly elastic impacts (i.e., conservation of energy).

14.49 For the game of billiards of Prob. 14.48, it is now assumed that $v_0 = 12$ ft/s, $v_C = 7$ ft/s, and $c = 3$ ft. Determine (a) the velocities $\mathbf{v}_A$ and $\mathbf{v}_B$ of balls A and B, (b) the point A' where ball A hits the side of the table.

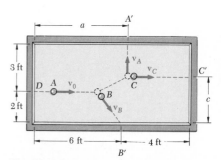

Fig. P14.48

* 14.10. Variable Systems of Particles. All the systems of particles considered so far consisted of well-defined particles. These systems did not gain or lose any particles during their motion. In a large number of engineering applications, however, it is necessary to consider *variable systems of particles,* i.e., systems which are continually gaining or losing particles, or doing both at the same time. Consider, for example, a hydraulic turbine. Its analysis involves the determination of the forces exerted by a stream of water on rotating blades, and we note that the particles of water in contact with the blades form an everchanging system which continually acquires and loses particles. Rockets furnish another example of variable systems, since their propulsion depends upon the continual ejection of fuel particles.

We recall that all the kinetics principles established so far were derived for constant systems of particles, which neither gain nor lose particles. We must therefore find a way to reduce the analysis of a variable system of particles to that of an auxiliary constant system. The procedure to follow is indicated in Secs. 14.11 and 14.12 for two broad categories of applications.

* 14.11. Steady Stream of Particles. Consider a steady stream of particles, such as a stream of water diverted by a fixed vane or a flow of air through a duct or through a blower. In order to determine the resultant of the forces exerted on the particles in contact with the vane, duct, or blower, we isolate these particles and denote by S the system thus defined (Fig. 14.9). We observe that S is a variable system of particles, since it continually gains particles flowing in and loses an equal number of particles flowing out. Therefore, the kinetics principles that have been established so far cannot be directly applied to S.

However, we may easily define an auxiliary system of particles which does remain constant for a short interval of time Δt. Consider at time t the system S *plus* the particles which will enter S during the interval of time Δt (Fig. 14.10a). Next, consider at time $t + \Delta t$ the system S *plus* the particles which have left S during the interval Δt (Fig. 14.10c). Clearly, *the same particles are involved in both cases,* and we may apply to these particles the principle of impulse and momentum. Since the total mass m of the system S remains constant, the particles entering the system and those leaving the system in the time Δt must have the same mass Δm. Denoting by $\mathbf{v}_A$ and $\mathbf{v}_B$, respectively, the velocities of the particles entering S at A and leaving S at B, we represent the momentum of the particles entering S by $(\Delta m)\mathbf{v}_A$ (Fig. 14.10a) and the momentum of the particles leaving S by $(\Delta m)\mathbf{v}_B$ (Fig. 14.10c). We also represent the momenta $m_i\mathbf{v}_i$ of the particles forming S and the impulses of the forces exerted on S by the appropriate vectors, and indicate by gray plus and equals signs that the system of the momenta and impulses in parts *a* and *b* of Fig. 14.10 is equipollent to the system of the momenta in part *c* of the same figure.

Since the resultant $\Sigma m_i\mathbf{v}_i$ of the momenta of the particles of S is found on both sides of the equals sign, it may be omitted. We conclude that *the system formed by the momentum $(\Delta m)\mathbf{v}_A$ of the particles entering S in the*

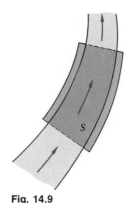

Fig. 14.9

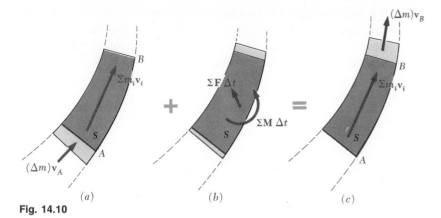

Fig. 14.10

time Δt *and the impulses of the forces exerted on S during that time is equipollent to the momentum* $(\Delta m)\mathbf{v}_B$ *of the particles leaving S in the same time* Δt. We may therefore write

$$(\Delta m)\mathbf{v}_A + \Sigma\mathbf{F}\,\Delta t = (\Delta m)\mathbf{v}_B \tag{14.38}$$

A similar equation may be obtained by taking the moments of the vectors involved (see Sample Prob. 14.5). Dividing all terms of Eq. (14.38) by Δt and letting Δt approach zero, we obtain at the limit

$$\Sigma\mathbf{F} = \frac{dm}{dt}(\mathbf{v}_B - \mathbf{v}_A) \tag{14.39}$$

where $\mathbf{v}_B - \mathbf{v}_A$ represents the difference between the *vectors* $\mathbf{v}_B$ and $\mathbf{v}_A$.

If SI units are used, dm/dt is expressed in kg/s and the velocities in m/s; we check that both members of Eq. (14.39) are expressed in the same units (newtons). If U.S. customary units are used, dm/dt must be expressed in slugs/s and the velocities in ft/s; we check again that both members of the equation are expressed in the same units (pounds).†

The principle we have established may be used to analyze a large number of engineering applications. Some of the most common are indicated below.

Fluid Stream Diverted by a Vane. If the vane is fixed, the method of analysis given above may be applied directly to find the force $\mathbf{F}$ exerted by the vane on the stream. We note that $\mathbf{F}$ is the only force which needs to be considered since the pressure in the stream is constant (atmospheric pressure). The force exerted by the stream on the vane will be equal and

---

† It is often convenient to express the mass rate of flow dm/dt as the product ρQ, where ρ is the density of the stream (mass per unit volume) and Q its volume rate of flow (volume per unit time). If SI units are used, ρ is expressed in kg/m³ (for instance, $\rho = 1000$ kg/m³ for water) and Q in m³/s. However, if U.S. customary units are used, ρ will generally have to be computed from the corresponding specific weight γ (weight per unit volume), $\rho = \gamma/g$. Since γ is expressed in lb/ft³ (for instance, $\gamma = 62.4$ lb/ft³ for water), ρ is obtained in slugs/ft³. The volume rate of flow Q is expressed in ft³/s.

opposite to **F**. If the vane moves with a constant velocity, the stream is not steady. However, it will appear steady to an observer moving with the vane. We should therefore choose a system of axes moving with the vane. Since this system of axes is not accelerated, Eq. (14.38) may still be used, but v_A and v_B must be replaced by the *relative velocities* of the stream with respect to the vane (see Sample Prob. 14.6).

Fluid Flowing through a Pipe. The force exerted by the fluid on a pipe transition such as a bend or a contraction may be determined by considering the system of particles S in contact with the transition. Since, in general, the pressure in the flow will vary, we should also consider the forces exerted on S by the adjoining portions of the fluid.

Jet Engine. In a jet engine, air enters with no velocity through the front of the engine and leaves through the rear with a high velocity. The energy required to accelerate the air particles is obtained by burning fuel. While the exhaust gases contain burned fuel, the mass of the fuel is small compared with the mass of the air flowing through the engine and usually may be neglected. Thus, the analysis of a jet engine reduces to that of an air stream. This stream may be considered as a steady stream if all velocities are measured with respect to the airplane. The air stream shall be assumed, therefore, to enter the engine with a velocity **v** of magnitude equal to the speed of the airplane and to leave with a velocity **u** equal to the relative velocity of the exhaust gases (Fig. 14.11). Since the intake and exhaust pressures are nearly atmospheric, the only external force which needs to be considered is the force exerted by the engine on the air stream. This force is equal and opposite to the thrust.†

Fan. We consider the system of particles S shown in Fig. 14.12. The velocity v_A of the particles entering the system is assumed equal to zero,

Fig. 14.11

---

† Note that if the airplane is accelerated, it cannot be used as a newtonian frame of reference. The same result will be obtained for the thrust, however, by using a reference frame at rest with respect to the atmosphere, since the air particles will then be observed to enter the engine with no velocity and to leave it with a velocity of magnitude $u - v$.

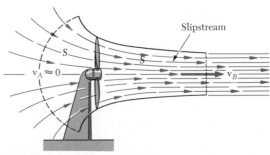

Fig. 14.12

and the velocity $\mathbf{v}_B$ of the particles leaving the system is the velocity of the *slipstream*. The rate of flow may be obtained by multiplying v_B by the cross-sectional area of the slipstream. Since the pressure all around S is atmospheric, the only external force acting on S is the thrust of the fan.

Airplane Propeller. In order to obtain a steady stream of air, velocities should be measured with respect to the airplane. Thus, the air particles will be assumed to enter the system with a velocity $\mathbf{v}$ of magnitude equal to the speed of the airplane and to leave with a velocity $\mathbf{u}$ equal to the relative velocity of the slipstream.

* 14.12. Systems Gaining or Losing Mass. We shall now analyze a different type of variable system of particles, namely, a system which gains mass by continually absorbing particles or loses mass by continually expelling particles. Consider the system S shown in Fig. 14.13. Its mass, equal to m at the instant t, increases by Δm in the interval of time Δt.

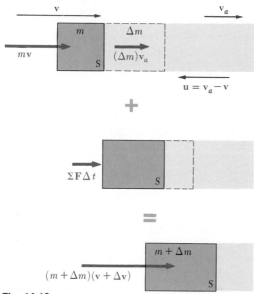

Fig. 14.13

In order to apply the principle of impulse and momentum to the analysis of this system, we must consider at time t the system S *plus* the particles of mass Δm which S absorbs during the time interval Δt. The velocity of S at time t is denoted by $\mathbf{v}$, and its velocity at time $t + \Delta t$ is denoted by $\mathbf{v} + \Delta\mathbf{v}$, while the absolute velocity of the particles which are absorbed is denoted by $\mathbf{v}_a$. Applying the principle of impulse and momentum, we write

$$m\mathbf{v} + (\Delta m)\mathbf{v}_a + \Sigma\mathbf{F}\,\Delta t = (m + \Delta m)(\mathbf{v} + \Delta\mathbf{v})$$

Solving for the sum $\Sigma\mathbf{F}\,\Delta t$ of the impulses of the external forces acting on S (excluding the forces exerted by the particles being absorbed), we have

$$\Sigma\mathbf{F}\,\Delta t = m\Delta\mathbf{v} + \Delta m(\mathbf{v} - \mathbf{v}_a) + (\Delta m)(\Delta\mathbf{v}) \tag{14.40}$$

Introducing the *relative velocity* $\mathbf{u}$ with respect to S of the particles which are absorbed, we write $\mathbf{u} = \mathbf{v}_a - \mathbf{v}$ and note, since $v_a < v$, that the relative velocity $\mathbf{u}$ is directed to the left, as shown in Fig. 14.13. Neglecting the last term in Eq. (14.40), which is of the second order, we write

$$\Sigma\mathbf{F}\,\Delta t = m\,\Delta\mathbf{v} - (\Delta m)\mathbf{u}$$

Dividing through by Δt and letting Δt approach zero, we have at the limit†

$$\Sigma\mathbf{F} = m\frac{d\mathbf{v}}{dt} - \frac{dm}{dt}\mathbf{u} \tag{14.41}$$

Rearranging the terms and recalling that $d\mathbf{v}/dt = \mathbf{a}$, where $\mathbf{a}$ is the acceleration of the system S, we write

$$\Sigma\mathbf{F} + \frac{dm}{dt}\mathbf{u} = m\mathbf{a} \tag{14.42}$$

which shows that the action on S of the particles being absorbed is equivalent to a thrust

$$\mathbf{P} = \frac{dm}{dt}\mathbf{u} \tag{14.43}$$

which tends to slow down the motion of S, since the relative velocity $\mathbf{u}$ of the particles is directed to the left. If SI units are used, dm/dt is expressed in kg/s, the relative velocity u in m/s, and the corresponding thrust in newtons. If U.S. customary units are used, dm/dt must be expressed in slugs/s and u in ft/s; the corresponding thrust will then be expressed in pounds.‡

The equations obtained may also be used to determine the motion of a system S losing mass. In this case, the rate of change of mass is negative, and the action on S of the particles being expelled is equivalent to a thrust in the direction of $-\mathbf{u}$, that is, in the direction opposite to that in which the particles are being expelled. A *rocket* represents a typical case of a system continually losing mass (see Sample Prob. 14.7).

---

† When the absolute velocity $\mathbf{v}_a$ of the particles absorbed is zero, we have $\mathbf{u} = -\mathbf{v}$, and formula (14.41) becomes

$$\Sigma\mathbf{F} = \frac{d}{dt}(m\mathbf{v})$$

Comparing the formula obtained to Eq. (12.3) of Sec. 12.3, we observe that Newton's second law may be applied to a system gaining mass. *provided that the particles absorbed are initially at rest.* It may also be applied to a system losing mass, *provided that the velocity of the particles expelled is zero* with respect to the frame of reference selected.

‡ See footnote on page 679.

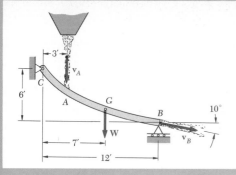

SAMPLE PROBLEM 14.6

Grain falls from a hopper onto a chute CB at the rate of 240 lb/s. It hits the chute at A with a velocity of 20 ft/s and leaves at B with a velocity of 15 ft/s, forming an angle of 10° with the horizontal. Knowing that the combined weight of the chute and of the grain it supports is a force $\mathbf{W}$ of magnitude 600 lb applied at G, determine the reaction at the roller-support B and the components of the reaction at the hinge C.

Solution. We apply the principle of impulse and momentum for the time interval Δt to the system consisting of the chute, the grain it supports, and the amount of grain which hits the chute in the interval Δt. Since the chute does not move, it has no momentum. We also note that the sum $\Sigma m_i \mathbf{v}_i$ of the momenta of the particles supported by the chute is the same at t and $t + \Delta t$ and thus may be omitted.

 $+$ 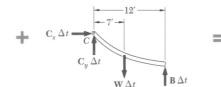 $=$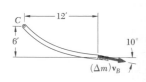

Since the system formed by the momentum $(\Delta m)\mathbf{v}_A$ and the impulses is equipollent to the momentum $(\Delta m)\mathbf{v}_B$, we write

$\xrightarrow{+}$ x components:
$$C_x \, \Delta t = (\Delta m)v_B \cos 10° \qquad (1)$$

$+\uparrow y$ components:
$$-(\Delta m)v_A + C_y \, \Delta t - W \, \Delta t + B \, \Delta t$$
$$= -(\Delta m)v_B \sin 10° \quad (2)$$

$+\uparrow$ moments about C:
$$-3(\Delta m)v_A - 7(W \, \Delta t) + 12(B \, \Delta t)$$
$$= 6(\Delta m)v_B \cos 10° - 12(\Delta m)v_B \sin 10° \quad (3)$$

Using the given data, $W = 600$ lb, $v_A = 20$ ft/s, $v_B = 15$ ft/s, $\Delta m/\Delta t = 240/32.2 = 7.45$ slugs/s, and solving Eq. (3) for B and Eq. (1) for C_x,

$$12B = 7(600) + 3(7.45)(20) + 6(7.45)(15)(\cos 10° - 2 \sin 10°)$$
$$12B = 5075 \qquad B = 423 \text{ lb} \qquad\qquad \mathbf{B} = 423 \text{ lb} \uparrow \quad \blacktriangleleft$$

$$C_x = (7.45)(15) \cos 10° = 110.1 \text{ lb} \qquad \mathbf{C}_x = 110.1 \text{ lb} \rightarrow \quad \blacktriangleleft$$

Substituting for B and solving Eq. (2) for C_y,

$$C_y = 600 - 423 + (7.45)(20 - 15 \sin 10°) = 307 \text{ lb}$$
$$\mathbf{C}_y = 307 \text{ lb} \uparrow \quad \blacktriangleleft$$

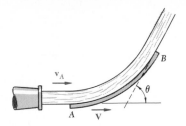

SAMPLE PROBLEM 14.7

A nozzle discharges a stream of water of cross-sectional area A with a velocity $\mathbf{v}_A$. The stream is deflected by a *single* blade which moves to the right with a constant vleocity $\mathbf{V}$. Assuming that the water moves along the blade at constant speed, determine (a) the components of the force $\mathbf{F}$ exerted by the blade on the stream, (b) the velocity $\mathbf{V}$ for which maximum power is developed.

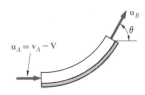

a. **Components of Force Exerted on Stream.** We choose a coordinate system which moves with the blade at a constant velocity $\mathbf{V}$. The particles of water strike the blade with a relative velocity $\mathbf{u}_A = \mathbf{v}_A - \mathbf{V}$ and leave the blade with a relative velocity $\mathbf{u}_B$. Since the particles move along the blade at a constant speed, the relative velocities $\mathbf{u}_A$ and $\mathbf{u}_B$ have the same magnitude u. Denoting the density of water by ρ, the mass of the particles striking the blade during the time interval Δt is $\Delta m = A\rho(v_A - V) \Delta t$; an equal mass of particles leaves the blade during Δt. We apply the principle of impulse and momentum to the system formed by the particles in contact with the blade and by those striking the blade in the time Δt.

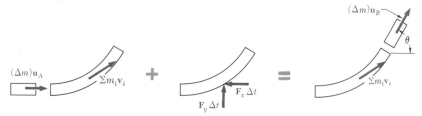

Recalling that $\mathbf{u}_A$ and $\mathbf{u}_B$ have the same magnitude u, and omitting the momentum $\Sigma m_i \mathbf{v}_i$ which appears on both sides, we write

$\xrightarrow{+}$ x components: $(\Delta m)u - F_x \Delta t = (\Delta m)u \cos \theta$

$+\uparrow y$ components: $+F_y \Delta t = (\Delta m)u \sin \theta$

Substituting $\Delta m = A\rho(v_A - V) \Delta t$ and $u = v_A - V$, we obtain

$$\mathbf{F}_x = A\rho(v_A - V)^2(1 - \cos \theta) \leftarrow \qquad \mathbf{F}_y = A\rho(v_A - V)^2 \sin \theta \uparrow \quad \blacktriangleleft$$

b. **Velocity of Blade for Maximum Power.** The power is obtained by multiplying the velocity V of the blade by the component F_x of the force exerted by the stream on the blade.

$$\text{Power} = F_x V = A\rho(v_A - V)^2(1 - \cos \theta)V$$

Differentiating the power with respect to V and setting the derivative equal to zero, we obtain

$$\frac{d(\text{power})}{dV} = A\rho(v_A^2 - 4v_A V + 3V^2)(1 - \cos \theta) = 0$$

$$V = v_A \qquad V = \tfrac{1}{3}v_A \qquad \text{For maximum power } \mathbf{V} = \tfrac{1}{3}v_A \rightarrow \quad \blacktriangleleft$$

Note. These results are valid only when a *single* blade deflects the stream. Different results are obtained when a series of blades deflects the stream, as in a Pelton-wheel turbine. (See Prob. 14.73.)

SAMPLE PROBLEM 14.8

A rocket of initial mass m_0 (including shell and fuel) is fired vertically at time $t = 0$. The fuel is consumed at a constant rate $q = dm/dt$ and is expelled at a constant speed u relative to the rocket. Derive an expression for the velocity of the rocket at time t, neglecting the resistance of the air.

Solution. At time t, the mass of the rocket shell and remaining fuel is $m = m_0 - qt$, and the velocity is **v**. During the time interval Δt, a mass of fuel $\Delta m = q\,\Delta t$ is expelled with a speed u relative to the rocket. Denoting by $\mathbf{v}_e$ the absolute velocity of the expelled fuel, we apply the principle of impulse and momentum between time t and time $t + \Delta t$.

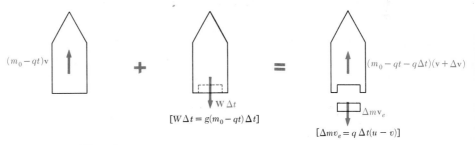

We write

$$(m_0 - qt)v - g(m_0 - qt)\,\Delta t = (m_0 - qt - q\,\Delta t)(v + \Delta v) - q\,\Delta t(u - v)$$

Dividing through by Δt, and letting Δt approach zero, we obtain

$$-g(m_0 - qt) = (m_0 - qt)\frac{dv}{dt} - qu$$

Separating variables and integrating from $t = 0$, $v = 0$ to $t = t$, $v = v$,

$$dv = \left(\frac{qu}{m_0 - qt} - g\right)dt \qquad \int_0^v dv = \int_0^t \left(\frac{qu}{m_0 - qt} - g\right)dt$$

$$v = [-u\ln(m_0 - qt) - gt]_0^t \qquad v = u\ln\frac{m_0}{m_0 - qt} - gt \quad \blacktriangleleft$$

Remark. The mass remaining at time t_f, after all the fuel has been expended, is equal to the mass of the rocket shell $m_s = m_0 - qt_f$, and the maximum velocity attained by the rocket is $v_m = u\ln(m_0/m_s) - gt_f$. Assuming that the fuel is expelled in a relatively short period of time, the term gt_f is small and we have $v_m \approx u\ln(m_0/m_s)$. In order to escape the gravitational field of the earth, a rocket must reach a velocity of 11.18 km/s. Assuming $u = 2200$ m/s and $v_m = 11.18$ km/s, we obtain $m_0/m_s = 161$. Thus, to project each kilogram of the rocket shell into space, it is necessary to consume more than 161 kg of fuel if a propellant yielding $u = 2200$ m/s is used.

685

Problems

Note. In the following problems use $\rho = 1000 \text{ kg/m}^3$ for the density of water in SI units, and $\gamma = 62.4 \text{ lb/ft}^3$ for its specific weight in U.S. customary units. (See footnote on page 679.)

14.50 A hose discharges water at a rate of 12 m³/min from the stern of a 30-Mg fireboat. If the velocity of the water stream is 35 m/s, determine the reaction on the boat.

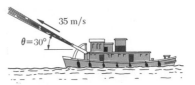

Fig. P14.50

14.51 Sand is discharged at the rate of m kg/s from a conveyor belt moving with a velocity $\mathbf{v}_0$. The sand is deflected by a plate at B so that it falls in a vertical stream. After falling a distance h, the sand is again deflected as shown by a curved plate at C. Neglecting the friction between the sand and the plates, determine the force required to hold each plate in the position shown.

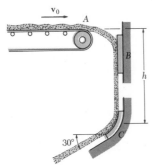

Fig. P14.51

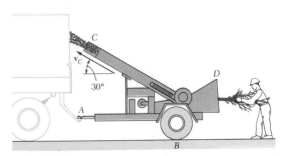

Fig. P14.52

14.52 Tree limbs and branches are being fed at D at a rate of 10 lb/s into a shredder which spews the resulting wood chips at C with a velocity of 60 ft/s. Determine the horizontal component of the force exerted by the shredder on the truck hitch at A.

14.53 A stream of water of cross-sectional area A and velocity $\mathbf{v}_1$ strikes a plate which is held motionless by a force $\mathbf{P}$. Determine the magnitude of $\mathbf{P}$, knowing that $A = 500 \text{ mm}^2$, $v_1 = 25 \text{ m/s}$, and $V = 0$.

14.54 A stream of water of cross-sectional area A and velocity $\mathbf{v}_1$ strikes a plate which moves to the right with a velocity $\mathbf{V}$. Determine the magnitude of $\mathbf{V}$, knowing that $A = 600 \text{ mm}^2$, $v_1 = 30 \text{ m/s}$, and $P = 400 \text{ N}$.

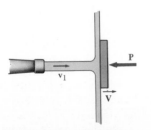

Fig. P14.53 and P14.54

14.55 Water flows in a continuous sheet from between two plates A and B with a velocity $\mathbf{v}$ of magnitude 90 ft/s. The stream is split into two parts by a smooth horizontal plate C. Knowing that the rates of flow in each of the two resulting streams are, respectively, $Q_1 = 30$ gal/min and $Q_2 = 150$ gal/min, determine (*a*) the angle θ, (*b*) the total force exerted by the stream on the plate $(1 \text{ ft}^3 = 7.48 \text{ gal})$.

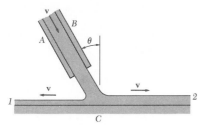

Fig. P14.55 and P14.56

14.56 Water flows in a continuous sheet from between two plates A and B with a velocity $\mathbf{v}$ of magnitude 120 ft/s. The stream is split into two parts by a smooth horizontal plate C. Determine the rates of flow Q_1 and Q_2 in each of the two resulting streams, knowing that $\theta = 25°$ and that the total force exerted by the streams on the plate is a 90-lb vertical force $(1 \text{ ft}^3 = 7.48 \text{ gal})$.

14.57 The nozzle shown discharges water at the rate of 250 gal/min. Knowing that at both A and B the stream of water moves with a velocity of magnitude 90 ft/s and neglecting the weight of the vane, determine the components of the reactions at C and D $(1 \text{ ft}^3 = 7.48 \text{ gal})$.

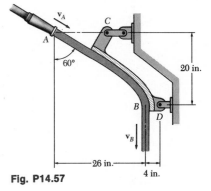

Fig. P14.57

14.58 The nozzle shown discharges water at the rate of $1.2 \text{ m}^3/\text{min}$. Knowing that at both A and B the water moves with a velocity of magnitude 50 m/s and neglecting the weight of the vane, determine the force-couple system which must be applied at C to hold the vane in place.

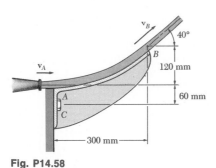

Fig. P14.58

14.59 Knowing that the blade AB of Sample Prob. 14.7 is in the shape of an arc of a circle, show that the resultant force $\mathbf{F}$ exerted by the blade on the stream is applied at the midpoint C of the arc AB. (*Hint.* First show that the line of action of $\mathbf{F}$ must pass through the center O of the circle.)

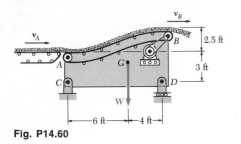

Fig. P14.60

14.60 The final component of a conveyor system receives sand at a rate of 240 lb/s at A and discharges it at B. The sand is moving horizontally at A and B with a velocity of magnitude $v_A = v_B = 15$ ft/s. Knowing that the combined weight of the component and of the sand it supports is $W = 900$ lb, determine the reactions at C and D.

14.61 Coal is being discharged from a first conveyor belt at a rate of 150 kg/s. It is received at A by a second belt which discharges it again at B. Knowing that $v_1 = 3$ m/s and $v_2 = 4.25$ m/s and that the second conveyor-belt assembly and the coal it supports have a total mass of 500 kg, determine the reactions at C and D.

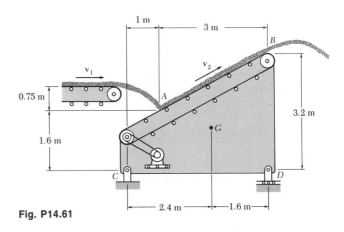

Fig. P14.61

14.62 Solve Prob. 14.61, assuming that the velocity of the second conveyor belt is decreased to $v_2 = 2.55$ m/s and that the mass of coal it supports is increased by 50 kg.

14.63 The total drag due to air friction on a jet airplane traveling at 900 km/h is 35 kN. Knowing that the exhaust velocity is 600 m/s relative to the airplane, determine the mass of air which must pass through the engine per second to maintain the speed of 900 km/h in level flight.

14.64 While cruising in level flight at a speed of 540 mi/h, a jet airplane scoops in air at a rate of 150 lb/s and discharges it with a velocity of 2000 ft/s relative to the airplane. Determine the total drag due to air friction on the airplane.

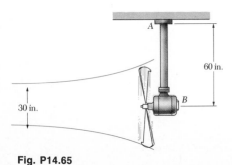

Fig. P14.65

14.65 For the ceiling-mounted fan shown, determine the maximum allowable air velocity in the slipstream if the bending moment in the supporting rod AB is not to exceed 80 lb·ft. Assume $\gamma = 0.076$ lb/ft^3 for air and neglect the approach velocity of the air.

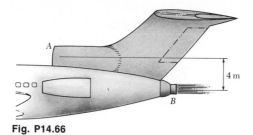

Fig. P14.66

14.66 The jet engine shown scoops in air at A at a rate of 90 kg/s and discharges it at B with a velocity of 750 m/s relative to the airplane. Determine the magnitude and line of action of the propulsive thrust developed by the engine when the speed of the airplane is (*a*) 500 km/h, (*b*) 1000 km/h.

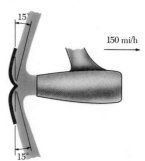

Fig. P14.67

14.67 In order to shorten the distance required for landing, a jet airplane is equipped with movable vanes which partially reverse the direction of the air discharged by each of its engines. Each engine scoops in air at a rate of 250 lb/s and discharges it with a velocity of 2000 ft/s relative to the engine. At an instant when the speed of the airplane is 150 mi/h, determine the reversed thrust provided by each of the engines.

Fig. P14.68

14.68 The helicopter shown has a mass of 12 Mg when empty and can produce a maximum downward air speed of 30 m/s in its 16-m-diameter slipstream. Assuming $\rho = 1.21$ kg/m³ for air, determine the maximum combined payload and fuel load the helicopter can carry while hovering in midair.

14.69 A jet airliner is cruising at a speed of 900 km/h with each of its three engines discharging air with a velocity of 800 m/s relative to the plane. Determine the speed of the airliner after it has lost the use of (*a*) one of its engines, (*b*) two of its engines. Assume that the drag due to air friction is proportional to the square of the speed and that the remaining engines keep operating at the same rate.

Fig. P14.69

Fig. P14.70

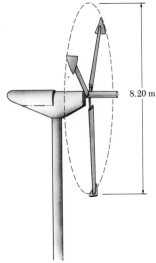

8.20 m

Fig. P14.71 and P14.72

14.70 A 35,000-lb jet airplane maintains a constant speed of 480 mi/h while climbing at an angle $\alpha = 15°$. The airplane scoops in air at a rate of 600 lb/s and discharges it with a velocity of 2250 ft/s relative to the airplane. If the pilot changes to a horizontal flight and the same engine conditions are maintained, determine (a) the initial acceleration of the plane, (b) the maximum horizontal speed attained. Assume that the drag due to air friction is proportional to the square of the speed.

14.71 The wind turbine-generator shown has an output-power rating of 20 kW for a wind speed of 45 km/h. For the given wind speed, determine (a) the kinetic energy of the air particles entering the 8.20-m-diameter circle per second, (b) the efficiency of this energy-conversion system. Assume $\rho = 1.21$ kg/m³ for air.

14.72 For a given wind speed, the wind turbine-generator shown produces 15 kW of electric power and, as an energy-conversion system, has an efficiency of 0.30. Determine (a) the kinetic energy of the air particles entering the 8.20-m-diameter circle per second, (b) the wind speed.

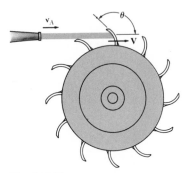

Fig. P14.73

14.73 In a Pelton-wheel turbine, a stream of water is deflected by a series of blades so that the rate at which water is deflected by the blades is equal to the rate at which water issues from the nozzle ($\Delta m/\Delta t = A\rho v_A$). Using the same notation as in Sample Prob. 14.7, (a) determine the velocity **V** of the blades for which maximum power is developed, (b) derive an expression for the maximum power, (c) derive an expression for the mechanical efficiency.

14.74 While cruising in level flight at a speed of 570 mi/h, a jet airplane scoops in air at a rate of 240 lb/s and discharges it with a velocity of 2200 ft/s relative to the airplane. Determine (a) the power actually used to propel the airplane, (b) the total power developed by the engine, (c) the mechanical efficiency of the airplane.

*14.75 The depth of water flowing in a rectangular channel of width b at a speed v_1 and depth d_1 increases to a depth d_2 at a *hydraulic jump*. Express the rate of flow Q in terms of b, d_1, and d_2.

Fig. P14.75

*14.76 Determine the rate of flow in the channel of Prob. 14.75, knowing that $d_1 = 1.2$ m, $d_2 = 1.5$ m, and the rectangular channel is 4 m wide.

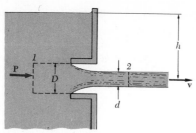

Fig. P14.77

14.77 A circular reentrant orifice (also called Borda's mouthpiece) of diameter D is placed at a depth h below the surface of a tank. Knowing that the speed of the issuing stream is $v = \sqrt{2gh}$ and assuming that the speed of approach v_1 is zero, show that the diameter of the stream is $d = D/\sqrt{2}$. (*Hint*. Consider the section of water indicated, and note that P is equal to the pressure at a depth h multiplied by the area of the orifice.)

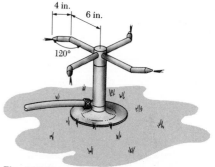

Fig. P14.78

14.78 A garden sprinkler has four rotating arms, each of which consists of two horizontal straight sections of pipe forming an angle of 120°. Each arm discharges water at a rate of 5 gal/min with a velocity of 60 ft/s relative to the arm. Knowing that the friction between the moving and stationary parts of the sprinkler is equivalent to a couple of magnitude $M = 0.275$ lb·ft, determine the constant rate at which the sprinkler rotates (1 ft³ = 7.48 gal).

14.79 The ends of a chain lie in piles at A and C. When given an initial speed v, the chain keeps moving freely at that speed over the pulley at B. Neglecting friction, determine the required value of h.

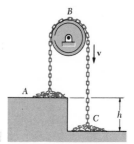

14.80 The ends of a chain lie in piles at A and C. When released from rest at time $t = 0$, the chain moves over the pulley at B, which has a negligible mass. Denoting by L the length of chain connecting the two piles and neglecting friction, determine the speed v of the chain at time t.

14.81 Determine the speed v of the chain of Prob. 14.80, after a length x of chain has been transferred from pile A to pile C.

Fig. P14.79 and P14.80

(1)   *(2)*

Fig. P14.82

14.82 A chain of length l and mass m falls through a small hole in a plate. Initially, when y is very small, the chain is at rest. In each case shown, determine (*a*) the acceleration of the first link A as a function of y, (*b*) the velocity of the chain as the last link passes through the hole. In case *1* assume that the individual links are at rest until they fall through the hole; in case *2* assume that at any instant all links have the same speed. Ignore the effect of friction.

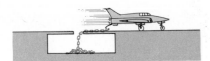

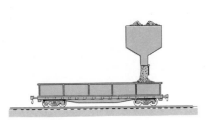

14.83 A possible method for reducing the speed of a training plane as it lands on an aircraft carrier consists in having the tail of the plane hook into the end of a heavy chain of length l which lies in a pile below deck. Denoting by m the mass of the plane and by v_0 its speed at touchdown, and assuming no other retarding force, determine (*a*) the required mass of the chain if the speed of the plane is to be reduced to βv_0, where $\beta < 1$, (*b*) the maximum value of the force exerted by the chain on the plane.

Fig. P14.83 and P14.84

14.84 As a 12,000-lb training plane lands on an aircraft carrier at a speed of 105 mi/h, its tail hooks into the end of a long chain which lies in a pile below deck. Knowing that the chain weighs 24 lb/ft and assuming no other retarding force, determine the maximum deceleration of the plane.

14.85 A railroad car of length L and mass m_0 when empty is moving freely on a horizontal track while being loaded with sand from a stationary chute at a rate $q = dm/dt$. Knowing that the car was approaching the chute at a speed v_0, determine after the car has cleared the chute (*a*) the mass of the car and its load, (*b*) the speed of the car.

Fig. P14.85

14.86 For the railroad car of Prob. 14.85, express in terms of t the magnitude of the horizontal force **P** which should be applied to the car to keep it moving, (*a*) at the constant speed v_0, (*b*) with a constant acceleration a, while being loaded. Assume that $t = 0$ when the loading operation begins.

14.87 The main propulsion system of a space shuttle consists of three identical rocket engines, each of which burns the hydrogen-oxygen propellant at the rate of 750 lb/s and ejects it with a relative velocity of 12,500 ft/s. Determine the total thrust provided by the three engines.

Fig. P14.87 and P14.88

14.88 The main propulsion system of a space shuttle consists of three identical rocket engines which provide a total thrust of 1200 kips. Determine the rate at which the hydrogen-oxygen propellant is burned by each of the three engines, knowing that it is ejected with a relative velocity of 12,500 ft/s.

14.89 A space vehicle describing a circular orbit at a speed of 24×10^3 km/h releases at its front end a capsule which has a gross mass of 600 kg, including 400 kg of fuel. If the fuel is consumed at the constant rate of 18 kg/s and ejected with a relative velocity of 3000 m/s, determine (*a*) the tangential acceleration of the capsule as the engine is fired, (*b*) the maximum speed attained by the capsule.

Fig. P14.89

14.90 A rocket has a mass of 1200 kg, including 1000 kg of fuel, which is consumed at the rate of 12.5 kg/s and ejected with a relative velocity of 4000 m/s. Knowing that the rocket is fired vertically from the ground, determine its acceleration (*a*) as it is fired, (*b*) as the last particle of fuel is being consumed.

14.91 A weather satellite of mass 5000 kg, including fuel, has been ejected from a space shuttle describing a low circular orbit around the earth. After the satellite has slowly drifted to a safe distance from the shuttle, its engine is fired to increase its velocity by 2430 m/s as a first step to its transfer to a geosynchronous orbit. Knowing that the fuel is ejected with a relative velocity of 4200 m/s, determine the mass of fuel consumed in this maneuver.

14.92 Determine the increase in velocity of the weather satellite of Prob. 14.91 after 1500 kg of fuel has been consumed.

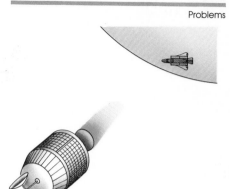

Fig. P14.91 and P14.92

14.93 A 1200-lb spacecraft is mounted on top of a rocket weighing 42,600 lb, including 40,000 lb of fuel. Knowing that the fuel is consumed at a rate of 500 lb/s and ejected with a relative velocity of 12,000 ft/s, determine the maximum speed imparted to the spacecraft when the rocket is fired vertically from the ground.

14.94 The rocket used to launch the 1200-lb spacecraft of Prob. 14.93 is redesigned to include two stages *A* and *B*, each weighing 21,300 lb, including 20,000 lb of fuel. The fuel is again consumed at a rate of 500 lb/s and ejected with a relative velocity of 12,000 ft/s. Knowing that when stage *A* expels its last particle of fuel, its casing is released and jettisoned, determine (*a*) the speed of the rocket at that instant, (*b*) the maximum speed imparted to the spacecraft.

14.95 Determine the altitude reached by the spacecraft of Prob. 14.93 when all the fuel of its launching rocket has been consumed.

14.96 For the spacecraft and the two-stage launching rocket of Prob. 14.94, determine the altitude at which (*a*) stage *A* of the rocket is released, (*b*) the fuel of both stages has been consumed.

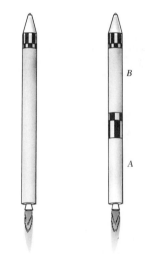

Fig. P14.93   **Fig. P14.94**

14.97 In Prob. 14.91, determine the distance separating the weather satellite from the space shuttle 80 s after its engine has been fired, knowing that the fuel is consumed at a rate of 18.75 kg/s.

14.98 For the rocket of Prob. 14.90, determine (*a*) the altitude at which all the fuel has been consumed, (*b*) the velocity of the rocket at that time.

14.99 In a rocket, the kinetic energy imparted to the consumed and ejected fuel is wasted as far as propelling the rocket is concerned. The useful power is equal to the product of the force available to propel the rocket and the speed of the rocket. If v is the speed of the rocket and u is the relative speed of the expelled fuel, show that the mechanical efficiency of the rocket is $\eta = 2uv/(u^2 + v^2)$. Explain why $\eta = 1$ when $u = v$.

14.100 In a jet airplane, the kinetic energy imparted to the exhaust gases is wasted as far as propelling the airplane is concerned. The useful power is equal to the product of the force available to propel the airplane and the speed of the airplane. If v is the speed of the airplane and u is the relative speed of the expelled gases, show that the mechanical efficiency of the airplane is $\eta = 2v/(u + v)$. Explain why $\eta = 1$ when $u = v$.

Review and Summary

In this chapter we analyzed the motion of *systems of particles*, i.e., the motion of a large number of particles considered together. In the first part of the chapter we considered systems consisting of well-defined particles, while in the second part we analyzed systems which are continually gaining or losing particles, or doing both at the same time.

Effective forces

We first defined the *effective force* of a particle P_i of a given system as the product $m_i\mathbf{a}_i$ of its mass m_i and its acceleration $\mathbf{a}_i$ with respect to a newtonian frame of reference centered at O [Sec. 14.2]. We then showed that *the system of the external forces acting on the particles and the system of the effective forces of the particles are equipollent*, i.e., both systems have the *same resultant* and the *same moment resultant* about O:

$$\sum_{i=1}^{n} \mathbf{F}_i = \sum_{i=1}^{n} m_i\mathbf{a}_i \qquad (14.4)$$

$$\sum_{i=1}^{n} (\mathbf{r}_i \times \mathbf{F}_i) = \sum_{i=1}^{n} (\mathbf{r}_i \times m_i\mathbf{a}_i) \qquad (14.5)$$

Linear and angular momentum of a system of particles

Defining the *linear momentum* $\mathbf{L}$ and the *angular momentum* $\mathbf{H}_O$ *about point O* of the system of particles [Sec. 14.3] as

$$\mathbf{L} = \sum_{i=1}^{n} m_i\mathbf{v}_i \qquad \mathbf{H}_O = \sum_{i=1}^{n} (\mathbf{r}_i \times m_i\mathbf{v}_i) \quad (14.6, 14.7)$$

we showed that Eqs. (14.4) and (14.5) may be replaced by the equations

$$\Sigma\mathbf{F} = \dot{\mathbf{L}} \qquad \Sigma\mathbf{M}_O = \dot{\mathbf{H}}_O \qquad (14.10, 14.11)$$

which express that *the resultant and the moment resultant about O of the external forces are, respectively, equal to the rates of change of the linear momentum and of the angular momentum about O of the system of particles.*

In Sec. 14.4, we defined the mass center of a system of particles as the point G whose position vector $\bar{\mathbf{r}}$ satisfies the equation

$$m\bar{\mathbf{r}} = \sum_{i=1}^{n} m_i \mathbf{r}_i \qquad (14.12)$$

where m represents the total mass $\sum_{i=1}^{n} m_i$ of the particles. Differentiating both members of Eq. (14.12) twice with respect to t, we obtained the relations

$$\mathbf{L} = m\bar{\mathbf{v}} \qquad \dot{\mathbf{L}} = m\bar{\mathbf{a}} \qquad (14.14, \ 14.15)$$

where $\bar{\mathbf{v}}$ and $\bar{\mathbf{a}}$ represent, respectively, the velocity and the acceleration of the mass center G. Substituting for $\dot{\mathbf{L}}$ from (14.15) into (14.10), we obtained the equation

$$\Sigma \mathbf{F} = m\bar{\mathbf{a}} \qquad (14.16)$$

from which we concluded that *the mass center of a system of particles moves as if the entire mass of the system and all the external forces were concentrated at that point* [Sample Prob. 14.1].

In Sec. 14.5 we considered the motion of the particles of a system with respect to a centroidal frame $Gx'y'z'$ attached to the mass center G of the system and in translation with respect to the newtonian frame $Oxyz$ (Fig. 14.14). We defined the *angular momentum of the system about its mass center G* as the sum of the moments about G of the momenta $m_i\mathbf{v}_i'$ of the particles in their motion relative to the frame $Gx'y'z'$. We also noted that the same result may be obtained by considering the moments about G of the momenta $m_i\mathbf{v}_i$ of the particles in their absolute motion. We wrote therefore

$$\mathbf{H}_G = \sum_{i=1}^{n} (\mathbf{r}_i' \times m_i\mathbf{v}_i) = \sum_{i=1}^{n} (\mathbf{r}_i' \times m_i\mathbf{v}_i') \qquad (14.24)$$

and derived the relation

$$\Sigma \mathbf{M}_G = \dot{\mathbf{H}}_G \qquad (14.23)$$

which expresses that *the moment resultant about G of the external forces is equal to the rate of change of the angular momentum about G of the system of particles*. As we shall see later, this relation is fundamental to the study of the motion of rigid bodies.

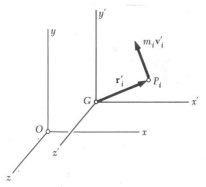

Fig. 14.14

When no external force acts on a system of particles [Sec. 14.6], it follows from Eqs. (14.10) and (14.11) that the linear momentum $\mathbf{L}$ and the angular momentum $\mathbf{H}_O$ of the system are conserved [Sample Probs. 14.2 and 14.3]. In problems involving central forces, the angular momentum of the system about the center of force O will also be conserved.

The kinetic energy T of a system of particles was defined as the sum of the kinetic energies of the particles [Sec. 14.7]:

Kinetic energy of a system of particles

$$T = \frac{1}{2} \sum_{i=1}^{n} m_i v_i^2 \tag{14.28}$$

Using the centroidal frame of reference $Gx'y'z'$ of Fig. 14.14, we noted that the kinetic energy of the system may also be obtained by adding the kinetic energy $\frac{1}{2}m\bar{v}^2$ associated with the motion of the mass center G and the kinetic energy of the system in its motion relative to the frame $Gx'y'z'$:

$$T = \tfrac{1}{2}m\bar{v}^2 + \frac{1}{2} \sum_{i=1}^{n} m_i v_i'^2 \tag{14.29}$$

Principle of work and energy

The *principle of work and energy* may be applied to a system of particles as well as to individual particles [Sec. 14.8]. We wrote

$$T_1 + U_{1 \rightarrow 2} = T_2 \tag{14.30}$$

and noted that $U_{1 \rightarrow 2}$ represents the work of *all* the forces acting on the particles of the system, internal as well as external.

Conservation of energy

If all the forces acting on the particles of the system are *conservative*, we may determine the potential energy V of the system and write

$$T_1 + V_1 = T_2 + V_2 \tag{14.31}$$

which expresses the *principle of conservation of energy* for a system of particles.

Principle of impulse and momentum

We saw in Sec. 14.9 that the *principle of impulse and momentum* for a system of particles may be expressed graphically as shown in Fig. 14.15. It states that the momenta of the particles at time t_1 and the impulses of the external forces from t_1 to t_2 form a system of vectors equipollent to the system of the momenta of the particles at time t_2.

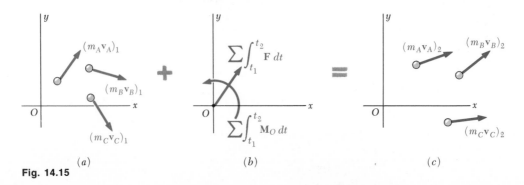

Fig. 14.15

If no external force acts on the particles of the system, the systems of momenta shown in parts *a* and *c* of Fig. 14.15 are equipollent and we have

$$\mathbf{L}_1 = \mathbf{L}_2 \qquad (\mathbf{H}_O)_1 = (\mathbf{H}_O)_2 \qquad (14.36,\ 14.37)$$

Many problems involving the motion of systems of particles may be solved by applying simultaneously the principle of impulse and momentum and the principle of conservation of energy [Sample Prob. 14.4] or by expressing that the linear momentum, angular momentum, and energy of the system are conserved [Sample Prob. 14.5].

Use of conservation principles in the solution of problems involving systems of particles

In the second part of the chapter, we considered *variable systems of particles*. First we considered a *steady stream of particles*, such as a stream of water diverted by a fixed vane or the flow of air through a jet engine [Sec. 14.11]. Applying the principle of impulse and momentum to a system S of particles during a time interval Δt, and including the particles which enter the system at A during that time interval and those (of the same mass Δm) which leave the system at B, we concluded that *the system formed by the momentum $(\Delta m)\mathbf{v}_A$ of the particles entering S in the time Δt and the impulses of the forces exerted on S during that time is equipollent to the momentum $(\Delta m)\mathbf{v}_B$ of the particles leaving S in the same time Δt* (Fig. 14.16). Equating the *x* components, *y* components, and moments about a fixed point of the vectors involved, we could obtain as many as three equations, which could be solved for the desired unknowns [Sample Probs. 14.6 and 14.7]. From this result, we could also derive the following expression for the resultant $\Sigma\mathbf{F}$ of the forces exerted on S,

Variable systems of particles
Steady stream of particles

$$\Sigma\mathbf{F} = \frac{dm}{dt}(\mathbf{v}_B - \mathbf{v}_A) \qquad (14.39)$$

where $\mathbf{v}_B - \mathbf{v}_A$ represents the difference between the *vectors* $\mathbf{v}_B$ and $\mathbf{v}_A$ and where dm/dt is the mass rate of flow of the stream (see footnote, page 679).

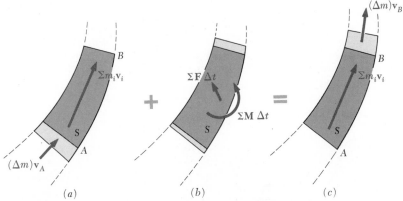

Fig. 14.16

Systems gaining or losing mass

Considering next a system of particles gaining mass by continually absorbing particles or losing mass by continually expelling particles [Sec. 14.12], as in the case of a rocket, we applied the principle of impulse and momentum to the system during a time interval Δt, being careful to include the particles gained or lost during that time interval [Sample Prob. 14.8]. We also noted that the action on a system S of the particles being *absorbed* by S was equivalent to a thrust

$$\mathbf{P} = \frac{dm}{dt}\mathbf{u} \qquad (14.43)$$

where dm/dt is the rate at which mass is being absorbed, and $\mathbf{u}$ the velocity of the particles *relative to* S. In the case of particles being *expelled* by S, the rate dm/dt is negative and the thrust $\mathbf{P}$ is exerted in a direction opposite to that in which the particles are being expelled.

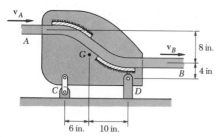

Fig. P14.101

Review Problems

14.101 A jet of water having a cross-sectional area $A = 1.2$ in^2 and moving with a velocity of magnitude $v_A = v_B = 60$ ft/s is deflected by the two vanes shown, which are welded to a vertical plate Knowing that the combined weight of the plate and vanes is 10 lb, determine the reactions at C and D.

14.102 A 1-oz bullet is fired with a velocity of 1600 ft/s into block A, which weighs 10 lb. The coefficient of kinetic friction between block A and the cart BC is 0.50. Knowing that the cart weighs 8 lb and can roll freely, determine (*a*) the final velocity of the cart and block, (*b*) the final position of the block on the cart.

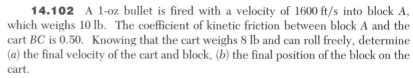

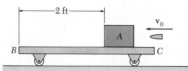

Fig. P14.102

14.103 A 500-lb space vehicle is traveling with a velocity $\mathbf{v}_0 = (1500$ ft/s$)\mathbf{i}$ when explosive charges separate it into three parts A, B, and C, weighing, respectively, 120 lb, 180 lb, and 200 lb. Knowing that immediately after the explosion, the velocity of part A is $\mathbf{v}_A = (2000$ ft/s$)\mathbf{i} - (200$ ft/s$)\mathbf{j} + (300$ ft/s$)\mathbf{k}$ and that the velocity of part B is $\mathbf{v}_B = (1000$ ft/s$)\mathbf{i} + (300$ ft/s$)\mathbf{j} - (500$ ft/s$)\mathbf{k}$, determine the corresponding velocity of part C.

14.104 The stream of water shown flows at a rate of 200 gal/min and moves with a velocity of magnitude 80 ft/s at both A and B. The vane is supported by a pin connection at C and by a load cell at D which can exert only a vertical force. Neglecting the weight of the vane, determine the reactions at C and D (1 ft^3 = 7.48 gal).

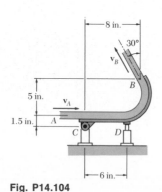

Fig. P14.104

14.105 A rotary power plow is used to remove snow from a level section of railroad track. The plow car is placed ahead of an engine which propels it at a constant speed of 15 mi/h. The plow clears 150 tons of snow per minute, projecting it in the direction shown with a velocity of 50 ft/s relative to the plow car. Neglecting rolling friction, determine (a) the magnitude of the force **P** exerted by the engine on the plow car, (b) the lateral force exerted on the plow car by the track.

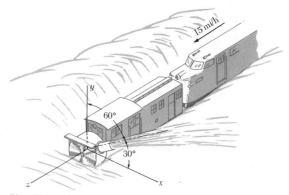

Fig. P14.105

14.106 The acceleration of a rocket is observed to be 30 m/s^2 at $t = 0$, as it is fired vertically from the ground, and 350 m/s^2 at $t = 80$ s. Knowing that the fuel is consumed at the rate of 10 kg/s, determine (a) the initial mass of the rocket, (b) the relative velocity with which the fuel is ejected.

14.107 A chain of length l and total mass m lies in a pile on the floor. If its end A is raised vertically at a constant speed v, determine (a) the force **P** applied to A at the time when half the chain is off the floor, (b) the reaction exerted by the floor at that time.

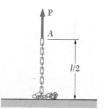

Fig. P14.107

14.108 Solve Prob. 14.107, assuming that end A of the chain is being *lowered* to the floor at a constant speed v.

14.109 Three identical freight cars are located on the same track. Car A is moving to the right with a velocity $\mathbf{v}_0$, while cars B and C are at rest. Determine the velocity of each car after all collisions have taken place, assuming that cars A and B bounce off each other with a coefficient of restitution $e = 1$ and that cars B and C (a) also bounce off each other with $e = 1$, (b) get automatically coupled as they hit each other, (c) were already tightly coupled when A hit B.

Fig. P14.109

14.110 Gravel falls with practically zero velocity onto a conveyor belt at the constant rate $q = dm/dt$. (a) Determine the magnitude of the force **P** required to maintain a constant belt speed v. (b) Show that the kinetic energy acquired by the gravel in a given time interval is equal to half the work done in that interval by the force **P**. Explain what happens to the other half of the work done by **P**.

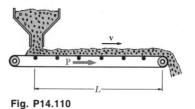

Fig. P14.110

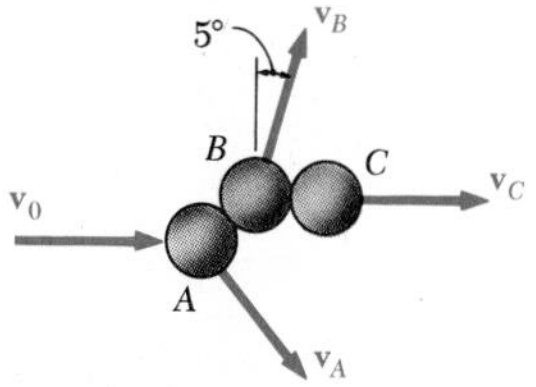

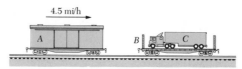

14.111 In a game of billiards, ball A is moving with a velocity $\mathbf{v}_0 = (3 \text{ m/s})\mathbf{i}$ when it strikes balls B and C which are at rest side by side. After the collision, B and C are observed to move in the directions shown with velocities of magnitude $v_B = 1.375$ m/s and $v_C = 1.700$ m/s, respectively. Determine (a) the magnitude and direction of the velocity $\mathbf{v}_A$, (b) the percentage of the initial energy lost in the collision.

Fig. P14.111

14.112 A 40-ton boxcar A is moving with a velocity of 4.5 mi/h in a railroad switchyard when it strikes, and is automatically coupled with, a 20-ton flatcar B which carries a 30-ton trailer truck C. Both the flatcar and the truck are at rest with their brakes released. Determine the velocity of the cars and of the truck (a) immediately after the coupling of the two cars, (b) immediately after the end of the flatcar hits the truck. Neglect friction and assume the impact between the flatcar and the truck to be perfectly plastic ($e = 0$).

Fig. P14.112

The following problems are designed to be solved with a computer.

14.C1 A 12-kg projectile is moving with a velocity of 40 m/s when it explodes into two fragments A and B, of mass 3 kg and 9 kg, respectively. Knowing that immediately after the explosion fragments A and B travel in the directions shown, write a computer program and use it to calculate the velocity of each fragment for values of θ_A from 30 to 120° at 5° intervals.

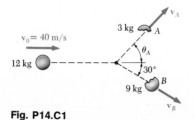

Fig. P14.C1

14.C2 Block *B* starts from rest and slides on the 25-lb wedge *A*, which is supported by a horizontal surface. Neglecting friction, write a computer program which can be used to calculate (*a*) the velocity of *B* relative to *A* after it has slid 3 ft down the inclined surface, (*b*) the corresponding velocity of *A*. Use this program to calculate the velocities for values of the weight of block *B* from 5 to 20 lb at 1-lb intervals.

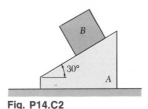

Fig. P14.C2

14.C3 A 35,000-lb jet airplane maintains a constant speed of 480 mi/h while climbing at an angle $\alpha_0 = 15°$. The airplane scoops in air at a rate of 600 lb/s and discharges it with a velocity of 2250 ft/s relative to the airplane. If the pilot changes the angle of climb α and the same engine conditions are maintained, write a computer program and use it to calculate for values of α from 0 to 20° at 1° intervals (*a*) the initial acceleration of the plane, (*b*) the maximum speed attained. Assume that the drag due to air friction is proportional to the square of the speed.

Fig. P14.C3

14.C4 The stream of water shown flows at the rate of 1.25 m³/min and moves with a velocity of magnitude 40 m/s at both *A* and *B*. The vane is supported by a pin connection at *C* and by a load cell at *D* which can exert only a vertical force. Neglecting the weight of the vane, write a computer program and use it to calculate the reactions at *C* and *D* for values of θ from 0 to 90° at 10° intervals.

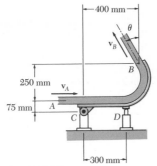

Fig. P14.C4

Kinematics of Rigid Bodies

15.1. Introduction. In this chapter, we shall study the kinematics of *rigid bodies*. We shall investigate the relations existing between the time, the positions, the velocities, and the accelerations of the various particles forming a rigid body. As we shall see, the various types of rigid-body motion may be conveniently grouped as follows:

1. *Translation.* A motion is said to be a translation if any straight line inside the body keeps the same direction during the motion. It may also be observed that in a translation all the particles forming the body move along parallel paths. If these paths are straight lines, the motion is said to be a *rectilinear translation* (Fig. 15.1); if the paths are curved lines, the motion is a *curvilinear translation* (Fig. 15.2).

2. *Rotation about a Fixed Axis.* In this motion, the particles forming the rigid body move in parallel planes along circles centered on the same fixed axis (Fig. 15.3). If this axis, called the *axis of rotation*, intersects the rigid body, the particles located on the axis have zero velocity and zero acceleration.

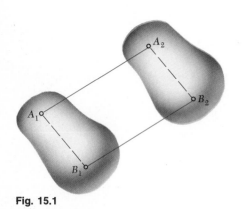

Fig. 15.1

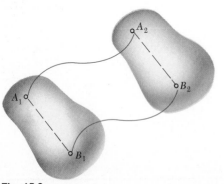

Fig. 15.2

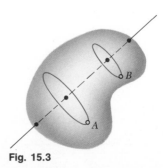

Fig. 15.3

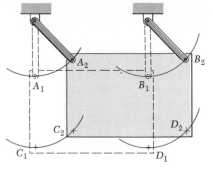

(*a*) Curvilinear translation

Fig. 15.4

(*b*) Rotation

Rotation should not be confused with certain types of curvilinear translation. For example, the plate shown in Fig. 15.4*a* is in curvilinear translation, with all its particles moving along *parallel* circles, while the plate shown in Fig. 15.4*b* is in rotation, with all its particles moving along *concentric* circles. In the first case, any given straight line drawn on the plate will maintain the same direction, whereas in the second case, point O remains fixed.

Because each particle moves in a given plane, the rotation of a body about a fixed axis is said to be a *plane motion*.

3. *General Plane Motion.* There are many other types of plane motion, i.e., motions in which all the particles of the body move in parallel planes. Any plane motion which is neither a rotation nor a translation is referred to as a general plane motion. Two examples of general plane motion are given in Fig. 15.5.

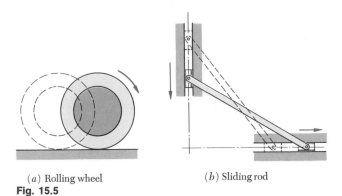

(*a*) Rolling wheel   (*b*) Sliding rod

Fig. 15.5

4. *Motion about a Fixed Point.* This is the three-dimensional motion of a rigid body attached at a fixed point O. An example of motion about a fixed point is provided by the motion of a top on a rough floor (Fig. 15.6).

5. *General Motion.* Any motion of a rigid body which does not fall in any of the categories above is referred to as a general motion.

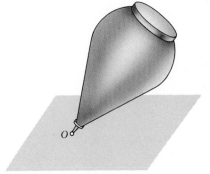

Fig. 15.6

After a brief discussion in Sec. 15.2 of the motion of translation, we shall consider in Sec. 15.3 the rotation of a rigid body about a fixed axis. We shall define the *angular velocity* and the *angular acceleration* of the body and learn to express the velocity and the acceleration of a given point of the body in terms of its position vector and the angular velocity and angular acceleration of the body.

The following sections are devoted to the study of the general plane motion of a rigid body and to its application to the analysis of mechanisms such as gears, connecting rods, and pin-connected linkages. Resolving the plane motion of a slab into a translation and a rotation (Secs. 15.5 and 15.6), we shall then express the velocity of a point B of the slab as the sum of the velocity of a reference point A and of the velocity of B relative to a frame of reference translating with A (i.e., moving with A but not rotating). The same approach is used later in Sec. 15.8 to express the acceleration of B in terms of the acceleration of A and of the acceleration of B relative to a frame translating with A.

An alternative method for the analysis of velocities in plane motion, based on the concept of *instantaneous center of rotation,* is given in Sec. 15.7; and still another method of analysis, based on the use of parametric expressions for the coordinates of a given point, is presented in Sec. 15.9.

The motion of a particle relative to a rotating frame of reference and the concept of *Coriolis acceleration* are discussed in Secs. 15.10 and 15.11, and the results obtained are applied to the analysis of the plane motion of mechanisms containing parts which slide on each other.

The remaining part of the chapter is devoted to the analysis of the three-dimensional motion of a rigid body, namely, the motion of a rigid body with a fixed point and the general motion of a rigid body. In Secs. 15.12 and 15.13, a fixed frame of reference or a frame of reference in translation will be used to carry out this analysis; whereas in Secs. 15.14 and 15.15, we shall consider the motion of the body relative to a rotating frame or to a frame in general motion and use again the concept of Coriolis acceleration.

15.2. Translation. Consider a rigid body in translation (either rectilinear or curvilinear translation), and let A and B be any two of its particles (Fig. 15.7a). Denoting, respectively, by $\mathbf{r}_A$ and $\mathbf{r}_B$ the position vectors of A and B with respect to a fixed frame of reference and by $\mathbf{r}_{B/A}$ the vector joining A and B, we write

$$\mathbf{r}_B = \mathbf{r}_A + \mathbf{r}_{B/A} \tag{15.1}$$

Let us differentiate this relation with respect to t. We note that from the very definition of a translation, the vector $\mathbf{r}_{B/A}$ must maintain a constant direction; its magnitude must also be constant, since A and B belong to the same rigid body. Thus, the derivative of $\mathbf{r}_{B/A}$ is zero and we have

$$\mathbf{v}_B = \mathbf{v}_A \tag{15.2}$$

Differentiating once more, we write

$$\mathbf{a}_B = \mathbf{a}_A \qquad (15.3)$$

Thus, *when a rigid body is in translation, all the points of the body have the same velocity and the same acceleration at any given instant* (Fig. 15.7*b* and *c*). In the case of curvilinear translation, the velocity and acceleration change in direction as well as in magnitude at every instant. In the case of rectilinear translation, all particles of the body move along parallel straight lines, and their velocity and acceleration keep the same direction during the entire motion.

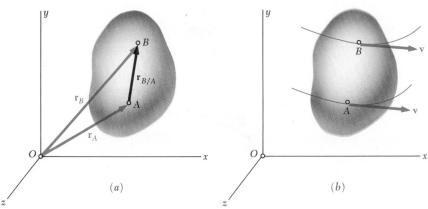

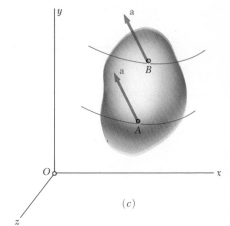

(a) (b) (c)

Fig. 15.7

15.3. Rotation about a Fixed Axis. Consider a rigid body which rotates about a fixed axis *AA'*. Let *P* be a point of the body and **r** its position vector with respect to a fixed frame of reference. For convenience, we shall assume that the frame is centered at point *O* on *AA'* and that the *z* axis coincides with *AA'* (Fig. 15.8). Let *B* be the projection of *P* on *AA'*; since *P* must remain at a constant distance from *B*, it will describe a circle of center *B* and of radius $r \sin \phi$, where ϕ denotes the angle formed by **r** and *AA'*.

The position of *P* and of the entire body is completely defined by the angle θ the line *BP* forms with the *zx* plane. The angle θ is known as the *angular coordinate* of the body. The angular coordinate is defined as positive when counterclockwise as viewed from *A'* and will be expressed in radians (rad) or, occasionally, in degrees (°) or revolutions (rev). We recall that

$$1 \text{ rev} = 2\pi \text{ rad} = 360°$$

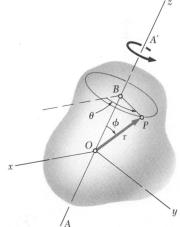

Fig. 15.8

We recall from Sec. 11.9 that the velocity $\mathbf{v} = d\mathbf{r}/dt$ of a particle *P* is a vector tangent to the path of *P* and of magnitude $v = ds/dt$. Observing that the length Δs of the arc described by *P* when the body rotates through $\Delta \theta$ is

Path length

$$\Delta s = (BP)\,\Delta\theta = (r\sin\phi)\,\Delta\theta$$

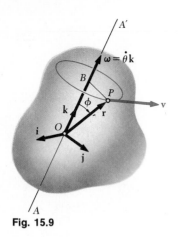

Fig. 15.9

and dividing both members by Δt, we obtain at the limit, as Δt approaches zero,

$$v = \frac{ds}{dt} = r\dot{\theta} \sin \phi \tag{15.4}$$

where $\dot{\theta}$ denotes the time derivative of θ. (Note that while the angle θ depends on the position of P within the body, the rate of change $\dot{\theta}$ is itself independent of P.) We conclude that the velocity $\mathbf{v}$ of P is a vector perpendicular to the plane containing AA' and $\mathbf{r}$, and of magnitude v defined by (15.4). But this is precisely the result we would obtain if we drew along AA' a vector $\boldsymbol{\omega} = \dot{\theta}\mathbf{k}$ and formed the vector product $\boldsymbol{\omega} \times \mathbf{r}$ (Fig. 15.9). We thus write

$$\mathbf{v} = \frac{d\mathbf{r}}{dt} = \boldsymbol{\omega} \times \mathbf{r} \tag{15.5}$$

The vector

$$\boldsymbol{\omega} = \omega\mathbf{k} = \dot{\theta}\mathbf{k} \tag{15.6}$$

is called the *angular velocity* of the body. It is directed along the axis of rotation, it is equal in magnitude to the rate of change $\dot{\theta}$ of the angular coordinate, and its sense may be obtained by the right-hand rule (Sec. 3.6) from the sense of rotation of the body.†

We shall now determine the acceleration $\mathbf{a}$ of the particle P. Differentiating (15.5) and recalling the rule for the differentiation of a vector product (Sec. 11.10), we write

$$\mathbf{a} = \frac{d\mathbf{v}}{dt} = \frac{d}{dt}(\boldsymbol{\omega} \times \mathbf{r})$$

$$= \frac{d\boldsymbol{\omega}}{dt} \times \mathbf{r} + \boldsymbol{\omega} \times \frac{d\mathbf{r}}{dt}$$

$$= \frac{d\boldsymbol{\omega}}{dt} \times \mathbf{r} + \boldsymbol{\omega} \times \mathbf{v} \tag{15.7}$$

The vector $d\boldsymbol{\omega}/dt$ is denoted by $\boldsymbol{\alpha}$ and called the *angular acceleration* of the body. Substituting also for $\mathbf{v}$ from (15.5), we have

$$\mathbf{a} = \boldsymbol{\alpha} \times \mathbf{r} + \boldsymbol{\omega} \times (\boldsymbol{\omega} \times \mathbf{r}) \tag{15.8}$$

Differentiating (15.6), and recalling that $\mathbf{k}$ is constant in magnitude and direction, we have

$$\boldsymbol{\alpha} = \alpha\mathbf{k} = \dot{\omega}\mathbf{k} = \ddot{\theta}\mathbf{k} \tag{15.9}$$

---

† It will be shown in Sec. 15.12 in the more general case of a rigid body rotating simultaneously about axes having different directions that angular velocities obey the parallelogram law of addition and, thus, are actually vector quantities.

Thus, the angular acceleration of a body rotating about a fixed axis is a vector directed along the axis of rotation, and equal in magnitude to the rate of change $\dot{\omega}$ of the angular velocity. Returning to (15.8), we note that the acceleration of P is the sum of two vectors. The first vector is equal to the vector product $\boldsymbol{\alpha} \times \mathbf{r}$; it is tangent to the circle described by P and represents, therefore, the tangential component of the acceleration. The second vector is equal to the *vector triple product* $\boldsymbol{\omega} \times (\boldsymbol{\omega} \times \mathbf{r})$ obtained by forming the vector product of $\boldsymbol{\omega}$ and $\boldsymbol{\omega} \times \mathbf{r}$; since $\boldsymbol{\omega} \times \mathbf{r}$ is tangent to the circle described by P, the vector triple product is directed toward the center B of the circle and represents, therefore, the normal component of the acceleration.

Rotation of a Representative Slab. The rotation of a rigid body about a fixed axis may be defined by the motion of a representative slab in a reference plane perpendicular to the axis of rotation. Let us choose the xy plane as the reference plane and assume that it coincides with the plane of the figure, with the z axis pointing out of the paper (Fig. 15.10). Recalling from (15.6) that $\boldsymbol{\omega} = \omega\mathbf{k}$, we note that a positive value of the scalar ω corresponds to a counterclockwise rotation of the representative slab, and a negative value to a clockwise rotation. Substituting $\omega\mathbf{k}$ for $\boldsymbol{\omega}$ into Eq. (15.5), we express the velocity of any given point P of the slab as

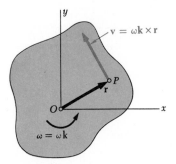

Fig. 15.10

$$\mathbf{v} = \omega\mathbf{k} \times \mathbf{r} \tag{15.10}$$

Since the vectors $\mathbf{k}$ and $\mathbf{r}$ are mutually perpendicular, the magnitude of the velocity $\mathbf{v}$ is

$$v = r\omega \tag{15.10'}$$

and its direction may be obtained by rotating $\mathbf{r}$ through $90°$ in the sense of rotation of the slab.

Substituting $\boldsymbol{\omega} = \omega\mathbf{k}$ and $\boldsymbol{\alpha} = \alpha\mathbf{k}$ into Eq. (15.8), and observing that cross-multiplying $\mathbf{r}$ twice by $\mathbf{k}$ results in a $180°$ rotation of the vector $\mathbf{r}$, we express the acceleration of point P as

$$\mathbf{a} = \alpha\mathbf{k} \times \mathbf{r} - \omega^2\mathbf{r} \tag{15.11}$$

Resolving $\mathbf{a}$ into tangential and normal components (Fig. 15.11), we write

$$\begin{array}{ll} \mathbf{a}_t = \alpha\mathbf{k} \times \mathbf{r} & a_t = r\alpha \\ \mathbf{a}_n = -\omega^2\mathbf{r} & a_n = r\omega^2 \end{array} \tag{15.11'}$$

The tangential component $\mathbf{a}_t$ points in the counterclockwise direction if the scalar α is positive, and in the clockwise direction if α is negative. The normal component $\mathbf{a}_n$ always points in the direction opposite to that of $\mathbf{r}$, that is, toward O.

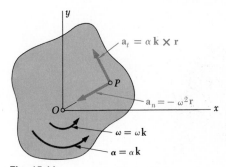

Fig. 15.11

15.4. Equations Defining the Rotation of a Rigid Body about a Fixed Axis. The motion of a rigid body rotating about a fixed axis AA' is said to be *known* when its angular coordinate θ may be expressed as a known function of t. In practice, however, the rotation of a rigid body is seldom defined by a relation between θ and t. More often, the conditions of motion will be specified by the type of angular acceleration that the body possesses. For example, α may be given as a function of t, or as a function of θ, or as a function of ω. Recalling the relations (15.6) and (15.9), we write

$$\omega = \frac{d\theta}{dt} \tag{15.12}$$

$$\alpha = \frac{d\omega}{dt} = \frac{d^2\theta}{dt^2} \tag{15.13}$$

or, solving (15.12) for dt and substituting into (15.13),

$$\alpha = \omega \frac{d\omega}{d\theta} \tag{15.14}$$

Since these equations are similar to those obtained in Chap. 11 for the rectilinear motion of a particle, their integration may be performed by following the procedure outlined in Sec. 11.3.

Two particular cases of rotation are frequently encountered:

1. *Uniform Rotation.* This case is characterized by the fact that the angular acceleration is zero. The angular velocity is thus constant, and the angular coordinate is given by the formula

$$\theta = \theta_0 + \omega t \tag{15.15}$$

2. *Uniformly Accelerated Rotation.* In this case, the angular acceleration is constant. The following formulas relating angular velocity, angular coordinate, and time may then be derived in a manner similar to that described in Sec. 11.5. The similitude between the formulas derived here and those obtained for the rectilinear uniformly accelerated motion of a particle is easily noted.

$$\begin{aligned} \omega &= \omega_0 + \alpha t \\ \theta &= \theta_0 + \omega_0 t + \tfrac{1}{2}\alpha t^2 \\ \omega^2 &= \omega_0^2 + 2\alpha(\theta - \theta_0) \end{aligned} \tag{15.16}$$

It should be emphasized that formula (15.15) may be used only when $\alpha = 0$, and formulas (15.16) only when $\alpha = $ constant. In any other case, the general formulas (15.12) to (15.14) should be used.

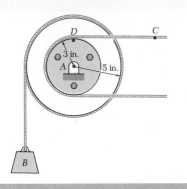

Load B is connected to a double pulley by one of the two inextensible cables shown. The motion of the pulley is controlled by cable C, which has a constant acceleration of 9 in./s^2 and an initial velocity of 12 in./s, both directed to the right. Determine (a) the number of revolutions executed by the pulley in 2 s, (b) the velocity and change in position of the load B after 2 s, and (c) the acceleration of point D on the rim of the inner pulley at $t = 0$.

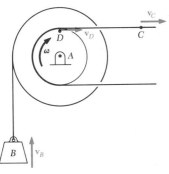

a. Motion of Pulley. Since the cable is inextensible, the velocity of point D is equal to the velocity of point C and the tangential component of the acceleration of D is equal to the acceleration of C.

$$(\mathbf{v}_D)_0 = (\mathbf{v}_C)_0 = 12 \text{ in./s} \rightarrow \qquad (\mathbf{a}_D)_t = \mathbf{a}_C = 9 \text{ in./s}^2 \rightarrow$$

Noting that the distance from D to the center of the pulley is 3 in., we write

$$(v_D)_0 = r\omega_0 \qquad 12 \text{ in./s} = (3 \text{ in.})\omega_0 \qquad \omega_0 = 4 \text{ rad/s} \downarrow$$

$$(a_D)_t = r\alpha \qquad 9 \text{ in./s}^2 = (3 \text{ in.})\alpha \qquad \alpha = 3 \text{ rad/s}^2 \downarrow$$

Using the equations of uniformly accelerated motion, we obtain, for $t = 2$ s,

$$\omega = \omega_0 + \alpha t = 4 \text{ rad/s} + (3 \text{ rad/s}^2)(2 \text{ s}) = 10 \text{ rad/s}$$
$$\omega = 10 \text{ rad/s} \downarrow$$
$$\theta = \omega_0 t + \tfrac{1}{2}\alpha t^2 = (4 \text{ rad/s})(2 \text{ s}) + \tfrac{1}{2}(3 \text{ rad/s}^2)(2 \text{ s})^2 = 14 \text{ rad}$$
$$\theta = 14 \text{ rad} \downarrow$$

$$\text{Number of revolutions} = (14 \text{ rad})\left(\frac{1 \text{ rev}}{2\pi \text{ rad}}\right) = 2.23 \text{ rev} \quad \blacktriangleleft$$

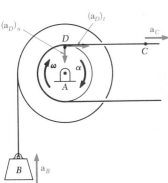

b. Motion of Load B. Using the following relations between linear and angular motion, with $r = 5$ in., we write

$$v_B = r\omega = (5 \text{ in.})(10 \text{ rad/s}) = 50 \text{ in./s} \qquad v_B = 50 \text{ in./s} \uparrow \quad \blacktriangleleft$$
$$\Delta y_B = r\theta = (5 \text{ in.})(14 \text{ rad}) = 70 \text{ in.} \qquad \Delta y_B = 70 \text{ in. upward} \quad \blacktriangleleft$$

c. Acceleration of Point D at $t = 0$. The tangential component of the acceleration is

$$(\mathbf{a}_D)_t = \mathbf{a}_C = 9 \text{ in./s}^2 \rightarrow$$

Since, at $t = 0$, $\omega_0 = 4$ rad/s, the normal component of the acceleration is

$$(a_D)_n = r_D\omega_0^2 = (3 \text{ in.})(4 \text{ rad/s})^2 = 48 \text{ in./s}^2 \qquad (\mathbf{a}_D)_n = 48 \text{ in./s}^2 \downarrow$$

The magnitude and direction of the total acceleration may be obtained by writing

$$\tan \phi = (48 \text{ in./s}^2)/(9 \text{ in./s}^2) \qquad \phi = 79.4°$$
$$a_D \sin 79.4° = 48 \text{ in./s}^2 \qquad a_D = 48.8 \text{ in./s}^2$$
$$a_D = 48.8 \text{ in./s}^2 \searrow 79.4° \quad \blacktriangleleft$$

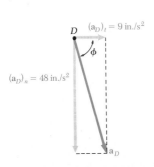

Problems

15.1 The motion of an oscillating crank is defined by the relation $\theta = \theta_0 \cos (2\pi t/T)$, where θ is expressed in radians and t in seconds. Knowing that $\theta_0 = 0.75$ rad and $T = 1.20$ s, determine the maximum angular velocity and the maximum angular acceleration of the crank.

15.2 The motion of a disk rotating in an oil bath is defined by the relation $\theta = \theta_0(1 - e^{-t/3})$, where θ is expressed in radians and t in seconds. Knowing that $\theta_0 = 1.20$ rad, determine the angular coordinate, velocity, and acceleration of the disk when (a) $t = 0$, (b) $t = 3$ s, (c) $t = \infty$.

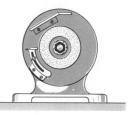

Fig. P15.3

15.3 A small grinding wheel is attached to the shaft of an electric motor which has a rated speed of 3600 rpm. When the power is turned on, the unit reaches its rated speed in 5 s, and when the power is turned off, the unit coasts to rest in 70 s. Assuming uniformly accelerated motion, determine the number of revolutions that the motor executes (a) in reaching its rated speed, (b) in coasting to rest.

15.4 During the starting phase of a computer, it is observed that a storage disk which was initially at rest executed 2.5 revolutions in 0.3 s. Assuming that the motion was uniformly accelerated, determine (a) the angular acceleration of the disk, (b) the final angular velocity of the disk.

15.5 As steam is slowly injected into a turbine, the angular acceleration of the rotor is observed to increase linearly with the time t. Knowing that the rotor starts from rest at $t = 0$ and that after 10 s the rotor has completed 20 revolutions, write the equations of motion for the rotor and determine (a) the angular velocity at $t = 20$ s, (b) the time required for the rotor to complete its first 40 revolutions.

15.6 The bent rod $ABCD$ rotates about a line joining points A and D with a constant angular velocity of 75 rad/s. Knowing that at the instant considered the velocity of corner C is upward, determine the velocity and acceleration of corner B.

15.7 In Prob. 15.6, determine the velocity and acceleration of corner B, assuming that the angular velocity is 75 rad/s and decreases at the rate of 600 rad/s^2.

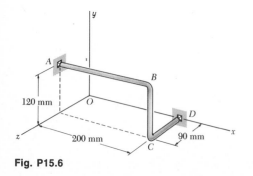

Fig. P15.6

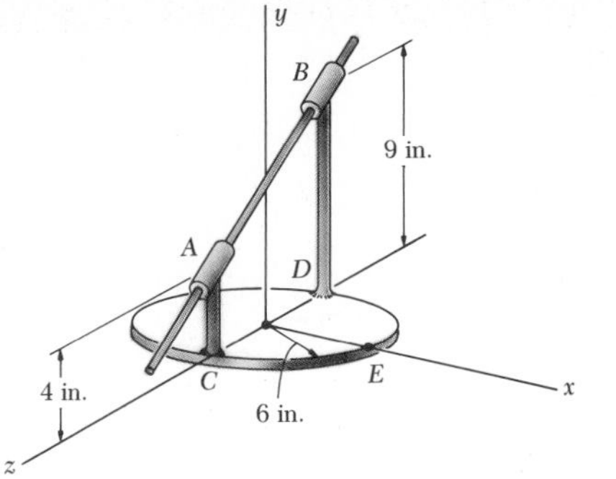

Fig. P15.8

15.8 A circular plate of 6-in. radius is supported by two bearings A and B as shown. The plate rotates about the rod joining A and B with a constant angular velocity of 26 rad/s. Knowing that, at the instant considered, the velocity of point C is directed to the right, determine the velocity and acceleration of point E.

15.9 The assembly shown consists of the straight rod ABC which passes through and is welded to the rectangular plate $DEFH$. The assembly rotates about the axis AC with a constant angular velocity of 9 rad/s. Knowing that the motion when viewed from C is counterclockwise, determine the velocity and acceleration of corner F.

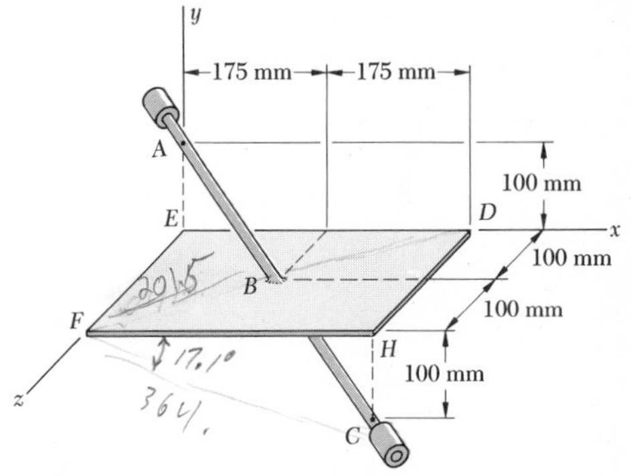

Fig. P15.9

15.10 In Prob. 15.9, determine the velocity and acceleration of corner H, assuming that the angular velocity is 9 rad/s and decreases at the rate of 18 rad/s^2.

15.11 In Prob. 15.8, determine the velocity and acceleration of point E, assuming that the angular velocity is 26 rad/s and increases at the rate of 65 rad/s^2.

15.12 The earth makes one complete revolution about the sun in 365.24 days. Assuming that the orbit of the earth is circular and has a radius of 93,000,000 mi, determine the velocity and acceleration of the earth.

15.13 The earth makes one complete revolution on its axis in 23.93 h. Knowing that the mean radius of the earth is 3960 mi, determine the linear velocity and acceleration of a point on the surface of the earth (*a*) at the equator, (*b*) at Philadelphia, latitude 40° north, (*c*) at the North Pole.

15.14 The belt shown moves without slipping over two idler pulleys. At the instant shown, the velocity of point C is 400 mm/s to the left and its acceleration is 600 mm/s^2 to the right. Determine (*a*) the angular velocity and angular acceleration of each pulley, (*b*) the total acceleration of point A, (*c*) the total acceleration of point B.

15.15 The belt shown moves without slipping over two idler pulleys. At the instant shown, the angular velocity of the 80-mm-radius pulley is 5 rad/s clockwise. Determine the angular acceleration of the 80-mm-radius pulley for which the total acceleration of point B is 3.5 m/s^2.

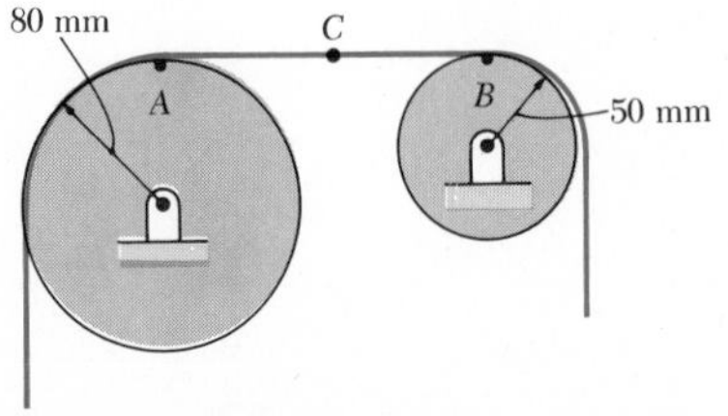

Fig. P15.14 and P15.15

15.16 It is known that the static-friction force between the small block *B* and the plate will be exceeded and that the block will start sliding on the plate when the total acceleration of the block reaches 4 m/s². If the plate starts from rest at $t = 0$ and is accelerated at the constant rate of 5 rad/s², determine the time *t* and the angular velocity of the plate when the block starts sliding, assuming $r = 250$ mm.

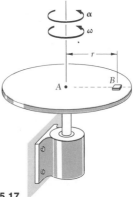

Fig. P15.16 and P15.17

15.17 A small block *B* rests on a horizontal plate which rotates about a fixed axis. The plate starts from rest at $t = 0$ and is accelerated at the constant rate of 0.4 rad/s². Knowing that $r = 250$ mm, determine the magnitude of the total acceleration of the block when (*a*) $t = 0$, (*b*) $t = 1$ s, (*c*) $t = 3$ s.

15.18 The sprocket wheel and chain shown are initially at rest. If the wheel has a uniform angular acceleration of 90 rad/s² counterclockwise, determine (*a*) the acceleration of point *A* of the chain, (*b*) the magnitude of the acceleration of point *B* of the wheel after 3 s.

15.19 The sprocket wheel shown is being operated at a speed of 600 rpm counterclockwise. When the power is turned off it is observed that the wheel and chain coast to rest in 4 s. Assuming uniformly decelerated motion, determine the magnitudes of the velocity and acceleration of point *B* of the wheel (*a*) immediately before the power is turned off, (*b*) 2.5 s later.

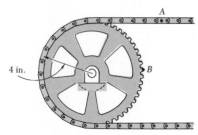

P15.18 and P15.19

15.20 Ring *C* has an inside radius of 55 mm and an outside radius of 60 mm and is positioned between two wheels *A* and *B*, each of 24-mm outside radius. Knowing that wheel *A* rotates with a constant angular velocity of 300 rpm and that no slipping occurs, determine (*a*) the angular velocity of ring *C* and of wheel *B*, (*b*) the acceleration of the points of *A* and *B* which are in contact with *C*.

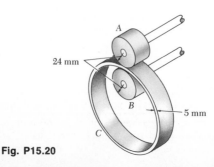

Fig. P15.20

15.21   A computer tape moves over two drums. During a 4-s interval the speed of the tape is increased uniformly from $v_0 = 2$ ft/s to $v_1 = 6$ ft/s. Knowing that the tape does not slip on the drums, determine (a) the angular acceleration of drum A, (b) the number of revolutions executed by drum A during the 4-s interval.

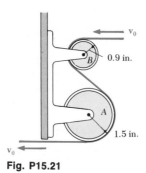

Fig. P15.21

15.22   Solve Prob. 15.21, considering drum B instead of drum A.

15.23   A mixing drum of 125-mm outside radius rests on two casters, each of 25-mm radius. The drum executes 15 revolutions during the time interval t, while its angular velocity is being increased uniformly from 20 to 50 rpm. Knowing that no slipping occurs between the drum and the casters, determine (a) the angular acceleration of the casters, (b) the time interval t.

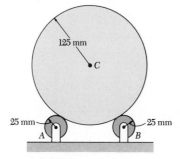

Fig. P15.23

15.24   A pulley and two loads are connected by inextensible cords as shown. Load A has a constant acceleration of 300 mm/s² and an initial velocity of 240 mm/s, both directed upward. Determine (a) the number of revolutions executed by the pulley in 3 s, (b) the velocity and position of load B after 3 s, (c) the acceleration of point D on the rim of the pulley at $t = 0$.

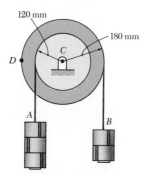

Fig. P15.24 and P15.25

15.25   A pulley and two loads are connected by inextensible cords as shown. The pulley starts from rest at $t = 0$ and is accelerated at the uniform rate of 2.4 rad/s² clockwise. At $t = 4$ s, determine the velocity and position (a) of load A, (b) of load B.

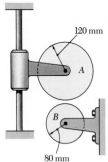
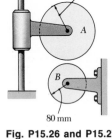

120 mm

A

B

80 mm

Fig. P15.26 and P15.27

15.26 Disk A is at rest when it is brought into contact with disk B, which is rotating freely at 500 rpm clockwise. After 8 s of slippage, during which each disk has a constant angular acceleration, disk B reaches a final angular velocity of 150 rpm clockwise. Determine the angular acceleration of each disk during the period of slippage.

15.27 and 15.28 A simple friction drive consists of two disks A and B. Initially, disk B has a clockwise angular velocity of 500 rpm, and disk A is at rest. It is known that disk B will coast to rest in 60 s. However, rather than waiting until both disks are at rest to bring them together, disk A is given a constant angular acceleration of 3 rad/s² counterclockwise. Determine (*a*) at what time the disks may be brought together if they are not to slip, (*b*) the angular velocity of each disk as contact is made.

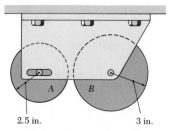
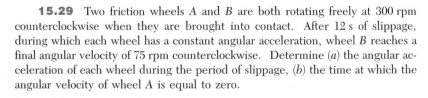

2.5 in.   3 in.

Fig. P15.28 and P15.29

15.29 Two friction wheels A and B are both rotating freely at 300 rpm counterclockwise when they are brought into contact. After 12 s of slippage, during which each wheel has a constant angular acceleration, wheel B reaches a final angular velocity of 75 rpm counterclockwise. Determine (*a*) the angular acceleration of each wheel during the period of slippage, (*b*) the time at which the angular velocity of wheel A is equal to zero.

*15.30 In a continuous printing process, paper is drawn into the presses at a constant speed v. Denoting by r the radius of paper on the roll at any given time and by b the thickness of the paper, derive an expression for the angular acceleration of the paper roll.

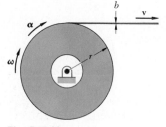

b

v

α

r

ω

Fig. P15.30

15.5. General Plane Motion. As indicated in Sec. 15.1, we understand by general plane motion a plane motion which is neither a translation nor a rotation. As we shall presently see, however, *a general plane motion may always be considered as the sum of a translation and a rotation.*

Consider, for example, a wheel rolling on a straight track (Fig. 15.12). Over a certain interval of time, two given points A and B will have moved, respectively, from A_1 to A_2 and from B_1 to B_2. The same result could be obtained through a translation which would bring A and B into A_2 and B_1' (the line AB remaining vertical), followed by a rotation about A bringing B into B_2. Although the original rolling motion differs from the combination of translation and rotation when these motions are taken in succession, the original motion may be completely duplicated by a combination of simultaneous translation and rotation.

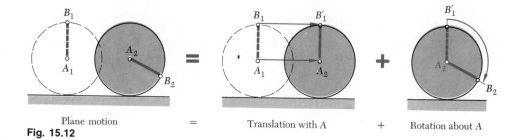

Plane motion   =   Translation with A +   Rotation about A

Fig. 15.12

Another example of plane motion is given in Fig. 15.13, which represents a rod whose extremities slide, respectively, along a horizontal and a vertical track. This motion may be replaced by a translation in a horizontal direction and a rotation about A (Fig. 15.13a) or by a translation in a vertical direction and a rotation about B (Fig. 15.13b).

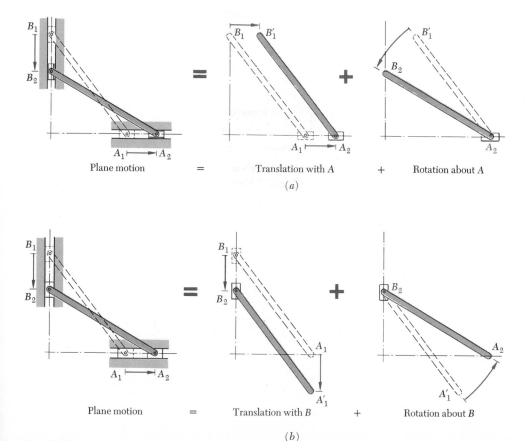

Plane motion   =   Translation with A +   Rotation about A

(a)

Plane motion   =   Translation with B +   Rotation about B

(b)

Fig. 15.13

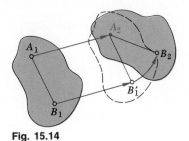

Fig. 15.14

In general, we shall consider a small displacement which brings two particles A and B of a representative slab, respectively, from A_1 and B_1 into A_2 and B_2 (Fig. 15.14). This displacement may be divided into two parts, one in which the particles move into A_2 and B_1' while the line AB maintains the same direction, the other in which B moves into B_2 while A remains fixed. The first part of the motion is clearly a translation and the second part a rotation about A.

Recalling from Sec. 11.12 the definition of the "relative motion" of a particle with respect to a moving frame of reference—as opposed to its "absolute motion" with respect to a fixed frame of reference—we may restate as follows the result obtained above: Given two particles A and B of a rigid slab in plane motion, the relative motion of B with respect to a frame attached to A and of fixed orientation is a rotation. To an observer moving with A but not rotating, particle B will appear to describe an arc of circle centered at A.

15.6. Absolute and Relative Velocity in Plane Motion.
We saw in the preceding section that any plane motion of a slab may be replaced by a translation defined by the motion of an arbitrary reference point A, and a simultaneous rotation about A. The absolute velocity $\mathbf{v}_B$ of a particle B of the slab is obtained from the relative-velocity formula derived in Sec. 11.12,

$$\mathbf{v}_B = \mathbf{v}_A + \mathbf{v}_{B/A} \qquad (15.17)$$

where the right-hand member represents a vector sum. The velocity $\mathbf{v}_A$ corresponds to the translation of the slab with A, while the relative velocity

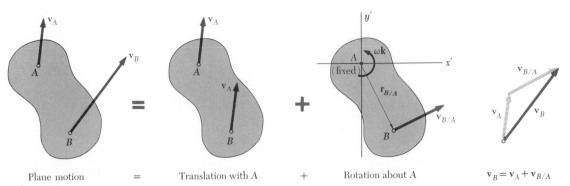

Plane motion $=$ Translation with A $+$ Rotation about A $\qquad \mathbf{v}_B = \mathbf{v}_A + \mathbf{v}_{B/A}$

Fig. 15.15

$v_{B/A}$ is associated with the rotation of the slab about A and is measured with respect to axes centered at A and of fixed orientation (Fig. 15.15). Denoting by $r_{B/A}$ the position vector of B relative to A, and by ωk the angular velocity of the slab with respect to axes of fixed orientation, we have from (15.10) and (15.10')

$$v_{B/A} = \omega k \times r_{B/A} \qquad v_{B/A} = r\omega \qquad (15.18)$$

where r is the distance from A to B. Substituting for $v_{B/A}$ from (15.18) into (15.17), we may also write

$$v_B = v_A + \omega k \times r_{B/A} \qquad (15.17')$$

As an example, we shall consider again the rod AB of Fig. 15.13. Assuming that the velocity v_A of end A is known, we propose to find the velocity v_B of end B and the angular velocity ω of the rod, in terms of the velocity v_A, the length l, and the angle θ. Choosing A as a reference point, we express that the given motion is equivalent to a translation with A and a simultaneous rotation about A (Fig. 15.16). The absolute velocity of B must therefore be equal to the vector sum

$$v_B = v_A + v_{B/A} \qquad (15.17)$$

We note that while the direction of $v_{B/A}$ is known, its magnitude $l\omega$ is unknown. However, this is compensated for by the fact that the direction of v_B is known. We may therefore complete the diagram of Fig. 15.16. Solving for the magnitudes v_B and ω, we write

$$v_B = v_A \tan\theta \qquad \omega = \frac{v_{B/A}}{l} = \frac{v_A}{l\cos\theta} \qquad (15.19)$$

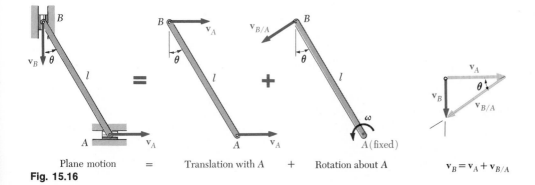

| Plane motion | = | Translation with A | + | Rotation about A | $v_B = v_A + v_{B/A}$ |

Fig. 15.16

The same result may be obtained by using B as a point of reference. Resolving the given motion into a translation with B and a simultaneous rotation about B (Fig. 15.17), we write the equation

$$\mathbf{v}_A = \mathbf{v}_B + \mathbf{v}_{A/B} \qquad (15.20)$$

which is represented graphically in Fig. 15.17. We note that $\mathbf{v}_{A/B}$ and $\mathbf{v}_{B/A}$ have the same magnitude $l\omega$ but opposite sense. The sense of the relative velocity depends, therefore, upon the point of reference which has been selected and should be carefully ascertained from the appropriate diagram (Fig. 15.16 or 15.17).

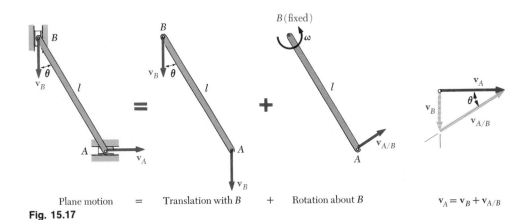

Plane motion   =   Translation with B +   Rotation about B $\mathbf{v}_A = \mathbf{v}_B + \mathbf{v}_{A/B}$

Fig. 15.17

Finally, we observe that the angular velocity ω of the rod in its rotation about B is the same as in its rotation about A. It is measured in both cases by the rate of change of the angle θ. This result is quite general; we should therefore bear in mind that *the angular velocity ω of a rigid body in plane motion is independent of the reference point.*

Most mechanisms consist not of one but of *several* moving parts. When the various parts of a mechanism are pin-connected, its analysis may be carried out by considering each part as a rigid body, while keeping in mind that the points where two parts are connected must have the same absolute velocity (see Sample Prob. 15.3). A similar analysis may be used when gears are involved, since the teeth in contact must also have the same absolute velocity. However, when a mechanism contains parts which slide on each other, the relative velocity of the parts in contact must be taken into account (see Secs. 15.10 and 15.11).

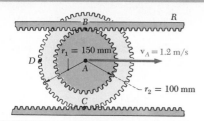

The double gear shown rolls on the stationary lower rack; the velocity of its center A is 1.2 m/s directed to the right. Determine (a) the angular velocity of the gear, (b) the velocities of the upper rack R and of point D of the gear.

a. **Angular Velocity of the Gear.** Since the gear rolls on the lower rack, its center A moves through a distance equal to the outer circumference $2\pi r_1$ for each full revolution of the gear. Noting that 1 rev = 2π rad, and that when A moves to the right ($x_A > 0$) the gear rotates clockwise ($\theta < 0$), we write

$$\frac{x_A}{2\pi r_1} = -\frac{\theta}{2\pi} \qquad x_A = -r_1\theta$$

Differentiating with respect to the time t and substituting the known values $v_A = 1.2$ m/s and $r_1 = 150$ mm = 0.150 m, we obtain

$$v_A = -r_1\omega \qquad 1.2 \text{ m/s} = -(0.150 \text{ m})\omega \qquad \omega = -8 \text{ rad/s}$$
$$\omega = \omega\mathbf{k} = -(8 \text{ rad/s})\mathbf{k} \quad \blacktriangleleft$$

where $\mathbf{k}$ is a unit vector pointing out of the paper.

b. **Velocities.** The rolling motion is resolved into two component motions: a translation with the center A and a rotation about the center A. In the translation, all points of the gear move with the same velocity $\mathbf{v}_A$. In the rotation, each point P of the gear moves about A with a relative velocity $\mathbf{v}_{P/A} = \omega\mathbf{k} \times \mathbf{r}_{P/A}$, where $\mathbf{r}_{P/A}$ is the position vector of P relative to A.

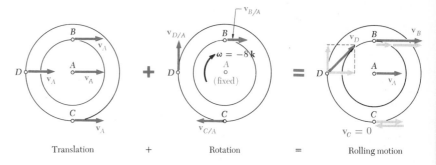

| Translation | + | Rotation | = | Rolling motion |

Velocity of Upper Rack. The velocity of the upper rack is equal to the velocity of point B; we write

$$\mathbf{v}_R = \mathbf{v}_B = \mathbf{v}_A + \mathbf{v}_{B/A} = \mathbf{v}_A + \omega\mathbf{k} \times \mathbf{r}_{B/A}$$
$$= (1.2 \text{ m/s})\mathbf{i} - (8 \text{ rad/s})\mathbf{k} \times (0.100 \text{ m})\mathbf{j}$$
$$= (1.2 \text{ m/s})\mathbf{i} + (0.8 \text{ m/s})\mathbf{i} = (2 \text{ m/s})\mathbf{i}$$
$$\mathbf{v}_R = 2 \text{ m/s} \rightarrow \quad \blacktriangleleft$$

Velocity of Point D:

$$\mathbf{v}_D = \mathbf{v}_A + \mathbf{v}_{D/A} = \mathbf{v}_A + \omega\mathbf{k} \times \mathbf{r}_{D/A}$$
$$= (1.2 \text{ m/s})\mathbf{i} - (8 \text{ rad/s})\mathbf{k} \times (-0.150 \text{ m})\mathbf{i}$$
$$= (1.2 \text{ m/s})\mathbf{i} + (1.2 \text{ m/s})\mathbf{j}$$
$$\mathbf{v}_D = 1.697 \text{ m/s} \measuredangle 45° \quad \blacktriangleleft$$

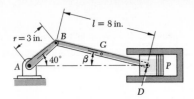

SAMPLE PROBLEM 15.3

In the engine system shown, the crank AB has a constant clockwise angular velocity of 2000 rpm. For the crank position indicated, determine (a) the angular velocity of the connecting rod BD, (b) the velocity of the piston P.

Motion of Crank AB. The crank AB rotates about point A. Expressing ω_{AB} in rad/s and writing $v_B = r\omega_{AB}$, we obtain

$$\omega_{AB} = \left(2000\,\frac{\text{rev}}{\text{min}}\right)\left(\frac{1\ \text{min}}{60\ \text{s}}\right)\left(\frac{2\pi\ \text{rad}}{1\ \text{rev}}\right) = 209.4\ \text{rad/s}$$

$$v_B = (AB)\omega_{AB} = (3\ \text{in.})(209.4\ \text{rad/s}) = 628.3\ \text{in./s}$$

$$\mathbf{v}_B = 628.3\ \text{in./s}\ \searrow 50°$$

Motion of Connecting Rod BD. We consider this motion as a general plane motion. Using the law of sines, we compute the angle β between the connecting rod and the horizontal:

$$\frac{\sin 40°}{8\ \text{in.}} = \frac{\sin \beta}{3\ \text{in.}} \qquad \beta = 13.95°$$

The velocity $\mathbf{v}_D$ of the point D where the rod is attached to the piston must be horizontal, while the velocity of point B is equal to the velocity $\mathbf{v}_B$ obtained above. Resolving the motion of BD into a translation with B and a rotation about B, we obtain

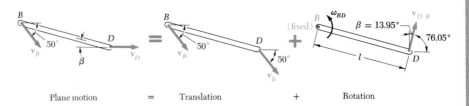

Plane motion   =   Translation   +   Rotation

Expressing the relation between the velocities $\mathbf{v}_D$, $\mathbf{v}_B$, and $\mathbf{v}_{D/B}$, we write

$$\mathbf{v}_D = \mathbf{v}_B + \mathbf{v}_{D/B}$$

We draw the vector diagram corresponding to this equation. Recalling that $\beta = 13.95°$, we determine the angles of the triangle and write

$$\frac{v_D}{\sin 53.95°} = \frac{v_{D/B}}{\sin 50°} = \frac{628.3\ \text{in./s}}{\sin 76.05°}$$

$$v_{D/B} = 495.9\ \text{in./s} \qquad \mathbf{v}_{D/B} = 495.9\ \text{in./s}\ \measuredangle 76.05°$$

$$v_D = 523.4\ \text{in./s} = 43.6\ \text{ft/s} \qquad \mathbf{v}_D = 43.6\ \text{ft/s} \rightarrow$$

$$\mathbf{v}_P = \mathbf{v}_D = 43.6\ \text{ft/s} \rightarrow \quad \blacktriangleleft$$

Since $v_{D/B} = l\omega_{BD}$, we have

$$495.9\ \text{in./s} = (8\ \text{in.})\omega_{BD} \qquad \omega_{BD} = 62.0\ \text{rad/s}\ \uparrow \quad \blacktriangleleft$$

Problems

15.31 An automobile travels to the right at a constant speed of 90 km/h. If the diameter of a wheel is 550 mm, determine the velocities of points B, C, D, and E on the rim of the wheel.

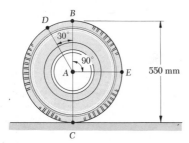

Fig. P15.31

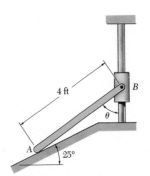

Fig. P15.32

15.32 Collar B moves upward with a constant velocity of 5 ft/s. At the instant when $\theta = 50°$, determine (a) the angular velocity of rod AB, (b) the velocity of end A of the rod.

15.33 Solve Prob. 15.32, assuming that $\theta = 30°$.

15.34 Collar A moves with a constant velocity of 900 mm/s to the right. At the instant when $\theta = 30°$, determine (a) the angular velocity of rod AB, (b) the velocity of collar B.

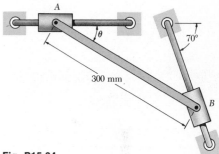

Fig. P15.34

15.35 The plate shown moves in the xy plane. Knowing that $(v_A)_x = 250$ mm/s, $(v_B)_y = -450$ mm/s, and $(v_C)_x = -500$ mm/s, determine (a) the angular velocity of the plate, (b) the velocity of point A.

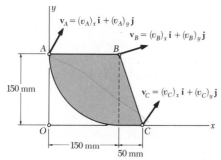

Fig. P15.35

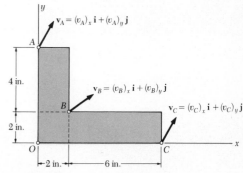

Fig. P15.36

15.36 The plate shown moves in the xy plane. Knowing that $(v_A)_x = 12$ in./s, $(v_B)_x = -4$ in./s, and $(v_C)_y = -24$ in./s, determine (a) the angular velocity of the plate, (b) the velocity of point B.

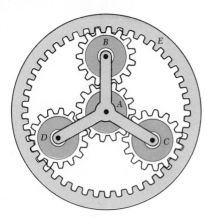

Fig. P15.39 and P15.40

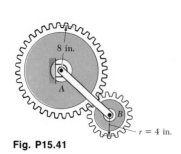

Fig. P15.41

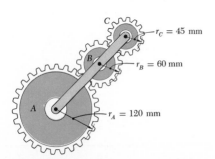

Fig. P15.43 and P15.44

15.37 In Prob. 15.36, determine (*a*) the velocity of point *A*, (*b*) the point of the plate with zero velocity.

15.38 In Prob. 15.35, determine (*a*) the velocity of point *B*, (*b*) the point of the plate with zero velocity.

15.39 In the planetary gear system shown, the radius of gears *A*, *B*, *C*, and *D* is *a* and the radius of the outer gear *E* is 3*a*. Knowing that the angular velocity of gear *A* is ω_A clockwise and that the outer gear *E* is stationary, determine (*a*) the angular velocity of each planetary gear, (*b*) the angular velocity of the spider connecting the planetary gears.

15.40 In the planteary gear system shown, the radius of gears *A*, *B*, *C*, and *D* is 30 mm and the radius of the outer gear *E* is 90 mm. Knowing that gear *E* has an angular velocity of 180 rpm clockwise and that the central gear *A* has an angular velocity of 240 rpm clockwise, determine (*a*) the angular velocity of each planetary gear, (*b*) the angular velocity of the spider connecting the planetary gears.

15.41 Gear *A* rotates clockwise with a constant angular velocity of 60 rpm. Determine the angular velocity of gear *B* if the angular velocity of arm *AB* is (*a*) 40 rpm counterclockwise, (*b*) 40 rpm clockwise.

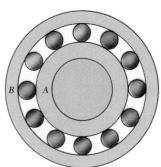

Fig. P15.42

15.42 In the simplified sketch of a ball bearing shown, the diameter of the inner race *A* is 2.5 in. and the diameter of each ball is 0.5 in. The outer race *B* is stationary while the inner race has an angular velocity of 3600 rpm. Determine (*a*) the speed of the center of each ball, (*b*) the angular velocity of each ball, (*c*) the number of times per minute each ball describes a complete circle.

15.43 Three gears *A*, *B*, and *C* are pinned at their centers to rod *ABC*. Knowing that end *A* of rod *ABC* is fixed and that gear *A* does not rotate, determine the angular velocity of gears *B* and *C* when rod *ABC* rotates clockwise with a constant angular velocity of 75 rpm.

15.44 Three gears *A*, *B*, and *C* are pinned at their centers to rod *ABC*. Knowing that end *C* of rod *ABC* is fixed and that gear *C* does not rotate, determine the angular velocity of gears *A* and *B* when rod *ABC* rotates clockwise with a constant angular velocity of 80 rpm.

15.45 The disk shown has a constant angular velocity of 500 rpm counterclockwise. Knowing that rod BD is 250 mm long, determine the angular velocity of rod BD and the velocity of collar D when (a) $\theta = 0$, (b) $\theta = 90°$.

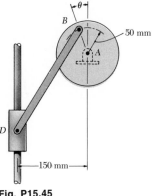

Fig. P15.45

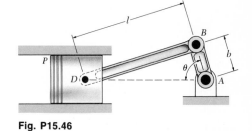

Fig. P15.46

15.46 In the engine system shown, $l = 8$ in. and $b = 3$ in.; crank AB rotates with a constant angular velocity of 2000 rpm clockwise. Determine the velocity of piston P and the angular velocity of the connecting rod for the position corresponding to (a) $\theta = 0$, (b) $\theta = 90°$, (c) $\theta = 180°$.

15.47 Solve Prob. 15.46 for the position corresponding to $\theta = 30°$.

15.48 Solve Prob. 15.45 for the position corresponding to $\theta = 60°$.

15.49 through 15.52 In the position shown, bar AB has a constant angular velocity of 3 rad/s counterclockwise. Determine the angular velocity of bars BD and DE.

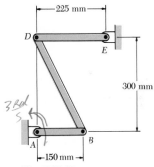

Fig. P15.49

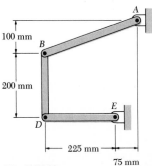

Fig. P15.50

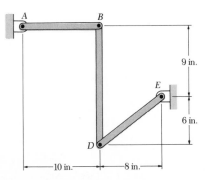

Fig. P15.51

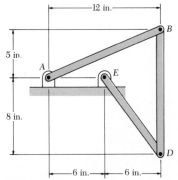

Fig. P15.52

15.53 The flanged wheel shown rolls to the right with a constant velocity of 1.5 m/s. Knowing that rod AB is 0.9 m long, determine the velocity of A and the angular velocity of the rod when (a) $\beta = 0$, (b) $\beta = 90°$.

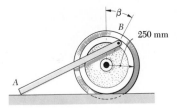

Fig. P15.53

15.54 Solve Prob. 15.53, assuming that $\beta = 30°$.

*15.55 For the gearing system shown, derive an expression for the angular velocity ω_C of gear C and show that ω_C is independent of the radius of gear B. Assume that point A is fixed and denote the angular velocities of rod ABC and gear A by ω_{ABC} and ω_A, respectively.

Fig. P15.55

15.7. Instantaneous Center of Rotation in Plane Motion.
Consider the general plane motion of a slab. We shall show that at any given instant the velocities of the various particles of the slab are the same as if the slab were rotating about a certain axis perpendicular to the plane of the slab, called the *instantaneous axis of rotation*. This axis intersects the plane of the slab at a point C, called the *instantaneous center of rotation* of the slab.

To prove our statement, we first recall that the plane motion of a slab may always be replaced by a translation defined by the motion of an arbitrary reference point A, and by a rotation about A. As far as the velocities are concerned, the translation is characterized by the velocity $\mathbf{v}_A$ of the reference point A and the rotation is characterized by the angular velocity ω of the slab (which is independent of the choice of A). Thus, the velocity $\mathbf{v}_A$ of point A and the angular velocity ω of the slab define completely the velocities of all the other particles of the slab (Fig. 15.18a). Now let us assume that $\mathbf{v}_A$ and ω are known and that they are both different from zero. (If $\mathbf{v}_A = 0$, point A is itself the instantaneous center of rotation, and if $\omega = 0$, all the particles have the same velocity $\mathbf{v}_A$.) These velocities could be obtained by letting the slab rotate with the angular velocity ω about a point C located on the perpendicular to $\mathbf{v}_A$ at a distance $r = v_A/\omega$ from A

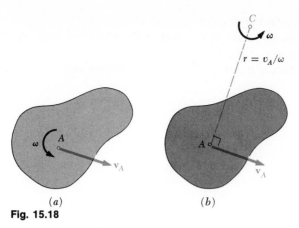

$r = v_A/\omega$

(a)   (b)

Fig. 15.18

as shown in Fig. 15.18b. We check that the velocity of A would be perpendicular to AC and that its magnitude would be $r\omega = (v_A/\omega)\omega = v_A$. Thus the velocities of all the other particles of the slab would be the same as originally defined. Therefore, *as far as the velocities are concerned, the slab seems to rotate about the instantaneous center C* at the instant considered.

The position of the instantaneous center may be defined in two other ways. If the directions of the velocities of two particles A and B of the slab are known and if they are different, the instantaneous center C is obtained by drawing the perpendicular to $\mathbf{v}_A$ through A and the perpendicular to $\mathbf{v}_B$ through B and determining the point in which these two lines intersect (Fig. 15.19a). If the velocities $\mathbf{v}_A$ and $\mathbf{v}_B$ of two particles A and B are perpendicular to the line AB and if their magnitudes are known, the instantaneous center may be found by intersecting the line AB with the line joining the extremities of the vectors $\mathbf{v}_A$ and $\mathbf{v}_B$ (Fig. 15.19b). Note that if $\mathbf{v}_A$ and $\mathbf{v}_B$ were parallel in Fig. 15.19a or if $\mathbf{v}_A$ and $\mathbf{v}_B$ had the same magnitude in Fig. 15.19b, the instantaneous center C would be at an infinite distance and ω would be zero; all points of the slab would have the same velocity.

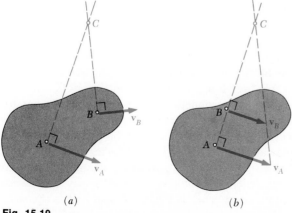

(a)   (b)

Fig. 15.19

To see how the concept of instantaneous center of rotation may be put to use, let us consider again the rod of Sec. 15.6. Drawing the perpendicular to $\mathbf{v}_A$ through A and the perpendicular to $\mathbf{v}_B$ through B (Fig. 15.20), we obtain the instantaneous center C. At the instant considered, the velocities

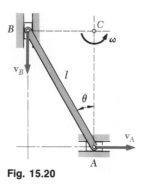

Fig. 15.20

of all the particles of the rod are thus the same as if the rod rotated about C. Now, if the magnitude v_A of the velocity of A is known, the magnitude ω of the angular velocity of the rod may be obtained by writing

$$\omega = \frac{v_A}{AC} = \frac{v_A}{l \cos \theta}$$

The magnitude of the velocity of B may then be obtained by writing

$$v_B = (BC)\omega = l \sin \theta \frac{v_A}{l \cos \theta} = v_A \tan \theta$$

Note that only *absolute* velocities are involved in the computation.

The instantaneous center of a slab in plane motion may be located either on the slab or outside the slab. If it is located on the slab, the particle C coinciding with the instantaneous center at a given instant t must have zero velocity at that instant. However, it should be noted that the instantaneous center of rotation is valid only at a given instant. Thus, the particle C of the slab which coincides with the instantaneous center at time t will generally not coincide with the instantaneous center at time $t + \Delta t$; while its velocity is zero at time t, it will probably be different from zero at time $t + \Delta t$. This means that, in general, the particle C *does not have zero acceleration* and, therefore, that the *accelerations* of the various particles of the slab *cannot* be determined as if the slab were rotating about C.

As the motion of the slab proceeds, the instantaneous center moves in space. But it was just pointed out that the position of the instantaneous center on the slab keeps changing. Thus, the instantaneous center describes one curve in space, called the *space centrode,* and another curve on the slab, called the *body centrode* (Fig. 15.21). It may be shown that at any instant, these two curves are tangent at C and that as the slab moves, the body centrode appears to *roll* on the space centrode.

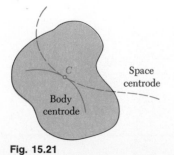

Space
centrode

Body
centrode

Fig. 15.21

Solve Sample Prob. 15.2, using the method of the instantaneous center of rotation.

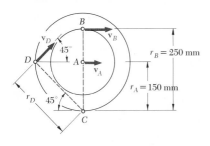

a. Angular Velocity of the Gear. Since the gear rolls on the stationary lower rack, the point of contact C of the gear with the rack has no velocity; point C is therefore the instantaneous center of rotation. We write

$$v_A = r_A \omega \qquad 1.2 \text{ m/s} = (0.150 \text{ m})\omega \qquad \omega = 8 \text{ rad/s} \downarrow \blacktriangleleft$$

b. Velocities. All points of the gear seem to rotate about the instantaneous center as far as velocities are concerned.

Velocity of Upper Rack. Recalling that $v_R = v_B$, we write

$$v_R = v_B = r_B \omega \qquad v_R = (0.250 \text{ m})(8 \text{ rad/s}) = 2 \text{ m/s}$$
$$\mathbf{v}_R = 2 \text{ m/s} \rightarrow \blacktriangleleft$$

Velocity of Point D. Since $r_D = (0.150 \text{ m})\sqrt{2} = 0.2121 \text{ m}$, we write

$$v_D = r_D \omega \qquad v_D = (0.2121 \text{ m})(8 \text{ rad/s}) = 1.697 \text{ m/s}$$
$$\mathbf{v}_D = 1.697 \text{ m/s} \measuredangle 45° \blacktriangleleft$$

Solve Sample Prob. 15.3, using the method of the instantaneous center of rotation.

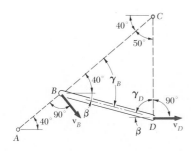

Motion of Crank AB. Referring to Sample Prob. 15.3, we obtain the velocity of point B; $\mathbf{v}_B = 628.3 \text{ in./s} \searrow 50°$.

Motion of the Connecting Rod BD. We first locate the instantaneous center C by drawing lines perpendicular to the absolute velocities $\mathbf{v}_B$ and $\mathbf{v}_D$. Recalling from Sample Prob. 15.3 that $\beta = 13.95°$ and that $BD = 8 \text{ in.}$, we solve the triangle BCD.

$$\gamma_B = 40° + \beta = 53.95° \qquad \gamma_D = 90° - \beta = 76.05°$$
$$\frac{BC}{\sin 76.05°} = \frac{CD}{\sin 53.95°} = \frac{8 \text{ in.}}{\sin 50°}$$
$$BC = 10.14 \text{ in.} \qquad CD = 8.44 \text{ in.}$$

Since the connecting rod BD seems to rotate about point C, we write

$$v_B = (BC)\omega_{BD}$$
$$628.3 \text{ in./s} = (10.14 \text{ in.})\omega_{BD}$$
$$\omega_{BD} = 62.0 \text{ rad/s} \uparrow \blacktriangleleft$$
$$v_D = (CD)\omega_{BD} = (8.44 \text{ in.})(62.0 \text{ rad/s})$$
$$= 523 \text{ in./s} = 43.6 \text{ ft/s}$$
$$\mathbf{v}_P = \mathbf{v}_D = 43.6 \text{ ft/s} \rightarrow \blacktriangleleft$$

Problems

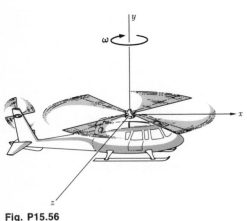

Fig. P15.56

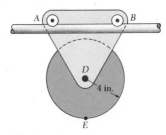

Fig. P15.57

15.56 A helicopter moves horizontally in the x direction at a speed of 120 mi/h. Knowing that the main blades rotate clockwise with an angular velocity of 180 rpm, determine the instantaneous axis of rotation of the main blades.

15.57 The trolley shown moves to the left along a horizontal pipe support at a speed of 15 in./s. Knowing that the 4-in.-radius disk has an angular velocity of 6 rad/s counterclockwise, determine (a) the instantaneous center of rotation of the disk, (b) the velocity of point E.

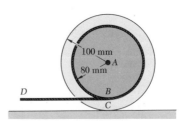

Fig. P15.58

15.58 A drum of radius 80 mm is mounted on a cylinder of radius 100 mm. A cord is wound around the drum, and its extremity D is pulled to the left with a constant velocity of 120 mm/s, causing the cylinder to roll without sliding. Determine (a) the angular velocity of the cylinder, (b) the velocity of the center of the cylinder, (c) the length of cord which is wound or unwound per second.

15.59 A roll of paper rests on a horizontal surface as shown. End D of the paper is then pulled to the right with a constant velocity of 195 mm/s and it is observed that the center A of the roll moves to the right with a constant velocity of 75 mm/s. Determine (a) the instantaneous axis of rotation of the roll of paper, (b) the velocity of point B, (c) the number of millimeters of paper unwound per second.

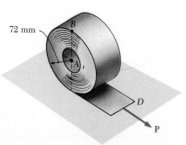

Fig. P15.59

15.60 Knowing that at the instant shown the angular velocity of rod *BE* is 4 rad/s counterclockwise, determine (*a*) the angular velocity of rod *AD*, (*b*) the velocity of collar *D*, (*c*) the velocity of point *A*.

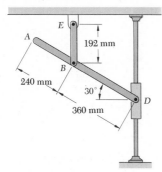

Fig. P15.60 and P15.61

15.61 Knowing that at the instant shown the velocity of collar *D* is 1.6 m/s upward, determine (*a*) the angular velocity of rod *AD*, (*b*) the velocity of point *B*, (*c*) the velocity of point *A*.

15.62 Knowing that at the instant shown the velocity of collar *D* is 4.8 in./s to the left, determine (*a*) the instantaneous center of rotation of rod *BE*, (*b*) the angular velocities of crank *AB* and rod *BE*, (*c*) the velocity of point *E*.

15.63 Knowing that at the instant shown the angular velocity of crank *AB* is 4.5 rad/s clockwise, determine (*a*) the angular velocity of rod *BE*, (*b*) the velocity of collar *D*, (*c*) the velocity of point *E*.

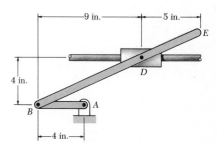

Fig. P15.62 and P15.63

15.64 Rod *BDE* is partially guided by a roller at *D* which moves in a vertical track. Knowing that at the instant shown the angular velocity of crank *AB* is 5 rad/s clockwise and that $\beta = 25°$, determine (*a*) the angular velocity of the rod, (*b*) the velocity of point *E*.

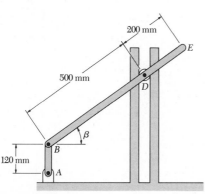

Fig. P15.64

15.65 Solve Prob. 15.64, assuming that $\beta = 40°$.

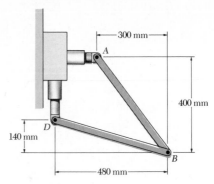

Fig. P15.66

15.66 Two links AB and BD, each 500 mm long, are connected at B and guided by hydraulic cylinders attached at A and D. Knowing that D is stationary and that the velocity of A is 750 mm/s to the right, determine at the instant shown (*a*) the angular velocity of each link, (*b*) the velocity of B.

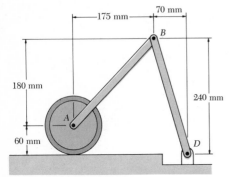

Fig. P15.67

15.67 A 60-mm-radius wheel is connected to a fixed support D by two links AB and BD. At the instant shown, the velocity of the center A of the wheel is 300 mm/s to the left. Determine (*a*) the angular velocity of each link, (*b*) the velocity of pin B.

15.68 Two slots have been cut in plate FG and the plate has been placed so that the slots fit two fixed pins A and B. Knowing that at the instant shown the angular velocity of crank DE is 6 rad/s clockwise, determine (*a*) the velocity of point F, (*b*) the velocity of point G.

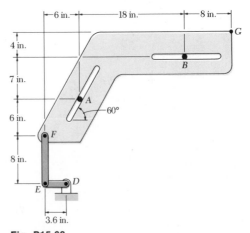

Fig. P15.68

15.69 In Prob. 15.68, determine the velocity of the point of the plate which is in contact with (*a*) pin A, (*b*) pin B.

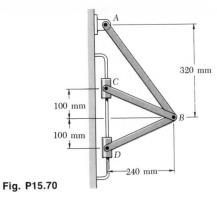

Fig. P15.70

15.70 Two collars C and D move along the vertical rod shown. Knowing that the velocity of collar C is 660 mm/s downward, determine (a) the velocity of collar D, (b) the angular velocity of member AB.

15.71 Two links AB and BD, each of length 30 in., are connected to three collars. At the instant shown $\beta = 55°$ and the velocity of collar A is 90 in./s to the right. Determine (a) the corresponding value of γ, (b) the velocity of collar D.

15.72 and 15.73 Two 500-mm rods are pin-connected at D as shown. Knowing that B moves to the left with a constant velocity of 360 mm/s, determine at the instant shown (a) the angular velocity of each rod, (b) the velocity of E.

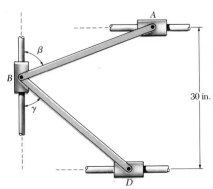

Fig. P15.71

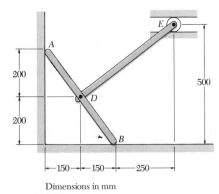

Dimensions in mm
Fig. P15.72

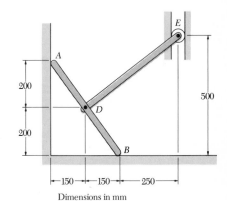

Dimensions in mm
Fig. P15.73

15.74 Describe the space centrode and the body centrode of link AB of Prob. 15.71 as collar B moves upward. (*Note.* The body centrode need not lie on a physical portion of the link.)

15.75 Describe the space centrode and the body centrode of the drum of Prob. 15.58 as the drum rolls on the horizontal surface.

15.76 Using the method of Sec. 15.7, solve Prob. 15.50.
15.77 Using the method of Sec. 15.7, solve Prob. 15.51.
15.78 Using the method of Sec. 15.7, solve Prob. 15.52.
15.79 Using the method of Sec. 15.7, solve Prob. 15.53.

It is quite evident that the determination of accelerations is considerably more involved than the determination of velocities. Yet in the example considered here, the extremities A and B of the rod were moving along straight tracks, and the diagrams drawn were relatively simple. If A and B had moved along curved tracks, the accelerations $\mathbf{a}_A$ and $\mathbf{a}_B$ should have been resolved into normal and tangential components and the solution of the problem would have involved six different vectors.

When a mechanism consists of several moving parts which are pin-connected, its analysis may be carried out by considering each part as a rigid body, while keeping in mind that the points where two parts are connected must have the same absolute acceleration (see Sample Prob. 15.7). In the case of meshed gears, the tangential components of the accelerations of the teeth in contact are equal, but their normal components are different.

* 15.9. Analysis of Plane Motion in Terms of a Parameter.

In the case of certain mechanisms, it is possible to express the coordinates x and y of all the significant points of the mechanism by means of simple analytic expressions containing a single parameter. It may be advantageous in such a case to determine directly the absolute velocity and the absolute acceleration of the various points of the mechanism, since the components of the velocity and of the acceleration of a given point may be obtained by differentiating the coordinates x and y of that point.

Let us consider again the rod AB whose extremities slide, respectively, in a horizontal and a vertical track (Fig. 15.25). The coordinates x_A and y_B of the extremities of the rod may be expressed in terms of the angle θ the rod forms with the vertical:

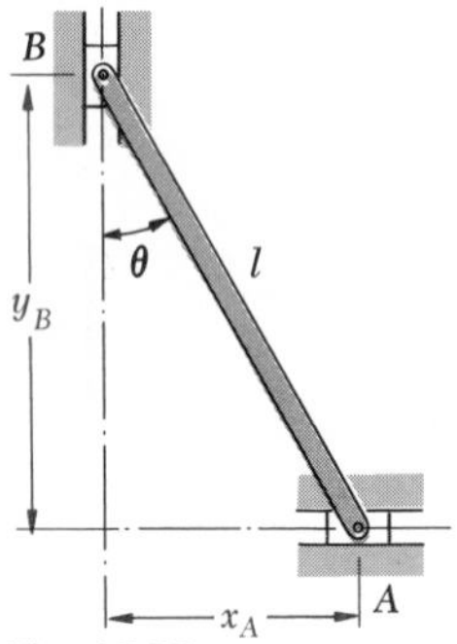

Fig. 15.25

$$x_A = l \sin \theta \qquad y_B = l \cos \theta \qquad (15.24)$$

Differentiating Eqs. (15.24) twice with respect to t, we write

$$v_A = \dot{x}_A = l\dot{\theta} \cos \theta$$
$$a_A = \ddot{x}_A = -l\dot{\theta}^2 \sin \theta + l\ddot{\theta} \cos \theta$$

$$v_B = \dot{y}_B = -l\dot{\theta} \sin \theta$$
$$a_B = \ddot{y}_B = -l\dot{\theta}^2 \cos \theta - l\ddot{\theta} \sin \theta$$

Recalling that $\dot{\theta} = \omega$ and $\ddot{\theta} = \alpha$, we obtain

$$v_A = l\omega \cos \theta \qquad\qquad v_B = -l\omega \sin \theta \qquad (15.25)$$

$$a_A = -l\omega^2 \sin \theta + l\alpha \cos \theta \qquad a_B = -l\omega^2 \cos \theta - l\alpha \sin \theta \qquad (15.26)$$

We note that a positive sign for v_A or a_A indicates that the velocity $\mathbf{v}_A$ or the acceleration $\mathbf{a}_A$ is directed to the right; a positive sign for v_B or a_B indicates that $\mathbf{v}_B$ or $\mathbf{a}_B$ is directed upward. Equations (15.25) may be used, for example, to determine v_B and ω when v_A and θ are known. Substituting for ω in (15.26), we may then determine a_B and α if a_A is known.

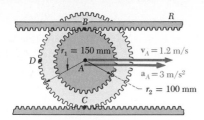

SAMPLE PROBLEM 15.6

The center of the double gear of Sample Prob. 15.2 has a velocity of 1.2 m/s to the right and an acceleration of 3 m/s² to the right. Recalling that the lower rack is stationary, determine (a) the angular acceleration of the gear, (b) the acceleration of points B, C, and D of the gear.

a. Angular Acceleration of the Gear. In Sample Prob. 15.2, we found that $x_A = -r_1\theta$ and $v_A = -r_1\omega$. Differentiating the latter with respect to time, we obtain $a_A = -r_1\alpha$.

$$v_A = -r_1\omega \qquad 1.2 \text{ m/s} = -(0.150 \text{ m})\omega \qquad \omega = -8 \text{ rad/s}$$
$$a_A = -r_1\alpha \qquad 3 \text{ m/s}^2 = -(0.150 \text{ m})\alpha \qquad \alpha = -20 \text{ rad/s}^2$$
$$\alpha = \alpha\mathbf{k} = -(20 \text{ rad/s}^2)\mathbf{k} \quad \blacktriangleleft$$

b. Accelerations. The rolling motion of the gear is resolved into a translation with A and a rotation about A.

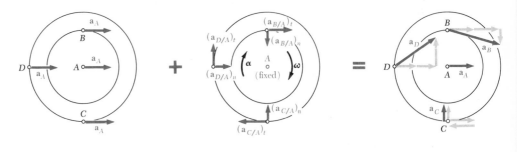

Translation $\qquad +$ $\qquad$ Rotation $\qquad =$ $\qquad$ Rolling motion

Acceleration of Point B. Adding vectorially the accelerations corresponding to the translation and to the rotation, we obtain

$$\mathbf{a}_B = \mathbf{a}_A + \mathbf{a}_{B/A} = \mathbf{a}_A + (\mathbf{a}_{B/A})_t + (\mathbf{a}_{B/A})_n$$
$$= \mathbf{a}_A + \alpha\mathbf{k} \times \mathbf{r}_{B/A} - \omega^2 \mathbf{r}_{B/A}$$
$$= (3 \text{ m/s}^2)\mathbf{i} - (20 \text{ rad/s}^2)\mathbf{k} \times (0.100 \text{ m})\mathbf{j} - (8 \text{ rad/s})^2(0.100 \text{ m})\mathbf{j}$$
$$= (3 \text{ m/s}^2)\mathbf{i} + (2 \text{ m/s}^2)\mathbf{i} - (6.40 \text{ m/s}^2)\mathbf{j}$$
$$a_B = 8.12 \text{ m/s}^2 \; \diagdown \; 52.0° \quad \blacktriangleleft$$

Acceleration of Point C
$$\mathbf{a}_C = \mathbf{a}_A + \mathbf{a}_{C/A} = \mathbf{a}_A + \alpha\mathbf{k} \times \mathbf{r}_{C/A} - \omega^2 \mathbf{r}_{C/A}$$
$$= (3 \text{ m/s}^2)\mathbf{i} - (20 \text{ rad/s}^2)\mathbf{k} \times (-0.150 \text{ m})\mathbf{j} - (8 \text{ rad/s})^2(-0.150 \text{ m})\mathbf{j}$$
$$= (3 \text{ m/s}^2)\mathbf{i} - (3 \text{ m/s}^2)\mathbf{i} + (9.60 \text{ m/s}^2)\mathbf{j}$$
$$a_C = 9.60 \text{ m/s}^2 \uparrow \quad \blacktriangleleft$$

Acceleration of Point D

$$\mathbf{a}_D = \mathbf{a}_A + \mathbf{a}_{D/A} = \mathbf{a}_A + \alpha\mathbf{k} \times \mathbf{r}_{D/A} - \omega^2 \mathbf{r}_{D/A}$$
$$= (3 \text{ m/s}^2)\mathbf{i} - (20 \text{ rad/s}^2)\mathbf{k} \times (-0.150 \text{ m})\mathbf{i} - (8 \text{ rad/s})^2(-0.150 \text{ m})\mathbf{i}$$
$$= (3 \text{ m/s}^2)\mathbf{i} + (3 \text{ m/s}^2)\mathbf{j} + (9.60 \text{ m/s}^2)\mathbf{i}$$
$$a_D = 12.95 \text{ m/s}^2 \; \measuredangle \; 13.4° \quad \blacktriangleleft$$

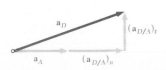

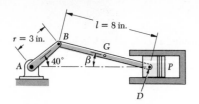

SAMPLE PROBLEM 15.7

Crank AB of the engine system of Sample Prob. 15.3 has a constant clockwise angular velocity of 2000 rpm. For the crank position shown, determine the angular acceleration of the connecting rod BD and the acceleration of point D.

Motion of Crank AB. Since the crank rotates about A with constant $\omega_{AB} = 2000$ rpm $= 209.4$ rad/s, we have $\alpha_{AB} = 0$. The acceleration of B is therefore directed toward A and has a magnitude

$$a_B = r\omega_{AB}^2 = (\tfrac{3}{12}\text{ ft})(209.4\text{ rad/s})^2 = 10{,}962\text{ ft/s}^2$$
$$\mathbf{a}_B = 10{,}962\text{ ft/s}^2 \; \diagup\!\!\!\!\diagdown\; 40°$$

Motion of the Connecting Rod BD. The angular velocity ω_{BD} and the value of β were obtained in Sample Prob. 15.3:

$$\omega_{BD} = 62.0\text{ rad/s} \uparrow \qquad \beta = 13.95°$$

The motion of BD is resolved into a translation with B and a rotation about B. The relative acceleration $\mathbf{a}_{D/B}$ is resolved into normal and tangential components:

$$(a_{D/B})_n = (BD)\omega_{BD}^2 = (\tfrac{8}{12}\text{ ft})(62.0\text{ rad/s})^2 = 2563\text{ ft/s}^2$$
$$(\mathbf{a}_{D/B})_n = 2563\text{ ft/s}^2 \; \diagdown\!\!\!\diagup\; 13.95°$$
$$(a_{D/B})_t = (BD)\alpha_{BD} = (\tfrac{8}{12})\alpha_{BD} = 0.6667\alpha_{BD}$$
$$(\mathbf{a}_{D/B})_t = 0.6667\alpha_{BD} \; \diagup\!\!\!\!\diagup\; 76.05°$$

While $(\mathbf{a}_{D/B})_t$ must be perpendicular to BD, its sense is not known.

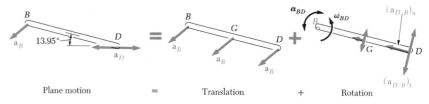

Plane motion = Translation + Rotation

Noting that the acceleration $\mathbf{a}_D$ must be horizontal, we write

$$\mathbf{a}_D = \mathbf{a}_B + \mathbf{a}_{D/B} = \mathbf{a}_B + (\mathbf{a}_{D/B})_n + (\mathbf{a}_{D/B})_t$$
$$[a_D \leftrightarrow] = [10{,}962 \; \diagup\!\!\!\!\diagdown\; 40°] + [2563 \; \diagdown\!\!\!\diagup\; 13.95°] + [0.6667\alpha_{BD} \; \diagup\!\!\!\!\diagup\; 76.05°]$$

Equating x and y components, we obtain the following scalar equations:

$\xrightarrow{+} x$ components:
$$-a_D = -10{,}962\cos 40° - 2563\cos 13.95° + 0.6667\alpha_{BD}\sin 13.95°$$
$+\uparrow y$ components:
$$0 = -10{,}962\sin 40° + 2563\sin 13.95° + 0.6667\alpha_{BD}\cos 13.95°$$

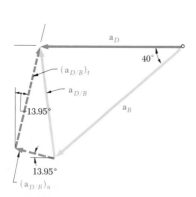

Solving the equations simultaneously, we obtain $\alpha_{BD} = +9940$ rad/s^2 and $a_D = +9290$ ft/s^2. The positive signs indicate that the senses shown on the vector polygon are correct; we write

$$\alpha_{BD} = 9940\text{ rad/s}^2 \uparrow \quad \blacktriangleleft$$
$$\mathbf{a}_D = 9290\text{ ft/s}^2 \leftarrow \quad \blacktriangleleft$$

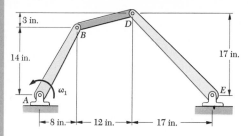

SAMPLE PROBLEM 15.8

The linkage $ABDE$ moves in the vertical plane. Knowing that in the position shown crank AB has a constant angular velocity ω_1 of 20 rad/s counterclockwise, determine the angular velocities and angular accelerations of the connecting rod BD and of the crank DE.

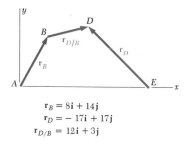

$$r_B = 8i + 14j$$
$$r_D = -17i + 17j$$
$$r_{D/B} = 12i + 3j$$

Solution. While this problem could be solved by the method used in Sample Prob. 15.7, we shall make full use of the vector approach in the present case. The position vectors r_B, r_D, and $r_{D/B}$ are chosen as shown in the sketch.

Velocities. Since the motion of each element of the linkage is contained in the plane of the figure, we have

$$\omega_{AB} = \omega_{AB}k = (20 \text{ rad/s})k \qquad \omega_{BD} = \omega_{BD}k \qquad \omega_{DE} = \omega_{DE}k$$

where k is a unit vector pointing out of the paper. We now write

$$v_D = v_B + v_{D/B}$$
$$\omega_{DE}k \times r_D = \omega_{AB}k \times r_B + \omega_{BD}k \times r_{D/B}$$
$$\omega_{DE}k \times (-17i + 17j) = 20k \times (8i + 14j) + \omega_{BD}k \times (12i + 3j)$$
$$-17\omega_{DE}j - 17\omega_{DE}i = 160j - 280i + 12\omega_{BD}j - 3\omega_{BD}i$$

Equating the coefficients of the unit vectors i and j, we obtain the following two scalar equations:

$$-17\omega_{DE} = -280 - 3\omega_{BD}$$
$$-17\omega_{DE} = +160 + 12\omega_{BD}$$
$$\omega_{BD} = -(29.33 \text{ rad/s})k \qquad \omega_{DE} = (11.29 \text{ rad/s})k \quad \blacktriangleleft$$

Accelerations. Noting that at the instant considered crank AB has a constant angular velocity, we write

$$\alpha_{AB} = 0 \qquad \alpha_{BD} = \alpha_{BD}k \qquad \alpha_{DE} = \alpha_{DE}k$$
$$a_D = a_B + a_{D/B} \tag{1}$$

Each term of Eq. (1) is evaluated separately:

$$a_D = \alpha_{DE}k \times r_D - \omega_{DE}^2 r_D$$
$$= \alpha_{DE}k \times (-17i + 17j) - (11.29)^2(-17i + 17j)$$
$$= -17\alpha_{DE}j - 17\alpha_{DE}i + 2170i - 2170j$$
$$a_B = \alpha_{AB}k \times r_B - \omega_{AB}^2 r_B = 0 - (20)^2(8i + 14j)$$
$$= -3200i - 5600j$$
$$a_{D/B} = \alpha_{BD}k \times r_{D/B} - \omega_{BD}^2 r_{D/B}$$
$$= \alpha_{BD}k \times (12i + 3j) - (29.33)^2(12i + 3j)$$
$$= 12\alpha_{BD}j - 3\alpha_{BD}i - 10,320i - 2580j$$

Substituting into Eq. (1) and equating the coefficients of i and j, we obtain

$$-17\alpha_{DE} + 3\alpha_{BD} = -15,690$$
$$-17\alpha_{DE} - 12\alpha_{BD} = -6010$$
$$\alpha_{BD} = -(645 \text{ rad/s}^2)k \qquad \alpha_{DE} = (809 \text{ rad/s}^2)k \quad \blacktriangleleft$$

Problems

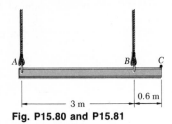

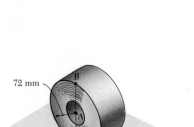

Fig. P15.80 and P15.81

15.80 A 3.6-m steel beam is lowered by means of two cables unwinding at the same speed from overhead cranes. As the beam approaches the ground, the crane operators apply brakes to slow down the unwinding motion. At the instant considered the deceleration of the cable attached at A is 4.2 m/s², while that of the cable attached at B is 1.5 m/s². Determine (a) the angular acceleration of the beam, (b) the acceleration of point C.

15.81 The acceleration of point C is 0.3 m/s² downward and the angular acceleration of the beam is 0.8 rad/s² clockwise. Knowing that the angular velocity of the beam is zero at the instant considered, determine the acceleration of each cable.

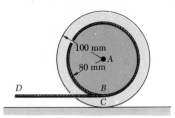

Fig. P15.82

15.82 A roll of paper is at rest on a horizontal surface. A force $\mathbf{P}$ applied to the paper produces the following accelerations: $\mathbf{a}_A = 300$ mm/s² and $\mathbf{a}_D = 780$ mm/s², both to the right. Determine (a) the angular acceleration of the roll, (b) the acceleration of point B.

15.83 In Prob. 15.82, determine the point on the vertical surface of the roll of paper which (a) has no acceleration, (b) has an acceleration of 500 mm/s² to the right.

Fig. P15.84

15.84 A drum of radius 80 mm is mounted on a cylinder of radius 100 mm. A cord is wound around the drum and is pulled in such a way that point D has a velocity of 75 mm/s and an acceleration of 400 mm/s², both directed to the left. Assuming that the cylinder rolls without slipping, determine the acceleration (a) of point A, (b) of point B, (c) of point C.

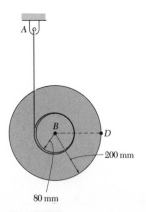

Fig. P15.85

15.85 At the instant shown the center B of the double pulley has a velocity of 0.6 m/s and an acceleration of 2.4 m/s², both directed downward. Knowing that the cord wrapped around the inner pulley is attached to a fixed support at A, determine the acceleration of point D.

15.86 An automobile travels to the right at a constant speed of 90 km/h. Knowing that the diameter of a wheel is 550 mm, determine the acceleration (*a*) of point *C*, (*b*) of point *D*.

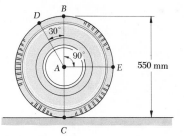

Fig. P15.86 and P15.87

15.87 An automobile is started from rest and has a constant acceleration of 2.4 m/s² to the right. Knowing that the wheel shown rolls without sliding, determine the speed of the automobile at which the magnitude of the acceleration of point *B* of the wheel will be 16 m/s².

15.88 The 5-in.-radius drum rolls without slipping on a portion of a belt which moves downward to the left with a constant velocity of 6 in./s. Knowing that at a given instant the velocity and acceleration of the center *A* of the drum are as shown, determine the acceleration of point *D*.

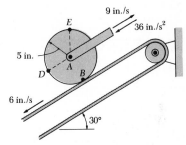

Fig. P15.88

15.89 In Prob. 15.88, determine the acceleration of point *E*.

15.90 The disk shown has a constant angular velocity of 500 rpm counterclockwise. Knowing that rod *BD* is 250 mm long, determine the acceleration of collar *D* when (*a*) $\theta = 90°$, (*b*) $\theta = 180°$.

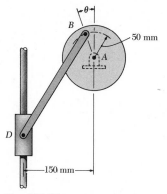

Fig. P15.90

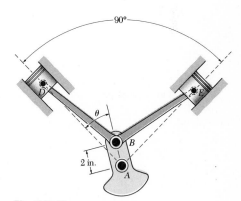

Fig. P15.91

15.91 In the two-cylinder air compressor shown the connecting rods *BD* and *BE* are each 7.5 in. long and crank *AB* rotates about the fixed point *A* with a constant angular velocity of 1500 rpm clockwise. Determine the acceleration of each piston when $\theta = 0$.

15.92 In Prob. 15.91, determine the acceleration of each piston when $\theta = 45°$.

15.93 Arm AB rotates with a constant angular velocity of 60 rpm clockwise. Knowing that gear A does not rotate, determine the acceleration of the tooth of gear B which is in contact with gear A.

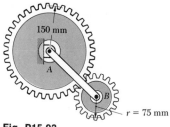

Fig. P15.93

15.94 The cross $BHDF$ is supported by two links AB and DE. Knowing that at the instant shown link AB rotates with a constant angular velocity of 4 rad/s clockwise, determine (a) the angular velocity of the cross, (b) the angular acceleration of the cross, (c) the acceleration of point H.

15.95 In Prob. 15.94, determine the acceleration (a) of point F, (b) of point G.

15.96 and 15.97 For the linkage indicated, determine the angular acceleration (a) of bar BD, (b) of bar DE.

> **15.96** Linkage of Prob. 15.52.
> **15.97** Linkage of Prob. 15.51.

15.98 End A of rod AB moves to the right with a constant velocity of 2 m/s. For the position shown, determine (a) the angular acceleration of rod AB, (b) the acceleration of the midpoint G of rod AB.

15.99 In the position shown, end A of rod AB has a velocity of 2.5 m/s and an acceleration of 1.5 m/s², both directed to the right. Determine (a) the angular acceleration of rod AB, (b) the acceleration of the midpoint G of rod AB.

15.100 In the position shown, end A of rod AB has a velocity of 2.5 ft/s and an acceleration of 1.8 ft/s², both directed to the left. Determine (a) the angular acceleration of rod AB, (b) the acceleration of the midpoint G of rod AB.

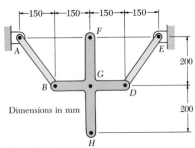

Fig. P15.94

Dimensions in mm

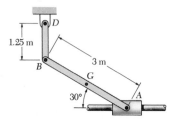

Fig. P15.98 and P15.99

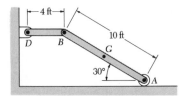

Fig. P15.100 and P15.101

15.101 End A of rod AB moves to the left with a constant velocity of 2.5 ft/s. For the position shown, determine (a) the angular acceleration of rod AB, (b) the acceleration of the midpoint G of rod AB.

15.102 Denoting by $\mathbf{r}_A$ the position vector of a point A of a rigid slab which moves in plane motion, show that (a) the position vector $\mathbf{r}_C$ of the instantaneous center of rotation is

$$\mathbf{r}_C = \mathbf{r}_A + \frac{\boldsymbol{\omega} \times \mathbf{v}_A}{\omega^2}$$

where $\boldsymbol{\omega}$ is the angular velocity of the slab and $\mathbf{v}_A$ the velocity of point A, (b) the acceleration of the instantaneous center of rotation is zero if, and only if,

$$\mathbf{a}_A = \frac{\alpha}{\omega}\mathbf{v}_A + \boldsymbol{\omega} \times \mathbf{v}_A$$

where $\boldsymbol{\alpha} = \alpha\mathbf{k}$ is the angular acceleration of the slab.

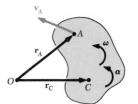

Fig. P15.102

*15.103 Rod AB slides with its ends in contact with the floor and the inclined plane. Using the method of Sec. 15.9, derive an expression for the angular velocity of the rod in terms of v_B, θ, l, and β.

*15.104 Derive an expression for the angular acceleration of rod AB in terms of v_B, θ, l, and β, knowing that the acceleration of point B is zero.

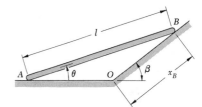

Fig. P15.103 and P15.104

*15.105 Crank AB rotates with a constant clockwise angular velocity ω, and $\theta = 0$ at $t = 0$. Using the method of Sec. 15.9, derive an expression for the velocity of piston P in terms of the time t.

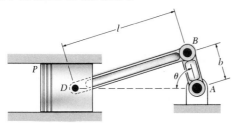

Fig. P15.105

*15.106 The drive disk of the Scotch crosshead mechanism shown has an angular velocity ω and an angular acceleration α, both directed clockwise. Using the method of Sec. 15.9 and assuming that $\beta = 0$, derive an expression for (a) the velocity of point A, (b) the acceleration of point A.

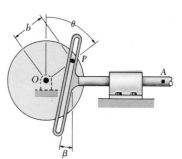

Fig. P15.106

*15.107 Solve Prob. 15.106 for an arbitrary value of angle β.

*15.108 Knowing that rod AB rotates with an angular velocity ω and an angular acceleration α, both counterclockwise, derive expressions for the velocity and acceleration of collar C.

*15.109 Knowing that collar C moves to the right with a constant velocity $\mathbf{v}_0$, derive expressions for the angular velocity and angular acceleration of rod AB in the position shown.

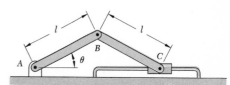

Fig. P15.108 and P15.109

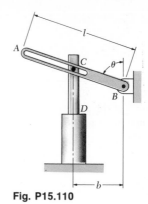

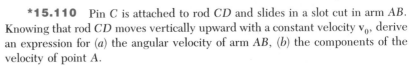

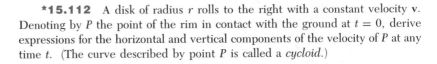

Fig. P15.110

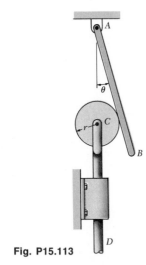

*15.110 Pin C is attached to rod CD and slides in a slot cut in arm AB. Knowing that rod CD moves vertically upward with a constant velocity v_0, derive an expression for (a) the angular velocity of arm AB, (b) the components of the velocity of point A.

*15.111 In Prob. 15.110, derive an expression for the angular acceleration of arm AB.

*15.112 A disk of radius r rolls to the right with a constant velocity v. Denoting by P the point of the rim in contact with the ground at $t = 0$, derive expressions for the horizontal and vertical components of the velocity of P at any time t. (The curve described by point P is called a *cycloid*.)

*15.113 The position of rod AB is controlled by a disk of radius r which is attached to yoke CD. Knowing that the yoke moves vertically upward with a constant velocity v_0, derive an expression for the angular velocity of rod AB.

Fig. P15.113

*15.114 In Prob. 15.113, derive an expression for the angular acceleration of rod AB.

15.10. Rate of Change of a Vector with Respect to a Rotating Frame.

We saw in Sec. 11.10 that the rate of change of a vector is the same with respect to a fixed frame and with respect to a frame in translation. In this section, we shall compare the rates of change of a vector $\mathbf{Q}$ with respect to a fixed frame and with respect to a rotating frame of reference.† We shall also learn to determine the rate of change of $\mathbf{Q}$ with respect to one frame of reference when $\mathbf{Q}$ is defined by its components in another frame.

Consider two frames of reference centered at O, a fixed frame $OXYZ$ and a frame $Oxyz$ which rotates about the fixed axis OA; let $\mathbf{\Omega}$ denote the angular velocity of the frame $Oxyz$ at a given instant (Fig. 15.26). Consider now a vector function $\mathbf{Q}(t)$ represented by the vector $\mathbf{Q}$ attached at O; as the time t varies, both the direction and the magnitude of $\mathbf{Q}$ change. Since

† It is recalled that the selection of a fixed frame of reference is arbitrary. Any frame may be designated as "fixed"; all others will then be considered as moving.

the variation of **Q** is viewed differently by an observer using $OXYZ$ as a frame of reference and by an observer using $Oxyz$, we should expect the rate of change of **Q** to depend upon the frame of reference which has been selected. Therefore, we shall denote by $(\dot{\mathbf{Q}})_{OXYZ}$ the rate of change of **Q** with respect to the fixed frame $OXYZ$, and by $(\dot{\mathbf{Q}})_{Oxyz}$ its rate of change with respect to the rotating frame of $Oxyz$. We propose to determine the relation existing between these two rates of change.

Let us first resolve the vector **Q** into components along the x, y, and z axes of the rotating frame. Denoting by **i**, **j**, and **k** the corresponding unit vectors, we write

$$\mathbf{Q} = Q_x\mathbf{i} + Q_y\mathbf{j} + Q_z\mathbf{k} \qquad (15.27)$$

Differentiating (15.27) with respect to t and considering the unit vectors **i**, **j**, **k** as fixed, we obtain the rate of change of **Q** *with respect to the rotating frame Oxyz:*

$$(\dot{\mathbf{Q}})_{Oxyz} = \dot{Q}_x\mathbf{i} + \dot{Q}_y\mathbf{j} + \dot{Q}_z\mathbf{k} \qquad (15.28)$$

To obtain the rate of change of **Q** *with respect to the fixed frame OXYZ*, we must consider the unit vectors **i**, **j**, **k** as variable when differentiating (15.27). We therefore write

$$(\dot{\mathbf{Q}})_{OXYZ} = \dot{Q}_x\mathbf{i} + \dot{Q}_y\mathbf{j} + \dot{Q}_z\mathbf{k} + Q_x\frac{d\mathbf{i}}{dt} + Q_y\frac{d\mathbf{j}}{dt} + Q_z\frac{d\mathbf{k}}{dt} \quad (15.29)$$

Recalling (15.28), we observe that the sum of the first three terms in the right-hand member of (15.29) represents the rate of change $(\dot{\mathbf{Q}})_{Oxyz}$. We note, on the other hand, that the rate of change $(\dot{\mathbf{Q}})_{OXYZ}$ would reduce to the last three terms in (15.29) if the vector **Q** were fixed within the frame $Oxyz$, since $(\dot{\mathbf{Q}})_{Oxyz}$ would then be zero. But in that case, $(\dot{\mathbf{Q}})_{OXYZ}$ would represent the velocity of a particle located at the tip of **Q** and belonging to a body rigidly attached to the frame $Oxyz$. Thus, the last three terms in (15.29) represent the velocity of that particle; since the frame $Oxyz$ has an angular velocity Ω with respect to $OXYZ$ at the instant considered, we write, by (15.5),

$$Q_x\frac{d\mathbf{i}}{dt} + Q_y\frac{d\mathbf{j}}{dt} + Q_z\frac{d\mathbf{k}}{dt} = \mathbf{\Omega} \times \mathbf{Q} \qquad (15.30)$$

Substituting from (15.28) and (15.30) into (15.29), we obtain the fundamental relation

$$(\dot{\mathbf{Q}})_{OXYZ} = (\dot{\mathbf{Q}})_{Oxyz} + \mathbf{\Omega} \times \mathbf{Q} \qquad (15.31)$$

We conclude that the rate of change of the vector **Q** with respect to the fixed frame $OXYZ$ is made of two parts: The first part represents the rate of change of **Q** with respect to the rotating frame $Oxyz$; the second part, $\mathbf{\Omega} \times \mathbf{Q}$, is induced by the rotation of the frame $Oxyz$.

The use of relation (15.31) simplifies the determination of the rate of change of a vector **Q** with respect to a fixed frame of reference $OXYZ$ when

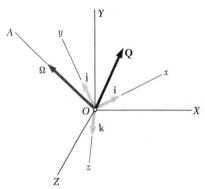

Fig. 15.26

the vector $\mathbf{Q}$ is defined by its components along the axes of a rotating frame $Oxyz$, since this relation does not require the separate computation of the derivatives of the unit vectors defining the orientation of the rotating frame.

15.11. Plane Motion of a Particle Relative to a Rotating Frame. Coriolis Acceleration.

Consider two frames of reference, both centered at O and both in the plane of the figure, a fixed frame OXY, and a rotating frame Oxy (Fig. 15.27). Let P be a particle moving in the plane of the figure. While the position vector $\mathbf{r}$ of P is the same in both frames, its rate of change depends upon the frame of reference which has been selected.

The absolute velocity $\mathbf{v}_P$ of the particle is defined as the velocity observed from the fixed frame OXY and is equal to the rate of change $(\dot{\mathbf{r}})_{OXY}$ of $\mathbf{r}$ with respect to that frame. We may, however, express $\mathbf{v}_P$ in terms of the rate of change $(\dot{\mathbf{r}})_{Oxy}$ observed from the rotating frame if we make use of Eq. (15.31). Denoting by $\boldsymbol{\Omega}$ the angular velocity of the frame Oxy with respect to OXY at the instant considered, we write

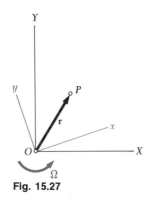

Fig. 15.27

$$\mathbf{v}_P = (\dot{\mathbf{r}})_{OXY} = \boldsymbol{\Omega} \times \mathbf{r} + (\dot{\mathbf{r}})_{Oxy} \tag{15.32}$$

But $(\dot{\mathbf{r}})_{Oxy}$ defines the velocity of the particle P relative to the rotating frame Oxy. Denoting the rotating frame by $\mathcal{F}$ for short, we shall represent the velocity $(\dot{\mathbf{r}})_{Oxy}$ of P relative to the rotating frame by the symbol $\mathbf{v}_{P/\mathcal{F}}$. If we imagine that a rigid slab has been attached to the rotating frame, $v_{P/\mathcal{F}}$ will represent the velocity of P along the path that it describes on that slab (Fig. 15.28). On the other hand, the term $\boldsymbol{\Omega} \times \mathbf{r}$ in (15.32) will represent the velocity $\mathbf{v}_{P'}$ of the point P' of the slab—or rotating frame—which coincides with P at the instant considered. Thus, we have

$$\mathbf{v}_P = \mathbf{v}_{P'} + \mathbf{v}_{P/\mathcal{F}} \tag{15.33}$$

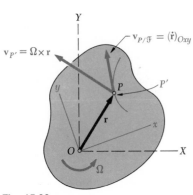

Fig. 15.28

where $\mathbf{v}_P$ = absolute velocity of particle P
$\mathbf{v}_{P'}$ = velocity of point P' of moving frame $\mathcal{F}$ coinciding with P
$\mathbf{v}_{P/\mathcal{F}}$ = velocity of P relative to moving frame $\mathcal{F}$

The absolute acceleration $\mathbf{a}_P$ of the particle is defined as the rate of change of $\mathbf{v}_P$ with respect to the fixed frame OXY. Computing the rates of change with respect to OXY of the terms in (15.32), we write

$$\mathbf{a}_P = \dot{\mathbf{v}}_P = \dot{\boldsymbol{\Omega}} \times \mathbf{r} + \boldsymbol{\Omega} \times \dot{\mathbf{r}} + \frac{d}{dt}[(\dot{\mathbf{r}})_{Oxy}] \tag{15.34}$$

where all derivatives are defined with respect to OXY, except where indicated otherwise. Referring to Eq. (15.31), we note that the last term in (15.34) may be expressed as

$$\frac{d}{dt}[(\dot{\mathbf{r}})_{Oxy}] = (\ddot{\mathbf{r}})_{Oxy} + \boldsymbol{\Omega} \times (\dot{\mathbf{r}})_{Oxy}$$

On the other hand, $\dot{\mathbf{r}}$ represents the velocity $\mathbf{v}_P$ and may be replaced by the right-hand member of Eq. (15.32). After completing these two substitutions into (15.34), we write

745

15.11. Plane Motion of a Particle Relative to a
Rotating Frame. Coriolis Acceleration

$$\mathbf{a}_P = \dot{\mathbf{\Omega}} \times \mathbf{r} + \mathbf{\Omega} \times (\mathbf{\Omega} \times \mathbf{r}) + 2\mathbf{\Omega} \times (\dot{\mathbf{r}})_{Oxy} + (\ddot{\mathbf{r}})_{Oxy} \qquad (15.35)$$

Referring to the expression (15.8) obtained in Sec. 15.3 for the acceleration of a particle in a rigid body rotating about a fixed axis, we note that the sum of the first two terms represents the acceleration $\mathbf{a}_{P'}$ of the point P' of the rotating frame which coincides with P at the instant considered. On the other hand, the last term defines the acceleration $\mathbf{a}_{P/\mathcal{F}}$ of P relative to the rotating frame. If it were not for the third term, which has not been accounted for, a relation similar to (15.33) could be written for the accelerations, and $\mathbf{a}_P$ could be expressed as the sum of $\mathbf{a}_{P'}$ and $\mathbf{a}_{P/\mathcal{F}}$. However, it is clear that *such a relation would be incorrect* and that we must include the additional term. This term, which we shall denote by $\mathbf{a}_c$, is called the *complementary acceleration*, or *Coriolis acceleration*, after the French mathematician de Coriolis (1792–1843). We write

$$\mathbf{a}_P = \mathbf{a}_{P'} + \mathbf{a}_{P/\mathcal{F}} + \mathbf{a}_c \qquad (15.36)$$

where $\mathbf{a}_P$ = absolute acceleration of particle P
$\mathbf{a}_{P'}$ = acceleration of point P' of moving frame $\mathcal{F}$ coinciding with P
$\mathbf{a}_{P/\mathcal{F}}$ = acceleration of P relative to moving frame $\mathcal{F}$
$\mathbf{a}_c = 2\mathbf{\Omega} \times (\dot{\mathbf{r}})_{Oxy} = 2\mathbf{\Omega} \times \mathbf{v}_{P/\mathcal{F}}$
= complementary, or Coriolis, acceleration†

We note that since point P' moves in a circle about the origin O, its acceleration $\mathbf{a}_{P'}$ has, in general, two components: a component $(\mathbf{a}_{P'})_t$ tangent to the circle, and a component $(\mathbf{a}_{P'})_n$ directed toward O. Similarly, the acceleration $\mathbf{a}_{P/\mathcal{F}}$ generally has two components: a component $(\mathbf{a}_{P/\mathcal{F}})_t$ tangent to the path that P describes on the rotating slab, and a component $(\mathbf{a}_{P/\mathcal{F}})_n$ directed toward the center of curvature of that path. We further note that since the vector $\mathbf{\Omega}$ is perpendicular to the plane of motion, and thus to $\mathbf{v}_{P/\mathcal{F}}$, the magnitude of the Coriolis acceleration $\mathbf{a}_c = 2\mathbf{\Omega} \times \mathbf{v}_{P/\mathcal{F}}$ is equal to $2\Omega v_{P/\mathcal{F}}$, and its direction may be obtained by rotating the vector $\mathbf{v}_{P/\mathcal{F}}$ through $90°$ in the sense of rotation of the moving frame (Fig. 15.29). The Coriolis acceleration reduces to zero when either $\mathbf{\Omega}$ or $\mathbf{v}_{P/\mathcal{F}}$ is zero.

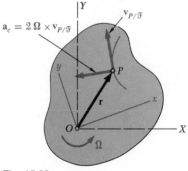

Fig. 15.29

---

† It is important to note the difference between Eq. (15.36) and Eq. (15.21) of Sec. 15.8. When we wrote

$$\mathbf{a}_B = \mathbf{a}_A + \mathbf{a}_{B/A} \qquad (15.21)$$

in Sec. 15.8, we were expressing the absolute acceleration of point B as the sum of its acceleration $\mathbf{a}_{B/A}$ relative to a *frame in translation* and of the acceleration $\mathbf{a}_A$ of a point of that frame. We are now trying to relate the absolute acceleration of point P to its acceleration $\mathbf{a}_{P/\mathcal{F}}$ relative to a *rotating frame* $\mathcal{F}$ and to the acceleration $\mathbf{a}_{P'}$ of the point P' of that frame which coincides with P; Eq. (15.36) shows that because the frame is rotating, it is necessary to include an additional term representing the Coriolis acceleration $\mathbf{a}_c$.

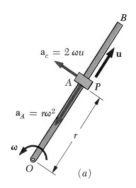

(a)

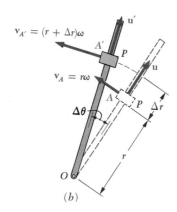

(b)

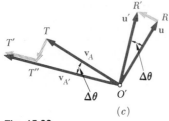

(c)

Fig. 15.30

The following example will help in understanding the physical meaning of the Coriolis acceleration. Consider a collar P which is made to slide at a constant relative speed u along a rod OB rotating at a constant angular velocity ω about O (Fig. 15.30a). According to formula (15.36), the absolute acceleration of P may be obtained by adding vectorially the acceleration $\mathbf{a}_A$ of the point A of the rod coinciding with P, the relative acceleration $\mathbf{a}_{P/OB}$ of P with respect to the rod, and the Coriolis acceleration $\mathbf{a}_c$. Since the angular velocity ω of the rod is constant, $\mathbf{a}_A$ reduces to its normal component $(\mathbf{a}_A)_n$ of magnitude $r\omega^2$; and since u is constant, the relative acceleration $\mathbf{a}_{P/OB}$ is zero. According to the definition given above, the Coriolis acceleration is a vector perpendicular to OB, of magnitude $2\omega u$, and directed as shown in the figure. The acceleration of the collar P consists, therefore, of the two vectors shown in Fig. 15.30a. Note that the result obtained may be checked by applying the relation (11.44).

To understand better the significance of the Coriolis acceleration, we shall consider the absolute velocity of P at time t and at time $t + \Delta t$ (Fig. 15.30b). At time t, the velocity may be resolved into its components $\mathbf{u}$ and $\mathbf{v}_A$, and at time $t + \Delta t$ into its components $\mathbf{u}'$ and $\mathbf{v}_{A'}$. Drawing these components from the same origin (Fig. 15.30c), we note that the change in velocity during the time Δt may be represented by the sum of three vectors, $\overrightarrow{RR'}$, $\overrightarrow{TT''}$, and $\overrightarrow{T''T'}$. The vector $\overrightarrow{TT''}$ measures the change in direction of the velocity $\mathbf{v}_A$, and the quotient $\overrightarrow{TT''}/\Delta t$ represents the acceleration $\mathbf{a}_A$ when Δt approaches zero. We check that the direction of $\overrightarrow{TT''}$ is that of $\mathbf{a}_A$ when Δt approaches zero and that

$$\lim_{\Delta t \to 0} \frac{TT''}{\Delta t} = \lim_{\Delta t \to 0} v_A \frac{\Delta \theta}{\Delta t} = r\omega\omega = r\omega^2 = a_A$$

The vector $\overrightarrow{RR'}$ measures the change in direction of $\mathbf{u}$ due to the rotation of the rod; the vector $\overrightarrow{T''T'}$ measures the change in magnitude of $\mathbf{v}_A$ due to the motion of P on the rod. The vectors $\overrightarrow{RR'}$ and $\overrightarrow{T''T'}$ result from the *combined effect* of the relative motion of P and of the rotation of the rod; they would vanish if *either* of these two motions stopped. We may easily verify that the sum of these two vectors defines the Coriolis acceleration. Their direction is that of $\mathbf{a}_c$ when Δt approaches zero, and since $RR' = u\,\Delta\theta$ and $T''T' = v_{A'} - v_A = (r + \Delta r)\omega - r\omega = \omega\,\Delta r$, we check that

$$\lim_{\Delta t \to 0} \left(\frac{RR'}{\Delta t} + \frac{T''T'}{\Delta t}\right) = \lim_{\Delta t \to 0} \left(u\frac{\Delta\theta}{\Delta t} + \omega\frac{\Delta r}{\Delta t}\right) = u\omega + \omega u = 2\omega u = a_c$$

Formulas (15.33) and (15.36) may be used to analyze the motion of mechanisms which contain parts sliding on each other. They make it possible, for example, to relate the absolute and relative motions of sliding pins and collars (see Sample Probs. 15.9 and 15.10). The concept of Coriolis acceleration is also very useful in the study of long-range projectiles and of other bodies whose motions are appreciably affected by the rotation of the earth. As was pointed out in Sec. 12.2, a system of axes attached to the earth does not truly constitute a newtonian frame of reference; such a system of axes should actually be considered as rotating. The formulas derived in this section will therefore facilitate the study of the motion of bodies with respect to axes attached to the earth.

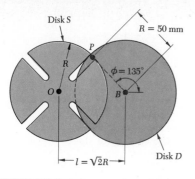

Disk S
$R = 50$ mm
P
R
$\phi = 135°$
O
B
$l = \sqrt{2}R$
Disk D

SAMPLE PROBLEM 15.9

The Geneva mechanism shown is used in many counting instruments and in other applications where an intermittent rotary motion is required. Disk D rotates with a constant counterclockwise angular velocity ω_D of 10 rad/s. A pin P is attached to disk D and slides along one of several slots cut in disk S. It is desirable that the angular velocity of disk S be zero as the pin enters and leaves each slot; in the case of four slots, this will occur if the distance between the centers of the disks is $l = \sqrt{2}\,R$.

At the instant when $\phi = 150°$, determine (a) the angular velocity of disk S, (b) the velocity of pin P relative to disk S.

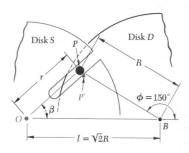

Disk S
P
Disk D
R
r
P'
$\phi = 150°$
β
O
B
$l = \sqrt{2}R$

Solution. We solve triangle OPB, which corresponds to the position $\phi = 150°$. Using the law of cosines, we write

$$r^2 = R^2 + l^2 - 2Rl \cos 30° = 0.551R^2 \qquad r = 0.742R = 37.1 \text{ mm}$$

From the law of sines,

$$\frac{\sin \beta}{R} = \frac{\sin 30°}{r} \qquad \sin \beta = \frac{\sin 30°}{0.742} \qquad \beta = 42.4°$$

Since pin P is attached to disk D, and since disk D rotates about point B, the magnitude of the absolute velocity of P is

$$v_P = R\omega_D = (50 \text{ mm})(10 \text{ rad/s}) = 500 \text{ mm/s}$$
$$\mathbf{v}_P = 500 \text{ mm/s} \measuredangle 60°$$

We consider now the motion of pin P along the slot in disk S. Denoting by P' the point of disk S which coincides with P at the instant considered and selecting a rotating frame $\mathcal{S}$ attached to disk S, we write

$$\mathbf{v}_P = \mathbf{v}_{P'} + \mathbf{v}_{P/\mathcal{S}}$$

Noting that $\mathbf{v}_{P'}$ is perpendicular to the radius OP and that $\mathbf{v}_{P/\mathcal{S}}$ is directed along the slot, we draw the velocity triangle corresponding to the equation above. From the triangle, we compute

$$\gamma = 90° - 42.4° - 30° = 17.6°$$
$$v_{P'} = v_P \sin \gamma = (500 \text{ mm/s}) \sin 17.6°$$
$$\mathbf{v}_{P'} = 151.2 \text{ mm/s} \measuredangle 42.4°$$
$$v_{P/\mathcal{S}} = v_P \cos \gamma = (500 \text{ mm/s}) \cos 17.6°$$
$$\mathbf{v}_{P/\mathcal{S}} = \mathbf{v}_{P/\mathcal{S}} = 477 \text{ mm/s} \measuredangle 42.4° \quad \blacktriangleleft$$

Since $\mathbf{v}_{P'}$ is perpendicular to the radius OP, we write

$$v_{P'} = r\omega_\mathcal{S} \qquad 151.2 \text{ mm/s} = (37.1 \text{ mm})\omega_\mathcal{S}$$
$$\omega_\mathcal{S} = \omega_\mathcal{S} = 4.08 \text{ rad/s} \; \downarrow \quad \blacktriangleleft$$

SAMPLE PROBLEM 15.10

In the Geneva mechanism of Sample Prob. 15.9, disk D rotates with a constant counterclockwise angular velocity ω_D of 10 rad/s. At the instant when $\phi = 150°$, determine the angular acceleration of disk S.

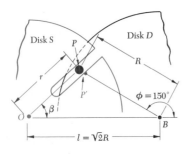

Solution. Referring to Sample Prob. 15.9, we obtain the angular velocity of the frame $\mathcal{S}$ attached to disk S and the velocity of the pin relative to $\mathcal{S}$:

$$\omega_\mathcal{S} = 4.08 \text{ rad/s} \downarrow$$

$$\beta = 42.4° \qquad \mathbf{v}_{P/\mathcal{S}} = 477 \text{ mm/s} \nearrow 42.4°$$

Since pin P moves with respect to the rotating frame $\mathcal{S}$, we write

$$\mathbf{a}_P = \mathbf{a}_{P'} + \mathbf{a}_{P/\mathcal{S}} + \mathbf{a}_c \qquad (1)$$

Each term of this vector equation is investigated separately.

Absolute Acceleration $\mathbf{a}_P$. Since disk D rotates with a constant angular velocity, the absolute acceleration $\mathbf{a}_P$ is directed toward B. We have

$$a_P = R\omega_D^2 = (50 \text{ mm})(10 \text{ rad/s})^2 = 5000 \text{ mm/s}^2$$

$$\mathbf{a}_P = 5000 \text{ mm/s}^2 \searrow 30°$$

Acceleration $\mathbf{a}_{P'}$ *of the Coinciding Point* P'. The acceleration $\mathbf{a}_{P'}$ of the point P' of the frame $\mathcal{S}$ which coincides with P at the instant considered is resolved into normal and tangential components. (We recall from Sample Prob. 15.9 that $r = 37.1$ mm.)

$$(a_{P'})_n = r\omega_\mathcal{S}^2 = (37.1 \text{ mm})(4.08 \text{ rad/s})^2 = 618 \text{ mm/s}^2$$
$$(\mathbf{a}_{P'})_n = 618 \text{ mm/s}^2 \nearrow 42.4°$$
$$(a_{P'})_t = r\alpha_\mathcal{S} = 37.1\alpha_\mathcal{S} \qquad (\mathbf{a}_{P'})_t = 37.1\alpha_\mathcal{S} \nwarrow 42.4°$$

Relative Acceleration $\mathbf{a}_{P/\mathcal{S}}$. Since the pin P moves in a straight slot cut in disk S, the relative acceleration $\mathbf{a}_{P/\mathcal{S}}$ must be parallel to the slot; i.e., its direction must be $\measuredangle 42.4°$.

Coriolis Acceleration $\mathbf{a}_c$. Rotating the relative velocity $\mathbf{v}_{P/\mathcal{S}}$ through 90° in the sense of $\omega_\mathcal{S}$, we obtain the direction of the Coriolis component of the acceleration: $\searform 42.4°$. We write

$$a_c = 2\omega_\mathcal{S} v_{P/\mathcal{S}} = 2(4.08 \text{ rad/s})(477 \text{ mm/s}) = 3890 \text{ mm/s}^2$$
$$\mathbf{a}_c = 3890 \text{ mm/s}^2 \searrow 42.4°$$

We rewrite Eq. (1) and substitute the accelerations found above:

$$\mathbf{a}_P = (\mathbf{a}_{P'})_n + (\mathbf{a}_{P'})_t + \mathbf{a}_{P/\mathcal{S}} + \mathbf{a}_c$$
$$[5000 \searrow 30°] = [618 \nearrow 42.4°] + [37.1\alpha_\mathcal{S} \nwarrow 42.4°]$$
$$+ [a_{P/\mathcal{S}} \measuredangle 42.4°] + [3890 \searrow 42.4°]$$

Equating components in a direction perpendicular to the slot,

$$5000 \cos 17.6° = 37.1\alpha_\mathcal{S} - 3890$$

$$\boldsymbol{\alpha}_\mathcal{S} = \alpha_\mathcal{S} = 233 \text{ rad/s}^2 \downarrow \quad \blacktriangleleft$$

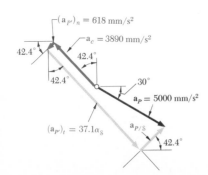

Problems

15.115 and 15.116 Two rotating rods are connected by slider block *P*. The rod attached at *A* rotates with a constant clockwise angular velocity ω_A. For the given data, determine for the position shown (*a*) the angular velocity of the rod attached at *B*, (*b*) the relative velocity of slider block *P* with respect to the rod on which it slides.

15.115 $b = 8$ in., $\omega_A = 6$ rad/s.
15.116 $b = 300$ mm, $\omega_A = 10$ rad/s.

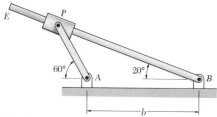

Fig. P15.115 and P15.117

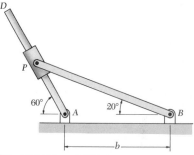

Fig. P15.116 and P15.118

15.117 and 15.118 Two rotating rods are connected by slider block *P*. The velocity v_0 of the slider block relative to the rod on which it slides is constant and is directed outward. For the given data, determine the angular velocity of each rod for the position shown.

15.117 $b = 300$ mm, $v_0 = 480$ mm/s.
15.118 $b = 8$ in., $v_0 = 9$ in./s.

15.119 The motion of pin *P* is guided by slots cut in rods *AE* and *BD*. Knowing that the rods rotate with the constant angular velocities $\omega_A = 4$ rad/s $\downarrow$ and $\omega_B = 5$ rad/s $\downarrow$, determine the velocity of pin *P* for the position shown.

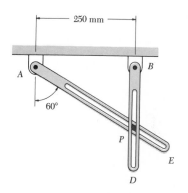

Fig. P15.119 and P15.120

15.120 The motion of pin *P* is guided by slots cut in rods *AE* and *BD*. Knowing that the rods rotate with the constant angular velocities $\omega_A = 4$ rad/s $\downarrow$ and $\omega_B = 5$ rad/s $\uparrow$, determine the velocity of pin *P* for the position shown.

15.121 Four pins slide in four separate slots cut in a circular plate as shown. When the plate is at rest, each pin has a velocity directed as shown and of the same constant magnitude *u*. If each pin maintains the same velocity in relation to the plate when the plate rotates about *O* with a constant *counterclockwise* angular velocity ω, determine the acceleration of each pin.

15.122 Solve Prob. 15.121, assuming that the plate rotates about *O* with a constant *clockwise* angular velocity ω.

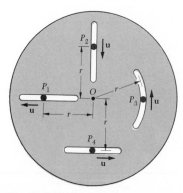

Fig. P15.121

15.123 At the instant shown the length of the boom is being decreased at the constant rate of 6 in./s and the boom is being lowered at the constant rate of 0.075 rad/s. Knowing that $\theta = 30°$, determine (a) the velocity of point B, (b) the acceleration of point B.

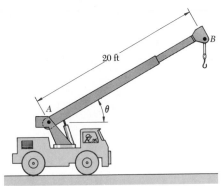

Fig. P15.123 and P15.124

15.124 At the instant shown the length of the boom is being increased at the constant rate of 6 in./s and the boom is being lowered at the constant rate of 0.075 rad/s. Knowing that $\theta = 30°$, determine (a) the velocity of point B, (b) the acceleration of point B.

15.125 The hollow cylinder AB rotates clockwise at the constant rate $\omega = 8$ rad/s about an axis at C. Knowing that in the position shown rod DE is being moved to the right at the constant speed of 2.5 m/s, determine (a) the acceleration of point E, (b) the acceleration of the point of the rod which coincides with point C.

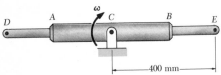

Fig. P15.125

15.126 In the automated welding setup shown, the position of the two welding tips G and H is controlled by the hydraulic cylinder D and rod BC. The cylinder is bolted to the vertical plate which at the instant shown rotates counter-clockwise about A with a constant angular velocity of 1.6 rad/s. Knowing that at the same instant the length EF of the welding assembly is increasing at the constant rate of 300 mm/s, determine (a) the velocity of tip H, (b) the acceleration of tip H.

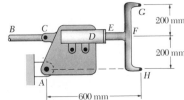

Fig. P15.126

15.127 In Prob. 15.126, determine (a) the velocity of tip G, (b) the acceleration of tip G.

15.128 The motion of blade D is controlled by the robot arm ABC. At the instant shown the arm is rotating clockwise at the constant rate $\omega = 1.8$ rad/s and the length of portion BC of the arm is being decreased at the constant rate of 250 mm/s. Determine (a) the velocity of D, (b) the acceleration of D.

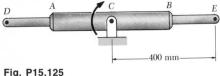

Fig. P15.128

15.129 The hydraulic cylinder CD is welded to an arm which rotates clockwise about A at the constant rate $\omega = 2.4$ rad/s. Knowing that in the position shown rod BE is being moved to the right at the constant rate of 15 in./s with respect to the cylinder, determine (a) the velocity of point B, (b) the acceleration of point B.

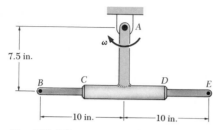

Fig. P15.129

15.130 In Prob. 15.129, determine (a) the velocity of point E, (b) the acceleration of point E.

15.131 The Geneva mechanism shown is used to provide an intermittent rotary motion of disk S. Disk D rotates with a constant counterclockwise angular velocity ω_D of 10 rad/s. A pin P is attached to disk D and may slide in one of the six equally spaced slots cut in disk S. It is desirable that the angular velocity of disk S be zero as the pin enters and leaves each of the six slots; this will occur if the distance between the centers of the disks and the radii of the disks are related as shown. Determine the angular velocity and angular acceleration of disk S at the instant when $\phi = 150°$.

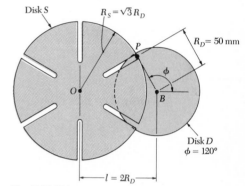

Fig. P15.131

15.132 In Prob. 15.131, determine the angular velocity and angular acceleration of disk S at the instant when $\phi = 135°$.

15.133 In Prob. 15.115, determine the angular acceleration of the rod attached at B.

15.134 In Prob. 15.118, determine the angular acceleration of the rod attached at B.

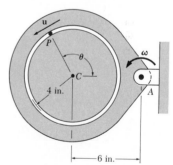

15.135 Pin P slides in the circular slot cut in the plate shown at a constant relative speed $u = 16$ in./s. Assuming that at the instant shown the angular velocity ω of the plate is 6 rad/s and is increasing at the rate of 20 rad/s², determine the acceleration of pin P when $\theta = 90°$.

Fig. P15.135

15.136 Solve Prob. 15.135 when $\theta = 120°$.

15.137 Rod AD is bent in the shape of an arc of circle of radius $b = 150$ mm. The position of the rod is controlled by pin B which slides in a horizontal slot and also slides along the rod. Knowing that at the instant shown pin B moves to the right at a constant speed of 75 mm/s, determine (a) the angular velocity of the rod, (b) the angular acceleration of the rod.

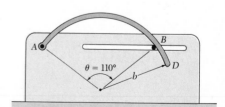

15.138 Solve Prob. 15.137 when $\theta = 90°$.

Fig. P15.137

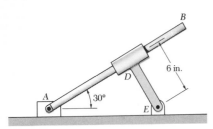

Fig. P15.141

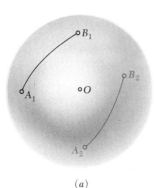

(a)

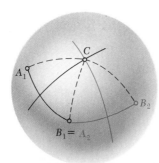

(b)

Fig. 15.31

*15.139 In Prob. 15.119, determine the acceleration of pin P.

*15.140 In Prob. 15.120, determine the acceleration of pin P.

*15.141 Rod AB passes through a collar which is welded to link DE. Knowing that at the instant shown block A moves to the right at a constant speed of 75 in./s, determine (a) the angular velocity of rod AB, (b) the velocity relative to the collar of the point of the rod in contact with the collar, (c) the acceleration of the point of the rod in contact with the collar. (*Hint*. Rod AB and link DE have the same $\boldsymbol{\omega}$ and the same $\boldsymbol{\alpha}$.)

*15.142 Solve Prob. 15.141, assuming that block A moves to the left at a constant speed of 75 in./s.

*15.12. Motion about a Fixed Point.

We have studied in Sec. 15.3 the motion of a rigid body constrained to rotate about a fixed axis. We shall now consider the more general case of the motion of a rigid body which has a fixed point O.

First, we shall prove that *the most general displacement of a rigid body with a fixed point O is equivalent to a rotation of the body about an axis through O.*† Instead of considering the rigid body itself, we may detach a sphere of center O from the body and analyze the motion of that sphere. Clearly, the motion of the sphere completely characterizes the motion of the given body. Since three points define the position of a solid in space, the center O and two points A and B on the surface of the sphere will define the position of the sphere and thus the position of the body. Let A_1 and B_1 characterize the position of the sphere at one instant, and A_2 and B_2 its position at a later instant (Fig. 15.31*a*). Since the sphere is rigid, the lengths of the arcs of great circle A_1B_1 and A_2B_2 must be equal, but except for this requirement, the positions of A_1, A_2, B_1, and B_2 are arbitrary. We propose to prove that the points A and B may be brought, respectively, from A_1 and B_1 into A_2 and B_2 by a single rotation of the sphere about an axis.

For convenience, and without loss of generality, we may select point B so that its initial position coincides with the final position of A; thus, $B_1 = A_2$ (Fig. 15.31*b*). We draw the arcs of great circle A_1A_2, A_2B_2 and the arcs bisecting, respectively, A_1A_2 and A_2B_2. Let C be the point of intersection of these last two arcs; we complete the construction by drawing A_1C, A_2C, and B_2C. As pointed out above, $A_1B_1 = A_2B_2$ on account of the rigidity of the sphere; on the other hand, since C is by construction equidistant from A_1, A_2, and B_2, we have $A_1C = A_2C = B_2C$. As a result, the spherical triangles A_1CA_2 and B_1CB_2 are congruent and the angles A_1CA_2 and B_1CB_2 are equal. Denoting by θ the common value of these angles, we conclude that the sphere may be brought from its initial position into its final position by a single rotation through θ about the axis OC.

† This is known as Euler's theorem.

It follows that the motion during a time interval Δt of a rigid body with a fixed point O may be considered as a rotation through $\Delta\theta$ about a certain axis. Drawing along that axis a vector of magnitude $\Delta\theta/\Delta t$ and letting Δt approach zero, we obtain at the limit the *instantaneous axis of rotation* and the angular velocity $\boldsymbol{\omega}$ of the body at the instant considered (Fig. 15.32). The velocity of a particle P of the body may then be obtained, as in Sec. 15.3, by forming the vector product of $\boldsymbol{\omega}$ and of the position vector $\mathbf{r}$ of the particle:

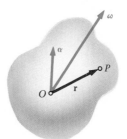

$$v = \frac{d\mathbf{r}}{dt} = \boldsymbol{\omega} \times \mathbf{r} \tag{15.37}$$

The acceleration of the particle is obtained by differentiating (15.37) with respect to t. As in Sec. 15.3 we have

Fig. 15.32

$$\mathbf{a} = \boldsymbol{\alpha} \times \mathbf{r} + \boldsymbol{\omega} \times (\boldsymbol{\omega} \times \mathbf{r}) \tag{15.38}$$

where the angular acceleration $\boldsymbol{\alpha}$ is defined as the derivative

$$\boldsymbol{\alpha} = \frac{d\boldsymbol{\omega}}{dt} \tag{15.39}$$

of the angular velocity $\boldsymbol{\omega}$.

In the case of the motion of a rigid body with a fixed point, the direction of $\boldsymbol{\omega}$ and of the instantaneous axis of rotation changes from one instant to the next. The angular acceleration $\boldsymbol{\alpha}$ therefore reflects the change in direction of $\boldsymbol{\omega}$ as well as its change in magnitude and, in general, *is not directed along the instantaneous axis of rotation.* While the particles of the body located on the instantaneous axis of rotation have zero velocity at the instant considered, they do not have zero acceleration. Also, the accelerations of the various particles of the body *cannot* be determined as if the body were rotating permanently about the instantaneous axis.

Recalling the definition of the velocity of a particle with position vector $\mathbf{r}$, we note that the angular acceleration $\boldsymbol{\alpha}$, as expressed in (15.39), represents the velocity of the tip of the vector $\boldsymbol{\omega}$. This property may be useful in the determination of the angular acceleration of a rigid body. For example, it follows that the vector $\boldsymbol{\alpha}$ is tangent to the curve described in space by the tip of the vector $\boldsymbol{\omega}$.

We should note that the vector $\boldsymbol{\omega}$ moves within the body, as well as in space. It thus generates two cones called, respectively, the *body cone* and the *space cone* (Fig. 15.33).† It may be shown that at any given instant, the two cones are tangent along the instantaneous axis of rotation and that as the body moves, the body cone appears to *roll* on the space cone.

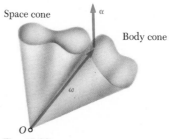

Space cone

Body cone

Fig. 15.33

---

†It is recalled that a *cone* is, by definition, a surface generated by a straight line passing through a fixed point. In general, the cones considered here *will not be circular cones.*

Before concluding our analysis of the motion of a rigid body with a fixed point, we should prove that angular velocities are actually vectors. As it was indicated in Sec. 2.3, some quantities, such as the *finite rotations* of a rigid body, have magnitude and direction but do not obey the parallelogram law of addition; these quantities cannot be considered as vectors. We shall see presently that angular velocities (and also *infinitesimal rotations*) *do obey* the parallelogram law and thus are truly vector quantities.

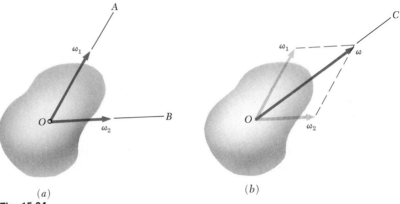

(a)   (b)

Fig. 15.34

Consider a rigid body with a fixed point O which at a given instant rotates simultaneously about the axes OA and OB with angular velocities ω_1 and ω_2 (Fig. 15.34a). We know that this motion must be equivalent at the instant considered to a single rotation of angular velocity ω. We propose to show that

$$\omega = \omega_1 + \omega_2 \tag{15.40}$$

i.e., that the resulting angular velocity may be obtained by adding ω_1 and ω_2 by the parallelogram law (Fig. 15.34b).

Consider a particle P of the body, defined by the position vector $\mathbf{r}$. Denoting, respectively, by $\mathbf{v}_1$, $\mathbf{v}_2$, and $\mathbf{v}$ the velocity of P when the body rotates about OA only, about OB only, and about both axes simultaneously, we write

$$\mathbf{v} = \omega \times \mathbf{r} \qquad \mathbf{v}_1 = \omega_1 \times \mathbf{r} \qquad \mathbf{v}_2 = \omega_2 \times \mathbf{r} \tag{15.41}$$

But the vectorial characters of *linear* velocities is well established (since they represent the derivatives of position vectors). We have therefore

$$\mathbf{v} = \mathbf{v}_1 + \mathbf{v}_2$$

where the plus sign indicates vector addition. Substituting from (15.41), we write

$$\omega \times \mathbf{r} = \omega_1 \times \mathbf{r} + \omega_2 \times \mathbf{r}$$
$$\omega \times \mathbf{r} = (\omega_1 + \omega_2) \times \mathbf{r}$$

where the plus sign still indicates vector addition. Since the relation obtained holds for an arbitrary $\mathbf{r}$, we conclude that (15.40) must be true.

*** 15.13. General Motion.** We shall now consider the most general motion of a rigid body in space. Let A and B be two particles of the body. We recall from Sec. 11.12 that the velocity of B with respect to the fixed frame of reference $OXYZ$ may be expressed as

$$\mathbf{v}_B = \mathbf{v}_A + \mathbf{v}_{B/A} \qquad (15.42)$$

where $\mathbf{v}_{B/A}$ is the velocity of B relative to a frame $AX'Y'Z'$ attached to A and of fixed orientation (Fig. 15.35). Since A is fixed in this frame, the motion of the body relative to $AX'Y'Z'$ is the motion of a body with a fixed point. The relative velocity $\mathbf{v}_{B/A}$ may, therefore, be obtained from (15.37), after $\mathbf{r}$ has been replaced by the position vector $\mathbf{r}_{B/A}$ of B relative to A. Substituting for $\mathbf{v}_{B/A}$ into (15.42), we write

$$\mathbf{v}_B = \mathbf{v}_A + \boldsymbol{\omega} \times \mathbf{r}_{B/A} \qquad (15.43)$$

where $\boldsymbol{\omega}$ is the angular velocity of the body at the instant considered.

The acceleration of B is obtained by a similar reasoning. We first write

$$\mathbf{a}_B = \mathbf{a}_A + \mathbf{a}_{B/A}$$

and, recalling Eq. (15.38),

$$\mathbf{a}_B = \mathbf{a}_A + \boldsymbol{\alpha} \times \mathbf{r}_{B/A} + \boldsymbol{\omega} \times (\boldsymbol{\omega} \times \mathbf{r}_{B/A}) \qquad (15.44)$$

where $\boldsymbol{\alpha}$ is the angular acceleration of the body at the instant considered.

Equations (15.43) and (15.44) show that *the most general motion of a rigid body is equivalent, at any given instant, to the sum of a translation,* in which all the particles of the body have the same velocity and acceleration as a reference particle A, *and of a motion in which particle A is assumed to be fixed.*†

It may easily be shown by solving (15.43) and (15.44) for $\mathbf{v}_A$ and $\mathbf{a}_A$ that the motion of the body with respect to a frame attached to B would be characterized by the same vectors $\boldsymbol{\omega}$ and $\boldsymbol{\alpha}$ as its motion relative to $AX'Y'Z'$. The angular velocity and angular acceleration of a rigid body at a given instant are thus independent of the choice of the reference point. On the other hand, one should keep in mind that whether it is attached to A or to B, the moving frame should maintain a fixed orientation; that is, it should remain parallel to the fixed reference frame $OXYZ$ throughout the motion of the rigid body. In many problems it is found more convenient to use a moving frame which is allowed to rotate as well as to translate. The use of such moving frames will be discussed in Secs. 15.14 and 15.15.

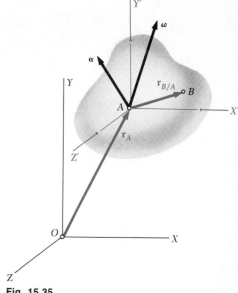

Fig. 15.35

---

†It is recalled from Sec. 15.12 that, in general, the vectors $\boldsymbol{\omega}$ and $\boldsymbol{\alpha}$ are not collinear, and that the accelerations of the particles of the body in their motion relative to the frame $AX'Y'Z'$ cannot be determined as if the body were rotating permanently about the instantaneous axis through A.

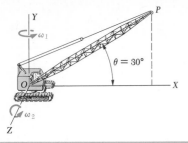

SAMPLE PROBLEM 15.11

The crane shown rotates with a constant angular velocity ω_1 of 0.30 rad/s. Simultaneously, the boom is being raised with a constant angular velocity ω_2 of 0.50 rad/s relative to the cab. Knowing that the length of the boom OP is $l = 12$ m, determine (a) the angular velocity ω of the boom, (b) the angular acceleration α of the boom, (c) the velocity $\mathbf{v}$ of the tip of the boom, (d) the acceleration $\mathbf{a}$ of the tip of the boom.

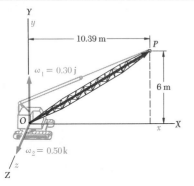

a. Angular Velocity of Boom. Adding the angular velocity ω_1 of the cab and the angular velocity ω_2 of the boom relative to the cab, we obtain the angular velocity ω of the boom at the instant considered:

$$\omega = \omega_1 + \omega_2 \qquad \omega = (0.30 \text{ rad/s})\mathbf{j} + (0.50 \text{ rad/s})\mathbf{k} \blacktriangleleft$$

b. Angular Acceleration of Boom. The angular acceleration α of the boom is obtained by differentiating ω. Since the vector ω_1 is constant in magnitude and direction, we have

$$\alpha = \dot{\omega} = \dot{\omega}_1 + \dot{\omega}_2 = 0 + \dot{\omega}_2$$

where the rate of change $\dot{\omega}_2$ is to be computed with respect to the fixed frame $OXYZ$. However, it is more convenient to use a frame $Oxyz$ attached to the cab and rotating with it, since the vector ω_2 also rotates with the cab and therefore has zero rate of change with respect to that frame. Using Eq. (15.31) with $\mathbf{Q} = \omega_2$ and $\mathbf{\Omega} = \omega_1$, we write

$$(\dot{\mathbf{Q}})_{OXYZ} = (\dot{\mathbf{Q}})_{Oxyz} + \mathbf{\Omega} \times \mathbf{Q}$$
$$(\dot{\omega}_2)_{OXYZ} = (\dot{\omega}_2)_{Oxyz} + \omega_1 \times \omega_2$$
$$\alpha = (\dot{\omega}_2)_{OXYZ} = 0 + (0.30 \text{ rad/s})\mathbf{j} \times (0.50 \text{ rad/s})\mathbf{k}$$
$$\alpha = (0.15 \text{ rad/s}^2)\mathbf{i} \blacktriangleleft$$

c. Velocity of Tip of Boom. Noting that the position vector of point P is $\mathbf{r} = (10.39 \text{ m})\mathbf{i} + (6 \text{ m})\mathbf{j}$ and using the expression found for ω in part a, we write

$$\mathbf{v} = \omega \times \mathbf{r} = \begin{vmatrix} \mathbf{i} & \mathbf{j} & \mathbf{k} \\ 0 & 0.30 \text{ rad/s} & 0.50 \text{ rad/s} \\ 10.39 \text{ m} & 6 \text{ m} & 0 \end{vmatrix}$$

$$\mathbf{v} = -(3 \text{ m/s})\mathbf{i} + (5.20 \text{ m/s})\mathbf{j} - (3.12 \text{ m/s})\mathbf{k} \blacktriangleleft$$

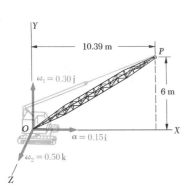

d. Acceleration of Tip of Boom. Recalling that $\mathbf{v} = \omega \times \mathbf{r}$, we write

$$\mathbf{a} = \alpha \times \mathbf{r} + \omega \times (\omega \times \mathbf{r}) = \alpha \times \mathbf{r} + \omega \times \mathbf{v}$$

$$\mathbf{a} = \begin{vmatrix} \mathbf{i} & \mathbf{j} & \mathbf{k} \\ 0.15 & 0 & 0 \\ 10.39 & 6 & 0 \end{vmatrix} + \begin{vmatrix} \mathbf{i} & \mathbf{j} & \mathbf{k} \\ 0 & 0.30 & 0.50 \\ -3 & 5.20 & -3.12 \end{vmatrix}$$

$$= 0.90\mathbf{k} - 0.94\mathbf{i} - 2.60\mathbf{i} - 1.50\mathbf{j} + 0.90\mathbf{k}$$

$$\mathbf{a} = -(3.54 \text{ m/s}^2)\mathbf{i} - (1.50 \text{ m/s}^2)\mathbf{j} + (1.80 \text{ m/s}^2)\mathbf{k} \blacktriangleleft$$

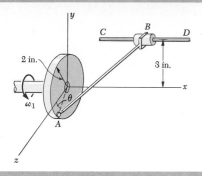

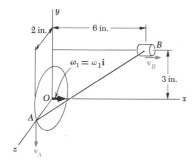

$\omega_1 = 12\,\mathbf{i}$
$\mathbf{r}_A = 2\mathbf{k}$
$\mathbf{r}_B = 6\mathbf{i} + 3\mathbf{j}$
$\mathbf{r}_{B/A} = 6\mathbf{i} + 3\mathbf{j} - 2\mathbf{k}$

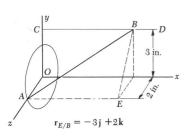

$\mathbf{r}_{E/B} = -3\mathbf{j} + 2\mathbf{k}$

SAMPLE PROBLEM 15.12

The rod AB, of length 7 in., is attached to the disk by a ball-and-socket connection and to the collar B by a clevis. The disk rotates in the yz plane at a constant rate $\omega_1 = 12$ rad/s, while the collar is free to slide along the horizontal rod CD. For the position $\theta = 0$, determine (a) the velocity of the collar, (b) the angular velocity of the rod.

a. **Velocity of Collar.** Since point A is attached to the disk and since collar B moves parallel to the x axis, we have

$$\mathbf{v}_A = \boldsymbol{\omega}_1 \times \mathbf{r}_A = 12\mathbf{i} \times 2\mathbf{k} = -24\mathbf{j} \qquad \mathbf{v}_B = v_B\mathbf{i}$$

Denoting by $\boldsymbol{\omega}$ the angular velocity of the rod, we write

$$\mathbf{v}_B = \mathbf{v}_A + \mathbf{v}_{B/A} = \mathbf{v}_A + \boldsymbol{\omega} \times \mathbf{r}_{B/A}$$

$$v_B\mathbf{i} = -24\mathbf{j} + \begin{vmatrix} \mathbf{i} & \mathbf{j} & \mathbf{k} \\ \omega_x & \omega_y & \omega_z \\ 6 & 3 & -2 \end{vmatrix}$$

$$v_B\mathbf{i} = -24\mathbf{j} + (-2\omega_y - 3\omega_z)\mathbf{i} + (6\omega_z + 2\omega_x)\mathbf{j} + (3\omega_x - 6\omega_y)\mathbf{k}$$

Equating the coefficients of the unit vectors, we obtain

$$v_B = \quad -2\omega_y \quad -3\omega_z \tag{1}$$
$$24 = 2\omega_x \qquad\qquad +6\omega_z \tag{2}$$
$$0 = 3\omega_x \quad -6\omega_y \tag{3}$$

Multiplying Eqs. (1), (2), (3), respectively, by 6, 3, -2 and adding, we write

$$6v_B + 72 = 0 \qquad v_B = -12 \qquad \mathbf{v}_B = -(12 \text{ in./s})\mathbf{i} \quad \blacktriangleleft$$

b. **Angular Velocity of Rod AB.** We note that the angular velocity cannot be determined from Eqs. (1), (2), and (3), since the determinant formed by the coefficients of ω_x, ω_y, and ω_z is zero. We must therefore obtain an additional equation by considering the constraint imposed by the clevis at B.

The collar-clevis connection at B permits rotation of AB about the rod CD and also about an axis perpendicular to the plane containing AB and CD. It prevents rotation of AB about the axis EB, which is perpendicular to CD and lies in the plane containing AB and CD. Thus the projection of $\boldsymbol{\omega}$ on $\mathbf{r}_{E/B}$ must be zero and we write†

$$\boldsymbol{\omega} \cdot \mathbf{r}_{E/B} = 0 \qquad (\omega_x\mathbf{i} + \omega_y\mathbf{j} + \omega_z\mathbf{k}) \cdot (-3\mathbf{j} + 2\mathbf{k}) = 0$$
$$-3\omega_y + 2\omega_z = 0 \tag{4}$$

Solving Eqs. (1) through (4) simultaneously, we obtain

$$v_B = -12 \qquad \omega_x = 3.69 \qquad \omega_y = 1.846 \qquad \omega_z = 2.77$$

$$\boldsymbol{\omega} = (3.69 \text{ rad/s})\mathbf{i} + (1.846 \text{ rad/s})\mathbf{j} + (2.77 \text{ rad/s})\mathbf{k} \quad \blacktriangleleft$$

†We could also note that the direction of EB is that of the vector triple product $\mathbf{r}_{B/C} \times (\mathbf{r}_{B/C} \times \mathbf{r}_{B/A})$ and write $\boldsymbol{\omega} \cdot [\mathbf{r}_{B/C} \times (\mathbf{r}_{B/C} \times \mathbf{r}_{B/A})] = 0$. This formulation would be particularly useful if the rod CD were skew.

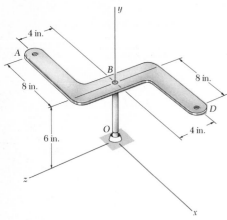

4 in.

A

8 in.

B

8 in.

D

O

4 in.

6 in.

y

z

x

Fig. P15.143

Problems

15.143 Plate ABD and rod OB are rigidly connected and rotate about the ball-and-socket joint O with an angular velocity $\boldsymbol{\omega} = \omega_x \mathbf{i} + \omega_y \mathbf{j} + \omega_z \mathbf{k}$. Knowing that $\mathbf{v}_A = (3\text{ in./s})\mathbf{i} + (14\text{ in./s})\mathbf{j} + (v_A)_z \mathbf{k}$ and $\omega_x = 1.5\text{ rad/s}$, determine (a) the angular velocity of the assembly, (b) the velocity of point D.

15.144 Solve Prob. 15.143, assuming that $\omega_x = -1.5\text{ rad/s}$.

15.145 The plate and rods shown are welded together to form an assembly which rotates about the ball-and-socket joint O with an angular velocity $\boldsymbol{\omega}$. Denoting the velocity of point A by $\mathbf{v}_A = (v_A)_x \mathbf{i} + (v_A)_y \mathbf{j} + (v_A)_z \mathbf{k}$, and knowing that $(v_A)_x = 10\text{ mm/s}$ and $(v_A)_z = 80\text{ mm/s}$, determine the velocity component $(v_A)_y$.

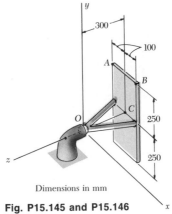

y

300

100

A

B

C

250

O

250

z

x

Dimensions in mm

Fig. P15.145 and P15.146

15.146 The plate and rods shown are welded together to form an assembly which rotates about the ball-and-socket joint O with an angular velocity $\boldsymbol{\omega}$. Knowing that $(v_A)_z = 500\text{ mm/s}$, $(v_B)_z = 100\text{ mm/s}$, and $\omega_z = 3\text{ rad/s}$, determine (a) the angular velocity of the assembly, (b) the velocities of points A and B.

15.147 The disk of a portable sander rotates at the constant rate $\omega_1 = 4400\text{ rpm}$ as shown. Determine the angular acceleration of the disk as a worker rotates the sander about the z axis with an angular velocity of 0.5 rad/s and an angular acceleration of 2.5 rad/s^2, both clockwise when viewed from the positive z axis.

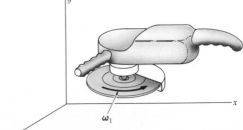

y

x

ω_1

Fig. P15.147

z

15.148 Knowing that the turbine rotor shown rotates at a constant rate $\omega_1 = 9000$ rpm, determine the angular acceleration of the rotor if the turbine housing has a constant angular velocity of 2.4 rad/s clockwise as viewed from (a) the positive y axis, (b) the positive z axis.

15.149 The fan of an automobile engine rotates about a horizontal axis at the rate of 2750 rpm in a clockwise sense when viewed from the rear of the engine. Knowing that the automobile is turning right along a path of radius 50 ft at a constant speed of 12 mi/h, determine the angular acceleration of the fan at the instant the automobile is moving due north.

15.150 Two disks A and B are mounted on an axle of length 2R and roll without sliding on a horizontal floor. Knowing that the axle rotates with a constant angular velocity ω_1, determine (a) the angular velocity of disk A, (b) the angular acceleration of disk A.

Fig. P15.148

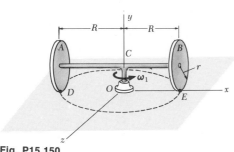

Fig. P15.150

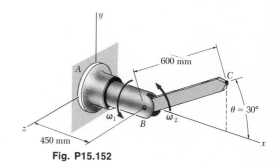

Fig. P15.152

15.151 In Prob. 15.150, determine (a) the angular velocity of disk B, (b) the angular acceleration of disk B.

15.152 The robotic component shown rotates with a constant angular velocity ω_1 of 3 rad/s about the x axis, while arm BC rotates about the z axis with an angular velocity ω_2 which, at the instant shown, has a magnitude $\omega_2 = 4$ rad/s and increases at the rate $\dot{\omega}_2 = 5$ rad/s². Determine the angular acceleration of arm BC.

15.153 In Prob. 15.152, determine (a) the velocity of point C, (b) the acceleration of point C.

15.154 A disk of radius r spins at the constant rate ω_2 about a horizontal axle held by a fork-ended vertical rod which rotates at the constant rate ω_1. For the position shown, determine (a) the angular acceleration of the disk, (b) the acceleration of point P on the rim of the disk if $\theta = 0$, (c) the acceleration of P if $\theta = 90°$.

15.155 A disk of radius r spins at the constant rate ω_2 about a horizontal axle held by a fork-ended vertical rod which rotates at the constant rate ω_1. For the position shown, determine the acceleration of point P for an arbitrary value of the angle θ.

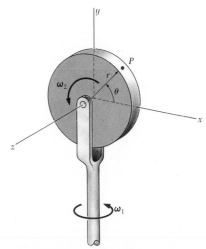

Fig. P15.154 and P15.155

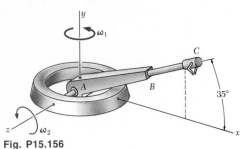

Fig. P15.156

15.156 At the instant shown, the robotic arm ABC is being rotated simultaneously at the constant rate $\omega_1 = 0.15$ rad/s about the y axis, and at the constant rate $\omega_2 = 0.25$ rad/s about the z axis. Knowing that the length of arm ABC is 40 in., determine (a) the angular acceleration of the arm, (b) the velocity of point C, (c) the acceleration of point C.

15.157 Solve Prob. 15.156, assuming that $\omega_1 = 0.25$ rad/s and $\omega_2 = 0.15$ rad/s.

15.158 Solve Prob. 15.156, assuming that at the instant shown the angular velocities ω_1 and ω_2 are both being decreased at the rate of 0.02 rad/s^2.

15.159 A disk of radius r is mounted on an axle of length $2r$. The axle is attached to a vertical shaft AD which rotates at the constant rate ω_1 and the disk rotates about the axle AB at the constant rate ω_2. Knowing that the angle θ remains constant and that the rim of the disk touches the y axis, determine (a) the angular acceleration of the disk, (b) the velocity of point C of the disk, (c) the acceleration of point C of the disk.

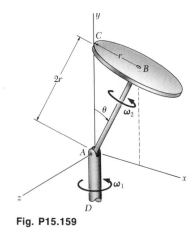

Fig. P15.159

15.160 Several rods are brazed together to form the robotic guide arm shown which is attached to a ball-and-socket joint at O. Rod OA slides in a straight inclined slot while rod OB slides in a slot parallel to the z axis. Knowing that at the instant shown $v_B = (180 \text{ mm/s})\mathbf{k}$, determine (a) the angular velocity of the guide arm, (b) the velocity of point A, (c) the velocity of point C.

15.161 In Prob. 15.160 the speed of point B is known to be constant. For the position shown, determine (a) the angular acceleration of the guide arm, (b) the acceleration of point C.

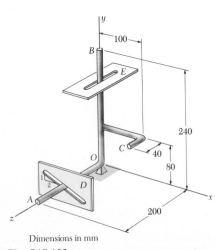

Dimensions in mm

Fig. P15.160

15.162 In the planetary gear system shown gears A and B are rigidly connected to each other and rotate as a unit about the inclined shaft. Gears C and D rotate with constant angular velocities of 30 rad/s and 20 rad/s, respectively (both counterclockwise when viewed from the right). Choosing the x axis to the right, the y axis upward, and the z axis pointing out of the plane of the figure, determine (*a*) the common angular velocity of gears A and B, (*b*) the angular velocity of shaft FH, which is rigidly attached to the inclined shaft.

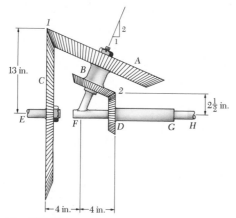

Fig. P15.162

15.163 In Prob. 15.162, determine (*a*) the common angular acceleration of gears A and B, (*b*) the acceleration of the tooth of gear A which is in contact with gear C at point *1*.

15.164 The small cone shown rolls without slipping on the inside surface of the large fixed cone. Denoting by ω_1 the constant angular velocity of the axis OB about the y axis, determine in terms of ω_1, β, and γ, (*a*) the rate of spin of the cone about the axis OB, (*b*) the total angular velocity of the cone, (*c*) the angular acceleration of the cone.

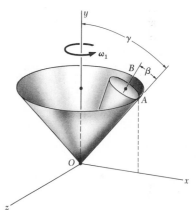

Fig. P15.164

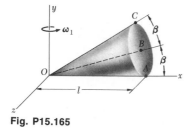

Fig. P15.165

15.165 The cone shown rolls on the zx plane with its apex at the origin of coordinates. Denoting by ω_1 the constant angular velocity of the axis OB of the cone about the y axis, determine (*a*) the rate of spin of the cone about the axis OB, (*b*) the total angular velocity of the cone, (*c*) the angular acceleration of the cone.

15.166 Rod BC, of length 21 in., is connected by ball-and-socket joints to collar C and to the rotating arm AB. Knowing that arm AB rotates in the xy plane at the constant rate $\omega_0 = 18$ rad/s, determine the velocity of collar C for the position shown.

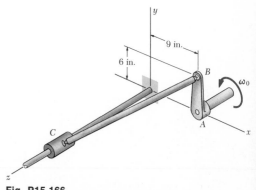

Fig. P15.166

15.167 Rod AB, of length 11 in., is connected by ball-and-socket joints to collars A and B, which slide along the two rods shown. Knowing that collar B moves downward at a constant speed of 54 in./s, determine the velocity of collar A when $c = 2$ in.

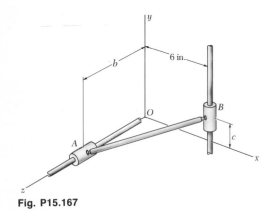

Fig. P15.167

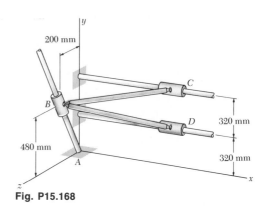

Fig. P15.168

15.168 Rods BC and BD are each 840 mm long and are connected by ball-and-socket joints to collars which may slide on the fixed rods shown. Knowing that collar B moves toward A at a constant speed of 390 mm/s, determine the velocity of collar C for the position shown.

15.169 In Prob. 15.168, determine the velocity of collar D.

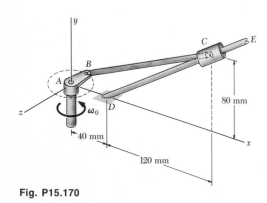

Fig. P15.170

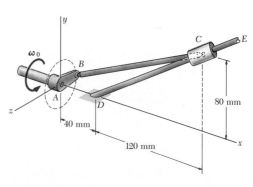

Fig. P15.171

15.170 and 15.171 Rod BC, of length 180 mm, is connected by ball-and-socket joints to the rotating arm AB and to collar C which may slide on the fixed rod DE. Arm AB is of length 20 mm and rotates at the constant rate $\omega_0 = 24$ rad/s. For the position shown determine the velocity of collar C.

15.172 Two shafts AC and EG, which lie in the vertical yz plane, are connected by a universal joint at D. Shaft AC rotates with a constant angular velocity ω_1 as shown. At a time when the arm of the crosspiece attached to shaft AC is vertical, determine the angular velocity of shaft EG.

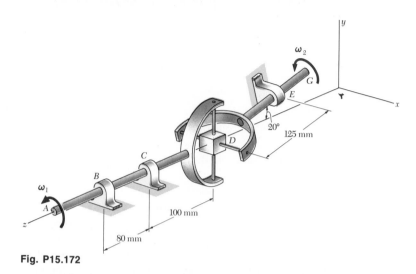

Fig. P15.172

15.173 Solve Prob. 15.172, assuming that the arm of the crosspiece attached to shaft AC is horizontal.

15.174 In Prob. 15.167, the ball-and-socket joint between the rod and collar B is replaced by the clevis connection shown. Determine (a) the angular velocity of the rod, (b) the velocity of collar A.

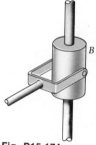

Fig. P15.174

15.175 In Prob. 15.168, the ball-and-socket joint between the rod and collar C is replaced by the clevis connection shown. Determine (a) the angular velocity of the rod, (b) the velocity of collar C.

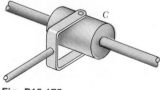

Fig. P15.175

*15.176 In Prob. 15.167, determine the acceleration of collar A when $c = 7$ in.

*15.177 In Prob. 15.167, determine the acceleration of collar A when $c = 2$ in.

*15.178 In Prob. 15.168, determine the acceleration of collar C.

*15.179 In Prob. 15.170, determine the acceleration of collar C.

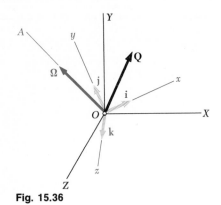

Fig. 15.36

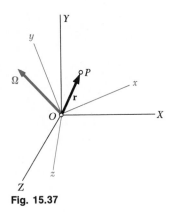

Fig. 15.37

*15.14. Three-Dimensional Motion of a Particle Relative to a Rotating Frame. Coriolis Acceleration. We saw in Sec. 15.10 that given a vector function $\mathbf{Q}(t)$ and two frames of reference centered at O—a fixed frame $OXYZ$ and a rotating frame $Oxyz$—the rates of change of $\mathbf{Q}$ with respect to the two frames satisfy the relation

$$(\dot{\mathbf{Q}})_{OXYZ} = (\dot{\mathbf{Q}})_{Oxyz} + \mathbf{\Omega} \times \mathbf{Q} \qquad (15.31)$$

We had assumed at the time that the frame $Oxyz$ was constrained to rotate about a fixed axis OA. However, the derivation given in Sec. 15.10 remains valid when the frame $Oxyz$ is constrained only to have a fixed point O. Under this more general assumption, the axis OA represents the *instantaneous* axis of rotation of the frame $Oxyz$ (Sec. 15.12) and the vector $\mathbf{\Omega}$, its angular velocity at the instant considered (Fig. 15.36).

We shall now consider the three-dimensional motion of a particle P relative to a rotating frame $Oxyz$ constrained to have a fixed origin O. Let $\mathbf{r}$ be the position vector of P at a given instant and $\mathbf{\Omega}$ be the angular velocity of the frame $Oxyz$ with respect to the fixed frame $OXYZ$ at the same instant (Fig. 15.37). The derivations given in Sec. 15.11 for the two-dimensional motion of a particle may readily be extended to the three-dimensional case, and we may express the absolute velocity $\mathbf{v}_P$ of P (i.e., its velocity with respect to the fixed frame $OXYZ$) as

$$\mathbf{v}_P = \mathbf{\Omega} \times \mathbf{r} + (\dot{\mathbf{r}})_{Oxyz} \qquad (15.45)$$

Denoting by $\mathcal{F}$ the rotating frame $Oxyz$, we write this relation in the alternative form

$$\mathbf{v}_P = \mathbf{v}_{P'} + \mathbf{v}_{P/\mathcal{F}} \qquad (15.46)$$

where $\mathbf{v}_P$ = absolute velocity of particle P
$\mathbf{v}_{P'}$ = velocity of point P' of moving frame $\mathcal{F}$ coinciding with P
$\mathbf{v}_{P/\mathcal{F}}$ = velocity of P relative to moving frame $\mathcal{F}$

The absolute acceleration $\mathbf{a}_P$ of P may be expressed as

$$\mathbf{a}_P = \dot{\mathbf{\Omega}} \times \mathbf{r} + \mathbf{\Omega} \times (\mathbf{\Omega} \times \mathbf{r}) + 2\mathbf{\Omega} \times (\dot{\mathbf{r}})_{Oxyz} + (\ddot{\mathbf{r}})_{Oxyz} \qquad (15.47)$$

We may also use the alternative form

$$\mathbf{a}_P = \mathbf{a}_{P'} + \mathbf{a}_{P/\mathcal{F}} + \mathbf{a}_c \qquad (15.48)$$

where $\mathbf{a}_P$ = absolute acceleration of particle P
$\qquad \mathbf{a}_{P'}$ = acceleration of point P' of moving frame $\mathcal{F}$ coinciding with P
$\qquad \mathbf{a}_{P/\mathcal{F}}$ = acceleration of P relative to moving frame $\mathcal{F}$
$\qquad \mathbf{a}_c = 2\mathbf{\Omega} \times (\dot{\mathbf{r}})_{Oxyz} = 2\mathbf{\Omega} \times \mathbf{v}_{P/\mathcal{F}}$
$\qquad\qquad$ = complementary, or Coriolis, acceleration†

We note that the Coriolis acceleration is perpendicular to the vectors $\mathbf{\Omega}$ and $\mathbf{v}_{P/\mathcal{F}}$. However, since these vectors are usually not perpendicular to each other, the magnitude of $\mathbf{a}_c$ is in general *not* equal to $2\Omega v_{P/\mathcal{F}}$, as was the case for the plane motion of a particle. We further note that the Coriolis acceleration reduces to zero when the vectors $\mathbf{\Omega}$ and $\mathbf{v}_{P/\mathcal{F}}$ are parallel, or when either of them is zero.

Rotating frames of reference are particularly useful in the study of the three-dimensional motion of rigid bodies. If a rigid body has a fixed point O, as was the case for the crane of Sample Prob. 15.11, we may use a frame $Oxyz$ which is neither fixed nor rigidly attached to the rigid body. Denoting by $\mathbf{\Omega}$ the angular velocity of the frame $Oxyz$, we then resolve the angular velocity $\boldsymbol{\omega}$ of the body into the components $\mathbf{\Omega}$ and $\boldsymbol{\omega}_{B/\mathcal{F}}$, where the second component represents the angular velocity of the body relative to the frame $Oxyz$ (see Sample Prob. 15.14). An appropriate choice of the rotating frame will often lead to a simpler analysis of the motion of the rigid body than would be possible with axes of fixed orientation. This is especially true in the case of the general three-dimensional motion of a rigid body, i.e., when the rigid body under consideration has no fixed point (see Sample Prob. 15.15).

*15.15. Frame of Reference in General Motion.

Consider a fixed frame of reference $OXYZ$ and a frame $Axyz$ which moves in a known, but arbitrary, fashion with respect to $OXYZ$ (Fig. 15.38). Let P be a particle moving in space. The position of P is defined at any instant by the vector $\mathbf{r}_P$ in the fixed frame, and by the vector $\mathbf{r}_{P/A}$ in the moving frame. Denoting by $\mathbf{r}_A$ the position vector of A in the fixed frame, we have

$$\mathbf{r}_P = \mathbf{r}_A + \mathbf{r}_{P/A} \qquad (15.49)$$

The absolute velocity $\mathbf{v}_P$ of the particle is obtained by writing

$$\mathbf{v}_P = \dot{\mathbf{r}}_P = \dot{\mathbf{r}}_A + \dot{\mathbf{r}}_{P/A} \qquad (15.50)$$

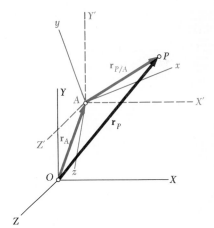

Fig. 15.38

where the derivatives are defined with respect to the fixed frame $OXYZ$. The first term in the right-hand member of (15.50) thus represents the velocity $\mathbf{v}_A$ of the origin A of the moving axes. On the other hand, since the rate of change of a vector is the same with respect to a fixed frame and with respect to a frame in translation (Sec. 11.10), the second term may be regarded as the velocity $\mathbf{v}_{P/A}$ of P relative to the frame $AX'Y'Z'$ of the same orientation as $OXYZ$ and the same origin as $Axyz$. We therefore have

$$\mathbf{v}_P = \mathbf{v}_A + \mathbf{v}_{P/A} \qquad (15.51)$$

† It is important to note the difference between Eq. (15.48) and Eq. (15.21) of Sec. 15.8. See the footnote on page 745.

But the velocity $\mathbf{v}_{P/A}$ of P relative to $AX'Y'Z'$ may be obtained from (15.45) by substituting $\mathbf{r}_{P/A}$ for $\mathbf{r}$ in that equation. We write

$$\mathbf{v}_P = \mathbf{v}_A + \mathbf{\Omega} \times \mathbf{r}_{P/A} + (\dot{\mathbf{r}}_{P/A})_{Axyz} \tag{15.52}$$

where $\mathbf{\Omega}$ is the angular velocity of the frame $Axyz$ at the instant considered.

The absolute acceleration $\mathbf{a}_P$ of the particle is obtained by differentiating (15.51) and writing

$$\mathbf{a}_P = \dot{\mathbf{v}}_P = \dot{\mathbf{v}}_A + \dot{\mathbf{v}}_{P/A} \tag{15.53}$$

where the derivatives are defined with respect to either of the frames $OXYZ$ or $AX'Y'Z'$. Thus, the first term in the right-hand member of (15.53) represents the acceleration $\mathbf{a}_A$ of the origin A of the moving axes and the second term, the acceleration $\mathbf{a}_{P/A}$ of P relative to the frame $AX'Y'Z'$. This acceleration may be obtained from (15.47) by substituting $\mathbf{r}_{P/A}$ for $\mathbf{r}$. We therefore write

$$\mathbf{a}_P = \mathbf{a}_A + \dot{\mathbf{\Omega}} \times \mathbf{r}_{P/A} + \mathbf{\Omega} \times (\mathbf{\Omega} \times \mathbf{r}_{P/A}) \\ + 2\mathbf{\Omega} \times (\dot{\mathbf{r}}_{P/A})_{Axyz} + (\ddot{\mathbf{r}}_{P/A})_{Axyz} \tag{15.54}$$

Formulas (15.52) and (15.54) make it possible to determine the velocity and acceleration of a given particle with respect to a fixed frame of reference, when the motion of the particle is known with respect to a moving frame. These formulas become more significant, and considerably easier to remember, if we note that the sum of the first two terms in (15.52) represents the velocity of the point P' of the moving frame which coincides with P at the instant considered, and that the sum of the first three terms in (15.54) represents the acceleration of the same point. Thus, the relations (15.46) and (15.48) of the preceding section are still valid in the case of a reference frame in general motion, and we write

$$\mathbf{v}_P = \mathbf{v}_{P'} + \mathbf{v}_{P/\mathcal{F}} \tag{15.46}$$
$$\mathbf{a}_P = \mathbf{a}_{P'} + \mathbf{a}_{P/\mathcal{F}} + \mathbf{a}_c \tag{15.48}$$

where the various vectors involved have been defined in Sec. 15.14.

We may note that if the moving reference frame $\mathcal{F}$ (or $Axyz$) is in translation, the velocity and acceleration of the point P' of the frame which coincides with P become, respectively, equal to the velocity and acceleration of the origin A of the frame. On the other hand, since the frame maintains a fixed orientation, $\mathbf{a}_c$ is zero, and the relations (15.46) and (15.48) reduce, respectively, to the relations (11.33) and (11.34) derived in Sec. 11.12.

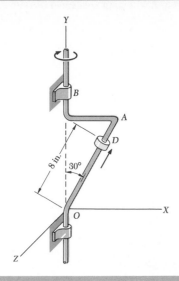

The bent rod OAB rotates about the vertical OB. At the instant considered, its angular velocity and angular acceleration are, respectively, 20 rad/s and 200 rad/s^2, both clockwise when viewed from the positive Y axis. The collar D moves along the rod and at the instant considered, $OD = 8$ in., and the velocity and acceleration of the collar relative to the rod are, respectively, 50 in./s and 600 in./s^2, both upward. Determine (a) the velocity of the collar, (b) the acceleration of the collar.

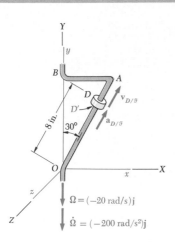

Frames of Reference. The frame $OXYZ$ is fixed. We attach the rotating frame $Oxyz$ to the bent rod. Its angular velocity and angular acceleration relative to $OXYZ$ are therefore $\mathbf{\Omega} = (-20 \text{ rad/s})\mathbf{j}$ and $\dot{\mathbf{\Omega}} = (-200 \text{ rad/s}^2)\mathbf{j}$, respectively. The position vector of D is

$$\mathbf{r} = (8 \text{ in.})(\sin 30°\mathbf{i} + \cos 30°\mathbf{j}) = (4 \text{ in.})\mathbf{i} + (6.93 \text{ in.})\mathbf{j}$$

a. Velocity $\mathbf{v}_D$. Denoting by D' the point of the rod which coincides with D and by $\mathcal{F}$ the rotating frame $Oxyz$, we write from Eq. (15.46)

$$\mathbf{v}_D = \mathbf{v}_{D'} + \mathbf{v}_{D/\mathcal{F}} \tag{1}$$

where

$$\mathbf{v}_{D'} = \mathbf{\Omega} \times \mathbf{r} = (-20 \text{ rad/s})\mathbf{j} \times [(4 \text{ in.})\mathbf{i} + (6.93 \text{ in.})\mathbf{j}] = (80 \text{ in./s})\mathbf{k}$$
$$\mathbf{v}_{D/\mathcal{F}} = (50 \text{ in./s})(\sin 30°\mathbf{i} + \cos 30°\mathbf{j}) = (25 \text{ in./s})\mathbf{i} + (43.3 \text{ in./s})\mathbf{j}$$

Substituting the values obtained for $\mathbf{v}_{D'}$ and $\mathbf{v}_{D/\mathcal{F}}$ into (1), we find

$$\mathbf{v}_D = (25 \text{ in./s})\mathbf{i} + (43.3 \text{ in./s})\mathbf{j} + (80 \text{ in./s})\mathbf{k} \quad \blacktriangleleft$$

b. Acceleration $\mathbf{a}_D$. From Eq. (15.48) we write

$$\mathbf{a}_D = \mathbf{a}_{D'} + \mathbf{a}_{D/\mathcal{F}} + \mathbf{a}_c \tag{2}$$

where

$$\begin{aligned}
\mathbf{a}_{D'} &= \dot{\mathbf{\Omega}} \times \mathbf{r} + \mathbf{\Omega} \times (\mathbf{\Omega} \times \mathbf{r}) \\
&= (-200 \text{ rad/s}^2)\mathbf{j} \times [(4 \text{ in.})\mathbf{i} + (6.93 \text{ in.})\mathbf{j}] - (20 \text{ rad/s})\mathbf{j} \times (80 \text{ in./s})\mathbf{k} \\
&= +(800 \text{ in./s}^2)\mathbf{k} - (1600 \text{ in./s}^2)\mathbf{i} \\
\mathbf{a}_{D/\mathcal{F}} &= (600 \text{ in./s}^2)(\sin 30°\mathbf{i} + \cos 30°\mathbf{j}) = (300 \text{ in./s}^2)\mathbf{i} + (520 \text{ in./s}^2)\mathbf{j} \\
\mathbf{a}_c &= 2\mathbf{\Omega} \times \mathbf{v}_{D/\mathcal{F}} \\
&= 2(-20 \text{ rad/s})\mathbf{j} \times [(25 \text{ in./s})\mathbf{i} + (43.3 \text{ in./s})\mathbf{j}] = (1000 \text{ in./s}^2)\mathbf{k}
\end{aligned}$$

Substituting the values obtained for $\mathbf{a}_{D'}$, $\mathbf{a}_{D/\mathcal{F}}$, and $\mathbf{a}_c$ into (2),

$$\mathbf{a}_D = -(1300 \text{ in./s}^2)\mathbf{i} + (520 \text{ in./s}^2)\mathbf{j} + (1800 \text{ in./s}^2)\mathbf{k} \quad \blacktriangleleft$$

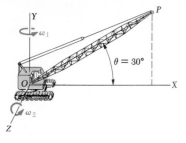

The crane shown rotates with a constant angular velocity ω_1 of 0.30 rad/s. Simultaneously, the boom is being raised with a constant angular velocity ω_2 of 0.50 rad/s relative to the cab. Knowing that the length of the boom OP is $l = 12$ m, determine (a) the velocity of the tip of the boom, (b) the acceleration of the tip of the boom.

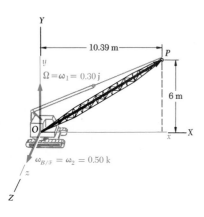

Frames of Reference. The frame $OXYZ$ is fixed. We attach the rotating frame $Oxyz$ to the cab. Its angular velocity with respect to the frame $OXYZ$ is therefore $\Omega = \omega_1 = (0.30 \text{ rad/s})\mathbf{j}$. The angular velocity of the boom relative to the cab and the rotating frame $Oxyz$ (or $\mathcal{F}$, for short) is $\omega_{B/\mathcal{F}} = \omega_2 = (0.50 \text{ rad/s})\mathbf{k}$.

a. Velocity $\mathbf{v}_P$. From Eq. (15.46) we write

$$\mathbf{v}_P = \mathbf{v}_{P'} + \mathbf{v}_{P/\mathcal{F}} \tag{1}$$

where $\mathbf{v}_{P'}$ is the velocity of the point P' of the rotating frame which coincides with P:

$$\mathbf{v}_{P'} = \Omega \times \mathbf{r} = (0.30 \text{ rad/s})\mathbf{j} \times [(10.39 \text{ m})\mathbf{i} + (6 \text{ m})\mathbf{j}] = -(3.12 \text{ m/s})\mathbf{k}$$

and where $\mathbf{v}_{P/\mathcal{F}}$ is the velocity of P relative to the rotating frame $Oxyz$. But the angular velocity of the boom relative to $Oxyz$ was found to be $\omega_{B/\mathcal{F}} = (0.50 \text{ rad/s})\mathbf{k}$. The velocity of its tip P relative to $Oxyz$ is therefore

$$\begin{aligned} \mathbf{v}_{P/\mathcal{F}} = \omega_{B/\mathcal{F}} \times \mathbf{r} &= (0.50 \text{ rad/s})\mathbf{k} \times [(10.39 \text{ m})\mathbf{i} + (6 \text{ m})\mathbf{j}] \\ &= -(3 \text{ m/s})\mathbf{i} + (5.20 \text{ m/s})\mathbf{j} \end{aligned}$$

Substituting the values obtained for $\mathbf{v}_{P'}$ and $\mathbf{v}_{P/\mathcal{F}}$ into (1), we find

$$\mathbf{v}_P = -(3 \text{ m/s})\mathbf{i} + (5.20 \text{ m/s})\mathbf{j} - (3.12 \text{ m/s})\mathbf{k} \qquad \blacktriangleleft$$

b. Acceleration $\mathbf{a}_P$. From Eq. (15.48) we write

$$\mathbf{a}_P = \mathbf{a}_{P'} + \mathbf{a}_{P/\mathcal{F}} + \mathbf{a}_c \tag{2}$$

Since Ω and $\omega_{B/\mathcal{F}}$ are both constant, we have

$$\mathbf{a}_{P'} = \Omega \times (\Omega \times \mathbf{r}) = (0.30 \text{ rad/s})\mathbf{j} \times (-3.12 \text{ m/s})\mathbf{k} = -(0.94 \text{ m/s}^2)\mathbf{i}$$

$$\begin{aligned} \mathbf{a}_{P/\mathcal{F}} = \omega_{B/\mathcal{F}} \times (\omega_{B/\mathcal{F}} \times \mathbf{r}) \\ = (0.50 \text{ rad/s})\mathbf{k} \times [-(3 \text{ m/s})\mathbf{i} + (5.20 \text{ m/s})\mathbf{j}] \\ = -(1.50 \text{ m/s}^2)\mathbf{j} - (2.60 \text{ m/s}^2)\mathbf{i} \end{aligned}$$

$$\begin{aligned} \mathbf{a}_c = 2\Omega \times \mathbf{v}_{P/\mathcal{F}} \\ = 2(0.30 \text{ rad/s})\mathbf{j} \times [-(3 \text{ m/s})\mathbf{i} + (5.20 \text{ m/s})\mathbf{j}] = (1.80 \text{ m/s}^2)\mathbf{k} \end{aligned}$$

Substituting for $\mathbf{a}_{P'}$, $\mathbf{a}_{P/\mathcal{F}}$, and $\mathbf{a}_c$ into (2), we find

$$\mathbf{a}_P = -(3.54 \text{ m/s}^2)\mathbf{i} - (1.50 \text{ m/s}^2)\mathbf{j} + (1.80 \text{ m/s}^2)\mathbf{k} \qquad \blacktriangleleft$$

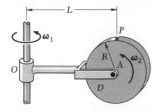

Disk D, of radius R, is pinned to end A of the arm OA of length L located in the plane of the disk. The arm rotates about a vertical axis through O at the constant rate ω_1, and the disk rotates about A at the constant rate ω_2. Determine (a) the velocity of point P located directly above A, (b) the acceleration of P, (c) the angular velocity and angular acceleration of the disk.

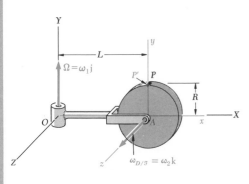

Frames of Reference. The frame $OXYZ$ is fixed. We attach the moving frame $Axyz$ to the arm OA. Its angular velocity with respect to the frame $OXYZ$ is therefore $\boldsymbol{\Omega} = \omega_1\mathbf{j}$. The angular velocity of disk D relative to the moving frame $Axyz$ (or $\mathfrak{F}$, for short) is $\boldsymbol{\omega}_{D/\mathfrak{F}} = \omega_2\mathbf{k}$. The position vector of P relative to O is $\mathbf{r} = L\mathbf{i} + R\mathbf{j}$ and its position vector relative to A is $\mathbf{r}_{P/A} = R\mathbf{j}$.

a. **Velocity** $\mathbf{v}_P$. Denoting by P' the point of the moving frame which coincides with P, we write from Eq. (15.46)

$$\mathbf{v}_P = \mathbf{v}_{P'} + \mathbf{v}_{P/\mathfrak{F}} \tag{1}$$

where $\mathbf{v}_{P'} = \boldsymbol{\Omega} \times \mathbf{r} = \omega_1\mathbf{j} \times (L\mathbf{i} + R\mathbf{j}) = -\omega_1 L\mathbf{k}$

$\mathbf{v}_{P/\mathfrak{F}} = \boldsymbol{\omega}_{D/\mathfrak{F}} \times \mathbf{r}_{P/A} = \omega_2\mathbf{k} \times R\mathbf{j} = -\omega_2 R\mathbf{i}$

Substituting the values obtained for $\mathbf{v}_{P'}$ and $\mathbf{v}_{P/\mathfrak{F}}$ into (1), we find

$$\mathbf{v}_P = -\omega_2 R\mathbf{i} - \omega_1 L\mathbf{k} \quad \blacktriangleleft$$

b. **Acceleration** $\mathbf{a}_P$. From Eq. (15.48) we write

$$\mathbf{a}_P = \mathbf{a}_{P'} + \mathbf{a}_{P/\mathfrak{F}} + \mathbf{a}_c \tag{2}$$

Since $\boldsymbol{\Omega}$ and $\boldsymbol{\omega}_{D/\mathfrak{F}}$ are both constant, we have

$$\mathbf{a}_{P'} = \boldsymbol{\Omega} \times (\boldsymbol{\Omega} \times \mathbf{r}) = \omega_1\mathbf{j} \times (-\omega_1 L\mathbf{k}) = -\omega_1^2 L\mathbf{i}$$
$$\mathbf{a}_{P/\mathfrak{F}} = \boldsymbol{\omega}_{D/\mathfrak{F}} \times (\boldsymbol{\omega}_{D/\mathfrak{F}} \times \mathbf{r}_{P/A}) = \omega_2\mathbf{k} \times (-\omega_2 R\mathbf{i}) = -\omega_2^2 R\mathbf{j}$$
$$\mathbf{a}_c = 2\boldsymbol{\Omega} \times \mathbf{v}_{P/\mathfrak{F}} = 2\omega_1\mathbf{j} \times (-\omega_2 R\mathbf{i}) = 2\omega_1\omega_2 R\mathbf{k}$$

Substituting the values obtained into (2), we find

$$\mathbf{a}_P = -\omega_1^2 L\mathbf{i} - \omega_2^2 R\mathbf{j} + 2\omega_1\omega_2 R\mathbf{k} \quad \blacktriangleleft$$

c. **Angular Velocity and Angular Acceleration of Disk.**

$$\boldsymbol{\omega} = \boldsymbol{\Omega} + \boldsymbol{\omega}_{D/\mathfrak{F}} \qquad \boldsymbol{\omega} = \omega_1\mathbf{j} + \omega_2\mathbf{k} \quad \blacktriangleleft$$

Using Eq. (15.31) with $\mathbf{Q} = \boldsymbol{\omega}$, we write

$$\boldsymbol{\alpha} = (\dot{\boldsymbol{\omega}})_{OXYZ} = (\dot{\boldsymbol{\omega}})_{Axyz} + \boldsymbol{\Omega} \times \boldsymbol{\omega}$$
$$= 0 + \omega_1\mathbf{j} \times (\omega_1\mathbf{j} + \omega_2\mathbf{k})$$
$$\boldsymbol{\alpha} = \omega_1\omega_2\mathbf{i} \quad \blacktriangleleft$$

Problems

15.180 Rod AB is welded to the 12-in.-radius plate which rotates at the constant rate $\omega_1 = 6$ rad/s. Knowing that collar D moves toward end B of the rod at a constant speed $u = 78$ in./s, determine, for the position shown, (a) the velocity of D, (b) the acceleration of D.

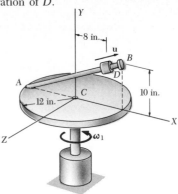

Fig. P15.180

15.181 The bent rod ABC rotates at the constant rate $\omega_1 = 4$ rad/s. Knowing that collar D moves downward along the rod at a constant relative speed $u = 65$ in./s, determine, for the position shown, (a) the velocity of D, (b) the acceleration of D.

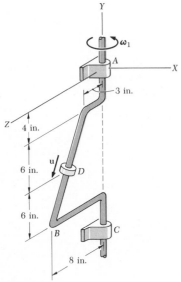

Fig. P15.181

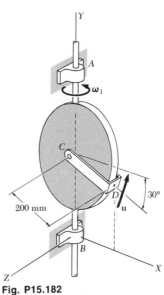

Fig. P15.182

15.182 The circular plate shown rotates about its vertical diameter at the constant rate $\omega_1 = 10$ rad/s. Knowing that in the position shown the disk lies in the XY plane and point D of strap CD moves upward at a constant relative speed $u = 1.5$ m/s, determine (a) the velocity of D, (b) the acceleration of D.

15.183 Manufactured items are spray-painted as they pass through the automated work station shown. Knowing that the bent pipe ACE rotates at the constant rate $\omega_1 = 0.4$ rad/s and that at point D the paint moves through the pipe at a constant relative speed $u = 150$ mm/s, determine, for the position shown, (a) the velocity of the paint at D, (b) the acceleration of the paint at D.

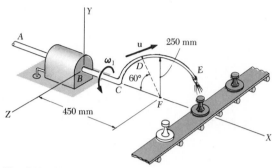

Fig. P15.183

15.184 Solve Prob. 15.180, assuming that, at the instant shown, the angular velocity ω_1 of the plate is 6 rad/s and is decreasing at the rate of 10 rad/s², while the relative speed u of the collar is 78 in./s and is decreasing at the rate of 208 in./s².

15.185 Solve Prob. 15.182, assuming that, at the instant shown, the angular velocity ω_1 of the plate is 10 rad/s and is decreasing at the rate of 25 rad/s², while the relative speed u of point D of strap CD is 1.5 m/s and is decreasing at the rate of 3 m/s².

15.186 Using the method of Sec. 15.14, solve Prob. 15.153.

15.187 Using the method of Sec. 15.14, solve Prob. 15.159.

15.188 Using the method of Sec. 15.14, solve Prob. 15.156.

15.189 Using the method of Sec. 15.14, solve Prob. 15.157.

15.190 The body AB and rod BC of the robotic component shown rotate at the constant rate $\omega_1 = 0.60$ rad/s about the Y axis. Simultaneously a wire-and-pulley control causes arm CD to rotate about C at the constant rate $\omega_2 = d\beta/dt = 0.45$ rad/s. Knowing that $\beta = 120°$, determine (a) the angular acceleration of arm CD, (b) the velocity of D, (c) the acceleration of D.

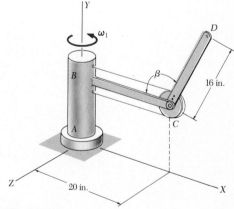

Fig. P15.190

15.191 Solve Prob. 15.190, assuming that $\beta = 60°$.

15.192 A disk of radius $r = 180$ mm rotates at the constant rate $\omega_2 = 8$ rad/s with respect to frame ABC, which itself rotates at the constant rate $\omega_1 = 6$ rad/s about the X axis. For the position shown, determine the velocity and acceleration of point D on the rim of the disk.

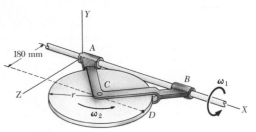

Fig. P15.192

15.193 Solve Prob. 15.192, assuming that $\omega_1 = 8$ rad/s and $\omega_2 = 6$ rad/s.

15.194 The remote manipulator system (RMS) shown is used to deploy payloads from the cargo bay of space shuttles. At the instant shown, the whole RMS is rotating at the constant rate $\omega_1 = 0.03$ rad/s about the axis AB. At the same time, portion BCD rotates as a rigid body at the constant rate $\omega_2 = d\beta/dt = 0.04$ rad/s about an axis through B parallel to the X axis. Knowing that $\beta = 30°$, determine (a) the angular acceleration of BCD, (b) the velocity of D, (c) the acceleration of D.

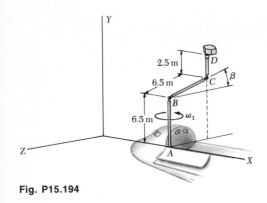

Fig. P15.194

15.195 The mechanism shown is used to raise a worker to the elevation of overhead electric and telephone wires. The entire mechanism rotates at the constant rate $\omega_1 = 0.15$ rad/s about the Y axis. The angle between arm AB and the horizontal is constant, while arm BC is being lowered at the constant rate $\omega_2 = d\beta/dt = -0.20$ rad/s. Knowing that AB and BC are each 4.5 m long, determine the acceleration of C at the instant shown.

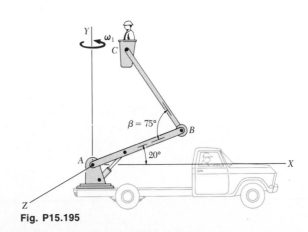

Fig. P15.195

15.196 A disk of 6-in. radius rotates at the constant rate $\omega_2 = 4$ rad/s with respect to arm ABC, which itself rotates at the constant rate $\omega_1 = 3$ rad/s about the Y axis. Determine (a) the angular acceleration of the disk, (b) the acceleration of point D on the rim of the disk.

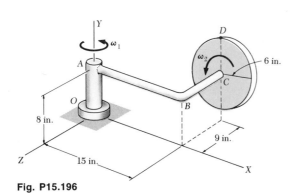

Fig. P15.196

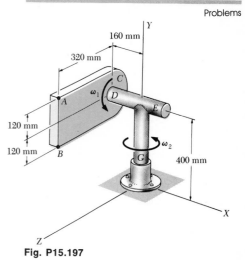

Fig. P15.197

15.197 In the portion of an industrial robot shown plate ABC rotates at the constant rate $\omega_1 = 2.4$ rad/s with respect to component DEG. At the same time, the entire unit rotates about the Y axis at the constant rate $\omega_2 = 2$ rad/s. Knowing that at the instant shown DE is parallel to the X axis and plate ABC is parallel to the YZ plane, determine (a) the velocity of point A, (b) the acceleration of point A.

15.198 In Prob. 15.197, determine (a) the velocity of point B, (b) the acceleration of point B.

15.199 Two disks, each of 5-in. radius, are welded to the 18-in. rod CD. The rod-and-disks unit rotates at the constant rate $\omega_2 = 3$ rad/s with respect to arm AB. Knowing that at the instant shown arm AB rotates about the Y axis at the constant rate $\omega_1 = 4$ rad/s, determine the velocity and acceleration of (a) point E, (b) point F.

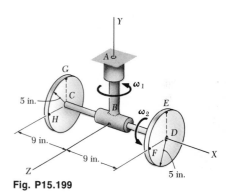

Fig. P15.199

15.200 In Prob. 15.199, determine the velocity and acceleration of (a) point G, (b) point H.

15.201 A square plate of side $2r$ is welded to a vertical shaft which rotates with a constant angular velocity ω_1. At the same time, rod AB of length r rotates about the center of the plate with a constant angular velocity ω_2 with respect to the plate. For the position of the plate shown, determine the acceleration of end B of the rod if (a) $\theta = 0$, (b) $\theta = 90°$, (c) $\theta = 180°$.

15.202 Solve Prob. 15.201, assuming $\omega_1 = 3$ rad/s, $\omega_2 = 2$ rad/s, and $r = 120$ mm.

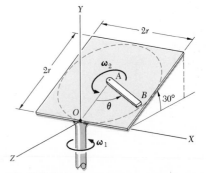

Fig. P15.201

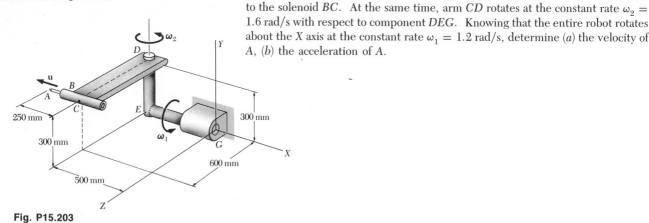

Fig. P15.203

15.203 The position of the stylus tip A is controlled by the robot shown. In the position shown the stylus moves at a constant speed $u = 180$ mm/s relative to the solenoid BC. At the same time, arm CD rotates at the constant rate $\omega_2 = 1.6$ rad/s with respect to component DEG. Knowing that the entire robot rotates about the X axis at the constant rate $\omega_1 = 1.2$ rad/s, determine (a) the velocity of A, (b) the acceleration of A.

Review and Summary

This chapter was devoted to the study of the kinematics of rigid bodies.

Rigid body in translation

We first considered the *translation* of a rigid body [Sec. 15.2] and observed that in such a motion, *all points of the body have the same velocity and the same acceleration at any given instant*.

Rigid body in rotation about a fixed axis

We next considered the *rotation* of a rigid body about a fixed axis [Sec. 15.3]. The position of the body is defined by the angle θ that the line BP, drawn from the axis of rotation to a point P of the body, forms with a fixed plane (Fig. 15.39). We found that the magnitude of the velocity of P is

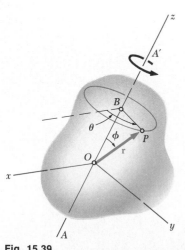

Fig. 15.39

$$v = \frac{ds}{dt} = r\dot{\theta} \sin \phi \qquad (15.4)$$

where $\dot{\theta}$ is the time derivative of θ. We then expressed the velocity of P as

$$\mathbf{v} = \frac{d\mathbf{r}}{dt} = \boldsymbol{\omega} \times \mathbf{r} \qquad (15.5)$$

where the vector

$$\boldsymbol{\omega} = \omega\mathbf{k} = \dot{\theta}\mathbf{k} \qquad (15.6)$$

is directed along the fixed axis of rotation and represents the *angular velocity* of the body.

Denoting by $\boldsymbol{\alpha}$ the derivative $d\boldsymbol{\omega}/dt$ of the angular velocity, we expressed the acceleration of P as

$$\mathbf{a} = \boldsymbol{\alpha} \times \mathbf{r} + \boldsymbol{\omega} \times (\boldsymbol{\omega} \times \mathbf{r}) \tag{15.8}$$

Differentiating (15.6), and recalling that $\mathbf{k}$ is constant in magnitude and direction, we found that

$$\boldsymbol{\alpha} = \alpha \mathbf{k} = \dot{\omega}\mathbf{k} = \ddot{\theta}\mathbf{k} \tag{15.9}$$

The vector $\boldsymbol{\alpha}$ represents the *angular acceleration* of the body and is directed along the fixed axis of rotation.

Next we considered the motion of a representative slab located in a plane perpendicular to the axis of rotation of the body (Fig. 15.40). Since

Rotation of a representative slab

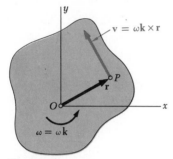

Fig. 15.40

the angular velocity is perpendicular to the slab, the velocity of a point P of the slab was expressed as

$$\mathbf{v} = \omega\mathbf{k} \times \mathbf{r} \tag{15.10}$$

where $\mathbf{v}$ is contained in the plane of the slab. Substituting $\boldsymbol{\omega} = \omega\mathbf{k}$ and $\boldsymbol{\alpha} = \alpha\mathbf{k}$ into (15.8), we found that the acceleration of P could be resolved into tangential and normal components (Fig. 15.41) respectively equal to

Tangential and normal components

$$\begin{array}{ll} \mathbf{a}_t = \alpha\mathbf{k} \times \mathbf{r} & a_t = r\alpha \\ \mathbf{a}_n = -\omega^2\mathbf{r} & a_n = r\omega^2 \end{array} \tag{15.11'}$$

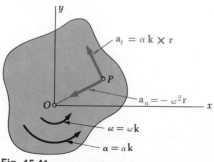

Fig. 15.41

Angular velocity and angular
acceleration of rotating slab

Recalling Eqs. (15.6) and (15.9), we obtained the following expressions for the *angular velocity* and the *angular acceleration* of the slab [Sec. 15.4]:

$$\omega = \frac{d\theta}{dt} \tag{15.12}$$

$$\alpha = \frac{d\omega}{dt} = \frac{d^2\theta}{dt^2} \tag{15.13}$$

or

$$\alpha = \omega \frac{d\omega}{d\theta} \tag{15.14}$$

We noted that these expressions are similar to those obtained in Chap. 11 for the rectilinear motion of a particle.

Two particular cases of rotation are frequently encountered: *uniform rotation* and *uniformly accelerated rotation*. Problems involving either of these motions may be solved by using equations similar to those used in Secs. 11.4 and 11.5 for the uniform rectilinear motion and the uniformly accelerated rectilinear motion of a particle, but where x, v, and a are replaced by θ, ω, and α, respectively [Sample Prob. 15.1].

Velocities in plane motion

The *most general plane motion* of a rigid slab may be considered as the *sum of a translation and a rotation* [Sec. 15.5]. For example, the slab shown in Fig. 15.42 may be assumed to translate with point A, while simultaneously rotating about A. It follows [Sec. 15.6] that the velocity of any point B of the slab may be expressed as

$$\mathbf{v}_B = \mathbf{v}_A + \mathbf{v}_{B/A} \tag{15.17}$$

where $\mathbf{v}_A$ is the velocity of A and $\mathbf{v}_{B/A}$ the relative velocity of B with respect to A or, more precisely, with respect to axes $x'y'$ translating with A. Denoting by $\mathbf{r}_{B/A}$ the position vector of B relative to A, we found that

$$\mathbf{v}_{B/A} = \omega\mathbf{k} \times \mathbf{r}_{B/A} \qquad v_{B/A} = r\omega \tag{15.18}$$

The fundamental equation (15.17) relating the absolute velocities of points A and B and the relative velocity of B with respect to A was ex-

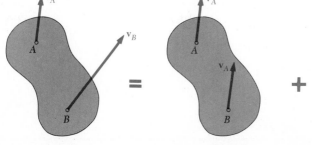

Fig. 15.42   Plane motion   =   Translation with A +   Rotation about A

$\mathbf{v}_B = \mathbf{v}_A + \mathbf{v}_{B/A}$

pressed in the form of a vector diagram and used to solve problems involving the motion of various types of mechanisms [Sample Probs. 15.2 and 15.3].

Another approach to the solution of problems involving the velocities of the points of a rigid slab in plane motion was presented in Sec. 15.7 and used in Sample Probs. 15.4 and 15.5. It is based on the determination of the *instantaneous center of rotation C* of the slab (Fig. 15.43).

Instantaneous center of rotation

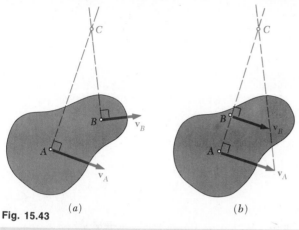

Fig. 15.43   *(a)*   *(b)*

The fact that any plane motion of a rigid slab may be considered as the sum of a translation of the slab with a reference point A and a rotation about A was used in Sec. 15.8 to relate the absolute accelerations of any two points A and B of the slab and the relative acceleration of B with respect to A. We had

Accelerations in plane motion

$$\mathbf{a}_B = \mathbf{a}_A + \mathbf{a}_{B/A} \qquad (15.21)$$

where $\mathbf{a}_{B/A}$ consisted of a *normal component* $(\mathbf{a}_{B/A})_n$ of magnitude $r\omega^2$ directed toward A, and a *tangential component* $(\mathbf{a}_{B/A})_t$ of magnitude $r\alpha$ perpendicular to the line AB (Fig. 15.44). The fundamental relation (15.21) was expressed in terms of vector diagrams or vector equations and used to determine the accelerations of given points of various mechanisms [Sample Probs. 15.6 through 15.8]. It should be noted that the instanta-

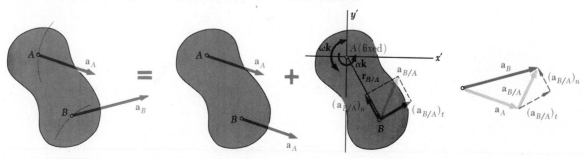

Fig. 15.44   Plane motion   =   Translation with A +   Rotation about A

Coordinates expressed in terms of a parameter

Rate of change of a vector with respect to a rotating frame

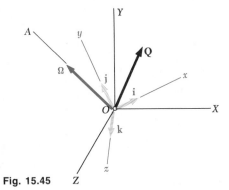

Fig. 15.45

Plane motion of a particle relative to a rotating frame

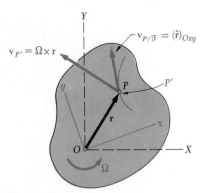

Fig. 15.46

neous center of rotation C considered in Sec. 15.7 cannot be used for the determination of accelerations, since point C, in general, does *not* have zero acceleration.

In the case of certain mechanisms, it is possible to express the coordinates x and y of all significant points of the mechanism by means of simple analytic expressions containing a *single parameter*. The components of the absolute velocity and acceleration of a given point may then be obtained by differentiating twice with respect to the time t the coordinates x and y of that point [Sec. 15.9].

While the rate of change of a vector is the same with respect to a fixed frame of reference and with respect to a frame in translation, the rate of change of a vector with respect to a rotating frame is different. Therefore, in order to study the motion of a particle relative to a rotating frame we first had to compare [Sec. 15.10] the rates of change of a general vector $\mathbf{Q}$ with respect to a fixed frame $OXYZ$ and with respect to a frame $Oxyz$ rotating with an angular velocity $\mathbf{\Omega}$ (Fig. 15.45). We obtained the fundamental relation

$$(\dot{\mathbf{Q}})_{OXYZ} = (\dot{\mathbf{Q}})_{Oxyz} + \mathbf{\Omega} \times \mathbf{Q} \tag{15.31}$$

and we concluded that the rate of change of the vector $\mathbf{Q}$ with respect to the fixed frame $OXYZ$ is made of two parts: The first part represents the rate of change of $\mathbf{Q}$ with respect to the rotating frame $Oxyz$; the second part, $\mathbf{\Omega} \times \mathbf{Q}$, is induced by the rotation of the frame $Oxyz$.

The next part of the chapter [Sec. 15.11] was devoted to the two-dimensional kinematic analysis of a particle P moving with respect to a frame $\mathcal{F}$ rotating with an angular velocity $\mathbf{\Omega}$ about a fixed axis (Fig. 15.46). We found that the absolute velocity of P could be expressed as

$$\mathbf{v}_P = \mathbf{v}_{P'} + \mathbf{v}_{P/\mathcal{F}} \tag{15.33}$$

where $\mathbf{v}_P$ = absolute velocity of particle P
$\quad\mathbf{v}_{P'}$ = velocity of point P' of moving frame $\mathcal{F}$ coinciding with P
$\quad\mathbf{v}_{P/\mathcal{F}}$ = velocity of P relative to moving frame $\mathcal{F}$

We noted that the same expression for $\mathbf{v}_P$ is obtained if the frame is in translation rather than in rotation. However, when the frame is in rotation, the expression for the acceleration of P is found to contain an additional term $\mathbf{a}_c$ called the *complementary acceleration* or *Coriolis acceleration*. We wrote

$$\mathbf{a}_P = \mathbf{a}_{P'} + \mathbf{a}_{P/\mathcal{F}} + \mathbf{a}_c \tag{15.36}$$

where $\mathbf{a}_P$ = absolute acceleration of particle P
$\quad\mathbf{a}_{P'}$ = acceleration of point P' of moving frame $\mathcal{F}$ coinciding with P
$\quad\mathbf{a}_{P/\mathcal{F}}$ = acceleration of P relative to moving frame $\mathcal{F}$
$\quad\mathbf{a}_c = 2\mathbf{\Omega} \times (\dot{\mathbf{r}})_{Oxy} = 2\mathbf{\Omega} \times \mathbf{v}_{P/\mathcal{F}}$
$\quad\quad$ = complementary, or Coriolis, acceleration

Since Ω and $v_{P/\mathscr{F}}$ are perpendicular to each other in the case of plane motion, the Coriolis acceleration was found to have a magnitude $a_c = 2\Omega v_{P/\mathscr{F}}$ and to point in the direction obtained by rotating the vector $v_{P/\mathscr{F}}$ through 90° in the sense of rotation of the moving frame. Formulas (15.33) and (15.36) may be used to analyze the motion of mechanisms which contain parts sliding on each other [Sample Probs. 15.9 and 15.10].

The last part of the chapter was devoted to the study of the kinematics of rigid bodies in three dimensions. We first considered the motion of a rigid body with a fixed point [Sec. 15.12]. After proving that the most general displacement of a rigid body with a fixed point O is equivalent to a rotation of the body about an axis through O, we were able to define the angular velocity $\boldsymbol{\omega}$ and the *instantaneous axis of rotation* of the body at a given instant. The velocity of a point P of the body (Fig. 15.47) could again be expressed as

$$\mathbf{v} = \frac{d\mathbf{r}}{dt} = \boldsymbol{\omega} \times \mathbf{r} \tag{15.37}$$

Differentiating this expression, we also wrote

$$\mathbf{a} = \boldsymbol{\alpha} \times \mathbf{r} + \boldsymbol{\omega} \times (\boldsymbol{\omega} \times \mathbf{r}) \tag{15.38}$$

However, since the direction of $\boldsymbol{\omega}$ changes from one instant to the next, the angular acceleration $\boldsymbol{\alpha}$ is, in general, not directed along the instantaneous axis of rotation [Sample Prob. 15.11].

It was shown in Sec. 15.13 that *the most general motion of a rigid body in space is equivalent, at any given instant, to the sum of a translation and a rotation.* Considering two particles A and B of the body, we found that

$$\mathbf{v}_B = \mathbf{v}_A + \mathbf{v}_{B/A} \tag{15.42}$$

where $\mathbf{v}_{B/A}$ is the velocity of B relative to a frame $AX'Y'Z'$ attached to A and of fixed orientation (Fig. 15.48). Denoting by $\mathbf{r}_{B/A}$ the position vector of B relative to A, we wrote

$$\mathbf{v}_B = \mathbf{v}_A + \boldsymbol{\omega} \times \mathbf{r}_{B/A} \tag{15.43}$$

where $\boldsymbol{\omega}$ is the angular velocity of the body at the instant considered [Sample Prob. 15.12]. The acceleration of B was obtained by a similar reasoning. We first wrote

$$\mathbf{a}_B = \mathbf{a}_A + \mathbf{a}_{B/A}$$

and, recalling Eq. (15.38),

$$\mathbf{a}_B = \mathbf{a}_A + \boldsymbol{\alpha} \times \mathbf{r}_{B/A} + \boldsymbol{\omega} \times (\boldsymbol{\omega} \times \mathbf{r}_{B/A}) \tag{15.44}$$

Motion of a rigid body with a fixed point

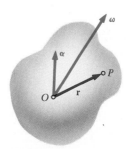

Fig. 15.47

General motion in space

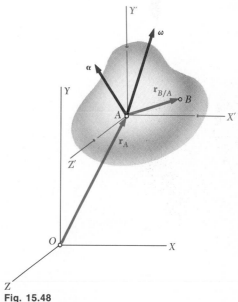

Fig. 15.48

Kinematics of Rigid Bodies

Three-dimensional motion of a particle relative to a rotating frame

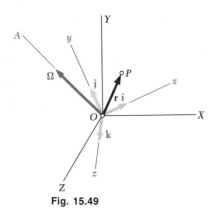

Fig. 15.49

Frame of reference in general motion

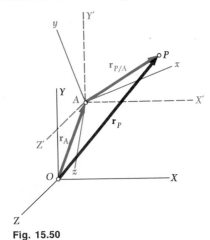

Fig. 15.50

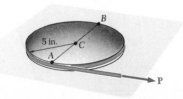

Fig. P15.204

In the final two sections of the chapter we considered the three-dimensional motion of a particle P relative to a frame $Oxyz$ rotating with an angular velocity $\mathbf{\Omega}$ with respect to a fixed frame $OXYZ$ (Fig. 15.49). In Sec. 15.14 we expressed the absolute velocity $\mathbf{v}_P$ of P as

$$\mathbf{v}_P = \mathbf{v}_{P'} + \mathbf{v}_{P/\mathcal{F}} \tag{15.46}$$

where $\mathbf{v}_P$ = absolute velocity of particle P
$\mathbf{v}_{P'}$ = velocity of point P' of moving frame $\mathcal{F}$ coinciding with P
$\mathbf{v}_{P/\mathcal{F}}$ = velocity of P relative to moving frame $\mathcal{F}$

The absolute acceleration $\mathbf{a}_P$ of P was then expressed as

$$\mathbf{a}_P = \mathbf{a}_{P'} + \mathbf{a}_{P/\mathcal{F}} + \mathbf{a}_c \tag{15.48}$$

where $\mathbf{a}_P$ = absolute acceleration of particle P
$\mathbf{a}_{P'}$ = acceleration of point P' of moving frame $\mathcal{F}$ coinciding with P
$\mathbf{a}_{P/\mathcal{F}}$ = acceleration of P relative to moving frame $\mathcal{F}$
$\mathbf{a}_c = 2\mathbf{\Omega} \times (\dot{\mathbf{r}})_{Oxyz} = 2\mathbf{\Omega} \times \mathbf{v}_{P/\mathcal{F}}$
= complementary, or Coriolis, acceleration

It was noted that the magnitude a_c of the Coriolis acceleration is not equal to $2\Omega v_{P/\mathcal{F}}$ [Sample Prob. 15.13] except in the special case when $\mathbf{\Omega}$ and $\mathbf{v}_{P/\mathcal{F}}$ are perpendicular to each other.

We also observed [Sec. 15.15] that Eqs. (15.46) and (15.48) remain valid when the frame $Axyz$ moves in a known, but arbitrary, fashion with respect to the fixed frame $OXYZ$ (Fig. 15.50) provided that the motion of A is included in the terms $\mathbf{v}_{P'}$ and $\mathbf{a}_{P'}$ representing the absolute velocity and acceleration of the coinciding point P'.

Rotating frames of reference are particularly useful in the study of the three-dimensional motion of rigid bodies. Indeed, there are many cases where an appropriate choice of the rotating frame will lead to a simpler analysis of the motion of the rigid body than would be possible with axes of fixed orientation [Sample Probs. 15.14 and 15.15].

Review Problems

15.204 A tape is wrapped around a 10-in.-diameter disk which is at rest on a horizontal table. A force $\mathbf{P}$ applied as shown produces the following accelerations: $\mathbf{a}_A = 45$ in./s² to the right, $\alpha = 6$ rad/s² counterclockwise as viewed from above. Determine (a) the acceleration of point B, (b) the point of the disk which has no acceleration.

15.205 Water flows through the curved pipe *OB*, which has a uniform radius of 18 in. and which rotates with a constant counterclockwise angular velocity of 150 rpm. If the velocity of the water relative to the pipe is 45 ft/s, determine the total acceleration of the particle of water *P*.

Fig. P15.205

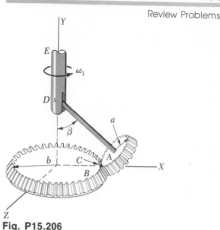

Fig. P15.206

15.206 Gear *A* rolls on the fixed gear *B* and rotates about axle *AD* which is rigidly attached at *D* to the vertical shaft *DE*. Knowing that shaft *DE* rotates with a constant angular velocity ω_1, determine (*a*) the rate of spin of gear *A* about axle *AD*, (*b*) the angular acceleration of gear *A*, (*c*) the acceleration of tooth *C* of gear *A*.

15.207 The flanged wheel rolls without slipping on the horizontal rail. If at a given instant the velocity and acceleration of the center of the wheel are as shown, determine the acceleration (*a*) of point *B*, (*b*) of point *C*, (*c*) of point *D*.

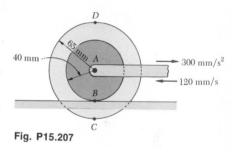

Fig. P15.207

15.208 A series of small machine components being moved by a conveyor belt pass over the 125-mm-radius idler pulley shown. At the instant shown, the velocity of point *A* is 350 mm/s to the left and its acceleration is 400 mm/s² to the right. Determine (*a*) the angular velocity and angular acceleration of the idler pulley, (*b*) the total acceleration of the machine component at *B*.

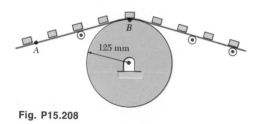

Fig. P15.208

15.209 Knowing that at the instant shown collar *D* moves downward with a constant velocity of 60 in./s, determine (*a*) the instantaneous center of rotation of link *BD*, (*b*) the angular velocities of crank *AB* and link *BD*, (*c*) the velocity of the midpoint of link *BD*.

15.210 Knowing that at the instant shown collar *D* moves downward with a constant velocity of 60 in./s, determine the angular acceleration of (*a*) link *BD*, (*b*) crank *AB*.

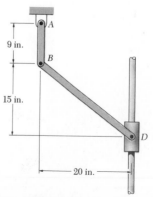

Fig. P15.209 and P15.210

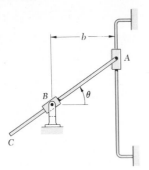

Fig. P15.211

15.211 Rod AC of length $2b$ is attached to a collar at A and passes through a pivoted collar at B. Knowing that collar A moves upward with a constant velocity $\mathbf{v}_A$, derive an expression for (a) the angular velocity of rod AC, (b) the velocity of the point of the rod in contact with the pivoted collar B, (c) the angular acceleration of rod AC.

15.212 The telescoping arm AB is used to raise a worker to the elevation of electric and telephone wires. Knowing that the length AB increases at the constant rate $dl/dt = 0.20$ m/s and that the arm rotates at the constant rate $\omega_1 = 0.25$ rad/s about a vertical axis, while the angle θ it forms with the horizontal maintains a constant value, determine the acceleration of point B when $l = 5$ m.

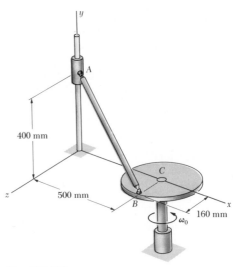

Fig. P15.213

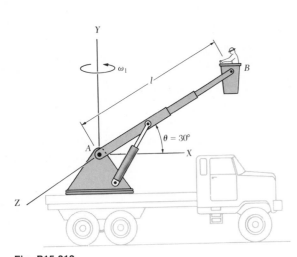

Fig. P15.212

15.213 Rod AB is connected by ball-and-socket joints to collar A and to the 320-mm-diameter rotating disk C. Knowing that the disk rotates counterclockwise in the zx plane at the constant rate $\omega_0 = 4$ rad/s, determine the velocity of collar A.

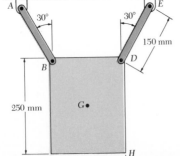

Fig. P15.214 and P15.215

15.214 A rectangular plate is supported by two 150-mm links as shown. Knowing that at the instant shown the angular velocity of link AB is 4 rad/s clockwise, determine (a) the angular velocity of the plate, (b) the velocity of the center of the plate, (c) the velocity of corner F.

15.215 Knowing that, at the instant shown, the angular velocity of link AB is 4 rad/s clockwise, determine (a) the angular velocity of the plate, (b) the points of the plate with a velocity equal to or less than 150 mm/s.

The following problems are designed to be solved with a computer.

783

Review Problems

15.C1 A 400-mm rod AB is guided by wheels at A and B which roll in the track shown. Knowing that B moves at a constant speed $v_B = 1.2$ m/s, write a computer program and use it to calculate the angular velocity of the rod and the velocity of A for values of θ from 0 to 150° at 10° intervals.

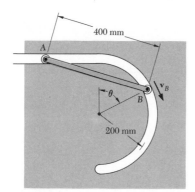

Fig. P15.C1

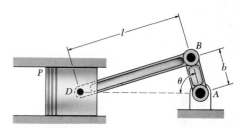

Fig. P15.C2

15.C2 In the engine system shown, $l = 8$ in. and $b = 3$ in. During a test of the system, crank AB is rotated with a constant angular velocity of 2000 rpm clockwise. Write a computer program and use it to calculate, for values of θ from 0 to 180° at 10° intervals, (a) the angular velocity and angular acceleration of the connecting rod BD, (b) the velocity and acceleration of the piston P.

15.C3 Write a computer program and use it to solve Sample Prob. 15.12 for values of θ from 0 to 360° at 30° intervals.

15.C4 Manufactured items are spray-painted as they pass through the automated work station shown. The bent pipe ACE rotates at the constant rate $\omega_1 = 0.4$ rad/s and the paint moves through the pipe at a constant relative speed $u = 150$ mm/s. Write a computer program and use it to calculate the velocity and the acceleration of the paint at D for values of θ from 0 to 150° at 10° intervals.

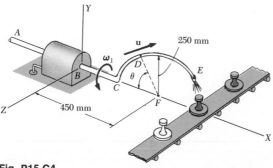

Fig. P15.C4

Plane Motion of Rigid Bodies: Forces and Accelerations

16.1. Introduction. In this chapter and in Chaps. 17 and 18, we shall study the *kinetics of rigid bodies,* i.e., the relations existing between the forces acting on a rigid body, the shape and mass of the body, and the motion produced. In Chaps. 12 and 13, we studied similar relations, assuming then that the body could be considered as a particle, i.e., that its mass could be concentrated in one point and that all forces acted at that point. We shall now take the shape of the body into account, as well as the exact location of the points of application of the forces. We shall also be concerned not only with the motion of the body as a whole but also with the motion of the body about its mass center.

Our approach will be to consider rigid bodies as made of large numbers of particles and to use the results obtained in Chap. 14 for the motion of systems of particles. In this chapter, we shall use specifically Eq. (14.16), $\Sigma \mathbf{F} = m\bar{\mathbf{a}}$, which relates the resultant of the external forces and the acceleration of the mass center G of the system of particles, and Eq. (14.23), $\Sigma \mathbf{M}_G = \dot{\mathbf{H}}_G$, which relates the moment resultant of the external forces and the angular momentum of the system of particles about G.

Except for Sec. 16.2, which applies to the most general case of the motion of a rigid body, the results derived in this chapter will be limited in two ways: (1) They will be restricted to the *plane motion* of rigid bodies, i.e., to a motion in which each particle of the body remains at a constant distance from a fixed reference plane. (2) The rigid bodies considered will consist only of plane slabs and of bodies which are symmetrical with respect to the reference plane.† The study of the plane motion of non-

---

† Or, more generally, bodies which have a principal centroidal axis of inertia perpendicular to the reference plane.

symmetrical three-dimensional bodies and, more generally, the motion of rigid bodies in three-dimensional space will be postponed until Chap. 18.

In Sec. 16.3, we define the angular momentum of a rigid body in plane motion and show that the rate of change of the angular momentum $\dot{\mathbf{H}}_G$ about the mass center is equal to the product $\bar{I}\alpha$ of the centroidal mass moment of inertia $\bar{I}$ and the angular acceleration α of the body. D'Alembert's principle, introduced in Sec. 16.4, is used to prove that the external forces acting on a rigid body are equivalent to a vector $m\bar{\mathbf{a}}$ attached at the mass center and a couple of moment $\bar{I}\alpha$.

In Sec. 16.5, we derive the principle of transmissibility using only the parallelogram law and Newton's laws of motion. The principle of transmissibility may therefore be removed from the list of axioms (Sec. 1.2) required for the study of the statics and dynamics of rigid bodies.

The concept of free-body-diagram equations is introduced in Sec. 16.6. Emphasis will be placed on the use of free-body-diagram equations in the solution of all problems involving the plane motion of rigid bodies.

After considering the plane motion of connected rigid bodies in Sec. 16.7, we shall be prepared to solve a variety of problems involving the translation, centroidal rotation, or unconstrained motion of rigid bodies. In Sec. 16.8 and in the remaining part of the chapter, we shall consider the solution of problems involving the noncentroidal rotation, rolling motion, or other partially constrained plane motions of rigid bodies.

16.2. Equations of Motion for a Rigid Body. Consider a rigid body acted upon by several external forces $\mathbf{F}_1$, $\mathbf{F}_2$, $\mathbf{F}_3$, etc. (Fig. 16.1). We may assume that the body is made of a large number n of particles of mass Δm_i $(i = 1, 2, \ldots, n)$ and apply the results obtained in Chap. 14 for a system of particles (Fig. 16.2). Considering first the motion of the mass center G of the body with respect to the newtonian frame of reference $Oxyz$, we recall Eq. (14.16) and write

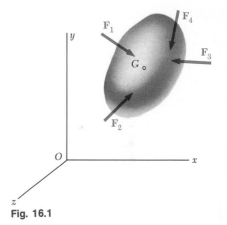

Fig. 16.1

$$\Sigma\mathbf{F} = m\bar{\mathbf{a}} \qquad (16.1)$$

where m is the mass of the body and $\bar{\mathbf{a}}$ the acceleration of the mass center G. Turning now to the motion of the body relative to the centroidal frame of reference $Gx'y'z'$, we recall Eq. (14.23) and write

$$\Sigma\mathbf{M}_G = \dot{\mathbf{H}}_G \qquad (16.2)$$

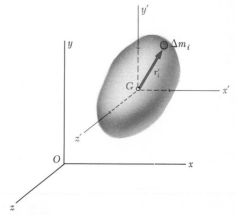

Fig. 16.2

where $\dot{\mathbf{H}}_G$ represents the rate of change of $\mathbf{H}_G$, the angular momentum about G of the system of particles forming the rigid body. In the following we shall simply refer to $\mathbf{H}_G$ as the *angular momentum of the rigid body about its mass center G*. Together Eqs. (16.1) and (16.2) express that the

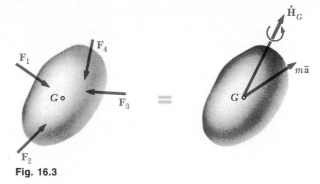

Fig. 16.3

system of the external forces is equipollent to the system consisting of the
vector $m\overline{\mathbf{a}}$ attached at G and the couple of moment $\dot{\mathbf{H}}_G$ (Fig. 16.3).†

Equations (16.1) and (16.2) apply in the most general case of the motion of a rigid body. In the rest of this chapter, however, we shall limit our
analysis to the *plane motion* of rigid bodies, i.e., to a motion in which each
particle remains at a constant distance from a fixed reference plane, and we
shall assume that the rigid bodies considered consist only of plane slabs and
of bodies which are symmetrical with respect to the reference plane. Further study of the plane motion of nonsymmetrical three-dimensional bodies
and of the motion of rigid bodies in three-dimensional space will be postponed until Chap. 18.

**16.3. Angular Momentum of a Rigid Body in Plane Mo-
tion.** Consider a rigid slab in plane motion. Assuming that the slab is
made of a large number n of particles P_i of mass Δm_i and recalling Eq.
(14.24) of Sec. 14.5, we note that the angular momentum $\mathbf{H}_G$ of the slab
about its mass center G may be computed by taking the moments about G
of the momenta of the particles of the slab in their motion with respect to
either of the frames Oxy or $Gx'y'$ (Fig. 16.4). Choosing the latter course,
we write

$$\mathbf{H}_G = \sum_{i=1}^{n} (\mathbf{r}'_i \times \mathbf{v}'_i \Delta m_i) \tag{16.3}$$

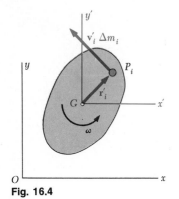

Fig. 16.4

where $\mathbf{r}'_i$ and $\mathbf{v}'_i \Delta m_i$ denote, respectively, the position vector and the linear
momentum of the particle P_i relative to the centroidal frame of reference
$Gx'y'$. But since the particle belongs to the slab, we have $\mathbf{v}'_i = \boldsymbol{\omega} \times \mathbf{r}'_i$,
where $\boldsymbol{\omega}$ is the angular velocity of the slab at the instant considered. We
write

$$\mathbf{H}_G = \sum_{i=1}^{n} [\mathbf{r}'_i \times (\boldsymbol{\omega} \times \mathbf{r}'_i)\Delta m_i]$$

† Since the systems involved act on a rigid body, we could conclude at this point, by
referring to Sec. 3.19, that the two systems are *equivalent* as well as equipollent and use
colored rather than gray equals signs in Fig. 16.3. However, by postponing this conclusion, we
shall be able to arrive at it independently (Secs. 16.4 and 18.5), thereby eliminating the
necessity of including the principle of transmissibility among the axioms of mechanics (Sec.
16.5).

Referring to Fig. 16.4, we easily verify that the expression obtained represents a vector of the same direction as $\boldsymbol{\omega}$ (that is, perpendicular to the slab) and of magnitude equal to $\omega \Sigma r_i'^2 \, \Delta m_i$. Recalling that the sum $\Sigma r_i'^2 \, \Delta m_i$ represents the moment of inertia $\bar{I}$ of the slab about a centroidal axis perpendicular to the slab, we conclude that the angular momentum $\mathbf{H}_G$ of the slab about its mass center is

787

16.4. Plane Motion of a Rigid Body. D'Alembert's
Principle

$$\mathbf{H}_G = \bar{I}\boldsymbol{\omega} \qquad (16.4)$$

Differentiating both members of Eq. (16.4) we obtain

$$\dot{\mathbf{H}}_G = \bar{I}\dot{\boldsymbol{\omega}} = \bar{I}\boldsymbol{\alpha} \qquad (16.5)$$

Thus the rate of change of the angular momentum of the slab is represented by a vector of the same direction as $\boldsymbol{\alpha}$ (that is, perpendicular to the slab) and of magnitude $\bar{I}\alpha$.

It should be kept in mind that the results obtained in this section have been derived for a rigid slab in plane motion. As we shall see in Chap. 18, they remain valid in the case of the plane motion of rigid bodies which are symmetrical with respect to the reference plane.† However, they do not apply in the case of nonsymmetrical bodies or in the case of three-dimensional motion.

16.4. Plane Motion of a Rigid Body. D'Alembert's Principle.
Consider a rigid slab of mass m moving under the action of several external forces $\mathbf{F}_1$, $\mathbf{F}_2$, $\mathbf{F}_3$, etc., contained in the plane of the slab (Fig. 16.5). Substituting for $\dot{\mathbf{H}}_G$ from Eq. (16.5) into Eq. (16.2) and writing the fundamental equations of motion (16.1) and (16.2) in scalar form, we have

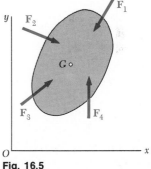

Fig. 16.5

$$\Sigma F_x = m\bar{a}_x \qquad \Sigma F_y = m\bar{a}_y \qquad \Sigma M_G = \bar{I}\alpha \qquad (16.6)$$

Equations (16.6) show that the acceleration of the mass center G of the slab and its angular acceleration $\boldsymbol{\alpha}$ may easily be obtained, once the resultant of the external forces acting on the slab and their moment resultant about G have been determined. Given appropriate initial conditions, the coordinates $\bar{x}$ and $\bar{y}$ of the mass center and the angular coordinate θ of the slab may then be obtained at any instant t by integration. Thus *the motion of the slab is completely defined by the resultant and moment resultant about G of the external forces acting on it.*

This property, which will be extended in Chap. 18 to the case of the three-dimensional motion of a rigid body, is characteristic of the motion of a rigid body. Indeed, as we saw in Chap. 14, the motion of a system of particles which are not rigidly connected will in general depend upon the

---

† Or, more generally, bodies which have a principal centroidal axis of inertia perpendicular to the reference plane.

specific external forces, as well as upon the internal forces, acting on the various particles.

Since the motion of a rigid body depends only upon the resultant and moment resultant of the external forces acting on it, it follows that *two systems of forces which are equipollent,* i.e., which have the same resultant and the same moment resultant, *are also equivalent;* that is, they have exactly the same effect on a given rigid body.†

Consider in particular the system of the external forces acting on a rigid body (Fig. 16.6a) and the system of the effective forces associated with the particles forming the rigid body (Fig. 16.6b). It was shown in Sec. 14.2 that the two systems thus defined are equipollent. But since the particles considered now form a rigid body, it follows from the discussion above that the two systems are also equivalent. We may thus state that *the external forces acting on a rigid body are equivalent to the effective forces of the various particles forming the body.* This statement is referred to as *d'Alembert's principle* after the French mathematician Jean le Rond d'Alembert (1717–1783), even though d'Alembert's original statement was written in a somewhat different form.

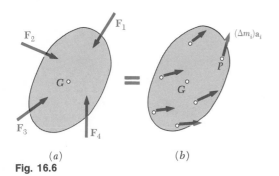

(a)   (b)

Fig. 16.6

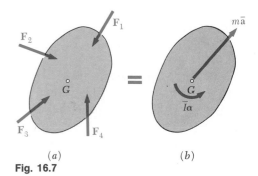

(a)   (b)

Fig. 16.7

The significance of d'Alembert's principle has been emphasized by the use of a colored equals sign in Fig. 16.6 and also in Fig. 16.7, where using results obtained earlier in this section, we have replaced the effective forces by a vector $m\bar{a}$ attached at the mass center G of the slab and a couple of moment $\bar{I}\alpha$.

---

† This result has already been derived in Sec. 3.19 from the principle of transmissibility (Sec. 3.3). The present derivation is independent of that principle, however, and will make possible its elimination from the axioms of mechanics (Sec. 16.5).

Translation. In the particular case of a body in translation, the angular acceleration of the body is identically equal to zero and its effective forces reduce to the vector $m\bar{\mathbf{a}}$ attached at G (Fig. 16.8). Thus, the resultant of the external forces acting on a rigid body in translation passes through the mass center of the body and is equal to $m\bar{\mathbf{a}}$.

789

16.4. Plane Motion of a Rigid Body. D'Alembert's Principle

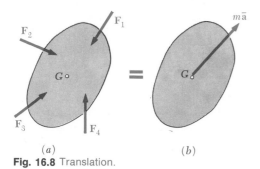

(a)   (b)

Fig. 16.8 Translation.

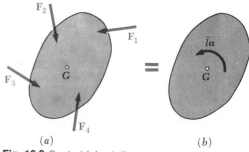

(a)   (b)

Fig. 16.9 Centroidal rotation.

Centroidal Rotation. When a slab, or, more generally, a body symmetrical with respect to the reference plane, rotates about a fixed axis perpendicular to the reference plane and passing through its mass center G, we say that the body is in *centroidal rotation*. Since the acceleration $\bar{\mathbf{a}}$ is identically equal to zero, the effective forces of the body reduce to the couple $\bar{I}\alpha$ (Fig. 16.9). Thus, the external forces acting on a body in centroidal rotation are equivalent to a couple of moment $\bar{I}\alpha$.

General Plane Motion. Comparing Fig. 16.7 with Figs. 16.8 and 16.9, we observe that from the point of view of *kinetics*, the most general plane motion of a rigid body symmetrical with respect to the reference plane may be replaced by the sum of a translation and a centroidal rotation. We should note that this statement is more restrictive than the similar statement made earlier from the point of view of *kinematics* (Sec. 15.5), since we now require that the mass center of the body be selected as the reference point.

Referring to Eqs. (16.6), we observe that the first two equations are identical with the equations of motion of a particle of mass m acted upon by the given forces $\mathbf{F}_1$, $\mathbf{F}_2$, $\mathbf{F}_3$, etc. We thus check that *the mass center G of a rigid body in plane motion moves as if the entire mass of the body were concentrated at that point, and as if all the external forces acted on it.* We recall that this result has already been obtained in Sec. 14.4 in the general case of a system of particles, the particles being not necessarily rigidly connected. We also note, as we did in Sec. 14.4, that the system of the external forces does not, in general, reduce to a single vector $m\bar{a}$ attached at G. Therefore, in the general case of the plane motion of a rigid body, *the resultant of the external forces acting on the body does not pass through the mass center of the body.*

Finally, we may observe that the last of Eqs. (16.6) would still be valid if the rigid body while subjected to the same applied forces were constrained to rotate about a fixed axis through G. Thus, *a rigid body in plane motion rotates about its mass center as if this point were fixed.*

* 16.5. A Remark on the Axioms of the Mechanics of Rigid Bodies.

The fact that two equipollent systems of external forces acting on a rigid body are also equivalent, i.e., have the same effect on that rigid body, has already been established in Sec. 3.19. But there it was derived from the *principle of transmissibility,* one of the axioms used in our study of the statics of rigid bodies. It should be noted that this axiom has not been used in the present chapter, because Newton's second and third laws of motion make its use unnecessary in the study of the dynamics of rigid bodies.

In fact, the principle of transmissibility may now be *derived* from the other axioms used in the study of mechanics. This principle stated, without proof (Sec. 3.3), that the conditions of equilibrium or motion of a rigid body remain unchanged if a force $\mathbf{F}$ acting at a given point of the rigid body is replaced by a force $\mathbf{F}'$ of the same magnitude and same direction, but acting at a different point, provided that the two forces have the same line of action. But since $\mathbf{F}$ and $\mathbf{F}'$ have the same moment about any given point, it is clear that they form two equipollent systems of external forces. Thus, we may now *prove,* as a result of what we established in the preceding section, that $\mathbf{F}$ and $\mathbf{F}'$ have the same effect on the rigid body (Fig. 3.3).

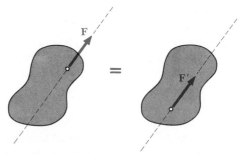

Fig. 3.3 (*repeated*)

The principle of transmissibility may therefore be removed from the list of axioms required for the study of the mechanics of rigid bodies. These axioms are reduced to the parallelogram law of addition of vectors and to Newton's laws of motion.

16.6. Solution of Problems Involving the Motion of a Rigid Body.

We saw in Sec. 16.4 that when a rigid body is in plane motion, there exists a fundamental relation between the forces $\mathbf{F}_1$, $\mathbf{F}_2$, $\mathbf{F}_3$, etc., acting on the body, the acceleration $\bar{\mathbf{a}}$ of its mass center, and the angular acceleration $\boldsymbol{\alpha}$ of the body. This relation, which is represented in Fig. 16.7 in the form of a *free-body-diagram equation,* may be used to determine the acceleration $\bar{\mathbf{a}}$ and the angular acceleration $\boldsymbol{\alpha}$ produced by a given system of forces acting on a rigid body or, conversely, to determine the forces which produce a given motion of the rigid body.

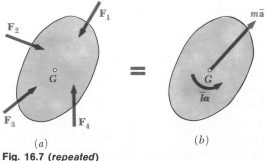

(a)　　　　　(b)

Fig. 16.7 (repeated)

While the three algebraic equations (16.6) may be used to solve problems of plane motion,† our experience in statics suggests that the solution of many problems involving rigid bodies could be simplified by an appropriate choice of the point about which the moments of the forces are computed. It is therefore preferable to remember the relation existing between the forces and the accelerations in the pictorial form shown in Fig. 16.7, and to derive from this fundamental relation the component or moment equations which fit best the solution of the problem under consideration.

The fundamental relation shown in Fig. 16.7 may be presented in an alternative form if we add to the external forces an inertia vector $-m\bar{\mathbf{a}}$ of sense opposite to that of $\bar{\mathbf{a}}$, attached at G, and an inertia couple $-\bar{I}\boldsymbol{\alpha}$ of moment equal in magnitude to $\bar{I}\alpha$ and of sense opposite to that of $\boldsymbol{\alpha}$ (Fig. 16.10). The system obtained is equivalent to zero, and the rigid body is said to be in dynamic equilibrium.

Whether the principle of equivalence of external and effective forces is directly applied, as in Fig. 16.7, or whether the concept of dynamic equilibrium is introduced, as in Fig. 16.10, the use of free-body-diagram equations showing vectorially the relationship existing between the forces applied on the rigid body and the resulting linear and angular accelerations presents considerable advantages over the blind application of formulas (16.6). These advantages may be summarized as follows:

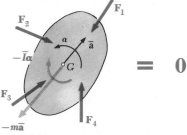

Fig. 16.10

---

† We recall that the last of Eqs. (16.6) is valid only in the case of the plane motion of a rigid body symmetrical with respect to the reference plane. In all other cases, the methods of Chap. 18 should be used.

1. A much clearer understanding of the effect of the forces on the motion of the body will result from the use of a pictorial representation.

2. This approach makes it possible to divide the solution of a dynamics problem into two parts: In the first part, the analysis of the kinematic and kinetic characteristics of the problem leads to the free-body diagrams of Fig. 16.7 or 16.10; in the second part, the diagram obtained is used to analyze by the methods of Chap. 3 the various forces and vectors involved.

3. A unified approach is provided for the analysis of the plane motion of a rigid body, regardless of the particular type of motion involved. While the kinematics of the various motions considered may vary from one case to the other, the approach to the kinetics of the motion is consistently the same. In every case we shall draw a diagram showing the external forces, the vector $m\bar{a}$ associated with the motion of G, and the couple $\bar{I}\alpha$ associated with the rotation of the body about G.

4. The resolution of the plane motion of a rigid body into a translation and a centroidal rotation, which is used here, is a basic concept which may be applied effectively throughout the study of mechanics. We shall use it again in Chap. 17 with the method of work and energy and the method of impulse and momentum.

5. As we shall see in Chap. 18, this approach may be extended to the study of the general three-dimensional motion of a rigid body. The motion of the body will again be resolved into a translation and a rotation about the mass center, and free-body-diagram equations will be used to indicate the relationship existing between the external forces and the rates of change of the linear and angular momentum of the body.

16.7. Systems of Rigid Bodies. The method described in the preceding section may also be used in problems involving the plane motion of several connected rigid bodies. A diagram similar to Fig. 16.7 or Fig. 16.10 may be drawn for each part of the system. The equations of motion obtained from these diagrams are solved simultaneously.

In some cases, as in Sample Prob. 16.3, a single diagram may be drawn for the entire system. This diagram should include all the external forces, as well as the vectors $m\bar{a}$ and the couples $\bar{I}\alpha$ associated with the various parts of the system. However, internal forces, such as the forces exerted by connecting cables, may be omitted since they occur in pairs of equal and opposite forces and are thus equipollent to zero. The equations obtained by expressing that the system of the external forces is equipollent to the system of the effective forces may be solved for the remaining unknowns.†

This second approach may not be used in problems involving more than three unknowns, since only three equations of motion are available when a single diagram is used. We shall not elaborate upon this point, since the discussion involved would be completely similar to that given in Sec. 6.11 in the case of the equilibrium of a system of rigid bodies.

---

† Note that we cannot speak of *equivalent* systems since we are not dealing with a single rigid body.

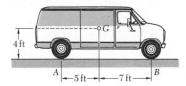

When the forward speed of the truck shown was 30 ft/s, the brakes were suddenly applied, causing all four wheels to stop rotating. It was observed that the truck skidded to rest in 20 ft. Determine the magnitude of the normal reaction and of the friction force at each wheel as the truck skidded to rest.

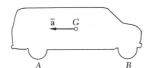

Kinematics of Motion. Choosing the positive sense to the right and using the equations of uniformly accelerated motion, we write

$$\bar{v}_0 = +30 \text{ ft/s} \qquad \bar{v}^2 = \bar{v}_0^2 + 2\bar{a}x \qquad 0 = (30)^2 + 2\bar{a}(20)$$
$$\bar{a} = -22.5 \text{ ft/s}^2 \qquad \bar{a} = 22.5 \text{ ft/s}^2 \leftarrow$$

Equations of Motion. The external forces consist of the weight $\mathbf{W}$ of the truck and of the normal reactions and friction forces at the wheels. (The vectors $\mathbf{N}_A$ and $\mathbf{F}_A$ represent the sum of the reactions at the rear wheels, while $\mathbf{N}_B$ and $\mathbf{F}_B$ represent the sum of the reactions at the front wheels.) Since the truck is in translation, the effective forces reduce to the vector $m\bar{a}$ attached at G. Three equations of motion are obtained by expressing that the system of the external forces is equivalent to the system of the effective forces.

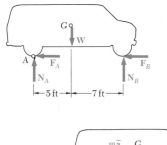

$$+\uparrow \Sigma F_y = \Sigma(F_y)_{\text{eff}}: \qquad N_A + N_B - W = 0$$

Since $F_A = \mu_k N_A$ and $F_B = \mu_k N_B$ where μ_k is the coefficient of kinetic friction, we find that

$$F_A + F_B = \mu_k(N_A + N_B) = \mu_k W$$

$$\xrightarrow{+} \Sigma F_x = \Sigma(F_x)_{\text{eff}}: \qquad -(F_A + F_B) = -m\bar{a}$$

$$-\mu_k W = -\frac{W}{32.2 \text{ ft/s}^2}(22.5 \text{ ft/s}^2)$$

$$\mu_k = 0.699$$

$$+\uparrow \Sigma M_A = \Sigma(M_A)_{\text{eff}}: \quad -W(5 \text{ ft}) + N_B(12 \text{ ft}) = m\bar{a}(4 \text{ ft})$$

$$-W(5 \text{ ft}) + N_B(12 \text{ ft}) = \frac{W}{32.2 \text{ ft/s}^2}(22.5 \text{ ft/s}^2)(4 \text{ ft})$$

$$N_B = 0.650W$$
$$F_B = \mu_k N_B = (0.699)(0.650W) \qquad F_B = 0.454W$$

$$+\uparrow \Sigma F_y = \Sigma(F_y)_{\text{eff}}: \qquad N_A + N_B - W = 0$$
$$N_A + 0.650W - W = 0$$
$$N_A = 0.350W$$
$$F_A = \mu_k N_A = (0.699)(0.350W) \qquad F_A = 0.245W$$

Reactions at Each Wheel. Recalling that the values computed above represent the sum of the reactions at the two front wheels or the two rear wheels, we obtain the magnitude of the reactions at each wheel by writing

$$N_{\text{front}} = \tfrac{1}{2}N_B = 0.325W \qquad N_{\text{rear}} = \tfrac{1}{2}N_A = 0.175W \quad \blacktriangleleft$$
$$F_{\text{front}} = \tfrac{1}{2}F_B = 0.227W \qquad F_{\text{rear}} = \tfrac{1}{2}F_A = 0.122W \quad \blacktriangleleft$$

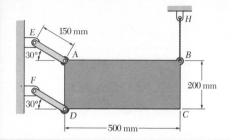

SAMPLE PROBLEM 16.2

The thin plate $ABCD$ has a mass of 8 kg and is held in the position shown by the wire BH and two links AE and DF. Neglecting the mass of the links, determine immediately after wire BH has been cut (a) the acceleration of the plate, (b) the force in each link.

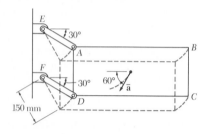

Kinematics of Motion. After wire BH has been cut, we observe that corners A and D move along parallel circles of radius 150 mm centered, respectively, at E and F. The motion of the plate is thus a curvilinear translation; the particles forming the plate move along parallel circles of radius 150 mm.

At the instant wire BH is cut, the velocity of the plate is zero. Thus the acceleration $\bar{\mathbf{a}}$ of the mass center G of the plate is tangent to the circular path which will be described by G.

Equations of Motion. The external forces consist of the weight $\mathbf{W}$ and the forces $\mathbf{F}_{AE}$ and $\mathbf{F}_{DF}$ exerted by the links. Since the plate is in translation, the effective forces reduce to the vector $m\bar{\mathbf{a}}$ attached at G and directed along the t axis. We shall now express that the system of the external forces is equivalent to the system of the effective forces.

a. Acceleration of the Plate.

$+\swarrow \Sigma F_t = \Sigma(F_t)_{\text{eff}}$:

$$W \cos 30° = m\bar{a}$$
$$mg \cos 30° = m\bar{a}$$
$$\bar{a} = g \cos 30° = (9.81 \text{ m/s}^2) \cos 30° \qquad (1)$$

$$\bar{\mathbf{a}} = 8.50 \text{ m/s}^2 \ \measuredangle\ 60° \quad \blacktriangleleft$$

b. Forces in Links AE and DF.

$+\nwarrow \Sigma F_n = \Sigma(F_n)_{\text{eff}}$: $\qquad F_{AE} + F_{DF} - W \sin 30° = 0 \qquad (2)$

$+\downarrow \Sigma M_G = \Sigma(M_G)_{\text{eff}}$:

$$(F_{AE} \sin 30°)(250 \text{ mm}) - (F_{AE} \cos 30°)(100 \text{ mm})$$
$$+(F_{DF} \sin 30°)(250 \text{ mm}) + (F_{DF} \cos 30°)(100 \text{ mm}) = 0$$
$$38.4F_{AE} + 211.6F_{DF} = 0$$
$$F_{DF} = -0.1815F_{AE} \qquad (3)$$

Substituting for F_{DF} from (3) into (2), we write

$$F_{AE} - 0.1815F_{AE} - W \sin 30° = 0$$
$$F_{AE} = 0.6109W$$
$$F_{DF} = -0.1815(0.6109W) = -0.1109W$$

Noting that $W = mg = (8 \text{ kg})(9.81 \text{ m/s}^2) = 78.48 \text{ N}$, we have

$$F_{AE} = 0.6109(78.48 \text{ N}) \qquad F_{AE} = 47.9 \text{ N T} \quad \blacktriangleleft$$
$$F_{DF} = -0.1109(78.48 \text{ N}) \qquad F_{DF} = 8.70 \text{ N C} \quad \blacktriangleleft$$

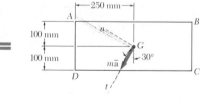

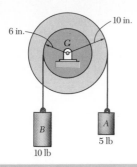

SAMPLE PROBLEM 16.3

A pulley weighing 12 lb and having a radius of gyration of 8 in. is connected to two cylinders as shown. Assuming no axle friction, determine the angular acceleration of the pulley and the acceleration of each cylinder.

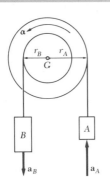

Sense of Motion. Although an arbitrary sense of motion may be assumed (since no friction forces are involved) and later checked by the sign of the answer, we may prefer first to determine the actual sense of rotation of the pulley. We first find the weight of cylinder B required to maintain the equilibrium of the pulley when it is acted upon by the 5-lb cylinder A. We write

$$+\uparrow\Sigma M_G = 0: \qquad W_B(6\text{ in.}) - (5\text{ lb})(10\text{ in.}) = 0 \qquad W_B = 8.33\text{ lb}$$

Since cylinder B actually weighs 10 lb, the pulley will rotate counterclockwise.

Kinematics of Motion. Assuming α counterclockwise and noting that $a_A = r_A\alpha$ and $a_B = r_B\alpha$, we obtain

$$\mathbf{a}_A = (\tfrac{10}{12}\text{ ft})\alpha \uparrow \qquad \mathbf{a}_B = (\tfrac{6}{12}\text{ ft})\alpha \downarrow$$

Equations of Motion. A single system consisting of the pulley and the two cylinders is considered. Forces external to this system consist of the weights of the pulley and the two cylinders and of the reaction at G. (The forces exerted by the cables on the pulley and on the cylinders are internal to the system considered and cancel out.) Since the motion of the pulley is a centroidal rotation and the motion of each cylinder is a translation, the effective forces reduce to the couple $\bar{I}\alpha$ and the two vectors ma_A and ma_B. The centroidal moment of inertia of the pulley is

$$\bar{I} = m\bar{k}^2 = \frac{W}{g}\bar{k}^2 = \frac{12\text{ lb}}{32.2\text{ ft/s}^2}(\tfrac{8}{12}\text{ ft})^2 = 0.1656\text{ lb}\cdot\text{ft}\cdot\text{s}^2$$

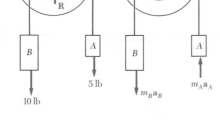

Since the system of the external forces is equipollent to the system of the effective forces, we write

$$+\uparrow\Sigma M_G = \Sigma(M_G)_{\text{eff}}:$$
$$(10\text{ lb})(\tfrac{6}{12}\text{ ft}) - (5\text{ lb})(\tfrac{10}{12}\text{ ft}) = +\bar{I}\alpha + m_B a_B(\tfrac{6}{12}\text{ ft}) + m_A a_A(\tfrac{10}{12}\text{ ft})$$
$$(10)(\tfrac{6}{12}) - (5)(\tfrac{10}{12}) = 0.1656\alpha + \tfrac{10}{32.2}(\tfrac{6}{12}\alpha)(\tfrac{6}{12}) + \tfrac{5}{32.2}(\tfrac{10}{12}\alpha)(\tfrac{10}{12})$$

$$\alpha = +2.374\text{ rad/s}^2 \qquad\qquad \boldsymbol{\alpha} = 2.37\text{ rad/s}^2 \uparrow \quad \blacktriangleleft$$
$$a_A = r_A\alpha = (\tfrac{10}{12}\text{ ft})(2.374\text{ rad/s}^2) \qquad \mathbf{a}_A = 1.978\text{ ft/s}^2 \uparrow \quad \blacktriangleleft$$
$$a_B = r_B\alpha = (\tfrac{6}{12}\text{ ft})(2.374\text{ rad/s}^2) \qquad \mathbf{a}_B = 1.187\text{ ft/s}^2 \downarrow \quad \blacktriangleleft$$

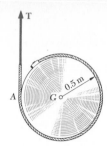

A cord is wrapped around a homogeneous disk of radius $r = 0.5$ m and mass $m = 15$ kg. If the cord is pulled upward with a force $\mathbf{T}$ of magnitude 180 N, determine (a) the acceleration of the center of the disk, (b) the angular acceleration of the disk, (c) the acceleration of the cord.

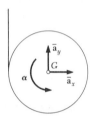

Equations of Motion. We assume that the components $\bar{\mathbf{a}}_x$ and $\bar{\mathbf{a}}_y$ of the acceleration of the center are directed, respectively, to the right and upward and that the angular acceleration of the disk is counterclockwise. The external forces acting on the disk consist of the weight $\mathbf{W}$ and the force $\mathbf{T}$ exerted by the cord. This system is equivalent to the system of the effective forces, which consists of a vector of components $m\bar{\mathbf{a}}_x$ and $m\bar{\mathbf{a}}_y$ attached at G and a couple $\bar{I}\alpha$. We write

$$\xrightarrow{+} \Sigma F_x = \Sigma(F_x)_{\text{eff}}: \qquad 0 = m\bar{a}_x \qquad\qquad \bar{\mathbf{a}}_x = 0 \quad\blacktriangleleft$$

$$+\uparrow\Sigma F_y = \Sigma(F_y)_{\text{eff}}: \qquad T - W = m\bar{a}_y$$

$$\bar{a}_y = \frac{T - W}{m}$$

Since $T = 180$ N, $m = 15$ kg, and $W = (15 \text{ kg})(9.81 \text{ m/s}^2) = 147.1$ N, we have

$$\bar{a}_y = \frac{180 \text{ N} - 147.1 \text{ N}}{15 \text{ kg}} = +2.19 \text{ m/s}^2 \qquad \bar{\mathbf{a}}_y = 2.19 \text{ m/s}^2 \uparrow \quad\blacktriangleleft$$

$$+\uparrow\Sigma M_G = \Sigma(M_G)_{\text{eff}}: \qquad -Tr = \bar{I}\alpha$$

$$-Tr = (\tfrac{1}{2}mr^2)\alpha$$

$$\alpha = -\frac{2T}{mr} = -\frac{2(180 \text{ N})}{(15 \text{ kg})(0.5 \text{ m})} = -48.0 \text{ rad/s}^2$$

$$\alpha = 48.0 \text{ rad/s}^2 \;\downarrow \quad\blacktriangleleft$$

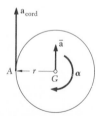

Acceleration of Cord. Since the acceleration of the cord is equal to the tangential component of the acceleration of point A on the disk, we write

$$\mathbf{a}_{\text{cord}} = (\mathbf{a}_A)_t = \bar{\mathbf{a}} + (\mathbf{a}_{A/G})_t$$

$$= [2.19 \text{ m/s}^2 \uparrow] + [(0.5 \text{ m})(48 \text{ rad/s}^2) \uparrow]$$

$$\mathbf{a}_{\text{cord}} = 26.2 \text{ m/s}^2 \uparrow \quad\blacktriangleleft$$

SAMPLE PROBLEM 16.5

A uniform sphere of mass m and radius r is projected along a rough horizontal surface with a linear velocity $\bar{v}_0$ and no angular velocity. Denoting by μ_k the coefficient of kinetic friction between the sphere and the floor, determine (a) the time t_1 at which the sphere will start rolling without sliding, (b) the linear velocity and angular velocity of the sphere at time t_1.

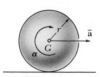

Equations of Motion. The positive sense is chosen to the right for $\bar{a}$ and clockwise for $\boldsymbol{\alpha}$. The external forces acting on the sphere consist of the weight $\mathbf{W}$, the normal reaction $\mathbf{N}$, and the friction force $\mathbf{F}$. Since the point of the sphere in contact with the surface is sliding to the right, the friction force $\mathbf{F}$ is directed to the left. While the sphere is sliding, the magnitude of the friction force is $F = \mu_k N$. The effective forces consist of the vector $m\bar{a}$ attached at G and the couple $\bar{I}\boldsymbol{\alpha}$. Expressing that the system of the external forces is equivalent to the system of the effective forces, we write

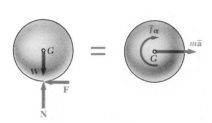

$$+\uparrow \Sigma F_y = \Sigma(F_y)_{\text{eff}}: \qquad N - W = 0$$
$$N = W = mg \qquad F = \mu_k N = \mu_k mg$$

$$\xrightarrow{+} \Sigma F_x = \Sigma(F_x)_{\text{eff}}: \qquad -F = m\bar{a} \qquad -\mu_k mg = m\bar{a} \qquad \bar{a} = -\mu_k g$$

$$+\downarrow \Sigma M_G = \Sigma(M_G)_{\text{eff}}: \qquad Fr = \bar{I}\alpha$$

Noting that $\bar{I} = \frac{2}{5}mr^2$ and substituting the value obtained for F, we write

$$(\mu_k mg)r = \tfrac{2}{5}mr^2\alpha \qquad \alpha = \frac{5}{2}\frac{\mu_k g}{r}$$

Kinematics of Motion. As long as the sphere both rotates and slides, its linear and angular motions are uniformly accelerated.

$$t = 0, \ \bar{v} = \bar{v}_0 \qquad \bar{v} = \bar{v}_0 + \bar{a}t = \bar{v}_0 - \mu_k gt \qquad (1)$$

$$t = 0, \ \omega_0 = 0 \qquad \omega = \omega_0 + \alpha t = 0 + \left(\frac{5}{2}\frac{\mu_k g}{r}\right)t \qquad (2)$$

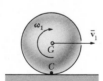

The sphere will start rolling without sliding when the velocity $\mathbf{v}_C$ of the point of contact C is zero. At that time, $t = t_1$, point C becomes the instantaneous center of rotation, and we have

$$\bar{v}_1 = r\omega_1 \qquad (3)$$

Substituting in (3) the values obtained for $\bar{v}_1$ and ω_1 by making $t = t_1$ in (1) and (2), respectively, we write

$$\bar{v}_0 - \mu_k gt_1 = r\left(\frac{5}{2}\frac{\mu_k g}{r}t_1\right) \qquad t_1 = \frac{2}{7}\frac{\bar{v}_0}{\mu_k g} \qquad \blacktriangleleft$$

Substituting for t_1 into (2), we have

$$\omega_1 = \frac{5}{2}\frac{\mu_k g}{r}t_1 = \frac{5}{2}\frac{\mu_k g}{r}\left(\frac{2}{7}\frac{\bar{v}_0}{\mu_k g}\right) \qquad \omega_1 = \frac{5}{7}\frac{\bar{v}_0}{r} \qquad \omega_1 = \frac{5}{7}\frac{\bar{v}_0}{r} \ \downarrow \qquad \blacktriangleleft$$

$$\bar{v}_1 = r\omega_1 = r\left(\frac{5}{7}\frac{\bar{v}_0}{r}\right) \qquad \bar{v}_1 = \tfrac{5}{7}\bar{v}_0 \qquad \mathbf{v}_1 = \tfrac{5}{7}\bar{v}_0 \rightarrow \qquad \blacktriangleleft$$

Problems

16.1 A 6-ft board is placed in a truck with one end resting against a block secured to the floor and the other leaning against a vertical partition. Determine the maximum allowable acceleration of the truck if the board is to remain in the position shown.

Fig. P16.1 and P16.2

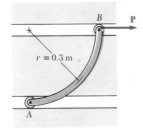

Fig. P16.3

16.2 A 6-ft board is placed in a truck with one end on the floor and the other leaning against a vertical partition. Knowing that the coefficient of static friction is 0.30 between the board and the floor and zero between the board and the partition, determine the maximum allowable deceleration of the truck if the board is to remain in the position shown.

16.3 The motion of the 2.5-kg rod AB is guided by two small wheels which roll freely in horizontal slots. If a force **P** of magnitude 8 N is applied at B, determine (*a*) the acceleration of the rod, (*b*) the reactions at A and B.

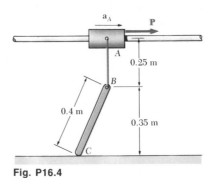

Fig. P16.4

16.4 A uniform rod BC of mass 4 kg is connected to a collar A by a 0.25-m cord AB. Neglecting the mass of the collar and cord, determine (*a*) the smallest constant acceleration $\mathbf{a}_A$ for which the cord and the rod lie in a straight line, (*b*) the corresponding tension in the cord.

16.5 Knowing that the coefficient of static friction between the tires and road is 0.75 for the car shown, determine the maximum possible acceleration on a level road, assuming (*a*) four-wheel drive, (*b*) conventional rear-wheel drive, (*c*) front-wheel drive.

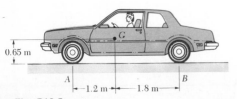

Fig. P16.5

16.6 Determine the distance through which the truck of Sample Prob. 16.1 will skid if (*a*) the rear-wheel brakes fail to operate, (*b*) the front-wheel brakes fail to operate.

16.7 A 25-kg rectangular panel is suspended from two skids *A* and *B* and is maintained in the position shown by a wire *CD*. Knowing that the coefficient of kinetic friction between each skid and the inclined track is $\mu_k = 0.20$, determine the normal component of the reaction at *B* immediately after wire *CD* has been cut.

16.8 In Prob. 16.7, determine (*a*) the maximum value of μ_k for which the skid at *B* remains in contact with the track after wire *CD* has been cut, (*b*) the corresponding acceleration of the panel.

16.9 A half section of pipe of weight *W* is pulled by a cable as shown. Denoting by μ_s and μ_k, respectively, the coefficients of static and kinetic friction at *A* and *B*, determine (*a*) the values of θ and *P* for which sliding and tipping of the pipe are both impending, (*b*) the acceleration of the pipe if *P* is then slightly increased, (*c*) the corresponding value of the normal component of the reaction at *A*.

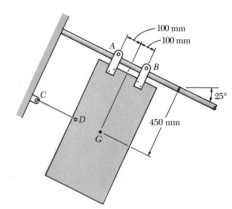

Fig. P16.7

16.10 Solve Prob. 16.9, assuming that $W = 200$ lb, $\mu_s = 0.50$, and $\mu_k = 0.40$.

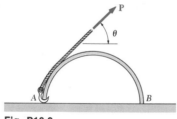

Fig. P16.9

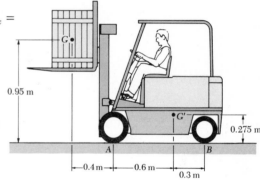

Fig. P16.11

16.11 A 2800-kg fork-lift truck carries a 1500-kg crate at the height shown. The truck is moving to the left when the brakes are applied, causing a deceleration of 3 m/s². Knowing that the coefficient of static friction between the crate and the fork lift is 0.50, determine (*a*) whether the crate will slide, (*b*) the vertical component of the reaction at each wheel.

16.12 In Prob. 16.11, determine the maximum allowable deceleration of the truck if the crate is not to slide forward and if the truck is not to tip forward.

16.13 An 80-lb cabinet is mounted on casters which allow it to move freely ($\mu = 0$) on the floor. If a 50-lb force is applied as shown, determine (*a*) the acceleration of the cabinet, (*b*) the range of values of *h* for which the cabinet will not tip.

16.14 Solve Prob. 16.13, assuming that the casters are locked and slide along the rough floor ($\mu_k = 0.25$).

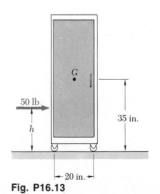

Fig. P16.13

16.15 A completely filled barrel and its contents have a combined mass of 125 kg. A 60-kg cylinder is connected to the barrel as shown. Knowing that the coefficients of friction between the barrel and the floor are $\mu_s = 0.35$ and $\mu_k = 0.30$, determine (a) the acceleration of the barrel, (b) the range of values of h for which the barrel will not tip.

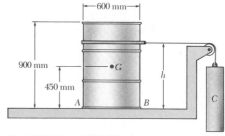

Fig. P16.15 and P16.16

16.16 A completely filled barrel and its contents have a combined mass of 160 kg. A cylinder C of mass m_C is connected to the barrel at a height $h = 600$ mm. Knowing that the coefficients of friction between the barrel and the floor are $\mu_s = 0.48$ and $\mu_k = 0.40$, determine the range of values of the mass of cylinder C for which the barrel will not tip.

16.17 Two uniform rods BD and EF, each of mass 2 kg, are welded together and attached to two links AB and CD. Neglecting the mass of the links, determine the force in each link immediately after the system has been released from rest with $\theta = 30°$.

16.18 Solve Prob. 16.17, assuming that the system has been released from rest with $\theta = 60°$.

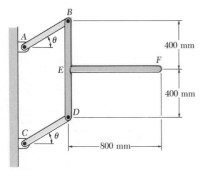

Fig. P16.17

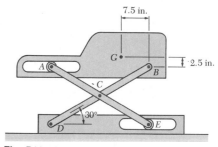

Fig. P16.19

16.19 Members ACE and DCB are each 30 in. long and are connected by a pin at C. The mass center of the 20-lb member AB is located at G. Determine (a) the acceleration of AB immediately after the system has been released from rest in the position shown, (b) the corresponding force exerted by roller A on member AB. Neglect the weight of members ACE and DCB.

16.20 The thin plate $ABCD$ weighs 16 lb and is held in position by the three inextensible wires AE, BF, and CH. Wire AE is then cut. Determine (a) the acceleration of the plate, (b) the tension in wires BF and CH immediately after wire AE has been cut.

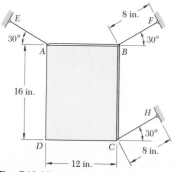

Fig. P16.20

16.21 A uniform semicircular plate of mass 8 kg is attached to two links *AB* and *DE*, each 250 mm long, and moves under its own weight. Neglecting the mass of the links and knowing that in the position shown the velocity of the plate is 1.2 m/s, determine the force in each link.

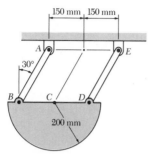

Fig. P16.21

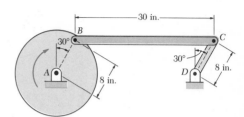

Fig. P16.22

16.22 The 15-lb rod *BC* connects a disk centered at *A* to crank *CD*. Knowing that the disk is made to rotate at the constant speed of 180 rpm, determine for the position shown the vertical components of the forces exerted on rod *BC* by the pins at *B* and *C*.

16.23 The T-shaped rod *ABCD* is guided by two pins which slide freely in parallel curved slots of radius 7.5 in. The rod weighs 6 lb, and its mass center is located at point *G*. Knowing that for the position shown the *vertical* component of the velocity of *D* is 4 ft/s upward and the *vertical* component of the acceleration of *D* is 15 ft/s² upward, determine the magnitude of the force **P**.

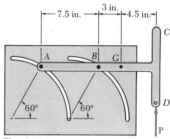

Fig. P16.23

16.24 Solve Prob. 16.23, knowing that for the position shown the *vertical* component of the velocity of *D* is 4 ft/s upward and the *vertical* component of the acceleration of *D* is zero.

*16.25 A 30-lb block is placed on a 5-lb platform *BD* which is held in the position shown by three wires. Determine the accelerations of the block and of the platform immediately after wire *AB* has been cut. Assume that the block (*a*) is rigidly attached to the platform, (*b*) can slide without friction on the platform.

*16.26 The coefficients of friction between the 30-lb block and the 5-lb platform *BD* are $\mu_s = 0.50$ and $\mu_k = 0.40$. Determine the accelerations of the block and of the platform immediately after wire *AB* has been cut.

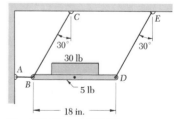

Fig. P16.25 and P16.26

*16.27 Draw the shear and bending-moment diagrams for the horizontal rod *EF* of Prob. 16.17.

*16.28 Draw the shear and bending-moment diagrams for the connecting rod *BC* of Prob. 16.22.

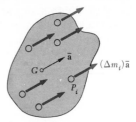

Fig. P16.29

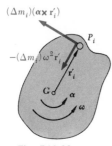

Fig. P16.30

16.29 For a rigid slab in translation, show that the system of the effective forces consists of vectors $(\Delta m_i)\bar{a}$ attached to the various particles of the slab, where $\bar{a}$ is the acceleration of the mass center G of the slab. Further show, by computing their sum and the sum of their moments about G, that the effective forces reduce to a single vector $m\bar{a}$ attached at G.

16.30 For a rigid slab in centroidal rotation, show that the system of the effective forces consists of vectors $-(\Delta m_i)\omega^2 r'_i$ and $(\Delta m_i)(\boldsymbol{\alpha} \times r'_i)$ attached to the various particles P_i of the slab, where $\boldsymbol{\omega}$ and $\boldsymbol{\alpha}$ are the angular velocity and angular acceleration of the slab, and where r'_i denotes the position vector of the particle P_i relative to the mass center G of the slab. Further show, by computing their sum and the sum of their moments about G, that the effective forces reduce to a couple $\bar{I}\boldsymbol{\alpha}$.

16.31 An electric motor is rotating at 1200 rpm when the load and power are cut off. The rotor weighs 180 lb and has a radius of gyration of 8 in. If the kinetic friction of the rotor produces a couple of magnitude 15 lb·in., how many revolutions will the rotor execute before stopping?

16.32 A turbine-generator unit is shut off when its rotor is rotating at 3600 rpm; it is observed that the rotor coasts to rest in 7.10 min. Knowing that the 1850-kg rotor has a radius of gyration of 234 mm, determine the average magnitude of the couple due to bearing friction.

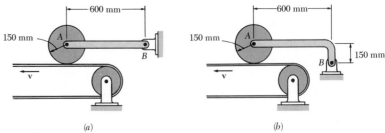

Fig. P16.33

16.33 The 6-kg disk is at rest when it is placed in contact with a conveyor belt moving at a constant speed. The link AB connecting the center of the disk to the support at B is of negligible weight. Knowing that the coefficient of kinetic friction between the disk and the belt is 0.30, determine for each of the arrangements shown the angular acceleration of the disk while slipping occurs.

16.34 Solve Prob. 16.33, assuming that the direction of motion of the conveyor belt is reversed.

16.35 The 10-in.-radius brake drum is attached to a larger flywheel which is not shown. The total mass moment of inertia of the flywheel and drum is 15 lb·ft·s² and the coefficient of kinetic friction between the drum and the brake shoe is 0.40. Knowing that the initial angular velocity is 240 rpm clockwise, determine the force which must be exerted by the hydraulic cylinder if the system is to stop in 60 revolutions.

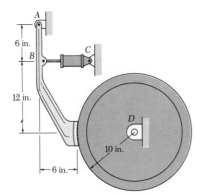

Fig. P16.35

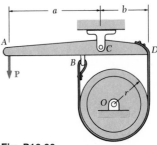

Fig. P16.36

16.36 The 100-mm-radius brake drum is attached to a flywheel which is not shown. The drum and flywheel together have a mass of 300 kg and a radius of gyration of 600 mm. The coefficient of kinetic friction between the brake band and the drum is 0.30. Knowing that a force **P** of magnitude 50 N is applied at A when the angular velocity is 180 rpm counterclockwise, determine the time required to stop the flywheel when $a = 200$ mm and $b = 160$ mm.

16.37 Solve Prob. 16.36, assuming that the coefficient of kinetic friction is 0.40 between the brake band and the drum.

16.38 Solve Prob. 16.35, assuming that the initial angular velocity of the flywheel is 240 rpm counterclockwise.

16.39 The flywheel shown consists of a 30-in.-diameter disk which weighs 240 lb. The coefficient of friction between the band and the flywheel is 0.35. If the initial angular velocity of the flywheel is 360 rpm clockwise, determine the magnitude of the force **P** required to stop the flywheel in 25 revolutions.

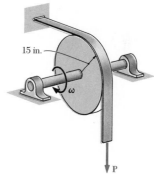

Fig. P16.39

16.40 Solve Prob. 16.39, assuming that the initial angular velocity of the flywheel is 360 rpm counterclockwise.

16.41 Two disks A and B, of mass $m_A = 2$ kg and $m_B = 4$ kg, are connected by a belt as shown. Assuming no slipping between the belt and the disks, determine the angular acceleration of each disk if a 2.70-N·m couple **M** is applied to disk A.

16.42 Solve Prob. 16.41, assuming that the 2.70-N·m couple **M** is applied to disk B.

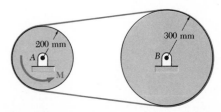

Fig. P16.41

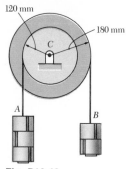

120 mm

180 mm

C

A

B

Fig. P16.43

r

A

Fig. P16.45

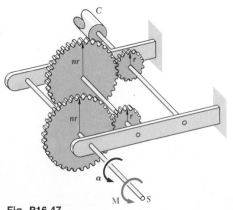

C

nr

r

nr

r

α

M   S

Fig. P16.47

16.43   The double pulley shown has a total mass of 6 kg and a centroidal radius of gyration of 135 mm. Five collars, each of mass 1.2 kg, are attached to cords A and B as shown. When the system is at rest and in equilibrium, one collar is removed from cord A. Neglecting friction, determine (a) the angular acceleration of the pulley, (b) the velocity of cord A at $t = 2.5$ s.

16.44   Solve Prob. 16.43, assuming that one collar is removed from cord B.

16.45   In order to determine the mass moment of inertia of a flywheel of radius $r = 1.5$ ft, a block A of weight 20 lb is attached to a cord wrapped around the rim of the flywheel. The block is released from rest and is observed to fall 12 ft in 4.50 s. To eliminate bearing friction from the computation, a second block of weight 40 lb is used and is observed to fall 12 ft in 2.80 s. Assuming that the magnitude of the couple due to bearing friction is constant, determine the mass moment of inertia of the flywheel.

16.46   Each of the double pulleys shown has a mass moment of inertia of 20 lb·ft·s² and is initially at rest. The outside radius is 18 in., and the inner radius is 9 in. Determine (a) the angular acceleration of each pulley, (b) the angular velocity of each pulley at $t = 3$ s, (c) the angular velocity of each pulley after point A on the cord has moved 10 ft.

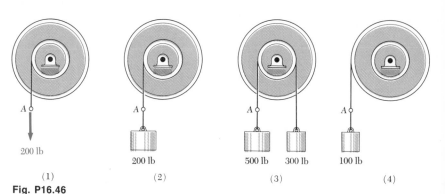

A

200 lb

(1)

A

200 lb

(2)

A

500 lb   300 lb

(3)

A

100 lb

(4)

Fig. P16.46

16.47   A coder C, used to record in digital form the rotation of a shaft S, is connected to the shaft by means of the gear train shown, which consists of four gears of the same thickness and of the same material. Two of the gears have a radius r and the other two a radius nr. Let I_R denote the ratio M/α of the magnitude M of the couple applied to shaft S and of the resulting angular acceleration α of S. (I_R is sometimes called the "reflected moment of inertia" of the coder and gear train.) Determine I_R in terms of the gear ratio n, the moment of inertia I_0 of the first gear, and the moment of inertia I_C of the coder. Neglect the moments of inertia of the shafts.

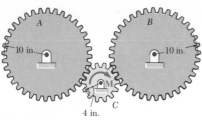

Fig. P16.48

16.48 Each of the gears A and B weighs 20 lb and has a radius of gyration of 7.5 in.; gear C weighs 5 lb and has a radius of gyration of 3 in. If a couple $\mathbf{M}$ of constant magnitude 50 lb·in. is applied to gear C, determine (a) the angular acceleration of gear A, (b) the tangential force which gear C exerts on gear A.

16.49 Disk A has a mass $m_A = 4$ kg, a radius $r_A = 300$ mm, and an initial angular velocity $\omega_0 = 300$ rpm clockwise. Disk B has a mass $m_B = 1.6$ kg, a radius $r_B = 180$ mm, and is at rest when it is brought into contact with disk A. Knowing that $\mu_k = 0.35$ between the disks and neglecting bearing friction, determine (a) the angular acceleration of each disk, (b) the reaction at the support C.

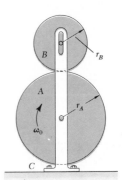

Fig. P16.49 and P16.50

16.50 Disk B is at rest when it is brought into contact with disk A, which has an initial angular velocity ω_0. (a) Show that the final angular velocities of the disks are independent of the coefficient of friction μ_k between the disks as long as $\mu_k \neq 0$. (b) Express the final angular velocity of disk A in terms of ω_0 and the ratio m_A/m_B of the masses of the two disks.

16.51 A belt of negligible mass passes between cylinders A and B and is pulled to the right with a force $\mathbf{P}$. Cylinders A and B weigh, respectively, 5 and 20 lb. The shaft of cylinder A is free to slide in a vertical slot and the coefficients of friction between the belt and each of the cylinders are $\mu_s = 0.50$ and $\mu_k = 0.40$. For $P = 3.60$ lb, determine (a) whether slipping occurs between the belt and either of the cylinders, (b) the angular acceleration of each cylinder.

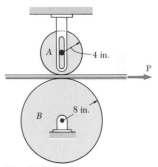

16.52 Solve Prob. 16.51 for $P = 2.00$ lb.

Fig. P16.51

16.53 A cylinder of radius r and mass m rests on two small casters A and B as shown. Initially, the cylinder is at rest and is set in motion by rotating caster B clockwise at high speed so that slipping occurs between the cylinder and caster B. Denoting by μ_k the coefficient of kinetic friction and neglecting the moment of inertia of the free caster A, derive an expression for the angular acceleration of the cylinder.

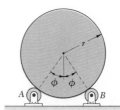

16.54 In Prob. 16.53, assume that no slipping can occur between caster B and the cylinder (such a case would exist if the cylinder and caster had gear teeth along their rims). Derive an expression for the maximum allowable counterclockwise acceleration α of the cylinder if it is not to lose contact with the caster at A.

Fig. P16.53

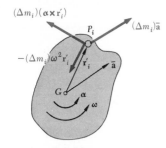

Fig. P16.56

16.55 Show that the system of the effective forces for a rigid slab in plane motion reduces to a single vector, and express the distance from the mass center G of the slab to the line of action of this vector in terms of the centroidal radius of gyration $\bar{k}$ of the slab, the magnitude $\bar{a}$ of the acceleration of G, and the angular acceleration α.

16.56 For a rigid slab in plane motion, show that the system of the effective forces consists of vectors $(\Delta m_i)\bar{\mathbf{a}}$, $-(\Delta m_i)\omega^2\mathbf{r}_i'$, and $(\Delta m_i)(\boldsymbol{\alpha} \times \mathbf{r}_i')$ attached to the various particles P_i of the slab, where $\bar{\mathbf{a}}$ is the acceleration of the mass center G of the slab, ω is the angular velocity of the slab, $\boldsymbol{\alpha}$ is its angular acceleration, and $\mathbf{r}_i'$ denotes the position vector of the particle P_i relative to G. Further show, by computing their sum and the sum of their moments about G, that the effective forces reduce to a vector $m\bar{\mathbf{a}}$ attached at G and a couple $\bar{I}\boldsymbol{\alpha}$.

16.57 The uniform slender rod AB of mass 1.50 kg is at rest on a frictionless horizontal surface. A force $\mathbf{P}$ of magnitude 2.70 N is applied at A in a horizontal direction perpendicular to the rod. Determine (a) the angular acceleration of the rod, (b) the acceleration of the center of the rod, (c) the point of the rod which has no acceleration.

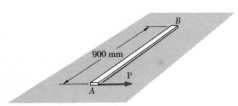

Fig. P16.57

16.58 (a) In Prob. 16.57, determine the point of the rod AB at which the force $\mathbf{P}$ should be applied if the acceleration of point B is to be zero. (b) Knowing that the magnitude of $\mathbf{P}$ is 2.70 N, determine the corresponding angular acceleration of the rod and the acceleration of the center of the rod.

16.59 A 10-in.-diameter disk weighing 15 lb rests on a frictionless horizontal surface. A force $\mathbf{P}$ of magnitude 3 lb is applied to a tape wrapped around the disk. Determine the acceleration of (a) point A, (b) point B.

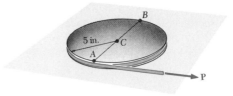

Fig. P16.59 and P16.60

16.60 A 10-in.-diameter disk weighing 15 lb rests on a frictionless horizontal surface. A force $\mathbf{P}$ of magnitude 12 oz is applied to a tape wrapped around the disk. Determine (a) the angular acceleration of the disk, (b) the acceleration of the center C of the disk, (c) the length of tape unwound after 2 s.

16.61 A 120-kg satellite has a radius of gyration of 600 mm with respect to the y axis and is symmetrical with respect to the zx plane. Its orientation is changed by firing four small rockets A, B, C, and D, each of which produces a 16.20-N thrust $\mathbf{T}$ directed as shown. Determine the angular acceleration of the satellite and the acceleration of its mass center G (a) when all four rockets are fired, (b) when all rockets except D are fired.

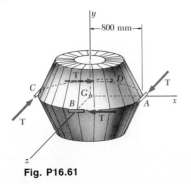

Fig. P16.61

16.62 Solve Prob. 16.61, assuming that only rocket A is fired.

16.63 The 240-kg crate shown is being lowered by means of two overhead cranes. Knowing that at the instant shown the deceleration of cable A is 7 m/s², while that of cable B is 1 m/s², determine the tension in each cable.

16.64 The 240-kg crate is being lowered by means of two overhead cranes. As the crate approaches the ground, the crane operators apply brakes to slow the motion. Determine the acceleration of each cable at that instant, knowing that $T_A = 1900$ N and $T_B = 1600$ N.

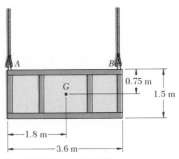

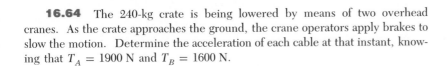

Fig. P16.63 and P16.64

16.65 For the disk and cord of Sample Prob. 16.4, determine the magnitude of the force **T** and the corresponding angular acceleration of the disk, knowing that the acceleration of the center of the disk is 0.8 m/s² (*a*) downward, (*b*) upward.

16.66 By pulling on the cord of a yo-yo just fast enough, a person manages to make the yo-yo spin counterclockwise, while remaining at a constant height above the floor. Denoting the weight of the yo-yo by W, the radius of the inner drum on which the cord is wound by r, and the radius of gyration of the yo-yo by $\bar{k}$, determine (*a*) the tension in the cord, (*b*) the angular acceleration of the yo-yo.

16.67 The steel roll shown weighs 2800 lb, has a centroidal radius of gyration of 6 in., and is lifted by two cables looped around its shaft. Knowing that for each cable $T_A = 690$ lb and $T_B = 730$ lb, determine (*a*) the angular acceleration of the roll, (*b*) the acceleration of its mass center.

Fig. P16.66

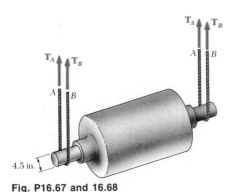

Fig. P16.67 and 16.68

16.68 The steel roll shown weighs 2800 lb, has a centroidal radius of gyration of 6 in., and is being lowered by two cables looped around its shaft. Knowing that at the instant shown the acceleration of the roll is 6 in./s² downward and that for each cable $T_A = 680$ lb, determine (*a*) the corresponding value of the tension T_B, (*b*) the angular acceleration of the roll.

16.69 through 16.72 A uniform slender bar *AB* of mass *m* is suspended from two springs as shown. If spring 2 breaks, determine at that instant (*a*) the angular acceleration of the bar, (*b*) the acceleration of point *A*, (*c*) the acceleration of point *B*.

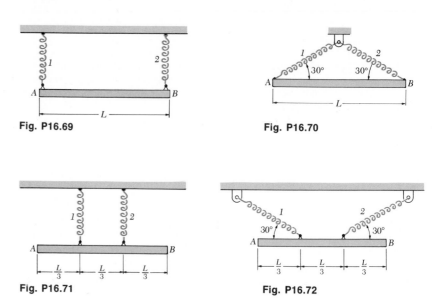

Fig. P16.69   Fig. P16.70

Fig. P16.71   Fig. P16.72

Fig. P16.73

16.73 A sphere of radius *r* and mass *m* is placed on a horizontal surface with no linear velocity but with a clockwise angular velocity ω_0. Denoting by μ_k the coefficient of kinetic friction between the sphere and the floor, determine (*a*) the time t_1 at which the sphere will start rolling without sliding, (*b*) the linear and angular velocities of the sphere at time t_1.

16.74 Solve Prob. 16.73, assuming that the sphere is replaced by a uniform disk of radius *r* and mass *m*.

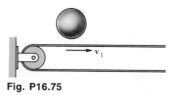

Fig. P16.75

16.75 A uniform sphere of radius *r* and mass *m* is placed with no initial velocity on a belt which moves to the right with a constant velocity v_1. Denoting by μ_k the coefficient of kinetic friction between the sphere and the belt, determine (*a*) the time t_1 at which the sphere will start rolling without sliding, (*b*) the linear velocity and the angular velocity of the sphere at time t_1.

16.76 Solve Prob. 16.75, assuming that the sphere is replaced by a wheel of radius *r*, mass *m*, and centroidal radius of gyration $\bar{k}$.

16.8. Constrained Plane Motion. Most engineering applications deal with rigid bodies which are moving under given constraints. Cranks, for example, must rotate about a fixed axis, wheels must roll without sliding, connecting rods must describe certain prescribed motions. In all such cases, definite relations exist between the components of the acceleration $\bar{a}$ of the mass center G of the body considered and its angular acceleration α; the corresponding motion is said to be a *constrained motion*.

The solution of a problem involving a constrained plane motion calls first for a *kinematic analysis* of the problem. Consider, for example, a slender rod AB of length l and mass m whose extremities are connected to blocks of negligible mass which slide along horizontal and vertical frictionless tracks. The rod is pulled by a force $\mathbf{P}$ applied at A (Fig. 16.11). We know from Sec. 15.8 that the acceleration $\bar{a}$ of the mass center G of the rod may be determined at any given instant from the position of the rod, its angular velocity, and its angular acceleration at that instant. Suppose, for instance, that the values of θ, ω, and α are known at a given instant and that we wish to determine the corresponding value of the force $\mathbf{P}$, as well as the reactions at A and B. We should first *determine the components $\bar{a}_x$ and $\bar{a}_y$ of the acceleration of the mass center G by the method of Sec. 15.8*. We next apply d'Alembert's principle (Fig. 16.12), using the expressions obtained for $\bar{a}_x$ and $\bar{a}_y$. The unknown forces $\mathbf{P}$, $\mathbf{N}_A$, and $\mathbf{N}_B$ may then be determined by writing and solving the appropriate equations.

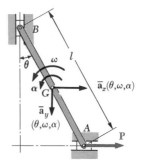

Fig. 16.11

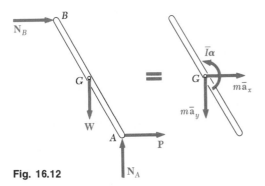

Fig. 16.12

Suppose now that the applied force $\mathbf{P}$, the angle θ, and the angular velocity ω of the rod are known at a given instant and that we wish to find the angular acceleration α of the rod and the components $\bar{a}_x$ and $\bar{a}_y$ of the acceleration of its mass center at that instant, as well as the reactions at A and B. The preliminary kinematic study of the problem will have for its object *to express the components $\bar{a}_x$ and $\bar{a}_y$ of the acceleration of G in terms of the angular acceleration α of the rod*. This will be done by first expressing the acceleration of a suitable reference point such as A in terms of the angular acceleration α. The components $\bar{a}_x$ and $\bar{a}_y$ of the acceleration of G may then be determined in terms of α, and the expressions obtained carried into Fig. 16.12. Three equations may then be derived in terms of α, N_A, and N_B, and solved for the three unknowns (see Sample Prob. 16.10). Note that the method of dynamic equilibrium may also be used to carry out the solution of the two types of problems we have considered (Fig. 16.13).

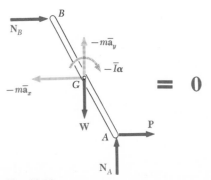

Fig. 16.13

When a mechanism consists of *several moving parts*, the approach just described may be used with each part of the mechanism. The procedure required to determine the various unknowns is then similar to the procedure followed in the case of the equilibrium of a system of connected rigid bodies (Sec. 6.11).

We have analyzed earlier two particular cases of constrained plane motion, the translation of a rigid body, in which the angular acceleration of the body is constrained to be zero, and the centroidal rotation, in which the acceleration $\bar{a}$ of the mass center of the body is constrained to be zero. Two other particular cases of constrained plane motion are of special interest, the *noncentroidal rotation* of a rigid body and the *rolling motion* of a disk or wheel. These two cases should be analyzed by one of the general methods described above. However, in view of the range of their applications, they deserve a few special comments.

Noncentroidal Rotation. This is the motion of a rigid body constrained to rotate about a fixed axis which does not pass through its mass center. Such a motion is called a *noncentroidal rotation*. The mass center G of the body moves along a circle of radius $\bar{r}$ centered at the point O, where the axis of rotation intersects the plane of reference (Fig. 16.14). Denoting, respectively, by ω and α the angular velocity and the angular acceleration of the line OG, we obtain the following expressions for the tangential and normal components of the acceleration of G:

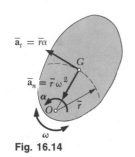

Fig. 16.14

$$\bar{a}_t = \bar{r}\alpha \qquad \bar{a}_n = \bar{r}\omega^2 \tag{16.7}$$

Since line OG belongs to the body, its angular velocity ω and its angular acceleration α also represent the angular velocity and the angular acceleration of the body in its motion relative to G. Equations (16.7) define, therefore, the kinematic relation existing between the motion of the mass center G and the motion of the body about G. They should be used to eliminate $\bar{a}_t$ and $\bar{a}_n$ from the equations obtained by applying d'Alembert's principle (Fig. 16.15) or the method of dynamic equilibrium (Fig. 16.16).

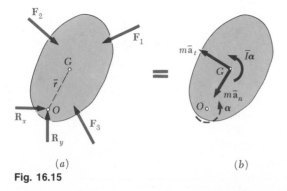

(a)   (b)

Fig. 16.15

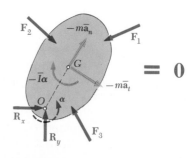

Fig. 16.16

An interesting relation may be obtained by equating the moments about the fixed point O of the forces and vectors shown, respectively, in parts a and b of Fig. 16.15. We write

$$+\uparrow\Sigma M_O = \bar{I}\alpha + (m\bar{r}\alpha)\bar{r} = (\bar{I} + m\bar{r}^2)\alpha$$

But according to the parallel-axis theorem, we have $\bar{I} + m\bar{r}^2 = I_O$, where I_O denotes the moment of inertia of the rigid body about the fixed axis. We therefore write

$$\Sigma M_O = I_O\alpha \tag{16.8}$$

Although formula (16.8) expresses an important relation between the sum of the moments of the external forces about the fixed point O and the product $I_O\alpha$, it should be clearly understood that this formula *does not mean* that the system of the external forces is equivalent to a couple of moment $I_O\alpha$. The system of the effective forces, and thus the system of the external forces, reduces to a couple only when O coincides with G—that is, *only when the rotation is centroidal* (Sec. 16.4). In the more general case of noncentroidal rotation, the system of the external forces does not reduce to a couple.

A particular case of noncentroidal rotation is of special interest—the case of *uniform rotation*, in which the angular velocity ω is constant. Since α is zero, the inertia couple in Fig. 16.16 vanishes and the inertia vector reduces to its normal component. This component (also called *centrifugal force*) represents the tendency of the rigid body to break away from the axis of rotation.

Rolling Motion. Another important case of plane motion is the motion of a disk or wheel rolling on a plane surface. If the disk is constrained to roll without sliding, the acceleration $\bar{a}$ of its mass center G and its angular acceleration α are not independent. Assuming that the disk is balanced, so that its mass center and its geometric center coincide, we first write that the distance $\bar{x}$ traveled by G during a rotation θ of the disk is $\bar{x} = r\theta$, where r is the radius of the disk. Differentiating this relation twice, we write

$$\bar{a} = r\alpha \tag{16.9}$$

Recalling that the system of the effective forces in plane motion reduces to a vector $m\bar{a}$ and a couple $\bar{I}\alpha$, we find that in the particular case of the rolling motion of a balanced disk, the effective forces reduce to a vector of magnitude $mr\alpha$ attached at G and to a couple of magnitude $\bar{I}\alpha$. We may thus express that the external forces are equivalent to the vector and couple shown in Fig. 16.17.

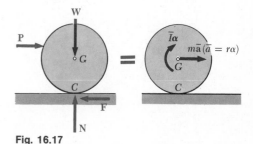

Fig. 16.17

When a disk *rolls without sliding*, there is no relative motion between the point of the disk in contact with the ground and the ground itself. As far as the computation of the friction force **F** is concerned, a rolling disk may thus be compared with a block at rest on a surface. The magnitude F of the friction force may have any value, as long as this value does not exceed the maximum value $F_m = \mu_s N$, where μ_s is the coefficient of static friction and N the magnitude of the normal force. In the case of a rolling disk, the magnitude F of the friction force should therefore be determined independently of N by solving the equation obtained from Fig. 16.17.

When *sliding is impending*, the friction force reaches its maximum value $F_m = \mu_s N$ and may be obtained from N.

When the disk *rotates and slides* at the same time, a relative motion exists between the point of the disk which is in contact with the ground and the ground itself, and the force of friction has the magnitude $F_k = \mu_k N$, where μ_k is the coefficient of kinetic friction. In this case, however, the motion of the mass center G of the disk and the rotation of the disk about G are independent, and $\bar{a}$ is not equal to $r\alpha$.

These three different cases may be summarized as follows:

Rolling, no sliding: $F \le \mu_s N$ $\bar{a} = r\alpha$

Rolling, sliding impending: $F = \mu_s N$ $\bar{a} = r\alpha$

Rotating and sliding: $F = \mu_k N$ $\bar{a}$ *and* α *independent*

When it is not known whether or not a disk slides, it should first be assumed that the disk rolls without sliding. If F is found smaller than, or equal to, $\mu_s N$, the assumption is proved correct. If F is found larger than $\mu_s N$, the assumption is incorrect and the problem should be started again, assuming rotating and sliding.

When a disk is *unbalanced*, i.e., when its mass center G does not coincide with its geometric center O, the relation (16.9) does not hold between $\bar{a}$ and α. However, a similar relation holds between the magnitude a_O of the acceleration of the geometric center and the angular acceleration α of an unbalanced disk which rolls without sliding. We have

$$a_O = r\alpha \tag{16.10}$$

To determine $\bar{a}$ in terms of the angular acceleration α and the angular velocity ω of the disk, we may use the relative-acceleration formula

$$\begin{aligned} \bar{\mathbf{a}} = \mathbf{a}_G &= \mathbf{a}_O + \mathbf{a}_{G/O} \\ &= \mathbf{a}_O + (\mathbf{a}_{G/O})_t + (\mathbf{a}_{G/O})_n \end{aligned} \tag{16.11}$$

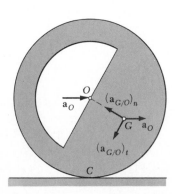

Fig. 16.18

where the three component accelerations obtained have the directions indicated in Fig. 16.18 and the magnitudes $a_O = r\alpha$, $(a_{G/O})_t = (OG)\alpha$, and $(a_{G/O})_n = (OG)\omega^2$.

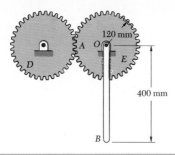

SAMPLE PROBLEM 16.6

The portion AOB of a mechanism consists of a 400-mm steel rod OB welded to a gear E of radius 120 mm which may rotate about a horizontal shaft O. It is actuated by a gear D and, at the instant shown, has a clockwise angular velocity of 8 rad/s and a counterclockwise angular acceleration of 40 rad/s². Knowing that rod OB has a mass of 3 kg and gear E a mass of 4 kg and a radius of gyration of 85 mm, determine (a) the tangential force exerted by gear D on gear E, (b) the components of the reaction at shaft O.

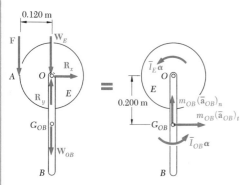

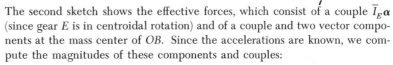

Solution. In determining the effective forces of the rigid body AOB we shall consider separately gear E and rod OB. Therefore, we shall first determine the components of the acceleration of the mass center G_{OB} of the rod:

$$(\bar{a}_{OB})_t = \bar{r}\alpha = (0.200 \text{ m})(40 \text{ rad/s}^2) = 8 \text{ m/s}^2$$
$$(\bar{a}_{OB})_n = \bar{r}\omega^2 = (0.200 \text{ m})(8 \text{ rad/s})^2 = 12.8 \text{ m/s}^2$$

Equations of Motion. Two sketches of the rigid body AOB have been drawn. The first shows the external forces consisting of the weight $\mathbf{W}_E$ of gear E, the weight $\mathbf{W}_{OB}$ of the rod OB, the force $\mathbf{F}$ exerted by gear D, and the components $\mathbf{R}_x$ and $\mathbf{R}_y$ of the reaction at O. The magnitudes of the weights are, respectively,

$$W_E = m_E g = (4 \text{ kg})(9.81 \text{ m/s}^2) = 39.2 \text{ N}$$
$$W_{OB} = m_{OB} g = (3 \text{ kg})(9.81 \text{ m/s}^2) = 29.4 \text{ N}$$

The second sketch shows the effective forces, which consist of a couple $\bar{I}_E \alpha$ (since gear E is in centroidal rotation) and of a couple and two vector components at the mass center of OB. Since the accelerations are known, we compute the magnitudes of these components and couples:

$$\bar{I}_E \alpha = m_E \bar{k}_E^2 \alpha = (4 \text{ kg})(0.085 \text{ m})^2(40 \text{ rad/s}^2) = 1.156 \text{ N} \cdot \text{m}$$
$$m_{OB}(\bar{a}_{OB})_t = (3 \text{ kg})(8 \text{ m/s}^2) = 24.0 \text{ N}$$
$$m_{OB}(\bar{a}_{OB})_n = (3 \text{ kg})(12.8 \text{ m/s}^2) = 38.4 \text{ N}$$
$$\bar{I}_{OB} \alpha = (\tfrac{1}{12} m_{OB} L^2)\alpha = \tfrac{1}{12}(3 \text{ kg})(0.400 \text{ m})^2(40 \text{ rad/s}^2) = 1.600 \text{ N} \cdot \text{m}$$

Expressing that the system of the external forces is equivalent to the system of the effective forces, we write the following equations:

$+\!\uparrow\Sigma M_O = \Sigma(M_O)_{\text{eff}}$:

$$F(0.120 \text{ m}) = \bar{I}_E \alpha + m_{OB}(\bar{a}_{OB})_t(0.200 \text{ m}) + \bar{I}_{OB}\alpha$$
$$F(0.120 \text{ m}) = 1.156 \text{ N} \cdot \text{m} + (24.0 \text{ N})(0.200 \text{ m}) + 1.600 \text{ N} \cdot \text{m}$$

$$F = 63.0 \text{ N} \qquad\qquad F = 63.0 \text{ N} \downarrow \quad \blacktriangleleft$$

$\xrightarrow{+} \Sigma F_x = \Sigma(F_x)_{\text{eff}}$:

$$R_x = m_{OB}(\bar{a}_{OB})_t$$
$$R_x = 24.0 \text{ N} \qquad\qquad R_x = 24.0 \text{ N} \rightarrow \quad \blacktriangleleft$$

$+\!\uparrow\Sigma F_y = \Sigma(F_y)_{\text{eff}}$:

$$R_y - F - W_E - W_{OB} = m_{OB}(\bar{a}_{OB})_n$$
$$R_y - 63.0 \text{ N} - 39.2 \text{ N} - 29.4 \text{ N} = 38.4 \text{ N}$$

$$R_y = 170.0 \text{ N} \qquad\qquad R_y = 170.0 \text{ N} \uparrow \quad \blacktriangleleft$$

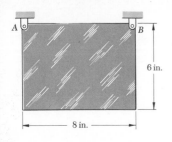

SAMPLE PROBLEM 16.7

A 6×8 in. rectangular plate weighing 60 lb is suspended from two pins A and B. If pin B is suddenly removed, determine (a) the angular acceleration of the plate, (b) the components of the reaction at pin A, immediately after pin B has been removed.

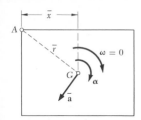

a. Angular Acceleration. We observe that as the plate rotates about point A, its mass center G describes a circle of radius $\bar{r}$ with center at A.

Since the plate is released from rest ($\omega = 0$), the normal component of the acceleration of G is zero. The magnitude of the acceleration $\bar{a}$ of the mass center G is thus $\bar{a} = \bar{r}\alpha$. We draw the diagram shown to express that the external forces are equivalent to the effective forces:

$$+\downarrow \Sigma M_A = \Sigma(M_A)_{\text{eff}}: \qquad W\bar{x} = (m\bar{a})\bar{r} + \bar{I}\alpha$$

Since $\bar{a} = \bar{r}\alpha$, we have

$$W\bar{x} = m(\bar{r}\alpha)\bar{r} + \bar{I}\alpha \qquad \alpha = \frac{W\bar{x}}{\dfrac{W}{g}\bar{r}^2 + \bar{I}} \qquad (1)$$

The centroidal moment of inertia of the plate is

$$\bar{I} = \frac{m}{12}(a^2 + b^2) = \frac{60\text{ lb}}{12(32.2\text{ ft/s}^2)}[(\tfrac{8}{12}\text{ ft})^2 + (\tfrac{6}{12}\text{ ft})^2]$$

$$= 0.1078\text{ lb}\cdot\text{ft}\cdot\text{s}^2$$

Substituting this value of $\bar{I}$ together with $W = 60$ lb, $\bar{r} = \tfrac{5}{12}$ ft, and $\bar{x} = \tfrac{4}{12}$ ft into Eq. (1), we obtain

$$\alpha = +46.4\text{ rad/s}^2 \qquad \boldsymbol{\alpha} = 46.4\text{ rad/s}^2 \downarrow \quad \blacktriangleleft$$

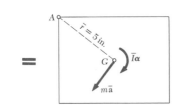

b. Reaction at A. Using the computed value of α, we determine the magnitude of the vector $m\bar{a}$ attached at G,

$$m\bar{a} = m\bar{r}\alpha = \frac{60\text{ lb}}{32.2\text{ ft/s}^2}(\tfrac{5}{12}\text{ ft})(46.4\text{ rad/s}^2) = 36.0\text{ lb}$$

Showing this result on the diagram, we write the equations of motion

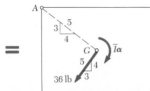

$$\xrightarrow{+} \Sigma F_x = \Sigma(F_x)_{\text{eff}}: \qquad A_x = -\tfrac{3}{5}(36\text{ lb})$$
$$= -21.6\text{ lb} \qquad\qquad \mathbf{A}_x = 21.6\text{ lb} \leftarrow \quad \blacktriangleleft$$

$$+\uparrow \Sigma F_y = \Sigma(F_y)_{\text{eff}}: \qquad A_y - 60\text{ lb} = -\tfrac{4}{5}(36\text{ lb})$$
$$A_y = +31.2\text{ lb} \qquad\qquad \mathbf{A}_y = 31.2\text{ lb} \uparrow \quad \blacktriangleleft$$

The couple $\bar{I}\boldsymbol{\alpha}$ is not involved in the last two equations; nevertheless, it should be indicated on the diagram.

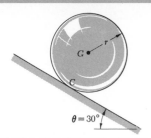

SAMPLE PROBLEM 16.8

A sphere of radius r and weight W is released with no initial velocity on the incline and rolls without slipping. Determine (a) the minimum value of the coefficient of static friction compatible with the rolling motion, (b) the velocity of the center G of the sphere after the sphere has rolled 10 ft, (c) the velocity of G if the sphere were to move 10 ft down a frictionless 30° incline.

$\theta = 30°$

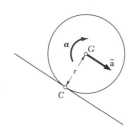

a. **Minimum μ_s for Rolling Motion.** The external forces **W**, **N**, and **F** form a system equivalent to the system of effective forces represented by the vector $m\bar{a}$ and the couple $\bar{I}\alpha$. Since the sphere rolls without sliding, we have $\bar{a} = r\alpha$.

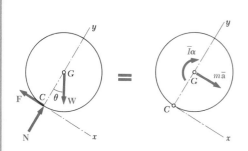

$+\downarrow \Sigma M_C = \Sigma(M_C)_{\text{eff}}$: $(W \sin \theta)r = (m\bar{a})r + \bar{I}\alpha$
$\qquad\qquad\qquad\qquad\qquad (W \sin \theta)r = (mr\alpha)r + \bar{I}\alpha$

Noting that $m = W/g$ and $\bar{I} = \frac{2}{5}mr^2$, we write

$$(W \sin \theta)r = \left(\frac{W}{g}r\alpha\right)r + \frac{2}{5}\frac{W}{g}r^2\alpha \qquad \alpha = +\frac{5g \sin \theta}{7r}$$

$$\bar{a} = r\alpha = \frac{5g \sin \theta}{7} = \frac{5(32.2 \text{ ft/s}^2) \sin 30°}{7} = 11.50 \text{ ft/s}^2$$

$+\searrow \Sigma F_x = \Sigma(F_x)_{\text{eff}}$: $\qquad W \sin \theta - F = m\bar{a}$

$$W \sin \theta - F = \frac{W}{g}\frac{5g \sin \theta}{7}$$

$F = +\frac{2}{7}W \sin \theta = \frac{2}{7}W \sin 30° \qquad \mathbf{F} = 0.143W \searrow 30°$

$+\nearrow \Sigma F_y = \Sigma(F_y)_{\text{eff}}$: $\qquad N - W \cos \theta = 0$
$\qquad\qquad\qquad N = W \cos \theta = 0.866W \qquad \mathbf{N} = 0.866W \measuredangle 60°$

$$\mu_s = \frac{F}{N} = \frac{0.143W}{0.866W} \qquad\qquad \mu_s = 0.165 \blacktriangleleft$$

b. **Velocity of Rolling Sphere.** We have uniformly accelerated motion:

$$\bar{v}_0 = 0 \qquad \bar{a} = 11.50 \text{ ft/s}^2 \qquad \bar{x} = 10 \text{ ft} \qquad \bar{x}_0 = 0$$
$$\bar{v}^2 = \bar{v}_0^2 + 2\bar{a}(\bar{x} - \bar{x}_0) \qquad \bar{v}^2 = 0 + 2(11.50 \text{ ft/s}^2)(10 \text{ ft})$$
$$\bar{v} = 15.17 \text{ ft/s} \qquad \bar{v} = 15.17 \text{ ft/s} \searrow 30° \blacktriangleleft$$

c. **Velocity of Sliding Sphere.** Assuming now no friction, we have $F = 0$ and obtain

$+\downarrow \Sigma M_G = \Sigma(M_G)_{\text{eff}}$: $\qquad 0 = \bar{I}\alpha \qquad \alpha = 0$

$+\searrow \Sigma F_x = \Sigma(F_x)_{\text{eff}}$: $\qquad W \sin 30° = m\bar{a} \qquad 0.50W = \frac{W}{g}\bar{a}$

$$\bar{a} = +16.1 \text{ ft/s}^2 \qquad \bar{\mathbf{a}} = 16.1 \text{ ft/s}^2 \searrow 30°$$

Substituting $\bar{a} = 16.1 \text{ ft/s}^2$ into the equations for uniformly accelerated motion, we obtain

$$\bar{v}^2 = \bar{v}_0^2 + 2\bar{a}(\bar{x} - \bar{x}_0) \qquad \bar{v}^2 = 0 + 2(16.1 \text{ ft/s}^2)(10 \text{ ft})$$
$$\bar{v} = 17.94 \text{ ft/s} \qquad \bar{v} = 17.94 \text{ ft/s} \searrow 30° \blacktriangleleft$$

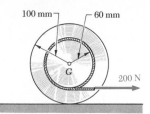

100 mm — 60 mm

G

200 N

A cord is wrapped around the inner drum of a wheel and pulled horizontally with a force of 200 N. The wheel has a mass of 50 kg and a radius of gyration of 70 mm. Knowing that $\mu_s = 0.20$ and $\mu_k = 0.15$, determine the acceleration of G and the angular acceleration of the wheel.

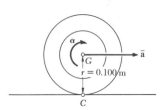

α

G

$\bar{a}$

$r = 0.100\ m$

C

a. **Assume Rolling without Sliding.** In this case, we have

$$\bar{a} = r\alpha = (0.100\ \text{m})\alpha$$

By comparing the friction force obtained with the maximum available friction force, we shall determine whether this assumption is justified. The moment of inertia of the wheel is

$$\bar{I} = m\bar{k}^2 = (50\ \text{kg})(0.070\ \text{m})^2 = 0.245\ \text{kg·m}^2$$

Equations of Motion

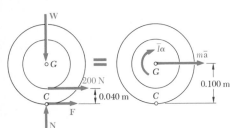

W

G = $\bar{I}\alpha$ G $m\bar{a}$

200 N 0.100 m

C C

0.040 m

F

N

$+\downarrow \Sigma M_C = \Sigma(M_C)_{\text{eff}}$: $(200\ \text{N})(0.040\ \text{m}) = m\bar{a}(0.100\ \text{m}) + \bar{I}\alpha$

$8.00\ \text{N·m} = (50\ \text{kg})(0.100\ \text{m})\alpha(0.100\ \text{m}) + (0.245\ \text{kg·m}^2)\alpha$

$\alpha = +10.74\ \text{rad/s}^2$

$\bar{a} = r\alpha = (0.100\ \text{m})(10.74\ \text{rad/s}^2) = 1.074\ \text{m/s}^2$

$\xrightarrow{+} \Sigma F_x = \Sigma(F_x)_{\text{eff}}$: $F + 200\ \text{N} = m\bar{a}$

$F + 200\ \text{N} = (50\ \text{kg})(1.074\ \text{m/s}^2)$

$F = -146.3\ \text{N}$ $\mathbf{F} = 146.3\ \text{N} \leftarrow$

$+\uparrow \Sigma F_y = \Sigma(F_y)_{\text{eff}}$:

$N - W = 0$ $N - W = mg = (50\ \text{kg})(9.81\ \text{m/s}^2) = 490.5\ \text{N}$

$\mathbf{N} = 490.5\ \text{N} \uparrow$

Maximum Available Friction Force

$$F_{\text{max}} = \mu_s N = 0.20(490.5\ \text{N}) = 98.1\ \text{N}$$

Since $F > F_{\text{max}}$, the assumed motion is impossible.

b. **Rotating and Sliding.** Since the wheel must rotate and slide at the same time, we draw a new diagram, where $\bar{a}$ and α are independent and where

$$F = F_k = \mu_k N = 0.15(490.5\ \text{N}) = 73.6\ \text{N}$$

From the computation of part *a*, it appears that $\mathbf{F}$ should be directed to the left. We write the following equations of motion:

W

0.060 m

G = $\bar{I}\alpha$ G $m\bar{a}$

200 N 0.100 m

C C

$F = 73.6\ \text{N}$

N

$\xrightarrow{+} \Sigma F_x = \Sigma(F_x)_{\text{eff}}$: $200\ \text{N} - 73.6\ \text{N} = (50\ \text{kg})\bar{a}$

$\bar{a} = +2.53\ \text{m/s}^2$ $\bar{\mathbf{a}} = 2.53\ \text{m/s}^2 \rightarrow$ ◄

$+\downarrow \Sigma M_G = \Sigma(M_G)_{\text{eff}}$:

$(73.6\ \text{N})(0.100\ \text{m}) - (200\ \text{N})(0.060\ \text{m}) = (0.245\ \text{kg·m}^2)\alpha$

$\alpha = -18.94\ \text{rad/s}^2$ $\boldsymbol{\alpha} = 18.94\ \text{rad/s}^2 \uparrow$ ◄

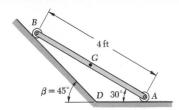

The extremities of a 4-ft rod, weighing 50 lb, may move freely and with no friction along two straight tracks as shown. If the rod is released with no velocity from the position shown, determine (a) the angular acceleration of the rod, (b) the reactions at A and B.

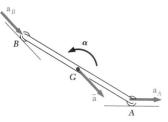

Kinematics of Motion. Since the motion is constrained, the acceleration of G must be related to the angular acceleration $\boldsymbol{\alpha}$. To obtain this relation, we shall first determine the magnitude of the acceleration $\mathbf{a}_A$ of point A in terms of α; assuming $\boldsymbol{\alpha}$ directed counterclockwise and noting that $a_{B/A} = 4\alpha$, we write

$$\mathbf{a}_B = \mathbf{a}_A + \mathbf{a}_{B/A}$$

$$[a_B \searrow 45°] = [a_A \rightarrow] + [4\alpha \nearrow 60°]$$

Noting that $\phi = 75°$ and using the law of sines, we obtain

$$a_A = 5.46\alpha \qquad a_B = 4.90\alpha$$

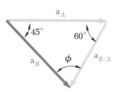

The acceleration of G is now obtained by writing

$$\bar{\mathbf{a}} = \mathbf{a}_G = \mathbf{a}_A + \mathbf{a}_{G/A}$$

$$\bar{\mathbf{a}} = [5.46\alpha \rightarrow] + [2\alpha \nearrow 60°]$$

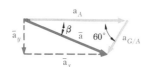

Resolving $\bar{\mathbf{a}}$ into x and y components, we obtain

$$\bar{a}_x = 5.46\alpha - 2\alpha \cos 60° = 4.46\alpha \qquad \bar{a}_x = 4.46\alpha \rightarrow$$
$$\bar{a}_y = -2\alpha \sin 60° = -1.732\alpha \qquad \bar{a}_y = 1.732\alpha \downarrow$$

Kinetics of Motion. We draw the two sketches shown to express that the system of external forces is equivalent to the system of effective forces represented by the vector of components $m\bar{a}_x$ and $m\bar{a}_y$ attached at G and the couple $\bar{I}\alpha$. We compute the following magnitudes:

$$\bar{I} = \tfrac{1}{12}ml^2 = \frac{1}{12}\frac{50 \text{ lb}}{32.2 \text{ ft/s}^2}(4 \text{ ft})^2 = 2.07 \text{ lb}\cdot\text{ft}\cdot\text{s}^2 \qquad \bar{I}\alpha = 2.07\alpha$$

$$m\bar{a}_x = \frac{50}{32.2}(4.46\alpha) = 6.93\alpha \qquad m\bar{a}_y = -\frac{50}{32.2}(1.732\alpha) = -2.69\alpha$$

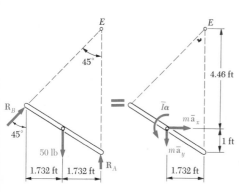

Equations of Motion

$+\uparrow \Sigma M_E = \Sigma(M_E)_{\text{eff}}$:
$$(50)(1.732) = (6.93\alpha)(4.46) + (2.69\alpha)(1.732) + 2.07\alpha$$
$$\alpha = +2.30 \text{ rad/s}^2 \qquad \alpha = 2.30 \text{ rad/s}^2 \uparrow \quad \blacktriangleleft$$

$\xrightarrow{+} \Sigma F_x = \Sigma(F_x)_{\text{eff}}$:
$$R_B \sin 45° = (6.93)(2.30) = 15.94$$
$$R_B = 22.5 \text{ lb} \qquad R_B = 22.5 \text{ lb} \measuredangle 45° \quad \blacktriangleleft$$

$+\uparrow \Sigma F_y = \Sigma(F_y)_{\text{eff}}$:
$$R_A + R_B \cos 45° - 50 = -(2.69)(2.30)$$
$$R_A = -6.19 - 15.94 + 50 = 27.9 \text{ lb} \qquad R_A = 27.9 \text{ lb} \uparrow \quad \blacktriangleleft$$

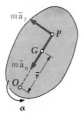

Fig. P16.77

Problems

16.77 Show that the couple $\bar{I}\alpha$ of Fig. 16.15 may be eliminated by attaching the vectors $m\bar{a}_t$ and $m\bar{a}_n$ at a point P called the *center of percussion*, located on line OG at a distance $GP = \bar{k}^2/\bar{r}$ from the mass center of the body.

16.78 A uniform slender rod, of length $L = 750$ mm and mass $m = 2$ kg, hangs freely from a hinge at A. If a force **P** of magnitude 12 N is applied at B horizontally to the left ($h = L$), determine (a) the angular acceleration of the rod, (b) the components of the reaction at A.

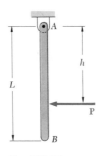

Fig. P16.78

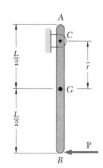

Fig. P16.79

16.79 A uniform slender rod, of length $L = 30$ in. and weight $W = 6$ lb, is supported as shown. A horizontal force **P** of magnitude 2.5 lb is applied at end B. For $\bar{r} = \frac{1}{4}L = 7.5$ in., determine (a) the angular acceleration of the rod, (b) the components of the reaction at C.

16.80 In Prob. 16.79, determine (a) the distance $\bar{r}$ for which the horizontal component of the reaction at C is zero, (b) the corresponding angular acceleration of the rod.

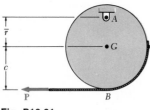

Fig. P16.81

16.81 A uniform disk, of radius $c = 160$ mm and mass $m = 6$ kg, hangs freely from a pin support at A. A force **P** of magnitude 20 N is applied as shown to a cord wrapped around the disk. For $\bar{r} = \frac{3}{4}c = 120$ mm, determine (a) the angular acceleration of the disk, (b) the components of the reaction at A.

16.82 In Prob. 16.81, determine (a) the distance $\bar{r}$ for which the horizontal component of the reaction at A is zero, (b) the corresponding angular acceleration of the disk.

16.83 In Prob. 16.78, determine (a) the distance h for which the horizontal component of the reaction at A is zero, (b) the corresponding angular acceleration of the rod.

16.84 A uniform slender rod of length l and mass m rotates about a vertical axis AA' with a constant angular velocity ω. Determine the tension in the rod at a distance x from the axis of rotation.

16.85 A large flywheel is mounted on a horizontal shaft and rotates at a constant rate of 1200 rpm. Experimental data show that the total force exerted by the flywheel on the shaft varies from 6000 lb upward to 9200 lb downward. Determine (a) the weight of the flywheel, (b) the distance from the center of the shaft to the mass center of the flywheel.

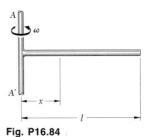

Fig. P16.84

16.86 A 120-mm-diameter hole is cut as shown in a thin disk of 600-mm diameter. The disk rotates in a horizontal plane about its geometric center A at the constant rate of 480 rpm. Knowing that the disk has a mass of 30 kg after the hole has been cut, determine the horizontal component of the force exerted by the shaft on the disk at A.

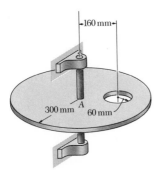

Fig. P16.86

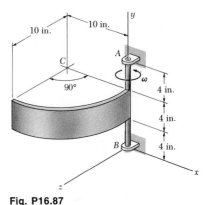

Fig. P16.87

16.87 A portion of a circular cylindrical shell forms a small vane which is welded to the vertical shaft AB. The vane and shaft rotate about the y axis with a constant angular velocity of 180 rpm counterclockwise. Knowing that the vane weighs 5 lb, determine the horizontal components of the reaction at A.

16.88 Centrifugal clutches of the type shown are used to control the operating speed of equipment such as movie cameras and dial telephones. Thin curved members AB and CD are connected by pins at A and C to the arm AC which may rotate about a fixed point O. Each of the members AB and CD has a mass of 4.5 g and a radius of 10 mm. As the clutch rotates counterclockwise, knobs H and K slide on the inside of a fixed cylindrical surface of radius 11 mm. Knowing that the coefficient of kinetic friction at H and K is 0.35 and that the clutch is to have a constant angular velocity of 3000 rpm, determine the couple $\mathbf{M}$ which must be applied to arm AC.

16.89 For the centrifugal clutch of Prob. 16.88, determine the constant angular velocity which will result from the application of a couple $\mathbf{M}$ of constant magnitude 30 N·mm.

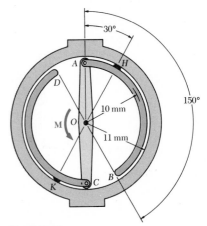

Fig. P16.88

16.90 and 16.91 A uniform beam of length L and weight W is supported as shown. If the cable suddenly breaks, determine (a) the acceleration of end B, (b) the reaction at the pin support.

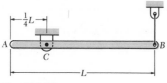

Fig. P16.90

Fig. P16.91

16.92 A uniform square plate of weight W is supported as shown. If the cable suddenly breaks, determine (a) the angular acceleration of the plate, (b) the acceleration of corner C, (c) the reaction at A.

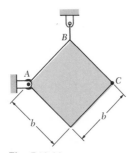

Fig. P16.92

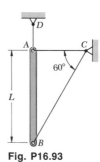

Fig. P16.93

16.93 A uniform slender rod AB of length L and weight W is supported as shown. If cable AD suddenly breaks, determine (a) the angular acceleration of the rod, (b) the tension in wires AC and BC.

16.94 Two uniform slender rods, AB of weight 6 lb and CD of weight 8 lb, are welded together to form a T-shaped assembly which swings freely about A in a vertical plane. Knowing that at the instant shown the assembly has a counterclockwise angular velocity of 8 rad/s, determine (a) the angular acceleration of the assembly, (b) the components of the reaction at A.

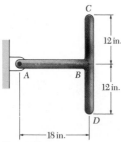

Fig. P16.94

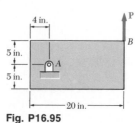

Fig. P16.95

16.95 A 12-lb uniform plate rotates about A in a vertical plane under the combined effect of gravity and of the vertical force $\mathbf{P}$. Knowing that at the instant shown the plate has an angular velocity of 20 rad/s and an angular acceleration of 30 rad/s² both counterclockwise, determine (a) the force $\mathbf{P}$, (b) the components of the reaction at A.

16.96 A 4-kg slender rod is welded to the edge of a 3-kg uniform disk as shown. The assembly rotates about A in a vertical plane under the combined effect of gravity and of the vertical force **P**. Knowing that at the instant shown the assembly has an angular velocity of 12 rad/s and an angular acceleration of 36 rad/s² both counterclockwise, determine (a) the force **P**, (b) the components of the reaction at A.

16.97 A 4-kg slender rod is welded to the edge of a 3-kg uniform disk as shown. The assembly swings freely about A in a vertical plane. Knowing that $P = 0$ and that in the position shown the assembly has an angular velocity of 15 rad/s counterclockwise, determine (a) the angular acceleration of the assembly, (b) the components of the reaction at A.

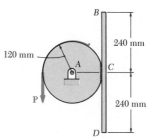

Fig. P16.96 and P16.97

16.98 Two uniform rods, ABC of mass 3 kg and DCE of mass 4 kg, are connected by a pin at C and by two cords BD and BE. The T-shaped assembly rotates in a vertical plane under the combined effect of gravity and of a couple **M** which is applied to rod ABC. Knowing that at the instant shown the tension is 8 N in cord BE and 2 N in cord BD, determine (a) the angular acceleration of the assembly, (b) the couple **M**.

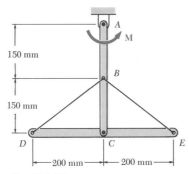

Fig. P16.98

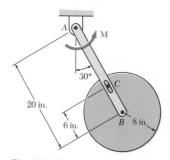

Fig. P16.99

16.99 A 20-lb uniform disk is attached to the 12-lb slender rod AB by means of frictionless pins at B and C. The assembly rotates in a vertical plane under the combined effect of gravity and of a couple **M** which is applied to rod AB. Knowing that at the instant shown the assembly has an angular velocity of 6 rad/s and an angular acceleration of 25 rad/s², both counterclockwise, determine (a) the couple **M**, (b) the force exerted by pin C on member AB.

16.100 Two slender rods, each of length l and mass m, are released from rest in the position shown. Knowing that a small frictionless knob at end B of rod AB bears on rod CD, determine immediately after release, (a) the acceleration of end C of rod CD, (b) the force exerted on knob B.

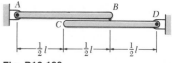

Fig. P16.100

16.101 Knowing that for the assembly of Prob. 16.98 the couple **M** is 7.55 N·m counterclockwise and the tension in cord BD is zero, determine (a) the angular acceleration of the assembly, (b) the tension in cord BE.

16.102 The uniform rod AB of mass m is released from rest when $\beta = 65°$. Assuming that the friction force between end A and the surface is large enough to prevent sliding, determine (a) the angular acceleration of the rod just after release, (b) the normal reaction and the friction force at A, (c) the minimum value of μ_s compatible with the described motion.

*16.103 Knowing that the coefficient of static friction between the rod and the floor is 0.35, determine the range of values of β for which the rod *will* slip immediately after being released from rest.

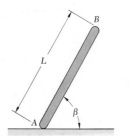

Fig. P16.102 and P16.103

16.104 Derive the equation $\Sigma M_C = I_C\alpha$ for the rolling disk of Fig. 16.17, where ΣM_C represents the sum of the moments of the external forces about the instantaneous center C and I_C the moment of inertia of the disk about C.

16.105 Show that in the case of an unbalanced disk, the equation derived in Prob. 16.104 is valid only when the mass center G, the geometric center O, and the instantaneous center C happen to lie in a straight line.

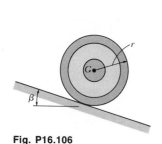

16.106 A wheel of radius r and centroidal radius of gyration $\overline{k}$ is released from rest on the incline and rolls without slipping. Derive an expression for the acceleration of the center of the wheel in terms of r, $\overline{k}$, β, and g.

Fig. P16.106

16.107 A flywheel is rigidly attached to a shaft of 60-mm radius which may roll along parallel rails as shown. When released from rest, the system rolls through a distance of 2.25 m in 15 s. Determine the centroidal radius of gyration of the system.

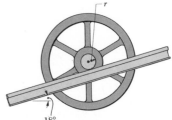

18°

Fig. P16.107 and P16.108

16.108 A flywheel of centroidal radius of gyration $\overline{k} = 500$ mm is rigidly attached to a shaft of radius $r = 40$ mm which may roll along parallel rails. Knowing that the system is released from rest, determine the distance through which it will roll in 20 s.

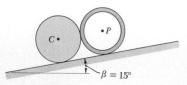

Fig. P16.109

16.109 A homogeneous cylinder C and a section of pipe P are in contact when they are released from rest. Knowing that both the cylinder and the pipe roll without slipping, determine the clear distance between them after 2.5 s.

16.110 through 16.113 A drum of 5-in. radius is attached to a disk of 10-in. radius. The disk and drum have a total weight of 16 lb and a radius of gyration of 7.5 in. A cord is attached as shown and pulled with a force **P** of magnitude 6 lb. Knowing that the disk rolls without sliding, determine (*a*) the angular acceleration of the disk and the acceleration of *G*, (*b*) the minimum value of the coefficient of static friction compatible with this motion.

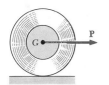

Fig. P16.110 and P16.114

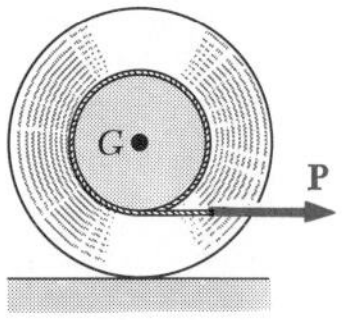

Fig. P16.111 and P16.115

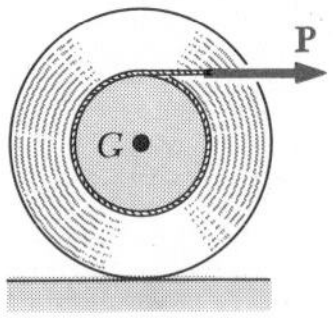

Fig. P16.112 and P16.116

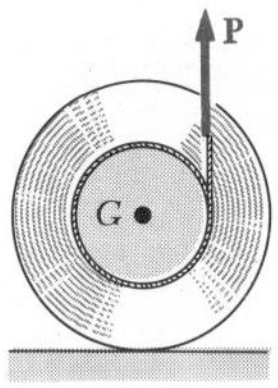

Fig. P16.113 and P16.117

16.114 through 16.117 A drum of 80-mm radius is attached to a disk of 160-mm radius. The disk and drum have a total mass of 5 kg and a radius of gyration of 120 mm. A cord is attached as shown and pulled with a force **P** of magnitude 18 N. Knowing that the coefficients of static and kinetic friction are $\mu_s = 0.20$ and $\mu_k = 0.15$, determine (*a*) whether or not the disk slides, (*b*) the angular acceleration of the disk and the acceleration of *G*.

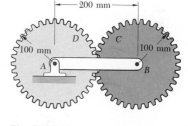

Fig. P16.118

16.118 and 16.119 Gear *C* has a mass of 5 kg and a centroidal radius of gyration of 75 mm. The uniform bar *AB* has a mass of 3 kg and gear *D* is stationary. If the system is released from rest in the position shown, determine (*a*) the angular acceleration of gear *C*, (*b*) the acceleration of point *B*.

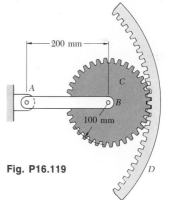

16.120 through 16.122 A bar of mass *m* is held as shown between four disks, each of mass *m'* and radius $r = 75$ mm. Determine the acceleration of the bar immediately after it has been released from rest, knowing that the normal forces exerted on the disks are sufficient to prevent any slipping and assuming that (*a*) $m = 5$ kg and $m' = 2$ kg, (*b*) the mass *m'* of the disks is negligible, (*c*) the mass *m* of the bar is negligible.

Fig. P16.119

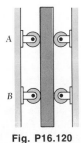

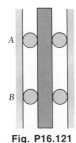

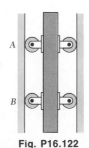

Fig. P16.120

Fig. P16.121

Fig. P16.122

16.123 A block B of mass m is attached to a cord wrapped around a cylinder of the same mass m and of radius r. The cylinder rolls without sliding on a horizontal surface. Determine the components of the accelerations of the center A of the cylinder and of the block B immediately after the system has been released from rest if (a) the block hangs freely, (b) the motion of the block is guided by a rigid member DAE, frictionless and of negligible mass, which is hinged to the cylinder at A.

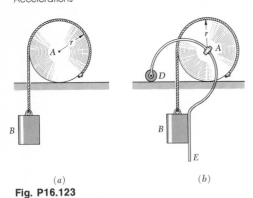

(a)   (b)

Fig. P16.123

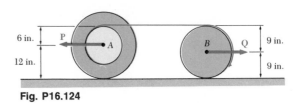

Fig. P16.124

16.124 The disk-and-drum assembly A has a total weight of 24 lb and a centroidal radius of gyration of 8 in. A cord is attached as shown to assembly A and to the 16-lb uniform disk B. Knowing that the disks roll without sliding, determine the acceleration of the center of each disk for $P = 0$ and $Q = 6$ lb.

16.125 Solve Prob. 16.124 for $P = 6$ lb and $Q = 0$.

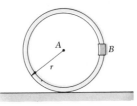

Fig. P16.126

16.126 A small block of mass m is attached at B to a hoop of mass m and radius r. The system is released from rest when B is directly above A and rolls without sliding. Knowing that at the instant shown the system has a clockwise angular velocity of magnitude $\omega = \sqrt{g/2r}$, determine (a) the angular acceleration of the hoop, (b) the acceleration of B.

16.127 The center of gravity G of a 4-lb unbalanced tracking wheel is located at a distance $r = 1.2$ in. from its geometric center B. The radius of the wheel is $R = 4$ in. and its centroidal radius of gyration is $\bar{k} = 3$ in. At the instant shown the center B of the wheel has a velocity of 2 ft/s and an acceleration of 5 ft/s², both directed to the left. Knowing that the wheel rolls without sliding and neglecting the weight of the driving yoke AB, determine the horizontal force $\mathbf{P}$ applied to the yoke.

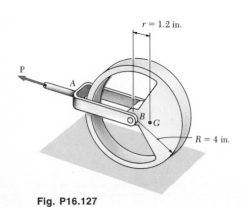

Fig. P16.127

16.128 A hemisphere of weight W and radius r is released from rest in the position shown. Determine (a) the minimum value of μ_s for which the hemisphere starts to roll without sliding, (b) the corresponding acceleration of point B. [*Hint*. Note that $OG = \frac{3}{8}r$ and that, by the parallel-axis theorem, $\bar{I} = \frac{2}{5}mr^2 - m(OG)^2$.]

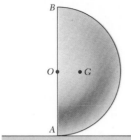

Fig. P16.128

16.129 Solve Prob. 16.128, considering a half cylinder instead of a hemisphere. [*Hint*. Note that $OG = 4r/3\pi$ and that, by the parallel-axis theorem, $\bar{I} = \frac{1}{2}mr^2 - m(OG)^2$.]

16.130 The ends of the 8-lb rod AB are attached to collars of negligible weight which slide without friction along fixed rods. If rod AB is released from rest in the position shown, determine immediately after release (a) the angular acceleration of the rod, (b) the reaction at A, (c) the reaction at B.

16.131 The motion of the 4-kg uniform rod ACB is guided by two blocks of negligible mass which slide without friction in the slots shown. If the rod is released from rest in the position shown, determine immediately after release (a) the angular acceleration of the rod, (b) the reaction at A.

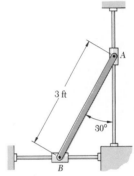

Fig. P16.130

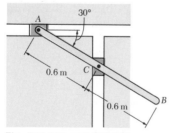

Fig. P16.131 and P16.132

16.132 The motion of the 4-kg uniform rod ACB is guided by two blocks of negligible mass which slide without friction in the slots shown. A horizontal force $\mathbf{P}$ is applied to block A, causing the rod to start from rest with a counterclockwise angular acceleration of 12 rad/s^2. Determine (a) the required force $\mathbf{P}$, (b) the corresponding reaction at A.

16.133 End B of the 15-lb uniform rod AB rests on a frictionless floor, while end A is attached to a horizontal cable AC. Knowing that at the instant shown the force $\mathbf{P}$ causes end B of the rod to start from rest with an acceleration of 9 ft/s^2 to the left, determine (a) the force $\mathbf{P}$, (b) the corresponding tension in cable AC.

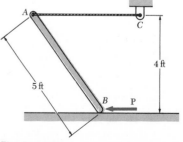

Fig. P16.133

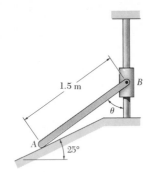

Fig. P16.134 and P16.135

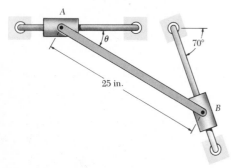

Fig. P16.136 and P16.137

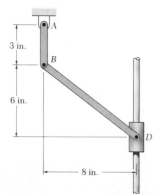

Fig. P16.140

16.134   End A of the 6-kg uniform rod AB rests on the inclined surface, while end B is attached to a collar of negligible mass which may slide along the vertical rod shown. Knowing that the rod is released from rest when $\theta = 35°$ and neglecting the effect of friction, determine immediately after release (*a*) the angular acceleration of the rod, (*b*) the reaction at B.

16.135   End A of the 6-kg uniform rod AB rests on the inclined surface, while end B is attached to a collar of negligible mass which may slide along the vertical rod shown. When the rod is at rest a vertical force $\mathbf{P}$ is applied at B, causing end B of the rod to start moving upward with an acceleration of 4 m/s². Knowing that $\theta = 35°$, determine the force $\mathbf{P}$.

16.136   The 4-lb uniform rod AB is attached to collars of negligible mass which may slide without friction along the fixed rods shown. Rod AB is at rest in the position $\theta = 25°$ when a horizontal force $\mathbf{P}$ is applied to collar A, causing it to start moving to the left with an acceleration of 12 ft/s². Determine (*a*) the force $\mathbf{P}$, (*b*) the reaction at B.

16.137   The 4-lb uniform rod AB is attached to collars of negligible mass which may slide without friction along the fixed rods shown. If rod AB is released from rest in the position $\theta = 25°$, determine immediately after release (*a*) the angular acceleration of the rod, (*b*) the reaction at B.

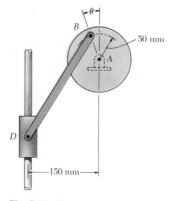

Fig. P16.138

16.138   The 250-mm uniform rod BD, of mass 5 kg, is connected as shown to disk A and to a collar of negligible mass, which may slide freely along a vertical rod. Knowing that disk A rotates counterclockwise at the constant rate of 500 rpm, determine the reaction at D when $\theta = 0$.

16.139   Solve Prob. 16.138 when $\theta = 90°$.

16.140   The 3-lb uniform rod BD is connected to crank AB and to a collar of negligible weight. A couple (not shown) is applied to crank AB causing it to rotate with an angular velocity of 12 rad/s counterclockwise and an angular acceleration of 80 rad/s² clockwise at the instant shown. Neglecting the effect of friction, determine the reaction at D.

16.141 In Prob. 16.140, determine the reaction at D, knowing that in the position shown crank AB has an angular velocity of 12 rad/s and an angular acceleration of 80 rad/s², both counterclockwise.

16.142 A driver starts his car with the door on the passenger's side wide open ($\theta = 0$). The door has a mass of 36 kg, a centroidal radius of gyration $\bar{k} = 300$ mm, and its mass center is located at a distance $\bar{r} = 525$ mm from its vertical axis of rotation. Knowing that the driver maintains a constant acceleration of 1.95 m/s², determine the angular velocity of the door as it slams shut ($\theta = 90°$).

16.143 For the car of Prob. 16.142, determine the smallest constant acceleration that the driver should maintain if the door is to close and to latch, knowing that it must hit the frame with an angular velocity of at least 1.8 rad/s for the latching mechanism to operate.

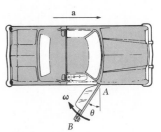

Fig. P16.142

16.144 In the engine system shown, $l = 8$ in., $b = 3$ in., the piston P weighs 4 lb, and the connecting rod BD is assumed to be a 5-lb uniform slender rod. During a test of the system, crank AB is made to rotate with a constant angular velocity of 2000 rpm clockwise with no force applied to the face of the piston. Determine the forces exerted on the connecting rod at B and D when $\theta = 0$. (Neglect the effect of the weight of the rod.)

16.145 Solve Prob. 16.144 when $\theta = 180°$.

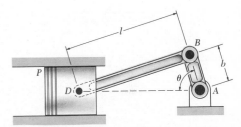

Fig. P16.144

16.146 and 16.147 Two rods AB and BC, of mass m' per unit length, are connected as shown to a disk which is made to rotate in a vertical plane at a constant angular velocity ω_0. For the position shown, determine the components of the forces exerted at A and B on rod AB.

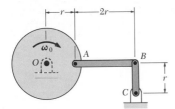

Fig. P16.146

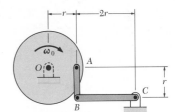

Fig. P16.147

*16.148 and *16.149 The 3-kg cylinder B and the 2-kg wedge A are held at rest in the position shown by cord C. Assuming that the cylinder rolls without sliding on the wedge and neglecting friction between the wedge and the ground, determine (a) the acceleration of the wedge, (b) the angular acceleration of the cylinder, immediately after cord C has been cut.

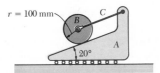

Fig. P16.148

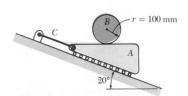

Fig. P16.149

*16.150 and *16.151 A uniform rod AB, of weight 30 lb and length 3 ft, is attached to the 40-lb cart C. Neglecting friction, determine immediately after the system has been released from rest, (a) the acceleration of the cart, (b) the angular acceleration of the rod.

Fig. P16.150

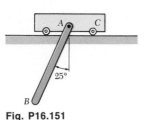

Fig. P16.151

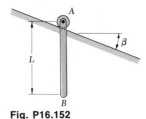

Fig. P16.152

*16.152 A uniform rod AB, of mass 4 kg and length $L = 1.5$ m, is released from rest in the position shown. Knowing that $\beta = 20°$, determine the values immediately after release of (a) the angular acceleration of the rod, (b) the acceleration of end A, (c) the reaction at A. Neglect the mass and friction of the roller at A.

*16.153 and *16.154 Each of the bars AB and BC is of length $L = 15$ in. and weight 4 lb. A horizontal force P of magnitude 3.5 lb is applied as shown. Determine the angular acceleration of each bar.

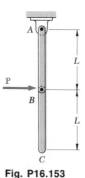

Fig. P16.153

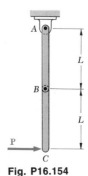

Fig. P16.154

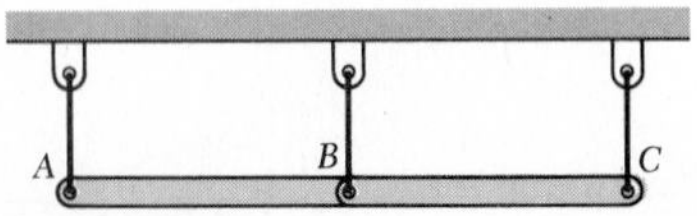

*16.155 Two identical uniform rods are connected by a pin at B and are held in a horizontal position by three wires as shown. If the wires attached at A and B are cut simultaneously, determine at that instant the acceleration of (a) point A, (b) point B.

Fig. P16.155

*16.156 (*a*) Determine the magnitude and the location of the maximum bending moment in the rod of Prob. 16.78. (*b*) Show that the answer to part *a* is independent of the mass of the rod.

*16.157 Draw the shear and bending-moment diagrams for the beam of Prob. 16.91 immediately after the cable at *B* breaks.

Review and Summary

In this chapter, we studied the *kinetics of rigid bodies,* i.e., the relations existing between the forces acting on a rigid body, the shape and mass of the body, and the motion produced. Except for the first two sections, which apply to the most general case of the motion of a rigid body, our analysis was restricted to the *plane motion of rigid slabs* and rigid bodies symmetrical with respect to the reference plane. The study of the plane motion of nonsymmetrical rigid bodies and of the motion of rigid bodies in three-dimensional space will be considered in Chap. 18.

We first recalled [Sec. 16.2] the two fundamental equations derived in Chap. 14 for the motion of a system of particles and observed that they apply in the most general case of the motion of a rigid body. The first equation defines the motion of the mass center *G* of the body; we have

$$\Sigma \mathbf{F} = m\overline{\mathbf{a}} \qquad (16.1)$$

where *m* is the mass of the body and $\overline{\mathbf{a}}$ the acceleration of *G*. The second is related to the motion of the body relative to a centroidal frame of reference; we wrote

$$\Sigma \mathbf{M}_G = \dot{\mathbf{H}}_G \qquad (16.2)$$

where $\dot{\mathbf{H}}_G$ is the rate of change of the angular momentum $\mathbf{H}_G$ of the body about its mass center *G*. Together, Eqs. (16.1) and (16.2) express that *the system of the external forces is equipollent to the system consisting of the vector $m\overline{\mathbf{a}}$ attached at G and the couple of moment $\dot{\mathbf{H}}_G$* (Fig. 16.19).

Fundamental equations of motion for a rigid body

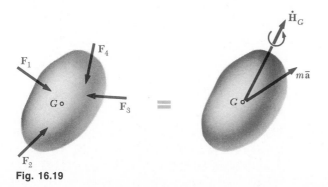

Fig. 16.19

Angular momentum in plane motion

Restricting our analysis at this point and for the rest of the chapter to the plane motion of rigid slabs and rigid bodies symmetrical with respect to the reference plane, we showed [Sec. 16.3] that the angular momentum of the body could be expressed as

$$\mathbf{H}_G = \bar{I}\boldsymbol{\omega} \qquad (16.4)$$

where $\bar{I}$ is the moment of inertia of the body about a centroidal axis perpendicular to the reference plane and $\boldsymbol{\omega}$ the angular velocity of the body. Differentiating both members of Eq. (16.4), we obtained

$$\dot{\mathbf{H}}_G = \bar{I}\dot{\boldsymbol{\omega}} = \bar{I}\boldsymbol{\alpha} \qquad (16.5)$$

which shows that, in the restricted case considered here, the rate of change of the angular momentum of the rigid body may be represented by a vector of the same direction as $\boldsymbol{\alpha}$ (i.e., perpendicular to the plane of reference) and of magnitude $\bar{I}\alpha$.

Equations for the plane motion of a rigid body

It follows from the above [Sec. 16.4] that the plane motion of a rigid slab or of a rigid body symmetrical with respect to the reference plane is defined by the three scalar equations

$$\Sigma F_x = m\bar{a}_x \qquad \Sigma F_y = m\bar{a}_y \qquad \Sigma M_G = \bar{I}\alpha \qquad (16.6)$$

D'Alembert's principle

It further follows that *the external forces acting on the rigid body are actually equivalent to the effective forces of the various particles forming the body*. This statement, known as *d'Alembert's principle*, may be expressed in the form of the vector diagram shown in Fig. 16.20, where the effective forces have been represented by a vector $m\bar{a}$ attached at G and a couple $\bar{I}\boldsymbol{\alpha}$. In the particular case of a slab in *translation*, the effective forces shown in part b of this figure reduce to the single vector $m\bar{\mathbf{a}}$, while in the particular case of a slab in *centroidal rotation*, they reduce to the single couple $\bar{I}\boldsymbol{\alpha}$; in any other case of plane motion, both the vector $m\bar{\mathbf{a}}$ and the couple $\bar{I}\boldsymbol{\alpha}$ should be included.

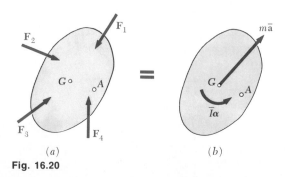

(a)   (b)

Fig. 16.20

Free-body diagram equation

Any problem involving the plane motion of a rigid slab may be solved by drawing a *free-body-diagram equation* similar to that of Fig. 16.20 [Sec. 16.6]. Three equations of motion may then be obtained by equating the x components, y components, or moments about an arbitrary point A, of the forces and vectors involved [Sample Probs. 16.1, 16.2, 16.4, and 16.5]. An alternative solution may also be obtained by adding to the external forces an *inertia vector* $-m\bar{a}$ of sense opposite to that of $\bar{a}$, attached at G, and an *inertia couple* $-\bar{I}\alpha$ of sense opposite to that of α. The system obtained in this way is equivalent to zero, and the slab is said to be in *dynamic equilibrium*.

Connected rigid bodies

The method described above may also be used to solve problems involving the plane motion of several connected rigid bodies [Sec. 16.7]. A free-body-diagram equation is drawn for each part of the system and the equations of motion obtained are solved simultaneously. In some cases, however, a single diagram may be drawn for the entire system, including all the external forces as well as the vectors $m\bar{a}$ and the couples $\bar{I}\alpha$ associated with the various parts of the system [Sample Prob. 16.3].

Constrained plane motion

In the second part of the chapter, we were concerned with rigid bodies *moving under given constraints* [Sec. 16.8]. While the kinetic analysis of the constrained plane motion of a rigid slab is the same as above, it must be supplemented by a *kinematic analysis* which has for its object to express the components $\bar{a}_x$ and $\bar{a}_y$ of the acceleration of the mass center G of the slab in terms of its angular acceleration α. Problems solved in this way included the *noncentroidal rotation* of rods and plates [Sample Probs. 16.6 and 16.7], the *rolling motion* of spheres and wheels [Sample Probs. 16.8 and 16.9], and the plane motion of *various types of linkages* [Sample Prob. 16.10].

Review Problems

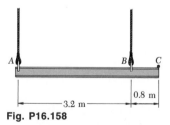

Fig. P16.158

16.158 A 4-m beam of mass 200 kg is lowered from a considerable height by means of two cables unwinding from overhead cranes. As the beam approaches the ground, the crane operators apply brakes to slow the unwinding motion. Determine the acceleration of each cable at that instant, knowing that $T_A = 1000$ N and $T_B = 1800$ N.

16.159 A uniform rod ABC weighs 16 lb and is connected to two collars of negligible weight which slide on horizontal, frictionless rods located in the same vertical plane. If a force **P** of magnitude 5 lb is applied at C, determine (*a*) the acceleration of the rod, (*b*) the reactions at B and C.

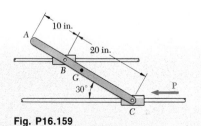

Fig. P16.159

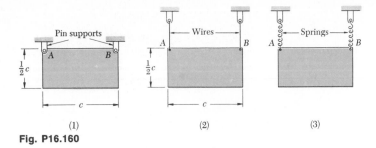

(1)   (2)   (3)

Fig. P16.160

16.160 A uniform plate of mass m is suspended in each of the ways shown. For each case determine immediately after the connection at B has been released (*a*) the angular acceleration of the plate, (*b*) the acceleration of its mass center.

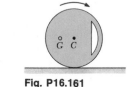

Fig. P16.161

16.161 The mass center G of a 5-kg wheel of radius $R = 300$ mm is located at a distance $r = 100$ mm from its geometric center C. The centroidal radius of gyration is $\bar{k} = 150$ mm. As the wheel rolls without sliding, its angular velocity varies and it is observed that $\omega = 8$ rad/s in the position shown. Determine the corresponding angular acceleration of the wheel.

16.162 Ends A and B of the 3-lb uniform rod slide without friction along the surfaces shown. When the rod is at rest a horizontal force $\mathbf{P}$ is applied at A, causing end A of the rod to start moving to the right with an acceleration of 8 ft/s². Determine (*a*) the force $\mathbf{P}$, (*b*) the corresponding reaction at B.

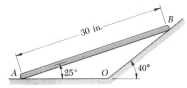

Fig. P16.162

Fig. P16.163

16.163 The rim of a flywheel has a mass of 1600 kg and a mean radius of 600 mm. As the flywheel rotates with a constant angular velocity of 360 rpm, radial forces are exerted on the rim by the spokes and internal forces are developed within the rim. Neglecting the weight of the spokes, determine (*a*) the internal forces in the rim, assuming the radial forces exerted by the spokes to be zero, (*b*) the radial force exerted by each spoke, assuming the tangential forces in the rim to be zero.

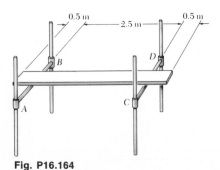

Fig. P16.164

16.164 A 3.5-m plank of mass 30 kg rests on two horizontal pipes AB and CD of a scaffolding. The pipes are 2.5 m apart, and the plank overhangs 0.5 m at each end. A 75-kg worker is standing on the plank when pipe CD suddenly breaks. Determine the initial acceleration of the worker, knowing that he was standing (*a*) in the middle of the plank, (*b*) just above pipe CD.

16.165 Cylindrical cans are transported from one elevation to another by the moving horizontal arms shown. Assuming that $\mu_s = 0.25$ between the cans and the arms, determine (a) the magnitude of the downward acceleration **a** for which the cans slide on the horizontal arms, (b) the smallest ratio h/d for which the cans tip before they slide.

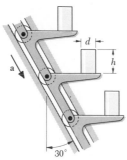

Fig. P16.165

16.166 Shortly after being fired, the experimental rocket shown weighs 25,000 lb and is moving upward with an acceleration of 45 ft/s². If at that instant rocket engine A fails, while rocket engine B continues to operate, determine (a) the acceleration of the mass center of the rocket, (b) the angular acceleration of the rocket. Assume that the rocket is a uniform slender rod 48 ft long.

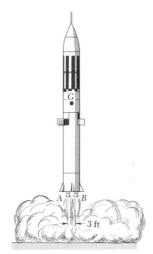

Fig. P16.166

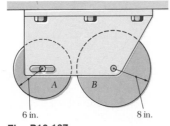

Fig. P16.167

16.167 Disk A weighs 5 lb and is initially at rest; disk B weighs 10 lb and has an initial angular velocity of 900 rpm clockwise. The disks are brought together by applying a horizontal force **P** of magnitude $P = 3$ lb to the axle of disk A. Knowing that $\mu_k = 0.25$ and neglecting bearing friction, determine (a) the angular acceleration of each disk, (b) the final angular velocity of each disk.

16.168 The motion of a 1.5-kg semicircular rod is guided by two blocks of negligible mass which slide without friction in the slots shown. A horizontal and variable force **P** is applied at B, causing B to move to the right at a constant speed of 5 m/s. For the position shown, determine (a) the force **P**, (b) the reaction at B.

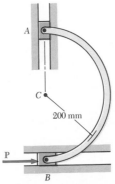

Fig. P16.168

16.169 A section of pipe rests on a plate. The plate is then given a constant acceleration **a** directed to the right. Assuming that the pipe rolls on the plate, determine (*a*) the acceleration of the pipe, (*b*) the distance through which the plate will move before the pipe reaches end *A*.

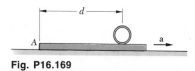

Fig. P16.169

The following problems are designed to be solved with a computer.

16.C1 A completely filled barrel and its contents have a combined mass of 125 kg. A cylinder is connected to the barrel as shown and it is known that the coefficients of friction between the barrel and the floor are $\mu_s = 0.35$ and $\mu_k = 0.30$. Write a computer program and use it to calculate the acceleration of the barrel and the range of values of *h* for which the barrel will not tip, for values of the mass of the cylinder from 25 to 75 kg at 5-kg intervals, and from 75 to 400 kg at 25-kg intervals.

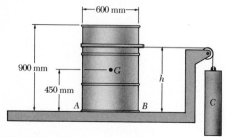

Fig. P16.C1

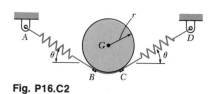

Fig. P16.C2

16.C2 A uniform disk is supported as shown by springs *AB* and *CD*. Write a computer program and use it to calculate, for values of θ from 30 to 90° at 5° intervals, the accelerations of the center *G* of the disk and of point *C* immediately after spring *AB* has broken.

16.C3 End *A* of the 6-kg uniform rod *AB* rests on the horizontal surface while end *B* is attached to a collar which is moved upward at a constant speed $v_B = 0.5$ m/s. Neglecting the effect of friction, write a computer program and use it to calculate the reaction at *A* for values of θ from 0 to 75° at 5° intervals.

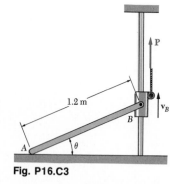

Fig. P16.C3

16.C4 In the engine system of Prob. 15.C2, piston *P* weighs 4 lb and the connecting rod *BD* is assumed to be a 5-lb uniform slender rod. Knowing that during the test of the system no force is applied to the face of the piston, write a computer program and use it to calculate the horizontal and vertical components of the forces exerted on the connecting rod at *B* and *D* for values of θ from 0 to 180° at 10° intervals. (Neglect the effect of the weight of the rod.)

Plane Motion of Rigid Bodies: Energy and Momentum Methods

17

17.1. Introduction. In this chapter the method of work and energy and the method of impulse and momentum will be used to analyze the plane motion of rigid bodies and of systems of rigid bodies.

The method of work and energy will be considered first. In Secs. 17.2 through 17.5, we shall define the work of a force and of a couple and then obtain an expression for the kinetic energy of a rigid body in plane motion. The principle of work and energy will then be used to solve problems involving displacements and velocities. In Sec. 17.6, the principle of conservation of energy will be applied to the solution of a variety of engineering problems.

In the second part of the chapter, we shall discuss in Secs. 17.8 and 17.9 the principle of impulse and momentum and apply it to the solution of problems involving velocities and time. In Sec. 17.10, the concept of conservation of angular momentum will be introduced and discussed.

In the last part of the chapter (Secs. 17.11 and 17.12), we shall consider problems involving the eccentric impact of rigid bodies. As was done in Chap. 13, where we considered the impact of particles, the coefficient of restitution between the colliding bodies will be used together with the principle of impulse and momentum in the solution of impact problems. It will also be shown that the method used is applicable not only when the colliding bodies move freely after the impact but also when the bodies are partially constrained in their motion.

17.2. Principle of Work and Energy for a Rigid Body.

The principle of work and energy will now be used to analyze the plane motion of rigid bodies. As was pointed out in Chap. 13, the method of work and energy is particularly well adapted to the solution of problems involving velocities and displacements. Its main advantage resides in the fact that the work of forces and the kinetic energy of particles are scalar quantities.

In order to apply the principle of work and energy to the analysis of the motion of a rigid body, we shall again assume that the rigid body is made of a large number n of particles of mass Δm_i. Recalling Eq. (14.30) of Sec. 14.8, we write

$$T_1 + U_{1 \to 2} = T_2 \tag{17.1}$$

where T_1, T_2 = initial and final values of total kinetic energy of the particles forming the rigid body

$U_{1 \to 2}$ = work of all forces acting on the various particles of the body

The total kinetic energy

$$T = \frac{1}{2} \sum_{i=1}^{n} \Delta m_i \, v_i^2 \tag{17.2}$$

is obtained by adding positive scalar quantities and is itself a positive scalar quantity. We shall see later how T may be determined for various types of motion of a rigid body.

The expression $U_{1 \to 2}$ in (17.1) represents the work of all the forces acting on the various particles of the body, whether these forces are internal or external. However, as we shall see presently, the total work of the internal forces holding together the particles of a rigid body is zero. Consider two particles A and B of a rigid body and the two equal and opposite forces $\mathbf{F}$ and $-\mathbf{F}$ they exert on each other (Fig. 17.1). While, in general, small displacements $d\mathbf{r}$ and $d\mathbf{r}'$ of the two particles are different, the components of these displacements along AB must be equal; otherwise, the particles would not remain at the same distance from each other, and the body would not be rigid. Therefore, the work of $\mathbf{F}$ is equal in magnitude and opposite in sign to the work of $-\mathbf{F}$, and their sum is zero. Thus, the total work of the internal forces acting on the particles of a rigid body is zero, and *the expression $U_{1 \to 2}$ in Eq. (17.1) reduces to the work of the external forces* acting on the body during the displacement considered.

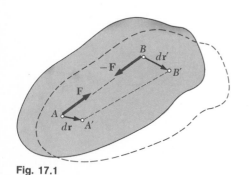

Fig. 17.1

17.3. Work of Forces Acting on a Rigid Body.

We saw in Sec. 13.2 that the work of a force $\mathbf{F}$ during a displacement of its point of application from A_1 to A_2 is

$$U_{1\to2} = \int_{A_1}^{A_2} \mathbf{F} \cdot d\mathbf{r} \qquad (17.3)$$

or

$$U_{1\to2} = \int_{s_1}^{s_2} (F \cos \alpha)\, ds \qquad (17.3')$$

where F is the magnitude of the force, α the angle it forms with the direction of motion of its point of application A, and s the variable of integration which measures the distance traveled by A along its path.

In computing the work of the external forces acting on a rigid body, it is often convenient to determine the work of a couple without considering separately the work of each of the two forces forming the couple. Consider the two forces $\mathbf{F}$ and $-\mathbf{F}$ forming a couple of moment $\mathbf{M}$ and acting on a rigid body (Fig. 17.2). Any small displacement of the rigid body bringing A and B, respectively, into A' and B'' may be divided into two parts, one in which points A and B undergo equal displacements $d\mathbf{r}_1$, the other in which A' remains fixed while B' moves into B'' through a displacement $d\mathbf{r}_2$ of magnitude $ds_2 = r\, d\theta$. In the first part of the motion, the work of $\mathbf{F}$ is equal in magnitude and opposite in sign to the work of $-\mathbf{F}$ and their sum is zero. In the second part of the motion, only force $\mathbf{F}$ works, and its work is $dU = F\, ds_2 = Fr\, d\theta$. But the product Fr is equal to the magnitude M of the moment of the couple. Thus, the work of a couple of moment $\mathbf{M}$ acting on a rigid body is

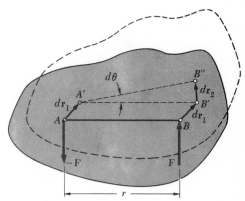

Fig. 17.2

$$dU = M\, d\theta \qquad (17.4)$$

where $d\theta$ is the small angle expressed in radians through which the body rotates. We again note that work should be expressed in units obtained by multiplying units of force by units of length. The work of the couple during a finite rotation of the rigid body is obtained by integrating both members of (17.4) from the initial value θ_1 of the angle θ to its final value θ_2. We write

$$U_{1\to2} = \int_{\theta_1}^{\theta_2} M\, d\theta \qquad (17.5)$$

When the moment $\mathbf{M}$ *of the couple is constant,* formula (17.5) reduces to

$$U_{1\to2} = M(\theta_2 - \theta_1) \qquad (17.6)$$

It was pointed out in Sec. 13.2 that a number of forces encountered in problems of kinetics *do no work.* They are forces applied to fixed points or acting in a direction perpendicular to the displacement of their point of application. Among the forces which do no work the following have been listed: the reaction at a frictionless pin when the body supported rotates about the pin, the reaction at a frictionless surface when the body in contact moves along the surface, the weight of a body when its center of gravity moves horizontally. We should also indicate now that *when a rigid body rolls without sliding on a fixed surface, the friction force* **F** *at the point of contact C does no work.* The velocity $\mathbf{v}_C$ of the point of contact C is zero, and the work of the friction force **F** during a small displacement of the rigid body is

$$dU = F \, ds_C = F(v_C \, dt) = 0$$

17.4. Kinetic Energy of a Rigid Body in Plane Motion.

Consider a rigid body of mass m in plane motion. We recall from Sec. 14.7 that, if the absolute velocity $\mathbf{v}_i$ of each particle P_i of the body is expressed as the sum of the velocity $\bar{\mathbf{v}}$ of the mass center G of the body and of the velocity $\mathbf{v}'_i$ of the particle relative to a frame $Gx'y'$ attached to G and of fixed orientation (Fig. 17.3), the kinetic energy of the system of particles forming the rigid body may be written in the form

$$T = \tfrac{1}{2}m\bar{v}^2 + \frac{1}{2}\sum_{i=1}^{n} \Delta m_i \, v_i'^2 \tag{17.7}$$

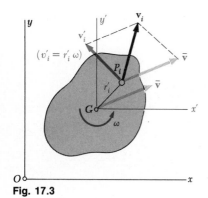

Fig. 17.3

But the magnitude v'_i of the relative velocity of P_i is equal to the product $r'_i\omega$ of the distance r'_i of P_i from the axis through G perpendicular to the plane of motion and of the magnitude ω of the angular velocity of the body at the instant considered. Substituting into (17.7), we have

$$T = \tfrac{1}{2}m\bar{v}^2 + \frac{1}{2}\left(\sum_{i=1}^{n} r_i'^2 \, \Delta m_i\right)\omega^2 \tag{17.8}$$

or, since the sum represents the moment of inertia $\bar{I}$ of the body about the axis through G,

$$T = \tfrac{1}{2}m\bar{v}^2 + \tfrac{1}{2}\bar{I}\omega^2 \tag{17.9}$$

We note that in the particular case of a body in translation ($\omega = 0$), the expression obtained reduces to $\tfrac{1}{2}m\bar{v}^2$, while in the case of a centroidal rotation ($\bar{v} = 0$), it reduces to $\tfrac{1}{2}\bar{I}\omega^2$. We conclude that the kinetic energy of a rigid body in plane motion may be separated into two parts: (1) the kinetic energy $\tfrac{1}{2}m\bar{v}^2$ associated with the motion of the mass center G of the body, and (2) the kinetic energy $\tfrac{1}{2}\bar{I}\omega^2$ associated with the rotation of the body about G.

Noncentroidal Rotation. The relation (17.9) is valid for any type of plane motion and may, therefore, be used to express the kinetic energy of a rigid body rotating with an angular velocity ω about a fixed axis through O (Fig. 17.4). In that case, however, the kinetic energy of the body may be expressed more directly by noting that the speed v_i of the particle P_i is equal to the product $r_i\omega$ of the distance r_i of P_i from the fixed axis and of the magnitude ω of the angular velocity of the body at the instant considered. Substituting into (17.2), we write

$$T = \frac{1}{2} \sum_{i=1}^{n} \Delta m_i \, (r_i\omega)^2 = \frac{1}{2} \left(\sum_{i=1}^{n} r_i^2 \, \Delta m_i \right) \omega^2$$

or, since the last sum represents the moment of inertia I_O of the body about the fixed axis through O,

$$T = \tfrac{1}{2} I_O \omega^2 \tag{17.10}$$

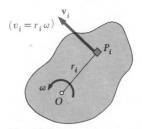

Fig. 17.4

We note that the results obtained are not limited to the motion of plane slabs or to the motion of bodies which are symmetrical with respect to the reference plane. They may be applied to the study of the plane motion of any rigid body, regardless of its shape. However, since Eq. (17.9) is applicable to any plane motion, we shall use it in the solution of sample problems, rather than Eq. (17.10) which may be used only for a noncentroidal rotation.

17.5. Systems of Rigid Bodies. When a problem involves several rigid bodies, each rigid body may be considered separately, and the principle of work and energy may be applied to each body. Adding the kinetic energies of all the particles and considering the work of all the forces involved, we may also write the equation of work and energy for the entire system. We have

$$T_1 + U_{1 \to 2} = T_2 \tag{17.11}$$

where T represents the arithmetic sum of the kinetic energies of the rigid bodies forming the system (all terms are positive) and $U_{1 \to 2}$ the work of all the forces acting on the various bodies, whether these forces are *internal* or *external* from the point of view of the system as a whole.

The method of work and energy is particularly useful in solving problems involving pin-connected members, or blocks and pulleys connected by inextensible cords, or meshed gears. In all these cases, the internal forces occur by pairs of equal and opposite forces, and the points of application of the forces in each pair *move through equal distances* during a small displacement of the system. As a result, the work of the internal forces is zero, and $U_{1 \to 2}$ reduces to the work of the *forces external to the system.*

17.6. Conservation of Energy. We saw in Sec. 13.6 that the work of conservative forces, such as the weight of a body or the force exerted by a spring, may be expressed as a change in potential energy. When a rigid body, or a system of rigid bodies, moves under the action of conservative forces, the principle of work and energy stated in Sec. 17.2 may be expressed in a modified form. Substituting for $U_{1\to2}$ from (13.19′) into (17.1), we write

$$T_1 + V_1 = T_2 + V_2 \qquad (17.12)$$

Formula (17.12) indicates that when a rigid body, or a system of rigid bodies, moves under the action of conservative forces, *the sum of the kinetic energy and of the potential energy of the system remains constant.* It should be noted that in the case of the plane motion of a rigid body, the kinetic energy of the body should include both the *translational* term $\frac{1}{2}m\bar{v}^2$ and the *rotational* term $\frac{1}{2}\bar{I}\omega^2$.

As an example of application of the principle of conservation of energy, we shall consider a slender rod AB, of length l and mass m, whose extremities are connected to blocks of negligible mass sliding along horizontal and vertical tracks. We assume that the rod is released with no initial velocity from a horizontal position (Fig. 17.5a), and we wish to determine its angular velocity after it has rotated through an angle θ (Fig. 17.5b).

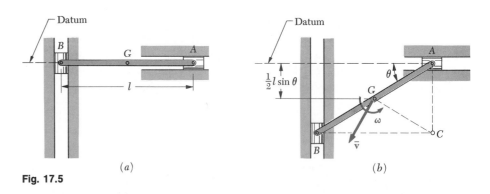

(a)   (b)

Fig. 17.5

Since the initial velocity is zero, we have $T_1 = 0$. Measuring the potential energy from the level of the horizontal track, we write $V_1 = 0$. After the rod has rotated through θ, the center of gravity G of the rod is at a distance $\frac{1}{2}l \sin \theta$ below the reference level and we have

$$V_2 = -\tfrac{1}{2}Wl \sin \theta = -\tfrac{1}{2}mgl \sin \theta$$

Observing that in this position, the instantaneous center of the rod is located at C, and that $CG = \frac{1}{2}l$, we write $\bar{v}_2 = \frac{1}{2}l\omega$ and obtain

$$T_2 = \tfrac{1}{2}m\bar{v}_2^2 + \tfrac{1}{2}\bar{I}\omega_2^2 = \tfrac{1}{2}m(\tfrac{1}{2}l\omega)^2 + \tfrac{1}{2}(\tfrac{1}{12}ml^2)\omega^2$$

$$= \frac{1}{2}\frac{ml^2}{3}\omega^2$$

Applying the principle of conservation of energy, we write

$$T_1 + V_1 = T_2 + V_2$$

$$0 = \frac{1}{2}\frac{ml^2}{3}\omega^2 - \tfrac{1}{2}mgl\sin\theta$$

$$\omega = \left(\frac{3g}{l}\sin\theta\right)^{1/2}$$

We recall that the advantages of the method of work and energy, as well as its shortcomings, were indicated in Sec. 13.4. In this connection, we wish to mention that the method of work and energy must be supplemented by the application of d'Alembert's principle when reactions at fixed axles, at rollers, or at sliding blocks are to be determined. For example, in order to compute the reactions at the extremities A and B of the rod of Fig. 17.5b, a diagram should be drawn to express that the system of the external forces applied to the rod is equivalent to the vector $m\bar{a}$ and the couple $\bar{I}\alpha$. The angular velocity ω of the rod, however, is determined by the method of work and energy before the equations of motion are solved for the reactions. The complete analysis of the motion of the rod and of the forces exerted on the rod requires, therefore, the combined use of the method of work and energy and of the principle of equivalence of the external and effective forces.

17.7. Power. *Power* was defined in Sec. 13.5 as the time rate at which work is done. In the case of a body acted upon by a force $\mathbf{F}$, and moving with a velocity $\mathbf{v}$, the power was expressed as follows:

$$\text{Power} = \frac{dU}{dt} = \mathbf{F}\cdot\mathbf{v} \qquad (13.13)$$

In the case of a rigid body rotating with an angular velocity ω and acted upon by a couple of moment $\mathbf{M}$ parallel to the axis of rotation, we have, by (17.4),

$$\text{Power} = \frac{dU}{dt} = \frac{M\,d\theta}{dt} = M\omega \qquad (17.13)$$

The various units used to measure power, such as the watt and the horsepower, were defined in Sec. 13.5.

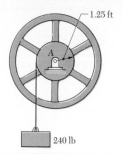

1.25 ft

240 lb

A 240-lb block is suspended from an inextensible cable which is wrapped around a drum of 1.25-ft radius rigidly attached to a flywheel. The drum and flywheel have a combined centroidal moment of inertia $\bar{I} = 10.5$ lb·ft·s². At the instant shown, the velocity of the block is 6 ft/s directed downward. Knowing that the bearing at A is poorly lubricated and that the bearing friction is equivalent to a couple **M** of magnitude 60 lb·ft, determine the velocity of the block after it has moved 4 ft downward.

Solution. We consider the system formed by the flywheel and the block. Since the cable is inextensible, the work done by the internal forces exerted by the cable cancels. The initial and final positions of the system and the external forces acting on the system are as shown.

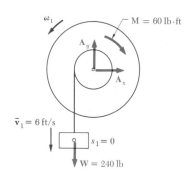

Kinetic Energy. *Position 1.*

Block: $\bar{v}_1 = 6$ ft/s

Flywheel: $\omega_1 = \dfrac{\bar{v}_1}{r} = \dfrac{6 \text{ ft/s}}{1.25 \text{ ft}} = 4.80$ rad/s

$$T_1 = \tfrac{1}{2}m\bar{v}_1^2 + \tfrac{1}{2}\bar{I}\omega_1^2$$
$$= \frac{1}{2}\frac{240 \text{ lb}}{32.2 \text{ ft/s}^2}(6 \text{ ft/s})^2 + \tfrac{1}{2}(10.5 \text{ lb·ft·s}^2)(4.80 \text{ rad/s})^2$$
$$= 255 \text{ ft·lb}$$

Position 2. Noting that $\omega_2 = \bar{v}_2/1.25$, we write

$$T_2 = \tfrac{1}{2}m\bar{v}_2^2 + \tfrac{1}{2}\bar{I}\omega_2^2$$
$$= \frac{1}{2}\frac{240}{32.2}(\bar{v}_2)^2 + (\tfrac{1}{2})(10.5)\left(\frac{\bar{v}_2}{1.25}\right)^2 = 7.09\bar{v}_2^2$$

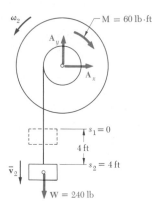

Work. During the motion, only the weight **W** of the block and the friction couple **M** do work. Noting that **W** does positive work and that the friction couple **M** does negative work, we write

$$s_1 = 0 \qquad s_2 = 4 \text{ ft}$$
$$\theta_1 = 0 \qquad \theta_2 = \frac{s_2}{r} = \frac{4 \text{ ft}}{1.25 \text{ ft}} = 3.20 \text{ rad}$$
$$U_{1\rightarrow2} = W(s_2 - s_1) - M(\theta_2 - \theta_1)$$
$$= (240 \text{ lb})(4 \text{ ft}) - (60 \text{ lb·ft})(3.20 \text{ rad})$$
$$= 768 \text{ ft·lb}$$

Principle of Work and Energy

$$T_1 + U_{1\rightarrow2} = T_2$$
$$255 \text{ ft·lb} + 768 \text{ ft·lb} = 7.09\bar{v}_2^2$$
$$\bar{v}_2 = 12.01 \text{ ft/s} \qquad \bar{v}_2 = 12.01 \text{ ft/s} \downarrow \quad \blacktriangleleft$$

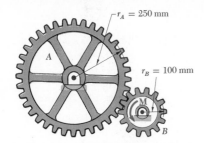

$r_A = 250$ mm

$r_B = 100$ mm

SAMPLE PROBLEM 17.2

Gear A has a mass of 10 kg and a radius of gyration of 200 mm, while gear B has a mass of 3 kg and a radius of gyration of 80 mm. The system is at rest when a couple **M** of magnitude 6 N·m is applied to gear B. Neglecting friction, determine (a) the number of revolutions executed by gear B before its angular velocity reaches 600 rpm, (b) the tangential force which gear B exerts on gear A.

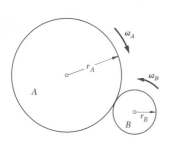

Motion of Entire System. Noting that the peripheral speeds of the gears are equal, we write

$$r_A\omega_A = r_B\omega_B \qquad \omega_A = \omega_B\frac{r_B}{r_A} = \omega_B\frac{100 \text{ mm}}{250 \text{ mm}} = 0.40\omega_B$$

For $\omega_B = 600$ rpm, we have

$$\omega_B = 62.8 \text{ rad/s} \qquad \omega_A = 0.40\omega_B = 25.1 \text{ rad/s}$$
$$\bar{I}_A = m_A\bar{k}_A^2 = (10 \text{ kg})(0.200 \text{ m})^2 = 0.400 \text{ kg·m}^2$$
$$\bar{I}_B = m_B\bar{k}_B^2 = (3 \text{ kg})(0.080 \text{ m})^2 = 0.0192 \text{ kg·m}^2$$

Kinetic Energy. Since the system is initially at rest, $T_1 = 0$. Adding the kinetic energies of the two gears when $\omega_B = 600$ rpm, we obtain

$$T_2 = \tfrac{1}{2}\bar{I}_A\omega_A^2 + \tfrac{1}{2}\bar{I}_B\omega_B^2$$
$$= \tfrac{1}{2}(0.400 \text{ kg·m}^2)(25.1 \text{ rad/s})^2 + \tfrac{1}{2}(0.0192 \text{ kg·m}^2)(62.8 \text{ rad/s})^2$$
$$= 163.9 \text{ J}$$

Work. Denoting by θ_B the angular displacement of gear B, we have

$$U_{1\to2} = M\theta_B = (6 \text{ N·m})(\theta_B \text{ rad}) = (6\,\theta_B) \text{ J}$$

Principle of Work and Energy

$$T_1 + U_{1\to2} = T_2$$
$$0 + (6\,\theta_B) \text{ J} = 163.9 \text{ J}$$
$$\theta_B = 27.32 \text{ rad} \qquad \theta_B = 4.35 \text{ rev} \quad \blacktriangleleft$$

Motion of Gear A. Kinetic Energy. Initially, gear A is at rest, $T_1 = 0$. When $\omega_B = 600$ rpm, the kinetic energy of gear A is

$$T_2 = \tfrac{1}{2}\bar{I}_A\omega_A^2 = \tfrac{1}{2}(0.400 \text{ kg·m}^2)(25.1 \text{ rad/s})^2 = 126.0 \text{ J}$$

Work. The forces acting on gear A are as shown. The tangential force **F** does work equal to the product of its magnitude and of the length $\theta_A r_A$ of the arc described by the point of contact. Since $\theta_A r_A = \theta_B r_B$, we have

$$U_{1\to2} = F(\theta_B r_B) = F(27.3 \text{ rad})(0.100 \text{ m}) = F(2.73 \text{ m})$$

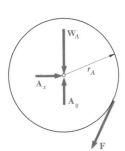

Principle of Work and Energy

$$T_1 + U_{1\to2} = T_2$$
$$0 + F(2.73 \text{ m}) = 126.0 \text{ J}$$
$$F = +46.2 \text{ N} \qquad F = 46.2 \text{ N} \nearrow \quad \blacktriangleleft$$

A sphere, a cylinder, and a hoop, each having the same mass and the same radius, are released from rest on an incline. Determine the velocity of each body after it has rolled through a distance corresponding to a change in elevation h.

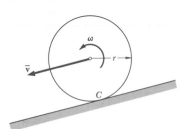

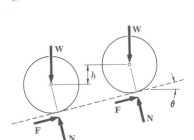

Solution. We shall first solve the problem in general terms and then find particular results for each body. We denote the mass by m, the centroidal moment of inertia by $\bar{I}$, the weight by W, and the radius by r.

Since each body rolls, the instantaneous center of rotation is located at C and we write

$$\omega = \frac{\bar{v}}{r}$$

Kinetic Energy

$$T_1 = 0$$
$$T_2 = \tfrac{1}{2}m\bar{v}^2 + \tfrac{1}{2}\bar{I}\omega^2$$
$$= \tfrac{1}{2}m\bar{v}^2 + \tfrac{1}{2}\bar{I}\left(\frac{\bar{v}}{r}\right)^2 = \tfrac{1}{2}\left(m + \frac{\bar{I}}{r^2}\right)\bar{v}^2$$

Work. Since the friction force $\mathbf{F}$ in rolling motion does no work,

$$U_{1\to2} = Wh$$

Principle of Work and Energy

$$T_1 + U_{1\to2} = T_2$$
$$0 + Wh = \tfrac{1}{2}\left(m + \frac{\bar{I}}{r^2}\right)\bar{v}^2 \qquad \bar{v}^2 = \frac{2Wh}{m + \bar{I}/r^2}$$

Noting that $W = mg$, we rearrange the result and obtain

$$\bar{v}^2 = \frac{2gh}{1 + \bar{I}/mr^2}$$

Velocities of Sphere, Cylinder, and Hoop. Introducing successively the particular expression for $\bar{I}$, we obtain

Sphere:	$\bar{I} = \tfrac{2}{5}mr^2$	$\bar{v} = 0.845\sqrt{2gh}$ ◄
Cylinder:	$\bar{I} = \tfrac{1}{2}mr^2$	$\bar{v} = 0.816\sqrt{2gh}$ ◄
Hoop:	$\bar{I} = mr^2$	$\bar{v} = 0.707\sqrt{2gh}$ ◄

Remark. We may compare the results with the velocity attained by a frictionless block sliding through the same distance. The solution is identical to the above solution except that $\omega = 0$; we find $\bar{v} = \sqrt{2gh}$.

Comparing the results, we note that the velocity of the body is independent of both its mass and radius. However, the velocity does depend upon the quotient $\bar{I}/mr^2 = \bar{k}^2/r^2$, which measures the ratio of the rotational kinetic energy to the translational kinetic energy. Thus the hoop, which has the largest $\bar{k}$ for a given radius r, attains the smallest velocity, while the sliding block, which does not rotate, attains the largest velocity.

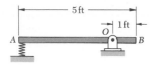

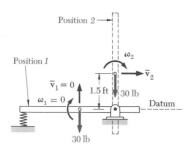

SAMPLE PROBLEMS 17.4

A 30-lb slender rod AB is 5 ft long and is pivoted about a point O which is 1 ft from end B. The other end is pressed against a spring of constant $k = 1800$ lb/in. until the spring is compressed 1 in. The rod is then in a horizontal position. If the rod is released from this position, determine its angular velocity and the reaction at the pivot O as the rod passes through a vertical position.

Position 1. Potential Energy. Since the spring is compressed 1 in., we have $x_1 = 1$ in.

$$V_e = \tfrac{1}{2}kx_1^2 = \tfrac{1}{2}(1800 \text{ lb/in.})(1 \text{ in.})^2 = 900 \text{ in} \cdot \text{lb}$$

Choosing the datum as shown, we have $V_g = 0$; therefore,

$$V_1 = V_e + V_g = 900 \text{ in} \cdot \text{lb} = 75 \text{ ft} \cdot \text{lb}$$

Kinetic Energy. Since the velocity in position 1 is zero, we have $T_1 = 0$.

Position 2. Potential Energy. The elongation of the spring is zero, and we have $V_e = 0$. Since the center of gravity of the rod is now 1.5 ft above the datum,

$$V_g = (30 \text{ lb})(+1.5 \text{ ft}) = 45 \text{ ft} \cdot \text{lb}$$
$$V_2 = V_e + V_g = 45 \text{ ft} \cdot \text{lb}$$

Kinetic Energy. Denoting by ω_2 the angular velocity of the rod in position 2, we note that the rod rotates about O and write $\bar{v}_2 = \bar{r}\omega_2 = 1.5\omega_2$.

$$\bar{I} = \tfrac{1}{12}ml^2 = \frac{1}{12}\frac{30 \text{ lb}}{32.2 \text{ ft/s}^2}(5 \text{ ft})^2 = 1.941 \text{ lb} \cdot \text{ft} \cdot \text{s}^2$$

$$T_2 = \tfrac{1}{2}m\bar{v}_2^2 + \tfrac{1}{2}\bar{I}\omega_2^2 = \frac{1}{2}\frac{30}{32.2}(1.5\omega_2)^2 + \tfrac{1}{2}(1.941)\omega_2^2 = 2.019\omega_2^2$$

Conservation of Energy

$$T_1 + V_1 = T_2 + V_2$$
$$0 + 75 \text{ ft} \cdot \text{lb} = 2.019\omega_2^2 + 45 \text{ ft} \cdot \text{lb}$$
$$\omega_2 = 3.86 \text{ rad/s} \downarrow \quad \blacktriangleleft$$

Reaction in Position 2. Since $\omega_2 = 3.86$ rad/s, the components of the acceleration of G as the rod passes through position 2 are

$$\bar{a}_n = \bar{r}\omega_2^2 = (1.5 \text{ ft})(3.86 \text{ rad/s})^2 = 22.3 \text{ ft/s}^2 \qquad \bar{a}_n = 22.3 \text{ ft/s}^2 \downarrow$$
$$\bar{a}_t = \bar{r}\alpha \qquad\qquad\qquad\qquad\qquad\qquad\qquad \bar{a}_t = \bar{r}\alpha \rightarrow$$

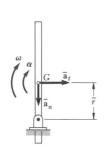

We express that the system of external forces is equivalent to the system of effective forces represented by the vector of components $m\bar{a}_t$ and $m\bar{a}_n$ attached at G and the couple $\bar{I}\alpha$.

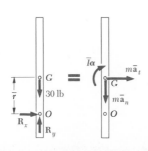

$$+\downarrow\Sigma M_O = \Sigma(M_O)_{\text{eff}}: \qquad 0 = \bar{I}\alpha + m(\bar{r}\alpha)\bar{r} \qquad \alpha = 0$$
$$\overset{+}{\rightarrow} \Sigma F_x = \Sigma(F_x)_{\text{eff}}: \qquad R_x = m(\bar{r}\alpha) \qquad R_x = 0$$
$$+\uparrow\Sigma F_y = \Sigma(F_y)_{\text{eff}}: \qquad R_y - 30 \text{ lb} = -m\bar{a}_n$$

$$R_y - 30 \text{ lb} = -\frac{30 \text{ lb}}{32.2 \text{ ft/s}^2}(22.3 \text{ ft/s}^2)$$

$$R_y = +9.22 \text{ lb} \qquad R = 9.22 \text{ lb} \uparrow \quad \blacktriangleleft$$

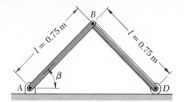

Each of the two slender rods shown is 0.75 m long and has a mass of 6 kg. If the system is released from rest when $\beta = 60°$, determine (a) the angular velocity of rod AB when $\beta = 20°$, (b) the velocity of point D at the same instant.

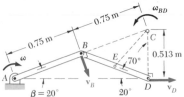

Kinematics of Motion When $\beta = 20°$. Since $\mathbf{v}_B$ is perpendicular to the rod AB and $\mathbf{v}_D$ is horizontal, the instantaneous center of rotation of rod BD is located at C. Considering the geometry of the figure, we obtain

$$BC = 0.75 \text{ m} \qquad CD = 2(0.75 \text{ m}) \sin 20° = 0.513 \text{ m}$$

Applying the law of cosines to triangle CDE, where E is located at the mass center of rod BD, we find $EC = 0.522$ m. Denoting by ω the angular velocity of rod AB, we have

$$\bar{v}_{AB} = (0.375 \text{ m})\omega \qquad \mathbf{\bar{v}}_{AB} = 0.375\omega \searrow$$
$$v_B = (0.75 \text{ m})\omega \qquad \mathbf{v}_B = 0.75\omega \searrow$$

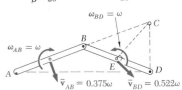

Since rod BD seems to rotate about point C, we may write

$$v_B = (BC)\omega_{BD} \qquad (0.75 \text{ m})\omega = (0.75 \text{ m})\omega_{BD} \qquad \boldsymbol{\omega}_{BD} = \omega \uparrow$$
$$\bar{v}_{BD} = (EC)\omega_{BD} = (0.522 \text{ m})\omega \qquad \mathbf{\bar{v}}_{BD} = 0.522\omega \searrow$$

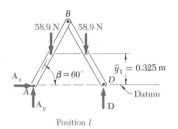

Position 1

Position 1. *Potential Energy.* Choosing the datum as shown, and observing that $W = (6 \text{ kg})(9.81 \text{ m/s}^2) = 58.86$ N, we have

$$V_1 = 2W\bar{y}_1 = 2(58.86 \text{ N})(0.325 \text{ m}) = 38.26 \text{ J}$$

Kinetic Energy. Since the system is at rest, $T_1 = 0$.

Position 2. *Potential Energy*

$$V_2 = 2W\bar{y}_2 = 2(58.86 \text{ N})(0.1283 \text{ m}) = 15.10 \text{ J}$$

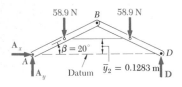

Position 2

Kinetic Energy

$$\bar{I}_{AB} = \bar{I}_{BD} = \tfrac{1}{12}ml^2 = \tfrac{1}{12}(6 \text{ kg})(0.75 \text{ m})^2 = 0.281 \text{ kg·m}^2$$
$$T_2 = \tfrac{1}{2}m\bar{v}_{AB}^2 + \tfrac{1}{2}\bar{I}_{AB}\omega_{AB}^2 + \tfrac{1}{2}m\bar{v}_{BD}^2 + \tfrac{1}{2}\bar{I}_{BD}\omega_{BD}^2$$
$$= \tfrac{1}{2}(6)(0.375\omega)^2 + \tfrac{1}{2}(0.281)\omega^2 + \tfrac{1}{2}(6)(0.522\omega)^2 + \tfrac{1}{2}(0.281)\omega^2$$
$$= 1.520\omega^2$$

Conservation of Energy

$$T_1 + V_1 = T_2 + V_2$$
$$0 + 38.26 \text{ J} = 1.520\omega^2 + 15.10 \text{ J}$$
$$\omega = 3.90 \text{ rad/s} \qquad \boldsymbol{\omega}_{AB} = 3.90 \text{ rad/s} \downarrow \quad \blacktriangleleft$$

Velocity of Point D

$$v_D = (CD)\omega = (0.513 \text{ m})(3.90 \text{ rad/s}) = 2.00 \text{ m/s}$$
$$\mathbf{v}_D = 2.00 \text{ m/s} \rightarrow \quad \blacktriangleleft$$

Problems

17.1 The rotor of a generator has an angular velocity of 3600 rpm when the generator is taken off line. The rotor, which weighs 300 lb and has a centroidal radius of gyration of 9 in., then coasts to rest. Knowing that the kinetic friction of the rotor produces a couple of magnitude 10 lb·in., determine the number of revolutions that the rotor executes before coming to rest.

17.2 Two disks of the same material are attached to a shaft as shown. Disk A is of radius r and has a thickness b, while disk B is of radius nr and thickness $3b$. A couple $\mathbf{M}$ of constant magnitude is applied when the system is at rest and is removed after the system has executed 2 revolutions. Determine the value of n which results in the largest final speed for a point on the rim of disk B.

17.3 Two disks of the same material are attached to a shaft as shown. Disk A has a mass of 15 kg and a radius $r = 125$ mm. Disk B is three times as thick as disk A. Knowing that a couple $\mathbf{M}$ of magnitude 20 N·m is to be applied to disk A when the system is at rest, determine the radius nr of disk B if the angular velocity of the system is to be 600 rpm after 4 revolutions.

Fig. P17.2 and P17.3

17.4 The flywheel of a punching machine weighs 900 lb and has a radius of gyration of 30 in. Each punching operation requires 2000 ft·lb of work. (*a*) Knowing that the speed of the flywheel is 200 rpm just before a punching, determine the speed immediately after the punching. (*b*) If a constant 25-lb·ft couple is applied to the shaft of the flywheel, determine the number of revolutions executed before the speed is again 200 rpm.

17.5 The flywheel of a small punch rotates at 300 rpm. It is known that 500 J of work must be done each time a hole is punched. It is desired that the speed of the flywheel after one punching be not less than 90 percent of the original speed of 300 rpm. (*a*) Determine the required moment of inertia of the flywheel. (*b*) If a constant 25-N·m couple is applied to the shaft of the flywheel, determine the number of revolutions required between two consecutive punchings, knowing that the initial velocity is to be 300 rpm at the start of each punching.

17.6 Disk A has a mass of 4 kg and a radius $r = 75$ mm; it is at rest when it is placed in contact with the belt, which moves at a constant speed $v = 18$ m/s. Knowing that $\mu_k = 0.25$ between the disk and the belt, determine the number of revolutions executed by the disk before it reaches a constant angular velocity.

17.7 A disk of uniform thickness and initially at rest is placed in contact with the belt, which moves with a constant velocity v. Denoting by μ_k the coefficient of kinetic friction between the disk and the belt, derive an expression for the number of revolutions executed by the disk before it reaches a constant angular velocity.

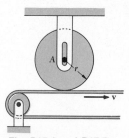

Fig. P17.6 and P17.7

17.8   The 10-in.-radius brake drum is attached to a larger flywheel which is not shown.   The total mass moment of inertia of the flywheel and drum is 16 lb·ft·s² and the coefficient of kinetic friction between the drum and the brake shoe is 0.40.   Knowing that the initial angular velocity is 240 rpm clockwise, determine the force which must be exerted by the hydraulic cylinder if the system is to stop in 75 revolutions.

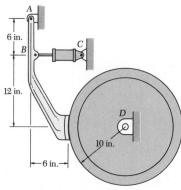

Fig. P17.8

17.9   Solve Prob. 17.8, assuming that the initial angular velocity of the flywheel is 240 rpm counterclockwise.

17.10   Each of the gears A and B weighs 20 lb and has a radius of gyration of 7.5 in.; gear C weighs 5 lb and has a radius of gyration of 3 in.   If a couple **M** of constant magnitude 50 lb·in. is applied to gear C, determine (a) the number of revolutions of gear C required for its angular velocity to increase from 450 rpm to 1800 rpm, (b) the corresponding tangential force acting on gear A.

17.11   Solve Prob. 17.10, assuming that the 50-lb·in. couple is applied to gear B.

17.12   The gear train shown consists of four gears of the same thickness and of the same material; two gears are of radius r, and the other two are of radius nr.   The system is at rest when the couple $\mathbf{M}_0$ is applied to shaft C. Denoting by I_0 the moment of inertia of a gear of radius r, determine the angular velocity of shaft A if the couple $\mathbf{M}_0$ is applied for one revolution of shaft C.

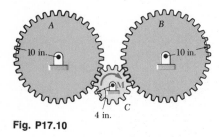

Fig. P17.10

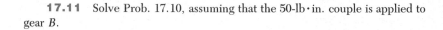

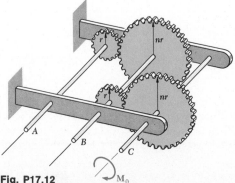

Fig. P17.12

17.13 The double pulley shown has a total mass of 6 kg and a centroidal radius of gyration of 140 mm. Five collars, each of mass 1.5 kg, are attached to cords A and B as shown. When the system is at rest and in equilibrium, one collar is removed from cord A. Knowing that the bearing friction is equivalent to a couple $\mathbf{M}$ of magnitude 0.5 N·m, determine the velocity of cord A after it has moved 900 mm.

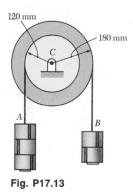

120 mm

180 mm

C

A

B

Fig. P17.13

17.14 Solve Prob. 17.13, assuming that one collar is removed from cord B.

17.15 The pulley shown has a mass of 5 kg and a radius of gyration of 150 mm. The 3-kg cylinder C is attached to a cord which is wrapped around the pulley as shown. A 1.8-kg collar B is then placed on the cylinder and the system is released from rest. After the cylinder has moved 300 mm, the collar is removed and the cylinder continues to move downward into a pit. Determine the velocity of cylinder C just before it strikes the bottom D of the pit.

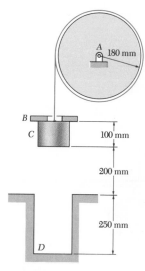

A 180 mm

B
C 100 mm

200 mm

250 mm

D

Fig. P17.15

A B

L

Fig. P17.16

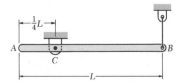

$\frac{1}{4}L$

A B
C

L

Fig. P17.17

17.16 and 17.17 A slender rod of length L and weight W is supported as shown. After the cable is cut the rod swings freely. (*a*) Determine the angular velocity of the rod as it first passes through a vertical position and the corresponding reaction at the pin support. (*b*) Solve part *a* for $W = 6$ lb and $L = 2.5$ ft.

17.18 A 3-kg slender rod rotates in a *vertical* plane about a pivot at B. A spring of constant $k = 300$ N/m and of unstretched length 120 mm is attached to the rod as shown. Knowing that in the position shown the rod has an angular velocity of 4 rad/s clockwise, determine the angular velocity of the rod after it has rotated through (*a*) 90°, (*b*) 180°.

17.19 Solve Prob. 17.18, assuming that the rod rotates in a *horizontal* plane.

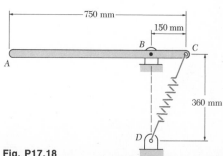

750 mm
150 mm

B C

A

360 mm

D

Fig. P17.18

17.20 A uniform sphere of radius r is placed at corner A and is given a slight clockwise motion. Assuming that the corner is sharp and becomes slightly embedded in the sphere, so that the coefficient of static friction at A is very large, determine (*a*) the angle β through which the sphere will have rotated when it loses contact with the corner, (*b*) the corresponding velocity of the center of the sphere.

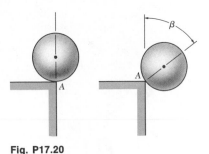

Fig. P17.20

17.21 Solve Prob. 17.20, assuming that the sphere is replaced by a uniform cylinder of radius r.

17.22 The 5-kg slender rod AB is welded to the 3-kg uniform disk which rotates about a pivot at A. A spring of constant 80 N/m is attached to the disk and is unstretched when rod AB is horizontal. Knowing that the assembly is released from rest in the position shown, determine its angular velocity after it has rotated through 90°.

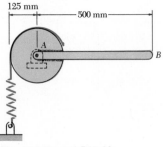

Fig. P17.22 and P17.23

17.23 The 5-kg slender rod AB is welded to the 3-kg uniform disk which rotates about a pivot at A. A spring of constant 80 N/m is attached to the disk and is unstretched when rod AB is horizontal. Determine the required angular velocity of the assembly when it is in the position shown, if its angular velocity is to be 9 rad/s after it has rotated through 90° clockwise.

17.24 A collar of weight 2 lb is rigidly attached at a distance $d = 12$ in. from the end of a uniform slender rod AB. The rod weighs 6 lb and is of length $L = 24$ in. Knowing that the rod is released from rest in the position shown, determine the angular velocity of the rod after it has rotated through 90°.

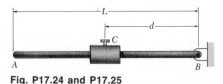

Fig. P17.24 and P17.25

17.25 A collar of weight 2 lb is rigidly attached to a slender rod AB of weight 6 lb and length $L = 24$ in. The rod is released from rest in the position shown. Determine the distance d for which the angular velocity of the rod is maximum after it has rotated through 90°.

17.26 A flywheel of centroidal radius of gyration $\overline{k} = 24$ in. is rigidly attached to a shaft of radius $r = 1.25$ in. which may roll along parallel rails. Knowing that the system is released from rest, determine the velocity of the center of the shaft after it has moved 8 ft.

17.27 A flywheel is rigidly attached to a 40-mm-radius shaft which rolls without sliding along parallel rails. The system is released from rest and attains a speed of 160 mm/s after moving 1.5 m along the rails. Determine the centroidal radius of gyration of the system.

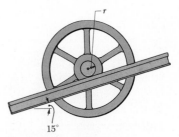

17.28 The mass center G of a 3-kg wheel of radius $R = 180$ mm is located at a distance $r = 60$ mm from its geometric center C. The centroidal radius of gyration of the wheel is $\overline{k} = 90$ mm. As the wheel rolls without sliding, its angular velocity is observed to vary. Knowing that $\omega = 8$ rad/s in the position shown, determine (a) the angular velocity of the wheel when the mass center G is directly above the geometric center C, (b) the reaction at the horizontal surface at the same instant.

Fig. P17.26 and P17.27

Fig. P17.28

17.29 A hemisphere of mass m and radius r is released from rest in the position shown. Assuming that the hemisphere rolls without sliding, determine (a) its angular velocity after it has rolled through 90°, (b) the normal reaction at the surface at the same instant. [*Hint.* Note that $GO = 3r/8$ and that, by the parallel-axis theorem, $\overline{I} = \frac{2}{5}mr^2 - m(GO)^2$.]

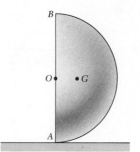

Fig. P17.29

17.30 A small sphere of mass m and radius r is released from rest at A and rolls without sliding on the curved surface to point B where it leaves the surface with a horizontal velocity. Knowing that $a = 1.5$ m and $b = 1.2$ m, determine (a) the speed of the sphere as it strikes the ground at C, (b) the corresponding distance c.

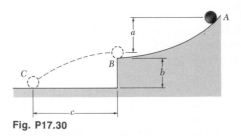

Fig. P17.30

17.31 Solve Prob. 17.30, assuming that the sphere is replaced by a thin-walled pipe of mass m and radius r.

17.32 and 17.33   Gear C has a mass of 3.2 kg and a centroidal radius of gyration of 60 mm. The uniform bar AB has a mass of 2.4 kg, and gear D is stationary. If the system is released from rest in the position shown, determine the velocity of point B after bar AB has rotated through 90°.

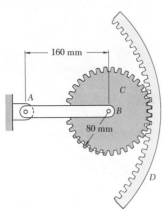

Fig. P17.32

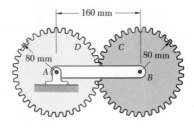

Fig. P17.33

17.34   The 9-kg rod AB is attached by pins to two 6-kg uniform disks as shown. The assembly rolls without sliding on a horizontal surface. If the assembly is released from rest when $\theta = 60°$, determine (a) the angular velocity of the disks when $\theta = 180°$, (b) the force exerted by the surface on each disk at that instant.

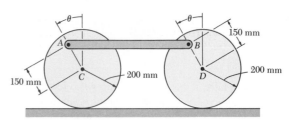

Fig. P17.34 and P17.35

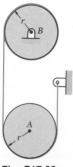

Fig. P17.36

17.35   The 9-kg rod AB is attached by pins to two 6-kg uniform disks as shown. The assembly rolls without sliding on a horizontal surface. Knowing that the velocity of rod AB is 840 mm/s to the left when $\theta = 0$, determine (a) the velocity of rod AB when $\theta = 180°$, (b) the force exerted by the surface on each disk at that instant.

17.36   Two uniform cylinders, each of weight $W = 21$ lb and radius $r = 6$ in., are connected by a belt as shown. If the system is released from rest, determine (a) the velocity of the center of cylinder A after it has moved through 4 ft, (b) the tension in the portion of the belt connecting the two cylinders.

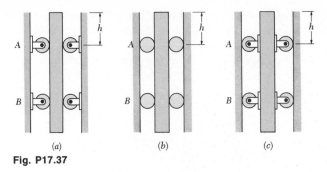

Fig. P17.37

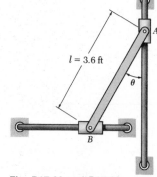

Fig. P17.38 and P17.39

17.37 A bar of weight $W = 10$ lb is held as shown between four disks each of weight $W' = 4$ lb and radius $r = 3$ in. Knowing that the forces exerted on the disks are sufficient to prevent slipping and that the bar is released from rest, for each of the cases shown determine the velocity of the bar after it has moved through the distance h.

17.38 The motion of the 6-lb slender bar AB is guided by collars of negligible weight which slide freely on the vertical and horizontal rods shown. Knowing that the bar is released from rest when $\theta = 15°$, determine the velocity of collars A and B when $\theta = 60°$.

17.39 The motion of the 5-lb slender bar AB is guided by collars of negligible weight which slide freely on the vertical and horizontal rods shown. Knowing that the bar is released from rest when $\theta = 30°$, determine the velocity of collar A when $\theta = 90°$.

17.40 The ends of a 12-kg rod AB are constrained to move along the slots shown. A spring of constant 120 N/m is attached to end A. Knowing that the rod is released from rest when $\theta = 0$ and that the initial tension in the spring is zero, determine the velocity of end A when (a) $\theta = 30°$, (b) $\theta = 60°$.

17.41 The ends of a 12-kg rod AB are constrained to move along the slots shown. A spring of constant 120 N/m is attached to end A in such a way that its tension is zero when $\theta = 0$. If the rod is released from rest when $\theta = 75°$, determine the angular velocity of the rod and the velocity of end A when $\theta = 30°$.

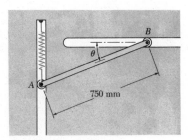

Fig. P17.40 and P17.41

17.42 The motion of the uniform rod AB is guided by small wheels of negligible mass which roll on the surface shown. If the rod is released from rest when $\theta = 0$, determine the maximum angular velocity and the corresponding velocity of end A during the ensuing motion.

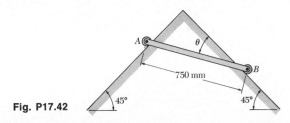

Fig. P17.42

17.43 The motion of a slender rod of length R is guided by pins at A and B which slide freely in slots cut in a vertical plate as shown. If end B is moved slightly to the left and then released, determine the angular velocity of the rod and the velocity of its mass center (*a*) at the instant when the velocity of end B is zero, (*b*) as end B passes through point D.

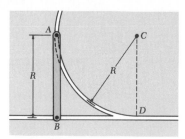

Fig. P17.43

17.44 The uniform rods AB and BC are of mass 3 kg and 8 kg, respectively, and collar C has a mass of 4 kg. If the system is released from rest in the position shown, determine the velocity of point B after rod AB has rotated through 90°.

17.45 The uniform rods AB and BC are of mass 3 kg and 8 kg, respectively, and collar C has a mass of 4 kg. Knowing that at the instant shown the velocity of collar C is 0.9 m/s downward, determine the velocity of point B after rod AB has rotated through 90°.

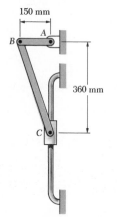

Fig. P17.44 and P17.45

17.46 Two uniform rods, each of mass m and length l, are connected to form the linkage shown. End D of rod BD may slide freely in the vertical slot, while end A of rod AB is attached to a fixed pin support. If the system is released from rest in the position shown, determine the velocity of end D at the instant when (*a*) ends A and D are at the same elevation, (*b*) rod AB is horizontal.

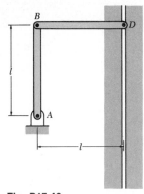

Fig. P17.46

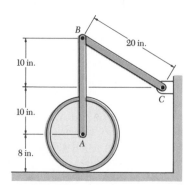

Fig. P17.47

17.47 Wheel A weighs 15 lb, has a centroidal radius of gyration of 6 in., and rolls without sliding on the horizontal surface. Each of the uniform rods AB and BC is 20 in. long and weighs 8 lb. If point A is moved slightly to the left and released, determine the velocity of point A as rod BC passes through a horizontal position.

17.48 The motor shown runs a machine attached to the shaft at A. The motor develops 7.5 hp and runs at a constant speed of 1200 rpm. Determine the magnitude of the couple exerted (a) by the shaft on pulley A, (b) by the motor on pulley B.

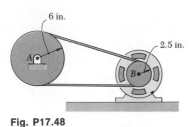

Fig. P17.48

17.49 Three shafts and four gears are used to form a gear train which will transmit 7.5 kW from the motor at A to a machine tool at F. (Bearings for the shafts are omitted in the sketch.) Knowing that the frequency of the motor is 30 Hz, determine the magnitude of the couple which is applied to shaft (a) AB, (b) CD, (c) EF.

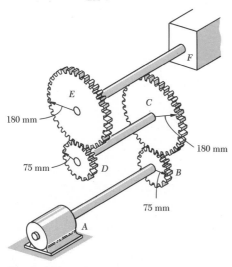

Fig. P17.49

17.50 Determine the magnitude of the couple which must be exerted by a motor to develop 400 W at a speed of (a) 3600 rpm, (b) 1200 rpm.

17.51 The experimental setup shown is used to measure the power output of a small turbine. When the turbine is operating at 200 rpm, the readings of the two spring scales are 10 and 22 lb, respectively. Determine the power being developed by the turbine.

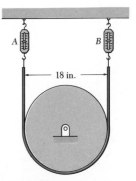

Fig. P17.51

17.8. Principle of Impulse and Momentum for the Plane Motion of a Rigid Body. We shall now apply the principle of impulse and momentum to the analysis of the plane motion of rigid bodies and of systems of rigid bodies. As was pointed out in Chap. 13, the method of impulse and momentum is particularly well adapted to the solution of problems involving time and velocities. Moreover, the principle of impulse and momentum provides the only practicable method for the solution of problems involving impulsive motion or impact (Secs. 17.11 and 17.12).

Considering again a rigid body as made of a large number of particles P_i, we recall from Sec. 14.9 that the system formed by the momenta of the particles at time t_1 and the system of the impulses of the external forces applied from t_1 to t_2 are together equipollent to the system formed by the momenta of the particles at time t_2. Since the vectors associated with a rigid body may be considered as sliding vectors, it follows (Sec. 3.19) that the systems of vectors shown in Fig. 17.6 are not only equipollent but truly *equivalent* in the sense that the vectors on the left-hand side of the equals

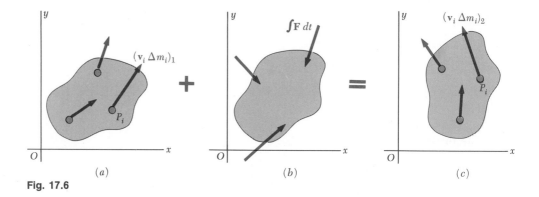

Fig. 17.6

sign may be transformed into the vectors on the right-hand side through the use of the fundamental operations listed in Sec. 3.13. We therefore write

$$\textbf{Syst Momenta}_1 + \textbf{Syst Ext Imp}_{1 \rightarrow 2} = \textbf{Syst Momenta}_2 \quad (17.14)$$

But the momenta $\mathbf{v}_i \, \Delta m_i$ of the particles may be reduced to a vector attached at G, equal to their sum

$$\mathbf{L} = \sum_{i=1}^{n} \mathbf{v}_i \, \Delta m_i$$

and to a couple of moment equal to the sum of their moments about G

$$\mathbf{H}_G = \sum_{i=1}^{n} \mathbf{r}_i' \times \mathbf{v}_i \, \Delta m_i$$

We recall from Sec. 14.3 that $\mathbf{L}$ and $\mathbf{H}_G$ define, respectively, the linear momentum and the angular momentum about G of the system of particles forming the rigid body. We also note from Eq. (14.14) that $\mathbf{L} = m\bar{\mathbf{v}}$. On the other hand, restricting the present analysis to the plane motion of a rigid slab or of a rigid body symmetrical with respect to the reference plane, we recall from Eq. (16.4) that $\mathbf{H}_G = \bar{I}\omega$. We thus conclude that the system of the momenta $\mathbf{v}_i\,\Delta m_i$ is equivalent to the *linear momentum vector* $m\bar{\mathbf{v}}$ attached at G and to the *angular momentum couple* $\bar{I}\omega$ (Fig. 17.7).

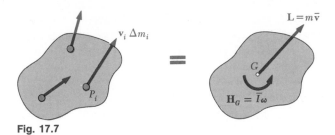

Fig. 17.7

Observing that the system of momenta reduces to the vector $m\bar{\mathbf{v}}$ in the particular case of a translation ($\omega = 0$) and to the couple $\bar{I}\omega$ in the particular case of a centroidal rotation ($\bar{\mathbf{v}} = 0$), we verify once more that the plane motion of a rigid body symmetrical with respect to the reference plane may be resolved into a translation with the mass center G and a rotation about G.

Replacing the system of momenta in parts a and c of Fig. 17.6 by the equivalent linear momentum vector and angular momentum couple, we obtain the three diagrams shown in Fig. 17.8. This figure expresses as a

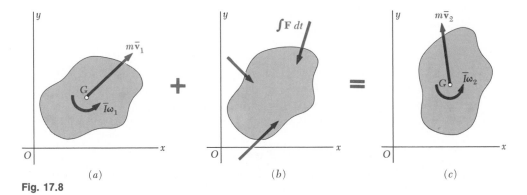

(a) (b) (c)

Fig. 17.8

free-body-diagram equation the fundamental relation (17.14) in the case of the plane motion of a rigid slab or of a rigid body symmetrical with respect to the reference plane.

Three equations of motion may be derived from Fig. 17.8. Two equations are obtained by summing and equating the x and y *components* of the momenta and impulses, and the third by summing and equating the *moments* of these vectors *about any given point*. The coordinate axes may be chosen fixed in space, or they may be allowed to move with the mass center of the body while maintaining a fixed direction. In either case, the point about which moments are taken should keep the same position relative to the coordinate axes during the interval of time considered.

In deriving the three equations of motion for a rigid body, care should be taken not to add indiscriminately linear and angular momenta. Confusion will be avoided if it is kept in mind that $m\bar{v}_x$ and $m\bar{v}_y$ represent the *components of a vector*, namely, the linear momentum vector $m\bar{v}$, while $\bar{I}\omega$ represents the *magnitude of a couple*, namely, the angular momentum couple $\bar{I}\omega$. Thus the quantity $\bar{I}\omega$ should be added only to the *moment* of the linear momentum $m\bar{v}$, never to this vector itself nor to its components. All quantities involved will then be expressed in the same units, namely N·m·s or lb·ft·s.

Noncentroidal Rotation. In this particular case of plane motion, the magnitude of the velocity of the mass center of the body is $\bar{v} = \bar{r}\omega$, where $\bar{r}$ represents the distance from the mass center to the fixed axis of rotation and ω the angular velocity of the body at the instant considered; the magnitude of the momentum vector attached at G is thus $m\bar{v} = m\bar{r}\omega$. Summing the moments about O of the momentum vector and momentum couple (Fig. 17.9) and using the parallel-axis theorem for moments of inertia, we find that the angular momentum $\mathbf{H}_O$ of the body about O has the magnitude†

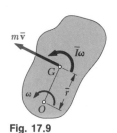

Fig. 17.9

$$\bar{I}\omega + (m\bar{r}\omega)\bar{r} = (\bar{I} + m\bar{r}^2)\omega = I_O\omega \tag{17.15}$$

Equating the moments about O of the momenta and impulses in (17.14), we write

$$I_O\omega_1 + \sum \int_{t_1}^{t_2} M_O \, dt = I_O\omega_2 \tag{17.16}$$

In the general case of plane motion of a rigid body symmetrical with respect to the reference plane, Eq. (17.16) may be used with respect to the instantaneous axis of rotation under certain conditions. It is recommended, however, that all problems of plane motion be solved by the general method described earlier in this section.

17.9. Systems of Rigid Bodies. The motion of several rigid bodies may be analyzed by applying the principle of impulse and momentum to each body separately (Sample Prob. 17.6).

However, in solving problems involving no more than three unknowns (including the impulses of unknown reactions), it is often found convenient

---

† Note that the sum $\mathbf{H}_A$ of the moments about an arbitrary point A of the momenta of the particles of a rigid slab is, in general, *not* equal to $I_A\omega$. (See Prob. 17.66.)

to apply the principle of impulse and momentum to the system as a whole. The momentum and impulse diagrams are drawn for the entire system of bodies. The diagrams of momenta should include a momentum vector, a momentum couple, or both, for each moving part of the system. Impulses of forces internal to the system may be omitted from the impulse diagram since they occur in pairs of equal and opposite vectors. Summing and equating successively the x components, y components, and moments of all vectors involved, one obtains three relations which express that the momenta at time t_1 and the impulses of the external forces form a system equipollent to the system of the momenta at time t_2.† Again, care should be taken not to add indiscriminately linear and angular momenta; each equation should be checked to make sure that consistent units have been used. This approach has been used in Sample Prob. 17.8 and, further on, in Sample Probs. 17.9 and 17.10.

17.10. Conservation of Angular Momentum. When no external force acts on a rigid body or a system of rigid bodies, the impulses of the external forces are zero and the system of the momenta at time t_1 is equipollent to the system of the momenta at time t_2. Summing and equating successively the x components, y components, and moments of the momenta at times t_1 and t_2, we conclude that the total linear momentum of the system is conserved in any direction and that its total angular momentum is conserved about any point.

There are many engineering applications, however, in which *the linear momentum is not conserved* yet *the angular momentum* $\mathbf{H}_O$ *of the system about a given point* O *is conserved:*

$$(\mathbf{H}_O)_1 = (\mathbf{H}_O)_2 \qquad (17.17)$$

Such cases occur when the lines of action of all external forces pass through O or, more generally, when the sum of the angular impulses of the external forces about O is zero.

Problems involving *conservation of angular momentum* about a point O may be solved by the general method of impulse and momentum, i.e., by drawing momentum and impulse diagrams as described in Secs. 17.8 and 17.9. Equation (17.17) is then obtained by summing and equating moments about O (Sample Prob. 17.8). As we shall see later in Sample Prob. 17.9, two additional equations may be written by summing and equating x and y components; these equations may be used to determine two unknown linear impulses, such as the impulses of the reaction components at a fixed point.

---

†Note that as in Sec. 16.7, we cannot speak of *equivalent* systems since we are not dealing with a single rigid body.

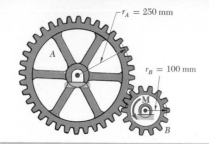

$r_A = 250$ mm

$r_B = 100$ mm

SAMPLE PROBLEM 17.6

Gear A has a mass of 10 kg and a radius of gyration of 200 mm, while gear B has a mass of 3 kg and a radius of gyration of 80 mm. The system is at rest when a couple **M** of magnitude 6 N·m is applied to gear B. Neglecting friction, determine (a) the time required for the angular velocity of gear B to reach 600 rpm, (b) the tangential force which gear B exerts on gear A. These gears have been previously considered in Sample Prob. 17.2.

Solution. We apply the principle of impulse and momentum to each gear separately. Since all forces and the couple are constant, their impulses are obtained by multiplying them by the unknown time t. We recall from Sample Prob. 17.2 that the centroidal moments of inertia and the final angular velocities are

$$\bar{I}_A = 0.400 \text{ kg·m}^2 \qquad \bar{I}_B = 0.0192 \text{ kg·m}^2$$
$$(\omega_A)_2 = 25.1 \text{ rad/s} \qquad (\omega_B)_2 = 62.8 \text{ rad/s}$$

Principle of Impulse and Momentum for Gear A. The systems of initial momenta, impulses, and final momenta are shown in three separate sketches.

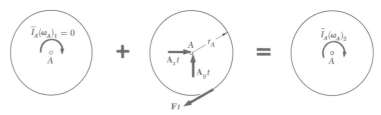

Syst Momenta₁ + Syst Ext Imp₁→₂ = Syst Momenta₂

$+\uparrow$ moments about A: $\qquad 0 - Ftr_A = -\bar{I}_A(\omega_A)_2$

$$Ft(0.250 \text{ m}) = (0.400 \text{ kg·m}^2)(25.1 \text{ rad/s})$$
$$Ft = 40.2 \text{ N·s}$$

Principle of Impulse and Momentum for Gear B.

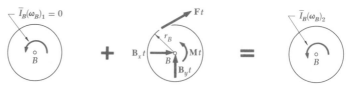

Syst Momenta₁ + Syst Ext Imp₁→₂ = Syst Momenta₂

$+\uparrow$ moments about B: $\qquad 0 + Mt - Ftr_B = \bar{I}_B(\omega_B)_2$

$$+(6 \text{ N·m})t - (40.2 \text{ N·s})(0.100 \text{ m}) = (0.0192 \text{ kg·m}^2)(62.8 \text{ rad/s})$$

$$t = 0.871 \text{ s} \quad \blacktriangleleft$$

Recalling that $Ft = 40.2$ N·s, we write

$$F(0.871 \text{ s}) = 40.2 \text{ N·s} \qquad F = +46.2 \text{ N}$$

Thus, the force exerted by gear B on gear A is $\qquad$ **F** = 46.2 N $\nearrow$ $\quad \blacktriangleleft$

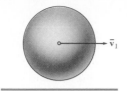

SAMPLE PROBLEM 17.7

A uniform sphere of mass m and radius r is projected along a rough horizontal surface with a linear velocity $\bar{\mathbf{v}}_1$ and no angular velocity. Denoting by μ_k the coefficient of kinetic friction between the sphere and the surface, determine (a) the time t_2 at which the sphere will start rolling without sliding, (b) the linear and angular velocities of the sphere at time t_2.

Solution. While the sphere is sliding relative to the surface, it is acted upon by the normal force $\mathbf{N}$, the friction force $\mathbf{F}$, and its weight $\mathbf{W}$ of magnitude $W = mg$.

Principle of Impulse and Momentum. We apply the principle of impulse and momentum to the sphere from the time $t_1 = 0$ when it is placed on the surface until the time $t_2 = t$ when it starts rolling without sliding.

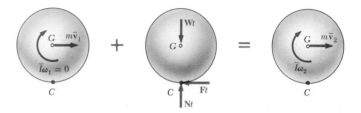

$$\textbf{Syst Momenta}_1 + \textbf{Syst Ext Imp}_{1 \to 2} = \textbf{Syst Momenta}_2$$

$+\uparrow y$ components:
$$Nt - Wt = 0 \tag{1}$$

$\xrightarrow{+} x$ components:
$$m\bar{v}_1 - Ft = m\bar{v}_2 \tag{2}$$

$+\downarrow$ moments about G:
$$Ftr = \bar{I}\omega_2 \tag{3}$$

From (1) we obtain $N = W = mg$. During the entire time interval considered, sliding occurs at point C and we have $F = \mu_k N = \mu_k mg$. Substituting for F into (2), we write

$$m\bar{v}_1 - \mu_k mgt = m\bar{v}_2 \qquad \bar{v}_2 = \bar{v}_1 - \mu_k gt \tag{4}$$

Substituting $F = \mu_k mg$ and $\bar{I} = \tfrac{2}{5}mr^2$ into (3),

$$\mu_k mgtr = \tfrac{2}{5}mr^2\omega_2 \qquad \omega_2 = \frac{5}{2}\frac{\mu_k g}{r}t \tag{5}$$

The sphere will start rolling without sliding when the velocity $\mathbf{v}_C$ of the point of contact is zero. At that time, point C becomes the instantaneous center of rotation, and we have $\bar{v}_2 = r\omega_2$. Substituting from (4) and (5), we write

$$\bar{v}_2 = r\omega_2 \qquad \bar{v}_1 - \mu_k gt = r\left(\frac{5}{2}\frac{\mu_k g}{r}t\right) \qquad t = \frac{2}{7}\frac{\bar{v}_1}{\mu_k g} \quad \blacktriangleleft$$

Substituting this expression for t into (5),

$$\omega_2 = \frac{5}{2}\frac{\mu_k g}{r}\left(\frac{2}{7}\frac{\bar{v}_1}{\mu_k g}\right) \qquad \omega_2 = \frac{5}{7}\frac{\bar{v}_1}{r} \qquad \omega_2 = \frac{5}{7}\frac{\bar{v}_1}{r}\,\downarrow \quad \blacktriangleleft$$

$$\bar{v}_2 = r\omega_2 \qquad \bar{v}_2 = r\left(\frac{5}{7}\frac{v_1}{r}\right) \qquad \bar{v}_2 = \tfrac{5}{7}\bar{v}_1 \rightarrow \quad \blacktriangleleft$$

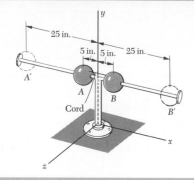

SAMPLE PROBLEM 17.8

Two solid spheres of radius 3 in., weighing 2 lb each, are mounted at A and B on the horizontal rod $A'B'$, which rotates freely about the vertical with a counterclockwise angular velocity of 6 rad/s. The spheres are held in position by a cord which is suddenly cut. Knowing that the centroidal moment of inertia of the rod and pivot is $\bar{I}_R = 0.25$ lb·ft·s², determine (a) the angular velocity of the rod after the spheres have moved to positions A' and B', (b) the energy lost due to the plastic impact of the spheres and the stops at A' and B'.

a. **Principle of Impulse and Momentum.** In order to determine the final angular velocity of the rod, we shall express that the initial momenta of the various parts of the system and the impulses of the external forces are together equipollent to the final momenta of the system.

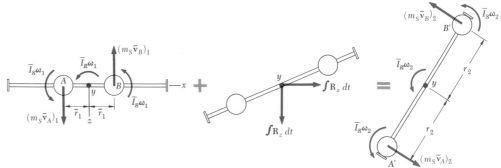

$$\text{Syst Momenta}_1 + \text{Syst Ext Imp}_{1\to2} = \text{Syst Momenta}_2$$

Observing that the external forces consist of the weights and the reaction at the pivot, which have no moment about the y axis, and noting that $\bar{v}_A = \bar{v}_B = \bar{r}\omega$, we write

$+\uparrow$moments about y axis:

$$2(m_S\bar{r}_1\omega_1)\bar{r}_1 + 2\bar{I}_S\omega_1 + \bar{I}_R\omega_1 = 2(m_S\bar{r}_2\omega_2)\bar{r}_2 + 2\bar{I}_S\omega_2 + \bar{I}_R\omega_2$$
$$(2m_S\bar{r}_1^2 + 2\bar{I}_S + \bar{I}_R)\omega_1 = (2m_S\bar{r}_2^2 + 2\bar{I}_S + \bar{I}_R)\omega_2 \qquad (1)$$

which expresses that *the angular momentum of the system about the y axis is conserved.* We now compute

$$\bar{I}_S = \tfrac{2}{5}m_S a^2 = \frac{2}{5}\left(\frac{2\text{ lb}}{32.2\text{ ft/s}^2}\right)(\tfrac{3}{12}\text{ ft})^2 = 0.00155\text{ lb·ft·s}^2$$

$$m_S\bar{r}_1^2 = \frac{2}{32.2}\left(\frac{5}{12}\right)^2 = 0.0108 \qquad m_S\bar{r}_2^2 = \frac{2}{32.2}\left(\frac{25}{12}\right)^2 = 0.2696$$

Substituting these values and $\bar{I}_R = 0.25$, $\omega_1 = 6$ rad/s into (1):

$$0.275(6\text{ rad/s}) = 0.792\omega_2 \qquad \omega_2 = 2.08\text{ rad/s}\uparrow \quad \blacktriangleleft$$

b. **Energy Lost.** The kinetic energy of the system at any instant is

$$T = 2(\tfrac{1}{2}m_S\bar{v}^2 + \tfrac{1}{2}\bar{I}_S\omega^2) + \tfrac{1}{2}\bar{I}_R\omega^2 = \tfrac{1}{2}(2m_S\bar{r}^2 + 2\bar{I}_S + \bar{I}_R)\omega^2$$

Recalling the numerical values found above, we have

$$T_1 = \tfrac{1}{2}(0.275)(6)^2 = 4.95\text{ ft·lb} \qquad T_2 = \tfrac{1}{2}(0.792)(2.08)^2 = 1.713\text{ ft·lb}$$
$$\Delta T = T_2 - T_1 = 1.71 - 4.95 \qquad \Delta T = -3.24\text{ ft·lb} \quad \blacktriangleleft$$

Problems

17.52 A small grinding wheel is attached to the shaft of an electric motor which has a rated speed of 3600 rpm. When the power is turned off, the unit coasts to rest in 70 s. The grinding wheel and rotor have a combined weight of 6 lb and a combined radius of gyration of 2 in. Determine the average magnitude of the couple due to kinetic friction in the bearings of the motor.

Fig. P17.52

17.53 A large flywheel has a mass of 2750 kg and a radius of gyration of 1.15 m. It is observed that 12.8 min is required for the flywheel to coast to rest from an angular velocity of 180 rpm. Determine the average magnitude of the couple due to kinetic friction in the bearings of the flywheel.

17.54 Three disks of the same thickness and same material are attached to a shaft as shown. Disks A and B each have a mass of 6 kg and a radius $r =$ 175 mm. A couple **M** of magnitude 15 N·m is applied to disk A when the system is at rest. Determine the radius nr of disk C if the angular velocity of the system is to be 1200 rpm after 4 s.

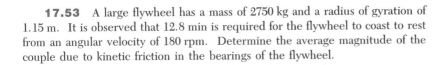

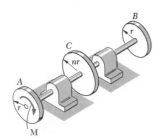

Fig. P17.54

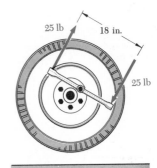

Fig. P17.55

17.55 A bolt located 2 in. from the center of an automobile wheel is tightened by applying the couple shown for 0.10 s. Assuming that the wheel is free to rotate and is initially at rest, determine the resulting angular velocity of the wheel. The wheel weighs 42 lb and has a radius of gyration of 10.8 in.

17.56 A sphere of radius r and weight W is placed in a corner with an initial counterclockwise angular velocity ω_0. Denoting by μ_k the coefficient of kinetic friction at A and B, derive an expression for the time required for the sphere to come to rest.

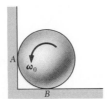

Fig. P17.56

17.57 A disk of constant thickness, initially at rest, is placed in contact with a belt which moves with a constant velocity **v**. The link AB connecting the center of the disk to the support at B has a negligible weight. Denoting by μ_k the coefficient of kinetic friction between the disk and the belt, derive an expression for the time required for the disk to reach a constant angular velocity.

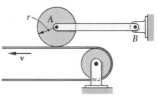

Fig. P17.57 and P17.58

17.58 Disk A has a mass of 2.5 kg and a radius $r = 75$ mm; it is placed in contact with the belt, which moves at a constant speed $v = 15$ m/s. The link AB connecting the center of the disk to the support at B has a negligible weight. Knowing that $\mu_k = 0.15$ between the disk and the belt, determine the time required for the disk to reach a constant angular velocity.

17.59 Each of the double pulleys shown has a centroidal mass moment of inertia of 0.25 kg·m², an inner radius of 100 mm, and an outer radius of 150 mm. Neglecting bearing friction, determine (*a*) the velocity of the cylinder 3 s after the system is released from rest, (*b*) the tension in the cord connecting the pulleys.

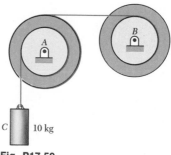

Fig. P17.59

17.60 Solve Prob. 17.59, assuming that the bearing friction at A and at B is equivalent to couples of magnitude 0.45 N·m.

17.61 Disk A is at rest when it is brought into contact with disk B, which has an initial angular velocity ω_0. Show that the final angular velocity of disk B depends only on ω_0 and the ratio of the masses m_A and m_B of the two disks.

17.62 Disk B has a mass $m_B = 4$ kg, a radius $r_B = 90$ mm, and an initial angular velocity $\omega_0 = 750$ rpm clockwise. Disk A has a mass $m_A = 6$ kg, a radius $r_A = 135$ mm, and is at rest when it is brought into contact with disk B. Neglecting friction in the bearings, determine (*a*) the final angular velocity of each disk, (*b*) the total impulse of the friction force which acted on disk A.

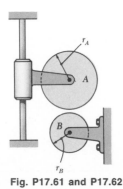

Fig. P17.61 and P17.62

17.63 A computer tape moves over the two drums shown. Drum A weighs 1.4 lb and has a radius of gyration of 0.75 in., while drum B weighs 3.5 lb and has a radius of gyration of 1.25 in. In the lower portion of the tape the tension is constant and equal to $T_A = 0.75$ lb. Knowing that the tape is initially at rest, determine (a) the required constant tension T_B if the velocity of the tape is to be $v = 10$ ft/s after 0.24 s, (b) the corresponding tension in the portion of tape between the drums.

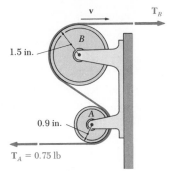

Fig. P17.63

17.64 Show that the system of momenta for a rigid slab in plane motion reduces to a single vector, and express the distance from the mass center G to the line of action of this vector in terms of the centroidal radius of gyration $\bar{k}$ of the slab, the magnitude $\bar{v}$ of the velocity of G, and the angular velocity ω.

17.65 Show that, when a rigid slab rotates about a fixed axis through O perpendicular to the slab, the system of momenta of its particles is equivalent to a single vector of magnitude $m\bar{r}\omega$, perpendicular to the line OG, and applied to a point P on this line, called the *center of percussion*, at a distance $GP = \bar{k}^2/\bar{r}$ from the mass center of the slab.

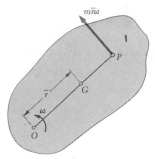

Fig. P17.65

17.66 Show that the sum $\mathbf{H}_A$ of the moments about a point A of the momenta of the particles of a rigid slab in plane motion is equal to $I_A\omega$, where ω is the angular velocity of the slab at the instant considered and I_A the moment of inertia of the slab about A, if and only if one of the following conditions is satisfied: (a) A is the mass center of the slab, (b) A is the instantaneous center of rotation, (c) the velocity of A is directed along a line joining point A and the mass center G.

17.67 Consider a rigid slab initially at rest and subjected to an impulsive force $\mathbf{F}$ contained in the plane of the slab. We define the *center of percussion P* as the point of intersection of the line of action of $\mathbf{F}$ with the perpendicular drawn from G. (a) Show that the instantaneous center of rotation C of the slab is located on line GP at a distance $GC = \bar{k}^2/GP$ on the opposite side of G. (b) Show that if the center of percussion were located at C the instantaneous center of rotation would be located at P.

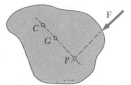

Fig. P17.67

17.68 A wheel of radius r and centroidal radius of gyration $\bar{k}$ is placed on an incline and released from rest at time $t = 0$. Assuming that the wheel rolls without slipping, determine (a) the velocity of the center at time t, (b) the coefficient of static friction required to prevent slipping.

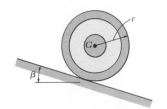

Fig. P17.68

17.69 A 15-kg uniform cylindrical roller, initially at rest, is acted upon by a 125-N force as shown. Assuming that the body rolls without slipping, determine (a) the velocity of the center G after 5 s, (b) the friction force required to prevent slipping.

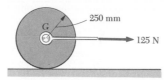

250 mm

G

125 N

Fig. P17.69

17.70 Cords are wrapped around a thin-walled pipe and a solid cylinder as shown. Knowing that the pipe and the cylinder are each released from rest at time $t = 0$, determine at time t the velocity of the center of (a) the pipe, (b) the cylinder.

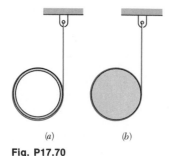

(a)   (b)

Fig. P17.70

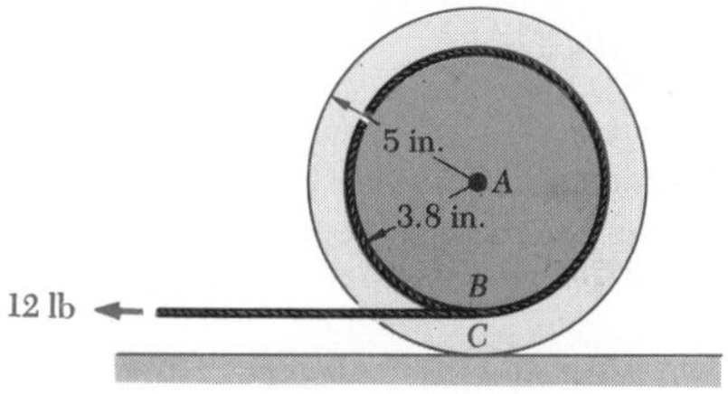

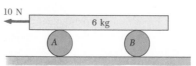

5 in.

A

3.8 in.

B

12 lb

C

Fig. P17.71

17.71 A drum of radius 3.8 in. is mounted on a cylinder of radius 5 in. The drum-cylinder unit weighs 18 lb and has a centroidal radius of gyration of 3 in. A cord is wrapped around the drum and pulled with a force of magnitude 12 lb. Knowing that the unit is initially at rest, determine (a) the velocity of center A of the wheel after 6 s, (b) the friction force required to prevent slipping.

17.72 and 17.73 The 6-kg carriage is supported as shown by two uniform disks, each having a mass of 4 kg and a radius of 75 mm. Knowing that the carriage is initially at rest, determine the velocity of the carriage 2.5 s after the 10-N force has been applied. Assume that the disks roll without sliding.

10 N

6 kg

A B

Fig. P17.72

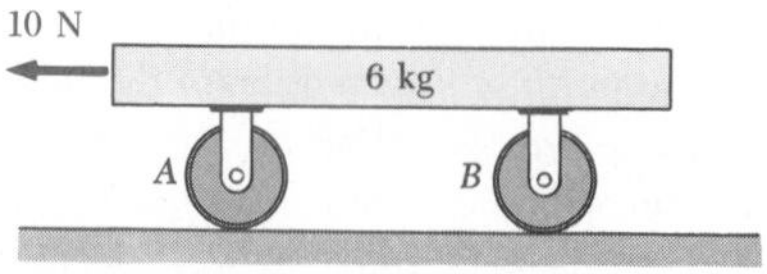

10 N

6 kg

A B

Fig. P17.73

P

Fig. P17.74

17.74 A 160-mm-diameter pipe of mass 6 kg rests on a 1.5-kg plate. The pipe and plate are initially at rest when a force $\mathbf{P}$ of magnitude 25 N is applied for 0.75 s. Knowing that $\mu_s = 0.25$ and $\mu_k = 0.20$ between the plate and both the pipe and the floor, determine (a) whether the pipe slides with respect to the plate, (b) the resulting velocities of the pipe and of the plate.

17.75 In the gear arrangement shown, gears A and C are attached to rod ABC, which is free to rotate about B, while the inner gear B is fixed. Knowing that the system is at rest, determine the magnitude of the couple $\mathbf{M}$ which must be applied to rod ABC, if 2.5 s later the angular velocity of the rod is to be 240 rpm clockwise. Gears A and C weigh 2.5 lb each and may be considered as disks of radius 2 in.; rod ABC weighs 4 lb.

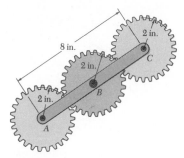

Fig. P17.75

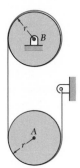

Fig. P17.76

17.76 Two uniform cylinders, each of weight $W = 21$ lb and radius $r = 5$ in., are connected by a belt as shown. If the system is released from rest, determine (*a*) the velocity of the center of cylinder A after 2.5 s, (*b*) the tension in the portion of the belt connecting the two cylinders.

17.77 A sphere of radius r and mass m is placed on a horizontal surface with no linear velocity but with a clockwise angular velocity ω_0. Denoting by μ_k the coefficient of kinetic friction between the sphere and the floor, determine (*a*) the time t_1 at which the sphere will start rolling without sliding, (*b*) the linear and angular velocities of the sphere at time t_1.

Fig. P17.77

Fig. P17.78

17.78 A sphere of mass m and radius r is projected along a rough horizontal surface with the initial velocities indicated. If the final velocity of the sphere is to be zero, express (*a*) the required ω_0 in terms of $\bar{v}_0$ and r, (*b*) the time required for the sphere to come to rest in terms of $\bar{v}_0$ and the coefficient of friction μ_k.

17.79 Disks A and B are made of the same material and are of the same thickness; they may rotate freely about the vertical shaft. Disk B is at rest when it is dropped onto disk A which is rotating with an angular velocity of 400 rpm. Knowing that the mass of disk A is 4 kg, determine (*a*) the final angular velocity of the disks, (*b*) the change in kinetic energy of the system.

17.80 In Prob. 17.79, show that if both disks are initially rotating, the change in kinetic energy ΔT of the system depends only upon the initial relative velocity $\omega_{B/A}$ of the disks, and derive an expression for ΔT in terms of $\omega_{B/A}$.

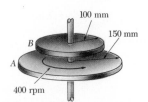

Fig. P17.79

17.81 Rod AB is of mass m and slides freely inside tube CD which is also of mass m. The angular velocity of the assembly was ω_1 when the rod was entirely inside the tube ($x = 0$). Neglecting the effect of friction, determine the angular velocity of the assembly when $x = \frac{2}{3}L$.

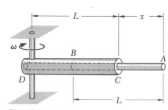

Fig. P17.81

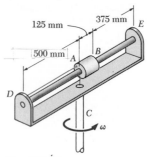

Fig. P17.82

17.82 A 1.6-kg tube AB may slide freely on rod DE which in turn may rotate freely in a horizontal plane. Initially the assembly is rotating with an angular velocity $\omega = 5$ rad/s and the tube is held in position by a cord. The moment of inertia of the rod and bracket about the vertical axis of rotation is 0.30 kg·m^2 and the centroidal moment of inertia of the tube about a vertical axis is 0.0025 kg·m^2. If the cord suddenly breaks, determine (a) the angular velocity of the assembly after the tube has moved to end E, (b) the energy lost during the plastic impact at E.

17.83 Two semicircular panels, each of radius r, are attached with hinges to a square plate as shown. The plate and the panels are made of the same material and have the same thickness. Knowing that when the panels are vertical the assembly is rotating with an angular velocity ω_0, determine the angular velocity of the assembly after the panels have come to rest in a horizontal position.

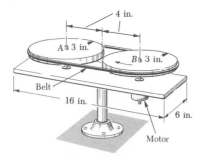

Fig. P17.83

17.84 Two 10-lb disks and a small motor are mounted on a 15-lb rectangular platform which is free to rotate about a central vertical spindle. The normal operating speed of the motor is 180 rpm. If the motor is started when the system is at rest, determine the angular velocity of all elements of the system after the motor has attained its normal operating speed. Neglect the mass of the motor and of the belt.

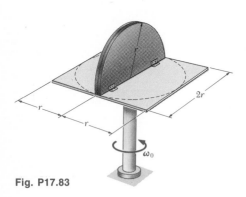

Fig. P17.84

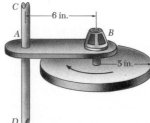

Fig. P17.85

17.85 A 10-lb disk is attached to the shaft of a motor mounted on arm AB which is free to rotate about the vertical axle CD. The arm-and-motor unit has a moment of inertia of 0.032 lb·ft·s^2 with respect to axle CD, and the normal operating speed of the motor is 360 rpm. Knowing that the system is initially at rest, determine the angular velocities of the arm and of the disk when the motor reaches a speed of 360 rpm.

Fig. P17.86

17.86 In the helicopter shown, a vertical tail propeller is used to prevent rotation of the cab as the speed of the main blades is changed. Assuming that the tail propeller is not operating, determine the final angular velocity of the cab after the speed of the main blades has been changed from 200 to 300 rpm. The speed of the main blades is measured relative to the cab, which has a centroidal moment of inertia of 1200 kg·m². Each of the four main blades is assumed to be a 5-m slender rod of mass 30 kg.

17.87 Assuming that the tail propeller in Prob. 17.86 is operating and that the angular velocity of the cab remains zero, determine the final horizontal velocity of the cab when the speed of the main blades is changed from 200 to 300 rpm. The cab has a mass of 720 kg and is initially at rest. Also determine the force exerted by the tail propeller if this change in speed takes place uniformly in 12 s.

17.88 Collar B has a mass of 3 kg and may slide freely on rod OA, which in turn may rotate freely in the horizontal plane. The assembly is rotating with an angular velocity $\omega = 1.8$ rad/s when a spring located between A and B is released, projecting the collar along the rod with an initial relative speed $v_r = 1.5$ m/s. Knowing that the moment of inertia about O of the rod and spring is 0.35 kg·m², determine (*a*) the minimum distance between the collar and point O in the ensuing motion, (*b*) the corresponding angular velocity of the assembly.

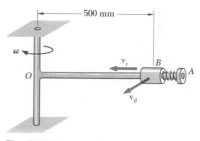

Fig. P17.88

17.89 In Prob. 17.88, determine the required magnitude of the initial relative velocity $\mathbf{v}_r$ if during the ensuing motion the minimum distance between collar B and point O is to be 300 mm.

17.90 In Prob. 17.82, determine the velocity of the tube relative to the rod as the tube strikes end E of the assembly.

17.91 In Prob. 17.81, determine the velocity of the rod relative to the tube when $x = \frac{2}{3}L$.

17.92 In Prob. 17.81, determine the velocity of the rod relative to the tube when $x = \frac{1}{2}L$.

*17.93 A uniform rod AB, of weight 15 lb and length 3.6 ft, is attached to the 25-lb cart C. Knowing that the system is released from rest in the position shown and neglecting friction, determine (*a*) the velocity of point B as rod AB passes through a vertical position, (*b*) the corresponding velocity of cart C.

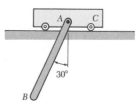

Fig. P17.93

*17.94 The 4-kg cylinder B and the 3-kg wedge A are at rest in the position shown. Cord C connecting the cylinder and the wedge is then cut and the cylinder rolls without sliding on the wedge. Neglecting friction between the wedge and the ground, determine (*a*) the angular velocity of the cylinder after it has rolled 180 mm down the wedge, (*b*) the corresponding velocity of the wedge.

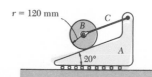

Fig. P17.94

870

Plane Motion of Rigid Bodies.
Energy and Momentum Methods

17.11. Impulsive Motion. We saw in Chap. 13 that the method of impulse and momentum is the only practicable method for the solution of problems involving impulsive motion. Now we shall also find that compared with the various problems considered in the preceding sections, problems involving impulsive motion are particularly well adapted to a solution by the method of impulse and momentum. The computation of linear impulses and angular impulses is quite simple, since the time interval considered being very short, the bodies involved may be assumed to occupy the same position during that time interval.

17.12. Eccentric Impact. In Secs. 13.13 and 13.14, we learned to solve problems of *central impact*, i.e., problems in which the mass centers of the two colliding bodies are located on the line of impact. We shall now analyze the *eccentric impact* of two rigid bodies. Consider two bodies which collide, and denote by $\mathbf{v}_A$ and $\mathbf{v}_B$ the velocities before impact *of the two points of contact A and B* (Fig. 17.10a). Under the impact, the two

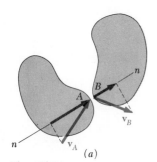

(a)

Fig. 17.10

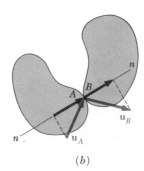

(b)

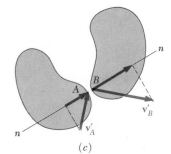

(c)

bodies will *deform,* and at the end of the period of deformation, the velocities $\mathbf{u}_A$ and $\mathbf{u}_B$ of A and B will have equal components along the line of impact nn (Fig. 17.10b). A period of *restitution* will then take place, at the end of which A and B will have velocities $\mathbf{v}'_A$ and $\mathbf{v}'_B$ (Fig. 17.10c). Assuming that the bodies are frictionless, we find that the forces they exert on each other are directed along the line of impact. Denoting, respectively, by $\int P\, dt$ and $\int R\, dt$ the magnitude of the impulse of one of these forces during the period of deformation and during the period of restitution, we recall that the coefficient of restitution e is defined as the ratio

$$e = \frac{\int R\, dt}{\int P\, dt} \tag{17.18}$$

We propose to show that the relation established in Sec. 13.13 between the relative velocities of two particles before and after impact also holds between the components along the line of impact of the relative velocities of the two points of contact A and B. We propose to show, therefore, that

$$(v'_B)_n - (v'_A)_n = e[(v_A)_n - (v_B)_n] \tag{17.19}$$

We shall first assume that the motion of each of the two colliding bodies of Fig. 17.10 is unconstrained. Thus the only impulsive forces exerted on the bodies during the impact are applied at A and B, respectively. Consider the body to which point A belongs and draw the three momentum and impulse diagrams corresponding to the period of deformation (Fig. 17.11). We denote by $\bar{v}$ and $\bar{u}$, respectively, the velocity of the mass center

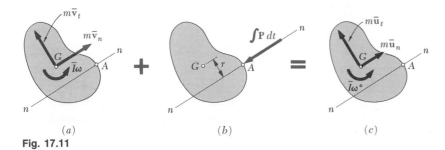

(a)   (b)   (c)

Fig. 17.11

at the beginning and at the end of the period of deformation, and by ω and $\omega°$ the angular velocity of the body at the same instants. Summing and equating the components of the momenta and impulses along the line of impact nn, we write

$$m\bar{v}_n - \int P\, dt = m\bar{u}_n \tag{17.20}$$

Summing and equating the moments about G of the momenta and impulses, we also write

$$\bar{I}\omega - r\int P\, dt = \bar{I}\omega° \tag{17.21}$$

where r represents the perpendicular distance from G to the line of impact. Considering now the period of restitution, we obtain in a similar way

$$m\bar{u}_n - \int R\, dt = m\bar{v}'_n \tag{17.22}$$

$$\bar{I}\omega° - r\int R\, dt = \bar{I}\omega' \tag{17.23}$$

where $\bar{v}'$ and ω' represent, respectively, the velocity of the mass center and the angular velocity of the body after impact. Solving (17.20) and (17.22) for the two impulses and substituting into (17.18), and then solving (17.21) and (17.23) for the same two impulses and substituting again into (17.18), we obtain the following two alternative expressions for the coefficient of restitution:

$$e = \frac{\bar{u}_n - \bar{v}'_n}{\bar{v}_n - \bar{u}_n} \qquad e = \frac{\omega° - \omega'}{\omega - \omega°} \qquad (17.24)$$

Multiplying by r the numerator and denominator of the second expression obtained for e, and adding respectively to the numerator and denominator of the first expression, we have

$$e = \frac{\bar{u}_n + r\omega° - (\bar{v}'_n + r\omega')}{\bar{v}_n + r\omega - (\bar{u}_n + r\omega°)} \qquad (17.25)$$

Observing that $\bar{v}_n + r\omega$ represents the component $(v_A)_n$ along nn of the velocity of the point of contact A and that, similarly, $\bar{u}_n + r\omega°$ and $\bar{v}'_n + r\omega'$ represent, respectively, the components $(u_A)_n$ and $(v'_A)_n$, we write

$$e = \frac{(u_A)_n - (v'_A)_n}{(v_A)_n - (u_A)_n} \qquad (17.26)$$

The analysis of the motion of the second body leads to a similar expression for e in terms of the components along nn of the successive velocities of point B. Recalling that $(u_A)_n = (u_B)_n$, and eliminating these two velocity components by a manipulation similar to the one used in Sec. 13.13, we obtain relation (17.19).

If one or both of the colliding bodies is constrained to rotate about a fixed point O, as in the case of a compound pendulum (Fig. 17.12a), an impulsive reaction will be exerted at O (Fig. 17.12b). We shall verify that while their derivation must be modified, Eqs. (17.26) and (17.19) remain valid. Applying formula (17.16) to the period of deformation and to the period of restitution, we write

$$I_O\omega - r\int P\,dt = I_O\omega° \qquad (17.27)$$
$$I_O\omega° - r\int R\,dt = I_O\omega' \qquad (17.28)$$

where r represents the perpendicular distance from the fixed point O to the line of impact. Solving (17.27) and (17.28) for the two impulses and substituting into (17.18), and then observing that $r\omega$, $r\omega°$, and $r\omega'$ represent the components along nn of the successive velocities of point A, we write

$$e = \frac{\omega° - \omega'}{\omega - \omega°} = \frac{r\omega° - r\omega'}{r\omega - r\omega°} = \frac{(u_A)_n - (v'_A)_n}{(v_A)_n - (u_A)_n}$$

and check that Eq. (17.26) still holds. Thus Eq. (17.19) remains valid when one or both of the colliding bodies is constrained to rotate about a fixed point O.

In order to determine the velocities of the two colliding bodies after impact, relation (17.19) should be used in conjunction with one or several other equations obtained by applying the principle of impulse and momentum (Sample Prob. 17.10).

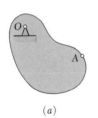

(a)

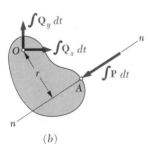

(b)

Fig. 17.12

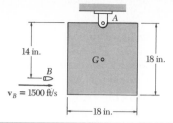

SAMPLE PROBLEM 17.9

A 0.05-lb bullet B is fired with a horizontal velocity of 1500 ft/s into the side of a 20-lb square panel suspended from a hinge at A. Knowing that the panel is initially at rest, determine (a) the angular velocity of the panel immediately after the bullet becomes embedded, (b) the impulsive reaction at A, assuming that the bullet becomes embedded in 0.0006 s.

Solution. *Principle of Impulse and Momentum.* We consider the bullet and the panel as a single system and express that the initial momenta of the bullet and panel and the impulses of the external forces are together equipollent to the final momenta of the system. Since the time interval $\Delta t = 0.0006$ s is very short, we neglect all nonimpulsive forces and consider only the external impulses $\mathbf{A}_x\,\Delta t$ and $\mathbf{A}_y\,\Delta t$.

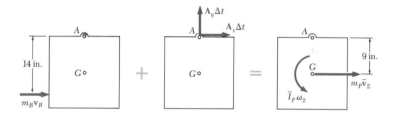

$$\text{Syst Momenta}_1 + \text{Syst Ext Imp}_{1\to2} = \text{Syst Momenta}_2$$

$+\!\uparrow$ moments about A: $\qquad m_B v_B(\tfrac{14}{12}\text{ ft}) + 0 = m_P \bar{v}_2(\tfrac{9}{12}\text{ ft}) + \bar{I}_P \omega_2 \qquad (1)$

$\xrightarrow{+}\; x$ components: $\qquad\qquad m_B v_B + A_x\,\Delta t = m_P \bar{v}_2 \qquad\qquad\quad (2)$

$+\!\uparrow y$ components: $\qquad\qquad\qquad 0 + A_y\,\Delta t = 0 \qquad\qquad\qquad\quad (3)$

The centroidal mass moment of inertia of the square panel is

$$\bar{I}_P = \tfrac{1}{6}m_P b^2 = \frac{1}{6}\left(\frac{20\text{ lb}}{32.2}\right)(\tfrac{18}{12}\text{ ft})^2 = 0.2329\text{ lb}\cdot\text{ft}\cdot\text{s}^2$$

Substituting this value as well as the given data into (1) and noting that

$$\bar{v}_2 = (\tfrac{9}{12}\text{ ft})\omega_2$$

we write

$$\left(\frac{0.05}{32.2}\right)(1500)(\tfrac{14}{12}) = 0.2329\omega_2 + \left(\frac{20}{32.2}\right)(\tfrac{9}{12}\omega_2)(\tfrac{9}{12})$$

$$\omega_2 = 4.67\text{ rad/s} \qquad\qquad \omega_2 = 4.67\text{ rad/s}\!\uparrow \quad \blacktriangleleft$$

$$\bar{v}_2 = (\tfrac{9}{12}\text{ ft})\omega_2 = (\tfrac{9}{12}\text{ ft})(4.67\text{ rad/s}) = 3.50\text{ ft/s}$$

Substituting $\bar{v}_2 = 3.50$ ft/s and $\Delta t = 0.0006$ s into Eq. (2), we have

$$\left(\frac{0.05}{32.2}\right)(1500) + A_x(0.0006) = \left(\frac{20}{32.2}\right)(3.50)$$

$$A_x = -259\text{ lb} \qquad\qquad \mathbf{A}_x = 259\text{ lb} \leftarrow \quad \blacktriangleleft$$

From Eq. (3), we find $\qquad\qquad A_y = 0 \qquad\qquad\qquad \mathbf{A}_y = 0 \quad \blacktriangleleft$

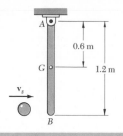

SAMPLE PROBLEM 17.10

A 2-kg sphere moving horizontally to the right with an initial velocity of 5 m/s strikes the lower end of an 8-kg rigid rod AB. The rod is suspended from a hinge at A and is initially at rest. Knowing that the coefficient of restitution between the rod and sphere is 0.80, determine the angular velocity of the rod and the velocity of the sphere immediately after the impact.

Principle of Impulse and Momentum. We consider the rod and sphere as a single system and express that the initial momenta of the rod and sphere and the impulses of the external forces are together equipollent to the final momenta of the system. We note that the only impulsive force external to the system is the impulsive reaction at A.

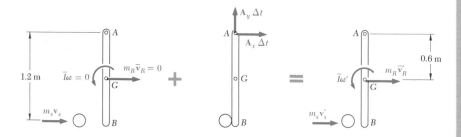

$$\text{Syst Momenta}_1 + \text{Syst Ext Imp}_{1\to2} = \text{Syst Momenta}_2$$

$+\uparrow$moments about A:

$$m_s v_s(1.2\text{ m}) = m_s v_s'(1.2\text{ m}) + m_R \bar{v}_R'(0.6\text{ m}) + \bar{I}\omega' \tag{1}$$

Since the rod rotates about A, we have $\bar{v}_R' = \bar{r}\omega' = (0.6\text{ m})\omega'$. Also,

$$\bar{I} = \tfrac{1}{12}mL^2 = \tfrac{1}{12}(8\text{ kg})(1.2\text{ m})^2 = 0.96\text{ kg}\cdot\text{m}^2$$

Substituting these values and the given data into Eq. (1), we have

$$(2\text{ kg})(5\text{ m/s})(1.2\text{ m}) = (2\text{ kg})v_s'(1.2\text{ m}) + (8\text{ kg})(0.6\text{ m})\omega'(0.6\text{ m}) + (0.96\text{ kg}\cdot\text{m}^2)\omega'$$
$$12 = 2.4v_s' + 3.84\omega' \tag{2}$$

Relative Velocities. Choosing positive to the right, we write

$$v_B' - v_s' = e(v_s - v_B)$$

Substituting $v_s = 5$ m/s, $v_B = 0$, and $e = 0.80$, we obtain

$$v_B' - v_s' = 0.80(5\text{ m/s}) \tag{3}$$

Again noting that the rod rotates about A, we write

$$v_B' = (1.2\text{ m})\omega' \tag{4}$$

Solving Eqs. (2) to (4) simultaneously, we obtain

$$\omega' = 3.21\text{ rad/s} \qquad \omega' = 3.21\text{ rad/s}\uparrow \quad \blacktriangleleft$$
$$v_s' = -0.143\text{ m/s} \qquad \mathbf{v}_s' = 0.143\text{ m/s} \leftarrow \quad \blacktriangleleft$$

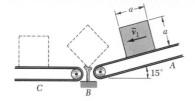

SAMPLE PROBLEM 17.11

A square package of side a and mass m moves down a conveyor belt A with a constant velocity $\bar{v}_1$. At the end of the conveyor belt, the corner of the package strikes a rigid support at B. Assuming that the impact at B is perfectly plastic, derive an expression for the smallest magnitude of the velocity $\bar{v}_1$ for which the package will rotate about B and reach conveyor belt C.

Principle of Impulse and Momentum. Since the impact between the package and the support is perfectly plastic, the package rotates about B during the impact. We apply the principle of impulse and momentum to the package and note that the only impulsive force external to the package is the impulsive reaction at B.

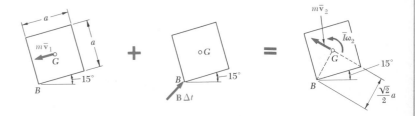

$$\text{Syst Momenta}_1 + \text{Syst Ext Imp}_{1\to2} = \text{Syst Momenta}_2$$

$+\uparrow$moments about B: $\qquad (m\bar{v}_1)(\tfrac{1}{2}a) + 0 = (m\bar{v}_2)(\tfrac{1}{2}\sqrt{2}a) + \bar{I}\omega_2$ (1)

Since the package rotates about B, we have $\bar{v}_2 = (GB)\omega_2 = \tfrac{1}{2}\sqrt{2}a\omega_2$. We substitute this expression, together with $\bar{I} = \tfrac{1}{6}ma^2$, into Eq. (1):

$$(m\bar{v}_1)(\tfrac{1}{2}a) = m(\tfrac{1}{2}\sqrt{2}a\omega_2)(\tfrac{1}{2}\sqrt{2}a) + \tfrac{1}{6}ma^2\omega_2 \qquad \bar{v}_1 = \tfrac{4}{3}a\omega_2 \quad (2)$$

Principle of Conservation of Energy. We apply the principle of conservation of energy between position 2 and position 3.

Position 2. $V_2 = Wh_2$. Recalling that $\bar{v}_2 = \tfrac{1}{2}\sqrt{2}a\omega_2$, we write

$$T_2 = \tfrac{1}{2}m\bar{v}_2^2 + \tfrac{1}{2}\bar{I}\omega_2^2 = \tfrac{1}{2}m(\tfrac{1}{2}\sqrt{2}a\omega_2)^2 + \tfrac{1}{2}(\tfrac{1}{6}ma^2)\omega_2^2 = \tfrac{1}{3}ma^2\omega_2^2$$

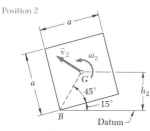

Position 2

$GB = \tfrac{1}{2}\sqrt{2}a = 0.707a$
$h_2 = GB \sin(45° + 15°)$
$\quad = 0.612a$

Position 3. Since the package must reach conveyor belt C, it must pass through position 3 where G is directly above B. Also, since we wish to determine the smallest velocity for which the package will reach this position, we choose $\bar{v}_3 = \omega_3 = 0$. Therefore $T_3 = 0$ and $V_3 = Wh_3$.

Conservation of Energy

$$T_2 + V_2 = T_3 + V_3$$
$$\tfrac{1}{3}ma^2\omega_2^2 + Wh_2 = 0 + Wh_3$$
$$\omega_2^2 = \frac{3W}{ma^2}(h_3 - h_2) = \frac{3g}{a^2}(h_3 - h_2) \quad (3)$$

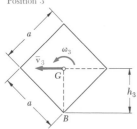

Position 3

$h_3 = GB = 0.707a$

Substituting the computed values of h_2 and h_3 into Eq. (3), we obtain

$$\omega_2^2 = \frac{3g}{a^2}(0.707a - 0.612a) = \frac{3g}{a^2}(0.095a) \qquad \omega_2 = \sqrt{0.285g/a}$$

$$\bar{v}_1 = \tfrac{4}{3}a\omega_2 = \tfrac{4}{3}a\sqrt{0.285g/a} \qquad \bar{v}_1 = 0.712\sqrt{ga} \quad \blacktriangleleft$$

Problems

17.95 A 30-g bullet is fired with a horizontal velocity of 400 m/s into a 4-kg wooden beam AB suspended from a pin support at A. Knowing that $h = 600$ mm and that the beam is initially at rest, determine (a) the velocity of the mass center G of the beam immediately after the bullet becomes embedded, (b) the impulsive reaction at A, assuming that the bullet becomes embedded in 1 ms.

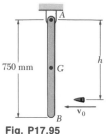

Fig. P17.95

17.96 In Prob. 17.95, determine (a) the required distance h if the impulsive reaction at A is to be zero, (b) the corresponding velocity of the mass center G of the beam after the bullet becomes embedded.

17.97 A 0.12-lb bullet is fired with a horizontal velocity $\mathbf{v}_0 = -(900 \text{ ft/s})\mathbf{i}$ into an 8-lb rectangular piece of timber suspended from a horizontal bar AB. Knowing that the timber is initially at rest and can rotate freely about AB, determine (a) the required distance h if the impulsive reaction at bar AB is to be zero, (b) the corresponding velocity of point D immediately after the bullet becomes embedded.

17.98 A 0.12-lb bullet is fired with a horizontal velocity $\mathbf{v}_0 = -(900 \text{ ft/s})\mathbf{i}$ into an 8-lb rectangular piece of timber suspended from a horizontal bar AB. Knowing that $h = 5$ in. and that the timber is initially at rest, determine (a) the velocity of point D immediately after the bullet becomes embedded, (b) the impulsive reaction at bar AB, assuming that the bullet becomes embedded in 0.002 s.

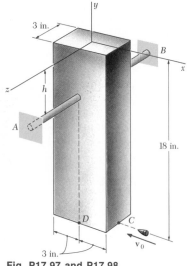

Fig. P17.97 and P17.98

17.99 An 8-kg wooden panel is suspended from a pin support at A and is initially at rest. A 2-kg metal sphere is released from rest at B and falls into a hemispherical cup C attached to the panel at a point located on its top edge. Assuming that the impact is perfectly plastic, determine the velocity of the mass center G of the panel immediately after the impact.

17.100 An 8-kg wooden panel is suspended from a pin support at A and is initially at rest. A 2-kg metal sphere is released from rest at B' and falls into a hemispherical cup C' attached to the panel at the same level as the mass center G. Assuming that the impact is perfectly plastic, determine the velocity of the mass center G of the panel immediately after the impact.

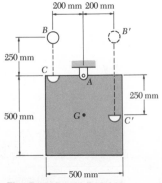

Fig. P17.99 and P17.100

17.101 A uniform rod AB of mass m is at rest on a frictionless horizontal surface when hook C of slider D engages a small pin at A. Knowing that the slider is pulled upward with a constant velocity $\mathbf{v}_0$, determine the impulse exerted on the rod (a) at A, (b) at B. Assume that the velocity of the slider is unchanged and that the impact is perfectly plastic.

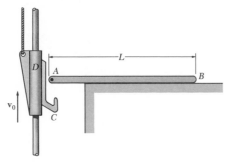

Fig. P17.101

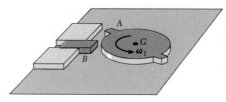

Fig. P17.102

17.102 A uniform disk of radius r and mass m is supported by a frictionless horizontal table. Initially the disk is spinning freely about its mass center G with a constant angular velocity ω_1. Suddenly latch B is moved to the right and is struck by a small stop A welded to the edge of the disk. Assuming that the impact of A and B is perfectly plastic, determine the angular velocity of the disk and the velocity of its mass center immediately after the impact.

17.103 A uniform slender rod AB is equipped at both ends with the hooks shown and is supported by a frictionless horizontal table. Initially the rod is hooked at A to a fixed pin C about which it rotates with a constant angular velocity ω_1. Suddenly end B of the rod hits and gets hooked to pin D, causing end A to be released. Determine the magnitude of the angular velocity ω_2 of the rod in its subsequent rotation about D.

Fig. P17.103

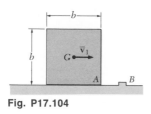

Fig. P17.104

17.104 A square block of mass m moves along a frictionless horizontal surface with a velocity $\overline{\mathbf{v}}_1$ and strikes a small obstruction at B. Assuming that the impact between corner A and obstruction B is perfectly plastic, determine the angular velocity of the block and the velocity of its mass center G immediately after the impact.

17.105 A slender rigid rod of mass m and with a centroidal radius of gyration $\overline{k}$ strikes a rigid knob at A with a vertical velocity of magnitude $\overline{v}_1$ and no angular velocity. Assuming that the impact is perfectly plastic, show that the magnitude of the velocity of G after the impact is

$$\overline{v}_2 = \overline{v}_1 \frac{r^2}{r^2 + \overline{k}^2}$$

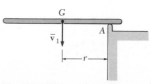

Fig. P17.105

17.106 A slender rod of mass m and length L is released from rest in the position shown and hits edge D. Assuming perfectly plastic impact at D, determine for $b = 0.6L$, (a) the angular velocity of the rod immediately after the impact, (b) the maximum angle through which the rod will rotate after the impact.

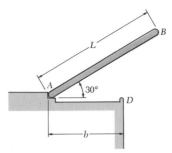

Fig. P17.106 and P17.107

17.107 A slender rod of mass m and length L is released from rest in the position shown and hits edge D. Assuming perfectly elastic impact ($e = 1$) at D, determine the distance b for which the rod will rebound with no angular velocity.

17.108 Solve Prob. 17.104, assuming that the impact between corner A and obstruction B is perfectly elastic.

17.109 A uniform slender rod of length L is dropped onto rigid supports at A and B. Immediately before striking A the velocity of the rod is $\bar{v}_1$. Since support B is slightly lower than support A, the rod strikes A before it strikes B. Assuming perfectly elastic impact at both A and B, determine the angular velocity of the rod and the velocity of its mass center immediately after the rod (a) strikes support A, (b) strikes support B, (c) again strikes support A.

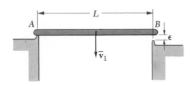

Fig. P17.109

17.110 A slender rod of length L forming an angle β with the vertical strikes a frictionless floor at A with a vertical velocity $\bar{v}_1$ and no angular velocity. Assuming that the impact at A is perfectly elastic, derive an expression for the angular velocity of the rod immediately after the impact.

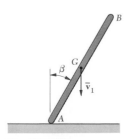

Fig. P17.110

17.111 Solve Prob. 17.102, assuming that the impact between the small stop A and latch B is perfectly elastic.

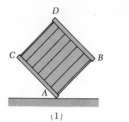
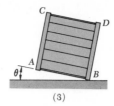

(1) (2) (3)

Fig. P17.112

17.112 A uniformly loaded square crate is released from rest with its corner D directly above A; it rotates about A until its corner B strikes the floor, and then rotates about B. The floor is sufficiently rough to prevent slipping and the impact at B is perfectly plastic. Denoting by ω_0 the angular velocity of the crate immediately before B strikes the floor, determine (a) the angular velocity of the crate immediately after B strikes the floor, (b) the fraction of the kinetic energy of the crate lost during the impact, (c) the angle θ through which the crate will rotate after B strikes the floor.

17.113 Rod AB, of mass 1.5 kg and length $L = 0.5$ m, is suspended from pin A which may move freely along a horizontal guide. If an impulse $\mathbf{Q}\,\Delta t = (1.8 \text{ N}\cdot\text{s})\mathbf{i}$ is applied at B, determine the maximum angle θ_m through which the rod will rotate during its subsequent motion.

17.114 For the rod of Prob. 17.113, determine the magnitude $Q\,\Delta t$ of the impulse for which the maximum angle through which the rod will rotate is (a) 90°, (b) 120°.

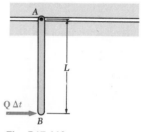

Fig. P17.113

17.115 A slender rod of length l and weight W is released in the position shown. It is observed that the rod rebounds to a horizontal position after striking the vertical surface. (a) Determine the coefficient of restitution between knob K and the surface. (b) Show that the same rebound may be expected for any position of knob K.

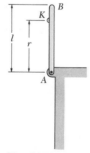

Fig. P17.115

17.116 A slender rod CD, of length L and mass m, is placed upright as shown on a frictionless horizontal surface. A second and identical rod AB is moving along the surface with a velocity $\bar{\mathbf{v}}_1$ when its end B strikes squarely end C of rod CD. Assuming that the impact is perfectly elastic, determine immediately after the impact, (a) the velocity of rod AB, (b) the angular velocity of rod CD, (c) the velocity of the mass center of rod CD.

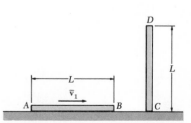

Fig. P17.116

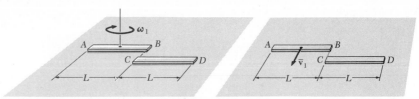

Fig. P17.117   Fig. P17.118

17.117 and 17.118 Two identical rods AB and CD, each of length L, may move freely on a frictionless horizontal surface. Rod AB is moving as indicated when end B strikes end C of rod CD, which is at rest. Knowing that at the instant of impact the rods are parallel and assuming perfectly elastic impact ($e = 1$), determine the angular velocity of each rod and the velocity of its mass center immediately after the impact.

17.117 Rod AB is rotating about its mass center with an angular velocity ω_1.

17.118 Rod AB is moving with a velocity $\bar{v}_1$ in a direction perpendicular to the longitudinal axes of the rods.

17.119 Block A of mass m is attached to a cord which is wrapped around a uniform disk of mass M. The block is released from rest and falls through a distance h before the cord becomes taut. Derive expressions for the velocity of the block and the angular velocity of the disk immediately after the impact. Assume that the impact is (a) perfectly plastic, (b) perfectly elastic.

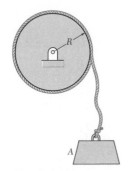

Fig. P17.119

17.120 Solve Prob. 17.119, assuming that $M = 8$ kg, $m = 3$ kg, $R = 250$ mm, and $h = 900$ mm.

17.121 A sphere of mass m is dropped onto the end of a slender rod BD of length L and mass $2m$. The rod is attached to a pin support at C which is located at a distance $c = \frac{1}{4}L$ from end B. Denoting by v_1 the velocity of the sphere just before it strikes the rod and assuming perfectly elastic impact, determine immediately after the impact (a) the velocity of the sphere, (b) the angular velocity of the rod.

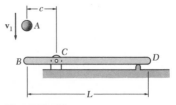

Fig. P17.121

17.122 Sphere A of mass m and radius r rolls without slipping with a velocity $\bar{v}_1$ on a horizontal plane. It hits squarely an identical sphere B which is at rest. Denoting by μ_k the coefficient of kinetic friction between the spheres and the plane, neglecting friction between the spheres, and assuming perfectly elastic impact ($e = 1$), determine (a) the linear and angular velocity of each sphere immediately after the impact, (b) the velocity of each sphere after it has started rolling uniformly. (c) Discuss the special case when $\mu_k = 0$.

Fig. P17.122

17.123 In a game of billiards, ball A is rolling without slipping with a velocity $\bar{\mathbf{v}}_0$ as it hits obliquely ball B which is at rest. Denoting by r the radius of each ball, by μ_k the coefficient of kinetic friction between the balls and the table, neglecting friction between the balls, and assuming perfectly elastic impact ($e = 1$), determine (a) the linear and angular velocity of each ball immediately after the impact, (b) the velocity of B after it has started rolling uniformly.

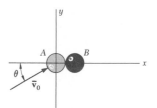

Fig. P17.123

17.124 In Prob. 17.123, determine the equation of the path described by the center of ball A while the ball is slipping.

17.125 For the billiard balls of Prob. 17.123, determine (a) the velocity of ball A after it has started rolling again without slipping, (b) the angle ϕ formed by the velocities of balls A and B after they have finished slipping.

17.126 A small rubber ball of radius r is thrown against a rough floor with a velocity $\bar{\mathbf{v}}_A$ of magnitude v_0 and a "backspin" $\boldsymbol{\omega}_A$ of magnitude ω_0. It is observed that the ball bounces from A to B, then from B to A, then from A to B, etc. Assuming perfectly elastic impact, determine the required magnitude ω_0 of the backspin in terms of v_0 and r.

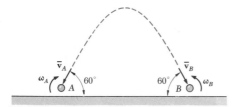

Fig. P17.126

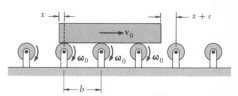

Fig. P17.127

17.127 A rectangular slab of mass m_S moves across a series of rollers, each of which is initially at rest and equivalent to a uniform disk of mass m_R. Since the length of the slab is slightly less than three times the distance b between two adjacent rollers, the slab leaves a roller just before it reaches another one. Each time a new roller enters into contact with the slab, slipping occurs between the roller and the slab for a short period of time (less than the time needed for the slab to move through the distance b). Denoting by $\mathbf{v}_0$ the velocity of the slab in the position shown, determine the velocity of the slab after it has moved (a) a distance b, (b) a distance nb.

17.128 Solve Prob. 17.127, assuming $v_0 = 3$ m/s, $m_S = 18$ kg, $m_R = 3$ kg, and $n = 8$.

*17.129 Each of the bars AB and BC is of length $L = 15$ in. and weight $W = 2.5$ lb. Determine the angular velocity of each bar immediately after the impulse $\mathbf{Q}\,\Delta t = (0.30\ \text{lb}\cdot\text{s})\mathbf{i}$ is applied at C.

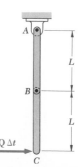

Fig. P17.129

Review and Summary

In this chapter we again considered the method of work and energy and the method of impulse and momentum. In the first part of the chapter we studied the method of work and energy and its application to the analysis of the motion of rigid bodies and systems of rigid bodies.

Principle of work and energy for a rigid body

In Sec. 17.2, we first expressed the principle of work and energy for a rigid body in the form

$$T_1 + U_{1\to2} = T_2 \tag{17.1}$$

where T_1 and T_2 represent the initial and final values of the kinetic energy of the rigid body and $U_{1\to2}$ the work of the *external forces* acting on the rigid body.

Work of a force or a couple

In Sec. 17.3, we recalled the expression found in Chap. 13 for the work of a force $\mathbf{F}$ applied at a point A, namely

$$U_{1\to2} = \int_{s_1}^{s_2} (F \cos \alpha)\, ds \tag{17.3'}$$

where F was the magnitude of the force, α the angle it formed with the direction of motion of A, and s the variable of integration measuring the distance traveled by A along its path. We also derived the expression for the *work of a couple of moment* $\mathbf{M}$ applied to a rigid body during a rotation in θ of the rigid body:

$$U_{1\to2} = \int_{\theta_1}^{\theta_2} M\, d\theta \tag{17.5}$$

Kinetic energy in plane motion

We then derived an expression for the kinetic energy of a rigid body in plane motion [Sec. 17.4]. We wrote

$$T = \tfrac{1}{2}m\bar{v}^2 + \tfrac{1}{2}\bar{I}\omega^2 \tag{17.9}$$

where $\bar{v}$ is the velocity of the mass center G of the body, ω the angular velocity of the body, and $\bar{I}$ its moment of inertia about an axis through G perpendicular to the plane of reference (Fig. 17.13) [Sample Prob. 17.3].

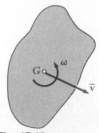

Fig. 17.13

We noted that the kinetic energy of a rigid body in plane motion may be separated into two parts: (1) the kinetic energy $\frac{1}{2}m\bar{v}^2$ associated with the motion of the mass center G of the body, and (2) the kinetic energy $\frac{1}{2}\bar{I}\omega^2$ associated with the rotation of the body about G.

Kinetic energy in rotation

For a rigid body rotating about a fixed axis through O with an angular velocity ω, we had

$$T = \tfrac{1}{2}I_O\omega^2 \qquad (17.10)$$

where I_O was the moment of inertia of the body about the fixed axis. We noted that the result obtained was not limited to the rotation of plane slabs or of bodies symmetrical with respect to the reference plane. It is valid regardless of the shape of the body or of the location of the axis of rotation.

Systems of rigid bodies

Equation (17.1) may be applied to the motion of systems of rigid bodies [Sec. 17.5] as long as all the forces acting on the various bodies involved—internal as well as external to the system—are included in the computation of $U_{1\rightarrow2}$. However, in the case of systems consisting of pin-connected members, or blocks and pulleys connected by inextensible cords, or meshed gears, the points of application of the internal forces move through equal distances and the work of these forces cancels out [Sample Probs. 17.1 and 17.2].

Conservation of energy

When a rigid body, or a system of rigid bodies, moves under the action of conservative forces, the principle of work and energy may be expressed in the form

$$T_1 + V_1 = T_2 + V_2 \qquad (17.12)$$

which is referred to as the *principle of conservation of energy* [Sec. 17.6]. This principle may be used to solve problems involving conservative forces such as the force of gravity or the force exerted by a spring [Sample Probs. 17.4 and 17.5]. However, when a reaction is to be determined, the principle of conservation of energy must be supplemented by the application of d'Alembert's principle [Sample Prob. 17.4].

Power

In Sec. 17.7, we extended the concept of power to a rotating body subjected to a couple, writing

$$\text{Power} = \frac{dU}{dt} = \frac{M\,d\theta}{dt} = M\omega \qquad (17.13)$$

where M is the magnitude of the couple and ω the angular velocity of the body.

Principle of impulse and momentum for
a rigid body

The middle part of the chapter was devoted to the method of impulse and momentum and its application to the solution of various types of problems involving the plane motion of rigid slabs and rigid bodies symmetrical with respect to the reference plane.

We first recalled the *principle of impulse and momentum* as it was derived in Sec. 14.9 for a system of particles and applied it to the *motion of a rigid body* [Sec. 17.8]. We wrote

$$\text{Syst Momenta}_1 + \text{Syst Ext Imp}_{1\to2} = \text{Syst Momenta}_2 \quad (17.14)$$

Next we showed that for a rigid slab or a rigid body symmetrical with respect to the reference plane, the system of the momenta of the particles forming the body is equivalent to a vector $m\bar{\mathbf{v}}$ attached at the mass center G of the body and a couple $\bar{I}\boldsymbol{\omega}$ (Fig. 17.14). The vector $m\bar{\mathbf{v}}$ is associated

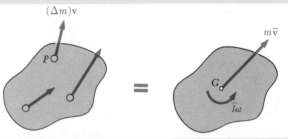

Fig. 17.14

with the translation of the body with G and represents the *linear momentum* of the body, while the couple $\bar{I}\boldsymbol{\omega}$ corresponds to the rotation of the body about G and represents the *angular momentum* of the body about an axis through G.

Equation (17.14) may be expressed graphically as shown in Fig. 17.15 by drawing three diagrams representing respectively the system of the initial momenta of the body, the impulses of the external forces acting on it, and the system of the final momenta of the body. Summing and

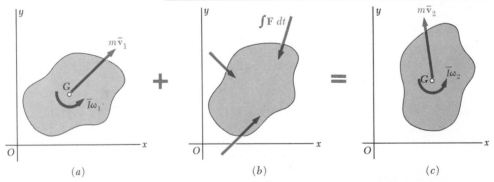

Fig. 17.15

equating respectively the *x components*, the *y components*, and the *moments about any given point* of the vectors shown in that figure, we obtain three equations of motion which may be solved for the desired unknowns [Sample Probs. 17.6 and 17.7].

In problems dealing with several connected rigid bodies [Sec. 17.9], each body may be considered separately [Sample Prob. 17.6] or, if no more than three unknowns are involved, the principle of impulse and momentum may be applied to the entire system, considering the impulses of the external forces only [Sample Prob. 17.8].

Conservation of angular momentum

When the lines of action of all the external forces acting on a system of rigid bodies pass through a given point O, the angular momentum of the system about O is conserved [Sec. 17.10]. It was suggested that problems involving conservation of angular momentum be solved by the general method described above [Sample Prob. 17.8].

Impulsive motion

The last part of the chapter was devoted to the *impulsive motion* and the *eccentric impact* of rigid bodies. In Sec. 17.11, we recalled that the method of impulse and momentum is the only practicable method for the solution of problems involving impulsive motion and that the computation of impulses in such problems is particularly simple [Sample Prob. 17.9].

Eccentric impact

In Sec. 17.12, we recalled that the eccentric impact of two rigid bodies is defined as an impact in which the mass centers of the colliding bodies are *not* located on the line of impact. It was shown that in such a situation a relation similar to that derived in Chap. 13 for the central impact of two particles and involving the coefficient of restitution e still holds, but that *the velocities of points A and B where contact occurs during the impact should be used*. We have

$$(v'_B)_n - (v'_A)_n = e[(v_A)_n - (v_B)_n] \qquad (17.19)$$

where $(v_A)_n$ and $(v_B)_n$ are the components along the line of impact of the velocities of A and B before the impact, and $(v'_A)_n$ and $(v'_B)_n$ their components after the impact (Fig. 17.16). Equation (17.19) is applicable not

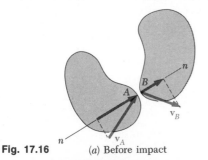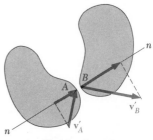

Fig. 17.16 (*a*) Before impact (*b*) After impact

only when the colliding bodies move freely after the impact but also when the bodies are partially constrained in their motion. It should be used in conjunction with one or several other equations obtained by applying the principle of impulse and momentum [Sample Prob. 17.10]. We also considered problems where the method of impulse and momentum and the method of work and energy may be combined [Sample Prob. 17.11].

Review Problems

17.130 A thin-walled spherical shell of mass m and radius r is used as a protective cover for a space shuttle experiment. In order to remove the cover, it is separated into two hemispherical shells which move to the final position shown. Knowing that during a test of the cover the initial angular velocity of the shell is $\boldsymbol{\omega} = \omega_0\mathbf{j}$, determine the angular velocity of the cover after it has reached its final position. (*Hint.* For a thin-walled shell, $I_{\text{diameter}} = \frac{2}{3}mr^2$.)

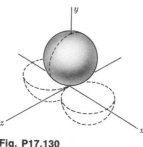

Fig. P17.130

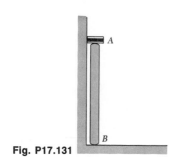

Fig. P17.131

17.131 A small matchbox is placed on top of rod AB. End B of the rod is given a slight horizontal push, causing it to slide on the horizontal floor. Assuming no friction and neglecting the weight of the matchbox, determine the angle θ through which the rod will have rotated when the matchbox loses contact with the rod.

17.132 A uniform square panel of side L and mass m is supported by a frictionless horizontal table. The panel is moving to the right with a velocity $\bar{\mathbf{v}}_1$ when a hook located at the midpoint E of side AB engages the fixed pin F. Assuming that the impact is perfectly plastic, determine the angular velocity of the panel in its subsequent rotation about F.

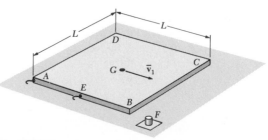

Fig. P17.132

17.133 Solve Prob. 17.132, assuming that the hook at E is removed and that a hook located at corner A engages the fixed pin F.

Fig. P17.134

17.134 The mass center G of a wheel of radius R is located at a distance r from its geometric center C. The centroidal radius of gyration of the wheel is denoted by $\bar{k}$. As the wheel rolls freely and without sliding on a horizontal plane, its angular velocity is observed to vary. Denoting by ω_1, ω_2, and ω_3, respectively, the angular velocity of the wheel when G is directly above C, level with C, and directly below C, show that ω_1, ω_2, and ω_3 satisfy the relation

$$\frac{\omega_2^2 - \omega_1^2}{\omega_3^2 - \omega_2^2} = \frac{g/R + \omega_1^2}{g/R + \omega_3^2}$$

17.135 A slender rod of length L is falling with a velocity $\bar{v}_1$ at the instant when the cords simultaneously become taut. Assuming that the impacts are perfectly plastic, determine the angular velocity of the rod and the velocity of its mass center immediately after the cords become taut.

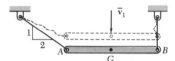

Fig. P17.135

17.136 A 25-g bullet is fired with a horizontal velocity of 400 m/s into the lower corner of a 3-kg square panel of side $b = 500$ mm. The panel is held by two vertical wires as shown. Determine immediately after the bullet becomes embedded, (a) the velocity of the mass center G, (b) the angular velocity of the panel.

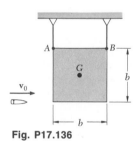

Fig. P17.136

17.137 A small 500-g ball may slide in a slender tube of length 1.2 m and of mass 1.5 kg which rotates freely about a vertical axis passing through its center C. If the angular velocity of the tube is 8 rad/s as the ball passes through C, determine the angular velocity of the tube (a) just before the ball leaves the tube, (b) just after the ball has left the tube.

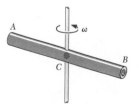

Fig. P17.137

17.138 Knowing that in Prob. 17.137 the speed of the ball is 1.8 m/s as it passes through C, determine the radial and transverse components of the velocity of the ball as it leaves the tube at B.

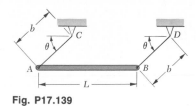

Fig. P17.139

17.139 A uniform rod of length L and weight W is attached to two wires, each of length b. The rod is released from rest when $\theta = 0$ and swings to the position $\theta = 90°$, at which time wire BD suddenly breaks. Determine the tension in wire AC (a) immediately before wire BD breaks, (b) immediately after wire BD breaks.

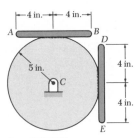

Fig. P17.140

17.140 Two 5-lb slender rods are welded to the edge of an 8-lb uniform disk as shown. Knowing that in the position shown the assembly has an angular velocity of 10 rad/s clockwise, determine the largest and the smallest angular velocity of the assembly as it rotates about pivot C.

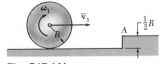

Fig. P17.141

17.141 A uniform sphere of radius R rolls to the right without slipping on a horizontal surface and strikes a step of height $\frac{1}{2}R$. Assuming that the impact is perfectly plastic and that no slipping occurs between the sphere and the corner of the step, determine (a) the angular velocity of the sphere and the velocity of its center immediately after the impact, (b) the smallest initial speed $\bar{v}_1$ for which the sphere will reach the upper level of the step.

The following problems are designed to be solved with a computer.

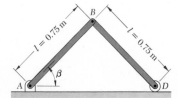

Fig. P17.C1

17.C1 Each of the two slender rods shown is 0.75 m long and has a mass of 6 kg. Knowing that the system is released from rest when $\beta = 60°$, write a computer program and use it to calculate the angular velocity of rod AB and the velocity of point D for values of β from 60 to 0° at 5° intervals.

Fig. P17.C2

17.C2 A half section of pipe of mass m and radius r is released from rest in the position shown and rolls without sliding. (*a*) Write a computer program to determine the angular velocity of the pipe after it has rolled through an angle θ. (*b*) Use this program to calculate the angular velocity of a half section of pipe of radius $r = 8$ in. for values of θ from 0 to 90° at 5° intervals. [Note that $GO = 2r/\pi$ and that by the parallel-axis theorem, $\bar{I} = mr^2 - m(GO)^2$.] (*c*) Expand the program to determine the approximate time required for the half section to roll through 180°, using 1° intervals. [*Hint:* The time Δt_i required for the half section to roll through an angle $\Delta\theta_i$ may be obtained by dividing $\Delta\theta_i$ expressed in radians by the average angular velocity $\frac{1}{2}(\omega_i + \omega_{i+1})$ of the half section over Δt_i if the angular acceleration of the half section is assumed to remain constant over Δt_i.]

17.C3 The 1.5-kg rod AB of length $L = 240$ mm slides freely inside the tube CD of mass m_t. Knowing that the angular velocity of the assembly was $\omega_1 = 8$ rad/s when the rod was entirely inside the tube ($x = 0$), write a computer program and use it to calculate the angular velocity of the assembly and the velocity of the rod relative to the tube for values of x from 40 to 200 mm at 40-mm intervals when (*a*) $m_t = 0.75$ kg, (*b*) $m_t = 1.5$ kg, (*c*) $m_t = 3$ kg.

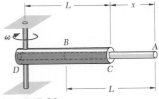

Fig. P17.C3

17.C4 A 0.8-kg sphere is dropped onto the end of a slender rod BD of length $L = 900$ mm and mass $m = 1.6$ kg. Knowing that the sphere strikes end B of the rod with a speed $v_1 = 10$ m/s and assuming perfectly elastic impact, write a computer program and use it to calculate immediately after the impact the velocity of the sphere and the angular velocity of the rod for values of x from 0 to 400 mm at 25-mm intervals.

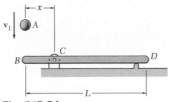

Fig. P17.C4

Kinetics of Rigid Bodies in Three Dimensions

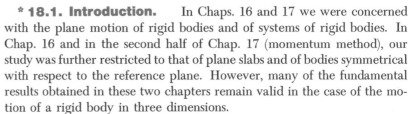

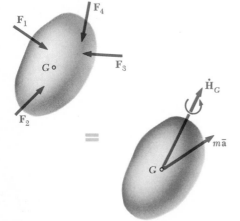

Fig. 18.1

* 18.1. Introduction.   In Chaps. 16 and 17 we were concerned with the plane motion of rigid bodies and of systems of rigid bodies. In Chap. 16 and in the second half of Chap. 17 (momentum method), our study was further restricted to that of plane slabs and of bodies symmetrical with respect to the reference plane. However, many of the fundamental results obtained in these two chapters remain valid in the case of the motion of a rigid body in three dimensions.

For example, the two fundamental equations

$$\Sigma \mathbf{F} = m\bar{\mathbf{a}} \qquad (18.1)$$

$$\Sigma \mathbf{M}_G = \dot{\mathbf{H}}_G \qquad (18.2)$$

on which the analysis of the plane motion of a rigid body was based, remain valid in the most general case of motion of a rigid body. As it was indicated in Sec. 16.2, these equations express that the system of the external forces is equipollent to the system consisting of the vector $m\bar{\mathbf{a}}$ attached at G and the couple of moment $\dot{\mathbf{H}}_G$ (Fig. 18.1). However, the relation $\mathbf{H}_G = \bar{I}\boldsymbol{\omega}$, which enabled us to determine the angular momentum of a rigid slab and which played an important part in the solution of problems involving the plane motion of slabs and bodies symmetrical with respect to the reference plane, ceases to be valid in the case of nonsymmetrical bodies or three-dimensional motion. It will thus be necessary for us to develop in Sec. 18.2 a more general method for computing the angular momentum $\mathbf{H}_G$ of a rigid body in three dimensions.

Similarly, the main feature of the impulse-momentum method discussed in Sec. 17.7, namely, the reduction of the momenta of the particles of a rigid body to a linear momentum vector $m\bar{\mathbf{v}}$ attached at the mass center G of the body and an angular momentum couple $\mathbf{H}_G$, remains valid. Here again, however, the relation $\mathbf{H}_G = \bar{I}\boldsymbol{\omega}$ will have to be discarded and replaced by the more general relation to be developed in Sec. 18.2.

We also note that the work-energy principle (Sec. 17.2) and the principle of conservation of energy (Sec. 17.6) still apply in the case of the motion of a rigid body in three dimensions. However, the expression obtained in Sec. 17.4 for the kinetic energy of a rigid body in plane motion will be replaced by a new expression to be developed in Sec. 18.4 for a rigid body in three-dimensional motion.

In the second part of the chapter, we shall first learn to determine the rate of change $\dot{\mathbf{H}}_G$ of the angular momentum $\mathbf{H}_G$ of a three-dimensional rigid body, using a rotating frame of reference with respect to which the moments and products of inertia of the body remain constant (Sec. 18.5). We shall then express Eqs. (18.1) and (18.2) in the form of free-body-diagram equations, which may be used to solve various problems involving the three-dimensional motion of rigid bodies (Secs. 18.6 through 18.8).

The last part of the chapter (Secs. 18.9 through 18.11) is devoted to the study of the motion of the gyroscope or, more generally, of an axisymmetrical body with a fixed point located on its axis of symmetry. In Sec. 18.10, we shall consider the particular case of the steady precession of a gyroscope and, in Sec. 18.11, the motion of an axisymmetrical body subjected to no force, except its own weight.

* 18.2. Angular Momentum of a Rigid Body in Three Dimensions.

We shall see in this section how the angular momentum $\mathbf{H}_G$ of the body about its mass center G may be determined from the angular velocity $\boldsymbol{\omega}$ of the body in the case of three-dimensional motion.

According to Eq. (14.24), the angular momentum of the body about G may be expressed as

$$\mathbf{H}_G = \sum_{i=1}^{n} (\mathbf{r}_i' \times \mathbf{v}_i' \, \Delta m_i) \tag{18.3}$$

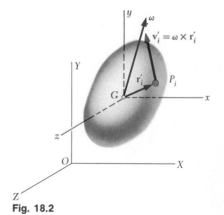

Fig. 18.2

where $\mathbf{r}_i'$ and $\mathbf{v}_i'$ denote, respectively, the position vector and the velocity of the particle P_i, of mass Δm_i, relative to the centroidal frame $Gxyz$ (Fig. 18.2). But $\mathbf{v}_i' = \boldsymbol{\omega} \times \mathbf{r}_i'$, where $\boldsymbol{\omega}$ is the angular velocity of the body at the instant considered. Substituting into (18.3), we have

$$\mathbf{H}_G = \sum_{i=1}^{n} [\mathbf{r}_i' \times (\boldsymbol{\omega} \times \mathbf{r}_i') \, \Delta m_i]$$

Recalling the rule for determining the rectangular components of a vector product (Sec. 3.5), we obtain the following expression for the x component of the angular momentum:

$$H_x = \sum_{i=1}^{n} [y_i(\boldsymbol{\omega} \times \mathbf{r}_i')_z - z_i(\boldsymbol{\omega} \times \mathbf{r}_i')_y] \Delta m_i$$

$$= \sum_{i=1}^{n} [y_i(\omega_x y_i - \omega_y x_i) - z_i(\omega_z x_i - \omega_x z_i)] \Delta m_i$$

$$= \omega_x \sum_i (y_i^2 + z_i^2) \Delta m_i - \omega_y \sum_i x_i y_i \Delta m_i - \omega_z \sum_i z_i x_i \Delta m_i$$

Replacing the sums by integrals in this expression and in the two similar expressions which are obtained for H_y and H_z, we have

$$H_x = \omega_x \int (y^2 + z^2)\, dm - \omega_y \int xy\, dm - \omega_z \int zx\, dm$$
$$H_y = -\omega_x \int xy\, dm + \omega_y \int (z^2 + x^2)\, dm - \omega_z \int yz\, dm \qquad (18.4)$$
$$H_z = -\omega_x \int zx\, dm - \omega_y \int yz\, dm + \omega_z \int (x^2 + y^2)\, dm$$

We note that the integrals containing squares represent the *centroidal mass moments of inertia* of the body about the x, y, and z axes, respectively (Sec. 9.11); we have

$$\bar{I}_x = \int (y^2 + z^2)\, dm \qquad \bar{I}_y = \int (z^2 + x^2)\, dm$$
$$\bar{I}_z = \int (x^2 + y^2)\, dm \qquad (18.5)$$

Similarly, the integrals containing products of coordinates represent the *centroidal mass products of inertia* of the body (Sec. 9.16); we have

$$\bar{I}_{xy} = \int xy\, dm \qquad \bar{I}_{yz} = \int yz\, dm \qquad \bar{I}_{zx} = \int zx\, dm \qquad (18.6)$$

Substituting from (18.5) and (18.6) into (18.4), we obtain the components of the angular momentum $\mathbf{H}_G$ of the body about its mass center G:

$$H_x = +\bar{I}_x \omega_x - \bar{I}_{xy}\omega_y - \bar{I}_{xz}\omega_z$$
$$H_y = -\bar{I}_{yx}\omega_x + \bar{I}_y \omega_y - \bar{I}_{yz}\omega_z \qquad (18.7)$$
$$H_z = -\bar{I}_{zx}\omega_x - \bar{I}_{zy}\omega_y + \bar{I}_z \omega_z$$

The relations (18.7) show that the operation which transforms the vector $\boldsymbol{\omega}$ into the vector $\mathbf{H}_G$ (Fig. 18.3) is characterized by the array of moments and products of inertia

$$\begin{pmatrix} \bar{I}_x & -\bar{I}_{xy} & -\bar{I}_{xz} \\ -\bar{I}_{yx} & \bar{I}_y & -\bar{I}_{yz} \\ -\bar{I}_{zx} & -\bar{I}_{zy} & \bar{I}_z \end{pmatrix} \qquad (18.8)$$

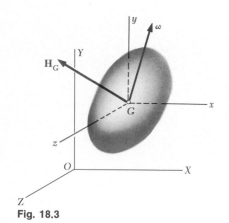

Fig. 18.3

The array (18.8) defines the *inertia tensor* of the body at its mass center G.† A new array of moments and products of inertia would be obtained if a different system of axes were used. The transformation characterized by this new array, however, would still be the same. Clearly, the angular momentum $\mathbf{H}_G$ corresponding to a given angular velocity $\boldsymbol{\omega}$ is independent of the choice of the coordinate axes. As was shown in Sec. 9.17, it is always possible to select a system of axes $Gx'y'z'$, called *principal axes of inertia,* with respect to which all the products of inertia of a given body are zero. The array (18.8) takes then the diagonalized form

$$\begin{pmatrix} \overline{I}_{x'} & 0 & 0 \\ 0 & \overline{I}_{y'} & 0 \\ 0 & 0 & \overline{I}_{z'} \end{pmatrix} \tag{18.9}$$

where $\overline{I}_{x'}, \overline{I}_{y'}, \overline{I}_{z'}$ represent the *principal centroidal moments of inertia* of the body, and the relations (18.7) reduce to

$$H_{x'} = \overline{I}_{x'}\omega_{x'} \qquad H_{y'} = \overline{I}_{y'}\omega_{y'} \qquad H_{z'} = \overline{I}_{z'}\omega_{z'} \tag{18.10}$$

We note that if the three principal centroidal moments of inertia $\overline{I}_{x'}$, $\overline{I}_{y'}, \overline{I}_{z'}$ are equal, the components $H_{x'}, H_{y'}, H_{z'}$ of the angular momentum about G are proportional to the components $\omega_{x'}, \omega_{y'}, \omega_{z'}$ of the angular velocity, and the vectors $\mathbf{H}_G$ and $\boldsymbol{\omega}$ are collinear. In general, however, the principal moments of inertia will be different, and the vectors $\mathbf{H}_G$ and $\boldsymbol{\omega}$ *will have different directions,* except when two of the three components of $\boldsymbol{\omega}$ happen to be zero, i.e., when $\boldsymbol{\omega}$ is directed along one of the coordinate axes. Thus, *the angular momentum $\mathbf{H}_G$ of a rigid body and its angular velocity $\boldsymbol{\omega}$ have the same direction if, and only if, $\boldsymbol{\omega}$ is directed along a principal axis of inertia.*‡ Since this condition is satisfied in the case of the plane motion of a rigid body symmetrical with respect to the reference plane, we were able in Secs. 16.3 and 17.8 to represent the angular momentum $\mathbf{H}_G$ of such a body by the vector $\overline{I}\boldsymbol{\omega}$. We must realize, however, that this result cannot be extended to the case of the plane motion of a nonsymmetrical body, or to the case of the three-dimensional motion of a

---

† Setting $\overline{I}_x = I_{11}, \overline{I}_y = I_{22}, \overline{I}_z = I_{33}$, and $-\overline{I}_{xy} = I_{12}, -\overline{I}_{xz} = I_{13}$, etc., we may write the inertia tensor (18.8) in the standard form

$$\begin{pmatrix} I_{11} & I_{12} & I_{13} \\ I_{21} & I_{22} & I_{23} \\ I_{31} & I_{32} & I_{33} \end{pmatrix}$$

Denoting by H_1, H_2, H_3 the components of the angular momentum $\mathbf{H}_G$ and by $\omega_1, \omega_2, \omega_3$ the components of the angular velocity $\boldsymbol{\omega}$, we may write the relations (18.7) in the form

$$H_i = \sum_j I_{ij}\omega_j$$

where i and j take the values 1, 2, 3. The quantities I_{ij} are said to be the *components* of the inertia tensor. Since $I_{ij} = I_{ji}$, the inertia tensor is a *symmetric tensor of the second order.*

‡ In the particular case when $\overline{I}_{x'} = \overline{I}_{y'} = \overline{I}_{z'}$, any line through G may be considered as a principal axis of inertia, and the vectors $\mathbf{H}_G$ and $\boldsymbol{\omega}$ are always collinear.

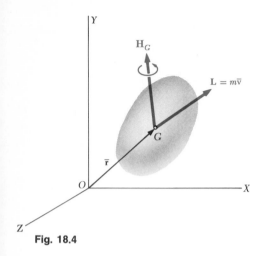

Fig. 18.4

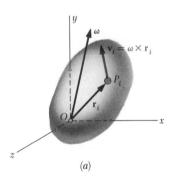

(a)

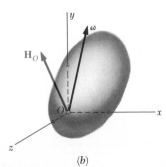

(b)

Fig. 18.5

rigid body. Except when ω happens to be directed along a principal axis of inertia, the angular momentum and angular velocity of a rigid body have different directions, and the relation (18.7) or (18.10) must be used to determine $\mathbf{H}_G$ from ω.

Reduction of the Momenta of the Particles of a Rigid Body to a Momentum Vector and a Couple at G. We saw in Sec. 17.8 that the system formed by the momenta of the various particles of a rigid body may be reduced to a vector $\mathbf{L}$ attached at the mass center G of the body, representing the linear momentum of the body, and to a couple $\mathbf{H}_G$, representing the angular momentum of the body about G (Fig. 18.4). We are now in a position to determine the vector $\mathbf{L}$ and the couple $\mathbf{H}_G$ in the most general case of three-dimensional motion of a rigid body. As in the case of the two-dimensional motion considered in Sec. 17.8, the linear momentum $\mathbf{L}$ of the body is equal to the product $m\overline{\mathbf{v}}$ of its mass m and of the velocity $\overline{\mathbf{v}}$ of its mass center G. The angular momentum $\mathbf{H}_G$, however, cannot be obtained any more by simply multiplying the angular velocity ω of the body by the scalar $\overline{I}$; it must now be obtained from the components of ω and from the centroidal moments and products of inertia of the body through the use of Eq. (18.7) or (18.10).

We should also note that once the linear momentum $m\overline{\mathbf{v}}$ and the angular momentum $\mathbf{H}_G$ of a rigid body have been determined, its angular momentum $\mathbf{H}_O$ about any given point O may be obtained by adding the moments about O of the vector $m\overline{\mathbf{v}}$ and of the couple $\mathbf{H}_G$. We write

$$\mathbf{H}_O = \overline{\mathbf{r}} \times m\overline{\mathbf{v}} + \mathbf{H}_G \qquad (18.11)$$

Angular Momentum of a Rigid Body Constrained to Rotate about a Fixed Point. In the particular case of a rigid body constrained to rotate in three-dimensional space about a fixed point O (Fig. 18.5a), it is sometimes convenient to determine the angular momentum $\mathbf{H}_O$ of the body about the fixed point O. While $\mathbf{H}_O$ could be obtained by first computing $\mathbf{H}_G$ as indicated above and then using Eq. (18.11), it is often advantageous to determine $\mathbf{H}_O$ directly from the angular velocity ω of the body and its moments and products of inertia with respect to a frame $Oxyz$ centered at the fixed point O. Recalling Eq. (14.7), we write

$$\mathbf{H}_O = \sum_{i=1}^{n} (\mathbf{r}_i \times \mathbf{v}_i \, \Delta m_i) \qquad (18.12)$$

where $\mathbf{r}_i$ and $\mathbf{v}_i$ denote, respectively, the position vector and the velocity of the particle P_i with respect to the fixed frame $Oxyz$. Substituting $\mathbf{v}_i = \omega \times \mathbf{r}_i$, and after manipulations similar to those used in the earlier part of this section, we find that the components of the angular momentum $\mathbf{H}_O$ (Fig. 18.5b) are given by the relations

$$
\begin{aligned}
H_x &= +I_x\,\omega_x - I_{xy}\omega_y - I_{xz}\omega_z \\
H_y &= -I_{yx}\omega_x + I_y\,\omega_y - I_{yz}\omega_z \\
H_z &= -I_{zx}\omega_x - I_{zy}\omega_y + I_z\,\omega_z
\end{aligned}
\qquad (18.13)
$$

where the moments of inertia I_x, I_y, I_z and the products of inertia I_{xy}, I_{yz}, I_{zx} are computed with respect to the frame $Oxyz$ centered at the fixed point O.

* 18.3. Application of the Principle of Impulse and Momentum to the Three-Dimensional Motion of a Rigid Body.

Before we can apply the fundamental equation (18.2) to the solution of problems involving the three-dimensional motion of a rigid body, we shall have to learn to compute the derivative of the vector $\mathbf{H}_G$. This will be done in Sec. 18.5. We may, however, immediately use the results obtained in the preceding section to solve problems by the impulse-momentum method.

Recalling that the system formed by the momenta of the particles of a center G of the body and an angular momentum couple $\mathbf{H}_G$, we represent center G of the body and an angular momentum couple $\mathbf{H}_G$, we represent graphically the fundamental relation

$$\textbf{Syst Momenta}_1 + \textbf{Syst Ext Imp}_{1 \rightarrow 2} = \textbf{Syst Momenta}_2 \qquad (17.4)$$

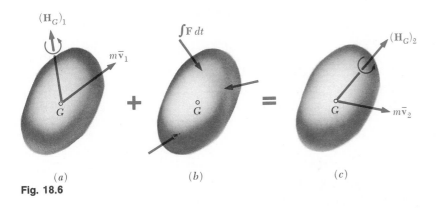

(a) (b) (c)

Fig. 18.6

by means of the three sketches shown in Fig. 18.6. To solve a given problem, we may use these sketches to write appropriate component and moment equations, keeping in mind that the components of the angular momentum $\mathbf{H}_G$ are related to the components of the angular velocity $\boldsymbol{\omega}$ by Eqs. (18.7) of the preceding section.

In solving problems dealing with the motion of a body rotating about a fixed point O, it will be convenient to eliminate the impulse of the reaction at O by writing an equation involving the moments of the momenta and impulses about O. We recall that the angular momentum $\mathbf{H}_O$ of the body about the fixed point O may be obtained either directly from Eqs. (18.13), or by first computing its linear momentum $m\bar{\mathbf{v}}$ and its angular momentum $\mathbf{H}_G$ and then using Eq. (18.11).

* 18.4. Kinetic Energy of a Rigid Body in Three Dimensions.

Consider a rigid body of mass m in three-dimensional motion. We recall from Sec. 14.6 that if the absolute velocity $\mathbf{v}_i$ of each particle P_i of the body is expressed as the sum of the velocity $\overline{\mathbf{v}}$ of the mass center G of the body and of the velocity $\mathbf{v}'_i$ of the particle relative to a frame $Gxyz$ attached to G and of fixed orientation (Fig. 18.7), the kinetic energy of the

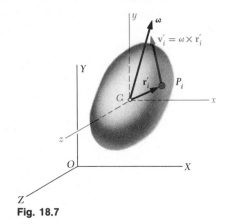

Fig. 18.7

system of particles forming the rigid body may be written in the form

$$T = \tfrac{1}{2}m\overline{v}^2 + \frac{1}{2}\sum_{i=1}^{n}\Delta m_i v_i'^2 \qquad (18.14)$$

where the last term represents the kinetic energy T' of the body relative to the centroidal frame $Gxyz$. Since $v'_i = |\mathbf{v}'_i| = |\boldsymbol{\omega} \times \mathbf{r}'_i|$, we write

$$T' = \frac{1}{2}\sum_{i=1}^{n}\Delta m_i v_i'^2 = \frac{1}{2}\sum_{i=1}^{n}|\boldsymbol{\omega} \times \mathbf{r}'_i|^2\,\Delta m_i$$

Expressing the square in terms of the rectangular components of the vector product, and replacing the sums by integrals, we have

$$
\begin{aligned}
T' &= \int[(\omega_x y - \omega_y x)^2 + (\omega_y z - \omega_z y)^2 + (\omega_z x - \omega_x z)^2]\,dm \\
&= \omega_x^2\int(y^2 + z^2)\,dm + \omega_y^2\int(z^2 + x^2)\,dm + \omega_z^2\int(x^2 + y^2)\,dm \\
&\quad - 2\omega_x\omega_y\int xy\,dm - 2\omega_y\omega_z\int yz\,dm - 2\omega_z\omega_x\int zx\,dm
\end{aligned}
$$

or, recalling the relations (18.5) and (18.6),

$$T' = \tfrac{1}{2}(\overline{I}_x\omega_x^2 + \overline{I}_y\omega_y^2 + \overline{I}_z\omega_z^2 - 2\overline{I}_{xy}\omega_x\omega_y - 2\overline{I}_{yz}\omega_y\omega_z - 2\overline{I}_{zx}\omega_z\omega_x) \qquad (18.15)$$

Substituting into (18.14) the expression (18.15) we have just obtained for the kinetic energy of the body relative to centroidal axes, we write

$$
\begin{aligned}
T = \tfrac{1}{2}m\overline{v}^2 + \tfrac{1}{2}(&\overline{I}_x\omega_x^2 + \overline{I}_y\omega_y^2 + \overline{I}_z\omega_z^2 - 2\overline{I}_{xy}\omega_x\omega_y \\
&- 2\overline{I}_{yz}\omega_y\omega_z - 2\overline{I}_{zx}\omega_z\omega_x) \qquad (18.16)
\end{aligned}
$$

If the axes of coordinates are chosen so that they coincide at the instant considered with the principal axes x', y', z' of the body, the relation obtained reduces to

$$T = \tfrac{1}{2}m\overline{v}^2 + \tfrac{1}{2}(\overline{I}_{x'}\omega_{x'}^2 + \overline{I}_{y'}\omega_{y'}^2 + \overline{I}_{z'}\omega_{z'}^2) \tag{18.17}$$

where $\overline{v}$ = velocity of mass center
        ω = angular velocity
        m = mass of rigid body
$\overline{I}_{x'}, \overline{I}_{y'}, \overline{I}_{z'}$ = principal centroidal moments of inertia

The results we have obtained enable us to extend to the three-dimensional motion of a rigid body the application of the principle of work and energy (Sec. 17.2) and of the principle of conservation of energy (Sec. 17.6).

Kinetic Energy of a Rigid Body with a Fixed Point.   In the particular case of a rigid body rotating in three-dimensional space about a fixed point O, the kinetic energy of the body may be expressed in terms of its moments and products of inertia with respect to axes attached at O (Fig. 18.8). Recalling the definition of the kinetic energy of a system of particles,

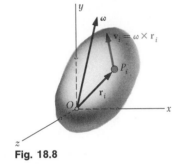

Fig. 18.8

and substituting $v_i = |\mathbf{v}_i| = |\boldsymbol{\omega} \times \mathbf{r}_i|$, we write

$$T = \frac{1}{2}\sum_{i=1}^{n}\Delta m_i v_i^2 = \frac{1}{2}\sum_{i=1}^{n}|\boldsymbol{\omega} \times \mathbf{r}_i|^2\,\Delta m_i \tag{18.18}$$

Manipulations similar to those used to derive Eq. (18.15) yield

$$T = \tfrac{1}{2}(I_x\omega_x^2 + I_y\omega_y^2 + I_z\omega_z^2 - 2I_{xy}\omega_x\omega_y - 2I_{yz}\omega_y\omega_z - 2I_{zx}\omega_z\omega_x)$$

$$\tag{18.19}$$

or, if the principal axes x', y', z' of the body at the origin O are chosen as coordinate axes,

$$T = \tfrac{1}{2}(I_{x'}\omega_{x'}^2 + I_{y'}\omega_{y'}^2 + I_{z'}\omega_{z'}^2) \tag{18.20}$$

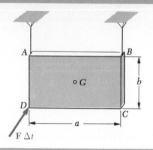

SAMPLE PROBLEM 18.1

A rectangular plate of mass m suspended from two wires at A and B is hit at D in a direction perpendicular to the plate. Denoting by $\mathbf{F}\,\Delta t$ the impulse applied at D, determine immediately after impact (*a*) the velocity of the mass center G, (*b*) the angular velocity of the plate.

Solution. We shall assume that the wires remain taut and thus that the components $\bar{v}_y$ of $\bar{\mathbf{v}}$ and ω_z of $\boldsymbol{\omega}$ are zero after impact. We therefore have

$$\bar{\mathbf{v}} = \bar{v}_x\mathbf{i} + \bar{v}_z\mathbf{k} \qquad \boldsymbol{\omega} = \omega_x\mathbf{i} + \omega_y\mathbf{j}$$

and since the x, y, z axes are principal axes of inertia,

$$\mathbf{H}_G = \bar{I}_x\omega_x\mathbf{i} + \bar{I}_y\omega_y\mathbf{j} \qquad \mathbf{H}_G = \tfrac{1}{12}mb^2\omega_x\mathbf{i} + \tfrac{1}{12}ma^2\omega_y\mathbf{j} \qquad (1)$$

Principle of Impulse and Momentum. Since the initial momenta are zero, the system of the impulses must be equivalent to the system of the final momenta:

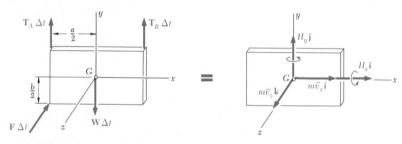

a. Velocity of Mass Center. Equating the components of the impulses and momenta in the x and z directions:

x components: $\qquad\qquad 0 = m\bar{v}_x \qquad \bar{v}_x = 0$
z components: $\qquad\qquad -F\,\Delta t = m\bar{v}_z \qquad \bar{v}_z = -F\,\Delta t/m$

$$\bar{\mathbf{v}} = \bar{v}_x\mathbf{i} + \bar{v}_z\mathbf{k} \qquad \bar{\mathbf{v}} = -(F\,\Delta t/m)\mathbf{k} \quad \blacktriangleleft$$

b. Angular Velocity. Equating the moments of the impulses and momenta about the x and y axes:

About x axis: $\qquad\qquad \tfrac{1}{2}bF\,\Delta t = H_x$
About y axis: $\qquad\qquad -\tfrac{1}{2}aF\,\Delta t = H_y$

$$\mathbf{H}_G = H_x\mathbf{i} + H_y\mathbf{j} \qquad \mathbf{H}_G = \tfrac{1}{2}bF\,\Delta t\mathbf{i} - \tfrac{1}{2}aF\,\Delta t\mathbf{j} \qquad (2)$$

Comparing Eqs. (1) and (2), we conclude that

$$\omega_x = 6F\,\Delta t/mb \qquad \omega_y = -6F\,\Delta t/ma$$
$$\boldsymbol{\omega} = \omega_x\mathbf{i} + \omega_y\mathbf{j} \qquad \boldsymbol{\omega} = (6F\,\Delta t/mab)(a\mathbf{i} - b\mathbf{j}) \quad \blacktriangleleft$$

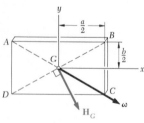

We note that $\boldsymbol{\omega}$ is directed along the diagonal AC.

Remark: Equating the y components of the impulses and momenta, and their moments about the z axis, we obtain two additional equations which yield $T_A = T_B = \tfrac{1}{2}W$. We thus verify that the wires remain taut and that our assumption was correct.

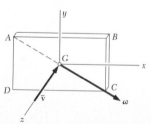

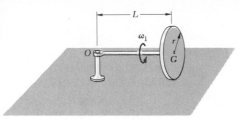

SAMPLE PROBLEM 18.2

A homogeneous disk of radius r and mass m is mounted on an axle OG of length L and negligible mass. The axle is pivoted at the fixed point O, and the disk is constrained to roll on a horizontal floor. Knowing that the disk rotates counterclockwise at the rate ω_1 about the axle OG, determine (a) the angular velocity of the disk, (b) its angular momentum about O, (c) its kinetic energy, (d) the vector and couple at G equivalent to the momenta of the particles of the disk.

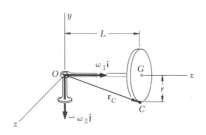

a. Angular Velocity. As the disk rotates about the axle OG it also rotates with the axle about the y axis at a rate ω_2 clockwise. The total angular velocity of the disk is therefore

$$\boldsymbol{\omega} = \omega_1\mathbf{i} - \omega_2\mathbf{j} \qquad (1)$$

To determine ω_2 we write that the velocity of C is zero:

$$\mathbf{v}_C = \boldsymbol{\omega} \times \mathbf{r}_C = 0$$
$$(\omega_1\mathbf{i} - \omega_2\mathbf{j}) \times (L\mathbf{i} - r\mathbf{j}) = 0$$
$$(L\omega_2 - r\omega_1)\mathbf{k} = 0 \qquad \omega_2 = r\omega_1/L$$

Substituting into (1) for ω_2: $\qquad\qquad \boldsymbol{\omega} = \omega_1\mathbf{i} - (r\omega_1/L)\mathbf{j}$ ◄

b. Angular Momentum About O. Assuming the axle to be part of the disk, we may consider the disk to have a fixed point at O. Since the x, y, and z axes are principal axes of inertia for the disk,

$$H_x = I_x\omega_x = (\tfrac{1}{2}mr^2)\omega_1$$
$$H_y = I_y\omega_y = (mL^2 + \tfrac{1}{4}mr^2)(-r\omega_1/L)$$
$$H_z = I_z\omega_z = (mL^2 + \tfrac{1}{4}mr^2)\,0 = 0$$
$$\mathbf{H}_O = \tfrac{1}{2}mr^2\omega_1\mathbf{i} - m(L^2 + \tfrac{1}{4}r^2)(r\omega_1/L)\mathbf{j} \ ◄$$

c. Kinetic Energy. Using the values obtained for the moments of inertia and the components of $\boldsymbol{\omega}$, we have

$$T = \tfrac{1}{2}(I_x\omega_x^2 + I_y\omega_y^2 + I_z\omega_z^2) = \tfrac{1}{2}[\tfrac{1}{2}mr^2\omega_1^2 + m(L^2 + \tfrac{1}{4}r^2)(-r\omega_1/L)^2]$$

$$T = \tfrac{1}{8}mr^2\left(6 + \frac{r^2}{L^2}\right)\omega_1^2 \ ◄$$

d. Momentum Vector and Couple at G. The linear momentum vector $m\bar{\mathbf{v}}$ and the angular momentum couple $\mathbf{H}_G$ are

$$m\bar{\mathbf{v}} = mr\omega_1\mathbf{k} \ ◄$$

and

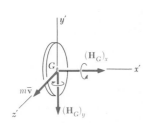

$$\mathbf{H}_G = \bar{I}_{x'}\omega_x\mathbf{i} + \bar{I}_{y'}\omega_y\mathbf{j} + \bar{I}_{z'}\omega_z\mathbf{k} = \tfrac{1}{2}mr^2\omega_1\mathbf{i} + \tfrac{1}{4}mr^2(-r\omega_1/L)\mathbf{j}$$

$$\mathbf{H}_G = \tfrac{1}{2}mr^2\omega_1\left(\mathbf{i} - \frac{r}{2L}\mathbf{j}\right) \ ◄$$

Problems

18.1 A thin rectangular plate of mass 6 kg rotates about its vertical diagonal AB with an angular velocity ω. Knowing that the z axis is perpendicular to the plate and that ω is constant and equal to 5 rad/s, determine the angular momentum of the plate about its mass center G.

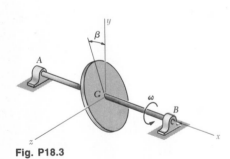

Fig. P18.1

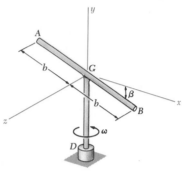

Fig. P18.2

18.2 A thin homogeneous rod AB of mass m and length $2b$ is welded at its midpoint G to a vertical shaft GD. Knowing that the shaft rotates with a constant angular velocity ω, determine the angular momentum of the rod about G.

18.3 A thin homogeneous disk of mass m and radius r is mounted on the horizontal axle AB. The plane of the disk forms an angle $\beta = 20°$ with the vertical. Knowing that the axle rotates with an angular velocity ω, determine the angle θ formed by the axle and the angular momentum of the disk about G.

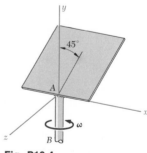

Fig. P18.4

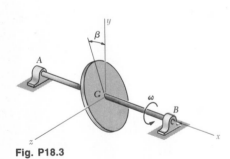

Fig. P18.3

18.4 A thin homogeneous square plate of mass m and side a is welded to a vertical shaft AB with which it forms an angle of 45°. Knowing that the shaft rotates with an angular velocity ω, determine the angular momentum of the plate about A.

18.5 A thin disk of weight $W = 10$ lb rotates at the constant rate $\omega_2 = 15$ rad/s with respect to arm ABC, which itself rotates at the constant rate $\omega_1 = 5$ rad/s about the y axis. Determine the angular momentum of the disk about its center C.

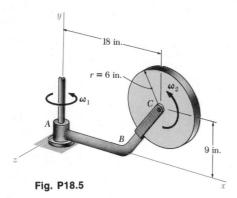

Fig. P18.5

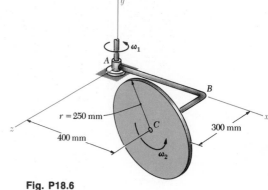

Fig. P18.6

18.6 A thin disk of mass $m = 5$ kg rotates at the constant rate $\omega_2 = 8$ rad/s with respect to the bent axle ABC, which itself rotates at the constant rate $\omega_1 = 3$ rad/s about the y axis. Determine the angular momentum of the disk about its center C.

18.7 The 25-kg projectile shown has a radius of gyration of 72 mm about its axis of symmetry Gx and of 300 mm about its transverse axis Gy. Its angular velocity ω may be resolved into two components: one component, directed along the axis of symmetry Gx, measures the *rate of spin* of the projectile, while the other, directed along GD, measures its *rate of precession*. Knowing that $\theta = 8°$ and that the angular momentum of the projectile about its mass center G is $\mathbf{H}_G = (600 \text{ g} \cdot \text{m}^2/\text{s})\mathbf{i} + (16 \text{ g} \cdot \text{m}^2/\text{s})\mathbf{j}$, determine (a) the rate of spin, (b) the rate of precession.

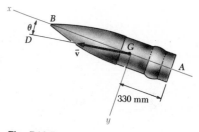

Fig. P18.7

18.8 Determine the angular momentum $\mathbf{H}_A$ of the projectile of Prob. 18.7 about the center A of its base, knowing that its mass center G has a velocity $\bar{\mathbf{v}}$ of 750 m/s. Give your answer in terms of components respectively parallel to the x and y axes shown and to a third axis z pointing toward you.

18.9 Determine the angular momentum of the disk of Prob. 18.6 about point A.

18.10 Determine the angular momentum of the disk of Prob. 18.5 about point A.

18.11 Determine the angular momentum $\mathbf{H}_O$ of the disk of Sample Prob. 18.2 from the expressions obtained for its linear momentum $m\bar{\mathbf{v}}$ and its angular momentum $\mathbf{H}_G$, using Eq. (18.11). Verify that the result obtained is the same that was obtained by direct computation.

18.12 Show that when a rigid body rotates about a fixed axis, its angular momentum is the same about any two points A and B on the fixed axis ($\mathbf{H}_A = \mathbf{H}_B$) if, and only if, the mass center G of the body is located on the fixed axis.

18.13 A homogeneous rod of mass m' per unit length is used to form the shaft shown. Knowing that the shaft rotates with a constant angular velocity ω, determine (a) the angular momentum of the shaft about G, (b) the angle formed by the angular momentum and the axis AB.

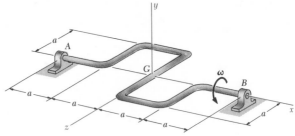

Fig. P18.13

18.14 Determine the angular momentum of the shaft of Prob. 18.13 about (a) point A, (b) point B.

18.15 Two triangular plates, each of mass 8 kg, are welded to a vertical shaft AB. Knowing that the system rotates at the constant rate $\omega = 6$ rad/s, determine its angular momentum about G.

18.16 Two L-shaped arms, each weighing 4 lb, are welded at the third points of the 2-ft shaft AB. Knowing that shaft AB rotates at the constant rate $\omega = 240$ rpm, determine (a) the angular momentum of the body about A, (b) the angle formed by the angular momentum and shaft AB.

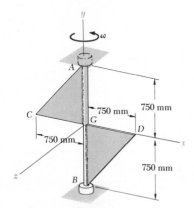

Fig. P18.15

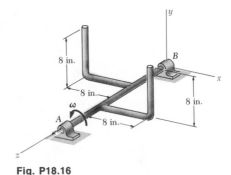

Fig. P18.16

18.17 For the body of Prob. 18.16, determine (a) the angular momentum about B, (b) the angle formed by the angular momentum and shaft BA.

18.18 A homogeneous wire, of mass 2.5 kg/m, is used to form the wire figure shown, which is suspended from point A. If an impulse $\mathbf{F}\,\Delta t = (3\text{ N}\cdot\text{s})\mathbf{j}$ is applied at point E of coordinates $x = 180$ mm, $y = -120$ mm, $z = 180$ mm, determine (a) the velocity of the mass center of the wire figure, (b) the angular velocity of the figure.

18.19 Solve Prob. 18.18, assuming that the impulse applied at point E is $\mathbf{F}\,\Delta t = -(3\text{ N}\cdot\text{s})\mathbf{k}$.

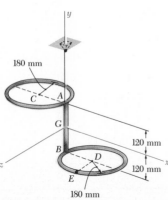

Fig. P18.18

18.20 A uniform rod of total mass m is bent into the shape shown and is suspended by a wire attached at its mass center G. The bent rod is hit at D in a direction perpendicular to the plane containing the rod (in the negative z direction). Denoting the corresponding impulse by $\mathbf{F}\,\Delta t$, determine immediately after the impact (a) the velocity of the mass center G, (b) the angular velocity of the rod.

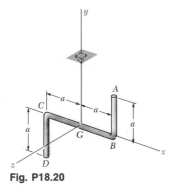

Fig. P18.20

18.21 Solve Prob. 18.20, assuming that the bent rod is hit at C.

18.22 A uniform rod of total mass m is bent into the shape shown and is suspended by a wire attached at B. The bent rod is hit at A in a vertical downward direction. Denoting the corresponding impulse by $\mathbf{F}\,\Delta t$, determine immediately after the impact (a) the velocity of the mass center G, (b) the angular velocity of the rod.

18.23 Solve Prob. 18.22, assuming that the bent rod is hit at C in the negative z direction.

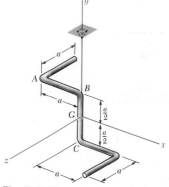

Fig. P18.22

18.24 Three slender homogeneous rods, each of mass m and length d, are welded together to form the assembly shown, which hangs from a wire attached at G. The assembly is hit at A in a vertical downward direction. Denoting the corresponding impulse by $\mathbf{F}\,\Delta t$, determine immediately after the impact (a) the velocity of the mass center G, (b) the angular velocity of the assembly.

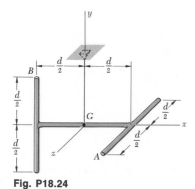

Fig. P18.24

18.25 Solve Prob. 18.24, assuming that the assembly is hit at B in a direction opposite to that of the z axis.

18.26 A circular plate of radius a and mass m supported by a ball-and-socket joint at A is rotating about the y axis with a constant angular velocity $\omega = \omega_0\mathbf{j}$ when an obstruction is suddenly introduced at B. Assuming the impact at B to be perfectly plastic, determine immediately after the impact (a) the angular velocity of the plate, (b) the velocity of its mass center G.

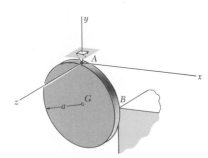

Fig. P18.26

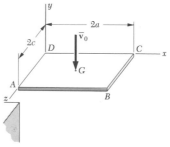

Fig. P18.27

18.27 A rectangular plate of mass m is falling with a velocity $\overline{\mathbf{v}}_0$ and no angular velocity when its corner A strikes an obstruction. Assuming the impact at A to be perfectly plastic, determine immediately after the impact (a) the angular velocity of the plate, (b) the velocity of its mass center G.

18.28 In order to analyze the twist performed by cats to land on their paws after being dropped upside down, the model shown may be used. It consists of two homogeneous cylinders, each of mass m, radius a, and length $3a$, mounted on a bent shaft. Small motors located inside the cylinders make it possible to spin each cylinder about the shaft at the rate ω_s after the assembly has been released from rest in position a. Observing that the entire assembly must rotate at the rate $\frac{1}{2}\omega_s$ about a horizontal axis through its mass center G if it is to complete a half turn about that axis when each cylinder has spun through 360° with respect to the shaft (position b), determine the required angle θ that each portion of the shaft should form with the horizontal. Neglect the mass of the shaft and of the motors. (*Hint.* Since no external force except its weight acts on the assembly, the angular momentum of the assembly about G must remain equal to zero.)

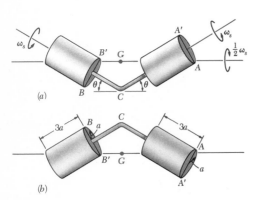

Fig. P18.28

18.29 A 1200-kg satellite designed to study the sun has an angular velocity $\boldsymbol{\omega}_0 = (0.050 \text{ rad/s})\mathbf{i} + (0.075 \text{ rad/s})\mathbf{k}$ when two small jets are activated at A and B in a direction parallel to the y axis. Knowing that the coordinate axes are principal centroidal axes, that the radii of gyration of the satellite are $\overline{k}_x = 1.120$ m, $\overline{k}_y = 1.200$ m, $\overline{k}_z = 0.900$ m, and that each jet produces a 50-N thrust, determine (a) the required operating time of each jet if the angular velocity of the satellite is to be reduced to zero, (b) the resulting change in the velocity of the mass center G.

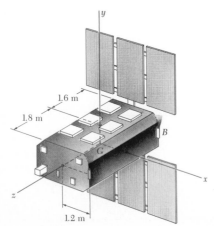

Fig. P18.29

18.30 If jet A in Prob. 18.29 is inoperative, determine (a) the required operating time of jet B to reduce to zero the x component of the angular velocity of the satellite, (b) the resulting final angular velocity, (c) the resulting change in the velocity of the mass center G.

18.31 An 800-lb geostationary satellite is spinning with an angular velocity $\omega_0 = (1.5 \text{ rad/s})\mathbf{j}$ when it is hit at A by a 6-oz meteorite traveling at a speed $v_0 = 4500$ ft/s relative to the satellite. Knowing that the radii of gyration of the satellite are $\bar{k}_x = \bar{k}_z = 28.8$ in. and $\bar{k}_y = 32.4$ in., and that immediately after the meteorite has become embedded the angular velocity of the satellite is observed to be $\omega = (0.6 \text{ rad/s})\mathbf{j} + (0.37 \text{ rad/s})\mathbf{k}$, determine the components of the velocity v_0 of the meteorite relative to the satellite.

18.32 An 800-lb geostationary satellite is spinning with an angular velocity $\omega_0 = (1.8 \text{ rad/s})\mathbf{j}$ when it is hit at B by a 6-oz meteorite traveling with a velocity $v_0 = -(3360 \text{ ft/s})\mathbf{i} + (3480 \text{ ft/s})\mathbf{j} - (2400 \text{ ft/s})\mathbf{k}$ relative to the satellite. Knowing that the radii of gyration of the satellite are $\bar{k}_x = \bar{k}_z = 28.8$ in. and $\bar{k}_y = 32.4$ in., and that immediately after the meteorite has become embedded the x component of the angular velocity ω of the satellite is observed to be $\omega_x = 0.36$ rad/s, determine (a) the distance b, (b) the y and z components of ω.

18.33 Denoting, respectively, by ω, $\mathbf{H}_O$, and T the angular velocity, the angular momentum, and the kinetic energy of a rigid body with a fixed point O, (a) prove that $\mathbf{H}_O \cdot \omega = 2T$; (b) show that the angle θ between ω and $\mathbf{H}_O$ will always be acute.

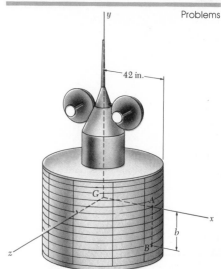

Fig. P18.31 and P18.32

18.34 Show that the kinetic energy of a rigid body with a fixed point O may be expressed as $T = \frac{1}{2}I_{OL}\omega^2$ where ω is the instantaneous angular velocity of the body and I_{OL} its moment of inertia about the line of action OL of ω. Derive this expression (a) from Eqs. (9.46) and (18.19), (b) by considering T as the sum of the kinetic energies of particles P_i describing circles of radius ρ_i about line OL.

18.35 Determine the kinetic energy of the plate of Prob. 18.1

18.36 Determine the kinetic energy of the plate of Prob. 18.4

18.37 Determine the kinetic energy of the disk of Prob. 18.3

18.38 Determine the kinetic energy of the disk of Prob. 18.6

18.39 Determine the kinetic energy of the disk of Prob. 18.5

18.40 Determine the kinetic energy of the body of Prob. 18.16

18.41 Determine the kinetic energy lost when the plate of Prob. 18.27 hits the obstruction.

18.42 Determine the kinetic energy lost when the plate of Prob. 18.26 hits the obstruction at B.

18.43 Determine the change in the kinetic energy of the satellite of Prob. 18.29 in its motion about its mass center due to the operation of the jets.

18.44 Determine the change in the kinetic energy of the satellite of Prob. 18.32 in its motion about its mass center due to the collision with the meteorite.

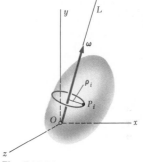

Fig. P18.34

* 18.5. Motion of a Rigid Body in Three Dimensions. As was indicated in Sec. 18.2, the fundamental equations

$$\Sigma \mathbf{F} = m\bar{\mathbf{a}} \tag{18.1}$$

$$\Sigma \mathbf{M}_G = \dot{\mathbf{H}}_G \tag{18.2}$$

remain valid in the most general case of the motion of a rigid body. Before Eq. (18.2) could be applied to the three-dimensional motion of a rigid body, however, it was necessary to derive Eqs. (18.7), which relate the components of the angular momentum $\mathbf{H}_G$ and those of the angular velocity $\boldsymbol{\omega}$. It still remains for us to find an effective and convenient way for computing the components of the derivative $\dot{\mathbf{H}}_G$ of the angular momentum.

Since $\mathbf{H}_G$ represents the angular momentum of the body in its motion relative to centroidal axes $GX'Y'Z'$ of fixed orientation (Fig. 18.9), and since $\dot{\mathbf{H}}_G$ represents the rate of change of $\mathbf{H}_G$ with respect to the same axes, it would seem natural to use components of $\boldsymbol{\omega}$ and $\mathbf{H}_G$ along the axes X', Y', Z' in writing the relations (18.7). But since the body rotates, its moments and products of inertia would change continually, and it would be necessary to determine their values as functions of the time. It is therefore more convenient to use axes x, y, z attached to the body, thus making sure that its moments and products of inertia will maintain the same values during the motion. This is permissible since, as indicated earlier, the transformation of $\boldsymbol{\omega}$ into $\mathbf{H}_G$ is independent of the system of coordinate axes which has been selected. The angular velocity $\boldsymbol{\omega}$, however, should still be *defined* with respect to the frame $GX'Y'Z'$ of fixed orientation. The vector $\boldsymbol{\omega}$ may then be *resolved* into components along the rotating $x, y,$ and z axes. Applying the relations (18.7), we obtain the *components* of the vector $\mathbf{H}_G$ along the rotating axes. The vector $\mathbf{H}_G$, however, represents the angular momentum about G of the body *in its motion relative to the frame* $GX'Y'Z'$.

Differentiating with respect to t the components of the angular momentum in (18.7), we define the rate of change of the vector $\mathbf{H}_G$ with respect to the rotating frame $Gxyz$:

$$(\dot{\mathbf{H}}_G)_{Gxyz} = \dot{H}_x \mathbf{i} + \dot{H}_y \mathbf{j} + \dot{H}_z \mathbf{k} \tag{18.21}$$

where $\mathbf{i}, \mathbf{j}, \mathbf{k}$ are the unit vectors along the rotating axes. Recalling from Sec. 15.10 that the rate of change $\dot{\mathbf{H}}_G$ of the vector $\mathbf{H}_G$ with respect to the frame $GX'Y'Z'$ may be obtained by adding to $(\dot{\mathbf{H}}_G)_{Gxyz}$ the vector product $\boldsymbol{\Omega} \times \mathbf{H}_G$, where $\boldsymbol{\Omega}$ denotes the angular velocity of the rotating frame, we write

$$\boxed{\dot{\mathbf{H}}_G = (\dot{\mathbf{H}}_G)_{Gxyz} + \boldsymbol{\Omega} \times \mathbf{H}_G} \tag{18.22}$$

where $\mathbf{H}_G$ = angular momentum of the body with respect to the frame $GX'Y'Z'$ of fixed orientation

$(\dot{\mathbf{H}}_G)_{Gxyz}$ = rate of change of $\mathbf{H}_G$ with respect to the rotating frame $Gxyz$, to be computed from the relations (18.7) and (18.21)

$\boldsymbol{\Omega}$ = angular velocity of the rotating frame $Gxyz$

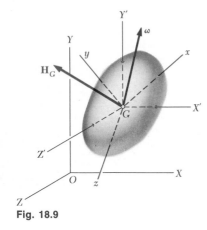

Fig. 18.9

Substituting for $\dot{\mathbf{H}}_G$ from (18.22) into (18.2), we have

907

18.6. Euler's Equations of Motion.
Extension of d'Alembert's Principle

$$\Sigma \mathbf{M}_G = (\dot{\mathbf{H}}_G)_{Gxyz} + \boldsymbol{\Omega} \times \mathbf{H}_G \qquad (18.23)$$

If the rotating frame is attached to the body, as has been assumed in this discussion, its angular velocity $\boldsymbol{\Omega}$ is identically equal to the angular velocity $\boldsymbol{\omega}$ of the body. There are many applications, however, where it is advantageous to use a frame of reference which is not actually attached to the body but rotates in an independent manner. For example, if the body considered is axisymmetrical, as in Sample Prob. 18.5 or Sec. 18.9, it is possible to select a frame of reference with respect to which the moments and products of inertia of the body remain constant, but which rotates less than the body itself.† As a result, simpler expressions may be obtained for the angular velocity $\boldsymbol{\omega}$ and the angular momentum $\mathbf{H}_G$ of the body than would have been possible if the frame of reference had actually been attached to the body. It is clear that in such cases the angular velocity $\boldsymbol{\Omega}$ of the rotating frame and the angular velocity $\boldsymbol{\omega}$ of the body are different.

*18.6. Euler's Equations of Motion. Extension of d'Alembert's Principle to the Motion of a Rigid Body in Three Dimensions.

If the x, y, and z axes are chosen to coincide with the principal axes of inertia of the body, the simplified relations (18.10) may be used to determine the components of the angular momentum $\mathbf{H}_G$. Omitting the primes from the subscripts, we write

$$\mathbf{H}_G = \bar{I}_x \omega_x \mathbf{i} + \bar{I}_y \omega_y \mathbf{j} + \bar{I}_z \omega_z \mathbf{k} \qquad (18.24)$$

where $\bar{I}_x$, $\bar{I}_y$, and $\bar{I}_z$ denote the principal centroidal moments of inertia of the body. Substituting for $\mathbf{H}_G$ from (18.24) into (18.23) and setting $\boldsymbol{\Omega} = \boldsymbol{\omega}$, we obtain the three scalar equations

$$\Sigma M_x = \bar{I}_x \dot{\omega}_x - (\bar{I}_y - \bar{I}_z)\omega_y \omega_z$$
$$\Sigma M_y = \bar{I}_y \dot{\omega}_y - (\bar{I}_z - \bar{I}_x)\omega_z \omega_x \qquad (18.25)$$
$$\Sigma M_z = \bar{I}_z \dot{\omega}_z - (\bar{I}_x - \bar{I}_y)\omega_x \omega_y$$

These equations, called *Euler's equations of motion* after the Swiss mathematician Leonhard Euler (1707–1783), may be used to analyze the motion of a rigid body about its mass center. In the following sections, however, we shall use Eq. (18.23) in preference to Eqs. (18.25), since the former is more general, and the compact vectorial form in which it is expressed is easier to remember.

Writing Eq. (18.1) in scalar form, we obtain the three additional equations

$$\Sigma F_x = m\bar{a}_x \qquad \Sigma F_y = m\bar{a}_y \qquad \Sigma F_z = m\bar{a}_z \qquad (18.26)$$

which together with Euler's equations form a system of six differential equations. Given appropriate initial conditions, these differential equa-

† More specifically, the frame of reference will have no spin (see Sec. 18.9).

tions have a unique solution. Thus, the motion of a rigid body in three dimensions is completely defined by the resultant and the moment resultant of the external forces acting on it. This result will be recognized as a generalization of a similar result obtained in Sec. 16.4 in the case of the plane motion of a rigid slab. It follows that in three as well as two dimensions, two systems of forces which are equipollent are also equivalent; that is, they have the same effect on a given rigid body.

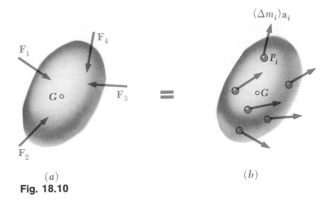

(a)   (b)

Fig. 18.10

Considering in particular the system of the external forces acting on a rigid body (Fig. 18.10a) and the system of the effective forces associated with the particles forming the rigid body (Fig. 18.10b), we may state that the two systems—which were shown in Sec. 14.2 to be equipollent—are also equivalent. This is the extension of d'Alembert's principle to the three-dimensional motion of a rigid body. Replacing the effective forces in Fig. 18.10b by an equivalent force-couple system, we verify that the system of the external forces acting on a rigid body in three-dimensional motion is equivalent to the system consisting of the vector $m\bar{a}$ attached at the mass center G of the body and the couple of moment $\dot{\mathbf{H}}_G$ (Fig. 18.11), where $\dot{\mathbf{H}}_G$ is obtained from the relations (18.7) and (18.22). Problems involving the three-dimensional motion of a rigid body may be solved by considering the free-body-diagram equation represented in Fig. 18.11 and writing appropriate scalar equations relating the components or moments of the external and effective forces (see Sample Prob. 18.3).

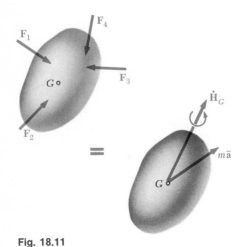

Fig. 18.11

* 18.7. Motion of a Rigid Body about a Fixed Point. When a rigid body is constrained to rotate about a fixed point O, it is desirable to write an equation involving the moments about O of the external and effective forces, since this equation will not contain the unknown reaction at O. While such an equation may be obtained from Fig. 18.11, it may be more convenient to write it by considering the rate of change of the angular momentum $\mathbf{H}_O$ of the body about the fixed point O (Fig. 18.12). Recalling Eq. (14.11), we write

$$\Sigma \mathbf{M}_O = \dot{\mathbf{H}}_O \tag{18.27}$$

where $\dot{\mathbf{H}}_O$ denotes the rate of change of the vector $\mathbf{H}_O$ with respect to the fixed frame $OXYZ$. A derivation similar to that used in Sec. 18.5 enables us to relate $\dot{\mathbf{H}}_O$ to the rate of change $(\dot{\mathbf{H}}_O)_{Oxyz}$ of $\mathbf{H}_O$ with respect to the rotating frame $Oxyz$. Substitution into (18.27) leads to the equation

$$\Sigma\mathbf{M}_O = (\dot{\mathbf{H}}_O)_{Oxyz} + \mathbf{\Omega} \times \mathbf{H}_O \qquad (18.28)$$

where $\Sigma\mathbf{M}_O$ = sum of the moments about O of the forces applied to the rigid body

$\mathbf{H}_O$ = angular momentum of the body with respect to the fixed frame $OXYZ$

$(\dot{\mathbf{H}}_O)_{Oxyz}$ = rate of change of $\mathbf{H}_O$ with respect to the rotating frame $Oxyz$, to be computed from the relations (18.13)

$\mathbf{\Omega}$ = angular velocity of the rotating frame $Oxyz$

If the rotating frame is attached to the body, its angular velocity $\mathbf{\Omega}$ is identically equal to the angular velocity $\boldsymbol{\omega}$ of the body. However, as indicated in the last paragraph of Sec. 18.5, there are many applications where it is advantageous to use a frame of reference which is not actually attached to the body but rotates in an independent manner.

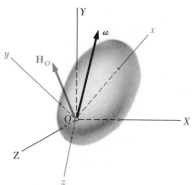

Fig. 18.12

* 18.8. Rotation of a Rigid Body about a Fixed Axis. We shall use Eq. (18.28), which was derived in the preceding section, to analyze the motion of a rigid body constrained to rotate about a fixed axis AB (Fig. 18.13). First, we note that the angular velocity of the body with respect to the fixed frame $OXYZ$ is represented by the vector $\boldsymbol{\omega}$ directed along the axis of rotation. Attaching the moving frame of reference $Oxyz$ to the body, with the z axis along AB, we have $\boldsymbol{\omega} = \omega\mathbf{k}$. Substituting $\omega_x = 0$, $\omega_y = 0$, $\omega_z = \omega$ into the relations (18.13), we obtain the components along the rotating axes of the angular momentum $\mathbf{H}_O$ of the body about O:

$$H_x = -I_{xz}\omega \qquad H_y = -I_{yz}\omega \qquad H_z = I_z\omega$$

Since the frame $Oxyz$ is attached to the body, we have $\mathbf{\Omega} = \boldsymbol{\omega}$ and Eq. (18.28) yields

$$\begin{aligned}
\Sigma\mathbf{M}_O &= (\dot{\mathbf{H}}_O)_{Oxyz} + \boldsymbol{\omega} \times \mathbf{H}_O \\
&= (-I_{xz}\mathbf{i} - I_{yz}\mathbf{j} + I_z\mathbf{k})\dot{\omega} + \omega\mathbf{k} \times (-I_{xz}\mathbf{i} - I_{yz}\mathbf{j} + I_z\mathbf{k})\omega \\
&= (-I_{xz}\mathbf{i} - I_{yz}\mathbf{j} + I_z\mathbf{k})\alpha + (-I_{xz}\mathbf{j} + I_{yz}\mathbf{i})\omega^2
\end{aligned}$$

The result obtained may be expressed by the three scalar equations

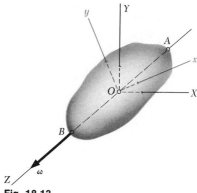

Fig. 18.13

$$\begin{aligned}
\Sigma M_x &= -I_{xz}\alpha + I_{yz}\omega^2 \\
\Sigma M_y &= -I_{yz}\alpha - I_{xz}\omega^2 \qquad (18.29) \\
\Sigma M_z &= I_z\alpha
\end{aligned}$$

When the forces applied to the body are known, the angular acceleration α may be obtained from the last of Eqs. (18.29). The angular velocity ω is then determined by integration and the values obtained for α and ω may be substituted into the first two equations (18.29). These equations plus the three equations (18.26) which define the motion of the mass center of the body may then be used to determine the reactions at the bearings A and B.

It should be noted that axes other than the ones shown in Fig. 18.13 may be selected to analyze the rotation of a rigid body about a fixed axis. In many cases, the principal axes of inertia of the body will be found more advantageous. It is therefore wise to revert to Eq. (18.28) and to select the system of axes which best fits the problem under consideration.

If the rotating body is symmetrical with respect to the xy plane, the products of inertia I_{xz} and I_{yz} are equal to zero and Eqs. (18.29) reduce to

$$\Sigma M_x = 0 \qquad \Sigma M_y = 0 \qquad \Sigma M_z = I_z \alpha \qquad (18.30)$$

which is in accord with the results obtained in Chap. 16. If, on the other hand, the products of inertia I_{xz} and I_{yz} are different from zero, the sum of the moments of the external forces about the x and y axes will also be different from zero, even when the body rotates at a constant rate ω. Indeed, in the latter case, Eqs. (18.29) yield

$$\Sigma M_x = I_{yz} \omega^2 \qquad \Sigma M_y = -I_{xz} \omega^2 \qquad \Sigma M_z = 0 \qquad (18.31)$$

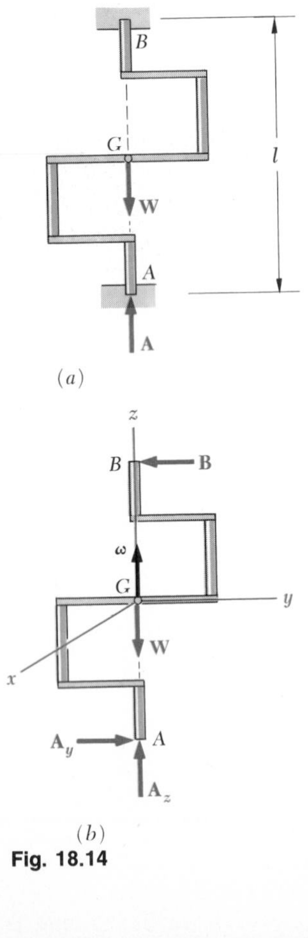

This last observation leads us to discuss the *balancing of rotating shafts*. Consider, for instance, the crankshaft shown in Fig. 18.14a, which is symmetrical about its mass center G. We first observe that when the crankshaft is at rest, it exerts no lateral thrust on its supports, since its center of gravity G is located directly above A. The shaft is said to be *statically balanced*. The reaction at A, often referred to as a *static reaction*, is vertical and its magnitude is equal to the weight W of the shaft. Let us now assume that the shaft rotates with a constant angular velocity ω. Attaching our frame of reference to the shaft, with its origin at G, the z axis along AB, and the y axis in the plane of symmetry of the shaft (Fig. 18.14b), we note that I_{xz} is zero and that I_{yz} is positive. According to Eqs. (18.31), the external forces must include a couple of moment $I_{yz} \omega^2 \mathbf{i}$. Since this couple is formed by the reaction at B and the horizontal component of the reaction at A, we have

$$\mathbf{A}_y = \frac{I_{yz} \omega^2}{l} \mathbf{j} \qquad \mathbf{B} = -\frac{I_{yz} \omega^2}{l} \mathbf{j} \qquad (18.32)$$

Since the bearing reactions are proportional to ω^2, the shaft will have a tendency to tear away from its bearings when rotating at high speeds. Moreover, since the bearing reactions $\mathbf{A}_y$ and $\mathbf{B}$, called *dynamic reactions*, are contained in the yz plane, they rotate with the shaft and cause the structure supporting it to vibrate. These undesirable effects will be avoided if by rearranging the distribution of mass around the shaft or by adding corrective masses, we let I_{yz} become equal to zero. The dynamic reactions $\mathbf{A}_y$ and $\mathbf{B}$ will vanish and the reactions at the bearings will reduce to the static reaction $\mathbf{A}_z$, the direction of which is fixed. The shaft will then be *dynamically as well as statically balanced*.

(a)

(b)

Fig. 18.14

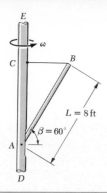

SAMPLE PROBLEM 18.3

A slender rod AB of length $L = 8$ ft and weight $W = 40$ lb is pinned at A to a vertical axle DE which rotates with a constant angular velocity ω of 15 rad/s. The rod is maintained in position by means of a horizontal wire BC attached to the axle and to the end B of the rod. Determine the tension in the wire and the reaction at A.

Solution. The effective forces reduce to the vector $m\bar{\mathbf{a}}$ attached at G and the couple $\dot{\mathbf{H}}_G$. Since G describes a horizontal circle of radius $\bar{r} = \frac{1}{2}L\cos\beta$ at the constant rate ω, we have

$$\bar{\mathbf{a}} = \mathbf{a}_n = -\bar{r}\omega^2\mathbf{I} = -(\tfrac{1}{2}L\cos\beta)\omega^2\mathbf{I} = -(450 \text{ ft/s}^2)\mathbf{I}$$
$$m\bar{\mathbf{a}} = \frac{40}{g}(-450\mathbf{I}) = -(559 \text{ lb})\mathbf{I}$$

Determination of $\dot{\mathbf{H}}_G$. We first compute the angular momentum $\mathbf{H}_G$. Using the principal centroidal axes of inertia x, y, z, we write

$$\bar{I}_x = \tfrac{1}{12}mL^2 \qquad\qquad \bar{I}_y = 0 \qquad\qquad \bar{I}_z = \tfrac{1}{12}mL^2$$
$$\omega_x = -\omega\cos\beta \qquad \omega_y = \omega\sin\beta \qquad \omega_z = 0$$
$$\mathbf{H}_G = \bar{I}_x\omega_x\mathbf{i} + \bar{I}_y\omega_y\mathbf{j} + \bar{I}_z\omega_z\mathbf{k}$$
$$\mathbf{H}_G = -\tfrac{1}{12}mL^2\omega\cos\beta\,\mathbf{i}$$

The rate of change $\dot{\mathbf{H}}_G$ of $\mathbf{H}_G$ with respect to axes of fixed orientation is obtained from Eq. (18.22). Observing that the rate of change $(\dot{\mathbf{H}}_G)_{Gxyz}$ of $\mathbf{H}_G$ with respect to the rotating frame $Gxyz$ is zero, and that the angular velocity $\boldsymbol{\Omega}$ of that frame is equal to the angular velocity $\boldsymbol{\omega}$ of the rod, we have

$$\dot{\mathbf{H}}_G = (\dot{\mathbf{H}}_G)_{Gxyz} + \boldsymbol{\omega}\times\mathbf{H}_G$$
$$\dot{\mathbf{H}}_G = 0 + (-\omega\cos\beta\,\mathbf{i} + \omega\sin\beta\,\mathbf{j})\times(-\tfrac{1}{12}mL^2\omega\cos\beta\,\mathbf{i})$$
$$\dot{\mathbf{H}}_G = \tfrac{1}{12}mL^2\omega^2\sin\beta\cos\beta\,\mathbf{k} = (645 \text{ lb}\cdot\text{ft})\mathbf{k}$$

Equations of Motion. Expressing that the system of the external forces is equivalent to the system of the effective forces, we write

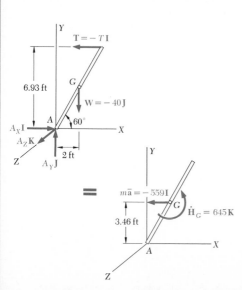

$$\Sigma\mathbf{M}_A = \Sigma(\mathbf{M}_A)_{\text{eff}}:$$
$$6.93\mathbf{J}\times(-T\mathbf{I}) + 2\mathbf{I}\times(-40\mathbf{J}) = 3.46\mathbf{J}\times(-559\mathbf{I}) + 645\,\mathbf{K}$$
$$(6.93T - 80)\mathbf{K} = (1934 + 645)\mathbf{K} \qquad T = 384 \text{ lb} \quad \blacktriangleleft$$

$$\Sigma\mathbf{F} = \Sigma\mathbf{F}_{\text{eff}}: \quad A_X\mathbf{I} + A_Y\mathbf{J} + A_Z\mathbf{K} - 384\mathbf{I} - 40\mathbf{J} = -559\mathbf{I}$$
$$\mathbf{A} = -(175 \text{ lb})\mathbf{I} + (40 \text{ lb})\mathbf{J} \quad \blacktriangleleft$$

Remark. The value of T could have been obtained from $\mathbf{H}_A$ and Eq. (18.28). However, the method used here also yields the reaction at A. Moreover, it draws attention to the effect of the asymmetry of the rod on the solution of the problem by clearly showing that both the vector $m\bar{\mathbf{a}}$ and the couple $\dot{\mathbf{H}}_G$ must be used to represent the effective forces.

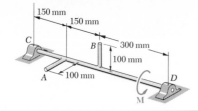

Two 100-mm rods A and B, each of mass 300 g, are welded to the shaft CD which is supported by bearings at C and D. If a couple $\mathbf{M}$ of magnitude equal to 6 N·m is applied to the shaft, determine the components of the dynamic reactions at C and D at the instant when the shaft has reached an angular velocity of 1200 rpm. Neglect the moment of inertia of the shaft itself.

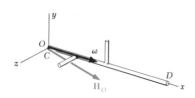

Angular Momentum about O. We attach to the body the frame of reference $Oxyz$ and note that the axes chosen are not principal axes of inertia for the body. Since the body rotates about the x axis, we have $\omega_x = \omega$ and $\omega_y = \omega_z = 0$. Substituting into Eqs. (18.13),

$$H_x = I_x\omega \qquad H_y = -I_{xy}\omega \qquad H_z = -I_{xz}\omega$$
$$\mathbf{H}_O = (I_x\mathbf{i} - I_{xy}\mathbf{j} - I_{xz}\mathbf{k})\omega$$

Moments of the External Forces about O. Since the frame of reference rotates with the angular velocity $\boldsymbol{\omega}$, Eq. (18.28) yields

$$\Sigma\mathbf{M}_O = (\dot{\mathbf{H}}_O)_{Oxyz} + \boldsymbol{\omega} \times \mathbf{H}_O$$
$$= (I_x\mathbf{i} - I_{xy}\mathbf{j} - I_{xz}\mathbf{k})\alpha + \omega\mathbf{i} \times (I_x\mathbf{i} - I_{xy}\mathbf{j} - I_{xz}\mathbf{k})\omega$$
$$= I_x\alpha\mathbf{i} - (I_{xy}\alpha - I_{xz}\omega^2)\mathbf{j} - (I_{xz}\alpha + I_{xy}\omega^2)\mathbf{k} \qquad (1)$$

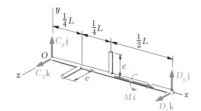

Dynamic Reaction at D. The external forces consist of the weights of the shaft and rods, the couple $\mathbf{M}$, the static reactions at C and D, and the dynamic reactions at C and D. Since the weights and static reactions are balanced, the external forces reduce to the couple $\mathbf{M}$ and the dynamic reactions $\mathbf{C}$ and $\mathbf{D}$ as shown in the figure. Taking moments about O, we have

$$\Sigma\mathbf{M}_O = L\mathbf{i} \times (D_y\mathbf{j} + D_z\mathbf{k}) + M\mathbf{i} = M\mathbf{i} - D_zL\mathbf{j} + D_yL\mathbf{k} \qquad (2)$$

Equating the coefficients of the unit vector $\mathbf{i}$ in (1) and (2),

$$M = I_x\alpha \qquad M = 2(\tfrac{1}{3}mc^2)\alpha \qquad \alpha = 3M/2mc^2$$

Equating the coefficients of $\mathbf{k}$ and $\mathbf{j}$ in (1) and (2):

$$D_y = -(I_{xz}\alpha + I_{xy}\omega^2)/L \qquad D_z = (I_{xy}\alpha - I_{xz}\omega^2)/L \qquad (3)$$

Using the parallel-axis theorem, and noting that the product of inertia of each rod is zero with respect to centroidal axes, we have

$$I_{xy} = \Sigma m\bar{x}\bar{y} = m(\tfrac{1}{2}L)(\tfrac{1}{2}c) = \tfrac{1}{4}mLc$$
$$I_{xz} = \Sigma m\bar{x}\bar{z} = m(\tfrac{1}{4}L)(\tfrac{1}{2}c) = \tfrac{1}{8}mLc$$

Substituting into (3) the values found for I_{xy}, I_{xz}, and α:

$$D_y = -\tfrac{3}{16}(M/c) - \tfrac{1}{4}mc\omega^2 \qquad D_z = \tfrac{3}{8}(M/c) - \tfrac{1}{8}mc\omega^2$$

Substituting $\omega = 1200$ rpm $= 125.7$ rad/s, $c = 0.100$ m, $M = 6$ N·m, and $m = 0.300$ kg, we have

$$D_y = -129.8 \text{ N} \qquad D_z = -36.8 \text{ N} \quad \blacktriangleleft$$

Dynamic Reaction at C. Using a frame of reference attached at D, we obtain equations similar to Eqs. (3), which yield

$$C_y = -152.2 \text{ N} \qquad C_z = -155.2 \text{ N} \quad \blacktriangleleft$$

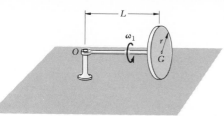

SAMPLE PROBLEM 18.5

A homogeneous disk of radius r and mass m is mounted on an axle OG of length L and negligible mass. The axle is pivoted at the fixed point O and the disk is constrained to roll on a horizontal floor. Knowing that the disk rotates counterclockwise at the constant rate ω_1 about the axle, determine (a) the force (assumed vertical) exerted by the floor on the disk, (b) the reaction at the pivot O.

Solution. The effective forces reduce to the vector $m\bar{\mathbf{a}}$ attached at G and the couple $\dot{\mathbf{H}}_G$. Recalling from Sample Prob. 18.2 that the axle rotates about the y axis at the rate $\omega_2 = r\omega_1/L$, we write

$$m\bar{\mathbf{a}} = -mL\omega_2^2\mathbf{i} = -mL(r\omega_1/L)^2\mathbf{i} = -(mr^2\omega_1^2/L)\mathbf{i} \qquad (1)$$

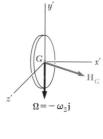

Determination of $\dot{\mathbf{H}}_G$. We recall from Sample Prob. 18.2 that the angular momentum of the disk about G is

$$\mathbf{H}_G = \tfrac{1}{2}mr^2\omega_1\left(\mathbf{i} - \frac{r}{2L}\mathbf{j}\right)$$

where $\mathbf{H}_G$ is resolved into components along the rotating axes x', y', z', with x' along OG and y' vertical. The rate of change $\dot{\mathbf{H}}_G$ of $\mathbf{H}_G$ with respect to axes of fixed orientation is obtained from Eq. (18.22). Noting that the rate of change $(\dot{\mathbf{H}}_G)_{Gx'y'z'}$ of $\mathbf{H}_G$ with respect to the rotating frame is zero, and that the angular velocity $\boldsymbol{\Omega}$ of that frame is

$$\boldsymbol{\Omega} = -\omega_2\mathbf{j} = -(r\omega_1/L)\mathbf{j}$$

we have

$$\dot{\mathbf{H}}_G = (\dot{\mathbf{H}}_G)_{Gx'y'z'} + \boldsymbol{\Omega}\times\mathbf{H}_G$$
$$= 0 - \frac{r\omega_1}{L}\mathbf{j}\times\tfrac{1}{2}mr^2\omega_1\left(\mathbf{i} - \frac{r}{2L}\mathbf{j}\right)$$
$$= \tfrac{1}{2}mr^2(r/L)\omega_1^2\mathbf{k} \qquad (2)$$

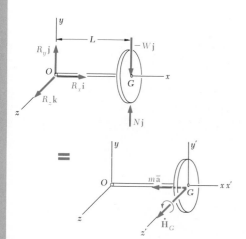

Equations of Motion. Expressing that the system of the external forces is equivalent to the system of the effective forces, we write

$$\Sigma\mathbf{M}_O = \Sigma(\mathbf{M}_O)_{\text{eff}}: \qquad L\mathbf{i}\times(N\mathbf{j} - W\mathbf{j}) = \dot{\mathbf{H}}_G$$
$$(N - W)L\mathbf{k} = \tfrac{1}{2}mr^2(r/L)\omega_1^2\mathbf{k}$$
$$N = W + \tfrac{1}{2}mr(r/L)^2\omega_1^2 \qquad \mathbf{N} = [W + \tfrac{1}{2}mr(r/L)^2\omega_1^2]\mathbf{j} \qquad (3) \quad \blacktriangleleft$$
$$\Sigma\mathbf{F} = \Sigma\mathbf{F}_{\text{eff}}: \qquad \mathbf{R} + N\mathbf{j} - W\mathbf{j} = m\bar{\mathbf{a}}$$

Substituting for N from (3), for $m\bar{\mathbf{a}}$ from (1), and solving for $\mathbf{R}$, we have

$$\mathbf{R} = -(mr^2\omega_1^2/L)\mathbf{i} - \tfrac{1}{2}mr(r/L)^2\omega_1^2\mathbf{j}$$

$$\mathbf{R} = -\frac{mr^2\omega_1^2}{L}\left(\mathbf{i} + \frac{r}{2L}\mathbf{j}\right) \quad \blacktriangleleft$$

Problems

18.45 Determine the rate of change $\dot{\mathbf{H}}_G$ of the angular momentum $\mathbf{H}_G$ of the plate of Prob. 18.1.

18.46 Determine the rate of change $\dot{\mathbf{H}}_G$ of the angular momentum $\mathbf{H}_G$ of the rod of Prob. 18.2.

18.47 Determine the rate of change $\dot{\mathbf{H}}_G$ of the angular momentum $\mathbf{H}_G$ of the disk of Prob. 18.3 for an arbitrary value of β, knowing that its angular velocity ω remains constant.

18.48 Determine the rate of change $\dot{\mathbf{H}}_A$ of the angular momentum $\mathbf{H}_A$ of the plate of Prob. 18.4, knowing that its angular velocity ω remains constant.

18.49 Determine the rate of change $\dot{\mathbf{H}}_C$ of the angular momentum $\mathbf{H}_C$ of the disk of Prob. 18.5.

18.50 Determine the rate of change $\dot{\mathbf{H}}_C$ of the angular momentum $\mathbf{H}_C$ of the disk of Prob. 18.6.

18.51 Determine the rate of change $\dot{\mathbf{H}}_G$ of the angular momentum $\mathbf{H}_G$ of the disk of Prob. 18.3 for an arbitrary value of β, knowing that the disk has an angular velocity $\omega = \omega\mathbf{i}$ and an angular acceleration $\alpha = \alpha\mathbf{i}$.

18.52 Determine the rate of change $\dot{\mathbf{H}}_A$ of the angular momentum $\mathbf{H}_A$ of the plate of Prob. 18.4, knowing that it has an angular velocity $\omega = \omega\mathbf{j}$ and an angular acceleration $\alpha = \alpha\mathbf{j}$.

18.53 When the 40-lb wheel shown is attached to a balancing machine and made to spin at a rate of 750 rpm, it is found that the forces exerted by the wheel on the machine are equivalent to a force-couple system consisting of a force $\mathbf{F} = (36.2 \text{ lb})\mathbf{j}$ applied at C and a couple $\mathbf{M}_C = -(10.85 \text{ lb} \cdot \text{ft})\mathbf{k}$, where the unit vectors form a right-handed triad which rotates with the wheel. (*a*) Determine the distance from the axis of rotation to the mass center of the wheel and the products of inertia I_{xy} and I_{zx}. (*b*) If only two corrective weights are to be used to balance the wheel both statically and dynamically, what should these weights be and at which of the points A, B, D, or E should they be placed?

18.54 After attaching the 40-lb wheel shown to a balancing machine and making it spin at a rate of 900 rpm, a mechanic has found that to balance the wheel both statically and dynamically, he should use two corrective weights, a 6-oz weight placed at A and a 2-oz weight placed at E. Using a right-handed frame of reference rotating with the wheel (with the z axis perpendicular to the plane of the figure), determine before the corrective weights have been attached (*a*) the distance from the axis of rotation to the mass center of the wheel and the products of inertia I_{xy} and I_{zx}, (*b*) the force-couple system at C equivalent to the forces exerted by the wheel on the machine.

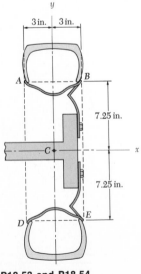

Fig. P18.53 and P18.54

18.55 A thin homogeneous triangular plate of mass 2.5 kg is welded to a light vertical axle supported by bearings at A and B. Knowing that the plate rotates at the constant rate $\omega = 8$ rad/s, determine the dynamic reactions at A and B.

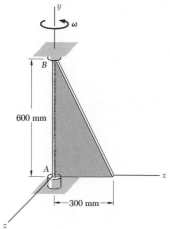

Fig. P18.55

18.56 A thin homogeneous rod of weight w per unit length is used to form the shaft shown. Knowing that the shaft rotates with a constant angular velocity ω, determine the dynamic reactions at A and B.

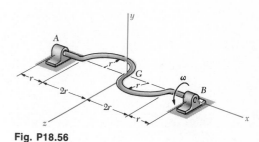

Fig. P18.56

18.57 Two L-shaped arms, each weighing 4 lb, are welded at the third points of a 2-ft horizontal shaft supported by bearings at A and B. The assembly is at rest ($\omega = 0$) when a couple of moment $\mathbf{M}_0 = (20 \text{ lb·in.})\mathbf{k}$ is applied to the shaft. Determine (a) the resulting angular acceleration of the shaft, (b) the dynamic reactions at A and B immediately after the couple has been applied.

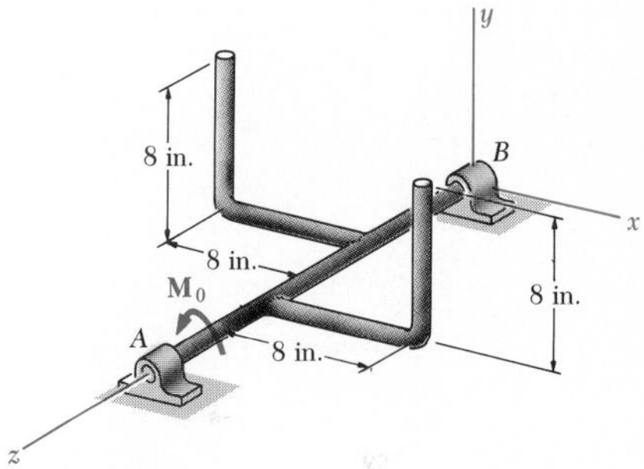

Fig. P18.57

18.58 Knowing that the plate of Prob. 18.55 is initially at rest ($\omega = 0$) when a couple of moment $\mathbf{M}_0 = (0.75 \text{ N·m})\mathbf{j}$ is applied to it, determine (a) the resulting angular acceleration of the plate, (b) the dynamic reactions at A and B immediately after the couple has been applied.

18.59 The sheet-metal component shown is of uniform thickness and has a mass of 600 g. It is attached to a light axle supported by bearings at A and B located 150 mm apart. The component is at rest when it is subjected to a couple $\mathbf{M}_0$ as shown. If the resulting angular acceleration is $\boldsymbol{\alpha} = (12 \text{ rad/s}^2)\mathbf{k}$, determine ($a$) the couple $\mathbf{M}_0$, (b) the dynamic reactions at A and B immediately after the couple has been applied.

18.60 Knowing that the shaft of Prob. 18.56 is initially at rest ($\omega = 0$), determine (a) the couple $\mathbf{M}_0$ required to impart to the shaft an acceleration $\boldsymbol{\alpha} = \alpha\mathbf{i}$, ($b$) the dynamic reactions at A and B immediately after the couple $\mathbf{M}_0$ has been applied.

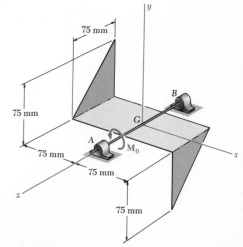

18.61 For the sheet-metal component of Prob. 18.59, determine (a) the angular velocity of the component 0.6 s after the couple $\mathbf{M}_0$ has been applied to it, (b) the magnitude of the dynamic reactions at A and B at that time.

Fig. P18.59

18.62 Two uniform rods CD and DE, each weighing 2.5 lb, are welded to shaft AB. Knowing that $\boldsymbol{\omega} = (12 \text{ rad/s})\mathbf{i}$ and $\boldsymbol{\alpha} = (45 \text{ rad/s}^2)\mathbf{i}$ at the instant shown, determine (a) the couple $\mathbf{M}$ applied to the shaft, (b) the dynamic reactions at A and B.

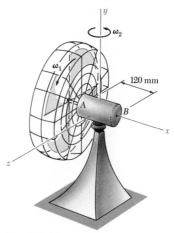

Fig. P18.63

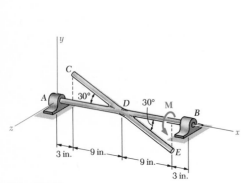

Fig. P18.62

18.63 The blade of an oscillating fan and the rotor of its motor have a total mass of 200 g and a combined radius of gyration of 75 mm. They are supported by bearings located at A and B, 120 mm apart, and have an angular velocity $\boldsymbol{\omega}_1$ of 2400 rpm at the high-speed setting. Determine the dynamic reactions at A and B when the motor casing has an angular velocity $\boldsymbol{\omega}_2$ of 0.5 rad/s as shown.

18.64 The essential structure of a certain type of aircraft turn indicator is shown. Springs AC and BD are initially stretched and exert equal vertical forces at A and B when the airplane is traveling in a straight path. Each spring has a constant of 600 N/m and the uniform disk has a mass of 250 g and spins at the rate of 12 000 rpm. Determine the angle through which the yoke will rotate when the pilot executes a horizontal turn of 800-m radius to the right at a speed of 720 km/h. Indicate whether point A will move up or down.

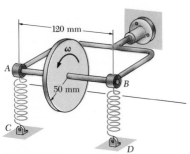

Fig. P18.64

18.65 Each wheel of an automobile weighs 50 lb and has a radius of gyration of 9 in. The automobile travels around an unbanked curve of radius 500 ft at a speed of 60 mi/h. Knowing that each wheel is of 23-in. diameter and that the transverse distance between the wheels is 60 in., determine the additional normal reaction at each outside wheel due to the rotation of the car.

18.66 A three-bladed airplane propeller weighs 300 lb and has a radius of gyration of 3 ft. Knowing that the propeller rotates at 1500 rpm, determine the magnitude of the couple applied by the propeller to its shaft when the airplane travels in a circular path of 1200-ft radius at 350 mi/h.

Fig. P18.67

18.67 A thin ring of radius a is attached by a collar at A to a vertical shaft which rotates with a constant angular velocity $\boldsymbol{\omega}$. Derive an expression (a) for the constant angle β that the plane of the ring forms with the vertical, (b) for the maximum value of ω for which the ring will remain vertical ($\beta = 0$).

18.68 A uniform square plate of side a and mass m is hinged at A and B to a clevis which rotates with a constant angular velocity $\boldsymbol{\omega}$. Determine (*a*) the value of ω for which angle β maintains the constant value $\beta = 40°$, (*b*) the range of values of ω for which the plate remains vertical ($\beta = 90°$).

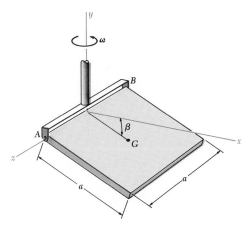

Fig. P18.68

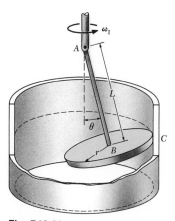

Fig. P18.69

18.69 The essential features of a type of crusher are shown. A disk of weight W is mounted on a shaft AB about which it can rotate freely. Shaft AB is attached by means of a clevis to a vertical shaft which is made to rotate at a constant angular velocity ω_1. The disk rolls on the inside of a vertical cylinder. (Only one half of the cylinder is shown.) Determine the minimum angular velocity ω_1 for which contact is maintained between the disk and the cylinder.

18.70 A slender rod AB of length L is attached to arm BC by a clevis at B and is made to rotate about the vertical with an angular velocity $\boldsymbol{\omega}$. Knowing that the rod forms an angle $\theta = 20°$ with the horizontal, derive an expression for the corresponding magnitude of the angular velocity ω.

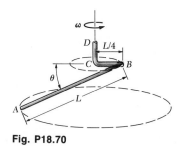

Fig. P18.70

18.71 Two disks, each of weight 10 lb and radius 12 in., spin as shown at 1200 rpm about rod AB, which is attached to shaft CD. The entire system is made to rotate about the z axis with an angular velocity $\boldsymbol{\Omega}$ of 60 rpm. (*a*) Determine the dynamic reactions at C and D as the system passes through the position shown. (*b*) Solve part *a* assuming that the direction of spin of disk B is reversed.

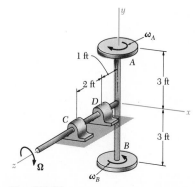

Fig. P18.71

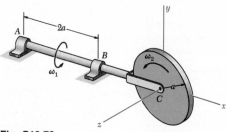

Fig. P18.72

18.72 A thin homogeneous disk of mass m and radius a is held by a fork-ended horizontal rod ABC. The disk and the rod rotate with the angular velocities shown. Assuming that both ω_1 and ω_2 are constant, determine the dynamic reactions at A and B.

18.73 A thin disk of mass $m = 5$ kg rotates with an angular velocity ω_2 with respect to the bent axle ABC, which itself rotates with an angular velocity ω_1 about the y axis. Knowing that $\omega_1 = 3$ rad/s and $\omega_2 = 8$ rad/s and that both are constant, determine the force-couple system representing the dynamic reaction at the support at A.

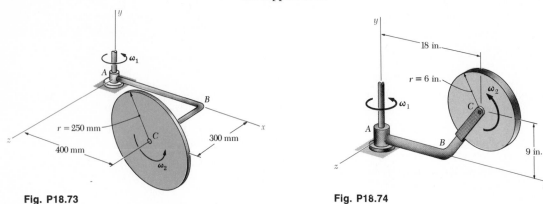

Fig. P18.73

Fig. P18.74

18.74 A thin disk of weight $W = 10$ lb rotates with an angular velocity ω_2 with respect to arm ABC, which itself rotates with an angular velocity ω_1 about the y axis. Knowing that $\omega_1 = 5$ rad/s and $\omega_2 = 15$ rad/s and that both are constant, determine the force-couple system representing the dynamic reaction at the support at A.

18.75 For the disk of Prob. 18.73, determine (a) the couple $M_1\mathbf{j}$ which should be applied to the bent axle ABC to give it an angular acceleration $\boldsymbol{\alpha}_1 = (7.5 \text{ rad/s}^2)\mathbf{j}$ when $\omega_1 = 3$ rad/s, knowing that the disk rotates at the constant rate $\omega_2 = 8$ rad/s, (b) the force-couple system representing the dynamic reaction at A at that instant. Assume that the bent axle ABC has a negligible mass.

18.76 For the disk of Prob. 18.73, determine (a) the couple $M_1\mathbf{j}$ which should be applied to the bent axle ABC to keep it rotating at the constant rate $\omega_1 = 3$ rad/s at an instant when the angular velocity ω_2 of the disk has a magnitude of 8 rad/s and decreases at the rate of 1.8 rad/s^2 due to axle friction at C, (b) the force-couple system representing the dynamic reaction at A at that instant.

18.77 For the disk of Prob. 18.72, determine (*a*) the couple $M_1\mathbf{i}$ which should be applied to rod AB to give it an angular acceleration $\alpha_1\mathbf{i}$ when the rod has an angular velocity $\omega_1\mathbf{i}$, knowing that the disk rotates at the constant rate ω_2, (*b*) the dynamic reactions at A and B at that instant.

18.78 For the disk of Prob. 18.74, determine (*a*) the couple $M_1\mathbf{j}$ which should be applied to arm ABC to give it an angular acceleration $\alpha_1 = -(7.5 \text{ rad/s}^2)\mathbf{j}$ when $\omega_1 = 5$ rad/s, knowing that the disk rotates at the constant rate $\omega_2 = 15$ rad/s, (*b*) the force-couple system representing the dynamic reaction at A at that instant. Assume that ABC has a negligible mass.

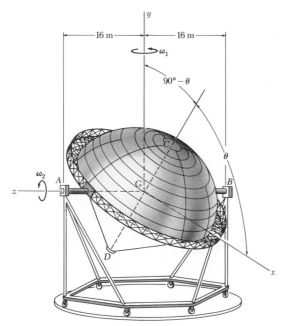

Fig. P18.79

*18.79 An experimental Fresnel-lens solar-energy concentrator may rotate about the horizontal axis AB which passes through its mass center G. It is supported at A and B by a steel framework which may rotate about the vertical y axis. The concentrator has a mass of 30 Mg, a radius of gyration of 12 m about its axis of symmetry CD, and a radius of gyration of 10 m about any transverse axis through G. Knowing that the angular velocities ω_1 amd ω_2 have constant magnitudes equal to 0.25 rad/s and 0.20 rad/s, respectively, determine for the position $\theta = 60°$ (*a*) the forces exerted on the concentrator at A and B, (*b*) the couple $M_2\mathbf{k}$ applied to the concentrator at that instant.

*18.80 A slender homogeneous rod AB of mass m and length L is made to rotate at the constant rate ω_2 about the horizontal z axis, while frame CD is made to rotate at the constant rate ω_1 about the vertical y axis. Express as a function of the angle θ (*a*) the couple $\mathbf{M}_1$ required to maintain the rotation of the frame, (*b*) the couple $\mathbf{M}_2$ required to maintain the rotation of the rod, (*c*) the dynamic reactions at the supports C and D.

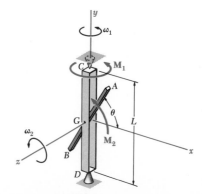

Fig. P18.80

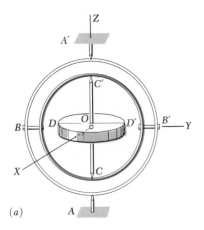

(a)

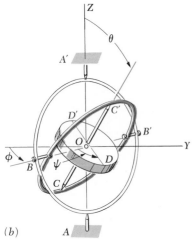

(b)

Fig. 18.15

A *gyroscope* consists essentially of a rotor which may spin freely about its geometric axis. When mounted in a Cardan's suspension (Fig. 18.15), a gyroscope may assume any orientation, but its mass center must remain fixed in space. In order to define the position of a gyroscope at a given instant, we shall select a fixed frame of reference $OXYZ$, with the origin O located at the mass center of the gyroscope and the Z axis directed along the line defined by the bearings A and A' of the outer gimbal, and we shall consider a reference position of the gyroscope in which the two gimbals and a given diameter DD' of the rotor are located in the fixed YZ plane (Fig. 18.15*a*). The gyroscope may be brought from this reference position into any arbitrary position (Fig. 18.15*b*) by means of the following steps: (1) a rotation of the outer gimbal through an angle ϕ about the axis AA', (2) a rotation of the inner gimbal through θ about BB', (3) a rotation of the rotor through ψ about CC'. The angles ϕ, θ, and ψ are called the *Eulerian angles;* they completely characterize the position of the gyroscope at any given instant. Their derivatives $\dot\phi$, $\dot\theta$, and $\dot\psi$ define, respectively, the rate of *precession*, the rate of *nutation*, and the rate of *spin* of the gyroscope at the instant considered.

In order to compute the components of the angular velocity and of the angular momentum of the gyroscope, we shall use a rotating system of axes $Oxyz$ *attached to the inner gimbal*, with the y axis along BB' and the z axis along CC' (Fig. 18.16). These axes are principal axes of inertia for the gyroscope but while they follow it in its precession and nutation, they do not spin. For that reason, they are more convenient to use than axes actually attached to the gyroscope. We shall now express the angular velocity $\boldsymbol{\omega}$ of the gyroscope with respect to the fixed frame of reference $OXYZ$ as the sum of three partial angular velocities corresponding, respectively, to the precession, the nutation, and the spin of the gyroscope. Denoting by $\mathbf{i}$, $\mathbf{j}$, $\mathbf{k}$ the unit vectors along the rotating axes, and by $\mathbf{K}$ the unit vector along the fixed Z axis, we have

$$\boldsymbol{\omega} = \dot\phi\mathbf{K} + \dot\theta\mathbf{j} + \dot\psi\mathbf{k} \tag{18.33}$$

Since the vector components obtained for $\boldsymbol{\omega}$ in (18.33) are not orthogonal (Fig. 18.16), we shall resolve the unit vector $\mathbf{K}$ into components along the x and z axes; we write

$$\mathbf{K} = -\sin\theta\,\mathbf{i} + \cos\theta\,\mathbf{k} \tag{18.34}$$

and, substituting for $\mathbf{K}$ into (18.33),

$$\boldsymbol{\omega} = -\dot\phi\sin\theta\,\mathbf{i} + \dot\theta\mathbf{j} + (\dot\psi + \dot\phi\cos\theta)\mathbf{k} \tag{18.35}$$

Since the coordinate axes are principal axes of inertia, the components of the angular momentum $\mathbf{H}_O$ may be obtained by multiplying the components of $\boldsymbol{\omega}$ by the moments of inertia of the rotor about the x, y, and z axes, respectively. Denoting by I the moment of inertia of the rotor about its spin axis, by I' its moment of inertia about a transverse axis through O, and neglecting the mass of the gimbals, we write

921

18.9. Motion of a Gyroscope. Eulerian Angles

$$\mathbf{H}_O = -I'\dot{\phi} \sin \theta \, \mathbf{i} + I'\dot{\theta}\mathbf{j} + I(\dot{\psi} + \dot{\phi} \cos \theta)\mathbf{k} \qquad (18.36)$$

Recalling that the rotating axes are attached to the inner gimbal, and thus do not spin, we express their angular velocity as the sum

$$\boldsymbol{\Omega} = \dot{\phi}\mathbf{K} + \dot{\theta}\mathbf{j} \qquad (18.37)$$

or, substituting for $\mathbf{K}$ from (18.34),

$$\boldsymbol{\Omega} = -\dot{\phi} \sin \theta \, \mathbf{i} + \dot{\theta}\mathbf{j} + \dot{\phi} \cos \theta \, \mathbf{k} \qquad (18.38)$$

Substituting for $\mathbf{H}_O$ and $\boldsymbol{\Omega}$ from (18.36) and (18.38) into the equation

$$\Sigma \mathbf{M}_O = (\dot{\mathbf{H}}_O)_{Oxyz} + \boldsymbol{\Omega} \times \mathbf{H}_O \qquad (18.28)$$

we obtain the three differential equations

$$\Sigma M_x = -I'(\ddot{\phi} \sin \theta + 2\dot{\theta}\dot{\phi} \cos \theta) + I\dot{\theta}(\dot{\psi} + \dot{\phi} \cos \theta)$$
$$\Sigma M_y = I'(\ddot{\theta} - \dot{\phi}^2 \sin \theta \cos \theta) + I\dot{\phi} \sin \theta(\dot{\psi} + \dot{\phi} \cos \theta) \qquad (18.39)$$
$$\Sigma M_z = I\frac{d}{dt}(\dot{\psi} + \dot{\phi} \cos \theta)$$

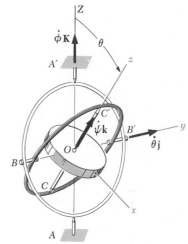

Fig. 18.16

The equations (18.39) define the motion of a gyroscope subjected to a given system of forces when the mass of its gimbals is neglected. They may also be used to define the motion of an *axisymmetrical body* (or body of revolution) attached at a point on its axis of symmetry, or the motion of an axisymmetrical body about its mass center. While the gimbals of the gyroscope helped us visualize the Eulerian angles, it is clear that these angles may be used to define the position of any rigid body with respect to axes centered at a point of the body, regardless of the way in which the body is actually supported.

Since the equations (18.39) are nonlinear, it will not be possible, in general, to express the Eulerian angles ϕ, θ, and ψ as analytical functions of the time t, and numerical methods of solution may have to be used. However, as we shall see in the following sections, there are several particular cases of interest which may be analyzed easily.

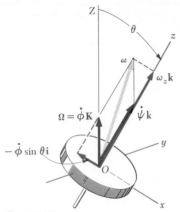

Fig. 18.17

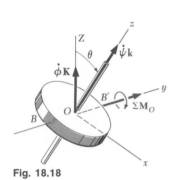

Fig. 18.18

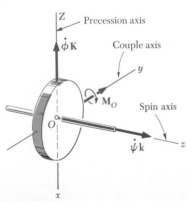

Fig. 18.19

*18.10. Steady Precession of a Gyroscope.** We shall consider in this section the particular case of gyroscopic motion in which the angle θ, the rate of precession $\dot{\phi}$, and the rate of spin $\dot{\psi}$ remain constant. We propose to determine the forces which must be applied to the gyroscope to maintain this motion, known as the *steady precession* of a gyroscope.

Instead of applying the general equations (18.39), we shall determine the sum of the moments of the required forces by computing the rate of change of the angular momentum of the gyroscope in the particular case considered. We first note that the angular velocity $\boldsymbol{\omega}$ of the gyroscope, its angular momentum $\mathbf{H}_O$, and the angular velocity $\boldsymbol{\Omega}$ of the rotating frame of reference (Fig. 18.17) reduce, respectively, to

$$\boldsymbol{\omega} = -\dot{\phi}\sin\theta\,\mathbf{i} + \omega_z\mathbf{k} \qquad (18.40)$$

$$\mathbf{H}_O = -I'\dot{\phi}\sin\theta\,\mathbf{i} + I\omega_z\mathbf{k} \qquad (18.41)$$

$$\boldsymbol{\Omega} = -\dot{\phi}\sin\theta\,\mathbf{i} + \dot{\phi}\cos\theta\,\mathbf{k} \qquad (18.42)$$

where $\omega_z = \dot{\psi} + \dot{\phi}\cos\theta =$ component along the spin axis of the total angular velocity of the gyroscope

Since θ, $\dot{\phi}$, and $\dot{\psi}$ are constant, the vector $\mathbf{H}_O$ is constant in magnitude and direction with respect to the rotating frame of reference, and its rate of change $(\dot{\mathbf{H}}_O)_{Oxyz}$ with respect to that frame is zero. Thus Eq. (18.28) reduces to

$$\Sigma\mathbf{M}_O = \boldsymbol{\Omega} \times \mathbf{H}_O \qquad (18.43)$$

which yields, after substitutions from (18.41) and (18.42),

$$\Sigma\mathbf{M}_O = (I\omega_z - I'\dot{\phi}\cos\theta)\dot{\phi}\sin\theta\,\mathbf{j} \qquad (18.44)$$

Since the mass center of the gyroscope is fixed in space, we have, by (18.1), $\Sigma\mathbf{F} = 0$; thus, the forces which must be applied to the gyroscope to maintain its steady precession reduce to a couple of moment equal to the right-hand member of Eq. (18.44). We note that *this couple should be applied about an axis perpendicular to the precession axis and to the spin axis of the gyroscope* (Fig. 18.18).

In the particular case when the precession axis and the spin axis are at a right angle to each other, we have $\theta = 90°$ and Eq. (18.44) reduces to

$$\Sigma\mathbf{M}_O = I\dot{\psi}\dot{\phi}\mathbf{j} \qquad (18.45)$$

Thus, if we apply to the gyroscope a couple $\mathbf{M}_O$ about an axis perpendicular to its axis of spin, the gyroscope will precess about an axis perpendicular to both the spin axis and the couple axis, in a sense such that the vectors representing, respectively, the spin, the couple, and the precession form a right-handed triad (Fig. 18.19).

Because of the relatively large couples required to change the orientation of their axles, gyroscopes are used as stabilizers in torpedoes and ships.

Spinning bullets and shells remain tangent to their trajectory because of gyroscopic action. And a bicycle is easier to keep balanced at high speeds because of the stabilizing effect of its spinning wheels. However, gyroscopic action is not always welcome and must be taken into account in the design of bearings supporting rotating shafts subjected to forced precession. The reactions exerted by its propellers on an airplane which changes its direction of flight must also be taken into consideration and compensated for whenever possible.

*18.11. Motion of an Axisymmetrical Body under No Force.

We shall consider in this section the motion about its mass center of an axisymmetrical body under no force, except its own weight. Examples of such a motion are furnished by projectiles, if air resistance is neglected, and by artificial satellites and space vehicles after burnout of their launching rockets.

Since the sum of the moments of the external forces about the mass center G of the body is zero, Eq. (18.2) yields $\dot{\mathbf{H}}_G = 0$. It follows that the angular momentum $\mathbf{H}_G$ of the body about G is constant. Thus, the direction of $\mathbf{H}_G$ is fixed in space and may be used to define the Z axis, or axis of precession (Fig. 18.20). Selecting a rotating system of axes $Gxyz$ with the z axis along the axis of symmetry of the body and the x axis in the plane defined by the Z and z axes, we have

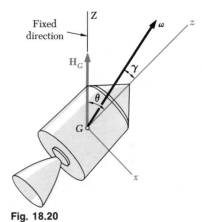

Fixed
direction

Fig. 18.20

$$H_x = -H_G \sin \theta \qquad H_y = 0 \qquad H_z = H_G \cos \theta \qquad (18.46)$$

where θ represents the angle formed by the Z and z axes, and H_G denotes the constant magnitude of the angular momentum of the body about G. Since the x, y, and z axes are principal axes of inertia for the body considered, we may write

$$H_x = I'\omega_x \qquad H_y = I'\omega_y \qquad H_z = I\omega_z \qquad (18.47)$$

where I denotes the moment of inertia of the body about its axis of symmetry, and I' its moment of inertia about a transverse axis through G. It follows from Eqs. (18.46) and (18.47) that

$$\omega_x = -\frac{H_G \sin \theta}{I'} \qquad \omega_y = 0 \qquad \omega_z = \frac{H_G \cos \theta}{I} \qquad (18.48)$$

The second of the relations obtained shows that the angular velocity ω has no component along the y axis, i.e., along an axis perpendicular to the Zz plane. Thus, the angle θ formed by the Z and z axes remains constant and *the body is in steady precession about the Z axis.*

Dividing the first and third of the relations (18.48) member by member, and observing from Fig. 18.21 that $-\omega_x/\omega_z = \tan \gamma$, we obtain the following relation between the angles γ and θ that the vectors ω and $\mathbf{H}_G$, respectively, form with the axis of symmetry of the body:

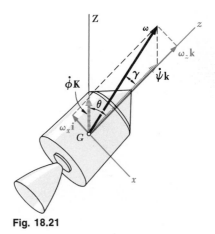

Fig. 18.21

$$\tan \gamma = \frac{I}{I'} \tan \theta \qquad (18.49)$$

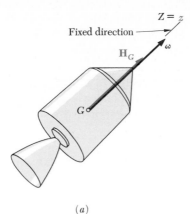

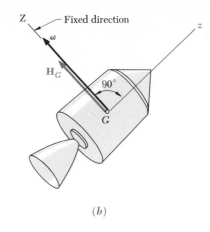

(a)   (b)

Fig. 18.22

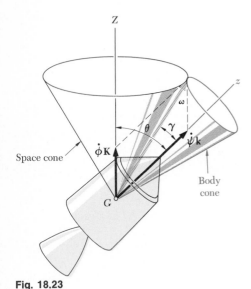

Fig. 18.23

There are two particular cases of motion of an axisymmetrical body under no force which involve no precession: (1) If the body is set to spin about its axis of symmetry, we have $\omega_x = 0$ and, by (18.47), $H_x = 0$; the vectors $\boldsymbol{\omega}$ and $\mathbf{H}_G$ have the same orientation and the body keeps spinning about its axis of symmetry (Fig. 18.22a). (2) If the body is set to spin about a transverse axis, we have $\omega_z = 0$ and, by (18.47), $H_z = 0$; again $\boldsymbol{\omega}$ and $\mathbf{H}_G$ have the same orientation and the body keeps spinning about the given transverse axis (Fig. 18.22b).

Considering now the general case represented in Fig. 18.21, we recall from Sec. 15.12 that the motion of a body about a fixed point—or about its mass center—may be represented by the motion of a body cone rolling on a space cone. In the case of steady precession, the two cones are circular, since the angles γ and $\theta - \gamma$ that the angular velocity $\boldsymbol{\omega}$ forms, respectively, with the axis of symmetry of the body and with the precession axis are constant. Two cases should be distinguished:

1. $I < I'$. This is the case of an elongated body, such as the space vehicle of Fig. 18.23. By (18.49) we have $\gamma < \theta$; the vector $\boldsymbol{\omega}$ lies inside the angle ZGz; the space cone and the body cone are tangent externally; the spin and the precession are both observed as counterclockwise from the positive z axis. The precession is said to be *direct*.

2. $I > I'$. This is the case of a flattened body, such as the satellite of Fig. 18.24. By (18.49) we have $\gamma > \theta$; since the vector $\boldsymbol{\omega}$ must lie outside the angle ZGz, the vector $\dot{\psi}\mathbf{k}$ has a sense opposite to that of the z axis; the space cone is inside the body cone; the precession and the spin have opposite senses; the precession is said to be *retrograde*.

Fig. 18.24

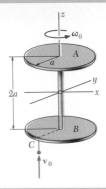

SAMPLE PROBLEM 18.6

A space satellite of mass m is known to be dynamically equivalent to two thin disks of equal mass. The disks are of radius $a = 800$ mm and are rigidly connected by a light rod of length $2a$. Initially the satellite is spinning freely about its axis of symmetry at the rate $\omega_0 = 60$ rpm. A meteorite, of mass $m_0 = m/1000$ and traveling with a velocity $\mathbf{v}_0$ of 2000 m/s relative to the satellite, strikes the satellite and becomes embedded at C. Determine (a) the angular velocity of the satellite immediately after impact, (b) the precession axis of the ensuing motion, (c) the rates of precession and spin of the ensuing motion.

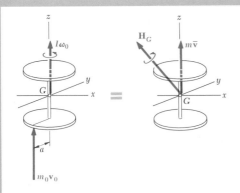

Solution. *Moments of Inertia.* We note that the axes shown are principal axes of inertia for the satellite and write

$$I = I_z = \tfrac{1}{2}ma^2 \qquad I' = I_x = I_y = 2[\tfrac{1}{4}(\tfrac{1}{2}m)a^2 + (\tfrac{1}{2}m)a^2] = \tfrac{5}{4}ma^2$$

Principle of Impulse and Momentum. We consider the satellite and the meteorite as a single system. Since no external force acts on this system, the momenta before and after impact are equipollent. Taking moments about G we write

$$-a\mathbf{j} \times m_0 v_0 \mathbf{k} + I\omega_0 \mathbf{k} = \mathbf{H}_G$$
$$\mathbf{H}_G = -m_0 v_0 a\mathbf{i} + I\omega_0 \mathbf{k} \qquad (1)$$

Angular Velocity after Impact. Substituting the values obtained for the components of $\mathbf{H}_G$ and for the moments of inertia into

$$H_x = I_x \omega_x \qquad H_y = I_y \omega_y \qquad H_z = I_z \omega_z$$

we write

$$-m_0 v_0 a = I'\omega_x = \tfrac{5}{4}ma^2 \omega_x \qquad 0 = I'\omega_y \qquad I\omega_0 = I\omega_z$$
$$\omega_x = -\frac{4}{5}\frac{m_0 v_0}{ma} \qquad \omega_y = 0 \qquad \omega_z = \omega_0 \qquad (2)$$

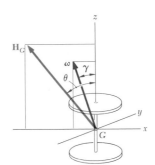

For the satellite considered we have $\omega_0 = 60$ rpm $= 6.283$ rad/s, $m_0/m = \frac{1}{1000}$, $a = 0.800$ m, and $v_0 = 2000$ m/s; we find

$$\omega_x = -2 \text{ rad/s} \qquad \omega_y = 0 \qquad \omega_z = 6.283 \text{ rad/s}$$
$$\omega = \sqrt{\omega_x^2 + \omega_z^2} = 6.594 \text{ rad/s} \qquad \tan\gamma = \frac{-\omega_x}{\omega_z} = +0.3183$$
$$\omega = 63.0 \text{ rpm} \qquad \gamma = 17.7° \quad \blacktriangleleft$$

Precession Axis. Since in free motion the direction of the angular momentum $\mathbf{H}_G$ is fixed in space, the satellite will precess about this direction. The angle θ formed by the precession axis and the z axis is

$$\tan\theta = \frac{-H_x}{H_z} = \frac{m_0 v_0 a}{I\omega_0} = \frac{2m_0 v_0}{ma\omega_0} = 0.796 \qquad \theta = 38.5° \quad \blacktriangleleft$$

Rates of Precession and Spin. We sketch the space and body cones for the free motion of the satellite. Using the law of sines, we compute the rates of precession and spin.

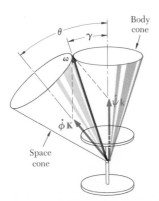

$$\frac{\omega}{\sin\theta} = \frac{\dot{\phi}}{\sin\gamma} = \frac{\dot{\psi}}{\sin(\theta - \gamma)}$$
$$\dot{\phi} = 30.8 \text{ rpm} \qquad \dot{\psi} = 35.9 \text{ rpm} \quad \blacktriangleleft$$

Problems

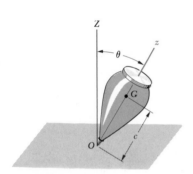

Fig. P18.81

18.81 A 2-kg disk of 150-mm diameter is attached to the end of a rod AB of negligible mass which is supported by a ball-and-socket joint at A. If the disk is observed to precess about the vertical in the sense indicated and at a constant rate of 36 rpm, determine the rate of spin $\dot{\psi}$ of the disk about AB when $\beta = 60°$.

18.82 Solve Prob. 18.81, assuming the same rate of precession and $\beta = 30°$.

18.83 The top shown weighs 3 oz and is supported at the fixed point O. The radii of gyration of the top with respect to its axis of symmetry and with respect to a transverse axis through O are 0.84 in. and 1.80 in., respectively. It is known that $c = 1.50$ in. and that the rate of spin of the top about its axis of symmetry is 1800 rpm. Determine the two possible rates of precession corresponding to $\theta = 30°$.

Fig. P18.83 and P18.84

18.84 The top shown is supported at the fixed point O and its moments of inertia about its axis of symmetry and about a transverse axis through O are denoted, respectively, by I and I'. (a) Show that the condition for steady precession of the top is

$$(I\omega_z - I'\dot{\phi} \cos \theta)\dot{\phi} = Wc$$

where $\dot{\phi}$ is the rate of precession and ω_z is the component of the angular velocity along the axis of symmetry of the top. (b) Show that if the rate of spin $\dot{\psi}$ of the top is very large compared with its rate of precession $\dot{\phi}$, the condition for steady precession is $I\dot{\psi}\dot{\phi} \approx Wc$. (c) Determine the percent error introduced when the slower of the two rates of precession obtained for the top of Prob. 18.83 is approximated by this last relation.

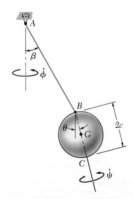

Fig. P18.85 and P18.86

18.85 A homogeneous sphere of radius c is attached as shown to a cord AB. The cord forms an angle β with the vertical and precesses at the constant rate $\dot{\phi}$, while the sphere spins at the constant rate $\dot{\psi}$ about its diameter BC. Determine the angle θ that BC forms with the vertical.

18.86 A homogeneous sphere of radius $c = 40$ mm is attached as shown to a cord AB. The cord forms an angle $\beta = 30°$ with the vertical and is observed to precess at the constant rate $\dot{\phi} = 5$ rad/s about the vertical through A. Determine the angle θ that the diameter BC forms with the vertical, knowing that the sphere (a) has no spin, (b) spins about its diameter BC at the rate $\dot{\psi} = 30$ rad/s, (c) spins about BC at the rate $\dot{\psi} = -30$ rad/s.

18.87 A homogeneous cone of height h and with a base of diameter $d < h$ is attached as shown to a cord AB. The cone spins about its axis BC at the constant rate $\dot{\psi}$ and precesses about the vertical through A at the constant rate $\dot{\phi}$. Determine the angle β for which the axis BC of the cone is aligned with the cord AB ($\theta = \beta$).

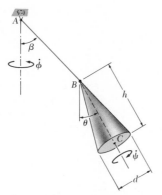

Fig. P18.87 and P18.88

18.88 A homogeneous cone of height $h = 12$ in. and base diameter $d = 6$ in. is attached as shown to a cord AB. Knowing that the angles that cord AB and the axis BC of the cone form with the vertical are, respectively, $\beta = 45°$ and $\theta = 30°$, and that the cone precesses at the constant rate $\dot{\phi} = 8$ rad/s in the sense indicated, determine (a) the rate of spin $\dot{\psi}$ of the cone about its axis BC, (b) the length of cord AB.

18.89 If the earth were a sphere, the gravitational attraction of the sun, moon, and planets would at all times be equivalent to a single force $\mathbf{R}$ acting at the mass center of the earth. However, the earth is actually an oblate spheroid and the gravitational system acting on the earth is equivalent to a force $\mathbf{R}$ and a couple $\mathbf{M}$. Knowing that the effect of the couple $\mathbf{M}$ is to cause the axis of the earth to precess about the axis GA at the rate of one revolution in 25,800 years, determine the average magnitude of the couple $\mathbf{M}$ applied to the earth. Assume that the average density of the earth is 5.51, that the average radius of the earth is 3960 mi, and that $\bar{I} = \frac{2}{5}mR^2$. (*Note.* This forced precession is known as the precession of the equinoxes and is not to be confused with the free precession discussed in Prob. 18.97.)

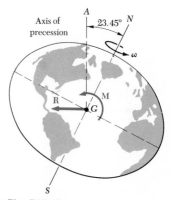

Fig. P18.89

18.90 A high-speed photographic record shows that a certain projectile was fired with a horizontal velocity $\bar{\mathbf{v}}$ of 600 m/s and with its axis of symmetry forming an angle $\beta = 4°$ with the horizontal. The rate of spin $\dot{\psi}$ of the projectile was 5000 rpm, and the atmospheric drag was equivalent to a force $\mathbf{D}$ of 150 N acting at the center of pressure C_P located at a distance $c = 120$ mm from G. (a) Knowing that the projectile has a mass of 18 kg and a radius of gyration of 40 mm with respect to its axis of symmetry, determine its approximate rate of steady precession. (b) If it is further known that the radius of gyration of the projectile with respect to a transverse axis through G is 160 mm, determine the exact values of the two possible rates of precession.

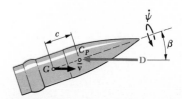

Fig. P18.90

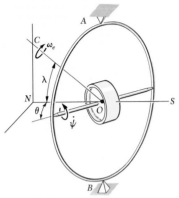

Fig. P18.91

18.91 The essential features of the gyrocompass are shown. The rotor spins at the rate $\dot{\psi}$ about an axis mounted in a single gimbal, which may rotate freely about the vertical axis AB. The angle formed by the axis of the rotor and the plane of the meridian is denoted by θ and the latitude of the position on the earth is denoted by λ. We note that line OC is parallel to the axis of the earth and we denote by $\boldsymbol{\omega}_e$ the angular velocity of the earth about its axis.

(a) Show that the equations of motion of the gyrocompass are

$$I'\ddot{\theta} + I\omega_z\omega_e \cos \lambda \sin \theta - I'\omega_e^2 \cos^2 \lambda \sin \theta \cos \theta = 0$$
$$I\dot{\omega}_z = 0$$

where ω_z is the component of the total angular velocity along the axis of the rotor, and I and I' are the moments of inertia of the rotor with respect to its axis of symmetry and a transverse axis through O, respectively.

(b) Neglecting the term containing ω_e^2, show that for small values of θ, we have

$$\ddot{\theta} + \frac{I\omega_z\omega_e \cos \lambda}{I'} \theta = 0$$

and that the axis of the gyrocompass oscillates about the north-south direction.

18.92 Show that for an axisymmetrical body under no force, the rates of precession and spin may be expressed, respectively, as

$$\dot{\phi} = \frac{H_G}{I'}$$

and

$$\dot{\psi} = \frac{H_G \cos \theta \, (I' - I)}{II'}$$

where H_G is the constant value of the angular momentum of the body.

18.93 (a) Show that for an axisymmetrical body under no force, the rate of precession may be expressed as

$$\dot{\phi} = \frac{I\omega_z}{I' \cos \theta}$$

where ω_z is the component of $\boldsymbol{\omega}$ along the axis of symmetry of the body. (b) Use this result to check that the condition (18.44) for steady precession is satisfied by an axisymmetrical body under no force.

18.94 Show that the angular velocity vector $\boldsymbol{\omega}$ of an axisymmetrical body under no force is observed from the body itself to rotate about the axis of symmetry at the constant rate

$$n = \frac{I' - I}{I'} \omega_z$$

where ω_z is the component of $\boldsymbol{\omega}$ along the axis of symmetry of the body.

18.95 For an axisymmetrical body under no force, prove (a) that the rate of retrograde precession can never be less than twice the rate of spin of the body about its axis of symmetry, (b) that in Fig. 18.24 the axis of symmetry of the body can never lie within the space cone.

18.96 A coin is tossed into the air. During the free motion the angle β between the plane of the coin and the horizontal is observed to be constant. (*a*) Derive an expression for the angle formed by the angular velocity of the coin and the vertical. (*b*) Denoting by $\dot{\psi}$ the rate of spin of the coin about its axis of symmetry, derive an expression for the rate of precession. (*c*) Solve parts *a* and *b* for the case $\beta = 20°$.

Fig. P18.96

18.97 Using the relation given in Prob. 18.94, determine the period of precession of the north pole of the earth about the axis of symmetry of the earth. The earth may be approximated by an oblate spheroid of axial moment of inertia I and of transverse moment of inertia $I' = 0.9967I$. (*Note*. Actual observations show a period of precession of the north pole of about 432.5 mean solar days; the difference between the observed and computed periods is due to the fact that the earth is not a perfectly rigid body. The free precession considered here should not be confused with the much slower precession of the equinoxes, which is a forced precession. See Prob. 18.89.)

18.98 The space station shown is known to precess about the fixed direction OC at the rate of one revolution per hour. Assuming that the station is dynamically equivalent to a homogeneous cylinder of length 30 m and radius 3 m, determine the rate of spin of the station about its axis of symmetry.

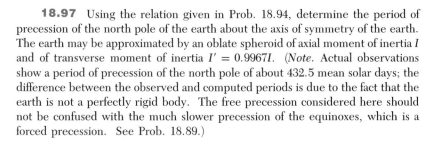

Fig. P18.98

18.99 The link connecting portions A and B of the space station of Prob. 18.98 may be severed to allow each portion to move freely. Each portion of the station is dynamically equivalent to a cylinder of length 15 m and radius 3 m. Knowing that when the link is severed, the station is oriented as shown, determine for portion B the axis of precession, the rate of precession, and the rate of spin about the axis of symmetry.

18.100 An 800-lb geostationary satellite is spinning with an angular velocity $\boldsymbol{\omega}_0 = (1.5 \text{ rad/s})\mathbf{j}$ when it is hit at B by a 6-oz meteorite traveling with a velocity $\mathbf{v}_0 = -(1600 \text{ ft/s})\mathbf{i} + (1300 \text{ ft/s})\mathbf{j} + (4000 \text{ ft/s})\mathbf{k}$ relative to the satellite. Knowing that $b = 20$ in. and that the radii of gyration of the satellite are $\bar{k}_x = \bar{k}_z = 28.8$ in. and $\bar{k}_y = 32.4$ in., determine the precession axis and the rates of precession and spin of the satellite after the impact.

18.101 Solve Prob. 18.100, assuming that the meteorite hits the satellite at A instead of B.

18.102 Solve Sample Prob. 18.6, assuming that the meteorite strikes the satellite at C with a velocity $\mathbf{v}_0 = +(2000 \text{ m/s})\mathbf{i}$.

18.103 After the motion determined in Sample Prob. 18.6 has been established, the rod connecting disks A and B of the satellite breaks, and disk A moves freely as a separate body. Knowing that the rod and the z axis coincide when the rod breaks, determine the precession axis, the rate of precession, and the rate of spin for the ensuing motion of disk A.

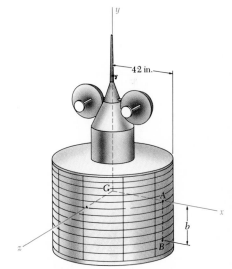

Fig. P18.100

18.104 The angular velocity vector of a football which has just been kicked is horizontal, and its axis of symmetry OC is oriented as shown. Knowing that the magnitude of the angular velocity is 180 rpm and that the ratio of the axial and transverse moments of inertia is $I/I' = \frac{1}{3}$, determine (*a*) the orientation of the axis of precession OA, (*b*) the rates of precession and spin.

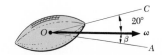

Fig. P18.104

18.105 The slender homogeneous rod AB of mass m and length L is free to rotate about a horizontal axle through its mass center G. The axle is supported by a frame of negligible mass which is free to rotate about the vertical CD. Knowing that, initially, $\theta = \theta_0$, $\dot{\theta} = 0$, and $\dot{\phi} = \dot{\phi}_0$, show that the rod will oscillate about the horizontal axle and determine (a) the range of values of angle θ during this motion, (b) the maximum value of $\dot{\theta}$, (c) the minimum value of $\dot{\phi}$.

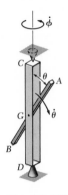

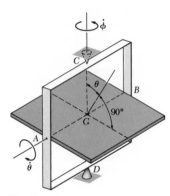

Fig. P18.105   **Fig. P18.106**

18.106 Solve Prob. 18.105, assuming that the rod is replaced by a thin homogeneous square plate of mass m and side a.

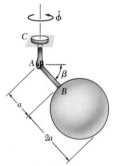

Fig. P18.107

18.107 A homogeneous sphere of mass m and radius a is welded to a rod AB of negligible mass, which is connected by a clevis to a vertical shaft AC. The rod and sphere can rotate freely about a horizontal axis at A, and shaft AC can rotate freely about a vertical axis. The system is released in the position $\beta = 0$ with an angular velocity $\dot{\phi}_0$ about the vertical axis and no angular velocity about the horizontal axis. Knowing that the largest value of β in the ensuing motion is $30°$, determine (a) the initial angular velocity $\dot{\phi}_0$, (b) the value of $\dot{\phi}$ when $\beta = 30°$.

*18.108 The top shown is supported at the fixed point O. We denote by ϕ, θ, and ψ the Eulerian angles defining the position of the top with respect to a fixed frame of reference. We shall consider the general motion of the top in which all Eulerian angles vary.

(a) Observing that $\Sigma M_Z = 0$ and $\Sigma M_z = 0$, and denoting by I and I', respectively, the moments of inertia of the top about its axis of symmetry and about a transverse axis through O, derive the two first-order differential equations of motion

$$I'\dot{\phi}\sin^2\theta + I(\dot{\psi} + \dot{\phi}\cos\theta)\cos\theta = \alpha \qquad (1)$$

$$I(\dot{\psi} + \dot{\phi}\cos\theta) = \beta \qquad (2)$$

where α and β are constants depending upon the initial conditions. These equations express that the angular momentum of the top is conserved about both the Z and z axes, i.e., that the rectangular component of $\mathbf{H}_O$ along each of these axes is constant.

(b) Use Eqs. (1) and (2) to show that the component ω_z of the angular velocity of the top is constant and that the rate of precession $\dot{\phi}$ depends upon the value of the angle of nutation θ.

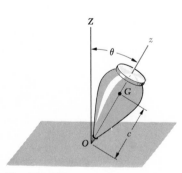

Fig. P18.108

*18.109 (*a*) Applying the principle of conservation of energy, derive a third differential equation for the general motion of the top of Prob. 18.108. (*b*) Eliminating the derivatives $\dot{\phi}$ and $\dot{\psi}$ from the equation obtained and from the two equations of Prob. 18.108, show that the rate of nutation $\dot{\theta}$ is defined by the differential equation $\dot{\theta}^2 = f(\theta)$, where

$$f(\theta) = \frac{1}{I'}\left(2E - \frac{\beta^2}{I} - 2mgc \cos\theta\right) - \left(\frac{\alpha - \beta\cos\theta}{I'\sin\theta}\right)^2$$

*18.110 A thin homogeneous disk is mounted on a light axle OA which is held by a ball-and-socket support at O. The disk is released in the position $\beta = 0$ with a rate of spin $\dot{\psi}_0$, clockwise as viewed from O, and with no precession or nutation. Knowing that the largest value of β in the ensuing motion is $\beta = 30°$, determine (*a*) the rate of spin $\dot{\psi}_0$ of the disk in its initial position, (*b*) the rates of precession and spin as the disk passes through its lowest position. (*Hint.* Use the principle of conservation of energy and the fact that the angular momentum of the body is conserved about both the Z and z axes; see Prob. 18.108, part *a*.)

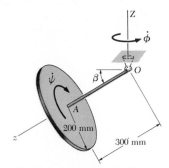

Fig. P18.110 and P18.111

*18.111 A thin homogeneous disk is mounted on a light axle OA which is held by a ball-and-socket support at O. The disk is released in the position $\beta = 0$ with a rate of spin $\dot{\psi} = .20$ rad/s, clockwise as viewed from O, and with no precession or nutation. Determine (*a*) the largest value of β in the ensuing motion, (*b*) the rates of precession and spin as the disk passes through its lowest position. (See hint of Prob. 18.110.)

*18.112 A homogeneous sphere of mass m and radius a is welded to a rod AB of negligible mass, which is held by a ball-and-socket support at A. The sphere is released in the position $\beta = 0$ with a rate of precession $\dot{\phi}_0 = \sqrt{17g/11a}$ and with no spin or nutation. Determine the largest value of β in the ensuing motion. (See hint of Prob. 18.110.)

*18.113 A homogeneous sphere of mass m and radius a is welded to a rod AB of length a and negligible mass, which is held by a ball-and-socket support at A. The sphere is released in the position $\beta = 0$ with a rate of precession $\dot{\phi} = \dot{\phi}_0$ and with no spin or nutation. Knowing that the largest value of β in the ensuing motion is 30°, determine (*a*) the rate of precession $\dot{\phi}_0$ of the sphere in its initial position, (*b*) the rates of precession and spin when $\beta = 30°$. (See hint of Prob. 18.110.)

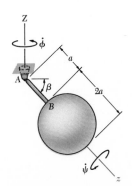

Fig. P18.112 and P18.113

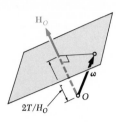

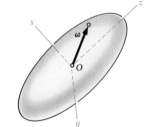

Fig. P18.114

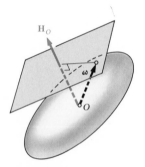

Fig. P18.115

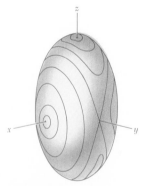

Fig. P18.117

*18.114   Consider a rigid body of arbitrary shape which is attached at its mass center O and subjected to no force other than its weight and the reaction of the support at O. (a) Prove that the angular momentum $\mathbf{H}_O$ of the body about the fixed point O is constant in magnitude and direction, that the kinetic energy T of the body is constant, and that the projection along $\mathbf{H}_O$ of the angular velocity $\boldsymbol{\omega}$ of the body is constant. (b) Show that the tip of the vector $\boldsymbol{\omega}$ describes a curve on a fixed plane in space (called the *invariable plane*), which is perpendicular to $\mathbf{H}_O$ and at a distance $2T/H_O$ from O. (c) Show that with respect to a frame of reference attached to the body and coinciding with its principal axes of inertia, the tip of the vector $\boldsymbol{\omega}$ appears to describe a curve on an ellipsoid of equation

$$I_x\omega_x^2 + I_y\omega_y^2 + I_z\omega_z^2 = 2T = \text{constant}$$

This ellipsoid (called the *Poinsot ellipsoid*) is rigidly attached to the body and is of the same shape as the ellipsoid of inertia, but of a different size.

*18.115   Referring to Prob. 18.114, (a) prove that the Poinsot ellipsoid is tangent to the invariable plane, (b) show that the motion of the rigid body must be such that the Poinsot ellipsoid appears to roll on the invariable plane. [*Hint.* In part *a*, show that the normal to the Poinsot ellipsoid at the tip of $\boldsymbol{\omega}$ is parallel to $\mathbf{H}_O$. It is recalled that the direction of the normal to a surface of equation $F(x, y, z) = \text{constant}$ at a point P is the same as that of **grad** F at point P.]

*18.116   Using the results obtained in Probs. 18.114 and 18.115, show that for an axisymmetrical body attached at its mass center O and under no force other than its weight and the reaction at O, the Poinsot ellipsoid is an ellipsoid of revolution and the space and body cones are both circular and are tangent to each other. Further show that (a) the two cones are tangent externally, and the precession is direct, when $I < I'$, where I and I' denote, respectively, the axial and transverse moment of inertia of the body, (b) the space cone is inside the body cone, and the precession is retrograde, when $I > I'$.

*18.117   Referring to Probs. 18.114 and 18.115, (a) show that the curve (called *polhode*) described by the tip of the vector $\boldsymbol{\omega}$ with respect to a frame of reference coinciding with the principal axes of inertia of the rigid body is defined by the equations

$$I_x\omega_x^2 + I_y\omega_y^2 + I_z\omega_z^2 = 2T = \text{constant} \tag{1}$$
$$I_x^2\omega_x^2 + I_y^2\omega_y^2 + I_z^2\omega_z^2 = H_O^2 = \text{constant} \tag{2}$$

and that this curve may, therefore, be obtained by intersecting the Poinsot ellipsoid with the ellipsoid defined by Eq. (2). (b) Further show, assuming $I_x > I_y > I_z$, that the polhodes obtained for various values of H_O have the shapes indicated in the figure. (c) Using the result obtained in part *b*, show that a rigid body under no force can rotate about a fixed centroidal axis if, and only if, that axis coincides with one of the principal axes of inertia of the body, and that the motion will be stable if the axis of rotation coincides with the major or minor axis of the Poinsot ellipsoid (z or x axis in the figure) and unstable if it coincides with the intermediate axis (y axis).

Review and Summary

This chapter was devoted to the kinetic analysis of the motion of rigid bodies in three dimensions.

We first noted [Sec. 18.1] that the two fundamental equations derived in Chap. 14 for the motion of a system of particles

Fundamental equations of motion for a rigid body

$$\Sigma \mathbf{F} = m\bar{\mathbf{a}} \qquad (18.1)$$

$$\Sigma \mathbf{M}_G = \dot{\mathbf{H}}_G \qquad (18.2)$$

provide the foundation of our analysis, just as they did in Chap. 16 in the case of the plane motion of rigid bodies. The computation of the angular momentum $\mathbf{H}_G$ of the body and of its derivative $\dot{\mathbf{H}}_G$, however, are now considerably more involved.

In Sec. 18.2, we saw that the rectangular components of the angular momentum $\mathbf{H}_G$ of a rigid body may be expressed as follows in terms of the components of its angular velocity $\boldsymbol{\omega}$ and of its centroidal moments and products of inertia:

Angular momentum of a rigid body in three dimensions

$$\begin{aligned} H_x &= +\bar{I}_x\,\omega_x - \bar{I}_{xy}\omega_y - \bar{I}_{xz}\omega_z \\ H_y &= -\bar{I}_{yx}\omega_x + \bar{I}_y\,\omega_y - \bar{I}_{yz}\omega_z \\ H_z &= -\bar{I}_{zx}\omega_x - \bar{I}_{zy}\omega_y + \bar{I}_z\,\omega_z \end{aligned} \qquad (18.7)$$

If *principal axes of inertia* $Gx'y'z'$ are used, these relations reduce to

$$H_{x'} = \bar{I}_{x'}\omega_{x'} \qquad H_{y'} = \bar{I}_{y'}\omega_{y'} \qquad H_{z'} = \bar{I}_{z'}\omega_{z'} \qquad (18.10)$$

We observed that, in general, *the angular momentum $\mathbf{H}_G$ and the angular velocity $\boldsymbol{\omega}$ do not have the same direction* (Fig. 18.25). They will, however, have the same direction if $\boldsymbol{\omega}$ is directed along one of the principal axes of inertia of the body.

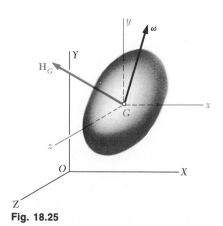

Fig. 18.25

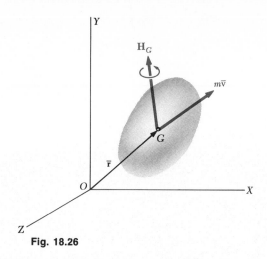

Fig. 18.26

Angular momentum about a given point

Recalling that the system of the momenta of the particles forming a rigid body may be reduced to the vector $m\bar{\mathbf{v}}$ attached at G and the couple $\mathbf{H}_G$ (Fig. 18.26), we noted that, once the linear momentum $m\bar{\mathbf{v}}$ and the angular momentum $\mathbf{H}_G$ of a rigid body have been determined, the angular momentum $\mathbf{H}_O$ of the body about any given point O may be obtained by writing

$$\mathbf{H}_O = \bar{\mathbf{r}} \times m\bar{\mathbf{v}} + \mathbf{H}_G \tag{18.11}$$

Rigid body with a fixed point

In the particular case of a rigid body *constrained to rotate about a fixed point O*, the components of the angular momentum $\mathbf{H}_O$ of the body about O may be obtained directly from the components of its angular velocity and from its moments and products of inertia with respect to axes through O. We wrote

$$\begin{aligned}
H_x &= +I_x\,\omega_x - I_{xy}\omega_y - I_{xz}\omega_z \\
H_y &= -I_{yx}\omega_x + I_y\,\omega_y - I_{yz}\omega_z \\
H_z &= -I_{zx}\omega_x - I_{zy}\omega_y + I_z\,\omega_z
\end{aligned} \tag{18.13}$$

Principle of impulse and momentum

The *principle of impulse and momentum* for a rigid body in three-dimensional motion [Sec. 18.3] is expressed by the same fundamental formula that was used in Chap. 17 for a rigid body in plane motion,

Syst Momenta$_1$ + Syst Ext Imp$_{1\to2}$ = Syst Momenta$_2$ (17.4)

but the systems of the initial and final momenta should now be represented as shown in Fig. 18.26, and $\mathbf{H}_G$ should be computed from the relations (18.7) or (18.10) [Sample Probs. 18.1 and 18.2].

The *kinetic energy* of a rigid body in three-dimensional motion may be divided into two parts [Sec. 18.4], one associated with the motion of its mass center G, and the other with its motion about G. Using principal centroidal axes x', y', z', we wrote

$$T = \tfrac{1}{2}m\bar{v}^2 + \tfrac{1}{2}(\bar{I}_{x'}\omega_{x'}^2 + \bar{I}_{y'}\omega_{y'}^2 + \bar{I}_{z'}\omega_{z'}^2) \qquad (18.17)$$

Kinetic energy of a rigid body in three dimensions

where
$\bar{v}$ = velocity of mass center
ω = angular velocity
m = mass of rigid body
$\bar{I}_{x'}, \bar{I}_{y'}, \bar{I}_{z'}$ = principal centroidal moments of inertia

We also noted that, in the case of a rigid body *constrained to rotate about a fixed point O*, the kinetic energy of the body may be expressed as

$$T = \tfrac{1}{2}(I_{x'}\omega_{x'}^2 + I_{y'}\omega_{y'}^2 + I_{z'}\omega_{z'}^2) \qquad (18.20)$$

where the x', y', and z' axes are the principal axes of inertia of the body at O. The results obtained in Sec. 18.4 make it possible to extend to the three-dimensional motion of a rigid body the application of the *principle of work and energy* and of the *principle of conservation of energy*.

The second part of the chapter was devoted to the application of the fundamental equations

$$\Sigma\mathbf{F} = m\bar{\mathbf{a}} \qquad (18.1)$$

$$\Sigma\mathbf{M}_G = \dot{\mathbf{H}}_G \qquad (18.2)$$

Using a rotating frame to write the equations of motion of a rigid body in space

to the motion of a rigid body in three dimensions. We first recalled [Sec. 18.5] that $\mathbf{H}_G$ represents the angular momentum of the body relative to a centroidal frame $GX'Y'Z'$ of fixed orientation (Fig. 18.27) and that $\dot{\mathbf{H}}_G$ in Eq. (18.2) represents the rate of change of $\mathbf{H}_G$ with respect to that frame. We noted that, as the body rotates, its moments and products of inertia with respect to the frame $GX'Y'Z'$ change continually. It is therefore more convenient to use a frame $Gxyz$ rotating with the body to resolve ω into components and to compute the moments and products of inertia which will be used to determine $\mathbf{H}_G$ from Eqs. (18.7) or (18.10). However, since $\dot{\mathbf{H}}_G$ in Eq. (18.2) represents the rate of change of $\mathbf{H}_G$ with respect to the frame $GX'Y'Z'$ of fixed orientation, we must use the method of Sec. 15.10 to determine its value. Recalling Eq. (15.31), we wrote

$$\dot{\mathbf{H}}_G = (\dot{\mathbf{H}}_G)_{Gxyz} + \mathbf{\Omega} \times \mathbf{H}_G \qquad (18.22)$$

where $\mathbf{H}_G$ = angular momentum of the body with respect to the frame $GX'Y'Z'$ of fixed orientation
$(\dot{\mathbf{H}}_G)_{Gxyz}$ = rate of change of $\mathbf{H}_G$ with respect to the rotating frame $Gxyz$, to be computed from the relations (18.7)
$\mathbf{\Omega}$ = angular velocity of the rotating frame $Gxyz$

Substituting for $\dot{\mathbf{H}}_G$ from (18.22) into (18.2), we obtained

$$\Sigma\mathbf{M}_G = (\dot{\mathbf{H}}_G)_{Gxyz} + \mathbf{\Omega} \times \mathbf{H}_G \qquad (18.23)$$

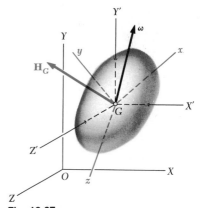

Fig. 18.27

If the rotating frame is actually attached to the body, its angular velocity $\boldsymbol{\Omega}$ is identically equal to the angular velocity $\boldsymbol{\omega}$ of the body. There are many applications, however, where it is advantageous to use a frame of reference which is not attached to the body but rotates in an independent manner [Sample Prob. 18.5].

Euler's equations of motion. D'Alembert's principle

Setting $\boldsymbol{\Omega} = \boldsymbol{\omega}$ in Eq. (18.23), using principal axes, and writing this equation in scalar form, we obtained *Euler's equations of motion* [Sec. 18.6]. A discussion of the solution of these equations and of the scalar equations corresponding to Eq. (18.1) led us to extend d'Alembert's principle to the three-dimensional motion of a rigid body and to conclude that the system of the external forces acting on the rigid body is not only equipollent, but actually *equivalent* to the effective forces of the body represented by the vector $m\overline{\mathbf{a}}$ and the couple $\dot{\mathbf{H}}_G$ (Fig. 18.28). Problems involving the three-dimensional motion of a rigid body may be solved by considering the free-body-diagram equation represented in Fig. 18.28 and writing appropriate scalar equations relating the components or moments of the external and effective forces [Sample Probs. 18.3 and 18.5].

Free-body-diagram equation

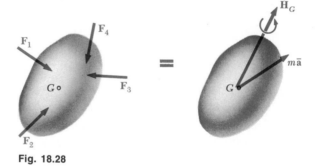

Fig. 18.28

Rigid body with a fixed point

In the case of a rigid body *constrained to rotate about a fixed point* O, an alternative method of solution may be used, involving the moments of the forces and the rate of change of the angular momentum about point O. We wrote [Sec 18.7]:

$$\Sigma \mathbf{M}_O = (\dot{\mathbf{H}}_O)_{Oxyz} + \boldsymbol{\Omega} \times \mathbf{H}_O \qquad (18.28)$$

where $\Sigma \mathbf{M}_O$ = sum of the moments about O of the forces applied to the rigid body

$\mathbf{H}_O$ = angular momentum of the body with respect to the fixed frame $OXYZ$

$(\dot{\mathbf{H}}_O)_{Oxyz}$ = rate of change of $\mathbf{H}_O$ with respect to the rotating frame $Oxyz$, to be computed from the relations (18.13)

$\boldsymbol{\Omega}$ = angular velocity of the rotating frame $Oxyz$

This approach may be used to solve certain problems involving the rotation of a rigid body about a fixed axis [Sec. 18.8], such as an unbalanced rotating shaft [Sample Prob. 18.4].

In the last part of the chapter, we considered the motion of *gyroscopes* and other *axisymmetrical bodies*. Introducing the *Eulerian angles* ϕ, θ, and ψ to define the position of a gyroscope (Fig. 18.29), we observed that their derivatives $\dot\phi$, $\dot\theta$, and $\dot\psi$ represent, respectively, the rates of *precession*, *nutation*, and *spin* of the gyroscope [Sec. 18.9]. Expressing the angular velocity ω in terms of these derivatives, we wrote

$$\omega = -\dot\phi \sin\theta\,\mathbf{i} + \dot\theta\,\mathbf{j} + (\dot\psi + \dot\phi\cos\theta)\mathbf{k} \qquad (18.35)$$

Motion of a gyroscope

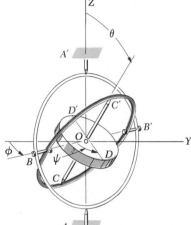

Fig. 18.29

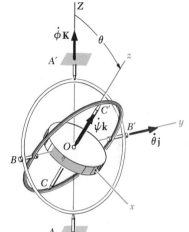

Fig. 18.30

where the unit vectors are associated with a frame $Oxyz$ attached to the inner gimbal of the gyroscope (Fig. 18.30) and rotate, therefore, with the angular velocity

$$\mathbf{\Omega} = -\dot\phi \sin\theta\,\mathbf{i} + \dot\theta\,\mathbf{j} + \dot\phi\cos\theta\,\mathbf{k} \qquad (18.38)$$

Denoting by I the moment of inertia of the gyroscope with respect to its spin axis z and by I' its moment of inertia with respect to a transverse axis through O, we wrote

$$\mathbf{H}_O = -I'\dot\phi \sin\theta\,\mathbf{i} + I'\dot\theta\,\mathbf{j} + I(\dot\psi + \dot\phi\cos\theta)\mathbf{k} \qquad (18.36)$$

Substituting for $\mathbf{H}_O$ and $\mathbf{\Omega}$ into Eq. (18.28) led us to the differential equations defining the motion of the gyroscope.

In the particular case of the *steady precession* of a gyroscope [Sec. 18.10], the angle θ, the rate of precession $\dot\phi$, and the rate of spin $\dot\psi$ remain constant. We saw that such a motion is possible only if the moments of the external forces about O satisfy the relation

$$\Sigma\mathbf{M}_O = (I\omega_z - I'\dot\phi\cos\theta)\dot\phi\sin\theta\,\mathbf{j} \qquad (18.44)$$

i.e., if the external forces reduce to a couple of moment equal to the right-hand member of Eq. (18.44) and applied *about an axis perpendicular to the precession axis and to the spin axis* (Fig. 18.31). The chapter ended with a discussion of the motion of an axisymmetrical body spinning and precessing *under no force* [Sec. 18.11; Sample Prob. 18.6].

Steady precession

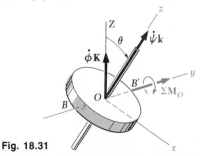

Fig. 18.31

Review Problems

18.118 Three 25-lb rotor disks are attached to a shaft which rotates at 720 rpm. Disk A is attached eccentrically so that its mass center is $\frac{1}{4}$ in. from the axis of rotation, while disks B and C are attached so that their mass centers coincide with the axis of rotation. Where should 2-lb weights be bolted to disks B and C to balance the system dynamically?

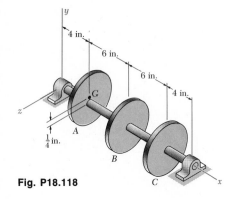

Fig. P18.118

Fig. P18.119

18.119 Two uniform rods AB and CD are welded together at B to form a T-shaped assembly of total mass m. The assembly is suspended from a ball-and-socket joint at A and is hit at C in a direction perpendicular to its plane (in the negative z direction). Denoting the corresponding impulse by **F** Δt, determine immediately after the impact (a) the angular velocity of the assembly, (b) its instantaneous axis of rotation.

18.120 A thin homogeneous disk of mass 800 g and radius 100 mm rotates at a constant rate $\omega_2 = 20$ rad/s with respect to arm ABC, which itself rotates at a constant rate $\omega_1 = 10$ rad/s about the x axis. Determine the angular momentum of the disk about point C.

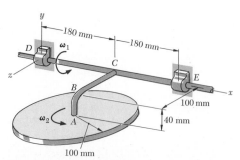

Fig. P18.120 and P18.121

18.121 A thin homogeneous disk of mass 800 g and radius 100 mm rotates at a constant rate $\omega_2 = 20$ rad/s with respect to arm ABC, which itself rotates at a constant rate $\omega_1 = 10$ rad/s about the x axis. For the position shown, determine the dynamic reactions at bearings D and E.

18.122 Gear A rolls on a fixed gear B and rotates about an axle AD of length $L = 500$ mm which is rigidly attached at D to a vertical shaft DE. Shaft DE is made to rotate with a constant angular velocity $\boldsymbol{\omega}_1$ of magnitude 4 rad/s. Assuming that gear A can be approximated by a thin disk of mass 2 kg and radius $a = 100$ mm, and that $\beta = 30°$, determine (a) the angular momentum of gear A about point D, (b) the kinetic energy of gear A.

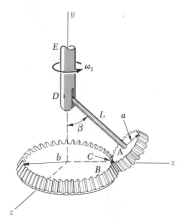

Fig. P18.122

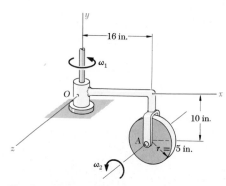

Fig. P18.123 and P18.124

18.123 A thin disk of weight $W = 8$ lb rotates at the constant rate $\omega_2 = 12$ rad/s with respect to arm OA, which itself rotates at the constant rate $\omega_1 = 4$ rad/s about the y axis. Determine the angular momentum of the disk about its center A.

18.124 A thin disk of weight $W = 8$ lb rotates with an angular velocity $\boldsymbol{\omega}_2$ with respect to arm OA, which itself rotates with an angular velocity $\boldsymbol{\omega}_1$ about the y axis. Determine (a) the couple $M_1\mathbf{j}$ which should be applied to arm OA to give it an angular acceleration $\boldsymbol{\alpha}_1 = (6\text{ rad/s}^2)\mathbf{j}$ when $\omega_1 = 4$ rad/s, knowing that the disk rotates at the constant rate $\omega_2 = 12$ rad/s, (b) the force-couple system representing the dynamic reaction at O at that instant. Assume that arm OA has a negligible mass.

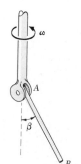

Fig. P18.125

18.125 A uniform rod AB of length l and mass m is attached to the pin of a clevis which rotates with a constant angular velocity $\boldsymbol{\omega}$. Determine (a) the constant angle β that the rod forms with the vertical, (b) the range of values of ω for which the rod remains vertical ($\beta = 0$).

18.126 A 240-kg satellite is spinning with an angular velocity $\boldsymbol{\omega}_0 = (1.5\text{ rad/s})\mathbf{i}$ when it is struck at A by a 30-g meteorite traveling with a velocity $\mathbf{v}_0 = -(576\text{ m/s})\mathbf{i} - (432\text{ m/s})\mathbf{j} + (960\text{ m/s})\mathbf{k}$ relative to the satellite. Knowing that the radii of gyration of the satellite are $\bar{k}_x = 300$ mm and $\bar{k}_y = \bar{k}_z = 400$ mm, determine the angular velocity of the satellite immediately after the meteorite has become embedded.

18.127 Determine the precession axis and the rates of precession and spin of the satellite of Prob. 18.126 after the impact.

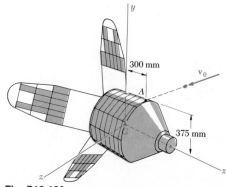

Fig. P18.126

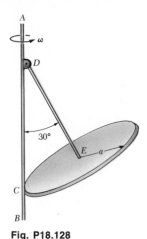

Fig. P18.128

18.128 A disk of mass m and radius a is rigidly attached to a rod DE of negligible mass. Rod DE is attached to a vertical shaft AB by a clevis at D and the disk leans against the shaft at C. Noting that when shaft AB is made to rotate, the same point of the disk will remain in contact with the shaft at C, determine the magnitude of the angular velocity ω for which the reaction at C will be zero.

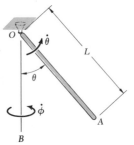

Fig. P18.129

18.129 A slender homogeneous rod OA of mass m and length L is supported by a ball-and-socket joint at O and may swing freely under its own weight. If the rod is held in a horizontal position ($\theta = 90°$) and given an initial angular velocity $\dot{\phi}_0 = \sqrt{8g/L}$ about the vertical OB, determine (a) the smallest value of θ in the ensuing motion, (b) the corresponding value of the angular velocity $\dot{\phi}$ of the rod about OB. (*Hint.* Apply the principle of conservation of energy and the principle of impulse and momentum, observing that since $\Sigma M_{OB} = 0$, the component of $\mathbf{H}_O$ along OB must be constant.)

The following problems are designed to be solved with a computer.

18.C1 A 20-g bullet is fired with an initial velocity $\mathbf{v}_0 = -(250 \text{ m/s})\mathbf{k}$ into an 8-kg circular plate which is suspended from a ball-and-socket joint at O. The bullet strikes point A and becomes embedded in the plate. Write a computer program and use it to calculate immediately after the impact the angular velocity of the plate and the instantaneous axis of rotation for values of θ from 0 to 180° at 10° intervals.

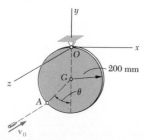

Fig. P18.C1

18.C2 An 800-lb geostationary satellite is spinning with an angular velocity $\omega_0 = (1.5 \text{ rad/s})\mathbf{j}$ when it is hit at B by a 6-oz meteorite traveling with a velocity $\mathbf{v}_0 = -(2800 \text{ ft/s})\mathbf{i} + (2900 \text{ ft/s})\mathbf{j} - (2000 \text{ ft/s})\mathbf{k}$ relative to the satellite. The radii of gyration of the satellite are $\bar{k}_x = \bar{k}_z = 28.8$ in. and $\bar{k}_y = 32.4$ in. Write a computer program and use it to calculate, immediately after the meteorite has become embedded at B, the rectangular components and the magnitude of the angular velocity of the satellite for values of b from 0 to 30 in. at 2-in. intervals. Expand the program written and use it to calculate, for the same values of b, the change in the kinetic energy associated with the rotation of the satellite.

18.C3 The top shown is supported at the fixed point O and is in steady precession about the Z axis. The top weighs 3 oz, its radii of gyration with respect to its axis of symmetry and with respect to a transverse axis through O are 0.84 in. and 1.80 in., respectively, and the distance from O to its mass center G is $c = 1.50$ in. Write a computer program to calculate, for given values of the angle θ and of the rate of spin $\dot{\psi}$, (a) the rate of precession $\dot{\phi}$ corresponding to the lower of the two values obtained by solving the equation given in part a of Prob. 18.84, (b) the approximate value of $\dot{\phi}$ obtained from the condition for steady precession, $I\dot{\psi}\dot{\phi} \approx Wc$, given in part b of Prob. 18.84, (c) the percentage error introduced by this approximation. Use the above program with values of θ equal to 15°, 30°, and 45°, and with values of $\dot{\psi}$ equal to 1050, 1200, 1500, 1800, 2400, and 6000 rpm.

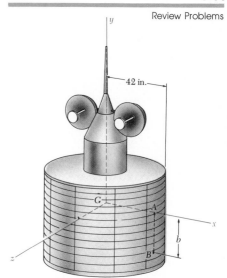

Fig. P18.C2

18.C4 The top shown is supported at the fixed point O and its radii of gyration about its axis of symmetry and about a transverse axis through O are denoted, respectively, by k and k'. The top is spinning at the rate $\dot{\psi}_0$ about its axis of symmetry, which forms an angle θ_0 with the vertical, as it is released with zero rate of precession ($\dot{\phi}_0 = 0$) and zero rate of nutation ($\dot{\theta} = 0$). The axis of the top is observed to drop and, as it drops, to start precessing. After reaching a maximum value θ_m, angle θ is observed to decrease until it has regained its initial value θ_0; the rates of precession and nutation are again zero. This motion, known as *cuspidal motion*, is repeated indefinitely, with point G oscillating between an upper level corresponding to $\theta = \theta_0$ and a lower level corresponding to $\theta = \theta_m$. (a) Referring to Prob. 18.109b, show that the rate of nutation $\dot{\theta}$ of the top is defined by the differential equation $\dot{\theta}^2 = f(\theta)$, where

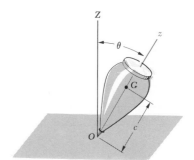

Fig. P18.C3 and P18.C4

$$f(\theta) = (\cos \theta_0 - \cos \theta)\left(\frac{2gc}{k'^2} - \frac{k^4}{k'^4}\dot{\psi}_0^2 \frac{\cos \theta_0 - \cos \theta}{\sin^2 \theta}\right)$$

and express $\dot{\psi}_0^2$ in terms of θ_0 and θ_m by making $\theta = \theta_m$ in the above equation and observing, since $\dot{\theta}_m = 0$, that $f(\theta_m) = 0$. (b) Write a computer program to calculate the period of oscillation of the axis of the top, observing that the time Δt_i corresponding to an increment $\Delta\theta_i$ of angle θ may be obtained by dividing $\Delta\theta_i$ by the average rate of nutation $\frac{1}{2}(\dot{\theta}_i + \dot{\theta}_{i+1})$ of the top over Δt_i if the second derivative $\ddot{\theta}$ of angle θ is assumed to remain constant over Δt_i. (c) Knowing that $c = 36$ mm, $k = 20$ mm, $k' = 45$ mm, $\theta_0 = 30°$, and $\theta_m = 40°$, determine the initial rate of spin $\dot{\psi}_0$ of the top and obtain an approximate value of the period of oscillation of its axis, using increments $\Delta\theta_i = 0.1°$.

Mechanical Vibrations

19.1. Introduction. A *mechanical vibration* is the motion of a particle or a body which oscillates about a position of equilibrium. Most vibrations in machines and structures are undesirable because of the increased stresses and energy losses which accompany them. They should therefore be eliminated or reduced as much as possible by appropriate design. The analysis of vibrations has become increasingly important in recent years owing to the current trend toward higher-speed machines and lighter structures. There is every reason to expect that this trend will continue and that an even greater need for vibration analysis will develop in the future.

The analysis of vibrations is a very extensive subject to which entire texts have been devoted. We shall therefore limit our present study to the simpler types of vibrations, namely, the vibrations of a body or a system of bodies with one degree of freedom.

A mechanical vibration generally results when a system is displaced from a position of stable equilibrium. The system tends to return to this position under the action of restoring forces (either elastic forces, as in the case of a mass attached to a spring, or gravitational forces, as in the case of a pendulum). But the system generally reaches its original position with a certain acquired velocity which carries it beyond that position. Since the process can be repeated indefinitely, the system keeps moving back and forth across its position of equilibrium. The time interval required for the system to complete a full cycle of motion is called the *period* of the vibration. The number of cycles per unit time defines the *frequency,* and the maximum displacement of the system from its position of equilibrium is called the *amplitude* of the vibration.

When the motion is maintained by the restoring forces only, the vibration is said to be a *free vibration* (Secs. 19.2 to 19.6). When a periodic force is applied to the system, the resulting motion is described as a *forced vibration* (Sec. 19.7). When the effects of friction may be neglected, the vibrations are said to be *undamped.* However, all vibrations are actually

942

damped to some degree. If a free vibration is only slightly damped, its amplitude slowly decreases until, after a certain time, the motion comes to a stop. But damping may be large enough to prevent any true vibration; the system then slowly regains its original position (Sec. 19.8). A damped forced vibration is maintained as long as the periodic force which produces the vibration is applied. The amplitude of the vibration, however, is affected by the magnitude of the damping forces (Sec. 19.9).

VIBRATIONS WITHOUT DAMPING

19.2. Free Vibrations of Particles. Simple Harmonic Motion. Consider a body of mass m attached to a spring of constant k (Fig. 19.1a). Since at the present time we are concerned only with the motion of its mass center, we shall refer to this body as a particle. When the particle is in static equilibrium, the forces acting on it are its weight **W** and the force **T** exerted by the spring, of magnitude $T = k\delta_{st}$, where δ_{st} denotes the elongation of the spring. We have, therefore,

$$W = k\delta_{st} \qquad (19.1)$$

Suppose now that the particle is displaced through a distance x_m from its equilibrium position and released with no initial velocity. If x_m has been chosen smaller than δ_{st}, the particle will move back and forth through its equilibrium position; a vibration of amplitude x_m has been generated. Note that the vibration may also be produced by imparting a certain initial velocity to the particle when it is in its equilibrium position $x = 0$ or, more generally, by starting the particle from any given position $x = x_0$ with a given initial velocity v_0.

To analyze the vibration, we shall consider the particle in a position P at some arbitrary time t (Fig. 19.1b). Denoting by x the displacement OP measured from the equilibrium position O (positive downward), we note that the forces acting on the particle are its weight **W** and the force **T** exerted by the spring which, in this position, has a magnitude $T = k(\delta_{st} + x)$. Recalling (19.1), we find that the magnitude of the resultant **F** of the two forces (positive downward) is

$$F = W - k(\delta_{st} + x) = -kx \qquad (19.2)$$

Thus the *resultant* of the forces exerted on the particle is proportional to the displacement *OP measured from the equilibrium position.* Recalling the sign convention, we note that **F** is always directed *toward* the equilibrium position O. Substituting for F into the fundamental equation $F = ma$ and recalling that a is the second derivative $\ddot{x}$ of x with respect to t, we write

$$m\ddot{x} + kx = 0 \qquad (19.3)$$

Note that the same sign convention should be used for the acceleration $\ddot{x}$ and for the displacement x, namely, positive downward.

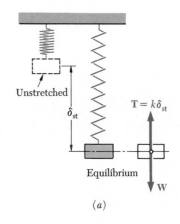

(a)

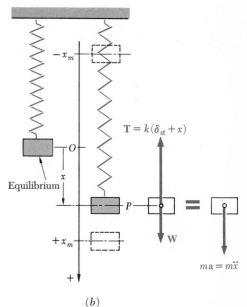

(b)

Fig. 19.1

Equation (19.3) is a linear differential equation of the second order. Setting

$$p^2 = \frac{k}{m} \tag{19.4}$$

we write (19.3) in the form

$$\ddot{x} + p^2 x = 0 \tag{19.5}$$

The motion defined by Eq. (19.5) is called *simple harmonic motion*. It is characterized by the fact that *the acceleration is proportional to the displacement and of opposite direction*. We note that each of the functions $x_1 = \sin pt$ and $x_2 = \cos pt$ satisfies (19.5). These functions, therefore, constitute two *particular solutions* of the differential equation (19.5). As we shall see presently, the *general solution* of (19.5) may be obtained by multiplying the two particular solutions by arbitrary constants A and B and adding. We write

$$x = A x_1 + B x_2 = A \sin pt + B \cos pt \tag{19.6}$$

Differentiating, we obtain successively the velocity and acceleration at time t:

$$v = \dot{x} = Ap \cos pt - Bp \sin pt \tag{19.7}$$

$$a = \ddot{x} = -Ap^2 \sin pt - Bp^2 \cos pt \tag{19.8}$$

Substituting from (19.6) and (19.8) into (19.5), we verify that the expression (19.6) provides a solution of the differential equation (19.5). Since this expression contains two arbitrary constants A and B, the solution obtained is the general solution of the differential equation. The values of the constants A and B depend upon the *initial conditions* of the motion. For example, we have $A = 0$ if the particle is displaced from its equilibrium position and released at $t = 0$ with no initial velocity, and we have $B = 0$ if P is started from O at $t = 0$ with a certain initial velocity. In general, substituting $t = 0$ and the initial values x_0 and v_0 of the displacement and velocity into (19.6) and (19.7), we find $A = v_0/p$ and $B = x_0$.

The expressions obtained for the displacement, velocity, and acceleration of a particle may be written in a more compact form if we observe that (19.6) expresses that the displacement $x = OP$ is the sum of the x components of two vectors $\mathbf{A}$ and $\mathbf{B}$, respectively, of magnitude A and B, directed as shown in Fig. 19.2a. As t varies, both vectors rotate clockwise; we also note that the magnitude of their resultant $\overrightarrow{OQ}$ is equal to the maximum displacement x_m. The simple harmonic motion of P along the x axis may thus be obtained by projecting on this axis the motion of a point Q describing an *auxiliary circle* of radius x_m with a *constant angular velocity p*.

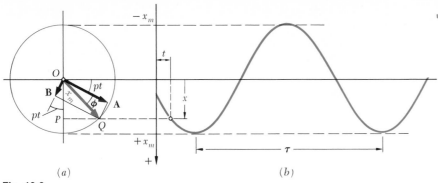

Fig. 19.2

Denoting by ϕ the angle formed by the vectors $\overrightarrow{OQ}$ and $\mathbf{A}$, we write

$$OP = OQ \sin(pt + \phi) \qquad (19.9)$$

which leads to new expressions for the displacement, velocity, and acceleration of P:

$$x = x_m \sin(pt + \phi) \qquad (19.10)$$

$$v = \dot{x} = x_m p \cos(pt + \phi) \qquad (19.11)$$

$$a = \ddot{x} = -x_m p^2 \sin(pt + \phi) \qquad (19.12)$$

The displacement-time curve is represented by a sine curve (Fig. 19.2b), and the maximum value x_m of the displacement is called the *amplitude* of the vibration. The angular velocity p of the point Q which describes the auxiliary circle is known as the *circular frequency* of the vibration and is measured in rad/s, while the angle ϕ which defines the initial position of Q on the circle is called the *phase angle*. We note from Fig. 19.2 that a full *cycle* has been described after the angle pt has increased by 2π rad. The corresponding value of t, denoted by τ, is called the *period* of the vibration and is measured in seconds. We have

$$\text{Period} = \tau = \frac{2\pi}{p} \qquad (19.13)$$

The number of cycles described per unit of time is denoted by f and is known as the *frequency* of the vibration. We write

$$\text{Frequency} = f = \frac{1}{\tau} = \frac{p}{2\pi} \qquad (19.14)$$

The unit of frequency is a frequency of 1 cycle per second, corresponding to a period of 1 s. In terms of base units the unit of frequency is thus $1/s$ or s^{-1}. It is called a *hertz* (Hz) in the SI system of units. It also follows from Eq. (19.14) that a frequency of $1\,s^{-1}$ or $1\,Hz$ corresponds to a circular frequency of 2π rad/s. In problems involving angular velocities expressed in revolutions per minute (rpm), we have $1\,\text{rpm} = \frac{1}{60}\,s^{-1} = \frac{1}{60}\,Hz$, or $1\,\text{rpm} = (2\pi/60)\,\text{rad/s}$.

Recalling that p was defined in (19.4) in terms of the constant k of the spring and the mass m of the particle, we observe that the period and the frequency are independent of the initial conditions and of the amplitude of the vibration. Note that τ and f depend on the *mass* rather than on the *weight* of the particle and thus are independent of the value of g.

The velocity-time and acceleration-time curves may be represented by sine curves of the same period as the displacement-time curve, but with different phase angles. From (19.11) and (19.12), we note that the maximum values of the magnitudes of the velocity and acceleration are

$$v_m = x_m p \qquad a_m = x_m p^2 \tag{19.15}$$

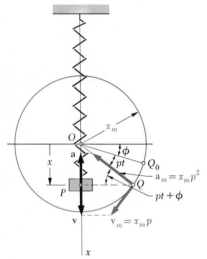

Fig. 19.3

Since the point Q describes the auxiliary circle, of radius x_m, at the constant angular velocity p, its velocity and acceleration are equal, respectively, to the expressions (19.15). Recalling Eqs. (19.11) and (19.12), we find, therefore, that the velocity and acceleration of P may be obtained at any instant by projecting on the x axis vectors of magnitudes $v_m = x_m p$ and $a_m = x_m p^2$ representing, respectively, the velocity and acceleration of Q at the same instant (Fig. 19.3).

The results obtained are not limited to the solution of the problem of a mass attached to a spring. They may be used to analyze the rectilinear motion of a particle *whenever the resultant* **F** *of the forces acting on the particle is proportional to the displacement x and directed toward O*. The fundamental equation of motion $F = ma$ may then be written in the form (19.5), which characterizes simple harmonic motion. Observing that the coefficient of x in (19.5) represents the square of the circular frequency p of the vibration, we easily obtain p and, after substitution into (19.13) and (19.14), the period τ and the frequency f of the vibration.

19.3. Simple Pendulum (Approximate Solution). Most of the vibrations encountered in engineering applications may be represented by a simple harmonic motion. Many others, although of a different type, may be *approximated* by a simple harmonic motion, provided that their amplitude remains small. Consider, for example, a *simple pendulum*, consisting of a bob of mass m attached to a cord of length l, which may oscillate in a vertical plane (Fig. 19.4a). At a given time t, the cord forms an angle θ with the vertical. The forces acting on the bob are its weight **W** and the force **T** exerted by the cord (Fig. 19.4b). Resolving the vector $m\mathbf{a}$ into tangential and normal components, with ma_t directed to the right, i.e., in

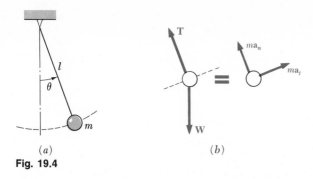

(a)

(b)

Fig. 19.4

the direction corresponding to increasing values of θ, and observing that $a_t = l\alpha = l\ddot{\theta}$, we write

$$\Sigma F_t = ma_t: \qquad\qquad -W\sin\theta = ml\ddot{\theta}$$

Noting that $W = mg$ and dividing through by ml, we obtain

$$\ddot{\theta} + \frac{g}{l}\sin\theta = 0 \qquad\qquad (19.16)$$

For oscillations of small amplitude, we may replace $\sin\theta$ by θ, expressed in radians, and write

$$\ddot{\theta} + \frac{g}{l}\theta = 0 \qquad\qquad (19.17)$$

Comparison with (19.5) shows that the equation obtained is that of a simple harmonic motion and that the circular frequency p of the oscillations is equal to $(g/l)^{1/2}$. Substitution into (19.13) yields the period of the small oscillations of a pendulum of length l:

$$\tau = \frac{2\pi}{p} = 2\pi\sqrt{\frac{l}{g}} \qquad\qquad (19.18)$$

* 19.4. Simple Pendulum (Exact Solution). Formula (19.18) is only approximate. To obtain an exact expression for the period of the oscillations of a simple pendulum, we must return to (19.16). Multiplying both terms by $2\dot{\theta}$ and integrating from an initial position corresponding to the maximum deflection, that is, $\theta = \theta_m$ and $\dot{\theta} = 0$, we write

$$\dot{\theta}^2 = \frac{2g}{l}(\cos\theta - \cos\theta_m)$$

or

$$\left(\frac{d\theta}{dt}\right)^2 = \frac{2g}{l}(\cos\theta - \cos\theta_m)$$

Replacing $\cos \theta$ by $1 - 2 \sin^2 (\theta/2)$ and $\cos \theta_m$ by a similar expression, solving for dt, and integrating over a quarter period from $t = 0$, $\theta = 0$ to $t = \tau/4$, $\theta = \theta_m$, we have

$$\tau = 2 \sqrt{\frac{l}{g}} \int_0^{\theta_m} \frac{d\theta}{\sqrt{\sin^2 (\theta_m/2) - \sin^2 (\theta/2)}}$$

The integral in the right-hand member is known as an *elliptic integral;* it cannot be expressed in terms of the usual algebraic or trigonometric functions. However, setting

$$\sin (\theta/2) = \sin (\theta_m/2) \sin \phi$$

we may write

$$\tau = 4 \sqrt{\frac{l}{g}} \int_0^{\pi/2} \frac{d\phi}{\sqrt{1 - \sin^2 (\theta_m/2) \sin^2 \phi}} \qquad (19.19)$$

where the integral obtained, commonly denoted by K, may be found in *tables of elliptic integrals* for various values of $\theta_m/2$.† In order to compare the result just obtained with that of the preceding section, we write (19.19) in the form

$$\tau = \frac{2K}{\pi} \left(2\pi \sqrt{\frac{l}{g}} \right) \qquad (19.20)$$

Formula (19.20) shows that the actual value of the period of a simple pendulum may be obtained by multiplying the approximate value (19.18) by the correction factor $2K/\pi$. Values of the correction factor are given in

Table 19.1 Correction Factor for the Period of a Simple Pendulum

θ_m	0°	10°	20°	30°	60°	90°	120°	150°	180°
K	1.571	1.574	1.583	1.598	1.686	1.854	2.157	2.768	∞
$2K/\pi$	1.000	1.002	1.008	1.017	1.073	1.180	1.373	1.762	∞

Table 19.1 for various values of the amplitude θ_m. We note that for ordinary engineering computations the correction factor may be omitted as long as the amplitude does not exceed 10°.

---

† See, for example, "Standard Mathematical Tables," Chemical Rubber Publishing Company, Cleveland, Ohio.

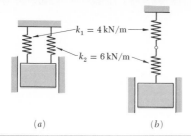

(a)   (b)

A 50-kg block moves between vertical guides as shown. The block is pulled 40 mm down from its equilibrium position and released. For each spring arrangement, determine the period of the vibration, the maximum velocity of the block, and the maximum acceleration of the block.

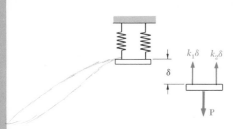

a. **Springs Attached in Parallel.** We first determine the constant k of a single spring equivalent to the two springs *by finding the magnitude of the force* **P** required to cause a given deflection δ. Since for a deflection δ the magnitudes of the forces exerted by the springs are, respectively, $k_1\delta$ and $k_2\delta$, we have

$$P = k_1\delta + k_2\delta = (k_1 + k_2)\delta$$

The constant k of the single equivalent spring is

$$k = \frac{P}{\delta} = k_1 + k_2 = 4 \text{ kN/m} + 6 \text{ kN/m} = 10 \text{ kN/m} = 10^4 \text{ N/m}$$

Period of Vibration: Since $m = 50$ kg, Eq. (19.4) yields

$$p^2 = \frac{k}{m} = \frac{10^4 \text{ N/m}}{50 \text{ kg}} \qquad p = 14.14 \text{ rad/s}$$

$$\tau = 2\pi/p \qquad\qquad\qquad \tau = 0.444 \text{ s} \quad \blacktriangleleft$$

Maximum Velocity: $\quad v_m = x_m p = (0.040 \text{ m})(14.14 \text{ rad/s})$

$$v_m = 0.566 \text{ m/s} \qquad \text{v}_m = 0.566 \text{ m/s} \updownarrow \quad \blacktriangleleft$$

Maximum Acceleration: $\quad a_m = x_m p^2 = (0.040 \text{ m})(14.14 \text{ rad/s})^2$

$$a_m = 8.00 \text{ m/s}^2 \qquad \text{a}_m = 8.00 \text{ m/s}^2 \updownarrow \quad \blacktriangleleft$$

b. **Springs Attached in Series.** We first determine the constant k of a single spring equivalent to the two springs *by finding the total elongation δ of the springs under a given static load* **P.** To facilitate the computation, a static load of magnitude $P = 12$ kN is used.

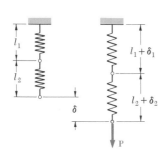

$$\delta = \delta_1 + \delta_2 = \frac{P}{k_1} + \frac{P}{k_2} = \frac{12 \text{ kN}}{4 \text{ kN/m}} + \frac{12 \text{ kN}}{6 \text{ kN/m}} = 5 \text{ m}$$

$$k = \frac{P}{\delta} = \frac{12 \text{ kN}}{5 \text{ m}} = 2.4 \text{ kN/m} = 2400 \text{ N/m}$$

Period of Vibration: $\quad p^2 = \frac{k}{m} = \frac{2400 \text{ N/m}}{50 \text{ kg}} \qquad p = 6.93 \text{ rad/s}$

$$\tau = \frac{2\pi}{p} \qquad\qquad\qquad \tau = 0.907 \text{ s} \quad \blacktriangleleft$$

Maximum Velocity: $\quad v_m = x_m p = (0.040 \text{ m})(6.93 \text{ rad/s})$

$$v_m = 0.277 \text{ m/s} \qquad \text{v}_m = 0.277 \text{ m/s} \updownarrow \quad \blacktriangleleft$$

Maximum Acceleration: $\quad a_m = x_m p^2 = (0.040 \text{ m})(6.93 \text{ rad/s})^2$

$$a_m = 1.920 \text{ m/s}^2 \qquad \text{a}_m = 1.920 \text{ m/s}^2 \updownarrow \quad \blacktriangleleft$$

Problems

19.1 Determine the maximum velocity and maximum acceleration of a particle which moves in simple harmonic motion with an amplitude of 3 mm and a frequency of 20 Hz.

19.2 A particle moves in simple harmonic motion. Knowing that the maximum velocity is 200 mm/s and the maximum acceleration is 4 m/s², determine the amplitude and frequency of the motion.

19.3 Determine the maximum velocity and maximum acceleration of a particle which moves in simple harmonic motion with an amplitude of 135 mm and a period of 1.2 s.

19.4 A 7.5-lb collar is attached to a spring of constant 5 lb/in. and may slide without friction on a horizontal rod. If the collar is moved 3 in. from its equilibrium position and released, determine the maximum velocity and maximum acceleration of the collar during the resulting motion.

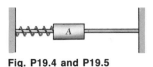

Fig. P19.4 and P19.5

19.5 A 4-lb collar is attached to a spring of constant 6 lb/in. and may slide without friction on a horizontal rod. The collar is at rest when it is struck with a mallet and given an initial velocity of 55 in./s. Determine the amplitude and the maximum acceleration of the collar during the resulting motion.

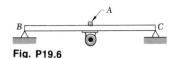

Fig. P19.6

19.6 A variable-speed motor is rigidly attached to beam BC. The rotor is slightly unbalanced and causes the beam to vibrate with a frequency equal to the motor speed. When the speed of the motor is less than 600 rpm or more than 1200 rpm, a small object placed at A is observed to remain in contact with the beam. For speeds between 600 and 1200 rpm the object is observed to "dance" and actually to lose contact with the beam. Determine the amplitude of the motion of A when the speed of the motor is (a) 600 rpm, (b) 1200 rpm. Give answers in both SI and U.S. customary units.

19.7 An instrument package B is placed on the shaking table C as shown. The table is made to move horizontally in simple harmonic motion with a frequency of 3 Hz. Knowing that the coefficient of static friction is $\mu_s = 0.40$ between the package and the table, determine the largest allowable amplitude of the motion if the package is not to slip on the table. Give answers in both SI and U.S. customary units.

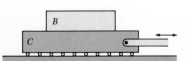

Fig. P19.7

19.8 A 2.5-kg collar rests on, but is not attached to, the spring shown. The collar is depressed 75 mm and released. If the ensuing motion is to be simple harmonic, determine (*a*) the largest permissible value of the spring constant *k*, (*b*) the corresponding position, velocity, and acceleration of the collar 0.12 s after it has been released.

19.9 A 5-kg collar is attached to a spring of constant $k = 900$ N/m as shown. If the collar is given a displacement of 50 mm downward from its equilibrium position and released, determine (*a*) the time required for the collar to move 75 mm upward, (*b*) the corresponding velocity and acceleration of the collar.

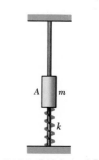

19.10 A 4-lb collar is attached to a spring of constant $k = 6$ lb/in. as shown. If the collar is given a displacement of 2.5 in. downward from its equilibrium position and released, determine (*a*) the time required for the collar to move 2 in. upward, (*b*) the corresponding velocity and acceleration of the collar.

Fig. P19.8, P19.9, and P19.10

19.11 In Prob. 19.10, determine the position, velocity, and acceleration of the collar 0.15 s after it has been released.

19.12 and 19.13 A 40-lb block is supported by the spring arrangement shown. If the block is moved vertically downward from its equilibrium position and released, determine (*a*) the period and frequency of the resulting motion, (*b*) the maximum velocity and acceleration of the block if the amplitude of the motion is 1.25 in.

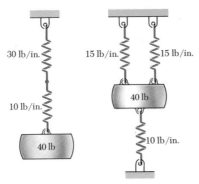

Fig. P19.12   **Fig. P19.13**

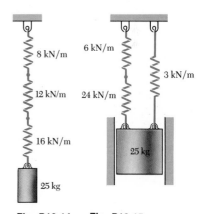

Fig. P19.14   **Fig. P19.15**

19.14 and 19.15 A 25-kg block is supported by the spring arrangement shown. If the block is moved vertically downward from its equilibrium position and released, determine (*a*) the period and frequency of the resulting motion, (*b*) the maximum velocity and acceleration of the block if the amplitude of the motion is 30 mm.

Fig. P19.16 and P19.17

19.16 The period of vibration of the system shown is observed to be 0.6 s. After cylinder *B* has been removed, the period is observed to be 0.5 s. Determine (*a*) the weight of cylinder *A*, (*b*) the constant of the spring.

19.17 The period of vibration of the system shown is observed to be 2.4 s. After cylinder *B* has been removed and replaced with a 4-lb cylinder, the period is observed to be 2.6 s. Determine (*a*) the weight of cylinder *A*, (*b*) the constant of the spring.

19.18 The period of vibration of the system shown is observed to be 0.8 s. If block *A* is removed, the period is observed to be 0.7 s. Determine (*a*) the mass of block *C*, (*b*) the period of vibration when both blocks *A* and *B* have been removed.

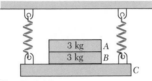

Fig. P19.18

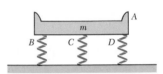

Fig. P19.19

19.19 A tray of mass *m* is attached to three springs as shown. The period of vibration of the tray is 0.5 s. After the center spring *C* has been removed, the period is observed to be 0.6 s. Knowing that the constant of spring *C* is 125 N/m, determine the mass *m* of the tray.

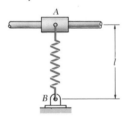

Fig. P19.20

19.20 A collar of mass *m* slides without friction on a horizontal rod and is attached to a spring *AB* of constant *k*. (*a*) If the unstretched length of the spring is just equal to *l*, show that the collar does not execute simple harmonic motion even when the amplitude of the oscillations is small. (*b*) If the unstretched length of the spring is less than *l*, show that the motion may be approximated by a simple harmonic motion for small oscillations.

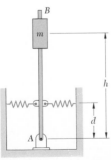

Fig. P19.21 and P19.22

19.21 Rod *AB* is attached to a hinge at *A* and to two springs, each of constant *k*. If $h = 700$ mm, $d = 300$ mm, and $m = 20$ kg, determine the value of *k* for which the period of small oscillations is (*a*) 1 s, (*b*) infinite. Neglect the mass of the rod and assume that each spring can act in either tension or compression.

19.22 If $h = 700$ mm and $d = 500$ mm and each spring has a constant $k = 600$ N/m, determine the mass *m* for which the period of small oscillations is (*a*) 0.50 s, (*b*) infinite. Neglect the mass of the rod and assume that each spring can act in either tension or compression.

19.23 Denoting by δ_{st} the static deflection of a beam under a given load, show that the frequency of vibration of the load is

$$f = \frac{1}{2\pi}\sqrt{\frac{g}{\delta_{st}}}$$

Neglect the mass of the beam, and assume that the load remains in contact with the beam.

Fig. P19.23

*19.24 Expanding the integrand in Eq. (19.19) of Sec. 19.4 into a series of even powers of $\sin \phi$ and integrating, show that the period of a simple pendulum of length l may be approximated by the formula

$$\tau = 2\pi\sqrt{\frac{l}{g}}\left(1 + \tfrac{1}{4}\sin^2\frac{\theta_m}{2}\right)$$

where θ_m is the amplitude of the oscillations.

*19.25 Using the formula given in Prob. 19.24, determine the amplitude θ_m for which the period of a simple pendulum is 1 percent longer than the period of the same pendulum for small oscillations.

*19.26 Using the data of Table 19.1, determine the period of a simple pendulum of length $l = 800$ mm (*a*) for small oscillations, (*b*) for oscillations of amplitude $\theta_m = 30°$, (*c*) for oscillations of amplitude $\theta_m = 90°$.

*19.27 Using a table of elliptic integrals, determine the period of a simple pendulum of length $l = 800$ mm if the amplitude of the oscillations is $\theta_m = 40°$.

19.5. Free Vibrations of Rigid Bodies.

The analysis of the vibrations of a rigid body or of a system of rigid bodies possessing a single degree of freedom is similar to the analysis of the vibrations of a particle. An appropriate variable, such as a distance x or an angle θ, is chosen to define the position of the body or system of bodies, and an equation relating this variable and its second derivative with respect to t is written. If the equation obtained is of the same form as (19.5), i.e., if we have

$$\ddot{x} + p^2 x = 0 \qquad \text{or} \qquad \ddot{\theta} + p^2\theta = 0 \qquad (19.21)$$

the vibration considered is a simple harmonic motion. The period and frequency of the vibration may then be obtained by identifying p and substituting into (19.13) and (19.14).

In general, a simple way to obtain one of Eqs. (19.21) is to express that the system of the external forces is equivalent to the system of the effective forces by drawing a diagram of the body for an arbitrary value of the variable and writing the appropriate equation of motion. We recall that our goal should be *the determination of the coefficient* of the variable x or

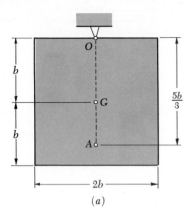

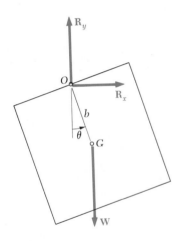

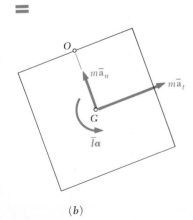

Fig. 19.5

θ, *not* the determination of the variable itself or of the derivative $\ddot{x}$ or $\ddot{\theta}$. Setting this coefficient equal to p^2, we obtain the circular frequency p, from which τ and f may be determined.

The method we have outlined may be used to analyze vibrations which are truly represented by a simple harmonic motion, or vibrations of small amplitude which can be *approximated* by a simple harmonic motion. As an example, we shall determine the period of the small oscillations of a square plate of side $2b$ which is suspended from the midpoint O of one side (Fig. 19.5a). We consider the plate in an arbitrary position defined by the angle θ that the line OG forms with the vertical and draw a diagram to express that the weight $\mathbf{W}$ of the plate and the components $\mathbf{R}_x$ and $\mathbf{R}_y$ of the reaction at O are equivalent to the vectors $m\mathbf{a}_t$ and $m\mathbf{a}_n$ and to the couple $\bar{I}\alpha$ (Fig. 19.5b). Since the angular velocity and angular acceleration of the plate are equal, respectively, to $\dot{\theta}$ and $\ddot{\theta}$, the magnitudes of the two vectors are, respectively, $mb\ddot{\theta}$ and $mb\dot{\theta}^2$, while the moment of the couple is $\bar{I}\ddot{\theta}$. In previous applications of this method (Chap. 16), we tried whenever possible to assume the correct sense for the acceleration. Here, however, we must assume the same positive sense for θ and $\ddot{\theta}$ in order to obtain an equation of the form (19.21). Consequently, the angular acceleration $\ddot{\theta}$ will be assumed positive counterclockwise, even though this assumption is obviously unrealistic. Equating moments about O, we write

$$+\uparrow \qquad\qquad -W(b \sin \theta) = (mb\ddot{\theta})b + \bar{I}\ddot{\theta}$$

Noting that $\bar{I} = \frac{1}{12}m[(2b)^2 + (2b)^2] = \frac{2}{3}mb^2$ and $W = mg$, we obtain

$$\ddot{\theta} + \frac{3}{5}\frac{g}{b}\sin \theta = 0 \tag{19.22}$$

For oscillations of small amplitude, we may replace $\sin \theta$ by θ, expressed in radians, and write

$$\ddot{\theta} + \frac{3}{5}\frac{g}{b}\theta = 0 \tag{19.23}$$

Comparison with (19.21) shows that the equation obtained is that of a simple harmonic motion and that the circular frequency p of the oscillations is equal to $(3g/5b)^{1/2}$. Substituting into (19.13), we find that the period of the oscillations is

$$\tau = \frac{2\pi}{p} = 2\pi\sqrt{\frac{5b}{3g}} \tag{19.24}$$

The result obtained is valid only for oscillations of small amplitude. A more accurate description of the motion of the plate is obtained by comparing Eqs. (19.16) and (19.22). We note that the two equations are identical if we choose l equal to $5b/3$. This means that the plate will oscillate as a simple pendulum of length $l = 5b/3$, and the results of Sec. 19.4 may be used to correct the value of the period given in (19.24). The point A of the plate located on line OG at a distance $l = 5b/3$ from O is defined as the *center of oscillation* corresponding to O (Fig. 19.5a).

SAMPLE PROBLEM 19.2

A cylinder of weight W and radius r is suspended from a looped cord as shown. One end of the cord is attached directly to a rigid support, while the other end is attached to a spring of constant k. Determine the period and frequency of vibration of the cylinder.

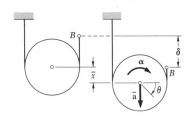

Kinematics of Motion. We express the linear displacement and the acceleration of the cylinder in terms of the angular displacement θ. Choosing the positive sense clockwise and measuring the displacements from the equilibrium position, we write

$$\bar{x} = r\theta \qquad \delta = 2\bar{x} = 2r\theta$$

$$\alpha = \ddot{\theta} \downarrow \qquad \bar{a} = r\alpha = r\ddot{\theta} \qquad \bar{\mathbf{a}} = r\ddot{\theta} \downarrow \qquad (1)$$

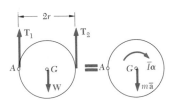

Equations of Motion. The system of external forces acting on the cylinder consists of the weight $\mathbf{W}$ and of the forces $\mathbf{T}_1$ and $\mathbf{T}_2$ exerted by the cord. We express that this system is equivalent to the system of effective forces represented by the vector $m\bar{\mathbf{a}}$ attached at G and the couple $\bar{I}\alpha$.

$$+\!\!\downarrow\!\Sigma M_A = \Sigma(M_A)_{\text{eff}}: \qquad Wr - T_2(2r) = m\bar{a}r + \bar{I}\alpha \qquad (2)$$

When the cylinder is in its position of equilibrium, the tension in the cord is $T_0 = \frac{1}{2}W$. We note that for an angular displacement θ, the magnitude of $\mathbf{T}_2$ is

$$T_2 = T_0 + k\delta = \tfrac{1}{2}W + k\delta = \tfrac{1}{2}W + k(2r\theta) \qquad (3)$$

Substituting from (1) and (3) into (2), and recalling that $\bar{I} = \frac{1}{2}mr^2$, we write

$$Wr - (\tfrac{1}{2}W + 2kr\theta)(2r) = m(r\ddot{\theta})r + \tfrac{1}{2}mr^2\ddot{\theta}$$

$$\ddot{\theta} + \frac{8}{3}\frac{k}{m}\theta = 0$$

The motion is seen to be simple harmonic, and we have

$$p^2 = \frac{8}{3}\frac{k}{m} \qquad p = \sqrt{\frac{8}{3}\frac{k}{m}}$$

$$\tau = \frac{2\pi}{p} \qquad \tau = 2\pi\sqrt{\frac{3}{8}\frac{m}{k}} \quad \blacktriangleleft$$

$$f = \frac{p}{2\pi} \qquad f = \frac{1}{2\pi}\sqrt{\frac{8}{3}\frac{k}{m}} \quad \blacktriangleleft$$

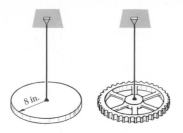

SAMPLE PROBLEM 19.3

A circular disk, weighing 20 lb and of radius 8 in., is suspended from a wire as shown. The disk is rotated (thus twisting the wire) and then released; the period of the torsional vibration is observed to be 1.13 s. A gear is then suspended from the same wire, and the period of torsional vibration for the gear is observed to be 1.93 s. Assuming that the moment of the couple exerted by the wire is proportional to the angle of twist, determine (*a*) the torsional spring constant of the wire, (*b*) the centroidal moment of inertia of the gear, (*c*) the maximum angular velocity reached by the gear if it is rotated through 90° and released.

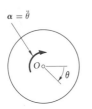

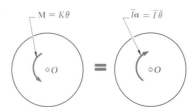

a. **Vibration of Disk.** Denoting by θ the angular displacement of the disk, we express that the magnitude of the couple exerted by the wire is $M = K\theta$, where K is the torsional spring constant of the wire. Since this couple must be equivalent to the couple $\bar{I}\alpha$ representing the effective forces of the disk, we write

$$+\uparrow\Sigma M_O = \Sigma(M_O)_{\text{eff}}: \qquad +K\theta = -\bar{I}\ddot{\theta}$$

$$\ddot{\theta} + \frac{K}{\bar{I}}\theta = 0$$

The motion is seen to be simple harmonic, and we have

$$p^2 = \frac{K}{\bar{I}} \qquad \tau = \frac{2\pi}{p} \qquad \tau = 2\pi\sqrt{\frac{\bar{I}}{K}} \tag{1}$$

For the disk, we have

$$\tau = 1.13 \text{ s} \qquad \bar{I} = \tfrac{1}{2}mr^2 = \frac{1}{2}\left(\frac{20 \text{ lb}}{32.2 \text{ ft/s}^2}\right)\left(\frac{8}{12}\text{ ft}\right)^2 = 0.138 \text{ lb}\cdot\text{ft}\cdot\text{s}^2$$

Substituting into (1), we obtain

$$1.13 = 2\pi\sqrt{\frac{0.138}{K}} \qquad K = 4.27 \text{ lb}\cdot\text{ft/rad} \blacktriangleleft$$

b. **Vibration of Gear.** Since the period of vibration of the gear is 1.93 s and $K = 4.27$ lb·ft/rad, Eq. (1) yields

$$1.93 = 2\pi\sqrt{\frac{\bar{I}}{4.27}} \qquad \bar{I}_{\text{gear}} = 0.403 \text{ lb}\cdot\text{ft}\cdot\text{s}^2 \blacktriangleleft$$

c. **Maximum Angular Velocity of the Gear.** Since the motion is simple harmonic, we have

$$\theta = \theta_m \sin pt \qquad \omega = \theta_m p \cos pt \qquad \omega_m = \theta_m p$$

Recalling that $\theta_m = 90° = 1.571$ rad and $\tau = 1.93$ s, we write

$$\omega_m = \theta_m p = \theta_m\left(\frac{2\pi}{\tau}\right) = (1.571 \text{ rad})\left(\frac{2\pi}{1.93 \text{ s}}\right)$$

$$\omega_m = 5.11 \text{ rad/s} \blacktriangleleft$$

Problems

19.28 The 4-kg uniform rod shown is attached to a spring of constant $k = 750$ N/m. If end A of the rod is depressed 40 mm and released, determine (*a*) the period of vibration, (*b*) the maximum velocity of end A.

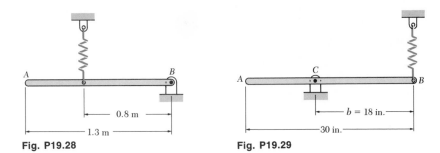

Fig. P19.28   **Fig. P19.29**

19.29 The uniform rod shown weighs 15 lb and is attached to a spring of constant $k = 4$ lb/in. If end B of the rod is depressed 0.4 in. and released, determine (*a*) the period of vibration, (*b*) the maximum velocity of end B.

19.30 A 15-lb slender rod AB is riveted to a 12-lb uniform disk as shown. A belt is attached to the rim of the disk and to a spring which holds the rod at rest in the position shown. If end A of the rod is moved 0.75 in. down and released, determine (*a*) the period of vibration, (*b*) the maximum velocity of end A.

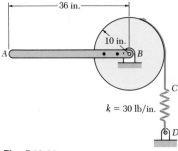

Fig. P19.30

19.31 A 5-kg uniform cylinder rolls without sliding on an incline and is attached to a spring AB as shown. If the center of the cylinder is moved 10 mm down the incline from its equilibrium position and released, determine (*a*) the period of vibration, (*b*) the maximum velocity of the center of the cylinder.

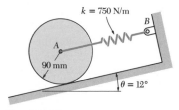

Fig. P19.31

19.32 A belt is placed around the rim of a 240-kg flywheel and attached as shown to two springs, each of constant $k = 15$ kN/m. If end C of the belt is pulled 40 mm down and released, the period of vibration of the flywheel is observed to be 0.5 s. Knowing that the initial tension in the belt is sufficient to prevent slipping, determine (*a*) the maximum angular velocity of the flywheel, (*b*) the centroidal radius of gyration of the flywheel.

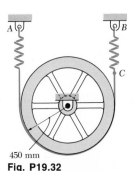

450 mm
Fig. P19.32

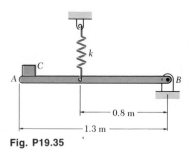

Fig. P19.35

19.33 Knowing that the centroidal radius of gyration of the flywheel in Prob. 19.32 is 380 mm, determine (*a*) the period of vibration, (*b*) the maximum angular velocity of the flywheel.

19.34 In Prob. 19.29, determine (*a*) the value of *b* for which the smallest period of vibration occurs, (*b*) the corresponding period of vibration.

19.35 The 4-kg uniform rod *AB* is attached as shown to a spring of constant $k = 750$ N/m. A small 0.5-kg block *C* is placed on the rod at *A*. (*a*) If end *A* of the rod is then moved down through a small distance δ_0 and released, determine the period of the vibration. (*b*) Determine the largest allowable value of δ_0 if block *C* is to remain at all times in contact with the rod.

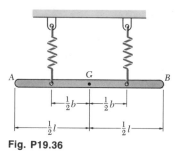

Fig. P19.36

19.36 A slender rod of mass *m* and length *l* is held by two springs, each of constant *k*. Determine the frequency of the resulting vibration if the rod is (*a*) given a small vertical displacement and released, (*b*) rotated through a small angle about a horizontal axis through *G* and released. (*c*) Determine the ratio *b/l* for which the frequencies found in parts *a* and *b* are equal.

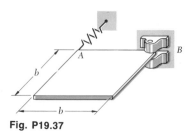

Fig. P19.37

19.37 A uniform square plate of mass *m* is supported in a horizontal plane by a vertical pin at *B* and is attached at *A* to a spring of constant *k*. If corner *A* is given a small displacement and released, determine the period of the resulting motion.

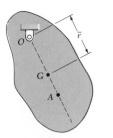

Fig. P19.38 and P19.39

19.38 A *compound pendulum* is defined as a rigid slab which oscillates about a fixed point *O*, called the center of suspension. Show that the period of oscillation of a compound pendulum is equal to the period of a simple pendulum of length *OA*, where the distance from *A* to the mass center *G* is $GA = \bar{k}^2/\bar{r}$. Point *A* is defined as the center of oscillation and coincides with the center of percussion defined in Prob. 17.65.

19.39 A rigid slab oscillates about a fixed point *O*. Show that the smallest period of oscillation occurs when the distance $\bar{r}$ from point *O* to the mass center *G* is equal to $\bar{k}$.

19.40 Show that if the compound pendulum of Prob. 19.38 is suspended from *A* instead of *O*, the period of oscillation is the same as before and the new center of oscillation is located at *O*.

19.41 A uniform bar of length l may oscillate about a hinge A located at a distance c from the mass center G of the bar. (*a*) Determine the frequency of small oscillations if $c = \frac{1}{4}l$. (*b*) Determine a second value of c for which the frequency of small oscillations is the same as that found in part *a*.

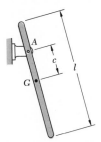

Fig. P19.41 and P19.42

19.42 A uniform bar of length l may oscillate about a hinge A located at a distance c from the mass center G of the bar. Determine (*a*) the distance c for which the period of oscillation is minimum, (*b*) the corresponding value of the period.

19.43 A uniform rectangular plate is suspended from a pin located at the midpoint of one edge as shown. Considering the dimension b constant, determine (*a*) the ratio c/b for which the period of oscillation of the plate is minimum, (*b*) the ratio c/b for which the period of oscillation of the plate is the same as the period of a simple pendulum of length c.

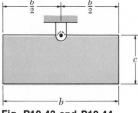

Fig. P19.43 and P19.44

19.44 A uniform rectangular plate is suspended from a pin located at the midpoint of one edge as shown. (*a*) Determine the period of small oscillations if $c = b$. (*b*) Considering the dimension b constant, determine a second value of c for which the period of oscillation is the same as that found in part *a*.

19.45 A thin homogeneous wire is bent into the shape of an equilateral triangle of side $l = 300$ mm. Determine the period of small oscillations if the wire (*a*) is suspended as shown, (*b*) is suspended from a pin located at the midpoint of one side.

Fig. P19.45

19.46 Two uniform rods, each of mass m and length l, are welded together to form the T-shaped assembly shown. Determine the frequency of small oscillations of the assembly.

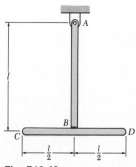

Fig. P19.46

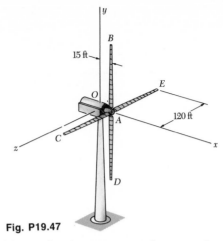

Fig. P19.47

19.47 Blade AB of the wind-turbine generator shown is to be temporarily removed. Motion of the turbine generator about the y axis is prevented, but the remaining three blades may oscillate as a unit about the x axis. Assuming that each blade is equivalent to a 120-ft slender rod, determine the period of small oscillations of the blades in the absence of wind.

Fig. P19.48 and P19.49

19.48 The period of small oscillations about A of a connecting rod is observed to be 1.03 s. Knowing that the distance r_a is 160 mm, determine the centroidal radius of gyration of the connecting rod.

19.49 A connecting rod is supported by a knife-edge at point A; the period of small oscillations is observed to be 0.87 s. The rod is then inverted and supported by a knife-edge at point B and the period of small oscillations is observed to be 0.78 s. Knowing that $r_a + r_b = 255$ mm, determine (*a*) the location of the mass center G, (*b*) the centroidal radius of gyration $\bar{k}$.

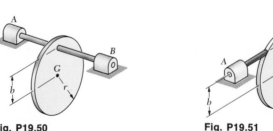

Fig. P19.50   **Fig. P19.51**

19.50 and 19.51 A thin disk of radius r may oscillate about an axis AB located as shown at a distance b from the mass center G. (*a*) Determine the period of small oscillations if $b = r$. (*b*) Determine a second value of b for which the period of oscillation is the same as that found in part a.

Fig. P19.52

19.52 A period of 4.1 s is observed for the angular oscillations of a 450-g gyroscope rotor suspended from a wire as shown. Knowing that a period of 6.2 s is obtained when a 50-mm-diameter steel sphere is suspended in the same fashion, determine the centroidal radius of gyration of the rotor. (Density of steel = 7850 kg/m³.)

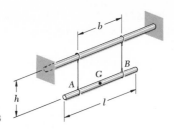

Fig. P19.53

19.53 A slender rod of length l is suspended from two vertical wires of length h, each located a distance $\frac{1}{2}b$ from the mass center G. Determine the period of oscillation when (a) the rod is rotated through a small angle about a vertical axis passing through G and released, (b) the rod is given a small horizontal translation along AB and released.

19.54 Solve Prob. 19.53, assuming that $h = 300$ mm, $b = 400$ mm, and $l = 600$ mm.

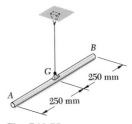
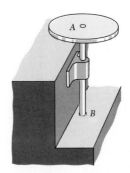

Fig. P19.55

19.55 A 3-kg slender rod is suspended from a steel wire which is known to have a torsional spring constant $K = 1.95$ N·m/rad. If the rod is rotated through 180° about the vertical and then released, determine (a) the period of oscillation, (b) the maximum velocity of end A of the rod.

19.56 A uniform disk of radius 12 in. and weighing 20 lb is attached to a vertical shaft which is rigidly held at B. It is known that the disk rotates through 5° when a 120-lb·in. static couple is applied to the disk. If the disk is rotated through 3° and then released, determine (a) the period of the resulting vibration, (b) the maximum velocity of a point on the rim of the disk.

19.57 A steel casting is rigidly bolted to the disk of Prob. 19.56. Knowing that the period of torsional vibration of the disk and casting is 0.41 s, determine the moment of inertia of the casting with respect to shaft AB.

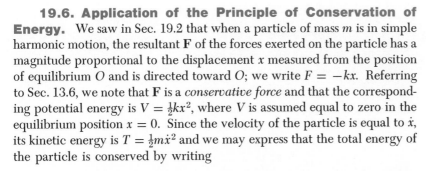

Fig. P19.56

19.6. Application of the Principle of Conservation of Energy.
We saw in Sec. 19.2 that when a particle of mass m is in simple harmonic motion, the resultant $\mathbf{F}$ of the forces exerted on the particle has a magnitude proportional to the displacement x measured from the position of equilibrium O and is directed toward O; we write $F = -kx$. Referring to Sec. 13.6, we note that $\mathbf{F}$ is a *conservative force* and that the corresponding potential energy is $V = \frac{1}{2}kx^2$, where V is assumed equal to zero in the equilibrium position $x = 0$. Since the velocity of the particle is equal to $\dot{x}$, its kinetic energy is $T = \frac{1}{2}m\dot{x}^2$ and we may express that the total energy of the particle is conserved by writing

$$T + V = \text{constant} \qquad \tfrac{1}{2}m\dot{x}^2 + \tfrac{1}{2}kx^2 = \text{constant}$$

Setting $p^2 = k/m$ as in (19.4), where p is the circular frequency of the vibration, we have

$$\dot{x}^2 + p^2x^2 = \text{constant} \qquad (19.25)$$

Equation (19.25) is characteristic of simple harmonic motion; it may be obtained directly from (19.5) by multiplying both terms by $2\dot{x}$ and integrating.

The principle of conservation of energy provides a convenient way for determining the period of vibration of a rigid body or of a system of rigid bodies possessing a single degree of freedom, once it has been established that the motion of the system is a simple harmonic motion, or that it may be approximated by a simple harmonic motion. Choosing an appropriate variable, such as a distance x or an angle θ, we consider two particular positions of the system:

1. *The displacement of the system is maximum;* we have $T_1 = 0$, and V_1 may be expressed in terms of the amplitude x_m or θ_m (choosing $V = 0$ in the equilibrium position).
2. *The system passes through its equilibrium position;* we have $V_2 = 0$, and T_2 may be expressed in terms of the maximum velocity $\dot{x}_m$ or $\dot{\theta}_m$.

We then express that the total energy of the system is conserved and write $T_1 + V_1 = T_2 + V_2$. Recalling from (19.15) that for simple harmonic motion the maximum velocity is equal to the product of the amplitude and of the circular frequency p, we find that the equation obtained may be solved for p.

As an example, we shall consider again the square plate of Sec. 19.5. In the position of maximum displacement (Fig. 19.6a), we have

$$T_1 = 0 \qquad V_1 = W(b - b\cos\theta_m) = Wb(1 - \cos\theta_m)$$

or, since $1 - \cos\theta_m = 2\sin^2(\theta_m/2) \approx 2(\theta_m/2)^2 = \theta_m^2/2$ for oscillations of small amplitude,

$$T_1 = 0 \qquad V_1 = \tfrac{1}{2}Wb\theta_m^2 \tag{19.26}$$

As the plate passes through its position of equilibrium (Fig. 19.6b), its velocity is maximum and we have

$$T_2 = \tfrac{1}{2}m\bar{v}_m^2 + \tfrac{1}{2}\bar{I}\omega_m^2 = \tfrac{1}{2}mb^2\dot{\theta}_m^2 + \tfrac{1}{2}\bar{I}\dot{\theta}_m^2 \qquad V_2 = 0$$

or, recalling from Sec. 19.5 that $\bar{I} = \tfrac{2}{3}mb^2$,

$$T_2 = \tfrac{1}{2}(\tfrac{5}{3}mb^2)\dot{\theta}_m^2 \qquad V_2 = 0 \tag{19.27}$$

Substituting from (19.26) and (19.27) into $T_1 + V_1 = T_2 + V_2$, and noting that the maximum velocity $\dot{\theta}_m$ is equal to the product $\theta_m p$, we write

$$\tfrac{1}{2}Wb\theta_m^2 = \tfrac{1}{2}(\tfrac{5}{3}mb^2)\theta_m^2 p^2 \tag{19.28}$$

which yields $p^2 = 3g/5b$ and

$$\tau = \frac{2\pi}{p} = 2\pi\sqrt{\frac{5b}{3g}} \tag{19.29}$$

as previously obtained.

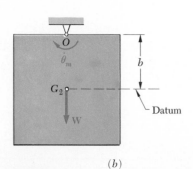

Fig. 19.6

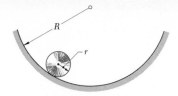

Determine the period of small oscillations of a cylinder of radius r which rolls without slipping inside a curved surface of radius R.

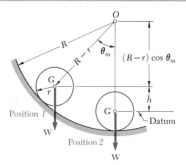

Position 1

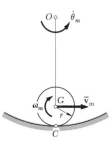

Position 2

Solution. We denote by θ the angle which line OG forms with the vertical. Since the cylinder rolls without slipping, we may apply the principle of conservation of energy between position 1, where $\theta = \theta_m$, and position 2, where $\theta = 0$.

Position 1. **Kinetic Energy.** Since the velocity of the cylinder is zero, we have $T_1 = 0$.

Potential Energy. Choosing a datum as shown and denoting by W the weight of the cylinder, we have

$$V_1 = Wh = W(R - r)(1 - \cos \theta)$$

Noting that for small oscillations $(1 - \cos \theta) = 2 \sin^2 (\theta/2) \approx \theta^2/2$, we have

$$V_1 = W(R - r)\frac{\theta_m^2}{2}$$

Position 2. Denoting by $\dot{\theta}_m$ the angular velocity of line OG as the cylinder passes through position 2, and observing that point C is the instantaneous center of rotation of the cylinder, we write

$$\bar{v}_m = (R - r)\dot{\theta}_m \qquad \omega_m = \frac{\bar{v}_m}{r} = \frac{R - r}{r}\dot{\theta}_m$$

Kinetic Energy

$$
\begin{aligned}
T_2 &= \tfrac{1}{2}m\bar{v}_m^2 + \tfrac{1}{2}\bar{I}\omega_m^2 \\
&= \tfrac{1}{2}m(R - r)^2\dot{\theta}_m^2 + \tfrac{1}{2}(\tfrac{1}{2}mr^2)\left(\frac{R - r}{r}\right)^2\dot{\theta}_m^2 \\
&= \tfrac{3}{4}m(R - r)^2\dot{\theta}_m^2
\end{aligned}
$$

Potential Energy $\qquad\qquad V_2 = 0$

Conservation of Energy

$$T_1 + V_1 = T_2 + V_2$$

$$0 + W(R - r)\frac{\theta_m^2}{2} = \tfrac{3}{4}m(R - r)^2\dot{\theta}_m^2 + 0$$

Since $\dot{\theta}_m = p\theta_m$ and $W = mg$, we write

$$mg(R - r)\frac{\theta_m^2}{2} = \tfrac{3}{4}m(R - r)^2(p\theta_m)^2 \qquad p^2 = \frac{2}{3}\frac{g}{R - r}$$

$$\tau = \frac{2\pi}{p} \qquad \tau = 2\pi\sqrt{\frac{3}{2}\frac{R - r}{g}} \quad \blacktriangleleft$$

Problems

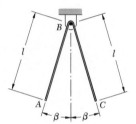

Fig. P19.58

Fig. P19.62 and P19.63

19.58 Using the method of Sec. 19.6, determine the period of a simple pendulum of length l.

19.59 The springs of an automobile are observed to expand 180 mm to an undeformed position as the body is lifted by several jacks. Assuming that each spring carries an equal portion of the weight of the automobile, determine the frequency of the free vertical vibrations of the body.

19.60 Using the method of Sec. 19.6, solve Prob. 19.4.

19.61 Using the method of Sec. 19.6, solve Prob. 19.5

19.62 A homogeneous wire of length $2l$ is bent as shown and allowed to oscillate about a frictionless pin at B. Denoting by τ_0 the period of small oscillations when $\beta = 0$, determine the angle β for which the period of small oscillations is $2\tau_0$.

19.63 Knowing that $l = 600$ mm and $\beta = 30°$, determine the period of small oscillations of the bent homogeneous wire shown.

19.64 A homogeneous wire is bent to form a square of side l which is supported by a ball-and-socket joint at A. Determine the period of small oscillations of the square (a) in the plane of the square, (b) in a direction perpendicular to the square.

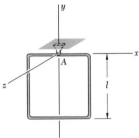

Fig. P19.64

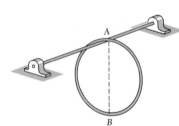

Fig. P19.65

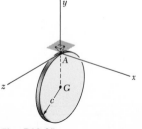

Fig. P19.66

19.65 A thin hoop of radius r and mass m is suspended from a rough rod as shown. Determine the frequency of small oscillations of the hoop (a) in the plane of the hoop, (b) in a direction perpendicular to the plane of the hoop. Assume that friction is sufficiently large to prevent slipping at A.

19.66 A uniform disk of radius c is supported by a ball-and-socket joint at A. Determine the frequency of small oscillations of the disk (a) in the plane of the disk, (b) in a direction perpendicular to the disk.

19.67 It is observed that when an 8-lb weight is attached to the rim of a 6-ft-diameter flywheel, the period of small oscillations of the flywheel is 22 s. Neglecting axle friction, determine the centroidal moment of inertia of the flywheel.

Fig. P19.67

19.68 Using the method of Sec. 19.6, solve Prob. 19.46.

19.69 A uniform rod ABC weighs 6 lb and is attached to two springs as shown. If end C is given a small displacement and released, determine the frequency of vibration of the rod.

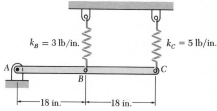

Fig. P19.69

19.70 For the rod of Prob. 19.69, determine the frequency of vibration of the rod if the springs are interchanged so that $k_B = 5$ lb/in. and $k_C = 3$ lb/in.

19.71 The 5-kg slender rod AB is welded to an 8-kg uniform disk. A spring of constant 450 N/m is attached to the disk and holds the rod at rest in the position shown. If end B of the rod is given a small displacement and released, determine the period of vibration of the rod.

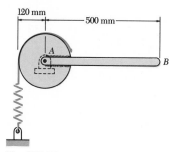

19.72 For the rod and disk of Prob. 19.71, determine the constant of the spring for which the period of vibration of the rod is 1.6 s.

Fig. P19.71

19.73 and 19.74 Two uniform rods AB and CD, each of length l and mass m, are attached to two gears as shown. Neglecting the mass of the gears, determine the period of small oscillations of the system.

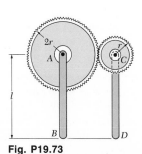

Fig. P19.73

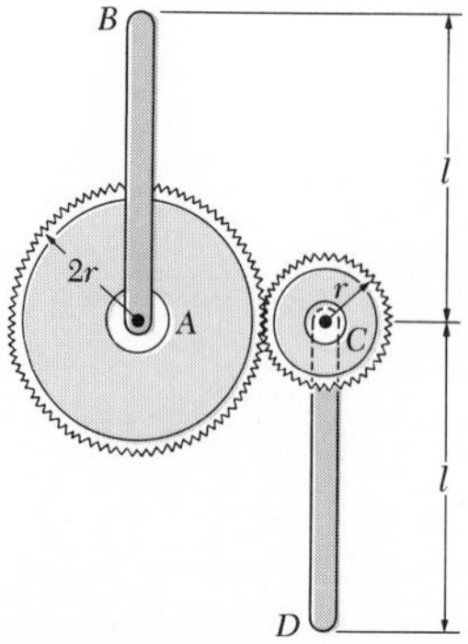

Fig. P19.74

19.75 Two 6-kg uniform disks are attached to the 9-kg rod AB as shown. Knowing that the constant of the spring is 5 kN/m and that the disks roll without sliding, determine the frequency of vibration of the system.

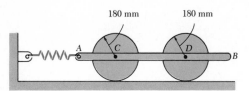

180 mm   180 mm

Fig. P19.75

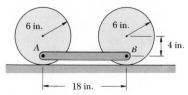

6 in.   6 in.

4 in.

18 in.

Fig. P19.76

19.76 The 20-lb rod AB is attached to two 8-lb disks as shown. Knowing that the disks roll without sliding, determine the frequency of small oscillations of the system.

19.77 Three identical rods are connected as shown. If $b = \frac{3}{4}l$, determine the frequency of small oscillations of the system.

19.78 Three identical rods are connected as shown. Determine (*a*) the distance b for which the frequency of oscillation is maximum, (*b*) the corresponding maximum frequency.

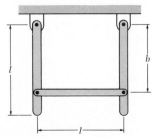

Fig. P19.77 and P19.78

19.79 The 3-lb rod AB is bolted to a 5-lb disk. Knowing that the disk rolls without sliding, determine the period of small oscillations of the system.

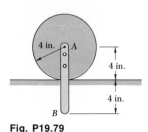

4 in.   A

4 in.

4 in.

B

Fig. P19.79

r

Fig. P19.80

19.80 A half section of pipe is placed on a horizontal surface, rotated through a small angle, and then released. Assuming that the pipe section rolls without sliding, determine the period of oscillation.

19.81 A uniform rod of length L is supported by a ball-and-socket joint at A and by a vertical wire CD. Derive an expression for the period of oscillation of the rod if end B is given a small horizontal displacement and then released.

19.82 Solve Prob. 19.81, assuming that $L = 900$ mm, $b = 750$ mm, and $h = 400$ mm.

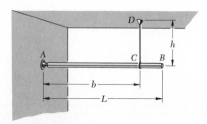

D

h

A

C   B

b

L

Fig. P19.81

19.83 A section of uniform pipe is suspended from two vertical cables attached at A and B. Determine the period of oscillation in terms of l_1 and l_2 when point B is given a small horizontal displacement to the right and released.

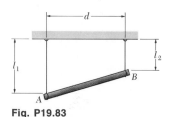

Fig. P19.83

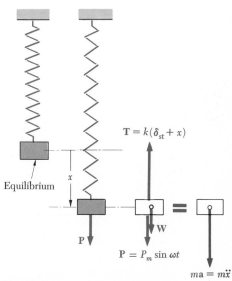

Fig. P19.84

*19.84 As a submerged body moves through a fluid, the particles of the fluid flow around the body and thus acquire kinetic energy. In the case of a sphere moving in an ideal fluid, the total kinetic energy acquired by the fluid is $\frac{1}{4}\rho V v^2$, where ρ is the mass density of the fluid, V the volume of the sphere, and v the velocity of the sphere. Consider a 1-lb hollow spherical shell of radius 3 in. which is held submerged in a tank of water by a spring of constant 3 lb/in. (a) Neglecting fluid friction, determine the period of vibration of the shell when it is displaced vertically and then released. (b) Solve part a, assuming that the tank is accelerated upward at the constant rate of 10 ft/s².

*19.85 A thin plate of length l rests on a half cylinder of radius r. Derive an expression for the period of small oscillations of the plate.

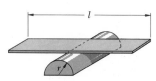

Fig. P19.85

19.7. Forced Vibrations. The most important vibrations from the point of view of engineering applications are the *forced vibrations* of a system. These vibrations occur when a system is subjected to a periodic force or when it is elastically connected to a support which has an alternating motion.

Consider first the case of a body of mass m suspended from a spring and subjected to a periodic force **P** of magnitude $P = P_m \sin \omega t$ (Fig. 19.7). This force may be an actual external force applied to the body, or it may be a centrifugal force produced by the rotation of some unbalanced part of the body (see Sample Prob. 19.5). Denoting by x the displacement of the body measured from its equilibrium position, we write the equation of motion,

$$+\downarrow \Sigma F = ma: \qquad P_m \sin \omega t + W - k(\delta_{st} + x) = m\ddot{x}$$

Recalling that $W = k\delta_{st}$, we have

$$m\ddot{x} + kx = P_m \sin \omega t \qquad (19.30)$$

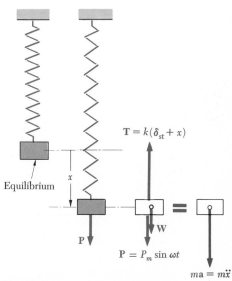

$$T = k(\delta_{st} + x)$$

Equilibrium

$P = P_m \sin \omega t$

$ma = m\ddot{x}$

Fig. 19.7

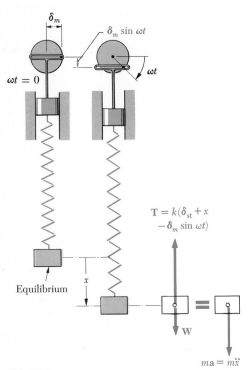

δ_m

$\delta_m \sin \omega t$

$\omega t = 0$

ωt

$T = k(\delta_{st} + x - \delta_m \sin \omega t)$

Equilibrium

x

$\mathbf{W}$

$ma = m\ddot{x}$

Fig. 19.8

Next we consider the case of a body of mass m suspended from a spring attached to a moving support whose displacement δ is equal to $\delta_m \sin \omega t$ (Fig. 19.8). Measuring the displacement x of the body from the position of static equilibrium corresponding to $\omega t = 0$, we find that the total elongation of the spring at time t is $\delta_{st} + x - \delta_m \sin \omega t$. The equation of motion is thus

$$+\downarrow \Sigma F = ma: \qquad W - k(\delta_{st} + x - \delta_m \sin \omega t) = m\ddot{x}$$

Recalling that $W = k\delta_{st}$, we have

$$m\ddot{x} + kx = k\delta_m \sin \omega t \qquad (19.31)$$

We note that Eqs. (19.30) and (19.31) are of the same form and that a solution of the first equation will satisfy the second if we set $P_m = k\delta_m$.

A differential equation such as (19.30) or (19.31), possessing a right-hand member different from zero, is said to be *nonhomogeneous*. Its general solution is obtained by adding a particular solution of the given equation to the general solution of the corresponding *homogeneous* equation (with right-hand member equal to zero). A *particular solution* of (19.30) or (19.31) may be obtained by trying a solution of the form

$$x_{part} = x_m \sin \omega t \qquad (19.32)$$

Substituting x_{part} for x into (19.30), we find

$$-m\omega^2 x_m \sin \omega t + kx_m \sin \omega t = P_m \sin \omega t$$

which may be solved for the amplitude,

$$x_m = \frac{P_m}{k - m\omega^2}$$

Recalling from (19.4) that $k/m = p^2$, where p is the circular frequency of the free vibration of the body, we write

$$x_m = \frac{P_m/k}{1 - (\omega/p)^2} \qquad (19.33)$$

Substituting from (19.32) into (19.31), we obtain in a similar way

$$x_m = \frac{\delta_m}{1 - (\omega/p)^2} \qquad (19.33')$$

The homogeneous equation corresponding to (19.30) or (19.31) is Eq. (19.3), defining the free vibration of the body. Its general solution, called the *complementary function*, was found in Sec. 19.2,

$$x_{\text{comp}} = A \sin pt + B \cos pt \tag{19.34}$$

Adding the particular solution (19.32) and the complementary function (19.34), we obtain the *general solution* of Eqs. (19.30) and (19.31):

$$x = A \sin pt + B \cos pt + x_m \sin \omega t \tag{19.35}$$

We note that the vibration obtained consists of two superposed vibrations. The first two terms in (19.35) represent a free vibration of the system. The frequency of this vibration, called the *natural frequency* of the system, depends only upon the constant k of the spring and the mass m of the body, and the constants A and B may be determined from the initial conditions. This free vibration is also called a *transient* vibration, since in actual practice, it will soon be damped out by friction forces (Sec. 19.9).

The last term in (19.35) represents the *steady-state* vibration produced and maintained by the impressed force or impressed support movement. Its frequency is the *forced frequency* imposed by this force or movement, and its amplitude x_m, defined by (19.33) or (19.33′), depends upon the *frequency ratio* ω/p. The ratio of the amplitude x_m of the steady-state vibration to the static deflection P_m/k caused by a force P_m, or to the amplitude δ_m of the support movement, is called the *magnification factor*. From (19.33) and (19.33′), we obtain

$$\text{Magnification factor} = \frac{x_m}{P_m/k} = \frac{x_m}{\delta_m} = \frac{1}{1 - (\omega/p)^2} \tag{19.36}$$

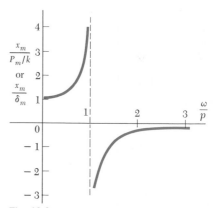

Fig. 19.9

The magnification factor has been plotted in Fig. 19.9 against the frequency ratio ω/p. We note that when $\omega = p$, the amplitude of the forced vibration becomes infinite. The impressed force or impressed support movement is said to be in *resonance* with the given system. Actually, the amplitude of the vibration remains finite because of damping forces (Sec. 19.9); nevertheless, such a situation should be avoided, and the forced frequency should not be chosen too close to the natural frequency of the system. We also note that for $\omega < p$ the coefficient of $\sin \omega t$ in (19.35) is positive, while for $\omega > p$ this coefficient is negative. In the first case the forced vibration is *in phase* with the impressed force or impressed support movement, while in the second case it is 180° *out of phase*.

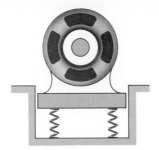

SAMPLE PROBLEM 19.5

A motor weighing 350 lb is supported by four springs, each having a constant of 750 lb/in. The unbalance of the rotor is equivalent to a weight of 1 oz located 6 in. from the axis of rotation. Knowing that the motor is constrained to move vertically, determine (a) the speed in rpm at which resonance will occur, (b) the amplitude of the vibration of the motor at a speed of 1200 rpm.

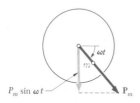

a. Resonance Speed. The resonance speed is equal to the circular frequency (in rpm) of the free vibration of the motor. The mass of the motor and the equivalent constant of the supporting springs are

$$m = \frac{350 \text{ lb}}{32.2 \text{ ft/s}^2} = 10.87 \text{ lb·s}^2/\text{ft}$$

$$k = 4(750 \text{ lb/in.}) = 3000 \text{ lb/in.} = 36{,}000 \text{ lb/ft}$$

$$p = \sqrt{\frac{k}{m}} = \sqrt{\frac{36{,}000}{10.87}} = 57.5 \text{ rad/s} = 549 \text{ rpm}$$

Resonance speed = 549 rpm ◄

b. Amplitude of Vibration at 1200 rpm. The angular velocity of the motor and the mass of the equivalent 1-oz weight are

$$\omega = 1200 \text{ rpm} = 125.7 \text{ rad/s}$$

$$m = (1 \text{ oz})\frac{1 \text{ lb}}{16 \text{ oz}}\frac{1}{32.2 \text{ ft/s}^2} = 0.00194 \text{ lb·s}^2/\text{ft}$$

The magnitude of the centrifugal force due to the unbalance of the rotor is

$$P_m = ma_n = mr\omega^2 = (0.00194 \text{ lb·s}^2/\text{ft})(\tfrac{6}{12} \text{ ft})(125.7 \text{ rad/s})^2 = 15.3 \text{ lb}$$

The static deflection that would be caused by a constant load P_m is

$$\frac{P_m}{k} = \frac{15.3 \text{ lb}}{3000 \text{ lb/in.}} = 0.00510 \text{ in.}$$

Substituting the value of P_m/k together with the known values of ω and p into Eq. (19.33), we obtain

$$x_m = \frac{P_m/k}{1 - (\omega/p)^2} = \frac{0.00510 \text{ in.}}{1 - (125.7/57.5)^2} = -0.00135 \text{ in.}$$

$x_m = 0.00135$ in. (out of phase) ◄

Note. Since $\omega > p$, the vibration is 180° out of phase with the centrifugal force due to the unbalance of the rotor. For example, when the unbalanced mass is directly below the axis of rotation, the position of the motor is $x_m = 0.00135$ in. above the position of equilibrium.

Problems

19.86 A 4-kg cylinder is suspended from a spring of constant 324 N/m and is acted upon by a vertical periodic force of magnitude $P = P_m \sin \omega t$, where $P_m = 15$ N. Determine the amplitude of the motion of the cylinder if (*a*) $\omega = 6$ rad/s, (*b*) $\omega = 12$ rad/s.

19.87 A cylinder of mass *m* is suspended from a spring of constant *k* and is acted upon by a vertical periodic force of magnitude $P = P_m \sin \omega t$. Determine the range of values of ω for which the amplitude of the vibration exceeds twice the static deflection caused by a constant force of magnitude P_m.

$P = P_m \sin \omega t$

Fig. P19.86 and P19.87

19.88 In Prob. 19.87, determine the range of values of ω for which the amplitude of the vibration is less than the static deflection caused by a constant force of magnitude P_m.

19.89 A simple pendulum of length *l* is suspended from a collar *C* which is forced to move horizontally according to the relation $x_C = \delta_m \sin \omega t$. Determine the range of values of ω for which the amplitude of the motion of the bob is less than δ_m. (Assume that δ_m is small compared with the length *l* of the pendulum.)

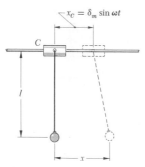

$x_C = \delta_m \sin \omega t$

C

l

x

Fig. P19.89

19.90 In Prob. 19.89, determine the range of values of ω for which the amplitude of the motion of the bob exceeds $2\delta_m$.

19.91 A 100-kg motor is supported by a light horizontal beam. The unbalance of the rotor is equivalent to a mass of 22 g located 180 mm from the axis of rotation. Knowing that the static deflection of the beam due to the weight of the motor is 5.5 mm, determine (*a*) the speed (in rpm) at which resonance will occur, (*b*) the amplitude of the steady-state vibration of the motor at a speed of 720 rpm.

Fig. P19.91

19.92 Solve Prob. 19.91, assuming that the 100-kg motor is supported by a nest of springs having a total constant of 45 kN/m.

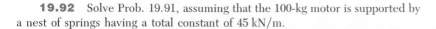

19.93 As the speed of a spring-supported motor is slowly increased from 300 to 400 rpm, the amplitude of the vibration due to the unbalance of the rotor is observed to decrease continuously from 0.075 to 0.040 in. Determine the speed at which resonance will occur.

19.94 For the spring-supported motor of Prob. 19.93, determine the speed of the motor for which the amplitude of the vibration is 0.125 in.

19.95 A motor of mass 18 kg is supported by four springs, each of constant 40 kN/m. The motor is constrained to move vertically, and the amplitude of its motion is observed to be 1.5 mm at a speed of 1200 rpm. Knowing that the mass of the rotor is 4 kg, determine the distance between the mass center of the rotor and the axis of the shaft.

19.96 In Prob. 19.95, determine the amplitude of the vertical motion of the motor at a speed of (*a*) 450 rpm, (*b*) 1600 rpm, (*c*) 900 rpm.

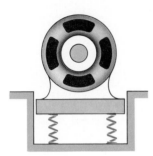

Fig. P19.95

19.97 Rod *AB* is rigidly attached to the frame of a motor running at a constant speed. When a collar of mass *m* is placed on the spring, it is observed to vibrate with an amplitude of 15 mm. When two collars, each of mass *m*, are placed on the spring, the amplitude is observed to be 18 mm. What amplitude of vibration should be expected when three collars, each of mass *m*, are placed on the spring? (Obtain two answers.)

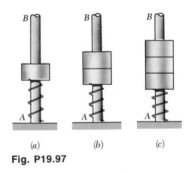

(*a*)   (*b*)   (*c*)

Fig. P19.97

19.98 Solve Prob. 19.97, assuming that the speed of the motor is changed and that one collar has an amplitude of 9 mm and two collars have an amplitude of 3 mm.

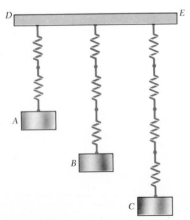

19.99 Three identical cylinders *A*, *B*, and *C* are suspended from a bar *DE* by two or more identical springs as shown. Bar *DE* is known to move vertically according to the relation $y = \delta_m \sin \omega t$. Knowing that the amplitudes of vibration of cylinders *A* and *B* are, respectively, 3 in. and 1.5 in., determine the expected amplitude of vibration of cylinder *C*. (Obtain two answers.)

19.100 Solve Prob. 19.99, assuming that the amplitudes of vibration of cylinders *A* and *B* are, respectively, 1.6 in. and 2.4 in.

Fig. P19.99

19.101 A variable-speed motor is rigidly attached to a beam *BC*. When the speed of the motor is less than 600 rpm or more than 1200 rpm, a small object placed at *A* is observed to remain in contact with the beam. For speeds between 600 and 1200 rpm the object is observed to "dance" and actually to lose contact with the beam. Determine the speed at which resonance will occur.

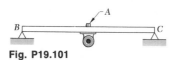

Fig. P19.101

19.102 A disk of mass *m* is attached to the midpoint of a vertical shaft *AB*, which revolves at a constant angular velocity ω. Denoting by *k* the spring constant of the system for a horizontal movement of the disk and by *e* the eccentricity of the disk with respect to the shaft, show that the deflection *r* of the center of the shaft may be written in the form

$$r = \frac{e(\omega/p)^2}{1 - (\omega/p)^2}$$

where $p = \sqrt{k/m}$.

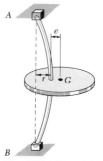

Fig. P19.102 and P19.103

19.103 A disk of mass 25 kg is attached with an eccentricity $e = 0.2$ mm to the midpoint of a vertical shaft *AB*, which revolves at a constant angular velocity ω. Knowing that a static force of 180 N will deflect the shaft 0.3 mm, determine (*a*) the angular velocity ω at which resonance occurs, (*b*) the deflection *r* of the shaft when $\omega = 1350$ rpm.

19.104 The amplitude of the motion of the pendulum bob shown is observed to be 30 mm when the amplitude of the motion of collar *C* is 10 mm. Knowing that the length of the pendulum is $l = 800$ mm, determine the two possible values of the frequency of the horizontal motion of collar *C*.

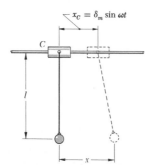

Fig. P19.104 and P19.105

19.105 A pendulum is suspended from collar *C* as shown. Knowing that the amplitude of the motion of the bob is 24 mm for $l = 600$ mm and 35 mm for $l = 400$ mm, determine the frequency and amplitude of the horizontal motion of collar *C*.

19.106 A small trailer weighing 600 lb with its load is supported by two springs, each of constant 100 lb/in. The trailer is pulled over a road, the surface of which may be approximated by a sine curve of amplitude $1\frac{1}{2}$ in. and of period 16 ft (i.e., the distance between two successive crests is 16 ft, and the vertical distance from a crest to a trough is 3 in.). Determine (*a*) the speed at which resonance will occur, (*b*) the amplitude of the vibration of the trailer at a speed of 40 mi/h.

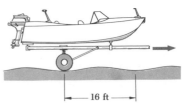

Fig. P19.106

19.107 Knowing that the amplitude of the vibration of the trailer of Prob. 19.106 is not to exceed $\frac{3}{4}$ in., determine the smallest speed at which the trailer can be pulled over the road.

DAMPED VIBRATIONS

* 19.8. Damped Free Vibrations. The vibrating systems considered in the first part of this chapter were assumed free of damping. Actually all vibrations are damped to some degree by friction forces. These forces may be caused by *dry friction*, or *Coulomb friction*, between rigid bodies, by *fluid friction* when a rigid body moves in a fluid, or by *internal friction* between the molecules of a seemingly elastic body.

A type of damping of special interest is the *viscous damping* caused by fluid friction at low and moderate speeds. Viscous damping is characterized by the fact that the friction force is *directly proportional to the speed* of the moving body. As an example, we shall consider again a body of mass m suspended from a spring of constant k, and we shall assume that the body is attached to the plunger of a dashpot (Fig. 19.10). The magnitude of the friction force exerted on the plunger by the surrounding fluid is equal to $c\dot{x}$, where the constant c, expressed in N·s/m or lb·s/ft and known as the *coefficient of viscous damping*, depends upon the physical properties of the fluid and the construction of the dashpot. The equation of motion is

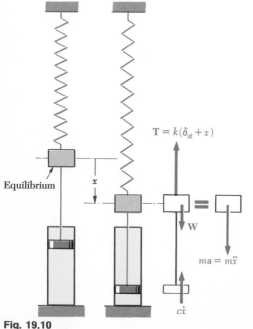

Fig. 19.10

$$+\downarrow\Sigma F = ma: \qquad W - k(\delta_{st} + x) - c\dot{x} = m\ddot{x}$$

Recalling that $W = k\delta_{st}$, we write

$$m\ddot{x} + c\dot{x} + kx = 0 \qquad (19.37)$$

Substituting $x = e^{\lambda t}$ into (19.37) and dividing through by $e^{\lambda t}$, we write the *characteristic equation*

$$m\lambda^2 + c\lambda + k = 0 \qquad (19.38)$$

and obtain the roots

$$\lambda = -\frac{c}{2m} \pm \sqrt{\left(\frac{c}{2m}\right)^2 - \frac{k}{m}} \qquad (19.39)$$

Defining the *critical damping coefficient* c_c as the value of c which makes the radical in (19.39) equal to zero, we write

$$\left(\frac{c_c}{2m}\right)^2 - \frac{k}{m} = 0 \qquad c_c = 2m\sqrt{\frac{k}{m}} = 2mp \qquad (19.40)$$

where p is the circular frequency of the system in the absence of damping. We may distinguish three different cases of damping, depending upon the value of the coefficient c.

1. *Heavy damping:* $c > c_c$. The roots λ_1 and λ_2 of the characteristic equation (19.38) are real and distinct, and the general solution of the differential equation (19.37) is

$$x = Ae^{\lambda_1 t} + Be^{\lambda_2 t} \qquad (19.41)$$

   This solution corresponds to a nonvibratory motion. Since λ_1 and λ_2 are both negative, x approaches zero as t increases indefinitely. However, the system actually regains its equilibrium position after a finite time.

2. *Critical damping:* $c = c_c$. The characteristic equation has a double root $\lambda = -c_c/2m = -p$, and the general solution of (19.37) is

$$x = (A + Bt)e^{-pt} \qquad (19.42)$$

   The motion obtained is again nonvibratory. Critically damped systems are of special interest in engineering applications since they regain their equilibrium position in the shortest possible time without oscillation.

3. *Light damping:* $c < c_c$. The roots of (19.38) are complex and conjugate, and the general solution of (19.37) is of the form

$$x = e^{-(c/2m)t}(A \sin qt + B \cos qt) \qquad (19.43)$$

where q is defined by the relation

$$q^2 = \frac{k}{m} - \left(\frac{c}{2m}\right)^2$$

Substituting $k/m = p^2$ and recalling (19.40), we write

$$q = p\sqrt{1 - \left(\frac{c}{c_c}\right)^2} \tag{19.44}$$

where the constant c/c_c is known as the *damping factor*. A substitution similar to the one used in Sec. 19.2 enables us to write the general solution of (19.37) in the form

$$x = x_m e^{-(c/2m)t} \sin{(qt + \phi)} \tag{19.45}$$

The motion defined by (19.45) is vibratory with diminishing amplitude (Fig. 19.11). Although this motion does not actually repeat itself, the time interval $\tau = 2\pi/q$, corresponding to two successive points where the curve (19.45) touches one of the limiting curves shown in Fig. 19.11, is commonly referred to as the period of the damped vibration. Recalling (19.44), we observe that τ is larger than the period of vibration of the corresponding undamped system.

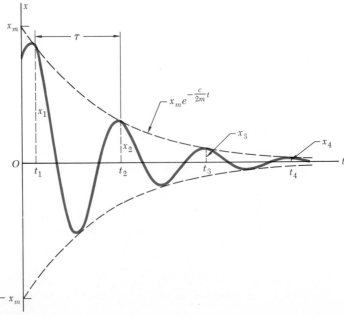

Fig. 19.11

* **19.9. Damped Forced Vibrations.** If the system considered in the preceding section is subjected to a periodic force **P** of magnitude $P = P_m \sin \omega t$, the equation of motion becomes

$$m\ddot{x} + c\dot{x} + kx = P_m \sin \omega t \qquad (19.46)$$

The general solution of (19.46) is obtained by adding a particular solution of (19.46) to the complementary function or general solution of the homogeneous equation (19.37). The complementary function is given by (19.41), (19.42), or (19.43), depending upon the type of damping considered. It represents a *transient* motion which is eventually damped out.

Our interest in this section is centered on the steady-state vibration represented by a particular solution of (19.46) of the form

$$x_{\text{part}} = x_m \sin (\omega t - \varphi) \qquad (19.47)$$

Substituting x_{part} for x into (19.46), we obtain

$$-m\omega^2 x_m \sin (\omega t - \varphi) + c\omega x_m \cos (\omega t - \varphi) + kx_m \sin (\omega t - \varphi) = P_m \sin \omega t$$

Making $\omega t - \varphi$ successively equal to 0 and to $\pi/2$, we write

$$c\omega x_m = P_m \sin \varphi \qquad (19.48)$$

$$(k - m\omega^2)x_m = P_m \cos \varphi \qquad (19.49)$$

Squaring both members of (19.48) and (19.49) and adding, we have

$$[(k - m\omega^2)^2 + (c\omega)^2]x_m^2 = P_m^2 \qquad (19.50)$$

Solving (19.50) for x_m and dividing (19.48) and (19.49) member by member, we obtain, respectively,

$$x_m = \frac{P_m}{\sqrt{(k - m\omega^2)^2 + (c\omega)^2}} \qquad \tan \varphi = \frac{c\omega}{k - m\omega^2} \qquad (19.51)$$

Recalling from (19.4) that $k/m = p^2$, where p is the circular frequency of the undamped free vibration, and from (19.40) that $2mp = c_c$, where c_c is the critical damping coefficient of the system, we write

$$\frac{x_m}{P_m/k} = \frac{x_m}{\delta_m} = \frac{1}{\sqrt{[1 - (\omega/p)^2]^2 + [2(c/c_c)(\omega/p)]^2}} \qquad (19.52)$$

$$\tan \varphi = \frac{2(c/c_c)(\omega/p)}{1 - (\omega/p)^2} \qquad (19.53)$$

Formula (19.52) expresses the magnification factor in terms of the frequency ratio ω/p and damping factor c/c_c. It may be used to determine the amplitude of the steady-state vibration produced by an impressed force of magnitude $P = P_m \sin \omega t$ or by an impressed support movement $\delta = \delta_m \sin \omega t$. Formula (19.53) defines in terms of the same parameters the *phase difference* φ between the impressed force or impressed support movement and the resulting steady-state vibration of the damped system. The magnification factor has been plotted against the frequency ratio in Fig. 19.12 for various values of the damping factor. We observe that the amplitude of a forced vibration may be kept small by choosing a large coefficient of viscous damping c or by keeping the natural and forced frequencies far apart.

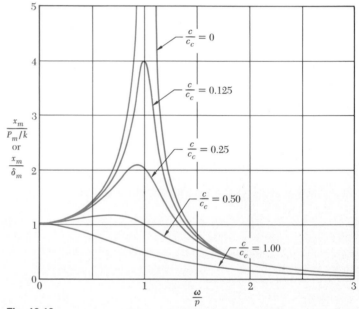

Fig. 19.12

---

* 19.10. Electrical Analogues. Oscillating electrical circuits are characterized by differential equations of the same type as those obtained in the preceding sections. Their analysis is therefore similar to that of a mechanical system, and the results obtained for a given vibrating system may be readily extended to the equivalent circuit. Conversely, any result obtained for an electrical circuit will also apply to the corresponding mechanical system.

Consider an electrical circuit consisting of an inductor of inductance L, a resistor of resistance R, and a capacitor of capacitance C, connected in series with a source of alternating voltage $E = E_m \sin \omega t$ (Fig. 19.13). It is recalled from elementary circuit theory† that if i denotes the current in the circuit and q the electric charge on the capacitor, the drop in potential is $L(di/dt)$ across the inductor, Ri across the resistor, and q/C across the capacitor. Expressing that the algebraic sum of the applied voltage and of the drops in potential around the circuit loop is zero, we write

$$E_m \sin \omega t - L\frac{di}{dt} - Ri - \frac{q}{C} = 0 \qquad (19.54)$$

Rearranging the terms and recalling that at any instant the current i is equal to the rate of change $\dot{q}$ of the charge q, we have

$$L\ddot{q} + R\dot{q} + \frac{1}{C}q = E_m \sin \omega t \qquad (19.55)$$

We verify that Eq. (19.55), which defines the oscillations of the electrical circuit of Fig. 19.13, is of the same type as Eq. (19.46), which characterizes the damped forced vibrations of the mechanical system of Fig. 19.10. By comparing the two equations, we may construct a table of the analogous mechanical and electrical expressions.

Table 19.2 may be used to extend to their electrical analogues the results obtained in the preceding sections for various mechanical systems. For instance, the amplitude i_m of the current in the circuit of Fig. 19.13 may be obtained by noting that it corresponds to the maximum value v_m of the velocity in the analogous mechanical system. Recalling that $v_m = \omega x_m$,

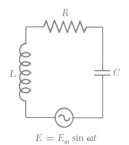

$E = E_m \sin \omega t$

Fig. 19.13

Table 19.2 Characteristics of a Mechanical System and of Its Electrical Analogue

Mechanical System		Electrical Circuit	
m	Mass	L	Inductance
c	Coefficient of viscous damping	R	Resistance
k	Spring constant	$1/C$	Reciprocal of capacitance
x	Displacement	q	Charge
v	Velocity	i	Current
P	Applied force	E	Applied voltage

†See A. E. Fitzgerald, D. E. Higginbotham, and A. Grabel, *Basic Electrical Engineering*, 5th ed., McGraw-Hill, New York, 1981; or J. W. Nilsson, *Electric Circuits*, Addison-Wesley, Reading, Mass., 1983.

substituting for x_m from Eq. (19.51), and replacing the constants of the mechanical system by the corresponding electrical expressions, we have

$$i_m = \frac{\omega E_m}{\sqrt{\left(\dfrac{1}{C} - L\omega^2\right)^2 + (R\omega)^2}}$$

$$i_m = \frac{E_m}{\sqrt{R^2 + \left(L\omega - \dfrac{1}{C\omega}\right)^2}} \qquad (19.56)$$

The radical in the expression obtained is known as the *impedance* of the electrical circuit.

The analogy between mechanical systems and electrical circuits holds for transient as well as steady-state oscillations. The oscillations of the circuit shown in Fig. 19.14, for instance, are analogous to the damped free vibrations of the system of Fig. 19.10. As far as the initial conditions are concerned, we may note that closing the switch S when the charge on the capacitor is $q = q_0$ is equivalent to releasing the mass of the mechanical system with no initial velocity from the position $x = x_0$. We should also observe that if a battery of constant voltage E is introduced in the electrical circuit of Fig. 19.14, closing the switch S will be equivalent to suddenly applying a force of constant magnitude P to the mass of the mechanical system of Fig. 19.10.

The discussion above would be of questionable value if its only result were to make it possible for mechanics students to analyze electrical circuits without learning the elements of circuit theory. It is hoped that this discussion will instead encourage students to apply to the solution of problems in mechanical vibrations the mathematical techniques they may learn in later courses in circuit theory. The chief value of the concept of electrical analogue, however, resides in its application to *experimental methods* for the determination of the characteristics of a given mechanical system. Indeed, an electrical circuit is much more easily constructed than is a mechanical model, and the fact that its characteristics may be modified by varying the inductance, resistance, or capacitance of its various components makes the use of the electrical analogue particularly convenient.

To determine the electrical analogue of a given mechanical system, we shall focus our attention on each moving mass in the system and observe which springs, dashpots, or external forces are applied directly to it. An equivalent electrical loop may then be constructed to match each of the mechanical units thus defined; the various loops obtained in that way will form together the desired circuit. Consider, for instance, the mechanical system of Fig. 19.15. We observe that the mass m_1 is acted upon by two springs of constants k_1 and k_2 and by two dashpots characterized by the coefficients of viscous damping c_1 and c_2. The electrical circuit should therefore include a loop consisting of an inductor of inductance L_1 propor-

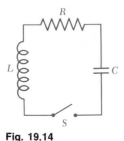

Fig. 19.14

tional to m_1, of two capacitors of capacitance C_1 and C_2 inversely proportional to k_1 and k_2, respectively, and of two resistors of resistance R_1 and R_2, proportional to c_1 and c_2, respectively. Since the mass m_2 is acted upon by the spring k_2 and the dashpot c_2, as well as by the force $P = P_m \sin \omega t$, the circuit should also include a loop containing the capacitor C_2, the resistor R_2, the new inductor L_2, and the voltage source $E = E_m \sin \omega t$ (Fig. 19.16).

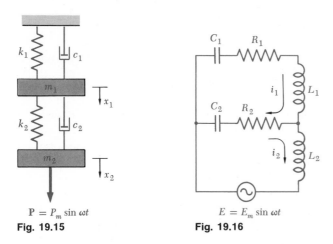

$$P = P_m \sin \omega t \qquad\qquad E = E_m \sin \omega t$$

Fig. 19.15   **Fig. 19.16**

To check that the mechanical system of Fig. 19.15 and the electrical circuit of Fig. 19.16 actually satisfy the same differential equations, we shall first derive the equations of motion for m_1 and m_2. Denoting, respectively, by x_1 and x_2 the displacements of m_1 and m_2 from their equilibrium positions, we observe that the elongation of the spring k_1 (measured from the equilibrium position) is equal to x_1, while the elongation of the spring k_2 is equal to the relative displacement $x_2 - x_1$ of m_2 with respect to m_1. The equations of motion for m_1 and m_2 are therefore

$$m_1\ddot{x}_1 + c_1\dot{x}_1 + c_2(\dot{x}_1 - \dot{x}_2) + k_1 x_1 + k_2(x_1 - x_2) = 0 \qquad (19.57)$$

$$m_2\ddot{x}_2 + c_2(\dot{x}_2 - \dot{x}_1) + k_2(x_2 - x_1) = P_m \sin \omega t \qquad (19.58)$$

Consider now the electrical circuit of Fig. 19.16; we denote, respectively, by i_1 and i_2 the current in the first and second loops, and by q_1 and q_2 the integrals $\int i_1\, dt$ and $\int i_2\, dt$. Noting that the charge on the capacitor C_1 is q_1, while the charge on C_2 is $q_1 - q_2$, we express that the sum of the potential differences in each loop is zero:

$$L_1\ddot{q}_1 + R_1\dot{q}_1 + R_2(\dot{q}_1 - \dot{q}_2) + \frac{q_1}{C_1} + \frac{q_1 - q_2}{C_2} = 0 \qquad (19.59)$$

$$L_2\ddot{q}_2 + R_2(\dot{q}_2 - \dot{q}_1) + \frac{q_2 - q_1}{C_2} = E_m \sin \omega t \qquad (19.60)$$

We easily check that Eqs. (19.59) and (19.60) reduce to (19.57) and (19.58), respectively, when the substitutions indicated in Table 19.2 are performed.

Problems

19.108 Show that in the case of heavy damping $(c > c_c)$, a body never passes through its position of equilibrium O (a) if it is released with no initial velocity from an arbitrary position or (b) if it is started from O with an arbitrary initial velocity.

19.109 Show that in the case of heavy damping $(c > c_c)$, a body released from an arbitrary position with an arbitrary initial velocity cannot pass more than once through its equilibrium position.

19.110 In the case of light damping, the displacements x_1, x_2, x_3, and so forth, shown in Fig. 19.11 may be assumed equal to the maximum displacements. Show that the ratio of any two successive maximum displacements x_n and x_{n+1} is a constant and that the natural logarithm of this ratio, called the *logarithmic decrement*, is

$$\ln \frac{x_n}{x_{n+1}} = \frac{2\pi(c/c_c)}{\sqrt{1 - (c/c_c)^2}}$$

19.111 In practice, it is often difficult to determine the logarithmic decrement of a system with light damping defined in Prob. 19.110 by measuring two successive maximum displacements. Show that the logarithmic decrement may also be expressed as $(1/k) \ln (x_n/x_{n+k})$, where k is the number of cycles between readings of the maximum displacement.

19.112 In a system with light damping $(c < c_c)$, the period of vibration is commonly defined as the time interval $\tau = 2\pi/q$ corresponding to two successive points where the displacement-time curve touches one of the limiting curves shown in Fig. 19.11. Show that the interval of time (a) between a maximum positive displacement and the following maximum negative displacement is $\frac{1}{2}\tau$, (b) between two successive zero displacements is $\frac{1}{2}\tau$, (c) between a maximum positive displacement and the following zero displacement is greater than $\frac{1}{4}\tau$.

19.113 Successive maximum displacements of a spring-mass-dashpot system similar to that shown in Fig. 19.10 are 64, 48, 36, and 27 mm. Knowing that $m = 15$ kg and $k = 1800$ N/m, determine (a) the damping factor c/c_c, (b) the value of the coefficient of viscous damping c. (*Hint.* See Probs. 19.110 and 19.111.)

19.114 The block shown is depressed 30 mm from its equilibrium position and released. Knowing that after 10 cycles the maximum displacement of the block is 18 mm, determine (a) the damping factor c/c_c, (b) the value of the coefficient of viscous damping. (*Hint.* See Probs. 19.110 and 19.111.)

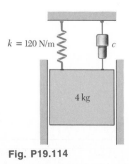

$k = 120$ N/m

c

4 kg

Fig. P19.114

19.115 The barrel of a field gun weighs 1200 lb and is returned into firing position after recoil by a recuperator of constant $k = 8000$ lb/ft. Determine (a) the value of the coefficient of damping of the recuperator which causes the barrel to return into firing position in the shortest possible time without oscillation, (b) the time needed for the barrel to move from its maximum-recoil position halfway back to its firing position.

19.116 Assuming that the barrel of the gun of Prob. 19.115 is modified, with a resulting increase in weight of 300 lb, determine (a) the constant k which should be used for the recuperator if the barrel is to remain critically damped, (b) the time needed for the modified barrel to move from its maximum-recoil position halfway back to its firing position.

19.117 In the case of the forced vibration of a system with a given damping factor c/c_c, determine the frequency ratio ω/p for which the amplitude of the vibration is maximum.

19.118 Show that for a small value of the damping factor c/c_c, (a) the maximum amplitude of a forced vibration occurs when $\omega \approx p$, (b) the corresponding value of the magnification factor is approximately $\frac{1}{2}(c_c/c)$.

19.119 A 35-lb motor is directly supported by a light horizontal beam which has a static deflection of 0.06 in. due to the weight of the motor. Knowing that the unbalance of the rotor is equivalent to a weight of 1 oz located 5 in. from the axis of rotation, determine the amplitude of vibration of the motor at a speed of 800 rpm, assuming (a) that no damping is present, (b) that the damping factor c/c_c is equal to 0.065.

19.120 A motor weighing 50 lb is supported by four springs, each having a constant of 1000 lb/in. The unbalance of the rotor is equivalent to a weight of 1 oz located 5 in. from the axis of rotation. Knowing that the motor is constrained to move vertically, determine the amplitude of the steady-state vibration of the motor at a speed of 1800 rpm, assuming (a) that no damping is present, (b) that the damping factor c/c_c is equal to 0.125.

19.121 Solve Prob. 19.95, assuming that a dashpot having a coefficient of damping $c = 350$ N·s/m has been connected to the motor and to the ground.

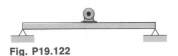

Fig. P19.122

19.122 A 50-kg motor is directly supported by a light horizontal beam which has a static deflection of 6 mm due to the weight of the motor. The unbalance of the rotor is equivalent to a mass of 100 g located 75 mm from the axis of rotation. Knowing that the amplitude of the vibration of the motor is 0.8 mm at a speed of 400 rpm, determine (a) the damping factor c/c_c, (b) the coefficient of damping c.

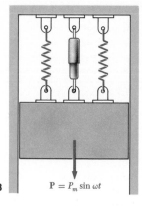

Fig. P19.123

$P = P_m \sin \omega t$

19.123 A machine element having a mass of 500 kg is supported by two springs, each having a constant of 48 kN/m. A periodic force of maximum magnitude equal to 150 N is applied to the element with a frequency of 2.4 Hz. Knowing that the coefficient of damping is 1600 N · s/m, determine the amplitude of the steady-state vibration of the element.

19.124 In Prob. 19.123, determine the required value of the coefficient of damping if the amplitude of the steady-state vibration of the element is to be 4 mm.

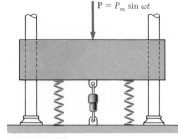

$P = P_m \sin \omega t$

Fig. P19.125

19.125 A platform of weight 200 lb, supported by two springs each of constant $k = 250$ lb/in., is subjected to a periodic force of maximum magnitude equal to 125 lb. Knowing that the coefficient of damping is 9 lb · s/in., determine (*a*) the natural frequency in rpm of the platform if there were no damping, (*b*) the frequency in rpm of the periodic force corresponding to the maximum value of the magnification factor, assuming damping, (*c*) the amplitude of the actual motion of the platform for each of the frequencies found in parts *a* and *b*.

19.126 Solve Prob. 19.125, assuming that the coefficient of damping is increased to 12 lb · s/in.

*19.127 The suspension of an automobile may be approximated by the simplified spring-and-dashpot system shown. (*a*) Write the differential equation defining the absolute motion of the mass *m* when the system moves at a speed *v* over a road of sinusoidal cross section as shown. (*b*) Derive an expression for the amplitude of the absolute motion of *m*.

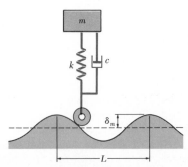

Fig. P19.127

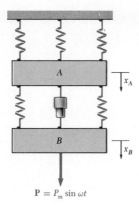

Fig. P19.128

*19.128 Two blocks A and B, each of mass m, are suspended as shown by means of five springs of the same constant k and are connected by a dashpot of coefficient of damping c. Block B is subjected to a force of magnitude $P = P_m \sin \omega t$. Write the differential equations defining the displacements x_A and x_B of the two blocks from their equilibrium positions.

19.129 Determine the range of values of the resistance R for which oscillations will take place in the circuit shown when switch S is closed.

19.130 Consider the circuit of Prob. 19.129 when the capacitor C is removed. If switch S is closed at time $t = 0$, determine (*a*) the final value of the current in the circuit, (*b*) the time t at which the current will have reached $(1 - 1/e)$ times its final value. (The desired value of t is known as the *time constant* of the circuit.)

19.131 and 19.132 Draw the electrical analogue of the mechanical system shown. (*Hint.* Draw the loops corresponding to the free bodies m and A.)

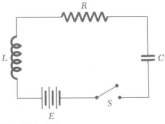

Fig. P19.129

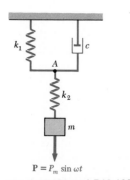

$P = P_m \sin \omega t$

Fig. P19.131 and P19.133

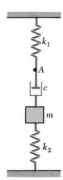

Fig. P19.132 and P19.134

19.133 and 19.134 Write the differential equations defining (*a*) the displacements of mass m and point A, (*b*) the currents in the corresponding loops of the electrical analogue.

Review and Summary

This chapter was devoted to the study of *mechanical vibrations*, i.e., to the analysis of the motion of particles and rigid bodies oscillating about a position of equilibrium. In the first part of the chapter [Secs. 19.2 through 19.7], we considered *vibrations without damping*, while the second part was devoted to *damped vibrations* [Secs. 19.8 through 19.10].

Free vibrations of a particle

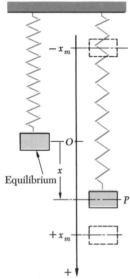

Fig. 19.17

In Sec. 19.2, we considered the *free vibrations of a particle*, i.e., the motion of a particle P subjected to a restoring force proportional to the displacement of the particle—such as the force exerted by a spring. If the displacement x of the particle P is measured from its equilibrium position O (Fig. 19.17), the resultant $\mathbf{F}$ of the forces acting on P (including its weight) has a magnitude kx and is directed toward O. Applying Newton's second law $F = ma$ and recalling that $a = \ddot{x}$, we wrote the differential equation

$$m\ddot{x} + kx = 0 \tag{19.3}$$

or, setting $p^2 = k/m$,

$$\ddot{x} + p^2 x = 0 \tag{19.5}$$

The motion defined by this equation is called a *simple harmonic motion*.
The solution of Eq. (19.5), which represents the displacement of the particle P, was expressed as

$$x = x_m \sin{(pt + \phi)} \tag{19.10}$$

where x_m = amplitude of the vibration
$p = \sqrt{k/m}$ = circular frequency
ϕ = phase angle

The *period of the vibration* (i.e., the time required for a full cycle) and its *frequency* (i.e., the number of cycles per second) were expressed as

$$\text{Period} = \tau = \frac{2\pi}{p} \tag{19.13}$$

$$\text{Frequency} = f = \frac{1}{\tau} = \frac{p}{2\pi} \tag{19.14}$$

The velocity and acceleration of the particle were obtained by differentiating Eq. (19.10), and their maximum values were found to be

$$v_m = x_m p \qquad a_m = x_m p^2 \tag{19.15}$$

Since all the above parameters depend directly upon the circular frequency p and thus upon the ratio k/m, it is essential in any given problem to calculate the value of the constant k; this may be done by determining the relation between the restoring force and the corresponding displacement of the particle [Sample Prob. 19.1].

It was also shown that the oscillatory motion of the particle P may be represented by the projection on the x axis of the motion of a point Q describing an auxiliary circle of radius x_m with the constant angular velocity p (Fig. 19.18). The instantaneous values of the velocity and accelera-

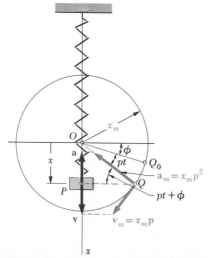

Fig. 19.18

tion of P may then be obtained by projecting on the x axis the vectors $\mathbf{v}_m$ and $\mathbf{a}_m$ representing, respectively, the velocity and acceleration of Q.

Simple pendulum

While the motion of a *simple pendulum* is not truly a simple harmonic motion, the formulas given above may be used with $p^2 = g/l$ to calculate the period and frequency of the *small oscillations* of a simple pendulum [Sec. 19.3]. Large-amplitude oscillations of a simple pendulum were discussed in Sec. 19.4.

Free vibrations of a rigid body

The *free vibrations of a rigid body* may be analyzed by choosing an appropriate variable, such as a distance x or an angle θ, to define the position of the body, drawing a diagram expressing the equivalence of the external and effective forces, and writing an equation relating the selected variable and its second derivative [Sec. 19.5]. If the equation obtained is of the form

$$\ddot{x} + p^2 x = 0 \quad \text{or} \quad \ddot{\theta} + p^2 \theta = 0 \quad (19.21)$$

the vibration considered is a simple harmonic motion and its period and frequency may be obtained *by identifying p* and substituting in (19.13) and (19.14) [Sample Probs. 19.2 and 19.3].

Using the principle of
conservation of energy

The *principle of conservation of energy* may be used as an alternative method for the determination of the period and frequency of the simple harmonic motion of a particle or rigid body [Sec. 19.6]. Choosing again an appropriate variable, such as θ, to define the position of the system, we express that the total energy of the system is conserved, $T_1 + V_1 = T_2 + V_2$, between the position of maximum displacement ($\theta_1 = \theta_m$) and the position of maximum velocity ($\dot{\theta}_2 = \dot{\theta}_m$). If the motion considered is simple harmonic, the two members of the equation obtained consist of homogeneous quadratic expressions in θ_m and $\dot{\theta}_m$, respectively.† Substituting $\dot{\theta}_m = \theta_m p$ in this equation, we may factor out θ_m^2 and solve for the circular frequency p [Sample Prob. 19.4].

Forced vibrations

In Sec. 19.7, we considered the *forced vibrations* of a mechanical system. These vibrations occur when the system is subjected to a periodic force (Fig. 19.19) or when it is elastically connected to a support which has an alternating motion (Fig. 19.20). In the first case, the motion of the system is defined by the differential equation

$$m\ddot{x} + kx = P_m \sin \omega t \qquad (19.30)$$

and in the second case by the differential equation

$$m\ddot{x} + kx = k\delta_m \sin \omega t \qquad (19.31)$$

The general solution of these equations is obtained by adding a particular

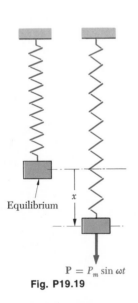

$$\mathbf{P} = P_m \sin \omega t$$

Fig. P19.19

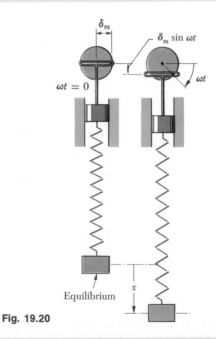

Fig. 19.20

†If the motion considered may only be *approximated* by a simple harmonic motion, such as for the small oscillations of a body under gravity, the potential energy must be approximated by a quadratic expression in θ_m.

solution of the form

$$x_{\text{part}} = x_m \sin \omega t \qquad (19.32)$$

to the general solution of the corresponding homogeneous equation. The particular solution (19.32) represents a *steady-state vibration* of the system, while the solution of the homogeneous equation represents a *transient free vibration* which may generally be neglected.

Dividing the amplitude x_m of the steady-state vibration by P_m/k in the case of a periodic force, or by δ_m in the case of an oscillating support, we defined the *magnification factor* of the vibration and found that

$$\text{Magnification factor} = \frac{x_m}{P_m/k} = \frac{x_m}{\delta_m} = \frac{1}{1 - (\omega/p)^2} \qquad (19.36)$$

According to Eq. (19.36), the amplitude x_m of the forced vibration *becomes infinite when $\omega = p$*, i.e., *when the forced frequency is equal to the natural frequency of the system*. The impressed force or impressed support movement is then said to be in *resonance* with the system [Sample Prob. 19.5]. Actually the amplitude of the vibration remains finite, due to damping forces.

In the last part of the chapter, we considered the *damped vibrations* of a mechanical system. First, we analyzed the *damped free vibrations* of a system with *viscous damping* [Sec. 19.8]. We found that the motion of such a system was defined by the differential equation

$$m\ddot{x} + c\dot{x} + kx = 0 \qquad (19.37)$$

where c is a constant called the *coefficient of viscous damping*. Defining the *critical damping coefficient* c_c as

$$c_c = 2m\sqrt{\frac{k}{m}} = 2mp \qquad (19.40)$$

where p is the circular frequency of the system in the absence of damping, we distinguished three different cases of damping, namely, (1) *heavy damping* when $c > c_c$, (2) *critical damping* when $c = c_c$, and (3) *light damping* when $c < c_c$. In the first two cases, the system when disturbed tends to regain its equilibrium position without any oscillation. In the third case, the motion is vibratory with diminishing amplitude.

In Sec. 19.9, we considered the *damped forced vibrations* of a mechanical system. These vibrations occur when a system with viscous damping is subjected to a periodic force **P** of magnitude $P = P_m \sin \omega t$ or when it is elastically connected to a support with an alternating motion $\delta = \delta_m \sin \omega t$. In the first case, the motion of the system was defined by the differential equation

$$m\ddot{x} + c\dot{x} + kx = P_m \sin \omega t \qquad (19.46)$$

and in the second case by a similar equation obtained by replacing P_m by $k\delta_m$ in (19.46).

Damped free vibrations

Damped forced vibrations

The *steady-state vibration* of the system is represented by a particular solution of Eq. (19.46) of the form

$$x_{part} = x_m \sin(\omega t - \varphi) \qquad (19.47)$$

Dividing the amplitude x_m of the steady-state vibration by P_m/k in the case of a periodic force, or by δ_m in the case of an oscillating support, we obtained the following expression for the magnification factor:

$$\frac{x_m}{P_m/k} = \frac{x_m}{\delta_m} = \frac{1}{\sqrt{[1 - (\omega/p)^2]^2 + [2(c/c_c)(\omega/p)]^2}} \qquad (19.52)$$

where $p = \sqrt{k/m}$ = natural circular frequency of undamped system
$c_c = 2mp$ = critical damping coefficient
c/c_c = damping factor

We also found that the *phase difference* φ between the impressed force or support movement and the resulting steady-state vibration of the damped system was defined by the relation

$$\tan \varphi = \frac{2(c/c_c)(\omega/p)}{1 - (\omega/p)^2} \qquad (19.53)$$

Electrical analogues

The chapter ended with a discussion of *electrical analogues* [Sec. 19.10], in which it was shown that the vibrations of mechanical systems and the oscillations of electrical circuits are defined by the same differential equations. Electrical analogues of mechanical systems may therefore be used to study or predict the behavior of these systems.

Review Problems

19.135 Disks A and B are of mass 5 kg and 14 kg, respectively, and a small block C of mass 1.5 kg is attached to the rim of disk B. Assuming that no slipping occurs between the disks, determine the period of small oscillations of the system.

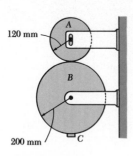

120 mm

200 mm

Fig. P19.135

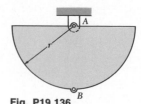

Fig. P19.136

19.136 A uniform semicircular disk of radius r is suspended from a hinge. Determine the period of small oscillations if the disk (*a*) is suspended from A as shown, (*b*) is suspended from point B.

19.137 A slender bar of length l is attached by a pin A to a collar of negligible mass. Determine the period of small oscillations of the bar, assuming that the coefficient of friction between the collar and the horizontal rod (*a*) is sufficient to prevent any movement of the collar, (*b*) is zero.

Fig. P19.137

19.138 A 150-kg electromagnet is at rest and is holding 100 kg of scrap steel when the current is turned off and the steel is dropped. Knowing that the cable and the supporting crane have a total stiffness equivalent to a spring of constant 200 kN/m, determine (*a*) the frequency, the amplitude, and the maximum velocity of the resulting motion, (*b*) the minimum tension which will occur in the cable during the motion, (*c*) the velocity of the magnet 0.03 s after the current is turned off.

Fig. P19.138

19.139 A 6-lb block is spring-mounted on the frame of a motor rotating at 1800 rpm. The rotor is unbalanced and the amplitude of the resulting motion of the block is measured and found to be 0.35 in. Knowing that the constant of the spring mounting is $k = 450$ lb/in., determine the amplitude of the motion of the motor.

19.140 A rod of mass m and length L rests on two pulleys A and B which rotate in the directions shown. Denoting by μ_k the coefficient of kinetic friction between the rod and the pulleys, determine the frequency of vibration if the rod is given a small displacement to the right and released.

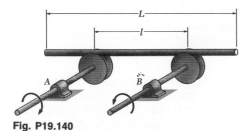

Fig. P19.140

19.141 A particle is placed with no initial velocity on a frictionless *plane* tangent to the surface of the earth. (*a*) Show that the particle will theoretically execute simple harmonic motion with a period of oscillation equal to that of a simple pendulum of length equal to the radius of the earth. (*b*) Compute the theoretical period of oscillation and show that it is equal to the periodic time of an earth satellite describing a low-altitude circular orbit. [*Hint*. See Eq. (12.44).]

Fig. P19.142

19.142 An automobile wheel-and-tire assembly of total weight 47 lb is attached to a mounting plate of negligible weight which is suspended from a steel wire. The torsional spring constant of the wire is known to be $K = 0.40$ lb·in./rad. The wheel is rotated through 90° about the vertical and then released. Knowing that the period of oscillation is observed to be 30 s, determine the centroidal mass moment of inertia and the centroidal radius of gyration of the wheel-and-tire assembly.

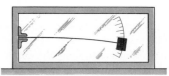

Fig. P19.143

19.143 A 15-kg beam is supported as shown by two uniform disks, each of mass 8 kg and radius 80 mm. Knowing that the disks roll without sliding, determine the period of vibration of the system if the beam is given a small displacement to the right and released.

19.144 Solve Prob. 19.143, assuming that the spring attached to the beam is removed.

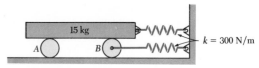

Fig. P19.145

19.145 A certain vibrometer used to measure vibration amplitudes consists essentially of a box containing a slender rod to which a mass m is attached; the natural frequency of the mass-rod system is known to be 5 Hz. When the box is rigidly attached to the casing of a motor rotating at 600 rpm, the mass is observed to vibrate with an amplitude of 0.06 in. relative to the box. Determine the amplitude of the vertical motion of the motor.

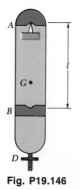

Fig. P19.146

19.146 If either a simple or a compound pendulum is used to determine experimentally the acceleration of gravity g, difficulties are encountered. In the simple pendulum, the string is not truly weightless, while in the compound pendulum, the exact location of the mass center is difficult to establish. In the case of a compound pendulum, the difficulty may be eliminated by using a reversible, or Kater, pendulum. Two knife-edges A and B are placed so that they are obviously not at the same distance from the mass center G, and the distance l is measured with great precision. The position of a counterweight D is then adjusted so that the period of oscillation τ is the same when either knife-edge is used. Show that the period τ obtained is equal to that of a true simple pendulum of length l and that $g = 4\pi^2 l/\tau^2$.

The following problems are designed to be solved with a computer.

19.C1 By expanding the integrand in Eq. (19.19) into a series of even powers of sin ϕ and integrating, it may be shown that the period of a simple pendulum of length l may be approximated by the expression

$$\tau = 2\pi\sqrt{\frac{l}{g}}\left[1 + \left(\frac{1}{2}\right)^2 c^2 + \left(\frac{1 \times 3}{2 \times 4}\right)^2 c^4 + \left(\frac{1 \times 3 \times 5}{2 \times 4 \times 6}\right)^2 c^6 + \cdots\right]$$

where $c = \sin\frac{1}{2}\theta_m$ and θ_m is the amplitude of the oscillations. Write a computer program and use it to calculate the sum of the series in brackets using successively 1, 2, 4, 8, and 16 terms for values of the amplitude θ_m from 30 to 120° at 30° intervals. Express the results with five significant figures.

19.C2 The 3-kg uniform rod AB is attached as shown to a spring of constant $k = 900$ N/m. A small block C of mass m_C is placed on the rod at end A which is then moved down through a small distance δ_0 and released. Write a computer program and use it to calculate the period of vibration and the largest allowable value of δ_0 for which block C will remain at all times in contact with the rod, for values of m_C from 0 to 2 kg at 0.5-kg intervals when (a) $b = 0.5$ m, (b) $b = 0.75$ m, (c) $b = 1$ m.

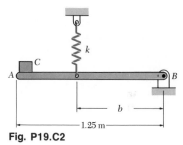

Fig. P19.C2

19.C3 For the motor of Sample Prob. 19.5, write a computer program and use it to determine the amplitude of the vibration and the maximum acceleration of the motor base for values of the motor speed from 300 to 1200 rpm at 50-rpm intervals.

19.C4 Solve Prob. 19.C3, assuming that a dashpot having a coefficient of damping $c = 75$ lb·s/ft has been connected to the motor base and to the ground. Also determine the phase difference between the motion of the motor base and that of the rotor for the various values of the motor speed considered.

Some Useful Definitions and Properties of Vector Algebra

The following definitions and properties of vector algebra were discussed fully in Chaps. 2 and 3 of *Vector Mechanics for Engineers: Statics.* They are summarized here for the convenience of the reader, with references to the appropriate sections of the *Statics* volume. Equation and illustration numbers are those used in the original presentation.

A.1. Addition of Vectors (Secs. 2.3 and 2.4). Vectors are defined as *mathematical expressions possessing magnitude and direction, which add according to the parallelogram law.* Thus the sum of two vectors **P** and **Q** is obtained by attaching the two vectors to the same point A and constructing a parallelogram, using **P** and **Q** as two sides of the parallelogram (Fig. 2.6). The diagonal that passes through A represents the sum of the vectors **P** and **Q**, and this sum is denoted by **P** + **Q**. Vector addition is *associative* and *commutative.*

The *negative vector* of a given vector **P** is defined as a vector having the same magnitude P and a direction opposite to that of **P** (Fig. 2.5); the negative of the vector **P** is denoted by −**P**. Clearly, we have

$$\mathbf{P} + (-\mathbf{P}) = 0$$

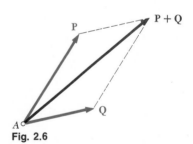

Fig. 2.6

Fig. 2.5

A.2. Product of a Scalar and a Vector (Sec. 2.4). The product $k\mathbf{P}$ of a scalar k and a vector $\mathbf{P}$ is defined as a vector having the same direction as $\mathbf{P}$ (if k is positive), or a direction opposite to that of $\mathbf{P}$ (if k is negative), and a magnitude equal to the product of the magnitude P and of the absolute value of k (Fig. 2.13).

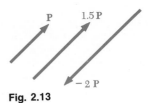

Fig. 2.13

A.3. Unit Vectors. Resolution of a Vector into Rectangular Components (Secs. 2.7 and 2.12). The vectors $\mathbf{i}$, $\mathbf{j}$, and $\mathbf{k}$, called *unit vectors*, are defined as vectors of magnitude 1, directed, respectively, along the positive x, y, and z axes (Fig. 2.32).

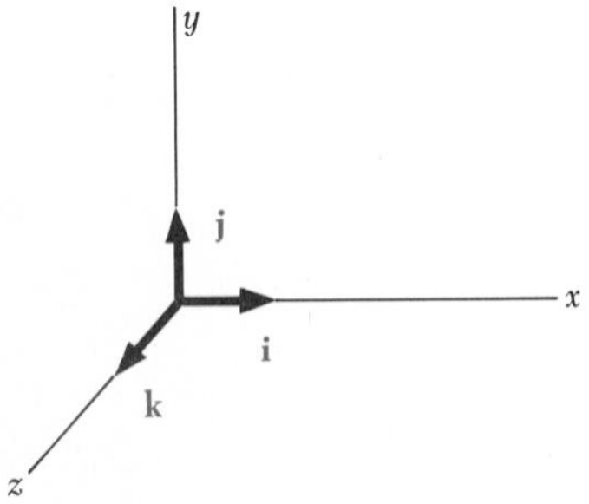

Denoting by F_x, F_y, and F_z the scalar components of a vector $\mathbf{F}$, we have (Fig. 2.33)

$$\mathbf{F} = F_x\mathbf{i} + F_y\mathbf{j} + F_z\mathbf{k} \tag{2.20}$$

In the particular case of a unit vector $\boldsymbol{\lambda}$ directed along a line forming angles θ_x, θ_y, and θ_z with the coordinate axes, we have

$$\boldsymbol{\lambda} = \cos\theta_x\mathbf{i} + \cos\theta_y\mathbf{j} + \cos\theta_z\mathbf{k} \tag{2.22}$$

Fig. 2.32

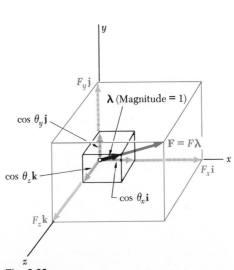

Fig. 2.33

A.4. Vector Product of Two Vectors (Secs. 3.4 and 3.5).

The vector product, or *cross product*, of two vectors **P** and **Q** is defined as the vector

$$V = P \times Q$$

which satisfies the following conditions:

1. The line of action of **V** is perpendicular to the plane containing **P** and **Q** (Fig. 3.6).
2. The magnitude of **V** is the product of the magnitudes of **P** and **Q** and of the sine of the angle θ formed by **P** and **Q** (the measure of which will always be 180° or less); we thus have

$$V = PQ \sin \theta \tag{3.1}$$

3. The sense of **V** is such that a person located at the tip of **V** will observe as counterclockwise the rotation through θ, which brings the vector **P** in line with the vector **Q**. Note that if **P** and **Q** do not have a common point of application, they should first be redrawn from the same point. The three vectors **P**, **Q**, and **V**—taken in that order—form a *right-handed triad*.

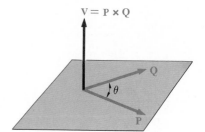

Fig. 3.6

Vector products are *distributive* but *not commutative*. We have

$$Q \times P = -(P \times Q) \tag{3.4}$$

Vector Products of Unit Vectors. It follows from the definition of the vector product of two vectors that

$$
\begin{array}{lll}
i \times i = 0 & j \times i = -k & k \times i = j \\
i \times j = k & j \times j = 0 & k \times j = -i \\
i \times k = -j & j \times k = i & k \times k = 0
\end{array} \tag{3.7}
$$

Rectangular Components of Vector Product. Resolving the vectors **P** and **Q** into rectangular components, we obtain the following expressions for the components of their vector product **V**:

$$
\begin{aligned}
V_x &= P_y Q_z - P_z Q_y \\
V_y &= P_z Q_x - P_x Q_z \\
V_z &= P_x Q_y - P_y Q_x
\end{aligned} \tag{3.9}
$$

In determinant form, we have

$$V = P \times Q = \begin{vmatrix} i & j & k \\ P_x & P_y & P_z \\ Q_x & Q_y & Q_z \end{vmatrix} \tag{3.10}$$

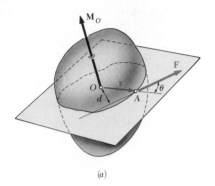

(a)

(b)

Fig. 3.12

A.5. Moment of a Force about a Point (Secs. 3.6 and 3.8).

The moment of a force $\mathbf{F}$ (or, more generally, of a vector $\mathbf{F}$) about a point O is defined as the vector product

$$\mathbf{M}_O = \mathbf{r} \times \mathbf{F} \tag{3.11}$$

where $\mathbf{r}$ denotes the *position vector* of the point of application A of $\mathbf{F}$ (Fig. 3.12a).

According to the definition of the vector product of two vectors given in Sec. A.4, the moment $\mathbf{M}_O$ must be perpendicular to the plane containing O and the force $\mathbf{F}$. Its magnitude is equal to

$$M_O = rF \sin \theta = Fd \tag{3.12}$$

where d is the perpendicular distance from O to the line of action of $\mathbf{F}$, and its sense is defined by the sense of the rotation which would bring the vector $\mathbf{r}$ in line with the vector $\mathbf{F}$; this rotation should be viewed as *counterclockwise* by an observer located at the tip of $\mathbf{M}_O$. Another way of defining the sense of $\mathbf{M}_O$ is furnished by the *right-hand rule*: Close your right hand and hold it so that your fingers are curled in the sense of the rotation that $\mathbf{F}$ would impart to the rigid body about a fixed axis directed along the line of action of $\mathbf{M}_O$; your thumb will indicate the sense of the moment $\mathbf{M}_O$ (Fig. 3.12b).

Rectangular Components of the Moment of a Force. Denoting by x, y, and z the coordinates of the point of application A of $\mathbf{F}$, we obtain the following expressions for the components of the moment $\mathbf{M}_O$ of $\mathbf{F}$:

$$\begin{aligned} M_x &= yF_z - zF_y \\ M_y &= zF_x - xF_z \\ M_z &= xF_y - yF_x \end{aligned} \tag{3.18}$$

In determinant form, we have

$$\mathbf{M}_O = \mathbf{r} \times \mathbf{F} = \begin{vmatrix} \mathbf{i} & \mathbf{j} & \mathbf{k} \\ x & y & z \\ F_x & F_y & F_z \end{vmatrix} \tag{3.19}$$

To compute the moment $\mathbf{M}_B$ about an arbitrary point B of a force $\mathbf{F}$ applied at A, we must use the vector $\mathbf{r}_{A/B} = \mathbf{r}_A - \mathbf{r}_B$ drawn from B to A instead of the vector $\mathbf{r}$. We write

$$\mathbf{M}_B = \mathbf{r}_{A/B} \times \mathbf{F} = (\mathbf{r}_A - \mathbf{r}_B) \times \mathbf{F} \tag{3.20}$$

or, using the determinant form,

$$\mathbf{M}_B = \begin{vmatrix} \mathbf{i} & \mathbf{j} & \mathbf{k} \\ x_{A/B} & y_{A/B} & z_{A/B} \\ F_x & F_y & F_z \end{vmatrix} \tag{3.21}$$

where $x_{A/B}$, $y_{A/B}$, $z_{A/B}$ are the components of the vector $\mathbf{r}_{A/B}$:

$$x_{A/B} = x_A - x_B \qquad y_{A/B} = y_A - y_B \qquad z_{A/B} = z_A - z_B$$

A.6. Scalar Product of Two Vectors (Sec. 3.9).

The scalar product, or *dot product*, of two vectors **P** and **Q** is defined as the product of the magnitudes of **P** and **Q** and of the cosine of the angle θ formed by **P** and **Q** (Fig. 3.19). The scalar product of **P** and **Q** is denoted by **P** · **Q**. We write

$$\mathbf{P} \cdot \mathbf{Q} = PQ \cos \theta \qquad (3.24)$$

Scalar products are *commutative* and *distributive*.

Scalar Products of Unit Vectors. It follows from the definition of the scalar product of two vectors that

$$\begin{aligned} \mathbf{i} \cdot \mathbf{i} = 1 && \mathbf{j} \cdot \mathbf{j} = 1 && \mathbf{k} \cdot \mathbf{k} = 1 \\ \mathbf{i} \cdot \mathbf{j} = 0 && \mathbf{j} \cdot \mathbf{k} = 0 && \mathbf{k} \cdot \mathbf{i} = 0 \end{aligned} \qquad (3.29)$$

Scalar Product Expressed in Terms of Rectangular Components. Resolving the vectors **P** and **Q** into rectangular components, we obtain

$$\mathbf{P} \cdot \mathbf{Q} = P_x Q_x + P_y Q_y + P_z Q_z \qquad (3.30)$$

Angle Formed by Two Vectors. It follows from (3.24) and (3.29) that

$$\cos \theta = \frac{\mathbf{P} \cdot \mathbf{Q}}{PQ} = \frac{P_x Q_x + P_y Q_y + P_z Q_z}{PQ} \qquad (3.32)$$

Projection of a Vector on a Given Axis. The projection of a vector **P** on the axis OL defined by the unit vector $\boldsymbol{\lambda}$ (Fig. 3.23) is

$$P_{OL} = OA = \mathbf{P} \cdot \boldsymbol{\lambda} \qquad (3.36)$$

A.7. Mixed Triple Product of Three Vectors (Sec. 3.10).

The mixed triple product of the three vectors **S**, **P**, and **Q** is defined as the scalar expression

$$\mathbf{S} \cdot (\mathbf{P} \times \mathbf{Q}) \qquad (3.38)$$

obtained by forming the scalar product of **S** with the vector product of **P** and **Q**. Mixed triple products are invariant under *circular permutations* but change sign under any other permutation:

$$\begin{aligned} \mathbf{S} \cdot (\mathbf{P} \times \mathbf{Q}) &= \mathbf{P} \cdot (\mathbf{Q} \times \mathbf{S}) = \mathbf{Q} \cdot (\mathbf{S} \times \mathbf{P}) \\ &= -\mathbf{S} \cdot (\mathbf{Q} \times \mathbf{P}) = -\mathbf{P} \cdot (\mathbf{S} \times \mathbf{Q}) = -\mathbf{Q} \cdot (\mathbf{P} \times \mathbf{S}) \end{aligned} \qquad (3.39)$$

Mixed Triple Product Expressed in Terms of Rectangular Components. The mixed triple product of **S**, **P**, and **Q** may be expressed in the form of a determinant

$$\mathbf{S} \cdot (\mathbf{P} \times \mathbf{Q}) = \begin{vmatrix} S_x & S_y & S_z \\ P_x & P_y & P_z \\ Q_x & Q_y & Q_z \end{vmatrix} \qquad (3.41)$$

The mixed triple product **S** · (**P** × **Q**) measures the volume of the parallelepiped having the vectors **S**, **P**, and **Q** for sides (Fig. 3.25).

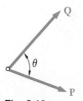

Fig. 3.19

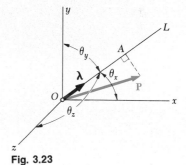

Fig. 3.23

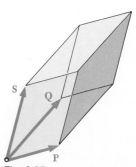

Fig. 3.25

moment M_{OL} of a force $\mathbf{F}$ (or, more generally, of a vector $\mathbf{F}$) about an axis
OL is defined as the projection OC on the axis OL of the moment $\mathbf{M}_O$ of $\mathbf{F}$
about O (Fig. 3.27). Denoting by $\boldsymbol{\lambda}$ the unit vector along OL, we have

$$M_{OL} = \boldsymbol{\lambda} \cdot \mathbf{M}_O = \boldsymbol{\lambda} \cdot (\mathbf{r} \times \mathbf{F}) \tag{3.42}$$

or, in determinant form,

$$M_{OL} = \begin{vmatrix} \lambda_x & \lambda_y & \lambda_z \\ x & y & z \\ F_x & F_y & F_z \end{vmatrix} \tag{3.43}$$

where $\lambda_x, \lambda_y, \lambda_z$ = direction cosines of axis OL
  x, y, z = coordinates of point of application of $\mathbf{F}$
  F_x, F_y, F_z = components of force $\mathbf{F}$

The moments of the force $\mathbf{F}$ about the three coordinate axes are given
by the expressions (3.18) obtained earlier for the rectangular components of
the moment $\mathbf{M}_O$ of $\mathbf{F}$ about O:

$$\begin{aligned} M_x &= yF_z - zF_y \\ M_y &= zF_x - xF_z \\ M_z &= xF_y - yF_x \end{aligned} \tag{3.18}$$

More generally, the moment of a force $\mathbf{F}$ applied at A about an axis
which does not pass through the origin is obtained by choosing an arbitrary
point B on the axis (Fig. 3.29) and determining the projection on the axis BL
of the moment $\mathbf{M}_B$ of $\mathbf{F}$ about B. We write

$$M_{BL} = \boldsymbol{\lambda} \cdot \mathbf{M}_B = \boldsymbol{\lambda} \cdot (\mathbf{r}_{A/B} \times \mathbf{F}) \tag{3.45}$$

where $\mathbf{r}_{A/B} = \mathbf{r}_A - \mathbf{r}_B$ represents the vector drawn from B to A. Expressing M_{BL} in the form of a determinant, we have

$$M_{BL} = \begin{vmatrix} \lambda_x & \lambda_y & \lambda_z \\ x_{A/B} & y_{A/B} & z_{A/B} \\ F_x & F_y & F_z \end{vmatrix} \tag{3.46}$$

where $\lambda_x, \lambda_y, \lambda_z$ = direction cosines of axis BL
  $x_{A/B} = x_A - x_B, \; y_{A/B} = y_A - y_B, \; z_{A/B} = z_A - z_B$
  F_x, F_y, F_z = components of force $\mathbf{F}$

It should be noted that the result obtained is independent of the choice of
the point B on the given axis; the same result would have been obtained if
point C had been chosen instead of B.

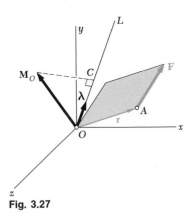

Fig. 3.27

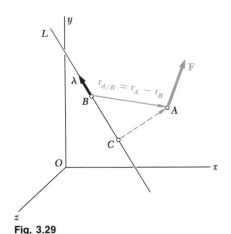

Fig. 3.29

Moments of Inertia
of Masses

MOMENTS OF INERTIA OF MASSES*

9.11. Moment of Inertia of a Mass. Consider a small mass Δm mounted on a rod of negligible mass which may rotate freely about an axis AA' (Fig. 9.20a). If a couple is applied to the system, the rod and mass, assumed initially at rest, will start rotating about AA'. The details of this motion will be studied later in dynamics. At present, we wish only to indicate that the time required for the system to reach a given speed of rotation is proportional to the mass Δm and to the square of the distance r. The product $r^2 \Delta m$ provides, therefore, a measure of the *inertia* of the system, i.e., of the resistance the system offers when we try to set it in motion. For this reason, the product $r^2 \Delta m$ is called the *moment of inertia* of the mass Δm with respect to the axis AA'.

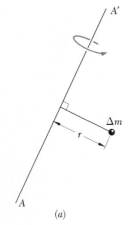

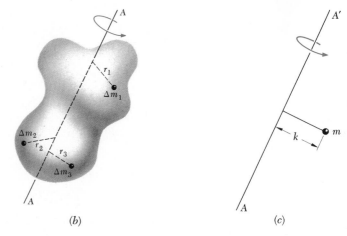

Fig. 9.20   (a)   (b)   (c)

Consider now a body of mass m which is to be rotated about an axis AA' (Fig. 9.20b). Dividing the body into elements of mass Δm_1, Δm_2, etc., we find that the resistance offered by the body is measured by the sum $r_1^2 \Delta m_1 + r_2^2 \Delta m_2 + \cdots$. This sum defines, therefore, the moment of inertia of the body with respect to the axis AA'. Increasing the number of elements, we find that the moment of inertia is equal, at the limit, to the integral

$$I = \int r^2 \, dm \qquad (9.28)$$

The *radius of gyration* k of the body with respect to the axis AA' is defined by the relation

$$I = k^2 m \qquad \text{or} \qquad k = \sqrt{\frac{I}{m}} \qquad (9.29)$$

The radius of gyration k represents, therefore, the distance at which the entire mass of the body should be concentrated if its moment of inertia with respect to AA' is to remain unchanged (Fig. 9.20c). Whether it is kept in its

---

*This repeats Secs. 9.11 through 9.17 of the volume on statics.

original shape (Fig. 9.20b) or whether it is concentrated as shown in Fig. 9.20c, the mass m will react in the same way to a rotation, or *gyration*, about AA'.

If SI units are used, the radius of gyration k is expressed in meters and the mass m in kilograms. The moment of inertia of a mass, therefore, will be expressed in kg·m². If U.S. customary units are used, the radius of gyration is expressed in feet and the mass in slugs, i.e., in lb·s²/ft. The moment of inertia of a mass, then, will be expressed in lb·ft·s².†

The moment of inertia of a body with respect to a coordinate axis may easily be expressed in terms of the coordinates x, y, z of the element of mass dm (Fig. 9.21). Noting, for example, that the square of the distance r from

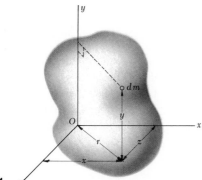

Fig. 9.21

the element dm to the y axis is $z^2 + x^2$, we express the moment of inertia of the body with respect to the y axis as

$$I_y = \int r^2 \, dm = \int (z^2 + x^2) \, dm$$

Similar expressions may be obtained for the moments of inertia with respect to the x and z axes. We write

$$I_x = \int (y^2 + z^2) \, dm$$
$$I_y = \int (z^2 + x^2) \, dm \qquad (9.30)$$
$$I_z = \int (x^2 + y^2) \, dm$$

† It should be kept in mind when converting the moment of inertia of a mass from U.S. customary units to SI units that the base unit pound used in the derived unit lb·ft·s² is a unit of force (*not* of mass) and should therefore be converted into newtons. We have

$$1 \text{ lb·ft·s}^2 = (4.45 \text{ N})(0.3048 \text{ m})(1 \text{ s})^2 = 1.356 \text{ N·m·s}^2$$

or, since N = kg·m/s²,

$$1 \text{ lb·ft·s}^2 = 1.356 \text{ kg·m}^2$$

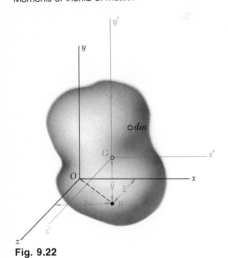

Fig. 9.22

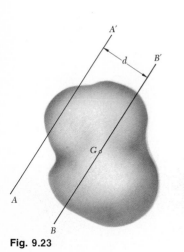

Fig. 9.23

9.12. Parallel-Axis Theorem.

Consider a body of mass m. Let $Oxyz$ be a system of rectangular coordinates with origin at an arbitrary point O, and $Gx'y'z'$ a system of parallel *centroidal axes*, i.e., a system with origin at the center of gravity G of the body† and with axes x', y', z', respectively, parallel to x, y, z (Fig. 9.22). Denoting by $\bar{x}$, $\bar{y}$, $\bar{z}$ the coordinates of G with respect to $Oxyz$, we write the following relations between the coordinates x, y, z of the element dm with respect to $Oxyz$ and its coordinates x', y', z' with respect to the centroidal axes $Gx'y'z'$:

$$x = x' + \bar{x} \qquad y = y' + \bar{y} \qquad z = z' + \bar{z} \qquad (9.31)$$

Referring to Eqs. (9.30), we may express the moment of inertia of the body with respect to the x axis as follows:

$$I_x = \int (y^2 + z^2)\, dm = \int [(y' + \bar{y})^2 + (z' + \bar{z})^2]\, dm$$
$$= \int (y'^2 + z'^2)\, dm + 2\bar{y} \int y'\, dm + 2\bar{z} \int z'\, dm + (\bar{y}^2 + \bar{z}^2) \int dm$$

The first integral in the expression obtained represents the moment of inertia $\bar{I}_{x'}$ of the body with respect to the centroidal axis x'; the second and third integrals represent the first moment of the body with respect to the $z'x'$ and $x'y'$ planes, respectively, and, since both planes contain G, the two integrals are zero; the last integral is equal to the total mass m of the body. We write, therefore,

$$I_x = \bar{I}_{x'} + m(\bar{y}^2 + \bar{z}^2) \qquad (9.32)$$

and, similarly,

$$I_y = \bar{I}_{y'} + m(\bar{z}^2 + \bar{x}^2) \qquad I_z = \bar{I}_{z'} + m(\bar{x}^2 + \bar{y}^2) \qquad (9.32')$$

We easily verify from Fig. 9.22 that the sum $\bar{z}^2 + \bar{x}^2$ represents the square of the distance OB between the y and y' axis. Similarly, $\bar{y}^2 + \bar{z}^2$ and $\bar{x}^2 + \bar{y}^2$ represent the squares of the distances between the x and x' axes, and the z and z' axes, respectively. Denoting by d the distance between an arbitrary axis AA' and a parallel centroidal axis BB' (Fig. 9.23), we may, therefore, write the following general relation between the moment of inertia I of the body with respect to AA' and its moment of inertia $\bar{I}$ with respect to BB':

$$I = \bar{I} + md^2 \qquad (9.33)$$

Expressing the moments of inertia in terms of the corresponding radii of gyration, we may also write

$$k^2 = \bar{k}^2 + d^2 \qquad (9.34)$$

where k and $\bar{k}$ represent the radii of gyration about AA' and BB', respectively.

---

†Note that the term centroidal is used to define an axis passing through the center of gravity G of the body, whether or not G coincides with the centroid of the volume of the body.

9.13. Moments of Inertia of Thin Plates. Consider a thin plate of uniform thickness t, made of a homogeneous material of density ρ (density = mass per unit volume). The mass moment of inertia of the plate with respect to an axis AA' *contained in the plane* of the plate (Fig. 9.24a) is

$$I_{AA',\text{mass}} = \int r^2 \, dm$$

Since $dm = \rho t \, dA$, we write

$$I_{AA',\text{mass}} = \rho t \int r^2 \, dA$$

But r represents the distance of the element of area dA to the axis AA'; the

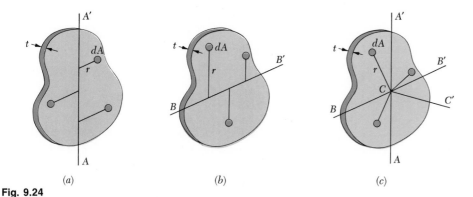

(a)   (b)   (c)

Fig. 9.24

integral is therefore equal to the moment of inertia of the area of the plate with respect to AA'. We have

$$I_{AA',\text{mass}} = \rho t I_{AA',\text{area}} \tag{9.35}$$

Similarly, we have with respect to an axis BB' perpendicular to AA' (Fig. 9.24b)

$$I_{BB',\text{mass}} = \rho t I_{BB',\text{area}} \tag{9.36}$$

Considering now the axis CC' *perpendicular* to the plate through the point of intersection C of AA' and BB' (Fig. 9.24c), we write

$$I_{CC',\text{mass}} = \rho t J_{C,\text{area}} \tag{9.37}$$

where J_C is the *polar* moment of inertia of the area of the plate with respect to point C.

Recalling the relation $J_C = I_{AA'} + I_{BB'}$ existing between polar and rectangular moments of inertia of an area, we write the following relation between the mass moments of inertia of a thin plate:

$$I_{CC'} = I_{AA'} + I_{BB'} \tag{9.38}$$

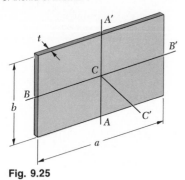

Fig. 9.25

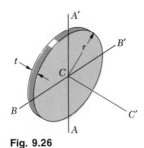

Fig. 9.26

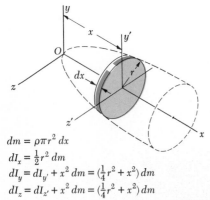

$dm = \rho \pi r^2\, dx$

$dI_x = \frac{1}{2} r^2\, dm$

$dI_y = dI_{y'} + x^2\, dm = (\frac{1}{4} r^2 + x^2)\, dm$

$dI_z = dI_{z'} + x^2\, dm = (\frac{1}{4} r^2 + x^2)\, dm$

Fig. 9.27 Determination of the moment of inertia of a body of revolution.

Rectangular Plate. In the case of a reactangular plate of sides a and b (Fig. 9.25), we obtain the following mass moments of inertia with respect to axes through the center of gravity of the plate:

$$I_{AA',\text{mass}} = \rho t I_{AA',\text{area}} = \rho t(\tfrac{1}{12}a^3 b)$$
$$I_{BB',\text{mass}} = \rho t I_{BB',\text{area}} = \rho t(\tfrac{1}{12}ab^3)$$

Observing that the product ρabt is equal to the mass m of the plate, we write the mass moments of inertia of a thin rectangular plate as follows:

$$I_{AA'} = \tfrac{1}{12}ma^2 \qquad I_{BB'} = \tfrac{1}{12}mb^2 \tag{9.39}$$
$$I_{CC'} = I_{AA'} + I_{BB'} = \tfrac{1}{12}m(a^2 + b^2) \tag{9.40}$$

Circular Plate. In the case of a circular plate, or disk, of radius r (Fig. 9.26), we write

$$I_{AA',\text{mass}} = \rho t I_{AA',\text{area}} = \rho t(\tfrac{1}{4}\pi r^4)$$

Observing that the product $\rho \pi r^2 t$ is equal to the mass m of the plate and that $I_{AA'} = I_{BB'}$, we write the mass moments of inertia of a circular plate as follows:

$$I_{AA'} = I_{BB'} = \tfrac{1}{4}mr^2 \tag{9.41}$$
$$I_{CC'} = I_{AA'} + I_{BB'} = \tfrac{1}{2}mr^2 \tag{9.42}$$

9.14. Determination of the Moment of Inertia of a Three-Dimensional Body by Integration.

The moment of inertia of a three-dimensional body is obtained by computing the integral $I = \int r^2\, dm$. If the body is made of a homogeneous material of density ρ, we have $dm = \rho\, dV$ and write $I = \rho \int r^2\, dV$. This integral depends only upon the shape of the body. Thus, in order to compute the moment of inertia of a three-dimensional body, it will generally be necessary to perform a triple, or at least a double, integration.

However, if the body possesses two planes of symmetry, it is usually possible to determine the body's moment of inertia through a single integration by choosing as an element of mass dm the mass of a thin slab perpendicular to the planes of symmetry. In the case of bodies of revolution, for example, the element of mass should be a thin disk (Fig. 9.27). Using formula (9.42), the moment of inertia of the disk with respect to the axis of revolution may be readily expressed as indicated in Fig. 9.27. Its moment of inertia with respect to each of the other two axes of coordinates will be obtained by using formula (9.41) and the parallel-axis theorem. Integration of the expressions obtained will yield the desired moment of inertia of the body of revolution.

9.15. Moments of Inertia of Composite Bodies.

The moments of inertia of a few common shapes are shown in Fig. 9.28. The moment of inertia with respect to a given axis of a body made of several of these simple shapes may be obtained by computing the moments of inertia of its component parts about the desired axis and adding them together. We should note, as we have already noted in the case of areas, that the radius of gyration of a composite body *cannot* be obtained by adding the radii of gyration of its component parts.

Slender rod	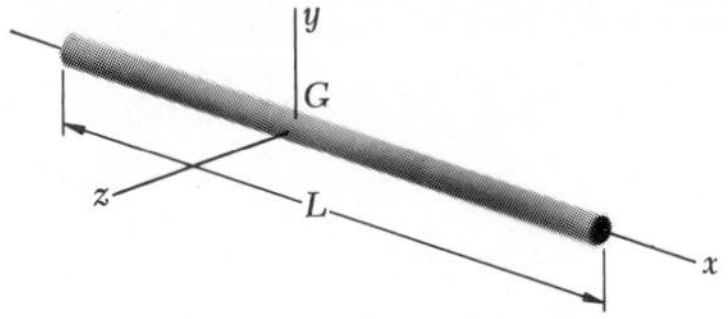	$I_y = I_z = \frac{1}{12}mL^2$
Thin rectangular plate	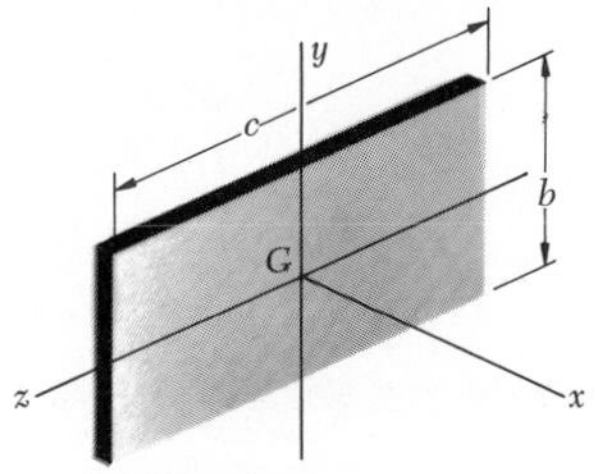	$I_x = \frac{1}{12}m(b^2 + c^2)$ $I_y = \frac{1}{12}mc^2$ $I_z = \frac{1}{12}mb^2$
Rectangular prism	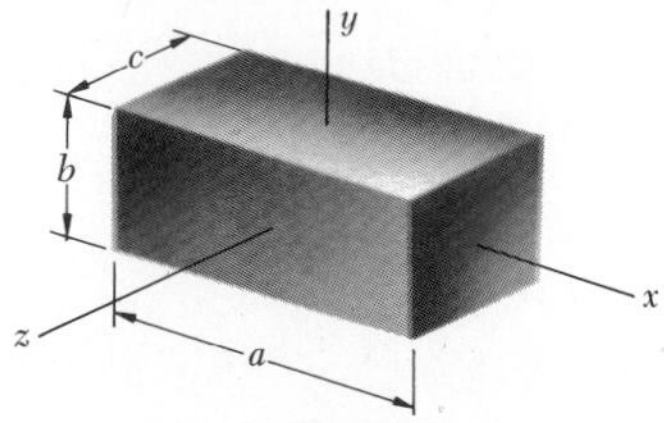	$I_x = \frac{1}{12}m(b^2 + c^2)$ $I_y = \frac{1}{12}m(c^2 + a^2)$ $I_z = \frac{1}{12}m(a^2 + b^2)$
Thin disk	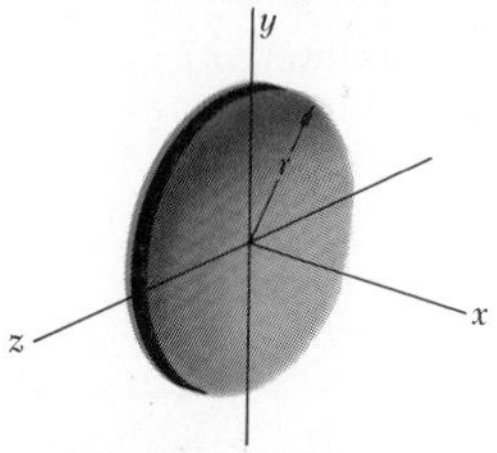	$I_x = \frac{1}{2}mr^2$ $I_y = I_z = \frac{1}{4}mr^2$
Circular cylinder	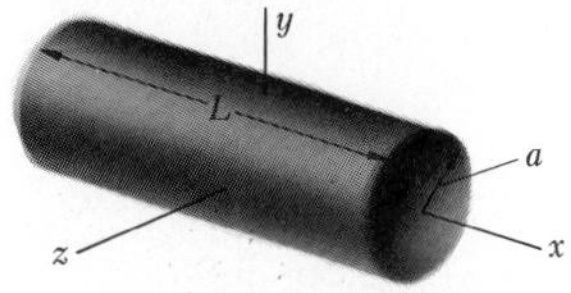	$I_x = \frac{1}{2}ma^2$ $I_y = I_z = \frac{1}{12}m(3a^2 + L^2)$
Circular cone	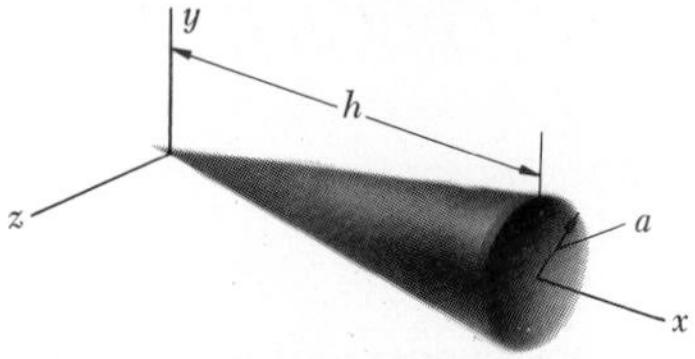	$I_x = \frac{3}{10}ma^2$ $I_y = I_z = \frac{3}{5}m(\frac{1}{4}a^2 + h^2)$
Sphere	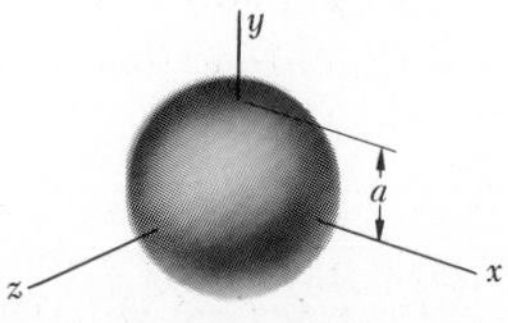	$I_x = I_y = I_z = \frac{2}{5}ma^2$

Fig. 9.28 Mass moments of inertia of common geometric shapes.

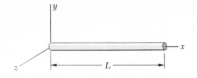

SAMPLE PROBLEM 9.9

Determine the mass moment of inertia of a slender rod of length L and mass m with respect to an axis perpendicular to the rod and passing through one end of the rod.

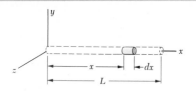

Solution. Choosing the differential element of mass shown, we write

$$dm = \frac{m}{L} dx$$

$$I_y = \int x^2 \, dm = \int_0^L x^2 \frac{m}{L} \, dx = \left[\frac{m}{L} \frac{x^3}{3} \right]_0^L \qquad I_y = \tfrac{1}{3} mL^2 \quad \blacktriangleleft$$

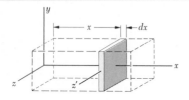

SAMPLE PROBLEM 9.10

Determine the mass moment of inertia of the homogeneous rectangular prism shown with respect to the z axis.

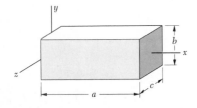

      Solution. We choose as a differential element of mass the thin slab shown for which

$$dm = \rho bc \, dx$$

Referring to Sec. 9.13, we find that the moment of inertia of the element with respect to the z' axis is

$$dI_{z'} = \tfrac{1}{12} b^2 \, dm$$

Applying the parallel-axis theorem, we obtain the mass moment of inertia of the slab with respect to the z axis.

$$dI_z = dI_{z'} + x^2 \, dm = \tfrac{1}{12} b^2 \, dm + x^2 \, dm = (\tfrac{1}{12} b^2 + x^2) \rho bc \, dx$$

Integrating from $x = 0$ to $x = a$, we obtain

$$I_z = \int dI_z = \int_0^a (\tfrac{1}{12} b^2 + x^2) \rho bc \, dx = \rho abc \, (\tfrac{1}{12} b^2 + \tfrac{1}{3} a^2)$$

Since the total mass of the prism is $m = \rho abc$, we may write

$$I_z = m(\tfrac{1}{12} b^2 + \tfrac{1}{3} a^2) \qquad I_z = \tfrac{1}{12} m(4a^2 + b^2) \quad \blacktriangleleft$$

We note that if the prism is slender, b is small compared to a and the expression for I_z reduces to $\tfrac{1}{3} ma^2$, which is the result obtained in Sample Prob. 9.9 when $L = a$.

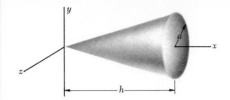

SAMPLE PROBLEM 9.11

Determine the mass moment of inertia of a right circular cone with respect to (a) its longitudinal axis, (b) an axis through the apex of the cone and perpendicular to its longitudinal axis, (c) an axis through the centroid of the cone and perpendicular to its longitudinal axis.

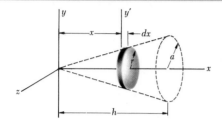

Solution. We choose the differential element of mass shown.

$$r = a\frac{x}{h} \qquad dm = \rho\pi r^2\, dx = \rho\pi\frac{a^2}{h^2}x^2\, dx$$

a. Moment of Inertia I_x. Using the expression derived in Sec. 9.13 for a thin disk, we compute the mass moment of inertia of the differential element with respect to the x axis.

$$dI_x = \tfrac{1}{2}r^2\, dm = \tfrac{1}{2}\left(a\frac{x}{h}\right)^2\left(\rho\pi\frac{a^2}{h^2}x^2\, dx\right) = \tfrac{1}{2}\rho\pi\frac{a^4}{h^4}x^4\, dx$$

Integrating from $x = 0$ to $x = h$, we obtain

$$I_x = \int dI_x = \int_0^h \tfrac{1}{2}\rho\pi\frac{a^4}{h^4}x^4\, dx = \tfrac{1}{2}\rho\pi\frac{a^4}{h^4}\frac{h^5}{5} = \tfrac{1}{10}\rho\pi a^4 h$$

Since the total mass of the cone is $m = \tfrac{1}{3}\rho\pi a^2 h$, we may write

$$I_x = \tfrac{1}{10}\rho\pi a^4 h = \tfrac{3}{10}a^2(\tfrac{1}{3}\rho\pi a^2 h) = \tfrac{3}{10}ma^2 \qquad I_x = \tfrac{3}{10}ma^2 \quad \blacktriangleleft$$

b. Moment of Inertia I_y. The same differential element will be used. Applying the parallel-axis theorem and using the expression derived in Sec. 9.13 for a thin disk, we write

$$dI_y = dI_{y'} + x^2\, dm = \tfrac{1}{4}r^2\, dm + x^2\, dm = (\tfrac{1}{4}r^2 + x^2)\, dm$$

Substituting the expressions for r and dm, we obtain

$$dI_y = \left(\frac{1}{4}\frac{a^2}{h^2}x^2 + x^2\right)\left(\rho\pi\frac{a^2}{h^2}x^2\, dx\right) = \rho\pi\frac{a^2}{h^2}\left(\frac{a^2}{4h^2} + 1\right)x^4\, dx$$

$$I_y = \int dI_y = \int_0^h \rho\pi\frac{a^2}{h^2}\left(\frac{a^2}{4h^2} + 1\right)x^4\, dx = \rho\pi\frac{a^2}{h^2}\left(\frac{a^2}{4h^2} + 1\right)\frac{h^5}{5}$$

Introducing the total mass of the cone m, we rewrite I_y as follows:

$$I_y = \tfrac{3}{5}(\tfrac{1}{4}a^2 + h^2)\tfrac{1}{3}\rho\pi a^2 h \qquad I_y = \tfrac{3}{5}m(\tfrac{1}{4}a^2 + h^2) \quad \blacktriangleleft$$

c. Moment of Inertia $\bar{I}_{y''}$. We apply the parallel-axis theorem and write

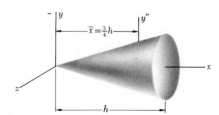

$$I_y = \bar{I}_{y''} + m\bar{x}^2$$

Solving for $\bar{I}_{y''}$ and recalling that $\bar{x} = \tfrac{3}{4}h$, we have

$$\bar{I}_{y''} = I_y - m\bar{x}^2 = \tfrac{3}{5}m(\tfrac{1}{4}a^2 + h^2) - m(\tfrac{3}{4}h)^2$$

$$\bar{I}_{y''} = \tfrac{3}{20}m(a^2 + \tfrac{1}{4}h^2) \quad \blacktriangleleft$$

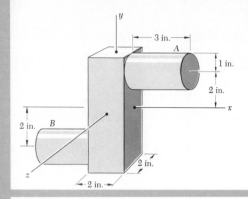

SAMPLE PROBLEM 9.12

A steel forging consists of a rectangular prism 6 × 2 × 2 in. and of two cylinders of diameter 2 in. and length 3 in., as shown. Determine the mass moments of inertia with respect to the coordinate axes. (Specific weight of steel = 490 lb/ft³.)

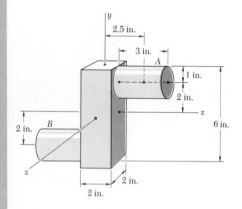

Computation of Masses
Prism

$$V = (2 \text{ in.})(2 \text{ in.})(6 \text{ in.}) = 24 \text{ in}^3$$

$$W = \frac{(24 \text{ in}^3)(490 \text{ lb/ft}^3)}{1728 \text{ in}^3/\text{ft}^3} = 6.81 \text{ lb}$$

$$m = \frac{6.81 \text{ lb}}{32.2 \text{ ft/s}^2} = 0.211 \text{ lb} \cdot \text{s}^2/\text{ft}$$

Each Cylinder

$$V = \pi(1 \text{ in.})^2(3 \text{ in.}) = 9.42 \text{ in}^3$$

$$W = \frac{(9.42 \text{ in}^3)(490 \text{ lb/ft}^3)}{1728 \text{ in}^3/\text{ft}^3} = 2.67 \text{ lb}$$

$$m = \frac{2.67 \text{ lb}}{32.2 \text{ ft/s}^2} = 0.0829 \text{ lb} \cdot \text{s}^2/\text{ft}$$

Mass Moments of Inertia. The mass moments of inertia of each component are computed from Fig. 9.28, using the parallel-axis theorem when necessary. Note that all lengths should be expressed in feet.

Prism

$I_x = I_z = \frac{1}{12}(0.211 \text{ lb} \cdot \text{s}^2/\text{ft})[(\frac{6}{12} \text{ ft})^2 + (\frac{2}{12} \text{ ft})^2] = 4.88 \times 10^{-3} \text{ lb} \cdot \text{ft} \cdot \text{s}^2$
$I_y = \frac{1}{12}(0.211 \text{ lb} \cdot \text{s}^2/\text{ft})[(\frac{2}{12} \text{ ft})^2 + (\frac{2}{12} \text{ ft})^2] = 0.977 \times 10^{-3} \text{ lb} \cdot \text{ft} \cdot \text{s}^2$

Each Cylinder

$I_x = \frac{1}{2}ma^2 + m\bar{y}^2 = \frac{1}{2}(0.0829 \text{ lb} \cdot \text{s}^2/\text{ft})(\frac{1}{12} \text{ ft})^2$
$\qquad\qquad + (0.0829 \text{ lb} \cdot \text{s}^2/\text{ft})(\frac{2}{12} \text{ ft})^2 = 2.59 \times 10^{-3} \text{ lb} \cdot \text{ft} \cdot \text{s}^2$
$I_y = \frac{1}{12}m(3a^2 + L^2) + m\bar{x}^2 = \frac{1}{12}(0.0829 \text{ lb} \cdot \text{s}^2/\text{ft})[3(\frac{1}{12} \text{ ft})^2 + (\frac{3}{12} \text{ ft})^2]$
$\qquad\qquad + (0.0829 \text{ lb} \cdot \text{s}^2/\text{ft})(\frac{2.5}{12} \text{ ft})^2 = 4.17 \times 10^{-3} \text{ lb} \cdot \text{ft} \cdot \text{s}^2$
$I_z = \frac{1}{12}m(3a^2 + L^2) + m(\bar{x}^2 + \bar{y}^2) = \frac{1}{12}(0.0829)[3(\frac{1}{12})^2 + (\frac{3}{12})^2]$
$\qquad\qquad + (0.0829)[(\frac{2.5}{12})^2 + (\frac{2}{12})^2] = 6.48 \times 10^{-3} \text{ lb} \cdot \text{ft} \cdot \text{s}^2$

Entire Body. Adding the values obtained,

$I_x = 4.88 \times 10^{-3} + 2(2.59 \times 10^{-3}) \qquad\qquad I_x = 10.06 \times 10^{-3} \text{ lb} \cdot \text{ft} \cdot \text{s}^2$ ◄
$I_y = 0.977 \times 10^{-3} + 2(4.17 \times 10^{-3}) \qquad\qquad I_y = 9.32 \times 10^{-3} \text{ lb} \cdot \text{ft} \cdot \text{s}^2$ ◄
$I_z = 4.88 \times 10^{-3} + 2(6.48 \times 10^{-3}) \qquad\qquad I_z = 17.84 \times 10^{-3} \text{ lb} \cdot \text{ft} \cdot \text{s}^2$ ◄

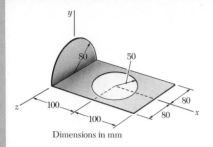

Dimensions in mm

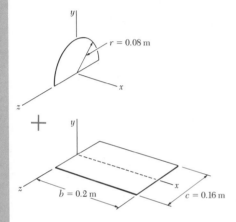

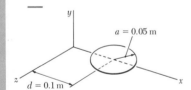

SAMPLE PROBLEM 9.13

A thin steel plate 4 mm thick is cut and bent to form the machine part shown. Knowing that the density of steel is 7850 kg/m^3, determine the mass moment of inertia of the machine part with respect to the coordinate axes.

Solution. We observe that the machine part consists of a semicircular plate, plus a rectangular plate, minus a circular plate.

Computation of Masses. *Semicircular Plate*

$$V_1 = \tfrac{1}{2}\pi r^2 t = \tfrac{1}{2}\pi(0.08 \text{ m})^2(0.004 \text{ m}) = 40.21 \times 10^{-6} \text{ m}^3$$
$$m_1 = \rho V_1 = (7.85 \times 10^3 \text{ kg/m}^3)(40.21 \times 10^{-6} \text{ m}^3) = 0.3156 \text{ kg}$$

Rectangular Plate

$$V_2 = (0.200 \text{ m})(0.160 \text{ m})(0.004 \text{ m}) = 128 \times 10^{-6} \text{ m}^3$$
$$m_2 = \rho V_2 = (7.85 \times 10^3 \text{ kg/m}^3)(128 \times 10^{-6} \text{ m}^3) = 1.005 \text{ kg}$$

Circular Plate

$$V_3 = \pi a^2 t = \pi(0.050 \text{ m})^2(0.004 \text{ m}) = 31.42 \times 10^{-6} \text{ m}^3$$
$$m_3 = \rho V_3 = (7.85 \times 10^3 \text{ kg/m}^3)(31.42 \times 10^{-6} \text{ m}^3) = 0.2466 \text{ kg}$$

Mass Moments of Inertia. Using the method presented in Sec. 9.13, we compute the mass moment of inertia of each component.

Semicircular Plate. From Fig. 9.28, we observe that for a circular plate of mass m and radius r

$$I_x = \tfrac{1}{2}mr^2 \qquad I_y = I_z = \tfrac{1}{4}mr^2$$

Because of symmetry, we note that for a semicircular plate

$$I_x = \tfrac{1}{2}(\tfrac{1}{2}mr^2) \qquad I_y = I_z = \tfrac{1}{2}(\tfrac{1}{4}mr^2)$$

Since the mass of the semicircular plate is $m_1 = \tfrac{1}{2}m$, we have

$$I_x = \tfrac{1}{2}m_1 r^2 = \tfrac{1}{2}(0.3156 \text{ kg})(0.08 \text{ m})^2 = 1.010 \times 10^{-3} \text{ kg·m}^2$$
$$I_y = I_z = \tfrac{1}{4}(\tfrac{1}{2}mr^2) = \tfrac{1}{4}m_1 r^2 = \tfrac{1}{4}(0.3156 \text{ kg})(0.08 \text{ m})^2 = 0.505 \times 10^{-3} \text{ kg·m}^2$$

Rectangular Plate

$$I_x = \tfrac{1}{12}m_2 c^2 = \tfrac{1}{12}(1.005 \text{ kg})(0.16 \text{ m})^2 = 2.144 \times 10^{-3} \text{ kg·m}^2$$
$$I_z = \tfrac{1}{3}m_2 b^2 = \tfrac{1}{3}(1.005 \text{ kg})(0.2 \text{ m})^2 = 13.400 \times 10^{-3} \text{ kg·m}^2$$
$$I_y = I_x + I_z = (2.144 + 13.400)(10^{-3}) = 15.544 \times 10^{-3} \text{ kg·m}^2$$

Circular Plate

$$I_x = \tfrac{1}{4}m_3 a^2 = \tfrac{1}{4}(0.2466 \text{ kg})(0.05 \text{ m})^2 = 0.154 \times 10^{-3} \text{ kg·m}^2$$
$$I_y = \tfrac{1}{2}m_3 a^2 + m_3 d^2$$
$$= \tfrac{1}{2}(0.2466 \text{ kg})(0.05 \text{ m})^2 + (0.2466 \text{ kg})(0.1 \text{ m})^2 = 2.774 \times 10^{-3} \text{ kg·m}^2$$
$$I_z = \tfrac{1}{4}m_3 a^2 + m_3 d^2 = \tfrac{1}{4}(0.2466 \text{ kg})(0.05 \text{ m})^2 + (0.2466 \text{ kg})(0.1 \text{ m})^2$$
$$= 2.620 \times 10^{-3} \text{ kg·m}^2$$

Entire Machine Part

$$I_x = (1.010 + 2.144 - 0.154)(10^{-3}) \text{ kg·m}^2 \qquad I_x = 3.00 \times 10^{-3} \text{ kg·m}^2 \quad \blacktriangleleft$$
$$I_y = (0.505 + 15.544 - 2.774)(10^{-3}) \text{ kg·m}^2 \qquad I_y = 13.28 \times 10^{-3} \text{ kg·m}^2 \quad \blacktriangleleft$$
$$I_z = (0.505 + 13.400 - 2.620)(10^{-3}) \text{ kg·m}^2 \qquad I_z = 11.29 \times 10^{-3} \text{ kg·m}^2 \quad \blacktriangleleft$$

Problems

9.88 A thin semicircular plate has a radius a and a mass m. Determine the mass moment of inertia of the plate with respect to (a) the centroidal axis BB', (b) the centroidal axis CC'.

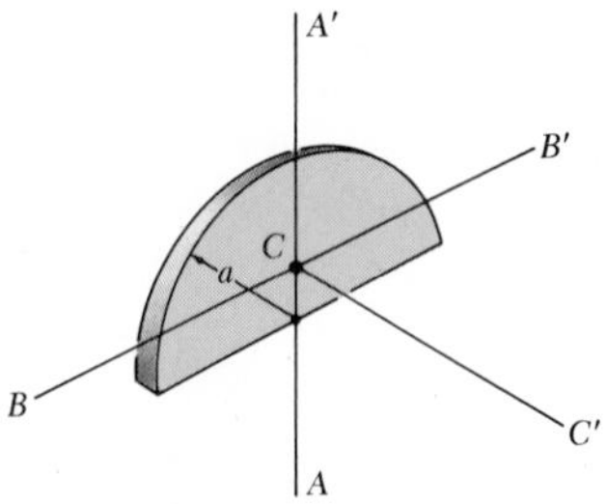

Fig. P9.88

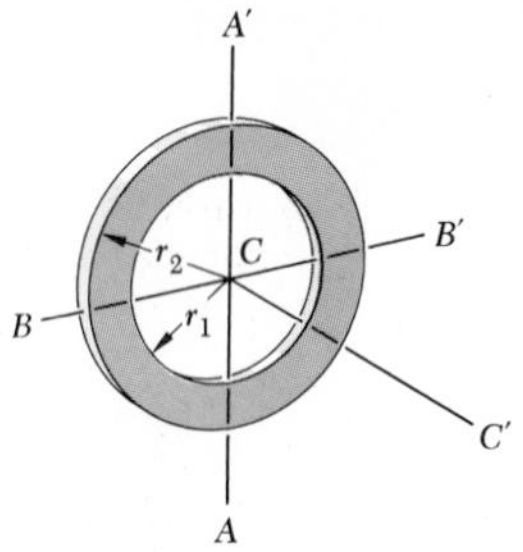

Fig. P9.89

9.89 Determine the mass moment of inertia of a ring of mass m, cut from a thin uniform plate, with respect to (a) the diameter AA' of the ring, (b) the axis CC' perpendicular to the plane of the ring.

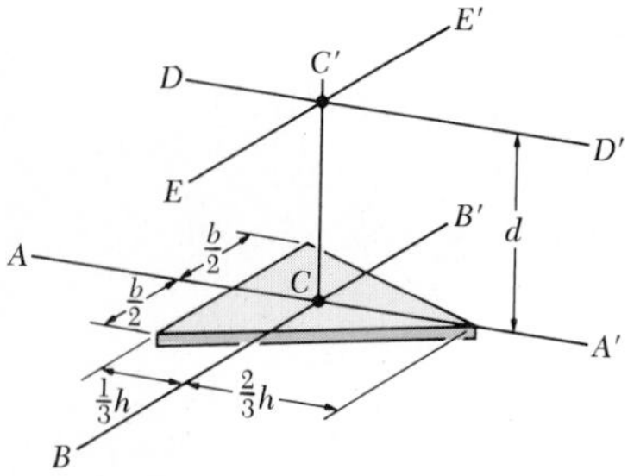

Fig. P9.90

9.90 A thin plate of mass m is cut in the shape of an isosceles triangle of base b and height h. Determine the mass moment of inertia of the plate with respect to (a) the centroidal axes AA' and BB' in the plane of the plate, (b) the centroidal axis CC' perpendicular to the plate.

9.91 Determine the mass moments of inertia of the plate of Prob. 9.90 with respect to axes DD' and EE' parallel to the centroidal axes AA' and BB' and located at a distance d from the plane of the plate.

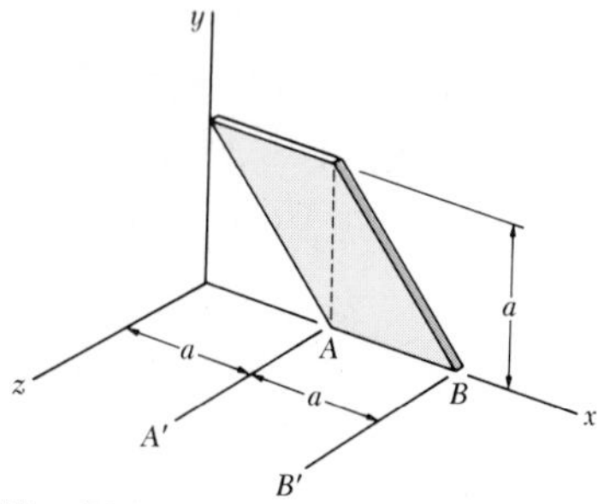

Fig. P9.92

9.92 A thin plate of mass m is cut in the shape of a parallelogram as shown. Determine the mass moment of inertia of the plate with respect to (a) the x axis, (b) the axis BB' perpendicular to the plate.

9.93 Determine the mass moment of inertia of the plate of Prob. 9.92 with respect to (a) the y axis, (b) the axis AA' perpendicular to the plate.

9.94 The area shown is revolved about the x axis to form a homogeneous solid of revolution of mass m. Express the mass moment of inertia of the solid with respect to the x axis in terms of m, a, and n.

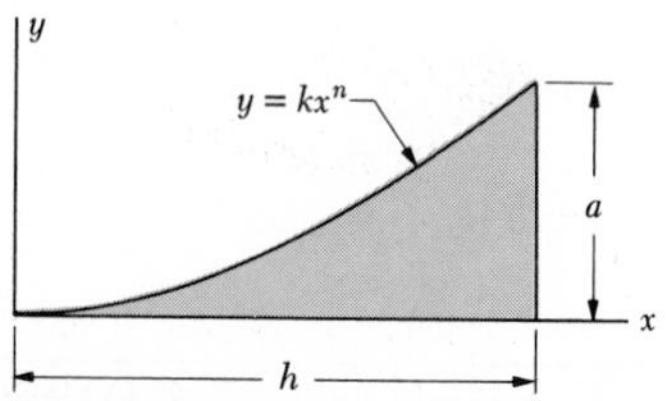

Fig. P9.94

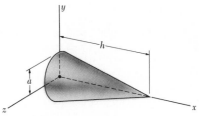

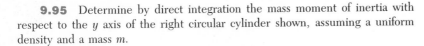

Fig. P9.95

9.95 Determine by direct integration the mass moment of inertia with respect to the y axis of the right circular cylinder shown, assuming a uniform density and a mass m.

9.96 Determine by direct integration the mass moment of inertia and the radius of gyration of the right circular cone with respect to the z axis, assuming a uniform density and a mass m.

Fig. P9.96

9.97 Determine by direct integration the mass moment of inertia and the radius of gyration with respect to the y axis of the paraboloid shown, assuming a uniform density and a mass m.

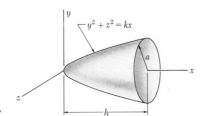

Fig. P9.97

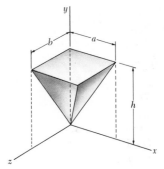

9.98 Determine by direct integration the mass moment of inertia with respect to the x axis of the pyramid shown, assuming a uniform density and a mass m.

9.99 Determine by direct integration the mass moment of inertia with respect to the y axis of the pyramid shown, assuming a uniform density and a mass m.

Fig. P9.98 and P9.99

9.100 A thin spherical dish of mass m is formed by passing a horizontal plane through a spherical shell of radius R and thickness t. Determine by direct integration the mass moment of inertia of the dish with respect to its vertical axis of symmetry. Check that the result obtained for $\phi = 90°$ (hemispherical shell) is the same as the answer to Prob. 9.104.

Fig. P9.100

9.101 A thin steel wire is bent into the shape of a circular arc as shown. Denoting by m' the mass per unit length of the wire, determine the mass moment of inertia of the wire with respect to each of the coordinate axes.

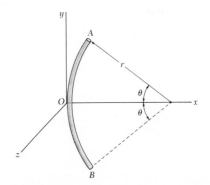

Fig. P9.101

9.102 and 9.103 Determine the mass moment of inertia and the radius of gyration of the steel flywheel shown with respect to its axis of rotation. (Density of steel = 7850 kg/m³; specific weight of steel = 490 lb/ft³.)

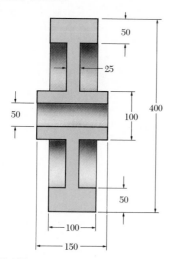

Fig. P9.102   Dimensions in mm

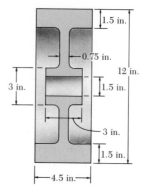

Fig. P9.103

9.104 Knowing that the thin hemispherical shell shown is of mass m and thickness t, determine the mass moment of inertia and the radius of gyration of the shell with respect to the x axis. (*Hint.* Consider the shell as formed by removing a hemisphere of radius r from a hemisphere of radius $r + t$; then neglect the terms containing t^2 and t^3 and keep those terms containing t.)

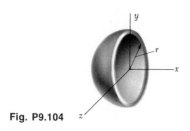

Fig. P9.104

9.105 The machine part shown is formed by machining a conical surface into a circular cylinder. For $b = \frac{1}{2}h$, determine the mass moment of inertia and the radius of gyration of the machine part with respect to the y axis.

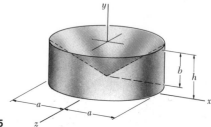

Fig. P9.105

9.106 For the 2-kg connecting rod shown, it has been experimentally determined that the mass moments of inertia of the rod with respect to the center-line axes of the bearings AA' and BB' are, respectively, $I_{AA'} = 78$ g·m² and $I_{BB'} = 41$ g·m². Knowing that $r_a + r_b = 290$ mm, determine (*a*) the location of the centroidal axis GG', (*b*) the radius of gyration with respect to axis GG'.

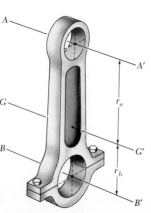

Fig. P9.106 and P9.107

9.107 Knowing that for the 3.25-kg connecting rod shown the mass moment of inertia with respect to axis AA' is 205 g·m², determine the mass moment of inertia with respect to axis BB', for $r_a = 225$ mm and $r_b = 140$ mm.

9.108 For the homogeneous ring shown, which is of density ρ, determine (a) the mass moment of inertia with respect to the axis BB', (b) the value of a_1 for which, given a_2 and h, $I_{BB'}$ is maximum, (c) the corresponding value of $I_{BB'}$.

9.109 Determine the mass moment of inertia and the radius of gyration of the steel ring shown with respect to the axis BB' with $a_1 = 3$ in., $a_2 = 5$ in., and $h = 4$ in. (Specific weight of steel = 490 lb/ft³.)

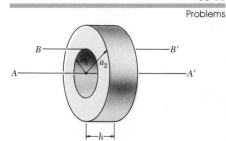

9.110 and 9.111 A section of sheet steel 2 mm thick is cut and bent into the machine component shown. Knowing that the density of steel is 7850 kg/m³, determine the mass moment of inertia of the component with respect to (a) the x axis, (b) the y axis, (c) the z axis.

Fig. P9.108 and P9.109

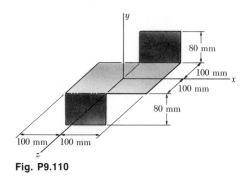

Fig. P9.110

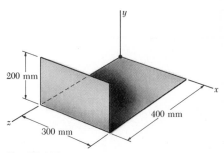

Fig. P9.111

9.112 A corner reflector for tracking by radar has two sides in the shape of a quarter circle of radius 15 in. and one side in the shape of a triangle. Each part of the reflector is formed from aluminum plate of uniform 0.05-in. thickness. Knowing that the specific weight of the aluminum used is 170 lb/ft³, determine the mass moment of inertia of the reflector with respect to each of the coordinate axes.

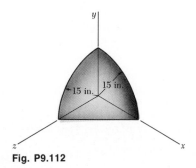

Fig. P9.112

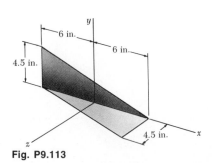

Fig. P9.113

9.113 A section of sheet steel 0.03 in. thick is cut and bent into the sheet-metal machine component shown. Knowing that the specific weight of steel is 490 lb/ft³, determine the mass moment of inertia of the component with respect to each of the coordinate axes.

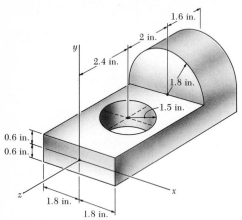

Fig. P9.115 and P9.116

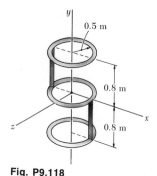

Fig. P9.118

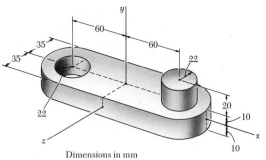

Dimensions in mm

Fig. P9.114 and P9.117

9.114 and 9.115 Determine the mass moment of inertia of the steel machine element shown with respect to the x axis. (Specific weight of steel = 490 lb/ft³; density of steel = 7850 kg/m³.)

9.116 and 9.117 Determine the mass moment of inertia of the steel machine element shown with respect to the y axis. (Specific weight of steel = 490 lb/ft³; density of steel = 7850 kg/m³.)

9.118 A homogeneous wire with a mass per unit length of 1.8 kg/m is used to form the figure shown. Determine the mass moment of inertia of the wire figure with respect to (*a*) the x axis, (*b*) the y axis.

9.119 In Prob. 9.118, determine the mass moment of inertia of the wire figure with respect to the z axis.

* 9.16. Moment of Inertia of a Body with Respect to an Arbitrary Axis through O. Mass Products of Inertia. We shall see in this section how the moment of inertia of a body may be determined with respect to an arbitrary axis OL through the origin (Fig. 9.29) if we have computed beforehand its moments of inertia with respect to the three coordinate axes, as well as certain other quantities to be defined below.

The moment of inertia of the body with respect to OL is represented by the integral $I_{OL} = \int p^2 \, dm$, where p denotes the perpendicular distance from the element of mass dm to the axis OL. But, denoting by $\boldsymbol{\lambda}$ the unit vector along OL and by $\mathbf{r}$ the position vector of the element dm, we observe that the perpendicular distance p is equal to the magnitude $r \sin \theta$ of the vector product $\boldsymbol{\lambda} \times \mathbf{r}$. We therefore write

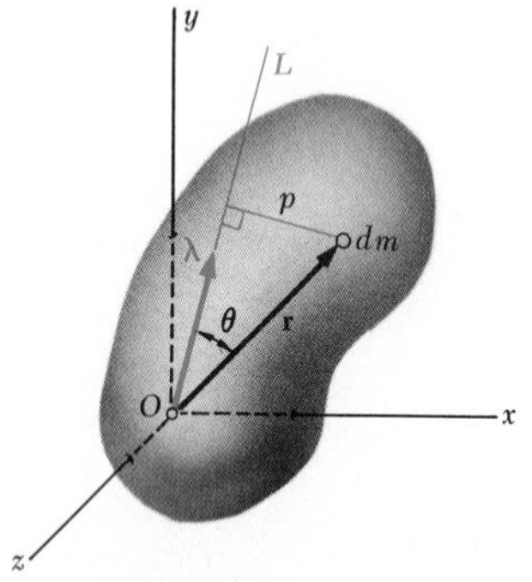

$$I_{OL} = \int p^2 \, dm = \int |\boldsymbol{\lambda} \times \mathbf{r}|^2 \, dm \qquad (9.43)$$

Expressing the square in terms of the rectangular components of the vector product, we have

$$I_{OL} = \int [(\lambda_x y - \lambda_y x)^2 + (\lambda_y z - \lambda_z y)^2 + (\lambda_z x - \lambda_x z)^2] \, dm$$

where the components λ_x, λ_y, λ_z of the unit vector $\boldsymbol{\lambda}$ represent the direction cosines of the axis OL, and the components x, y, z of $\mathbf{r}$ represent the coordinates of the element of mass dm. Expanding the squares in the expression obtained and rearranging the terms, we write

Fig. 9.29

$$I_{OL} = \lambda_x^2 \int (y^2 + z^2)\, dm + \lambda_y^2 \int (z^2 + x^2)\, dm + \lambda_z^2 \int (x^2 + y^2)\, dm$$
$$- 2\lambda_x \lambda_y \int xy\, dm - 2\lambda_y \lambda_z \int yz\, dm - 2\lambda_z \lambda_x \int zx\, dm \qquad (9.44)$$

Referring to Eqs. (9.30), we note that the first three integrals in (9.44) represent, respectively, the moments of inertia I_x, I_y, and I_z of the body with respect to the coordinate axes. The last three integrals in (9.44), which involve products of coordinates, are called the *products of inertia* of the body with respect to the x and y axes, the y and z axes, and the z and x axes, respectively. We write

$$I_{xy} = \int xy\, dm \qquad I_{yz} = \int yz\, dm \qquad I_{zx} = \int zx\, dm \qquad (9.45)$$

Substituting for the various integrals from (9.30) and (9.45) into (9.44), we have

$$I_{OL} = I_x \lambda_x^2 + I_y \lambda_y^2 + I_z \lambda_z^2 - 2I_{xy}\lambda_x\lambda_y - 2I_{yz}\lambda_y\lambda_z - 2I_{zx}\lambda_z\lambda_x \qquad (9.46)$$

We note that the definition of the products of inertia of a mass given in Eqs. (9.45) is an extension of the definition of the product of inertia of an area (Sec. 9.8). Mass products of inertia reduce to zero under the same conditions of symmetry as products of inertia of areas, and the parallel-axis theorem for mass products of inertia is expressed by relations similar to the formula derived for the product of inertia of an area. Substituting for x, y, z from Eqs. (9.31) into Eqs. (9.45), we verify that

$$\begin{aligned} I_{xy} &= \bar{I}_{x'y'} + m\bar{x}\bar{y} \\ I_{yz} &= \bar{I}_{y'z'} + m\bar{y}\bar{z} \\ I_{zx} &= \bar{I}_{z'x'} + m\bar{z}\bar{x} \end{aligned} \qquad (9.47)$$

where $\bar{x}$, $\bar{y}$, $\bar{z}$ are the coordinates of the center of gravity G of the body, and $\bar{I}_{x'y'}$, $\bar{I}_{y'z'}$, $\bar{I}_{z'x'}$ denote the products of inertia with respect to the centroidal axes x', y', z' (Fig. 9.22).

*9.17. Ellipsoid of Inertia. Principal Axes of Inertia.

Let us assume that the moment of inertia of the body considered in the preceding section has been determined with respect to a large number of axes OL through the fixed point O, and that a point Q has been plotted on each axis OL at a distance $OQ = 1/\sqrt{I_{OL}}$ from O. The locus of the points Q thus obtained forms a surface (Fig. 9.30). The equation of that surface may be obtained by substituting $1/(OQ)^2$ for I_{OL} in (9.46) and multiplying both sides of the equation by $(OQ)^2$. Observing that

$$(OQ)\lambda_x = x \qquad (OQ)\lambda_y = y \qquad (OQ)\lambda_z = z$$

where x, y, z denote the rectangular coordinates of a point Q of the surface, we write

$$I_x x^2 + I_y y^2 + I_z z^2 - 2I_{xy}xy - 2I_{yz}yz - 2I_{zx}zx = 1 \qquad (9.48)$$

The equation obtained is that of a *quadric*. Since the moment of inertia I_{OL} is different from zero for every axis OL, no point Q may be at an infinite

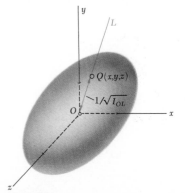

Fig. 9.30

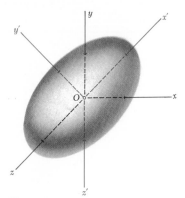

Fig. 9.31

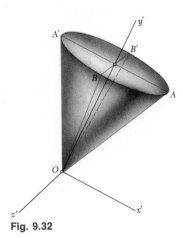

Fig. 9.32

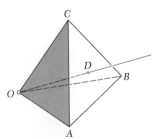

Fig. 9.33

distance from O. Thus, the quadric obtained is an *ellipsoid*. This ellipsoid, which defines the moment of inertia of the body with respect to any axis through O, is known as the *ellipsoid of inertia* of the body at O.

We observe that if the axes in Fig. 9.30 are rotated, the coefficients of the equation defining the ellipsoid change, since these are equal to the moments and products of inertia of the body with respect to the rotated coordinate axes. However, the *ellipsoid itself remains unaffected*, since its shape depends only upon the distribution of mass in the body considered. Suppose now that we choose as coordinate axes the principal axes x', y', z' of the ellipsoid of inertia (Fig. 9.31). The equation of the ellipsoid with respect to these coordinate axes will be of the form

$$I_{x'}x'^2 + I_{y'}y'^2 + I_{z'}z'^2 = 1 \qquad (9.49)$$

which does not contain any product of coordinates. Thus, the products of inertia of the body with respect to the x', y', z' axes are zero. The x', y', z' axes are known as the *principal axes of inertia* of the body at O, and the coefficients $I_{x'}$, $I_{y'}$, $I_{z'}$ as the *principal moments of inertia* of the body at O. Note that, given a body of arbitrary shape and a point O, it is always possible to find axes which are the principal axes of inertia of the body at O, that is, axes with respect to which the products of inertia of the body are zero. Indeed, no matter how odd or irregular the shape of the body may be, the moments of inertia of the body with respect to axes through O will define an ellipsoid, and this ellipsoid will have principal axes which, by definition, are the principal axes of the body at O.

If the principal axes of inertia x', y', z' are used as coordinate axes, the expression obtained in Eq. (9.46) for the moment of inertia of a body with respect to an arbitrary axis through O reduces to

$$I_{OL} = I_{x'}\lambda_{x'}^2 + I_{y'}\lambda_{y'}^2 + I_{z'}\lambda_{z'}^2 \qquad (9.50)$$

While the determination of the principal axes of inertia of a body of arbitrary shape is somewhat involved and would require solving a cubic equation, there are many cases when these axes may be spotted immediately. Consider, for instance, the homogeneous cone of elliptical base shown in Fig. 9.32; this cone possesses two mutually perpendicular planes of symmetry OAA' and OBB'. We check from the definition (9.45) that if the $x'y'$ and $y'z'$ planes are chosen to coincide with the two planes of symmetry, all the products of inertia are zero. The x', y', and z' axes thus selected are therefore the principal axes of inertia of the cone at O. In the case of the homogeneous regular tetrahedron $OABC$ shown in Fig. 9.33, the line joining the corner O to the center D of the opposite face is a principal axis of inertia at O and any line through O perpendicular to OD is also a principal axis of inertia at O. This property may be recognized if we observe that a rotation through $120°$ about OD leaves the shape and the mass distribution of the tetrahedron unchanged. It follows that the ellipsoid of inertia at O also remains unchanged under this rotation. The ellipsoid, therefore, is of revolution about OD, and the line OD, as well as any perpendicular line through O, must be a principal axis of the ellipsoid.

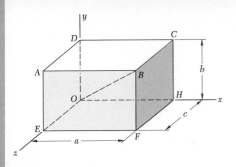

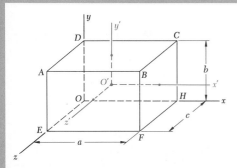

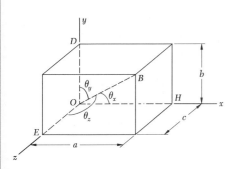

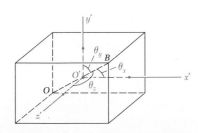

SAMPLE PROBLEM 9.14

Consider a rectangular prism of mass m and sides a, b, c. Determine (a) the mass moments and products of inertia of the prism with respect to the coordinate axes shown, (b) its moment of inertia with respect to the diagonal OB.

a. **Moments and Products of Inertia with Respect to the Coordinate Axes.** *Moments of Inertia.* Introducing the centroidal axes x', y', z', with respect to which the moments of inertia are given in Fig. 9.28, we apply the parallel-axis theorem:

$$I_x = \bar{I}_{x'} + m(\bar{y}^2 + \bar{z}^2) = \tfrac{1}{12}m(b^2 + c^2) + m(\tfrac{1}{4}b^2 + \tfrac{1}{4}c^2)$$
$$I_x = \tfrac{1}{3}m(b^2 + c^2) \quad \blacktriangleleft$$

Similarly,
$$I_y = \tfrac{1}{3}m(c^2 + a^2) \qquad I_z = \tfrac{1}{3}m(a^2 + b^2) \quad \blacktriangleleft$$

Products of Inertia. Because of symmetry, the products of inertia with respect to the centroidal axes x', y', z' are zero and these axes are principal axes of inertia. Using the parallel-axis theorem, we have

$$I_{xy} = \bar{I}_{x'y'} + m\bar{x}\bar{y} = 0 + m(\tfrac{1}{2}a)(\tfrac{1}{2}b) \qquad I_{xy} = \tfrac{1}{4}mab \quad \blacktriangleleft$$

Similarly,
$$I_{yz} = \tfrac{1}{4}mbc \qquad I_{zx} = \tfrac{1}{4}mca \quad \blacktriangleleft$$

b. **Moment of Inertia with Respect to OB.** We recall Eq. (9.46):

$$I_{OB} = I_x\lambda_x^2 + I_y\lambda_y^2 + I_z\lambda_z^2 - 2I_{xy}\lambda_x\lambda_y - 2I_{yz}\lambda_y\lambda_z - 2I_{zx}\lambda_z\lambda_x$$

where the direction cosines of OB are

$$\lambda_x = \cos\theta_x = \frac{OH}{OB} = \frac{a}{(a^2 + b^2 + c^2)^{1/2}}$$

$$\lambda_y = \frac{b}{(a^2 + b^2 + c^2)^{1/2}} \qquad \lambda_z = \frac{c}{(a^2 + b^2 + c^2)^{1/2}}$$

Substituting the values obtained for the moments and products of inertia and for the direction cosines:

$$I_{OB} = \frac{1}{a^2 + b^2 + c^2}\left[\tfrac{1}{3}m(b^2 + c^2)a^2 + \tfrac{1}{3}m(c^2 + a^2)b^2 + \tfrac{1}{3}m(a^2 + b^2)c^2\right.$$
$$\left. - \tfrac{1}{2}ma^2b^2 - \tfrac{1}{2}mb^2c^2 - \tfrac{1}{2}mc^2a^2\right]$$
$$I_{OB} = \frac{m}{6}\frac{a^2b^2 + b^2c^2 + c^2a^2}{a^2 + b^2 + c^2} \quad \blacktriangleleft$$

Alternate Solution. The moment of inertia I_{OB} may be obtained directly from the principal moments of inertia $\bar{I}_{x'}$, $\bar{I}_{y'}$, $\bar{I}_{z'}$, since the line OB passes through the centroid O'. Since the x', y', z' axes are principal axes of inertia, we use Eq. (9.50) and write

$$I_{OB} = \bar{I}_{x'}\lambda_x^2 + \bar{I}_{y'}\lambda_y^2 + \bar{I}_{z'}\lambda_z^2$$
$$= \frac{1}{a^2 + b^2 + c^2}\left[\frac{m}{12}(b^2 + c^2)a^2 + \frac{m}{12}(c^2 + a^2)b^2 + \frac{m}{12}(a^2 + b^2)c^2\right]$$
$$I_{OB} = \frac{m}{6}\frac{a^2b^2 + b^2c^2 + c^2a^2}{a^2 + b^2 + c^2} \quad \blacktriangleleft$$

Problems

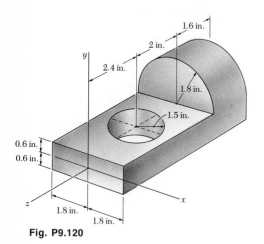

Fig. P9.120

9.120 and 9.121 Determine the mass products of inertia I_{xy}, I_{yz}, and I_{zx} of the steel machine element shown. (Specific weight of steel = 490 lb/ft³; density of steel = 7850 kg/m³.)

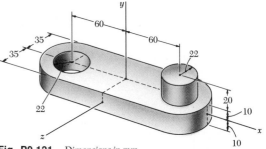

Fig. P9.121   Dimensions in mm

9.122 A section of sheet steel, 2 mm thick, is cut and bent into the machine component shown. Knowing that the density of steel is 7850 kg/m³, determine the mass products of inertia I_{xy}, I_{yz}, and I_{zx} of the component.

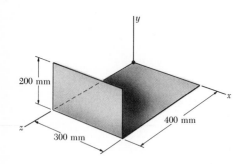

Fig. P9.122

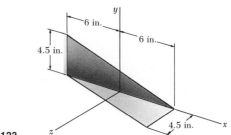

Fig. P9.123

9.123 A section of sheet steel, 0.03 in. thick, is cut and bent into the sheet-metal machine component shown. Knowing that the specific weight of steel is 490 lb/ft³, determine the mass products of inertia I_{xy}, I_{yz}, and I_{zx} of the component. (*Hint.* See results of Sample Prob. 9.6.)

9.124 For the corner reflector of Prob. 9.112, determine the mass products of inertia I_{xy}, I_{yz}, and I_{zx}. (*Hint.* See results of Sample Prob. 9.6 and Prob. 9.56.)

9.125 A homogeneous wire with a mass per unit length of 1.8 kg/m is used to form the figure shown. Determine the mass products of inertia I_{xy}, I_{yz}, and I_{zx} of the wire figure.

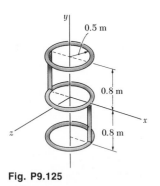

Fig. P9.125

9.126 Complete the derivation of Eqs. (9.47), which express the parallel-axis theorem for mass products of inertia.

9.127 For the homogeneous tetrahedron of mass m which is shown, (a) determine by direct integration the mass product of inertia I_{zx}, (b) deduce I_{yz} and I_{xy} from the result obtained in part a.

9.128 Determine the mass moment of inertia of the right circular cone of Sample Prob. 9.11 with respect to a generator of the cone.

9.129 Determine the mass moment of inertia of the rectangular prism of Sample Prob. 9.14 with respect to the diagonal OF of its base.

9.130 Determine the mass moment of inertia of the corner reflector of Probs. 9.112 and 9.124 with respect to the axis through the origin which forms equal angles with the x, y, and z axes.

9.131 Determine the mass moment of inertia of the machine part of Sample Prob. 9.13 with respect to an axis through the origin characterized by the unit vector $\boldsymbol{\lambda} = \frac{2}{3}\mathbf{i} + \frac{1}{3}\mathbf{j} + \frac{2}{3}\mathbf{k}$.

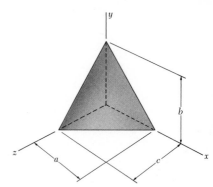

Fig. P9.127

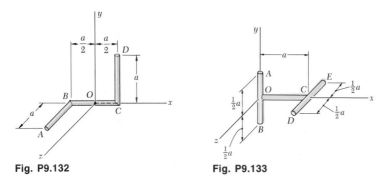

Fig. P9.132 **Fig. P9.133**

9.132 and 9.133 Three uniform rods, each of mass m, are welded together as shown. Determine (a) the mass moments of inertia and the mass products of inertia with respect to the coordinate axes, (b) the mass moment of inertia with respect to a line joining the origin O and point D.

9.134 The thin bent plate shown is of uniform density and mass m. Determine its mass moment of inertia with respect to a line joining points B and C.

9.135 The thin bent plate shown is of uniform density and mass m. Determine its mass moment of inertia with respect to a line joining the origin O and point A.

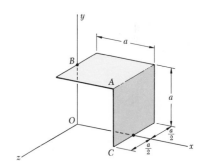

Fig. P9.134 and P9.135

9.136 Determine the value of the ratio a/h for which the ellipsoid of inertia of the right circular cone of Sample Prob. 9.11 is a sphere when computed (a) at the apex of the cone, (b) at the centroid of the cone.

9.137 Consider a homogeneous circular cylinder of radius a and length L. Determine the value of the ratio a/L for which the ellipsoid of inertia of the cylinder is a sphere when computed (a) at the centroid of the cylinder, (b) at the center of one of its bases.

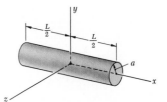

Fig. P9.137

9.138 Given an arbitrary solid and three rectangular axes x, y, and z, prove that the mass moment of inertia of the solid with respect to any one of the three axes cannot be larger than the sum of the moments of inertia of the solid with respect to the other two axes; i.e., prove that the inequality $I_x \leq I_y + I_z$ is satisfied, as well as two similar inequalities. Further prove that, if the solid is homogeneous and of revolution, and if x is the axis of revolution and y a transverse axis, then $I_y \geq \frac{1}{2}I_x$.

9.139 Consider a cube of mass m and side a. (*a*) Show that the ellipsoid of inertia at the center of the cube is a sphere, and use this property to determine the mass moment of inertia of the cube with respect to one of its diagonals. (*b*) Show that the ellipsoid of inertia at one of the corners of the cube is an ellipsoid of revolution, and determine the principal moments of inertia of the cube at that point.

9.140 Given a homogeneous solid of mass m and of arbitrary shape, and three rectangular axes x, y, and z of origin O, prove that the sum $I_x + I_y + I_z$ of the mass moments of inertia of the solid cannot be smaller than the similar sum computed for a sphere of the same mass and same material centered at O. Further prove, using the result of Prob. 9.138, that if the solid is of revolution and if x is the axis of revolution, then its moment of inertia I_y about a transverse axis y cannot be smaller than $\frac{3}{10}ma^2$, where a is the radius of the sphere of the same mass and same material.

Review and Summary

Moments of inertia of masses

Appendix B was devoted to the determination of *moments of inertia of masses*, which are encountered in dynamics in problems involving the rotation of a rigid body about an axis. The mass moment of inertia of a body with respect to an axis AA' (Fig. 9.39) was defined as

$$I = \int r^2 \, dm \tag{9.28}$$

where r is the distance from AA' to the element of mass [Sec. 9.11], and its *radius of gyration* as

$$k = \sqrt{\frac{I}{m}} \tag{9.29}$$

The moments of inertia of a body with respect to the coordinates axes were expressed as

$$I_x = \int (y^2 + z^2) \, dm$$

$$I_y = \int (z^2 + x^2) \, dm \tag{9.30}$$

$$I_z = \int (x^2 + y^2) \, dm$$

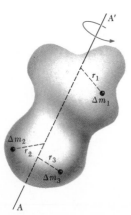

Fig. 9.39

We saw that the *parallel-axis theorem* also applies to mass moments of inertia [Sec. 9.12]. Thus, the moment of inertia I of a body with respect to an arbitrary axis AA' (Fig. 9.40), may be expressed as

$$I = \bar{I} + md^2 \tag{9.33}$$

where $\bar{I}$ is the moment of inertia of the body with respect to a centroidal axis BB' parallel to the axis AA', m the mass of the body, and d the distance between the two axes.

Parallel-axis theorem

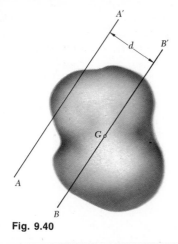

Fig. 9.40

The moments of inertia of *thin plates* may be readily obtained from the moments of inertia of their areas [Sec. 9.13]. We found that for a *rectangular plate* the moments of inertia with respect to the axes shown (Fig. 9.41) are

Moments of inertia of thin plates

$$I_{AA'} = \tfrac{1}{12}ma^2 \qquad I_{BB'} = \tfrac{1}{12}mb^2 \tag{9.39}$$
$$I_{CC'} = I_{AA'} + I_{BB'} = \tfrac{1}{12}m(a^2 + b^2) \tag{9.40}$$

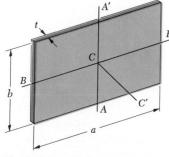

Fig. 9.41

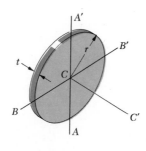

Fig. 9.42

while for a *circular plate* (Fig. 9.42) they are

$$I_{AA'} = I_{BB'} = \tfrac{1}{4}mr^2 \tag{9.41}$$
$$I_{CC'} = I_{AA'} + I_{BB'} = \tfrac{1}{2}mr^2 \tag{9.42}$$

Composite bodies

When a body possesses *two planes of symmetry*, it is usually possible to determine its moment of inertia with respect to a given axis through a *single integration* by considering an element of mass dm in the shape of a thin plate [Sample Probs. 9.10 and 9.11]. On the other hand, when a body is made of *several common geometric shapes*, its moment of inertia with respect to a given axis may be obtained from the values given in Fig. 9.28 and the application of the parallel-axis theorem [Sample Probs. 9.12 and 9.13].

Moment of inertia with respect to an arbitrary axis

In the second part of the appendix, we learned to determine the moment of inertia of a body *with respect to an arbitrary axis OL* drawn through the origin O [Sec. 9.16]. Denoting by λ_x, λ_y, λ_z the components of the unit vector $\boldsymbol{\lambda}$ along OL (Fig. 9.43) and introducing the *products of inertia*

$$I_{xy} = \int xy \, dm \qquad I_{yz} = \int yz \, dm \qquad I_{zx} = \int zx \, dm \quad (9.45)$$

we found that the moment of inertia of the body with respect to OL could be expressed as

$$I_{OL} = I_x \lambda_x^2 + I_y \lambda_y^2 + I_z \lambda_z^2 - 2I_{xy} \lambda_x \lambda_y - 2I_{yz} \lambda_y \lambda_z - 2I_{zx} \lambda_z \lambda_x \quad (9.46)$$

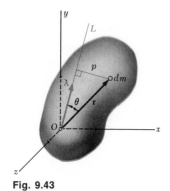

Fig. 9.43

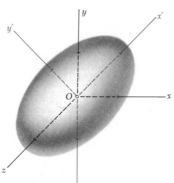

Fig. 9.44

Ellipsoid of inertia

By plotting a point Q along each axis OL at a distance $OQ = 1/\sqrt{I_{OL}}$ from O [Sec. 9.17], we obtained the surface of an ellipsoid, known as the *ellipsoid of inertia* of the body at point O. The principal axes x', y', z' of this ellipsoid (Fig. 9.44), are the *principal axes of inertia* of the body; that is, the products of inertia $I_{x'y'}$, $I_{y'z'}$, $I_{z'x'}$ of the body with respect to these axes are all zero. There are many situations where the principal axes of inertia of a body may be spotted from properties of symmetry of the body. Choosing these axes as coordinate axes, we may then express I_{OL} as

$$I_{OL} = I_{x'} \lambda_{x'}^2 + I_{y'} \lambda_{y'}^2 + I_{z'} \lambda_{z'}^2 \quad (9.50)$$

where $I_{x'}$, $I_{y'}$, $I_{z'}$ are the *principal moments of inertia* of the body at O.

How to Use the Interactive Tutorial Software

With Software Designed for an IBM PC™ or Compatible Computer:

C.1. Introduction. Two diskettes containing interactive software have been placed in the pocket inside the rear cover of this text. This software includes 12 "Tutorials," each consisting of a problem in dynamics for which students are more likely to require additional help. The tutorials were designed by the authors of this text, programmed by EnginComp Software, Inc., and should be easy to use and understand.

C.2. Equipment. This software is designed for an IBM PC or compatible computer with at least 256K of memory and DOS 2.0 and higher. Your IBM PC compatible must also have a graphics display card equivalent to the IBM color/graphics adapter.

C.3. Getting Started. In order to start the Supplementary Tutorial program, your system must be activated and the DOS version you are using must be loaded into the computer. Insert your tutorial disk into the A drive and type "A:". The computer should then prompt you with "A⟩". Simply type "TUTORIAL" following this prompt to enter the program.

The program will display a title page, and at the bottom of the screen it will type "PRESS ANY KEY TO CONTINUE". Throughout the tutorials, all inputs required and key strokes will be specified at the bottom of the screen in a similar fashion.

After striking any key, the program will display the main menu, from which all the tutorials can be accessed. The cursor can be moved by using the [↓] and [↑] arrow keys or by typing the number corresponding to your choice. Selections are made by striking the [F10] key when your choice is highlighted.

C.4. Inside the Tutorials. Once you have made your choice of tutorial on the main menu and pressed [F10], the program will show you the tutorial title page on the bottom half of the screen, and a "Help" screen on the top half. This "Help" screen lists all the special function keys and an explanation of their action. It can be displayed at any time during the program by simply striking [F1].

In a similar fashion, you can exit the tutorial at any time by striking the [Esc] key. The program will then ask you to strike [F10] if you want to exit to the main menu, or any other key to continue. This prevents accidental termination of a tutorial.

The title page of each tutorial is followed by the statement of the problem, and you will be asked to select part of the data involved. This is done by entering each item and striking the [F10] key after all the data has been entered.

In several instances, and after giving a wrong answer, you will be invited to strike the [F2] key to obtain an explanatory comment before trying again or proceeding with the solution.

You may review your progress at any time by striking the [Home] key and then [F10]. This will return you to the beginning of the tutorial. Striking [F10] repeatedly will then enable you to review rapidly your solution up to the point that you had reached.

In the course of a tutorial you will sometimes encounter calculations requiring the use of a hand calculator. You will then be asked whether you wish to perform these calculations yourself, thus getting additional practical experience, or whether you wish your "Tutor"—i.e., the computer—to do the calculations for you, saving you time and effort.

At the end of each tutorial, you will be asked if you wish to try again the tutorial with some different data. Entering a "Y" (for yes) will take you to the beginning of the tutorial, while entering an "N" (for no) will take you back to the main menu.

Sometimes it is convenient to have a hard copy of tutorial screens. In order to do this, you must have a printer that can handle text and graphics information. Also, you must run the DOS Graphics program before starting the tutorials. In order to get a hard copy, simply strike the [Shift] [Prt Sc] keys.

C.5. Keeping Score of Your Correct Answers.
The computer will keep score of your correct answers in the upper right-hand corner of the screen. The use of the [F9] key to supply the correct answer will count against your score, while the use of the [F10] key to review your progress will have no effect on your score at all. Your score will not be affected either, if you choose to have your "Tutor" perform the calculations for you when given that choice. However, if you choose to do the calculations yourself, your score will reflect the accuracy of your computations.

C.6. Exiting the Tutorials.
To exit the tutorials, simply get back to the main menu, select the option "Exit Dynamics Tutorials," using the arrow keys or striking the [Esc] key, and then press [F10]. Always exit the tutorials before turning off your computer.

C.7. Introduction. A diskette containing interactive software has been placed in the pocket inside the rear cover of this text. This software includes 12 "Tutorials," each consisting of a problem in dynamics for which students are more likely to require additional help. All the tutorials were designed by the authors of this text, programmed by EnginComp Software, Inc., and should be easy to use and understand.

C.8. Equipment. This software is supplied on a double-sided 800K disk and is designed for a Macintosh computer with at least 512K of RAM, i.e., a Mac 512E, Mac Plus, Mac SE, or Mac II.

C.9. Getting Started. In order to start the Supplementary Tutorial program, your system must be activated with a Startup Disk, either a floppy or hard disk with a System Folder on it. (A system version of 4.1 or higher is recommended.) Note that the Dynamics Tutorials disk does *not* contain a System Folder and *cannot* be used as a Startup Disk.

After starting up and inserting your tutorial disk in the computer, you will notice a disk icon in the upper-right-hand corner of the screen labeled "Vector Mechanics." Double clicking (moving the pointer over the disk icon and clicking the mouse button twice) will open the disk's directory window. Double clicking the icon labeled "Dynamics Tutorials" will launch the program by displaying the title page.

Clicking "Tutorials" in the lower-right-hand corner of the title page will bring a message at the bottom of the screen asking you to choose a lesson from the tutorials menu. Clicking on the heading "Tutorials" will cause the full menu to drop down. Moving the pointer down until the desired tutorial has been reached and releasing the mouse button will display the statement of the problem.

C.10. Inside the Tutorials. After selecting a tutorial from the Tutorials menu and clicking "Continue" at the statement of the problem, you will see an introductory figure and be asked to select part of the data involved. As you proceed through the tutorial you will be prompted to input more data, answer questions, or supply the missing elements of formulas and equations.

If you answer incorrectly, the bottom prompt will notify you when the correct answer and/or an explanatory comment are available in the Help menu. Note that all selections in this menu have shortcut command-key equivalents; "Answer," for instance, may be chosen by holding down the Cloverleaf (command) key and striking the [A] key.

You may review your progress at any time by selecting "Restart" from the Help menu. This will return you to the beginning of the tutorial. Striking [Enter] repeatedly will then enable you to rapidly review your solution up to the point that you have reached.

In the course of a tutorial you will sometimes encounter calculations requiring the use of a hand calculator. You will then be asked whether you wish to perform these calculations yourself, thus getting additional practical experience, or whether you wish your "Tutor"—i.e., the computer—to do the calculations for you, saving you time and effort.

At the end of each tutorial, you will be asked if you wish to try the tutorial again with some different data. Entering [Y] (for yes) will take you to the beginning of the tutorial, while entering [N] (for no) will take you back to the title screen.

Sometimes it is convenient to have a hard copy of tutorial screens. In order to do this, you must have an Imagewriter (I or II) printer connected to your Macintosh, turned on, and selected. Then choose "Print Screen" from the File menu.

C.11. Keeping Score of Your Correct Answers. The computer will keep score of your correct answers in the upper-right-hand corner of the screen. The use of "Answer" to supply the correct answer will count against your score, while the use of "Restart" to review your progress will have no effect on your score at all. Your score will not be affected either, if you choose to have your "Tutor" perform the calculations for you when given that choice. However, if you choose to do the calculations yourself, your score will reflect the accuracy of your computations.

C.12. Exiting the Tutorials. To exit the tutorials, choose "Quit" from the File menu. You will be able to select whether you wish to return to the title screen to begin another tutorial or completely quit the tutorial program. You may quit from either the title screen or from within a tutorial. Always exit the tutorial program before turning off your computer.

Index

Watt (unit), 582
Wedges, 334
Weight, 4, 529, 551
Wheel friction, 344–345
Wheels, 150, 151, 344, 811
Work:
    of a couple, 428–429, 442,
        837

Work (*Cont.*):
    and energy, principle of (*see*
        Principle, of work and energy)
    of a force, 427, 441, 574–577
    of force exerted by spring, 443, 576–
        577
    of forces on a rigid body, 428, 430,
        837

Work (*Cont.*):
    of gravitational force, 577
    input and output, 432–433
    virtual, 429
    of a weight, 442–443, 576
Wrench, 104–105

Zero-force member, 230

Answers to Even-Numbered Problems †

CHAPTER 11

11.2 $x = 14$ m, $v = 9$ m/s, $a = 12$ m/s^2.

11.4 $t = 0$, $x = 10$ ft, $a = -12$ ft/s^2;
$t = 2$ s, $x = 2$ ft, $a = 12$ ft/s^2.

11.6 (a) 1 s, 3 s. (b) $x = 25$ m, $a = 18$ m/s^2, 28 m.

11.8 $v = -8$ ft/s, $x = 64$ ft, 80 ft.

11.10 (a) 3 s. (b) 13 m, -28 m/s. (c) 32.5 m.

11.12 (a) 24 m^3/s^2. (b) 11.49 m/s.

11.14 (a) 25 s^{-2}. (b) 11.18 in./s.

11.16 (a) 15 m/s. (b) 14.74 m/s. (c) 15.25 m/s.

11.18 (a) 75 in. (b) Infinite. (c) 11.51 s.

11.20 (a) 240 ft. (b) Infinite.

11.22 (a) 6.73 m. (b) 4.64 m/s.

11.24 (a) 400 m. (b) -0.160 m/s^2. (c) 14.38 s.

11.26 (a) 2.04 km. (b) 211 km. (c) Infinite.

11.28 (a) $v = k\dfrac{T}{\pi}\left(1 - \cos\dfrac{\pi t}{T}\right)$,

$x = k\dfrac{T^2}{\pi^2}\left(\dfrac{\pi t}{T} - \sin\dfrac{\pi t}{T}\right)$.

(b) $2kT/\pi$. (c) $2kT^2/\pi$. (d) kT/π.

11.30 (a) 9.62 m/s ↑. (b) 29.6 m/s ↓.

11.32 (a) 13 ft/s. (b) 67 ft/s. (c) 220 ft.

11.34 (a) 12.36 s. (b) 60.4 km/h.

11.36 11.60 s, 50.4 m.

11.38 (a) 15.05 s, 734 ft.
(b) $\mathbf{v}_A = 42.5$ mi/h →, $\mathbf{v}_B = 23.7$ mi/h →.

11.40 (a) 10 m/s ↑. (b) 5 m/s ↑. (c) 15 m/s ↑.
(d) 10 m/s ↑.

11.42 (a) 27 in./s →. (b) 36 in./s →. (c) 9 in./s →.
(d) 18 in./s ←.

11.44 (a) $\mathbf{a}_A = 50$ mm/s^2 ←, $\mathbf{a}_B = 25$ mm/s^2 →.
(b) 800 mm →, 200 mm/s →.

11.46 (a) 5 mm/s^2 ↓, 10 mm/s ↑. (b) 90 mm ↑.

11.48 (a) 2.5 s. (b) 7.5 in. ↓.

11.50 $\mathbf{v}_A = 150$ mm/s ↑. $\mathbf{v}_B = 50$ mm/s ↓,
$\mathbf{v}_C = 250$ mm/s ↓.

11.52 (b) 20 ft/s, 74 ft, 176 ft.

11.54 (a) 54 m. (b) 18 s, 30 s.

11.56 36 s.

11.58 (a) 1125 ft/s^2, -81.5 ft/s^2. (b) 2.70 in. (c) 0.296 s.

11.60 (a) ± 0.714 m/s^2. (b) 18 km/h.

11.62 (a) 12.5 s, 187.5 m. (b) 20 s, 300m.

11.64 8.54 s, 58.3 mi/h.

11.66 104.4 km/h.

11.68 (a) 50 m/s, 1194 m. (b) 59.25 m/s.

11.70 (a) 10 s. (b) 5 ft/s.

11.72 (a) 45 mi/h. (b) 239 ft.

11.74 (a) 9.71 s. (b) 130.2 m.

11.78 (a) 46 m. (b) 54 m.

11.80 (a) 11.66 in./s ↗ 31.0°; 7.21 in./s^2 ↘ 56.3°.
(b) 8 in./s ←; 10 in./s^2 ∠ 36.9°.
(c) 8.49 in./s ∠ 45.0°; 20.9 in./s^2 ∠ 17.7°.

11.82 (a) 5.37 m/s. (b) 2.80 s; $x = -7.56$ m,
$y = 5.52$ m; ↘ 63.4°.

11.86 $v = \sqrt{c^2 + R^2p^2}$; $a = \sqrt{Rp^2}$.

11.88 (a) 16.17 m/s. (b) 32.7 m.

11.90 (a) 57.9 ft/s. (b) 55.6 ft/s $\leq v_0 \leq$ 70.9 ft/s.

11.92 16.83 ft/s $\leq v_0 \leq$ 28.4 ft/s.

11.94 $d_B = 5.47$ m, $d_C = 7.38$ m.

11.96 11.52 m/s $\leq v_0 \leq$ 13.29 m/s.

11.98 23.8 ft/s.

11.100 47.4°.

11.102 (a) 14.61 m. (b) 65.6°.

11.104 29.5°.

11.106 (a) 9.54 ft/s ↘ 27.0°. (b) 11.14 ft/s ↘ 6.5°.
(c) 111.4 ft ↘ 6.5°.

11.108 48.7° west of south.

11.110 (a) 7.01 in./s ↗ 60.6°. (b) 11.69 in./s^2 ↗ 60.6°.

11.112 (a) 52.5 mm/s^2 ∠ 51.4°. (b) 157.6 mm/s ∠ 51.4°.

11.114 20.1 ft/s ↘ 85.1°.

11.116 17.49 km/s ∠ 59.0°.

11.118 21.0 ft/s ↘ 82.8°.

11.120 1815 ft.

† Partial answers are provided or referenced for *all* problems designed to be solved with a computer.

11.122 89.7 ft/s.

11.124 (a) 1.694 m/s^2. (b) 1.406 m/s^2.

11.126 13.80 ft/s^2.

11.128 (a) 45.0 km/h $\measuredangle$ $26.4°$. (b) 2.70 m/s^2 $\measuredangle$ $5.6°$.

11.130 (a) 23.7 ft. (b) 13.03 ft.

11.132 (a) 3810 m. (b) 2480 m.

11.134 $\rho = R + c^2/Rp^2$.

11.136 881 mi.

11.138 84.4 minutes.

11.140 (a) $(60 \text{ in./s})\mathbf{e}_r + (160 \text{ in./s})\mathbf{e}_\theta$.
(b) $-(640 \text{ in./s}^2)\mathbf{e}_r + (640 \text{ in./s}^2)\mathbf{e}_\theta$. (c) 0.

11.142 (a) $v = 2b\omega$; $a = 4b\omega^2$.
(b) $\rho = b$; path is circle of radius b.

11.144 (a) $\mathbf{v} = 2\pi b\mathbf{e}_r$; $\mathbf{a} = -\frac{3}{2}\pi^2 b\mathbf{e}_\theta$.
(b) $\mathbf{v} = -\pi b\mathbf{e}_r + \pi b\mathbf{e}_\theta$; $\mathbf{a} = -\frac{1}{2}\pi^2 b\mathbf{e}_r - \pi^2 b\mathbf{e}_\theta$.

11.146 $v = b\sec^2\theta\,\dot\theta$.

11.148 $a = b\sec^2\theta\,[\ddot\theta + 2\dot\theta^2\tan\theta]$.

11.150 306 m/s $\uparrow$.

11.152 $v = (b/\theta^2)\sqrt{1 + \theta^2}\,\dot\theta$.

11.154 $a = (b\omega^2/\theta^3)\sqrt{4 + \theta^4}$.

11.156 (a) $\mathbf{v} = bk\mathbf{e}_\theta$; $\mathbf{a} = -\frac{1}{2}bk^2\mathbf{e}_r$.
(b) $\mathbf{v} = 2bk\mathbf{e}_r + 2bk\mathbf{e}_\theta$; $\mathbf{a} = 2bk^2\mathbf{e}_r + 4bk^2\mathbf{e}_\theta$.

11.158 $v = 2\pi\sqrt{A^2 + n^2B^2\cos^2 2\pi nt}$;
$a = 4\pi^2\sqrt{A^2 + n^4B^2\sin^2 2\pi nt}$.

11.160 $v = h\tan\beta\sqrt{(1 + 4\pi^2 t^2)} + \cot^2\beta$;
$a = 4\pi h\tan\beta\sqrt{1 + \pi^2 t^2}$.

11.162 $v = \pi A\sqrt{4 + \sin^2 2\pi t}$;
$a = 2\pi^2 A\sqrt{4 + \cos^2 2\pi t}$.

11.164 $\tan^{-1}(Rp/c)$.

11.166 2.44 m/s^2.

11.168 (a) $\mathbf{v} = -(8\pi \text{ in./s})\mathbf{i}$; $\mathbf{a} = (8\pi^2 \text{ in./s}^2)\mathbf{j}$.

11.170 60.3 km $\measuredangle$ $17.0°$.

11.172 13.59 m.

11.174 (a) $\mathbf{v} = (18.85 \text{ in./s})\mathbf{e}_\theta$; $\mathbf{a} = -(296 \text{ in./s}^2)\mathbf{e}_r$.
(b) $\mathbf{v} = -(37.7 \text{ in./s})\mathbf{e}_r$; $\mathbf{a} = -(237 \text{ in./s}^2)\mathbf{e}_\theta$.

11.176 (a) 40 mm/s $\measuredangle$ $70°$. (b) 358 mm $\measuredangle$ $46.6°$.

11.C1 $v_0 = 250$ m/s: $h = 1432$ m, $v_f = 122.2$ m/s.

11.C2 $v_0 = 6$ m/s: Strikes ground 3.76 m from firefigher; $v_0 = 10$ m/s: Strikes building 5.02 m from ground; $v_0 = 13$ m/s: Strikes roof 6.74 m from B; $v_0 = 15$ m/s: Strikes ground 20.5 m from firefighter.

11.C3 (a) Range $= 44,000$ ft; $t = 30$ s: $x = 27,600$ ft, $y = 8650$ ft.
(b) Range $= 41,300$ ft; $t = 30$ s: $x = 26,700$ ft, $y = 8220$ ft.
(c) Range $= 33,500$ ft; $t = 30$ s: $x = 24,000$ ft, $y = 6750$ ft.
(d) Range $= 27,400$ ft; $t = 30$ s: $x = 21,300$ ft, $y = 5340$ ft.

11.C4 38.4 s.

CHAPTER 12

12.2 (a) 3.24 N. (b) 2 kg.

12.4 3.63×10^6 kg·m/s.

12.6 (a) 5.00 s. (b) 0.612.

12.8 217 ft.

12.10 (a) $\mathbf{a}_A = 2.89$ ft/s^2 $\nearrow$; $\mathbf{a}_B = 1.446$ ft/s^2 $\searrow$.
(b) 124.0 lb.

12.12 (a) 469 m. (b) 13.33 kN C.

12.14 (a) 2.21 m/s^2 $\nearrow$. (b) 192.3 N.

12.16 $\mathbf{a}_A = \mathbf{a}_B = 4.65$ ft/s$^2 \rightarrow$. (b) 3.33 lb $\rightarrow$.

12.18 169.1 ft.

12.20 $\mathbf{a}_1 = 12.25$ m/s^2 $\nearrow$; $\mathbf{a}_2 = 3.87$ m/s^2 $\swarrow$.

12.22 $(mv_0/T_0)\ln 2$.

12.24 45.8 s.

12.26 (a) 25.8 ft/s $\downarrow$. (b) 0.

12.28 (a) 33.6 N. (b) $\mathbf{a}_A = 4.76$ m/s$^2 \rightarrow$;
$\mathbf{a}_B = 3.08$ m/s$^2 \downarrow$; $\mathbf{a}_C = 1.401$ m/s$^2 \leftarrow$.

12.30 (a) 22.4 ft/s^2 $\measuredangle$ $30°$. (b) 19.40 ft/s$^2 \rightarrow$.

12.32 (a) 2.49 m/s$^2 \leftarrow$. (b) 117.6 N.

12.34 (a) 14.24 ft/s. (b) $66.4°$.

12.36 3.47 m/s.

12.38 3.01 m/s $\leq v \leq 3.96$ m/s.

12.40 9.64 ft/s $\leq v \leq 15.85$ ft/s.

12.42 (a) $0.742W$. (b) $0.940W$.

12.44 (a) 1148 m. (b) 282 N; 1189 N.

12.46 $v_A = 86.5$ mi/h; $v_B = 135.1$ mi/h.

12.48 $24.1° \leq \theta \leq 155.9°$.

12.50 $v_{\min} = \sqrt{gr\tan(\theta - \phi_s)}$; $v_{\max} = \sqrt{gr\tan(\theta + \phi_s)}$

12.52 92.2 mi/h.

12.54 $0°$; $69.6°$; $180°$.

12.56 (a) Does not slide; 0.611 N $\measuredangle$ $75°$.
(b) Slides up; 0.957 N $\measuredangle$ $40°$.

12.58 0.512 s.

12.60 $10.1°$; $146.5°$.

12.62 $\sqrt{eV/mv_0^2}$.

12.64 (a) $F_r = 0.745$ lb, $F_\theta = 0$.
(b) $F_r = -3.98$ lb, $F_\theta = 3.98$ lb.

12.66 (a) $F_r = -118.4$ N, $F_\theta = 0$.
(b) $F_r = -39.5$ N, $F_\theta = 79.0$ N.

12.68 (a) 64.0 N. (b) 15.36 N.

12.70 (a) $v_r = -2.40$ m/s, $v_\theta = 3.60$ m/s.
(b) $a_r = -54.0$ m/s^2, $a_\theta = -57.6$ m/s^2.
(c) $F_\theta = -11.52$ N.

12.74 (b) $2\,mv_0^2/r_0$.

12.78 (a) 35.77×10^3 km, $22,230$ mi.
(b) 3070 m/s, $10,090$ ft/s.

12.80 (a) 3880 mi/h. (b) 317 mi/h.

12.82 5400 m/s.

12.84 (a) 0, 0. (b) $+300$ in./s^2. (c) $+6.00$ in./s.

12.86 (a) 0.450 m/s. (b) $a_r = +2.50$ m/s^2, $a_\theta = 0$.
(c) $+2.84$ m/s^2.

12.92 1.90×10^{27} kg.

12.94 18.82 km/s.

12.98 (a) 17,120 mi/h. (b) 12,720 mi.

12.100 (a) 159 m/s. (b) 70.1 m/s. (c) 2370 m/s.

12.102 3 days 23 h 29 min.

12.104 (a) 1 h 32 min 10 s. (b) 44 min 27 s.

12.106 19 h 17 min.

12.110 78.6°.

12.116 0.491 m/s^2.

12.118 (a) $F_r = -(6.71 \text{ lb}) \cos \theta$, $F_\theta = -(6.71 \text{ lb}) \sin \theta$.
(b) $\mathbf{P} = 0$; $\mathbf{Q} = 6.71$ lb ⬈ 2θ.

12.120 (a) $\mathbf{a}_A = 0$; $\mathbf{a}_B = 1.591$ m/s^2 ⬋.
(b) $\mathbf{a}_A = \mathbf{a}_B = 0.644$ m/s^2 ⬋.

12.122 (a) $a_t = g \cos \theta$; $v = \sqrt{2gl \sin \theta}$. (b) 41.8°.

12.124 (a) $\mathbf{a}_A = 6.90$ ft/s^2 ⬈; $\mathbf{a}_B = 3.45$ ft/s^2 ↓.
(b) 178.6 lb.

12.126 (a) 15,260 mi. (b) 7000 ft/s.

12.C1 $\mu = 0.05$: $\mathbf{a}_A = 2.76$ ft/s^2 →,
$\mathbf{a}_{B/A} = 17.16$ ft/s^2 ⬈ 30°;
$\mu = 0.20$: $\mathbf{a}_A = 0$, $\mathbf{a}_{B/A} = 10.52$ ft/s^2 ⬈ 30°;
$\mu = 0.60$: $\mathbf{a}_A = \mathbf{a}_{B/A} = 0$.

12.C2 $0 \le \theta \le 8.2°$; $52.1° \le \theta \le 84.7°$;
$176.3° \le \theta \le 180.0°$.

12.C3 33.8°.

12.C4 (a) $\theta = 60°$: 2.49×10^3 km, 13.98 min.;
$\theta = 110°$: 10.89×10^3 km, 45.3 min.;
periodic time = 1172 min.
(b) $\theta = 60°$: 2.79×10^3 km, 13.64 min.;
$\theta = 110°$: 14.51×10^3 km, 51.0 min.
(c) $\theta = 60°$: 3.94×10^3 km, 12.12 min.;
$\theta = 110°$: 59.3×10^3 km, 119.8 min.

CHAPTER 13

13.2 (a) 1350 J; 45.9 m. (b) 1350 J; 278 m.

13.4 2.84 m.

13.6 20.3 ft.

13.8 (a) 57.8 m. (b) 154 N →.

13.10 (a) 279 ft. (b) $F_{AB} = 19.38$ kips C;
$F_{BC} = 8.62$ kips C.

13.12 (a) 10.10 ft/s ⬈. (b) 121.2 lb.

13.14 (a) 5.89 ft/s ⬈. (b) 123.9 lb.

13.16 (a) 913 mm/s →. (b) 456 mm/s ←. (c) 9.38 N.

13.18 (a) 17.68 J. (b) $\mathbf{F}_A = 125.6$ N ↑; $\mathbf{F}_B = 137.4$ N ↑.

13.20 (a) 1.098 m/s ↑. (b) $\mathbf{F}_A = 218$ N ↑; $\mathbf{F}_D = 479$ N ↑.
(c) 391 J.

13.22 $v_1 = 3.43$ m/s; $v_2 = 4.72$ m/s.

13.24 $v_1 = 1.5$ m/s; $v_2 = 2.29$ m/s.

13.26 2.51 lb.

13.28 509 mm.

13.30 (a) 2.67 ft/s. (b) 20 lb.

13.32 (a) 88.5 ft. (b) 230 ft/s^2.

13.34 36.6 mm ↓; 86.3 N ↑.

13.36 451 mm.

13.38 (a) 11.18 km/s. (b) 2.37 km/s.

13.40 15,141 ft/s.

13.42 41.8°.

13.44 $\mathbf{a}_n = 64.4$ ft/s^2 ↑; $\mathbf{a}_t = 24.2$ ft/s^2 ←.

13.46 (a) 2.80 m/s; 331 mm. (b) $h = 533$ mm;
$d = 98.9$ mm.

13.48 (a) 572 W. (b) 4.71 s.

13.50 (a) 72.7 hp. (b) 342 hp.

13.52 For $t \le 50$ s: $P = (169.3$ kW/s$)t$.
At $t = 50$ s: $P = 8.46$ MW.
For $t > 50$ s: $P = 2.21$ MW.

13.54 3.02 ft/s^2 ↑.

13.56 (a) 375 kW. (b) 5.79 km/h.

13.58 3.19 m/s ↔.

13.60 (a) 11.04 ft/s. (b) 13.04 ft/s.

13.62 (a) 5.12 m/s. (b) 4.20 m/s.

13.64 (a) 1.981 m/s. (b) 221 mm.

13.66 14.45 in.

13.68 (a) 2.99 m/s. (b) 2.83 m/s.

13.70 $\mu_s = 0.364$; $\mu_k = 0.251$.

13.72 (a) 5.21 lb ⬋ 30°. (b) 7.08 lb ↑.

13.74 (b) 3.74 m/s ←.

13.76 (a) $\cot \phi = 0.243(12 - y)$.
(b) 60.0 lb; $\theta_x = 85.7°$, $\theta_y = 71.6°$, $\theta_z = 161.1°$.

13.78 (b) $V = -(x^2 + y^2 + z^2)^{-\frac{1}{2}}$.

13.80 (b) $V = -\frac{1}{4}(x^2 + y^2)^2$.

13.82 (a) $V = mgR[1 - (R/r)]$. (b) $T = \frac{1}{2}mgR^2/r$.
(c) $E = mgR[1 - (R/2r)]$.

13.84 (a) 2.81 MJ/kg. (b) 1.330 MJ/kg.

13.88 (a) 8.59 m/s. (b) 262 mm.

13.90 15.10 m/s.

13.92 (a) 25.3 in. (b) 7.58 ft/s.

13.96 (a) 252 m/s. (b) 8.39 km/s.

13.98 (a) 7960 ft/s. (b) 4820 ft/s.

13.100 3450 m/s.

13.102 21.3×10^3 ft/s.

13.104 63.8°.

13.108 $r_{max} = (1 + \sin \alpha)r_0$; $r_{min} = (1 - \sin \alpha)r_0$.

13.112 4 min 20 s.

13.114 6.37 s.

13.116 (a) 11.42 s. (b) $\mathbf{v} = -(125.5 \text{ m/s})\mathbf{j} - (194.5 \text{ m/s})\mathbf{k}$.

13.118 (a) 5.64 s. (b) $F_{AB} = 19,380$ lb T;
$F_{BC} = 8610$ lb T.

13.120 (a) 0.549 s. (b) 56.8 N.

13.122 (a) $\mathbf{v}_A = \mathbf{v}_B = 4.81$ m/s ⬋. (b) 3.32 N ⬈.

13.124 (a) 3.55 m/s ←; 1.686 s. (b) 3.40 s.

13.126 (a) 7 s. (b) 10.99 ft/s →. (c) 13.49 s.

13.128 4550 lb.

13.130 5.21W.

13.132 (a) 1.929 km/h →. (b) 44.6 kN.

13.134 (a) $v_{max} = 0.438$ m/s →. (b) 1.607 m →.
(c) 11.25 m ←.

13.136 (a) 7.79 ft/s →. (b) 83.6 kips ↑.

13.138 (a) 223 lb·s. (b) 159.2 lb·s.

13.140 (a) $v_A = 3.3$ ft/s →; $v_B = 6.5$ ft/s →.
(b) 0.1677 ft·lb.

13.142 0.875.

13.144 $v_A = 0.0938$ m/s →; $v_B = 0.281$ m/s →;
$v_C = 1.125$ m/s →.

13.146 (a) $v_A = 0.507\,v_0$ ⭸ $50.2°$; $v_B = 0.779\,v_0$ ⭶ $30°$.

13.148 (a) $\cos\theta = 2r/d$. (b) $ABC > \cos^{-1}(2r/d)$;
$ACB > \sin^{-1}(2r/d)$.

13.150 $0.728 \le e \le 0.762$.

13.152 (a) $22.5°$; $2h$. (b) $15.7°$; $0.610h$.

13.154 (a) 0.80. (b) 8 in.

13.156 2.80 s.

13.158 $v_B = 0.795$ ft/s →; $v_A = 4.58$ ft/s ⭶ $48.8°$.

13.160 $v_A = 2.68$ m/s ⭸ $45.7°$; $v_B = 0.624$ m/s ←.

13.162 $\theta_A = 7.3°$; $\theta_B = 25.8°$.

13.164 $d_A = 245$ mm ←; $d_B = 1280$ mm ←.

13.166 (a) 898 mm. (b) 0.442%.

13.168 1.868 in.

13.170 (a) 0.728. (b) 1.095 m.

13.172 96.0 mm.

13.174 (a) 2088 ft·lb. (b) 10 mi/h.

13.176 $v_r = 2.03$ m/s; $v_\theta = 0.225$ m/s.

13.178 (a) 7.5 J. (b) 13.5 J. (c) Part b includes both the
work done by the child and the work done by
the elevator motor in maintaining the constant
velocity.

13.180 95.7 kips.

13.182 155.7 s.

13.184 (a) $\frac{1}{2}L$. (b) $\frac{1}{4}L$.

13.186 (a) 18.75 s. (b) 3.33 kN T.

13.C1 (a) *200 mm from B:* 1.530 m/s;
300 mm from B: 1.655 m/s.
(b) 0.335 s.

13.C2 $v_0^2 = 3gr$: $x = -0.649r$; $v_0^2 = 4gr$: $x = 0.296r$.

13.C3 $\phi_0 = 75°$ and $105°$: $h_{min} = 1504$ km,
$h_{max} = 17\,390$ km;
$\phi_0 = 90°$: $h_{min} = 2400$ km, $h_{max} = 16\,490$ km.

13.C4 $\theta_0 = 0.15°$: (a) $v_A' = 0.1448$ m/s, $\theta_A' = 45.0°$ ⭷ ;
(b) $v_A' = 0.226$ m/s, $\theta_A' = 26.3°$ ⭷;
(c) $v_A' = 0.609$ m/s, $\theta_A' = 8.6°$ ⭷.
$\theta_0 = 4°$: (a) $v_A' = 2.791$ m/s, $\theta_A' = 48.3°$ ⭷;
(b) $v_A' = 2.794$ m/s, $\theta_A' = 46.8°$ ⭷;
(c) $v_A' = 2.82$ m/s, $\theta_A' = 41.0°$ ⭷.

CHAPTER 14

14.2 (a) 6.02 ft/s →. (b) 1286 ft/s →.

14.4 (a) 12.00 km/h →. (b) 3.43 km/h →.

14.6 (a) $v_A = v_B = v_C = 0.50$ km/h →.
(b) $v_A = v_B = 0.25$ km/h →; $v_C = 1.00$ km/h →.

14.8 (a) $v_x = 8.33$ ft/s, $v_z = 7.25$ ft/s.
(b) $-(4.51$ ft·lb·s$)\mathbf{k}$.

14.10 (a) $(3$ m$)\mathbf{i} + (1.5$ m$)\mathbf{j} + (1.5$ m$)\mathbf{k}$.
(b) $(17$ kg·m/s$)\mathbf{i} + (19$ kg·m/s$)\mathbf{j} - (5$ kg·m/s$)\mathbf{k}$.
(c) $(2$ kg·m²/s$)\mathbf{i} + (24.5$ kg·m²/s$)\mathbf{j} +$
$(25.5$ kg·m²/s$)\mathbf{k}$.

14.12 (a) 75.0 mi/h. (b) 1.300 s.

14.14 $x = 255$ m, $y = 7.93$ m, $z = -7.00$ m.

14.16 116.7 ft/s ⭸ $38.2°$.

14.18 $v_A = 919$ m/s; $v_B = 717$ m/s; $v_C = 619$ m/s.

14.20 185.6 km/h east, 70.9 km/h south, 49.2 km/h
down.

14.26 (a) 1293 ft·lb. (b) 832 ft·lb.

14.28 (a) 39.1 kJ. (b) 88.0 kJ.

14.30 (b) $E_A = 125.2 \times 10^3$ ft·lb;
$E_B = 219.1 \times 10^3$ ft·lb.

14.32 $v_A = 33.4$ ft/s ⭶ $32.2°$; $v_B = 18.71$ ft/s ⭸ $39.4°$.

14.34 (a) and (c) $v_A = v_B = \frac{1}{2}v_0$ →. (b) $v_A = v_0$ →;
$v_B = 0$.

14.36 $v_A = 1.710$ m/s; $v_B = 1.607$ m/s; $v_C = 4.42$ m/s.

14.38 $v_A = 0.533\,v_0$ ⭸ $30°$; $v_B = 0.814\,v_0$ ⭶ $19.1°$;
$v_C = 0.231\,v_0$ ←.

14.40 (a) $\frac{1}{2}v_0$. (b) $\frac{3}{4}v_0$. (c) $\frac{1}{8}$.

14.42 $x = 181.7$ mm, $z = 139.4$ mm.

14.44 (a) 0.400 m/s →. (b) 120.0 mm. (c) 6.86 rad/s.

14.46 $v_A = 2.56$ m/s ↑; $v_B = 4.24$ m/s ⭸ $31.9°$.
(c) 2.34 m.

14.48 (a) $v_B = 4.47$ ft/s ⭸ $63.4°$; $v_C = 8.00$ ft/s →.
(c) $c = 3.00$ ft.

14.50 7.00 kN ⭸ $30°$.

14.52 16.14 lb →.

14.54 4.18 m/s.

14.56 $Q_1 = 55.3$ gal/min; $Q_2 = 136.3$ gal/min.

14.58 $\mathbf{C}_x = 234$ N ←, $\mathbf{C}_y = 643$ N ↑,
$\mathbf{M}_C = 114.9$ N·m↾.

14.60 $\mathbf{C} = 388$ lb ↑; $\mathbf{D} = 512$ lb ↑.

14.62 $\mathbf{C}_x = 112.5$ N ←, $\mathbf{C}_y = 2680$ N ↑; $\mathbf{D} = 3470$ N ↑.

14.64 5.63 kips.

14.66 (a) 55.0 kN, 0.909 m below B.
(b) 42.5 kN, 2.35 m below B.

14.68 10.32 Mg.

14.70 (a) 8.33 ft/s². (b) 558 mi/h.

14.72 (a) 50.0 kJ/s. (b) 41.8 km/h.

14.74 (a) $15,450$ hp. (b) $28,060$ hp. (c) 0.551.

14.76 19.53 m³/s.

14.78 412 rpm.

14.80 $v = \sqrt{gh}\,\dfrac{e^{(2\sqrt{gh}/L)t} - 1}{e^{(2\sqrt{gh}/L)t} + 1}$

14.82 *Case 1:* (a) $g/3$. (b) $\sqrt{2gl/3}$.
Case 2: (a) gy/l. (b) $\sqrt{gl}$.

14.84 47.4 ft/s^2.

14.86 (a) qv_0. (b) $m_0 a + q(v_0 + 2at)$.

14.88 1030 lb/s.

14.90 (a) $31.9 \text{ m/s}^2 \uparrow$. (b) $240 \text{ m/s}^2 \uparrow$.

14.92 1498 m/s.

14.94 (a) 6030 ft/s. (b) $31,100 \text{ ft/s}$.

14.96 (a) $105,800 \text{ ft}$. (b) $669,500 \text{ ft}$.

14.98 (a) 173.9 km. (b) 6.38 km/s.

14.102 (a) 5.54 ft/s. (b) $0.641 \text{ ft from } B$.

14.104 $\mathbf{C}_x = 103.6 \text{ lb} \leftarrow$, $\mathbf{C}_y = 74.6 \text{ lb} \downarrow$; $\mathbf{D} = 134.5 \text{ lb} \uparrow$.

14.106 (a) 900 kg. (b) 3580 m/s.

14.108 (a) $\frac{1}{2}mg$. (b) $\frac{1}{2}mg + (mv^2/l)$.

14.110 (a) $P = qv$.

14.112 (a) $\mathbf{v}_A = \mathbf{v}_B = 3.00 \text{ mi/h} \rightarrow$; $\mathbf{v}_C = 0$. (b) $\mathbf{v}_A = \mathbf{v}_B = \mathbf{v}_C = 2.00 \text{ mi/h} \rightarrow$.

14.C1 $\theta_A = 50°$: $\mathbf{v}_A = 81.2 \text{ m/s} \measuredangle 50°$, $\mathbf{v}_B = 41.5 \text{ m/s} \measuredangle 30°$; $\theta_A = 105°$: $\mathbf{v}_A = 113.1 \text{ m/s} \measuredangle 75°$, $\mathbf{v}_B = 72.9 \text{ m/s} \measuredangle 30°$.

14.C2 $W_B = 8 \text{ lb}$: $\mathbf{v}_{B/A} = 10.87 \text{ ft/s} \measuredangle 30°$, $\mathbf{v}_A = 2.28 \text{ ft/s} \rightarrow$. $W_B = 16 \text{ lb}$: $\mathbf{v}_{B/A} = 11.69 \text{ ft/s} \measuredangle 30°$. $\mathbf{v}_A = 3.95 \text{ ft/s} \rightarrow$.

14.C3 $\alpha = 6°$: $a = 4.97 \text{ ft/s}^2$, $v = 528 \text{ mi/h}$; $\alpha = 12°$: $a = 1.639 \text{ ft/s}^2$, $v = 496 \text{ mi/h}$; $\alpha = 18°$: $a = -1.616 \text{ ft/s}^2$, $v = 464 \text{ mi/h}$.

14.C4 $\theta = 30°$: $\mathbf{C} = 1540 \text{ N} \measuredangle 35.8°$, $\mathbf{D} = 1622 \text{ N} \uparrow$; $\theta = 60°$: $\mathbf{C} = 1922 \text{ N} \measuredangle 36.0°$, $\mathbf{D} = 1546 \text{ N} \uparrow$.

CHAPTER 15

15.2 (a) 1.2 rad; 0.40 rad/s; -0.1333 rad/s^2. (b) 0.759 rad; 1.472 rad/s; -0.0491 rad/s^2. (c) 1.20 rad; 0; 0.

15.4 (a) 349 rad/s^2. (b) 1000 rpm.

15.6 $\mathbf{v}_B = (5.4 \text{ m/s})\mathbf{j} - (7.2 \text{ m/s})\mathbf{k}$; $\mathbf{a}_B = -(405 \text{ m/s}^2)\mathbf{i} - (432 \text{ m/s}^2)\mathbf{j} - (324 \text{ m/s}^2)\mathbf{k}$.

15.8 $\mathbf{v}_E = (156 \text{ in./s})\mathbf{i} + (144 \text{ in./s})\mathbf{j} + (60 \text{ in./s})\mathbf{k}$; $\mathbf{a}_E = -(338 \text{ ft/s}^2)\mathbf{i} + (312 \text{ ft/s}^2)\mathbf{j} + (130 \text{ ft/s}^2)\mathbf{k}$.

15.10 $\mathbf{v}_H = -(0.4 \text{ m/s})\mathbf{i} + (0.7 \text{ m/s})\mathbf{k}$; $\mathbf{a}_H = -(2 \text{ m/s}^2)\mathbf{i} - (6.5 \text{ m/s}^2)\mathbf{j} - (3 \text{ m/s}^2)\mathbf{k}$.

15.12 $66,700 \text{ mi/h}$; 0.01947 ft/s^2.

15.14 (a) Pulley A: $5 \text{ rad/s} \uparrow$, $7.5 \text{ rad/s}^2 \downarrow$; Pulley B: $8 \text{ rad/s} \uparrow$, $12 \text{ rad/s}^2 \downarrow$. (b) $2.09 \text{ m/s}^2 \measuredangle 73.3°$. (c) $3.26 \text{ m/s}^2 \measuredangle 79.4°$.

15.16 0.780 s; 3.90 rad/s.

15.18 (a) $30 \text{ ft/s}^2 \leftarrow$. (b) $24,300 \text{ ft/s}^2$.

15.20 (a) $\omega_C = 120 \text{ rpm}$; $\omega_B = 275 \text{ rpm}$. (b) $\mathbf{a}_A = 23.7 \text{ m/s}^2 \uparrow$; $\mathbf{a}_B = 19.90 \text{ m/s}^2 \downarrow$.

15.22 (a) $13.33 \text{ rad/s}^2 \uparrow$. (b) $34.0 \text{ rev} \uparrow$.

15.24 (a) 2.75 rev. (b) $1.710 \text{ m/s} \downarrow$; $3.11 \text{ m} \downarrow$. (c) $849 \text{ mm/s}^2 \measuredangle 32.0°$.

15.26 $\boldsymbol{\alpha}_A = 1.309 \text{ rad/s}^2 \uparrow$; $\boldsymbol{\alpha}_B = 4.58 \text{ rad/s}^2 \uparrow$.

15.28 (a) 15.52 s. (b) $\omega_A = 445 \text{ rpm} \uparrow$; $\omega_B = 371 \text{ rpm} \downarrow$.

15.30 $\boldsymbol{\alpha} = bv^2/(2\pi r^3) \downarrow$.

15.32 (a) $1.173 \text{ rad/s} \uparrow$. (b) $3.33 \text{ ft/s} \measuredangle 25°$.

15.34 (a) $3.68 \text{ rad/s} \downarrow$. (b) $1.017 \text{ m/s} \measuredangle 70°$.

15.36 (a) $4 \text{ rad/s} \downarrow$. (b) $-(4 \text{ in./s})\mathbf{i}$.

15.38 (a) $(250 \text{ mm/s})\mathbf{i} - (450 \text{ mm/s})\mathbf{j}$. (b) $x = 60 \text{ mm}$, $y = 100 \text{ mm}$.

15.40 (a) $150 \text{ rpm} \downarrow$. (b) $195 \text{ rpm} \downarrow$.

15.42 (a) 236 in./s. (b) 9000 rpm. (c) 1502.

15.44 $\omega_A = 50 \text{ rpm} \downarrow$; $\omega_B = 140 \text{ rpm} \downarrow$.

15.46 (a) $\mathbf{v}_P = 0$; $\omega_{BD} = 750 \text{ rpm} \uparrow$. (b) $\mathbf{v}_P = 52.4 \text{ ft/s} \rightarrow$; $\omega_{BD} = 0$. (c) $\mathbf{v}_P = 0$; $\omega_{BD} = 750 \text{ rpm} \downarrow$.

15.48 $\omega_{BD} = 5.79 \text{ rad/s} \uparrow$; $\mathbf{v}_D = 2.88 \text{ m/s} \downarrow$.

15.50 $\omega_{BD} = 1.5 \text{ rad/s} \downarrow$; $\omega_{DE} = 4 \text{ rad/s} \uparrow$.

15.52 $\omega_{BD} = 4.85 \text{ rad/s} \uparrow$; $\omega_{DE} = 6 \text{ rad/s} \uparrow$.

15.54 $\mathbf{v}_A = 2.34 \text{ m/s} \rightarrow$; $\omega = 0.974 \text{ rad/s} \downarrow$.

15.56 Vertical line intersecting zx plane at $x = 0$, $z = 9.34 \text{ ft}$.

15.58 (a) $6 \text{ rad/s} \uparrow$. (b) $600 \text{ mm/s} \leftarrow$. (c) 480 mm/s, wound.

15.60 (a) $4.27 \text{ rad/s} \downarrow$. (b) $1.330 \text{ m/s} \downarrow$. (c) $1.557 \text{ m/s} \measuredangle 34.7°$.

15.62 (a) $4 \text{ in. below } D$. (b) $\omega_{AB} = 2.7 \text{ rad/s} \uparrow$; $\omega_{BE} = 1.2 \text{ rad/s} \uparrow$. (c) $9.58 \text{ in./s} \measuredangle 38.8°$.

15.64 (a) $2.84 \text{ rad/s} \uparrow$. (b) $1.817 \text{ m/s} \measuredangle 82.4°$.

15.66 (a) $\omega_{AB} = 2.4 \text{ rad/s} \downarrow$; $\omega_{BD} = 1.5 \text{ rad/s} \downarrow$. (c) $0.75 \text{ m/s} \measuredangle 73.7°$.

15.68 (a) $22.0 \text{ in./s} \measuredangle 79.6°$. (b) $20.6 \text{ in./s} \measuredangle 20.5°$.

15.70 (a) $1260 \text{ mm/s} \uparrow$. (b) $1.25 \text{ rad/s} \uparrow$.

15.72 (a) $\omega_{AB} = 0.9 \text{ rad/s} \downarrow$; $\omega_{BD} = 0.338 \text{ rad/s} \downarrow$. (b) $78.8 \text{ mm/s} \leftarrow$.

15.74 Space centrode: Circle of 30-in. radius centered at intersection of pipes on which A and B move. Body centrode: Circle of 15-in. radius with center on rod AB at point equidistant from A and B.

15.80 (a) $0.9 \text{ rad/s}^2 \downarrow$. (b) $0.96 \text{ m/s}^2 \uparrow$.

15.82 (a) $6.67 \text{ rad/s}^2 \uparrow$. (b) $180 \text{ mm/s}^2 \leftarrow$.

15.84 (a) $2.00 \text{ m/s}^2 \leftarrow$. (b) $1.194 \text{ m/s}^2 \measuredangle 70.4°$. (c) $1.406 \text{ m/s}^2 \uparrow$.

15.86 (a) $2270 \text{ m/s}^2 \uparrow$. (b) $2270 \text{ m/s}^2 \measuredangle 60°$.

15.88 $37.1 \text{ in./s}^2 \measuredangle 46.0°$.

15.90 (a) $59.8 \text{ m/s}^2 \uparrow$. (b) $190.6 \text{ m/s}^2 \uparrow$.

15.92 $\mathbf{a}_D = 2930 \text{ ft/s}^2 \measuredangle 45°$; $\mathbf{a}_E = 2930 \text{ ft/s}^2 \measuredangle 45°$.

15.94 (a) $4 \text{ rad/s} \uparrow$. (b) $24 \text{ rad/s}^2 \uparrow$. (c) $10 \text{ m/s}^2 \uparrow$.

15.96 (a) $11.10 \text{ rad/s}^2 \downarrow$. (b) $4.53 \text{ rad/s}^2 \downarrow$.

15.98 (a) $1.232 \text{ rad/s}^2 \downarrow$. (b) $1.848 \text{ m/s}^2 \measuredangle 60°$.

15.100 (a) $1.011 \text{ rad/s}^2 \uparrow$. (b) $5.96 \text{ ft/s}^2 \measuredangle 57.0°$.

15.104 $\alpha = (v_B \sin \beta)^2 \sin \theta / (l^2 \cos^3 \theta)$

15.106 (a) $v_A = b\omega \cos \theta$. (b) $a_A = b\alpha \cos \theta - b\omega^2 \sin \theta$.

15.108 $v_C = -2l\omega \sin \theta$. (b) $a_C = -2l\alpha \sin \theta - 2l\omega^2 \cos \theta$.

15.110 (a) $\omega = (v_0/b) \sin^2 \theta$.
(b) $(v_A)_x = (v_0 l/b) \sin^2 \theta \cos \theta$.
$(v_A)_y = (v_0 l/b) \sin^3 \theta$.

15.112 $v_x = v[1 - \cos(vt/r)]$, $v_y = v \sin(vt/r)$.

15.114 $\alpha = (v_0/r)^2 (1 + \cos^2 \theta) \tan^3 \theta$.

15.116 (a) 5.16 rad/s $\downarrow$. (b) 1.339 m/s $\nwarrow$ 60°.

15.118 $\boldsymbol{\omega}_A = 2.52$ rad/s $\downarrow$; $\boldsymbol{\omega}_B = 1.299$ rad/s $\downarrow$.

15.120 1893 mm/s $\nwarrow$ 67.6°.

15.122 $\mathbf{a}_1 = r\omega^2\mathbf{i} + 2u\omega\mathbf{j}$; $\mathbf{a}_2 = -2u\omega\mathbf{i} - r\omega^2\mathbf{j}$;
$\mathbf{a}_3 = (-r\omega^2 - u^2/r + 2u\omega)\mathbf{i}$; $\mathbf{a}_4 = (r\omega^2 - 2u\omega)\mathbf{j}$.

15.124 (a) 18.97 in./s $\nwarrow$ 41.6°. (b) 1.622 in./s² $\nearrow$ 63.7°.

15.126 (a) 1006 mm/s $\angle$ 72.6°.
(b) 1811 mm/s² $\nwarrow$ 32.0°.

15.128 (a) 809 mm/s $\nearrow$ 35.5°. (b) 1.723 m/s² $\nwarrow$ 67.6°.

15.130 (a) 24.2 in./s $\nearrow$ 82.9°. (b) 64.4 in./s² $\nearrow$ 26.6°.

15.132 1.907 rad/s $\downarrow$; 90.0 rad/s² $\downarrow$.

15.134 1.891 rad/s² $\uparrow$.

15.136 614 in./s² $\nwarrow$ 55.6°.

15.138 (a) 0.354 rad/s $\uparrow$. (b) 0.125 rad/s² $\uparrow$.

15.140 29.7 m/s² $\nwarrow$ 53.9°.

15.142 (a) 3.61 rad/s $\downarrow$. (b) 86.6 in./s $\nearrow$ 30°.
(c) 563 in./s² $\nwarrow$ 41.6°.

15.144 (a) $-(1.5 \text{ rad/s})\mathbf{i} - (0.75 \text{ rad/s})\mathbf{j} - (1 \text{ rad/s})\mathbf{k}$.
(b) $(9 \text{ in./s})\mathbf{i} - (14 \text{ in./s})\mathbf{j} - (3 \text{ in./s})\mathbf{k}$.

15.146 (a) $(1.2 \text{ rad/s})\mathbf{i} + (2 \text{ rad/s})\mathbf{j} + (3 \text{ rad/s})\mathbf{k}$.
(b) $\mathbf{v}_A = -(1350 \text{ mm/s})\mathbf{i} + (60 \text{ mm/s})\mathbf{j} +$
$(500 \text{ mm/s})\mathbf{k}$; $\mathbf{v}_B = -(1350 \text{ mm/s})\mathbf{i} +$
$(660 \text{ mm/s})\mathbf{j} + (100 \text{ mm/s})\mathbf{k}$.

15.148 (a) $-(2260 \text{ rad/s}^2)\mathbf{k}$. (b) $(2260 \text{ rad/s}^2)\mathbf{j}$.

15.150 (a) $(R\omega_1/r)\mathbf{i} + \omega_1\mathbf{j}$. (b) $-(R\omega_1^2/r)\mathbf{k}$.

15.152 $-(12 \text{ rad/s}^2)\mathbf{j} + (5 \text{ rad/s}^2)\mathbf{k}$

15.154 (a) $\omega_1\omega_2\mathbf{i}$. (b) $-r(\omega_1^2 + \omega_2^2)\mathbf{i}$.
(c) $-r\omega_2^2\mathbf{j} + 2r\omega_1\omega_2\mathbf{k}$.

15.156 (a) $(0.0375 \text{ rad/s}^2)\mathbf{i}$.
(b) $-(5.74 \text{ in./s})\mathbf{i} + (8.19 \text{ in./s})\mathbf{j} - (4.91 \text{ in./s})\mathbf{k}$.
(c) $-(2.79 \text{ in./s}^2)\mathbf{i} - (1.434 \text{ in./s}^2)\mathbf{j} +$
$(1.721 \text{ in./s}^2)\mathbf{k}$.

15.158 (a) $(0.0375 \text{ rad/s}^2)\mathbf{i} - (0.02 \text{ rad/s}^2)\mathbf{j} -$
$(0.02 \text{ rad/s}^2)\mathbf{k}$.
(b) $-(5.74 \text{ in./s})\mathbf{i} + (8.19 \text{ in./s})\mathbf{j} - (4.91 \text{ in./s})\mathbf{k}$.
(c) $-(2.33 \text{ in./s}^2)\mathbf{i} - (2.09 \text{ in./s}^2)\mathbf{j} + (2.38 \text{ in./s}^2)\mathbf{k}$.

15.160 (a) $(0.75 \text{ rad/s})\mathbf{i} + (1.5 \text{ rad/s})\mathbf{j}$.
(b) $(300 \text{ mm/s})\mathbf{i} - (150 \text{ mm/s})\mathbf{j}$.
(c) $(60 \text{ mm/s})\mathbf{i} - (30 \text{ mm/s})\mathbf{j} - (90 \text{ mm/s})\mathbf{k}$.

15.162 (a) $(28.4 \text{ rad/s})\mathbf{i} + (5.24 \text{ rad/s})\mathbf{j}$. (b) $(25.8 \text{ rad/s})\mathbf{i}$.

15.164 (a) $\dfrac{\sin \gamma}{\sin \beta} \omega_1$. (b) $\dfrac{\sin (\gamma - \beta)}{\sin \beta} \omega_1 \, \measuredangle \, \gamma$.
(c) $\dfrac{\sin (\gamma - \beta) \sin \gamma}{\sin \beta} \omega_1^2 \, \mathbf{k}$.

15.166 $(54 \text{ in./s})\mathbf{k}$.

15.168 $-(34.5 \text{ mm/s})\mathbf{i}$.

15.170 $-(360 \text{ mm/s})\mathbf{i} - (240 \text{ mm/s})\mathbf{j}$.

15.172 $\omega_1/\cos 20°$.

15.174 (a) $-(2.77 \text{ rad/s})\mathbf{i} + (0.615 \text{ rad/s})\mathbf{j} - (4.15 \text{ rad/s})\mathbf{k}$.
(b) $(12 \text{ in./s})\mathbf{k}$.

15.176 $-(1148 \text{ in./s}^2)\mathbf{k}$.

15.178 $-(191.6 \text{ mm/s}^2)\mathbf{i}$.

15.180 (a) $(72 \text{ in./s})\mathbf{i} + (30 \text{ in./s})\mathbf{j} - (48 \text{ in./s})\mathbf{k}$.
(b) $-(288 \text{ in./s}^2)\mathbf{i} - (864 \text{ in./s}^2)\mathbf{k}$.

15.182 (a) $(0.75 \text{ m/s})\mathbf{i} + (1.299 \text{ m/s})\mathbf{j} - (1.732 \text{ m/s})\mathbf{k}$.
(b) $-(27.1 \text{ m/s}^2)\mathbf{i} + (5.63 \text{ m/s}^2)\mathbf{j} - (15.00 \text{ m/s}^2)\mathbf{k}$.

15.184 (a) $(72 \text{ in./s})\mathbf{i} + (30 \text{ in./s})\mathbf{j} - (48 \text{ in./s})\mathbf{k}$.
(b) $-(480 \text{ in./s}^2)\mathbf{i} - (80 \text{ in./s}^2)\mathbf{j} - (784 \text{ in./s}^2)\mathbf{k}$.

15.186 (a) $-(1.200 \text{ m/s})\mathbf{i} + (2.078 \text{ m/s})\mathbf{j} + (0.900 \text{ m/s})\mathbf{k}$.
(b) $-(9.81 \text{ m/s}^2)\mathbf{i} - (4.90 \text{ m/s}^2)\mathbf{j} + (12.47 \text{ m/s}^2)\mathbf{k}$.

15.190 (a) $-(0.27 \text{ rad/s}^2)\mathbf{i}$.
(b) $(6.24 \text{ in./s})\mathbf{i} - (3.6 \text{ in./s})\mathbf{j} - (16.8 \text{ in./s})\mathbf{k}$.
(c) $-(11.70 \text{ in./s}^2)\mathbf{i} - (2.81 \text{ in./s}^2)\mathbf{j} -$
$(7.48 \text{ in./s}^2)\mathbf{k}$.

15.192 $\mathbf{v}_D = -(1.08 \text{ m/s})\mathbf{j} - (1.44 \text{ m/s})\mathbf{k}$;
$\mathbf{a}_D = -(11.52 \text{ m/s}^2)\mathbf{i} + (17.28 \text{ m/s}^2)\mathbf{j} -$
$(6.48 \text{ m/s}^2)\mathbf{k}$.

15.194 (a) $-(1.2 \times 10^{-3} \text{ rad/s}^2)\mathbf{k}$.
(b) $-(168.9 \text{ mm/s})\mathbf{i} + (225 \text{ mm/s})\mathbf{j} + (230 \text{ mm/s})\mathbf{k}$.
(c) $(13.80 \text{ mm/s}^2)\mathbf{i} - (9.20 \text{ mm/s}^2)\mathbf{j} +$
$(14.07 \text{ mm/s}^2)\mathbf{k}$.

15.196 (a) $(12 \text{ rad/s}^2)\mathbf{i}$.
(b) $-(135 \text{ in./s}^2)\mathbf{i} - (96 \text{ in./s}^2)\mathbf{j} + (225 \text{ in./s}^2)\mathbf{k}$.

15.198 (a) $(640 \text{ mm/s})\mathbf{i} - (768 \text{ mm/s})\mathbf{j} + (32 \text{ mm/s})\mathbf{k}$.
(b) $-(0.512 \text{ m/s}^2)\mathbf{i} + (0.691 \text{ m/s}^2)\mathbf{j} - (3.12 \text{ m/s}^2)\mathbf{k}$.

15.200 (a) $\mathbf{v}_G = (51 \text{ in./s})\mathbf{k}$; $\mathbf{a}_G = (264 \text{ in./s}^2)\mathbf{i} -$
$(45 \text{ in./s}^2)\mathbf{j}$.
(b) $\mathbf{v}_H = (20 \text{ in./s})\mathbf{i} - (15 \text{ in./s})\mathbf{j} + (36 \text{ in./s})\mathbf{k}$.
$\mathbf{a}_H = (144 \text{ in./s}^2)\mathbf{i} - (125 \text{ in./s}^2)\mathbf{k}$.

15.202 (a) $(0.24 \text{ m/s}^2)\mathbf{j} - (1.856 \text{ m/s}^2)\mathbf{k}$.
(b) $-(2.81 \text{ m/s}^2)\mathbf{i} + (0.935 \text{ m/s}^2)\mathbf{j}$.
(c) $-(0.24 \text{ m/s}^2)\mathbf{j} + (3.73 \text{ m/s}^2)\mathbf{k}$.

15.204 (a) 15 in./s² $\leftarrow$.
(b) On AB 7.5 in. from A.

15.206 (a) $(b/a)\omega_1$.
(b) $(b/a)\omega_1^2 \sin \beta \, \mathbf{k}$.
(c) $(b/a)\omega_1^2[(a + b \cos \beta)\mathbf{i} + (b \sin \beta)\mathbf{j}]$.

15.208 (a) $\boldsymbol{\omega} = 2.8$ rad/s $\uparrow$; $\boldsymbol{\alpha} = 3.2$ rad/s² $\downarrow$.
(b) 1.058 m/s² $\nwarrow$ 67.8°.

15.210 (a) 18 rad/s² $\downarrow$. (b) 50 rad/s² $\uparrow$.

15.212 $-(271 \text{ mm/s}^2)\mathbf{i} - (86.6 \text{ mm/s}^2)\mathbf{k}$.

15.214 (a) 3 rad/s $\uparrow$. (b) 144.6 mm/s $\leftarrow$.
(c) 378 mm/s $\nwarrow$ 52.5°.

15.C1 $\theta = 50°$: $\boldsymbol{\omega} = 2.34$ rad/s $\downarrow$, $v_A = 0.938$ m/s $\rightarrow$;
$\theta = 130°$: $\boldsymbol{\omega} = 4.03$ rad/s $\downarrow$,
$v_A = 0.552$ m/s $\rightarrow$.

15.C2 $\theta = 60°$: (a) 41.5 rad/s $\nwarrow$, 14,470 rad/s^2 $\downarrow$;
(b) 54.3 ft/s $\rightarrow$, 3440 ft/s^2 $\rightarrow$.
$\theta = 120°$: (a) 41.5 rad/s $\downarrow$, 14,470 rad/s^2 $\downarrow$;
(b) 36.4 ft/s $\rightarrow$, 7530 ft/s^2 $\leftarrow$.

15.C3 $\theta = 30°$: $\mathbf{v}_B = -(11.38 \text{ in./s})\mathbf{i}$,
$\boldsymbol{\omega} = +(4.42 \text{ rad/s})\mathbf{i} + (1.038 \text{ rad/s})\mathbf{j} +$
$(2.40 \text{ rad/s})\mathbf{k}$.
$\theta = 240°$: $\mathbf{v}_B = (5.29 \text{ in./s})\mathbf{i}$,
$\boldsymbol{\omega} = -(5.50 \text{ rad/s})\mathbf{i} + (2.03 \text{ rad/s})\mathbf{j} -$
$(2.57 \text{ rad/s})\mathbf{k}$.

15.C4 $\theta = 50°$: $\mathbf{v}_D = (114.9 \text{ mm/s})\mathbf{i} + (96.4 \text{ mm/s})\mathbf{j} +$
$(76.6 \text{ mm/s})\mathbf{k}$, $\mathbf{a}_D = (57.9 \text{ mm/s}^2)\mathbf{i} -$
$(99.6 \text{ mm/s}^2)\mathbf{j} + (77.1 \text{ mm/s}^2)\mathbf{k}$;
$\theta = 120°$: $\mathbf{v}_D = (129.9 \text{ mm/s})\mathbf{i} - (75.0 \text{ mm/s})\mathbf{j} +$
$86.6 \text{ mm/s})\mathbf{k}$, $\mathbf{a}_D = -(45.0 \text{ mm/s}^2)\mathbf{i} -$
$(112.6 \text{ mm/s}^2)\mathbf{j} - (60.0 \text{ mm/s}^2)\mathbf{k}$.

CHAPTER 16

16.2 26.2 ft/s^2.
16.4 (a) 4.09 m/s^2 $\rightarrow$. (b) 42.5 N.
16.6 (a) 36.8 ft. (b) 42.3 ft.
16.8 (a) 0.222. (b) 2.17 m/s^2 $\searrow$ 25°.
16.10 (a) $\theta = 60.0°$; $P = 107.2$ lb. (b) 1.726 ft/s^2.
(c) 3.41 lb.
16.12 4.83 m/s^2 $\rightarrow$.
16.14 (a) 12.08 ft/s^2 $\rightarrow$. (b) 5.00 in. $\leq h \leq$ 37.0 in.
16.16 $m_C \leq 213$ kg.
16.18 $F_{AB} = 2.28$ N C; $F_{CD} = 31.7$ N C.
16.20 (a) 27.9 ft/s^2 $\searrow$ 60°. (b) $T_{BF} = 5.73$ lb;
$T_{CH} = 2.27$ lb.
16.22 $\mathbf{B}_y = 40.3$ lb $\downarrow$; $\mathbf{C}_y = 40.3$ lb $\downarrow$.
16.24 72.1 lb.
16.26 $(\mathbf{a}_b)_x = 6.61$ ft/s^2 $\rightarrow$, $(\mathbf{a}_b)_y = 15.67$ ft/s^2 $\downarrow$;
$\mathbf{a}_p = 31.3$ ft/s^2 $\searrow$ 30°.
16.28 $V_B = -40.3$ lb; $|M|_{\text{max}} = 25.2$ lb·ft.
16.32 89.6 N·m.
16.34 (a) 39.2 rad/s^2 $\nwarrow$. (b) 36.5 rad/s^2 $\nwarrow$.
16.36 74.5 s.
16.38 128.2 lb.
16.40 49.8 lb.
16.42 $\boldsymbol{\alpha}_A = 15.00$ rad/s^2 $\nwarrow$; $\boldsymbol{\alpha}_B = 10.00$ rad/s^2 $\nwarrow$.
16.44 (a) 10.59 rad/s^2 $\nwarrow$. (b) 3.18 m/s $\downarrow$.
16.46 (1) (a) 7.50 rad/s^2 $\nwarrow$; (b) 22.5 rad/s $\nwarrow$;
(c) 14.14 rad/s $\nwarrow$.
(2) (a) 6.38 rad/s^2 $\nwarrow$; (b) 19.15 rad/s $\nwarrow$;
(c) 13.05 rad/s $\nwarrow$.
(3) (a) 4.41 rad/s^2 $\nwarrow$; (b) 13.24 rad/s $\nwarrow$;
(c) 10.85 rad/s $\nwarrow$.
(4) (a) 5.56 rad/s^2 $\nwarrow$; (b) 16.67 rad/s $\nwarrow$;
(c) 8.61 rad/s $\nwarrow$.

16.48 (a) 19.08 rad/s^2 $\nwarrow$. (b) 5.56 lb $\nearrow$.
16.50 (b) $\boldsymbol{\omega}_A = \omega_0/(1 + m_B/m_A)$ $\downarrow$.
16.52 (a) No slipping.
(b) $\boldsymbol{\alpha}_A = 15.46$ rad/s^2 $\nwarrow$; $\boldsymbol{\alpha}_B = 7.73$ rad/s^2 $\downarrow$.
16.54 $\boldsymbol{\alpha} = (2g/r) \sin \phi$ $\nwarrow$.
16.58 (a) 300 mm from A.
(b) $\boldsymbol{\alpha} = 4.00$ rad/s^2 $\nwarrow$; $\overline{a} = 1.800$ m/s^2 $\rightarrow$.
16.60 (a) 7.73 rad/s^2 $\nwarrow$. (b) 1.610 ft/s^2 $\rightarrow$. (c) 77.3 in.
16.62 $\boldsymbol{\alpha} = -(0.300 \text{ rad/s}^2)\mathbf{j}$; $\overline{a} = (0.1350 \text{ m/s}^2)\mathbf{k}$.
16.64 $\mathbf{a}_A = 7.97$ m/s^2 $\uparrow$; $\mathbf{a}_B = 1.578$ m/s^2 $\uparrow$.
16.66 (a) $T = W$. (b) $\boldsymbol{\alpha} = rg/\overline{k}^2$ $\nwarrow$.
16.68 (a) 698 lb. (b) 0.315 rad/s^2 $\nwarrow$.
16.70 (a) $3g/L$ $\downarrow$. (b) $1.323g$ $\measuredangle$ 49.1°.
(c) $2.18g$ $\searrow$ 66.6°.
16.72 (a) g/L $\downarrow$. (b) $0.866g$ $\leftarrow$. (c) $1.323g$ $\nearrow$ 49.1°.
16.74 (a) $t_1 = \omega_0 r/3\mu_k g$.
(b) $\mathbf{v}_1 = \frac{1}{3}\omega_0 r \rightarrow$; $\boldsymbol{\omega}_1 = \frac{1}{3}\omega_0$ $\downarrow$.
16.76 (a) $t_1 = \dfrac{v_1}{\mu_k g} \dfrac{\overline{k}^2}{r^2 + \overline{k}^2}$.
(b) $v_1 \dfrac{\overline{k}^2}{r^2 + \overline{k}^2} \rightarrow$; $\dfrac{v_1 r}{r^2 + \overline{k}^2}$ $\nwarrow$.
16.78 (a) 24.0 rad/s^2 $\downarrow$.
(b) $\mathbf{A}_x = 6.00$ N $\leftarrow$; $\mathbf{A}_y = 19.62$ N $\uparrow$.
16.80 (a) 5.00 in. (b) 32.2 rad/s^2 $\downarrow$.
16.82 (a) 80 mm. (b) 41.7 rad/s^2 $\downarrow$.
16.84 $T = \dfrac{1}{2} \dfrac{m}{l} \omega^2(l^2 - x^2)$.
16.86 505 N $\rightarrow$.
16.88 41.9 N·mm $\nwarrow$.
16.90 (a) $\frac{9}{7}g$ $\downarrow$. (b) $\frac{4}{7}W$ $\uparrow$.
16.92 (a) $1.061g/b$ $\downarrow$. (b) $\frac{3}{2}g$ $\downarrow$. (c) $\frac{1}{4}W$ $\uparrow$.
16.94 (a) 21.1 rad/s^2 $\downarrow$.
(b) $\mathbf{A}_x = 32.8$ lb $\leftarrow$, $\mathbf{A}_y = 3.18$ lb $\uparrow$.
16.96 (a) 86.0 N $\downarrow$. (b) $\mathbf{A}_x = 69.1$ N $\leftarrow$, $\mathbf{A}_y = 172.0$ N $\uparrow$.
16.98 (a) 13.50 rad/s^2 $\nwarrow$. (b) 6.79 N·m $\nwarrow$.
16.100 (a) $1.8g$ $\downarrow$. (b) $0.2mg$ $\uparrow$.
16.102 (a) $0.634g/L$ $\downarrow$. (b) $\mathbf{N} = 0.866mg$ $\uparrow$,
$\mathbf{F} = 0.287mg \rightarrow$. (c) 0.332.
16.106 $\overline{a} = \dfrac{r^2}{r^2 + \overline{k}^2} g \sin \beta$.
16.108 3.86 m.
16.110 (a) $\boldsymbol{\alpha} = 9.27$ rad/s^2 $\downarrow$; $\overline{\mathbf{a}} = 7.73$ ft/s^2 $\rightarrow$. (b) 0.135.
16.112 (a) $\boldsymbol{\alpha} = 13.91$ rad/s^2 $\downarrow$; $\overline{\mathbf{a}} = 11.59$ ft/s^2 $\rightarrow$.
(b) 0.015.
16.114 (a) Does not slide. (b) $\boldsymbol{\alpha} = 14.40$ rad/s^2 $\downarrow$;
$\overline{\mathbf{a}} = 2.30$ m/s^2 $\rightarrow$.
16.116 (a) Does not slide. (b) $\boldsymbol{\alpha} = 21.6$ rad/s^2 $\downarrow$;
$\overline{\mathbf{a}} = 3.46$ m/s^2 $\rightarrow$.
16.118 (a) 72.4 rad/s^2 $\downarrow$. (b) 7.24 m/s^2 $\downarrow$.
16.120 (a) $5g/9$ $\downarrow$. (b) g $\downarrow$. (c) 0.
16.122 (a) $13g/17$ $\downarrow$. (b) g $\downarrow$. (c) $2g/3$ $\downarrow$.

17.108 $\omega_2 = \frac{3}{2}(\bar{v}_1/b)\ \downdownarrows$; $\bar{v}_2 = 0.791\ \bar{v}_1 \searrow 71.6°$.

17.110 $\omega = \dfrac{\bar{v}_1}{L}\ \dfrac{12 \sin \beta}{3 \sin^2 \beta + 1}\ \downdownarrows$.

17.112 (a) $\frac{1}{4}\omega_0\ \downdownarrows$. (b) $\frac{13}{16}$. (c) $1.5°$.

17.114 (a) 1.918 N·s. (b) 2.35 N·s.

17.116 (a) $0.6\bar{v}_1 \rightarrow$. (b) $2.4v_1/L\ \uparrow$. (c) $0.4\bar{v}_1 \rightarrow$.

17.118 $\omega_{AB} = \omega_{CD} = \frac{3}{2}(\bar{v}_1/L)\ \uparrow$; $\bar{v}_{AB} = \frac{3}{4}\bar{v}_1\ \downarrow$; $\bar{v}_{CD} = \frac{1}{4}\bar{v}_1\ \downarrow$.

17.120 (a) 1.801 m/s $\downarrow$; 7.20 rad/s $\downdownarrows$.
(b) 0.6 m/s $\uparrow$; 14.41 rad/s $\downdownarrows$.

17.122 (a) $\omega_A = \omega_1\ \downdownarrows$, $\omega_B = 0$; $\bar{v}_A = 0$, $\bar{v}_B = \bar{v}_1 \rightarrow$.
(b) $\bar{v}_A = 2\bar{v}_1/7 \rightarrow$; $\bar{v}_B = 5\bar{v}_1/7 \rightarrow$.
(c) Final motion is motion of part a.

17.124 $y^2 = [(2\bar{v}_0^2 \sin^2 \theta)/\mu_k g]x$.

17.126 $5v_0/4r$.

17.128 (a) 2.8 m/s $\rightarrow$. (b) 1.727 m/s $\rightarrow$.

17.130 $2\omega_0/5$.

17.132 $6\bar{v}_1/5L\ \downdownarrows$.

17.136 (a) 3.89 m/s $\measuredangle\ 31.0°$. (b) 8 rad/s $\uparrow$.

17.138 $v_r = 3.84$ m/s; $v_\theta = 2.40$ m/s.

17.140 14.68 rad/s $\downdownarrows$; 8.95 rad/s $\downdownarrows$.

17.C1 $\theta = 30°$: $\omega_{AB} = 2.86$ rad/s $\downdownarrows$, $v_D = 2.15$ m/s $\rightarrow$;
$\theta = 10°$: $\omega_{AB} = 4.99$ rad/s $\downdownarrows$. $v_D = 1.300$ m/s $\rightarrow$.

17.C2 (b) $\theta = 30°$: $\omega = 4.75$ rad/s $\uparrow$;
$\theta = 60°$: $\omega = 7.70$ rad/s $\uparrow$.
(c) $t = 0.785$ s.

17.C3 $x = 120$ mm: (a) 3.20 rad/s, 1.052 m/s;

17.C4 $x = 150$ mm: 8.67 m/s $\uparrow$, 8.89 rad/s $\uparrow$;
$x = 300$ mm: 3.33 m/s $\uparrow$, 22.2 rad/s $\uparrow$.

CHAPTER 18

18.2 $\frac{1}{6}mb^2\omega(\sin 2\beta\ \mathbf{i} + 2\cos^2 \beta\ \mathbf{j})$.

18.4 $\frac{1}{12}ma^2\omega(3\ \mathbf{j} + 2\ \mathbf{k})$.

18.6 $(0.234$ kg·m^2/s$)\mathbf{j} + (1.250$ kg·m^2/s$)\mathbf{k}$.

18.8 $(0.600$ kg·m^2/s$)\mathbf{i} + (0.016$ kg·m^2/s$)\mathbf{j} + (861$ kg·m^2/s$)\mathbf{k}$.

18.10 $-(1.747$ ft·lb·s$)\mathbf{i} + (3.59$ ft·lb·s$)\mathbf{j} + (0.582$ ft·lb·s$)\mathbf{k}$.

18.14 (a) $\frac{2}{3}m'a^3\omega(5\mathbf{i} - 3\mathbf{k})$. (b) $\frac{2}{3}m'a^3\omega(5\mathbf{i} - 3\mathbf{k})$.

18.16 (a) $-(1.041$ ft·lb·s$)\mathbf{i} + (1.041$ ft·lb·s$)\mathbf{j} + (2.31$ ft·lb·s$)\mathbf{k}$.
(b) $147.5°$.

18.18 (a) $(0.480$ m/s$)\mathbf{j}$.
(b) $-(3.99$ rad/s$)\mathbf{i} + (1.331$ rad/s$)\mathbf{j} + (1.504$ rad/s$)\mathbf{k}$.

18.20 (a) $-(F\ \Delta t/m)\mathbf{k}$. (b) $(12F\ \Delta t/7ma)(5\mathbf{i} + \mathbf{j})$.

18.22 (a) 0. (b) $(F\ \Delta t/ma)(2.5\mathbf{i} - 1.454\mathbf{j} + 2.194\mathbf{k})$.

18.24 (a) 0. (b) $(3F\ \Delta t/md)(\mathbf{i} - \frac{1}{4}\mathbf{k})$.

18.26 (a) $\frac{1}{6}\omega_0(-\mathbf{i} + \mathbf{j})$. (b) $\frac{1}{6}\omega_0 a\mathbf{k}$.

18.28 $42.9°$.

18.30 (a) 941 ms. (b) $(0.01693$ rad/s$)\mathbf{k}$.
(c) $-(39.2$ mm/s$)\mathbf{j}$.

18.32 (a) 22.1 in. (b) $\omega_y = 2.34$ rad/s, $\omega_z = 0.487$ rad/s.

18.36 $\frac{1}{8}ma^2\omega^2$.

18.38 10.98 J.

18.40 29.1 ft·lb.

18.42 $(5/48)ma^2\omega_0^2$.

18.44 $+229$ ft·lb.

18.46 $-\frac{1}{6}mb^2\omega^2 \sin 2\beta\ \mathbf{k}$.

18.48 $\frac{1}{6}ma^2\omega^2\mathbf{i}$.

18.50 $(3.75$ N·m$)\mathbf{i}$.

18.52 $\frac{1}{12}ma^2(2\omega^2\mathbf{i} + 3\alpha\mathbf{j} + 2\alpha\mathbf{k})$.

18.54 (a) 0.0453 in. below axis;
$I_{xy} = -2.35 \times 10^{-3}$ lb·ft·s^2, $I_{xz} = 0$.
(b) $\mathbf{F} = -(41.8$ lb$)\mathbf{j}$; $\mathbf{M}_C = (20.8$ lb·ft$)\mathbf{k}$.

18.56 $\mathbf{A} = (2w/3g)r^2\omega^2\mathbf{k}$; $\mathbf{B} = -(2w/3g)r^2\omega^2\mathbf{k}$.

18.58 (a) $(20.0$ rad/s$^2)\mathbf{j}$.
(b) $\mathbf{A} = -(3.75$ N$)\mathbf{k}$; $\mathbf{B} = -(1.250$ N$)\mathbf{k}$.

18.60 (a) $\pi(w/g)r^3\alpha\mathbf{i}$.
(b) $\mathbf{A} = (2w/3g)r^2\alpha\mathbf{j}$; $\mathbf{B} = -(2w/3g)r^2\alpha\mathbf{j}$.

18.62 (a) $(0.437$ lb·ft$)\mathbf{i}$.
(b) $\mathbf{A} = -(1.210$ lb$)\mathbf{j} + (0.378$ lb$)\mathbf{k}$;
$\mathbf{B} = (1.210$ lb$)\mathbf{j} - (0.378$ lb$)\mathbf{k}$.

18.64 $1.30°$; A will move up.

18.66 5630 lb·ft.

18.68 (a) $1.528\sqrt{g/a}$. (b) $0 \le \omega \le 1.225\sqrt{g/a}$.

18.70 $\omega = 2.70\sqrt{g/L}$.

18.72 $\mathbf{A} = -\frac{1}{4}ma\omega_1\omega_2\mathbf{k}$; $\mathbf{B} = \frac{1}{4}ma\omega_1\omega_2\mathbf{k}$.

18.74 $\mathbf{A} = -(11.65$ lb$)\mathbf{i}$; $\mathbf{M}_A = (2.91$ lb·ft$)\mathbf{i} + (8.73$ lb·ft$)\mathbf{k}$.

18.76 (a) $\mathbf{M}_1\mathbf{j} = 0$. (b) $\mathbf{A} = -(18.00$ N$)\mathbf{i} - (13.50$ N$)\mathbf{k}$;
$\mathbf{M}_A = (3.75$ N·m$)\mathbf{i} - (0.281$ N·m$)\mathbf{k}$.

18.78 (a) $-(5.39$ lb·ft$)\mathbf{j}$.
(b) $\mathbf{A} = -(11.65$ lb$)\mathbf{i} + (3.49$ lb$)\mathbf{k}$;
$\mathbf{M}_A = (5.53$ lb·ft$)\mathbf{i} + (8.73$ lb·ft$)\mathbf{k}$.

18.80 (a) $\mathbf{M}_1 = -\frac{1}{12}mL^2\omega_1\omega_2 \sin 2\theta\ \mathbf{j}$.
(b) $\mathbf{M}_2 = \frac{1}{24}mL^2\omega_1^2 \sin 2\theta\ \mathbf{k}$.
(c) $\mathbf{C} = \frac{1}{6}mL\omega_1\omega_2 \sin^2 \theta\ \mathbf{k}$; $\mathbf{D} = -\frac{1}{6}mL\omega_1\omega_2 \sin^2 \theta\ \mathbf{k}$.

18.82 1326 rpm.

18.84 (c) -7.8%.

18.86 (a) $30.0°$. (b) $24.9°$. (c) $37.4°$.

18.88 (a) 128.3 rad/s. (b) 2.17 in.

18.90 (a) 11.40 rpm. (b) 11.82 rpm and 322 rpm.

18.96 (c) $16.05°$; $-2.13\dot{\psi}$.

18.98 14.01 rev/h.

18.100 Precession axis: $\theta_x = 125.0°$, $\theta_y = 36.6°$, $\theta_z = 80.7°$; precession, 0.946 rad/s (retrograde); spin, 0.1593 rad/s.

18.102 (a) 109.4 rpm;
$\gamma_x = 90°$, $\gamma_y = 100.05°$, $\gamma_z = 10.05°$.
(b) $\theta_x = 90°$, $\theta_y = 113.9°$, $\theta_z = 23.9°$.
(c) Precession, 47.1 rpm; spin, 64.6 rpm.

18.104 (a) $\beta = 27.5°$.
(b) Precession, 83.5 rpm; spin, 112.8 rpm.

18.106 (a) $-\theta_0 \leq \theta \leq \theta_0$.
(b) $\dot{\phi}_0 \sin \theta_0 [(1 + \cos^2 \theta_0)/2]^{\frac{1}{2}}$.
(c) $\frac{1}{2}\dot{\phi}_0 (1 + \cos^2 \theta_0)$.

18.110 (a) 47.0 rad/s.
(b) Precession, 6.26 rad/s; spin, 50.1 rad/s.

18.112 27.5°.

18.118 On B: $6\frac{1}{4}$ in. below shaft; on C: $3\frac{1}{8}$ in. above.

18.120 $(112.8 \text{ g·m}^2/\text{s})\mathbf{i} + (80 \text{ g·m}^2/\text{s})\mathbf{j}$.

18.122 (a) $(0.825 \text{ kg·m}^2/\text{s})\mathbf{i} + (0.592 \text{ kg·m}^2/\text{s})\mathbf{j}$.
(b) 1.510 J.

18.124 (a) $(2.71 \text{ lb·ft})\mathbf{j}$.
(b) $\mathbf{F} = -(5.30 \text{ lb})\mathbf{i} - (1.988 \text{ lb})\mathbf{k}$;
$\mathbf{M}_O = (2.69 \text{ lb·ft})\mathbf{i} - (4.42 \text{ lb·ft})\mathbf{k}$.

18.126 $(2.00 \text{ rad/s})\mathbf{i} - (0.225 \text{ rad/s})\mathbf{j} + (0.0675 \text{ rad/s})\mathbf{k}$.

18.128 $\sqrt{8g/11a}$.

18.C1 $\theta = 60°$: $\boldsymbol{\omega} = (3.75 \text{ rad/s})\mathbf{i} - (10.83 \text{ rad/s})\mathbf{j}$, axis through O forming 70.9° ⦨ with x axis;
$\theta = 150°$: $\boldsymbol{\omega} = (0.335 \text{ rad/s})\mathbf{i} - (6.25 \text{ rad/s})\mathbf{j}$, axis through O forming 86.9° ⦨ with x axis.

18.C2 $b = 8$ in.: $\omega_x = 0.1085$ rad/s, $\omega_y = 1.950$ rad/s, $\omega_z = 0.674$ rad/s, $\omega = 2.07$ rad/s, $\Delta T = +174.0$ ft·lb;
$b = 24$ in.: $\omega_x = 0.326$ rad/s, $\omega_y = 1.950$ rad/s, $\omega_z = 0.370$ rad/s, $\omega = 2.01$ rad/s, $\Delta T = +158.0$ ft·lb.

18.C3 $\dot{\psi} = 1200$ rpm, $\theta = 30°$: (a) 78.3 rpm, (b) 62.4 rpm, (c) −20.3%.
$\dot{\psi} = 2400$ rpm, $\theta = 30°$: (a) 32.59 rpm, (b) 31.21 rpm, (c) −4.2%.

18.C4 $\dot{\psi}_0 = 1835$ rpm; $\tau = 0.211$ s.

CHAPTER 19

19.2 10 mm; 3.18 Hz.

19.4 4.01 ft/s; 64.4 ft/s^2.

19.6 (a) 2.48 mm; 0.0979 in. (b) 0.621 mm; 0.0245 in.

19.8 (a) 327 N/m. (b) 14.78 mm below equilibrium position; 841 mm/s ↑; 1.933 m/s^2 ↑.

19.10 (a) 0.0569 s. (b) 4.91 ft/s ↑; 24.2 ft/s^2 ↑.

19.12 (a) 0.738 s; 1.355 Hz. (b) 0.887 ft/s; 7.55 ft/s^2.

19.14 (a) 0.517 s; 1.934 Hz. (b) 0.365 m/s; 4.43 m/s^2.

19.16 (a) 6.82 lb. (b) 33.4 lb/ft.

19.18 (a) 6.80 kg. (b) 0.583 s.

19.22 (a) 3.56 kg. (b) 43.7 kg.

19.26 (a) 1.794 s. (b) 1.825 s. (c) 2.12 s.

19.28 (a) 0.430 s. (b) 0.584 m/s.

19.30 (a) 0.491 s. (b) 9.60 in./s.

19.32 (a) 1.117 rad/s. (b) 400 mm.

19.34 (a) 20 in. (b) 0.309 s.

19.36 (a) $f = (1/2\pi)\sqrt{2k/m}$. (b) $f = (1/2\pi)\sqrt{6b^2k/l^2m}$.
(c) $1/\sqrt{3}$.

19.42 (a) $l/\sqrt{12}$. (b) $4.77\sqrt{l/g}$.

19.44 (a) $\tau = 2\pi\sqrt{5b/6g}$. (b) $b/4$.

19.46 $f = (1/2\pi)\sqrt{18g/17l}$.

19.48 128.8 mm.

19.50 (a) $\tau = 2\pi\sqrt{3r/2g}$. (b) $r/2$.

19.52 11.17 mm.

19.54 (a) 0.952 s. (b) 1.099 s.

19.56 (a) 0.327 s. (b) 12.07 in./s.

19.62 75.5°.

19.64 (a) $\tau = 2\pi\sqrt{7l/6g}$. (b) $\tau = 2\pi\sqrt{5l/6g}$.

19.66 (a) $f = (1/2\pi)\sqrt{2g/3c}$. (b) $f = (1/2\pi)\sqrt{4g/5c}$.

19.70 4.56 Hz.

19.72 508 N/m.

19.74 $\tau = 2\pi\sqrt{10l/9g}$.

19.76 0.911 Hz.

19.78 (a) $0.291\ l$. (b) $f = 0.209\sqrt{g/l}$.

19.80 $\tau = 2\pi\sqrt{(\pi - 2)r/g}$.

19.82 1.135 s.

19.84 (a) and (b) 0.322 s.

19.86 (a) 83.3 mm. (b) 59.5 mm.

19.88 $\omega > \sqrt{2k/m}$.

19.90 $\sqrt{g/2l} < \omega < \sqrt{3g/2l}$.

19.92 (a) 203 rpm. (b) 43.0 μm.

19.94 274 rpm.

19.96 (a) 219 μm. (b) 960 μm. (c) Infinite.

19.98 1.286 mm or 1.800 mm.

19.100 0.686 in. or 4.80 in.

19.104 0.644 Hz; 0.455 Hz.

19.106 (a) 19.70 mi/h. (b) 0.481 in.

19.114 (a) 8.13×10^{-3}. (b) 0.356 N·s/m.

19.116 (a) 6400 lb/ft. (b) 0.1432 s.

19.120 (a) 0.0481 in. (b) 0.0234 in.

19.122 (a) 0.0905. (b) 366 N·s/m.

19.124 2.19 kN·s/m.

19.126 (a) 297 rpm. (b) 252 rpm. (c) 0.335 in.; 0.361 in.

19.128 $m\dfrac{d^2x_A}{dt^2} + 5kx_A - 2kx_B - c\dfrac{d}{dt}(x_B - x_A) = 0$

$m\dfrac{d^2x_B}{dt^2} - 2kx_A + 2kx_B + c\dfrac{d}{dt}(x_B - x_A) =$
$P_m \sin \omega t$.

19.130 (a) E/R. (b) L/R.

19.134 (a) $m\dfrac{d^2x_m}{dt^2} + k_2x_m + c\dfrac{d}{dt}(x_m - x_A) = 0$;

$c\dfrac{d}{dt}(x_A - x_m) + k_1x_A = 0$.

(b) $L\dfrac{d^2q_m}{dt^2} + \dfrac{q_m}{C_2} + R\dfrac{d}{dt}(q_m - q_A) = 0$;

$R\dfrac{d}{dt}(q_A - q_m) + \dfrac{q_A}{C_1} = 0$.

19.136 (a) $\tau = 6.82\sqrt{r/g}$. (b) $\tau = 6.68\sqrt{r/g}$.

19.138 (a) 5.81 Hz; 4.91 mm; 179.1 mm/s. (b) 491 N.
(c) 159.2 mm/s ↑.

19.140 $f = (1/2\pi)\sqrt{2\mu_k g/l}$.

19.142 0.760 lb·ft·s²; 8.66 in.

19.144 3.32 s.

19.C1 $\theta = 60°$: 1 term, 1.0625; 2 terms, 1.0713;
4 terms, 1.0731; 8 terms, 1.0732;
16 terms, 1.0732.

19.C2 $m_C = 0.5\ kg$: (a) 0.641 s, 102.2 mm;
(b) 0.428 s, 45.4 mm; (c) 0.321 s, 25.6 mm.
$m_C = 1.5\ kg$: (a) 0.828 s, 170.3 mm;
(b) 0.552 s, 75.7 mm; (c) 0.414 s, 42.6 mm.

19.C3 $\omega = 500\ rpm$: $x_m = 0.00515$ in. (in phase);
$a_m = 1.176$ ft/s².
$\omega = 1000\ rpm$: $x_m = 0.001535$ in. (out of phase);
$a_m = 1.403$ ft/s².

19.C4 $\omega = 500\ rpm$: $x_m = 0.00435$ in.;
$a_m = 0.994$ ft/s²; $\varphi = 32.3°$.
$\omega = 1000\ rpm$: $x_m = 0.001528$ in.;
$a_m = 1.397$ ft/s²; $\varphi = -5.39°$.

APPENDIX B

9.88 (a) $0.0699\ ma^2$. (b) $0.320\ ma^2$.

9.90 (a) $I_{AA'} = mb^2/24$; $I_{BB'} = mh^2/18$.
(b) $m(3b^2 + 4h^2)/72$.

9.92 (a) $ma^2/3$. (b) $3ma^2/2$.

9.94 $\frac{1}{2}ma^2(2n + 1)/(4n + 1)$.

9.96 $m(3a^2 + 2h^2)/20$; $\sqrt{(3a^2 + 2h^2)/20}$.

9.98 $m(b^2 + 3h^2)/5$.

9.100 $\frac{1}{3}mR^2(1 - \cos\phi)(2 + \cos\phi)$

9.102 1.514 kg·m²; 155.7 mm.

9.104 $2mr^2/3$; $0.816r$.

9.106 (a) $r_a = 176.9$ mm. (b) 87.8 mm.

9.108 (a) $\frac{1}{2}\rho\pi h(a_2^4 + 2a_2^2 a_1^2 - 3a_1^4)$. (b) $a_2/\sqrt{3}$.
(c) $\frac{2}{3}\rho\pi h a_2^4$.

9.110 (a) 5.14×10^{-3} kg·m². (b) 7.54×10^{-3} kg·m².
(c) 3.47×10^{-3} kg·m².

9.112 $I_x = I_z = 36.1 \times 10^{-3}$ lb·ft·s²;
$I_y = 30.0 \times 10^{-3}$ lb·ft·s².

9.114 0.883 g·m².

9.116 31.1×10^{-3} lb·ft·s².

9.118 (a) 9.97 kg·m². (b) 4.96 kg·m².

9.120 $I_{xy} = I_{zx} = 0$; $I_{yz} = -3.53 \times 10^{-3}$ lb·ft·s².

9.122 $I_{xy} = 14.13$ g·m²; $I_{yz} = 37.7$ g·m²;
$I_{zx} = 113.0$ g·m².

9.124 $I_{xy} = I_{yz} = 6.71 \times 10^{-3}$ lb·ft·s²;
$I_{zx} = 2.24 \times 10^{-3}$ lb·ft·s².

9.128 $(3ma^2/20)(a^2 + 6h^2)/(a^2 + h^2)$.

9.130 23.6×10^{-3} lb·ft·s².

9.132 (a) $I_{xy} = \frac{1}{4}ma^2$; $I_{yz} = 0$; $I_{zx} = -\frac{1}{4}ma^2$.
(b) $2ma^2/3$.

9.134 $17ma^2/54$.

9.136 (a) 2. (b) $\frac{1}{2}$.

SI Prefixes

Multiplication Factor	Prefix†	Symbol
$1\ 000\ 000\ 000\ 000 = 10^{12}$	tera	T
$1\ 000\ 000\ 000 = 10^{9}$	giga	G
$1\ 000\ 000 = 10^{6}$	mega	M
$1\ 000 = 10^{3}$	kilo	k
$100 = 10^{2}$	hecto‡	h
$10 = 10^{1}$	deka‡	da
$0.1 = 10^{-1}$	deci‡	d
$0.01 = 10^{-2}$	centi‡	c
$0.001 = 10^{-3}$	milli	m
$0.000\ 001 = 10^{-6}$	micro	μ
$0.000\ 000\ 001 = 10^{-9}$	nano	n
$0.000\ 000\ 000\ 001 = 10^{-12}$	pico	p
$0.000\ 000\ 000\ 000\ 001 = 10^{-15}$	femto	f
$0.000\ 000\ 000\ 000\ 000\ 001 = 10^{-18}$	atto	a

† The first syllable of every prefix is accented so that the prefix will retain its identity. Thus, the preferred pronunciation of kilometer places the accent on the first syllable, not the second.

‡ The use of these prefixes should be avoided, except for the measurement of areas and volumes and for the nontechnical use of centimeter, as for body and clothing measurements.

Principal SI Units Used in Mechanics

Quantity	Unit	Symbol	Formula
Acceleration	Meter per second squared	. . .	m/s^2
Angle	Radian	rad	†
Angular acceleration	Radian per second squared	. . .	rad/s^2
Angular velocity	Radian per second	. . .	rad/s
Area	Square meter	. . .	m^2
Density	Kilogram per cubic meter	. . .	kg/m^3
Energy	Joule	J	$N \cdot m$
Force	Newton	N	$kg \cdot m/s^2$
Frequency	Hertz	Hz	s^{-1}
Impulse	Newton-second	. . .	$kg \cdot m/s$
Length	Meter	m	‡
Mass	Kilogram	kg	‡
Moment of a force	Newton-meter	. . .	$N \cdot m$
Power	Watt	W	J/s
Pressure	Pascal	Pa	N/m^2
Stress	Pascal	Pa	N/m^2
Time	Second	s	‡
Velocity	Meter per second	. . .	m/s
Volume, solids	Cubic meter	. . .	m^3
Liquids	Liter	L	$10^{-3}\ m^3$
Work	Joule	J	$N \cdot m$

† Supplementary unit (1 revolution $= 2\pi$ rad $= 360°$).

‡ Base unit.

U.S. Customary Units and Their SI Equivalents

Quantity	U.S. Customary Unit	SI Equivalent
Acceleration	ft/s^2	0.3048 m/s^2
	$in./s^2$	0.0254 m/s^2
Area	ft^2	0.0929 m^2
	in^2	645.2 mm^2
Energy	$ft \cdot lb$	1.356 J
Force	kip	4.448 kN
	lb	4.448 N
	oz	0.2780 N
Impulse	$lb \cdot s$	$4.448 \text{ N} \cdot \text{s}$
Length	ft	0.3048 m
	in.	25.40 mm
	mi	1.609 km
Mass	oz mass	28.35 g
	lb mass	0.4536 kg
	slug	14.59 kg
	ton	907.2 kg
Moment of a force	$lb \cdot ft$	$1.356 \text{ N} \cdot \text{m}$
	$lb \cdot in.$	$0.1130 \text{ N} \cdot \text{m}$
Moment of inertia		
Of an area	in^4	$0.4162 \times 10^6 \text{ mm}^4$
Of a mass	$lb \cdot ft \cdot s^2$	$1.356 \text{ kg} \cdot \text{m}^2$
Momentum	$lb \cdot s$	$4.448 \text{ kg} \cdot \text{m/s}$
Power	$ft \cdot lb/s$	1.356 W
	hp	745.7 W
Pressure or stress	lb/ft^2	47.88 Pa
	lb/in^2 (psi)	6.895 kPa
Velocity	ft/s	0.3048 m/s
	$in./s$	0.0254 m/s
	mi/h (mph)	0.4470 m/s
	mi/h (mph)	1.609 km/h
Volume	ft^3	0.02832 m^3
	in^3	16.39 cm^3
Liquids	gal	3.785 L
	qt	0.9464 L
Work	$ft \cdot lb$	1.356 J